KINETICS OF MATERIALS

KINETICS OF MATERIALS

Robert W. Balluffi   Samuel M. Allen

W. Craig Carter

With Editorial Assistance from Rachel A. Kemper

Department of Materials Science and Engineering
Massachusetts Institute of Technology
Cambridge, Massachusetts

WILEY-INTERSCIENCE

A JOHN WILEY & SONS, INC., PUBLICATION

Published by John Wiley & Sons, Inc., Hoboken, New Jersey.
Published simultaneously in Canada.

For general information on our other products and services please contact our Customer Care Department with the U.S. at 877-762-2974, outside the U.S. at 317-572-3993 or fax 317-572-4002.

Wiley also publishes its books in a variety of electronic formats. Some content that appears in print, however, may not be available in electronic format.

Library of Congress Cataloging-in-Publication Data:

Balluffi, Robert W., 1924–
    Kinetics of Materials / Robert W. Balluffi, Samuel M. Allen, W. Craig Carter;
  edited by Rachel A. Kemper;
        p. cm.
    Includes bibliographical references and index.

    ISBN 13 978-0-471-24689-3 ISBN-10 0-471-24689-1

    1.Materials–Mechanical Properties.   2. Materials science.

  I. Allen, Samuel M. II. Carter, W. Craig. III. Kemper, Rachel A. IV. Title.
TA404.8.B35 2005
620.1'1292–dc22   2005047793

15 14 13 12

CONTENTS

PART II   MOTION OF DISLOCATIONS AND INTERFACES

PREFACE

This textbook has evolved from part of the first-year graduate curriculum in the Department of Materials Science and Engineering at the Massachusetts Institute of Technology (MIT). This curriculum includes four required semester-long subjects—"Materials at Equilibrium," "Mechanical Properties of Materials," "Electrical, Optical, and Magnetic Properties of Materials," and "Kinetic Processes in Materials." Together, these subjects introduce the essential building blocks of materials science and engineering at the beginning of graduate work and establish a foundation for more specialized topics.

Because the entire scope of kinetics of materials is far too great for a semester-length class or a textbook of reasonable length, we cover a range of selected topics representing the basic processes which bring about changes in the size, shape, composition, and atomistic structures of materials. The subject matter was selected with the criterion that structure is all-important in determining the properties (and applications) of materials. Topics concerned with fluid flow and kinetics, which are often important in the processing of materials, have not been included and may be found in standard texts such as those by Bird, Stewart, and Lightfoot [1] and Poirier and Geiger [2]. The major topics included in this book are:

I. Motion of atoms and molecules by diffusion

II. Motion of dislocations and interfaces

III. Morphological evolution due to capillary and applied mechanical forces

IV. Phase transformations

The various topics are generally introduced in order of increasing complexity. The text starts with diffusion, a description of the elementary manner in which atoms and molecules move around in solids and liquids. Next, the progressively more complex problems of describing the motion of dislocations and interfaces are addressed. Finally, treatments of still more complex kinetic phenomena—such as morphological evolution and phase transformations—are given, based to a large extent on topics treated in the earlier parts of the text.

The diffusional transport essential to many of these phenomena is driven by a wide variety of forces. The concept of a basic diffusion potential, which encompasses all of these forces, is therefore introduced early on and then used systematically in the analysis of the many kinetic processes that are considered.

We have striven to develop the subject in a systematic manner designed to provide readers with an appreciation of its analytic foundations and, in many cases, the approximations commonly employed in the field. We provide many extensive derivations of important results to help remove any mystery about their origins. Most attention is paid throughout to kinetic phenomena in crystalline materials; this reflects the interests and biases of the authors. However, selected phenomena in noncrystalline materials are also discussed and, in many cases, the principles involved apply across the board. We hope that with the knowledge gained from this book, students will be equipped to tackle topics that we have not addressed. The book therefore fills a significant gap, as no other currently available text covers a similarly wide range of topics.

The prerequisites for effective use of this book are a typical undergraduate knowledge of the structure of materials (including crystal imperfections), vector calculus and differential equations, elementary elasticity theory, and a somewhat deeper knowledge of classical thermodynamics and statistical mechanics. At MIT the latter prerequisite is met by requiring students to take "Materials at Equilibrium" before tackling "Kinetic Processes in Materials." To facilitate acquisition of prerequisites, we have included important background material in abbreviated form in Appendices. We have provided a list of our most frequently used symbols, which we have tried to keep in correspondence with general usage in the field. Also included are many exercises (with solutions) that amplify and extend the text.

Bibliography

1. B.R. Bird, W.E. Stewart, and N. Lightfoot. *Transport Phenomena.* John Wiley & Sons, New York, 2nd edition, 2002.

2. D.R. Poirier and G.H. Geiger. *Transport Phenomena in Materials Processing.* The Minerals, Metals and Materials Society, Warrendale, PA, 1994.

ACKNOWLEDGMENTS

We wish to acknowledge generous assistance from many friends and colleagues, especially Dr. John W. Cahn, Dr. Rowland M. Cannon, Prof. Adrian P. Sutton, Prof. Kenneth C. Russell, Prof. Donald R. Sadoway, Dr. Dominique Chatain, Prof. David N. Seidman, and Prof. Krystyn J. Van Vliet. Prof. David T. Wu graciously provided an unpublished draft of his theoretical developments in three-dimensional grain growth which we have incorporated into Chapter 15. We frequently consulted Prof. Paul Shewmon's valuable textbooks on diffusion, and he kindly gave us permission to adapt and reprint Exercise 3.4.

Scores of students have used draft versions of this book in their study of kinetics and many have provided thoughtful criticism that has been valuable in making improvements.

Particular thanks are due Catherine M. Bishop, Valerie LeBlanc, Nicolas Mounet, Gilbert Nessim, Nathaniel J. Quitoriano, Joel C. Williams, and Yi Zhang for their careful reading and suggestions. Ellen J. Siem provided illustrations from her Surface Evolver calculations. Scanning electron microscopy expertise was contributed by Jorge Feuchtwanger. Professors Alex King and Hans-Eckart Exner and Dr. Markus Döblinger furnished unpublished micrographs. Angela M. Locknar expended considerable effort securing hard-to-locate bibliographic sources. Andrew Standeven's care in drafting the bulk of the illustrations is appreciated. Jenna Picceri's and Geraldine Sarno's proofreading skills and work on gathering permissions are gratefully acknowledged. Finally, we wish to thank our editor, Rachel A. Kemper, for her invaluable assistance at all stages of the preparation of this work.

We are fortunate to have so many friends and colleagues who donated their time to help us correct and clarify the text. Although we have striven to remove them all, the remaining errors are the responsibility of the authors.

This textbook has evolved over eight years, during which our extended families have provided support, patience, indulgence, and sympathy. We thank you with all of our hearts.

NOTATION

Notation	Definition
$\vec{a}$	Vector a, the column vector $\vec{a}$
$\hat{a}$	Unit vector a
$\underline{A}$, $[A_{ij}]$	Matrix A, matrix A in component form
$\boldsymbol{A}$	Tensor A of rank two or greater
$\vec{a} \cdot \vec{b}$	Scalar, inner or dot product of $\vec{a}$ and $\vec{b}$
$\vec{a} \times \vec{b}$	Vector, outer or cross product of $\vec{a}$ and $\vec{b}$
$\vec{a}^T$, $\underline{A}^T$	Transpose of $\vec{a}$ or $\underline{A}$
$\mathcal{A}, A, a$	Total amount of A, amount of A per mole or per atom as deduced from context, density of A
$\langle a \rangle$	Average value of a
∇a	Gradient of scalar field a
$\nabla \cdot \vec{A}$	Divergence of vector field $\vec{A}$
$\nabla \cdot \nabla a \equiv \nabla^2 a$	Laplacian of scalar field a
δ_{ij}	Kronecker delta, $\delta_{ij} = 1$ for $i = j$; $\delta_{ij} = 0$ if $i \neq j$
$\mathcal{L}\{a\}$ or $\hat{a}$	Laplace transform of a
$Ca_K^{\bullet}$	Kröger–Vink notation for Ca on K-site with positive effective charge
V'_{Ag}	Kröger–Vink notation for vacancy on Ag-site with negative effective charge
$S_O^{\times}$	Kröger–Vink notation for S on O-site with zero effective charge

SYMBOLS—ROMAN

Symbol	Definition	Units
A	Area	m^2
a, b, c	Lattice constants	m
$\vec{b}, b$	Burgers vector, magnitude of Burgers vector	m
$\vec{b}$	Specific magnetic moment	$\mathrm{A\,m}^{-1}$
c, c_i	Concentration of molecules or atoms, concentration of species i	$\mathrm{m}^{-3}, d = 3$ $\mathrm{m}^{-2}, d = 2$ $\mathrm{m}^{-1}, d = 1$
$D, \boldsymbol{D}$	Mass diffusivity, diffusivity tensor	$\mathrm{m}^2\,\mathrm{s}^{-1}$
D^{XL}	Bulk diffusivity in crystalline material free of line or planar imperfections	$\mathrm{m}^2\,\mathrm{s}^{-1}$
D^B	Boundary diffusivity	$\mathrm{m}^2\,\mathrm{s}^{-1}$
D^D	Dislocation diffusivity	$\mathrm{m}^2\,\mathrm{s}^{-1}$
D^L	Liquid diffusivity	$\mathrm{m}^2\,\mathrm{s}^{-1}$
D^S	Surface diffusivity	$\mathrm{m}^2\,\mathrm{s}^{-1}$
$\widetilde{D}$	Chemical interdiffusivity	$\mathrm{m}^2\,\mathrm{s}^{-1}$
$^{\star}D$	Self-diffusivity in pure material	$\mathrm{m}^2\,\mathrm{s}^{-1}$
$^{\star}D_i$	Self-diffusivity of component i in multicomponent system	$\mathrm{m}^2\,\mathrm{s}^{-1}$
D_i	Intrinsic diffusivity of component i in multicomponent system	$\mathrm{m}^2\,\mathrm{s}^{-1}$
d	Spatial dimensionality	–
E	Activation energy	$\mathrm{J\,atom}^{-1}$
E	Young's elastic modulus	$\mathrm{Pa} = \mathrm{J\,m}^{-3}$
$\vec{E}$	Electric field vector	$\mathrm{V\,m}^{-1}$
$\mathbf{f}$	Correlation factor for atomic jumps in diffusion	—

SYMBOLS—ROMAN

Symbol	Definition	Units
$\mathcal{F}, F, f$	Helmholtz energy, Helmholtz energy per mole (or particle), Helmholtz energy density	$J, J\,mol^{-1}, J\,m^{-3}$
$\vec{F}, \vec{f}$	Force, force per unit length	$N, N\,m^{-1}$
$\mathcal{G}, G, g$	Gibbs energy, Gibbs energy per mole (or particle), Gibbs energy density	$J, J\,mol^{-1}, J\,m^{-3}$
$\mathcal{H}, H, h$	Enthalpy, enthalpy per mole (or particle), enthalpy density	$J, J\,mol^{-1}, J\,m^{-3}$
h	Planck constant	6.626×10^{-34} J s
I_q, I_i	Current of electrical charge, current of species i	$C\ s^{-1}, s^{-1}$
$\hat{\imath}, \hat{\jmath}, \hat{k}$	Unit vectors parallel to Cartesian coordinates x, y, z	—
$\vec{J}, \vec{J_i}$	Flux, flux of species i	$m^{-2}\,s^{-1}$
J	Nucleation rate	$m^{-3}\,s^{-1}$
K	Thermal conductivity	$J\,m^{-1}\,s^{-1}\,K^{-1}$
K	Rate constant	various
k	Boltzmann constant	$1.38 \times 10^{-23} J\,K^{-1}$
$L_{\alpha\beta}$	Onsager coupling coefficient (or tensor)	$m^{-2}\,s^{-1}\,N^{-1}$
$M, \boldsymbol{M}$	Mobility, mobility tensor	various
M_i°	Atomic or molecular weight of species i	$kg\,N_{\circ}^{-1}$
m	Mass	kg
N	Number	—
$\mathcal{N}$	Total number of atoms or molecules in subsystem	—
N_c	Number of components in a solution	—
$N_{\circ}$	Avogadro's number	6.023×10^{23}

SYMBOLS—ROMAN

Symbol	Definition	Units
n	Number per unit volume (concentration)	m^{-3}
$\hat{n}$	Unit normal vector at interface	—
n_d	Instantaneous diffusion-source strength	$m^{-2}, d = 3$ $m^{-1}, d = 2$ number, $d = 1$
P	Pressure	$Pa = J\,m^{-3}$
p	Probability	—
$\vec{p}$	Momentum	$kg\,m\,s^{-1}$
Q	Heat	J
q	Electrical charge	C
R	Radius	m
$\vec{r}$	Position vector relative to origin	m
r, θ, z	Cylindrical coordinates	—
r, θ, ϕ	Spherical coordinates	—
$\mathcal{S}, S, s$	Entropy, entropy per mole (or particle), entropy density	$J\,K^{-1}, J\,K^{-1}mol^{-1}, J\,K^{-1}m^{-3}$
T	Absolute temperature	K
T_m	Absolute melting temperature	K
t	Time	s
$\mathcal{U}, U, u$	Internal energy, internal energy per mole (or particle), internal energy density	$J, J\,mol^{-1}, J\,m^{-3}$
$\vec{u}$	Displacement field	m

SYMBOLS—ROMAN

Symbol	Definition	Units
V	Volume	m^3
$\vec{v}, v$	Velocity, speed	m s^{-1}
v	Specific volume	—
W, w	Work, work per unit volume	J
X_i	Composition variable: mole, atomic, or number fraction of component i	—
x, y, z	Cartesian orthogonal coordinates	m
x_1, x_2, x_3	General coordinates	—
z, z_c	Coordination number, effective coordination number for critical nucleus	—
Z	Partition function	—
Z	Zeldovich factor	—

SYMBOLS—GREEK

Symbol	Definition	Units
Γ', Γ	Atomic or molecular jump frequency for a particular jump, total jump frequency of particle in material	s^{-1}
γ, $\gamma(\hat{n})$	Surface or interfacial tension, work to produce unit interfacial area at constant stress and temperature at orientation $\hat{n}$	$J\,m^{-2}$
γ_i	Activity coefficient of component i	various
δ	Effective thickness of grain boundary or surface layer; diameter of dislocation core	m
η	Diffusion scaling factor, $x/\sqrt{4Dt}$	—
$\hat{\zeta}$	Unit vector tangent to dislocation	—
ε, $\boldsymbol{\varepsilon}$, ε_{ij}	Component of strain, strain tensor, strain tensor in component form	—
κ, κ_1, κ_2	Mean curvature; principal curvatures	m^{-1}
κ_γ	Weighted mean curvature	$J\,m^{-3}$
κ	Thermal diffusivity	$m^2\,s^{-1}$
λ	Wavelength	m
Λ	Elastic-energy shape factor	—
μ	Elastic shear modulus	$Pa = J\,m^{-3}$
μ, μ_i	Chemical potential, chemical potential of species i	J
μ_i^α, μ_i°	Chemical potential of species i in phase α, chemical potential of species i in reference state	J
ν	Frequency	s^{-1}

SYMBOLS—GREEK

Symbol	Definition	Units
ν	Poisson's ratio	—
$\vec{\xi}$	Capillarity vector	$\mathrm{J\,m^{-2}}$
ρ	Density	$\mathrm{kg\,m^{-3}}$
ρ	Electrical conductivity	$\mathrm{C\,V^{-1}m^{-1}\,s^{-1}}$
$\sigma,\,\boldsymbol{\sigma},\,\sigma_{ij}$	Stress, stress tensor, component of stress tensor	$\mathrm{Pa = J\,m^{-3}}$
$\dot{\sigma}$	Rate of entropy production per unit volume	$\mathrm{J\,m^{-3}\,s^{-1}\,K^{-1}}$
τ	Characteristic time	s
Φ_i	Diffusion potential for species i	J
ϕ	Electrical potential	$\mathrm{J\,C^{-1}}$
χ	Site fraction	—
$\Omega,\,\Omega_i,\,\langle\Omega\rangle$	Atomic volume, atomic volume of component i, average atomic volume	$\mathrm{m^3}$
ω	Angular frequency	$\mathrm{s^{-1}}$

CHAPTER 1

INTRODUCTION

Kinetics of Materials is the study of the rates at which various processes occur in materials—knowledge of which is fundamental to materials science and engineering. Many processes are of interest, including changes of size, shape, composition, and structure. In all cases, the system must be out of equilibrium during these processes if they are to occur at a finite rate. Because the departure from equilibrium may be large or small and because the range of phenomena is so broad, the study of kinetics is necessarily complex. This complexity is reduced by introducing approximations such as the assumption of local equilibrium in certain regions of a system, linear kinetics, or mean-field behavior. In much of this book we employ these approximations.

Ultimately, a knowledge of kinetics is valuable because it leads to prediction of the rates of materials processes of practical importance. Analyses of the kinetics of such processes are included here as an alternative to a purely theoretical approach. Some examples of these processes with well-developed kinetic models are the rates of diffusion of a chemical species through a material, conduction of heat during casting, grain growth, vapor deposition, sintering of powders, solidification, and diffusional creep.

The *mechanisms* by which materials change are of prime importance in determining the kinetics. Materials science and engineering emphasizes the role of a material's microstructure. Structure and mechanisms are the yarn from which materials science is woven [1]. Understanding kinetic processes in, for example, crystalline materials relies as much on a thorough familiarity with vacancies, interstitials, grain

Kinetics of Materials. By Robert W. Balluffi, Samuel M. Allen, and W. Craig Carter.
Copyright © 2005 John Wiley & Sons, Inc.

boundaries, and other crystal imperfections as it does on basic mathematics and physics. Extensive discussion of mechanisms is therefore a feature of this book.

We stress rigorous analysis, where possible, and try to build a foundation for understanding kinetics in preparation for concepts and phenomena that fall beyond the scope of this book. Also, in laying a foundation, we have selected basic topics that we feel will be part of the materials science and engineering curriculum for many years, no matter how technical applications of materials change. A comprehensive reading of this book and an effort at solving the exercises should provide the requisite tools for understanding most of the major aspects of kinetic processes in materials.

1.1   THERMODYNAMICS AND KINETICS

In the study of materials science, two broad topics are traditionally distinguished: thermodynamics and kinetics. *Thermodynamics* is the study of equilibrium states in which state variables of a system do not change with time, and *kinetics* is the study of the rates at which systems that are out of equilibrium change under the influence of various forces. The presence of the word *dynamics* in the term *thermodynamics* is therefore misleading but is retained for historical reasons.

In many cases, the study of kinetics concerns itself with the paths and rates adopted by systems approaching equilibrium. Thermodynamics provides invaluable information about the final state of a system, thus providing a basic reference state for any kinetic theory. Kinetic processes in a large system are typically rapid over short length scales, so that equilibrium is nearly satisfied locally; at the same time, longer-length-scale kinetic processes result in a slower approach to global equilibrium. Therefore, much of the machinery of thermodynamics can be applied locally under an assumption of local equilibrium. It is clear, therefore, that the subject of thermodynamics is closely intertwined with kinetics.

1.1.1   Classical Thermodynamics and Constructions of Kinetic Theories

Thermodynamics grew out of studies of systems that exchange energy. Joule and Kelvin established the relationship between work and the flow of heat which resulted in a statement of the first law of thermodynamics. In Clausius's treatise, *The Mechanical Theory of Heat*, the law of energy conservation was supplemented with a second law that defined entropy, a function that can only increase as an isolated system approaches equilibrium [2]. Poincaré coined the term *thermodynamiques* to refer to the new insights that developed from the first and second laws. Development of thermodynamics in the nineteenth century was devoted to practical considerations of work, energy supply, and efficiency of engines. At the end of the nineteenth century, J. Willard Gibbs transformed thermodynamics into the subject of phase stability, chemical equilibrium, and graphical constructions for analyzing equilibrium that is familiar to students of materials science. Gibbs used the first and second laws rigorously, but focused on the medium that stores energy during a work cycle. From Gibbs's careful and rigorous derivations of equilibrium conditions of matter, the modern subjects of chemical and material thermodynamics were born [3]. Modern theories of statistical and continuum thermodynamics—

which comprise the fundamental tools for the science of materials processes—derive from Gibbs's definitive works.

Thermodynamics is precise, but is strictly applicable to phenomena that are unachievable in finite systems in finite amounts of time. It provides concise descriptions of systems at equilibrium by specifying constant values for a small number of intensive parameters.

Two fundamental results from classical thermodynamics that form the basis for kinetic theories in materials are:

1. *If an extensive quantity can be exchanged between two bodies, a condition necessary for equilibrium is that the conjugate potential, which is an intensive quantity, must have the same value throughout both bodies.*

   This can be generalized to adjoining regions in materials. The equilibrium condition, which disallows spatial variations in a potential (e.g., the gradient in chemical potential or pressure), cannot exist in the presence of active physical processes that allow the potential's conjugate extensive density (composition or volume/mole) to change. This implies that a small set of homogeneous potentials can be specified for a heterogeneous system at equilibrium—and therefore the number of parameters required to characterize an equilibrium system is relatively small. For a system that is *not* at equilibrium, *any variation* of potential is permitted. There are an infinite number of ways that a potential [e.g., $\mu_i(x, y, z)$] can differ from its equilibrium value. Thus, the task of describing and analyzing nonequilibrium systems—the subject of the kinetics of materials—is more complex than describing equilibrium systems.

   With this complexity, construction of applicable kinetic theories and techniques requires approximations that must strike a balance between oversimplification and physical reality. Students will benefit from a solid understanding of which approximations are being made, why they are being made, and the fundamental physical principles on which they are founded.

2. *If a closed system is in equilibrium with reservoirs maintaining constant potentials (e.g., P and T), that system has a free-energy function [e.g., $G(P,T)$] that is minimized at equilibrium. Therefore, a necessary condition for equilibrium is that any variation in G must be nonnegative: $(\delta G)_{P,T} \geq 0$.*

   This leads to classical geometrical constructions of thermodynamics, including the common-tangent construction illustrated in Figs. 1.1 and 1.2. For closed systems that are not at equilibrium, a function $G(P,T)$ exists for the entire system—but only as a limiting value for the asymptotic approach to equilibrium. Away from equilibrium, the various parts of a system generally have gradients in potentials and there is no guarantee of the existence of an integrable local free-energy density. The total free energy must decrease to a minimum value at equilibrium. However, there is no recipe for calculating such a total free energy from the constituent parts of a nonequilibrium system. A quandary arises: general statements regarding the approach to equilibrium that are based on thermodynamic functions necessarily involve extrapolation away from equilibrium conditions. However, useful models and theories can be developed from approximate expressions for functions having minima that coincide with the equilibrium thermodynamic quantities and from assumptions of local equilibrium states. This approach is consistent

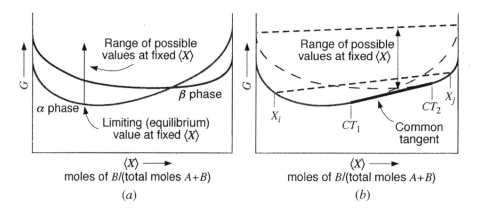

Figure 1.1:   (a) Curves of the minimum free energy of homogeneous α and β phases as a function of average (overall) composition, $\langle X \rangle$, at constant P and T. G is the free energy of 1 mole of solution. Under nonequilibrium conditions, free energies may be larger than those given by the curves, as the vertical arrow indicates. **(b)** Common-tangent construction showing a minimum free-energy curve for a system that may contain α phase, β phase, or both coexisting. The curve consists of a segment on the left which extends to the first point of common tangency, CT_1, a straight line segment between two points of common tangency, CT_1 and CT_2, and a further segment to the right of CT_2. The system at equilibrium consists of a homogeneous α phase up to composition CT_1, a mixture of coexisting α and β phases between CT_1 and CT_2, and a homogeneous β phase beyond CT_2. As in (a), an infinite number of higher free-energy states is possible for the system under nonequilibrium conditions. A subset of these correspond to linear mixtures of homogeneous α and β phases whose free energies are given by the lower dashed line in (b), where X_i and X_j are the α and β phase compositions, respectively. These free energies are plotted, and the energies that can be obtained from such mixtures are bound from above by the dashed line representing a mixture of pure A and pure B. However, in general, energies of the nonequilibrium system are not bound, as indicated in Fig. 1.2.

with the laws of thermodynamics and provides an insightful and organized theoretical foundation for kinetic theories.

Another approach is to build kinetic theories empirically and thus guarantee agreement between theory and experiment. Such theories often can successfully be extended to predict observations of new phenomena. Confidence in such predictions is increased by a thorough understanding of the atomic mechanisms of the system on which the primary observation is made and of the system to which predictions will be applied.

1.1.2   Averaging

Although it may be possible to use computation to simulate atomic motions and atomistic evolution, successful implementation of such a scheme would eliminate the need for much of this book if the computation could be performed in a reasonable amount of time. It is possible to construct interatomic potentials and forces between atoms that approximate real systems in a limited number of atomic configurations. Applying Newton's laws (or quantum mechanics, if required) to calculate the particle motions, the approximate behavior of large numbers of interacting par-

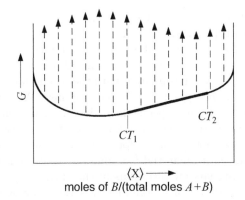

Figure 1.2:   Representation of all possible values of system molar free energy in Fig. 1.1. $\langle X \rangle$ is the average mole fraction of component B.

ticles can be simulated. At the time of this writing, believable approximations that simulate tens of millions of particles for microseconds can be performed by patient researchers with access to state-of-the-art computational facilities. Such calculations have been used to construct thermodynamic data as a foundation on which to build kinetic approximations. However, simulations for systems with sizes and time scales of technological interest do not appear feasible in any current and credible long-range forecast.

Just as statistical mechanics overcomes difficulties arising from large numbers of interacting particles by constructing rigorous methods of averaging, kinetic theory also uses averaging. However, the application of these methods to kinetically evolving systems is precluded because many of the fundamental assumptions of statistical mechanics (e.g., the ergodic hypothesis) do not apply.

Many theories developed in this book are expressed by equations or results involving continuous functions: for example, the spatially variable concentration $c(\vec{r})$. Materials systems are fundamentally discrete and do not have an inherent continuous structure from which continuous functions can be constructed. Whereas the composition at a particular point can be understood both intuitively and as an abstract quantity, a rigorous mathematical definition of a suitable composition function is not straightforward. Moreover, using a continuous position vector $\vec{r}$ in conjunction with a crystalline system having discrete atomic positions may lead to confusion.

The abstract conception of a continuum and the mathematics required to describe it and its variations are discussed below.

1.2   IRREVERSIBLE THERMODYNAMICS AND KINETICS

Irreversible thermodynamics originated in 1931 when Onsager presented a unified approach to irreversible processes [4]. In this book we explore some of Onsager's ideas, but it is worth remarking that his theory applies to systems that are *near* equilibrium.[1] Perhaps *zeroth- and first-order thermodynamics* would be

---

[1] *Near* is unfortunately a rather vague word when applied to the state of a system. Systems that are close to detailed balance where forward processes are almost balanced by backward processes,

more descriptive—but it really doesn't matter as long as the proper application of fundamental principles is retained.

Consider a material or system that is not at equilibrium. Its extensive state variables (total entropy; number of moles of chemical component, i; total magnetization; volume; etc.) will change consistent with the second law of thermodynamics (i.e., with an increase of entropy of all affected systems). At equilibrium, the values of the intensive variables are specified; for instance, if a chemical component is free to move from one part of the material to another and there are no barriers to diffusion, the chemical potential, μ_i, for each chemical component, i, must be uniform throughout the entire material.[2] So one way that a material can be out of equilibrium is if there are spatial variations in the chemical potential: $\mu_i(x, y, z)$. However, a chemical potential of a component is the amount of reversible work needed to add an infinitesimal amount of that component to a system at equilibrium. Can a chemical potential be defined when the system is not at equilibrium? This cannot be done rigorously, but based on decades of development of kinetic models for processes, it is useful to extend the concept of the chemical potential to systems close to, but not at, equilibrium.

Temperature is another quantity defined under equilibrium conditions and for which some doubt may arise regarding its applicability to nonequilibrium systems. Consider a bar of material with ends at different temperatures, as in Fig. 1.3. Suppose that the system has reached a steady state—the amount of heat absorbed by the bar at the hot end is equal to the amount of heat given off at the cold end. The temperature can be thought of as a continuous function, $T(x)$, which is sketched above the bar in Fig. 1.3. An imaginary thermometer placed along the bar would be expected to indicate the plotted temperatures as it moves from point to point. The thermometer in this case is *in local equilibrium* with an infinitesimal region of the bar. What kind of thermometer could perform such a measurement? In order not to affect the measurement, it must have a negligible heat capacity and be unable to conduct any significant amount of heat from the bar. Physically, no such

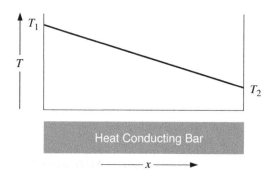

Figure 1.3:   Representation of a one-dimensional thermal gradient.

such as during diffusion, may be regarded as near equilibrium. Quantification of "nearness" has theoretical utility and is a topic of current research [5].

[2]Uniform chemical potential at equilibrium assumes that the component conveys no other work terms, such as charge in an electric field. If other energy-storage mechanisms are associated with a component, a generalized potential (the diffusion potential, developed in Section 2.2.3) will be uniform at equilibrium.

thermometer can exist—nor can a real material be divided infinitesimally. However, this does not mean that one's intuition about the existence of such a function $T(x)$ is wrong; it is reasonable to take a *continuum limit* (see Section 1.3.3) of such an idealized measurement and refer to the temperature at a point.

1.3 MATHEMATICAL BACKGROUND

A few basic physical and mathematical concepts are essential to the study of kinetics, and several of these concepts are introduced below using a mathematical language suited to a discussion of kinetics.

1.3.1 Fields

A field, $f(\vec{r})$, associates a physical quantity with a position, $\vec{r} = (x, y, z)$.[3] A field may be time-dependent: for example, $f(\vec{r}, t)$. The simplest case is a scalar field where the physical quantity can be described with one value at each point. For example, $T(\vec{r}, t)$ can represent the spatial and time-dependent temperature and $\rho(\vec{r}, t)$ the density.[4]

A vector field, such as force, $\vec{F}(\vec{r}, t)$ or flux, $\vec{J}(\vec{r}, t)$, requires specification of a magnitude and a direction in reference to a fixed frame. A rank-two tensor field such as stress, $\boldsymbol{\sigma}(\vec{r}, t)$, relates a vector field to another vector often attached to the material in question: for example, $\boldsymbol{\sigma} = \vec{F}(\vec{r}, t)/\vec{A}$, where $\vec{F}(\vec{r}, t)$ is the force exerted by the stress, $\boldsymbol{\sigma}$, on a virtual area embedded in the material and represented by the vector $\vec{A} = A\hat{n}$, where $\hat{n}$ is the unit normal to the area and A is the magnitude of the area.

Every sufficiently smooth scalar field has an associated natural vector field, which is the gradient field giving the direction and the magnitude of the steepest rate of ascent of the physical quantity associated with the field.[5]

1.3.2 Variations

Consider a *stationary* scalar field such as concentration, $c(\vec{r})$ (see Fig. 1.4), and the rate at which the values of c change as the position is moved with velocity $\vec{v}$ [suppose that an insect is walking on the surface of Fig. 1.4 with velocity $\vec{v}(x, y)$]. The value of c will change with time, t, according to $c(\vec{r} + \vec{v}t)$:

$$c(\vec{r} + \vec{v}t) = c(\vec{r}) + \nabla c \cdot \vec{v}|_{t=0}\, t + \cdots \tag{1.1}$$

where ∇c, the *gradient* of c, is the three-dimensional vector field defined by

$$\nabla c(\vec{r}) = \left(\frac{\partial c}{\partial x}, \frac{\partial c}{\partial y}, \frac{\partial c}{\partial z}\right) = \frac{\partial c}{\partial x}\hat{i} + \frac{\partial c}{\partial y}\hat{j} + \frac{\partial c}{\partial z}\hat{k} \tag{1.2}$$

∇c points in the direction of maximum rate of increase of the scalar field $c(\vec{r})$; the magnitude of the gradient vector is equal to this rate of increase. The instantaneous

---

[3]Here, Cartesian coordinates represent points. Other coordinate systems are employed when appropriate.
[4]However, the definition of each of these quantities depends on the choice of averaging of a physical quantity (e.g., kinetic energy or mass) at a point $\vec{r}$.
[5]The associated natural vector field exists as long as there is a definition of distance (a norm).

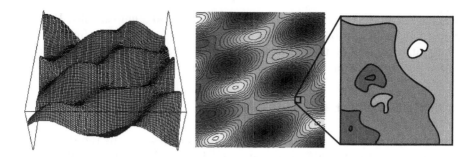

Figure 1.4: Representations of a two-dimensional scalar field are at the left and middle. A familiar example of a scalar field is the altitude of a point as a function of its longitude and latitude—a topographical map, as in the middle figure. It is understood in topographical maps that local averaging is performed. Details in the figure on the right may exist at "microscopic" scales that can be ignored for "macroscopic" model applications.

rate of change of c with respect to t is therefore

$$\frac{dc}{dt} = \nabla c \cdot \vec{v} \tag{1.3}$$

Equation 1.3 can be generalized further by considering a *time-dependent* field $c(\vec{r}, t)$; the instantaneous rate of change of c with velocity $\vec{v}(\vec{r})$ is then

$$\frac{dc}{dt} = \nabla c \cdot \vec{v} + \frac{\partial c}{\partial t} \tag{1.4}$$

Another type of derivative, the divergence of a vector field, is defined in Section 1.3.5.

1.3.3 Continuum Limits and Coarse Graining

Within the small volume of material shown at $\vec{r}$ in Fig. 1.5, a certain quantity of species i is expected. This specifies a concentration for that particular small box; this concentration will be in local equilibrium with some diffusion potential. However, materials are comprised of discrete atoms (molecules), which complicates the definition of local concentration when the volume sampled becomes comparable to the mean distance between atoms being counted. In Fig. 1.5, for the *physical*

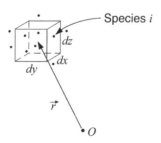

Figure 1.5: Infinitesimal volume, ΔV, with dimensions dx, dy, and dz located at position $\vec{r}$ with respect to the origin at O.

limit of small volume $\Delta V = dx\, dy\, dz$, the expectation of finding N atoms of species i in that volume vanishes as ΔV goes to zero.

Suppose that the atoms are distributed in space as in Fig. 1.5.[6] Consider the behavior of the concentration of i—defined by (number of atoms of type i)/(volume)— as the volume shrinks toward the point where $c(\vec{r})$ is evaluated as in Fig. 1.6. Apparently, the limiting value used intuitively to define the concentration $c(\vec{r})$ is

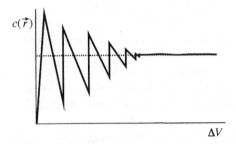

Figure 1.6:   Behavior of the concentration at a point $c(\vec{r})$ as the volume $\Delta V \to 0$.

not a well-defined limit of the function $c(\vec{r}, \Delta V \to 0)$. This conceptual difficulty can be removed by defining a *local convolution function* such as in Fig. 1.7. A continuum limit for the concentration of particles, $c(\vec{r})$, can be defined with a convolution function $\xi(\vec{r} - \vec{r'})$, which specifies, at a position $\vec{r}$, the weight to assign to a particle located at $\vec{r'}$:

$$c(\vec{r}) = \frac{\sum_{i=1}^{\mathcal{N}} \xi(\vec{r} - \vec{r}_i)}{\int_{\mathcal{V}} \xi(\vec{r})\, d^3 r} \tag{1.5}$$

This definition has the correct global behavior for large volumes $\mathcal{V}$ because

$$\int_{\mathcal{V}} c(r)\, d^3 r = \sum_{i=1}^{\mathcal{N}} \frac{\int_{\mathcal{V}} \xi(\vec{r} - \vec{r}_i)\, d^3 r}{\int_{\mathcal{V}} \xi(\vec{r})\, d^3 r} \approx \sum_{i=1}^{\mathcal{N}} \frac{\int_{\mathcal{V}} \xi(\vec{r})\, d^3 r}{\int_{\mathcal{V}} \xi(\vec{r})\, d^3 r} = \mathcal{N} \tag{1.6}$$

where it is assumed that the interference of convolution with the boundary of the domain $\mathcal{V}$ is negligible. Furthermore, the definition, Eq. 1.5, has the correct local behavior: suppose that a volume ΔV (with spatial dimensions large compared to

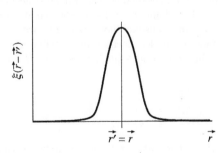

Figure 1.7:   The convolution function $\xi(\vec{r} - \vec{r'})$ accomplishes coarse graining of an object located at $\vec{r} = \vec{r'}$.

[6]Nicolas Mounet contributed significantly to the development of coarse graining in this section.

the width of the convolution function) contains a single isolated particle (i.e., the particle in ΔV is "far" from all others). Also, let the particle's i index be 1, with its position $\vec{r_1}$ at the center of ΔV; then

$$
\int_{\Delta V} c(r)\, d^3 r = \frac{\int_{\Delta V} \xi(\vec{r} - \vec{r_1})\, d^3 r + \sum_{i=2}^{\mathcal{N}} \int_{\Delta V} \xi(\vec{r} - \vec{r_i})\, d^3 r}{\int_{\mathcal{V}} \xi(\vec{r})\, d^3 r}
$$
$$
\approx \frac{\int_{\Delta V} \xi(\vec{r} - \vec{r_1})\, d^3 r}{\int_{\mathcal{V}} \xi(\vec{r})\, d^3 r} \approx 1 \tag{1.7}
$$

Defined by Eq. 1.5, $c(\vec{r})$ becomes a *coarse-grained* representation of the discrete particle positions.

In one dimension, an exemplary choice for a convolution function is $\xi(x - x_i) = \exp[(x - x_i)^2/B^2]$, where B is the characteristic coarse-grained length. With this choice, the coarse-grained one-dimensional concentration is

$$
c(x) = \frac{\sum_{i=1}^{\mathcal{N}} e^{(x - x_i)^2/B^2}}{\sqrt{\pi}\, B} \tag{1.8}
$$

Examples with different characteristic coarse-grain lengths are shown in Fig. 1.8.

In this book, it is assumed that the continuum limits exist and coarse-grained functions can be obtained that do not depend significantly on the choice of ξ.

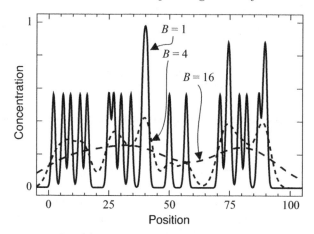

Figure 1.8: Example of one-dimensional coarse-grained concentrations of discrete data. Twenty-two atoms were placed randomly on a discrete lattice at positions x_i, $0 < x_i < 100$. The concentration curves are continuous and have areas that are approximately equal to the number of atoms in the random sample. Each atom contributes a unit area to the coarse-grained $c(x)$. Broader convolution functions (higher values of B) produce greater degrees of coarse graining.

1.3.4   Fluxes

A flux of i, $\vec{J_i}(\vec{r})$, describes the rate at which i flows through a unit area fixed with respect to a specified coordinate system. Let $\Delta \vec{A}$ be an oriented area, equal to $\hat{n} \Delta A = (A_x, A_y, A_z)$ in a Cartesian system.[7] If $\dot{M_i}$ is a smooth function that

---

[7] $\Delta A \equiv |\Delta \vec{A}|$ and $\hat{n} = \Delta \vec{A}/|\Delta \vec{A}|$

defines the rate at which i flows through area $\Delta \vec{A}$,

$$\dot{M}_i(\Delta \vec{A}) \propto |\Delta \vec{A}| \tag{1.9}$$

The proportionality factor must be a vector field $\vec{J}_i$:

$$\dot{M}_i(\Delta \vec{A}) = \vec{J}_i \cdot \Delta \vec{A} \tag{1.10}$$

This defines the local flux $\vec{J}_i(\vec{r})$ as the continuum limit of

$$\frac{\dot{M}_i(\Delta \vec{A})}{\Delta A} = \vec{J}_i(\vec{r}) \cdot \hat{n} \tag{1.11}$$

1.3.5 Accumulation

The amount of i that accumulates in a volume $\Delta V = dx\,dy\,dz$ (with outward-oriented normals) in a Cartesian system during the time interval δt is

$$\Delta M_i = (i \text{ that flowed in during } \delta t) - (i \text{ that flowed out during } \delta t) \\ + (i \text{ produced inside during } \delta t) \tag{1.12}$$

An expression for the accumulation can be written with the aid of Fig. 1.9, generalized to include the y and z components of the flux $\vec{J}$:

$$\begin{aligned} \delta M_i = &- \vec{J}_i(x + dx/2, 0, 0) \cdot \hat{i}\, dy\, dz\, \delta t + \vec{J}_i(x - dx/2, 0, 0) \cdot \hat{i}\, dy\, dz\, \delta t \\ &- \vec{J}_i(0, y + dy/2, 0) \cdot \hat{j}\, dz\, dx\, \delta t + \vec{J}_i(0, y - dy/2, 0) \cdot \hat{j}\, dz\, dx\, \delta t \\ &- \vec{J}_i(0, 0, z + dz/2) \cdot \hat{k}\, dx\, dy\, \delta t + \vec{J}_i(0, 0, z - dz/2) \cdot \hat{k}\, dx\, dy\, \delta t \\ &+ \dot{\rho}_i(\vec{r})\, \Delta V\, \delta t \end{aligned} \tag{1.13}$$

where $\dot{\rho}_i(\vec{r})$ is the rate of production of the density of i in ΔV. Expanding to first order in dx, dy, dz, subtracting, and using the continuum limit yields

$$\frac{\partial c_i}{\partial t} = -\nabla \cdot \vec{J}_i + \dot{\rho}_i \tag{1.14}$$

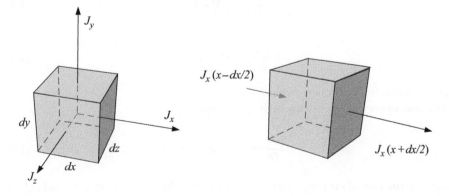

Figure 1.9: Accumulation of an extensive quantity arising from a divergence of its flux $\vec{J} = (J_x, J_y, J_z)$.

where the quantity $\nabla \cdot \vec{J_i}$ is the *divergence* of $\vec{J_i}$, which for a general flux $\vec{J}$ is

$$\nabla \cdot \vec{J} = \frac{\partial(\vec{J} \cdot \hat{\imath})}{\partial x} + \frac{\partial(\vec{J} \cdot \hat{\jmath})}{\partial y} + \frac{\partial(\vec{J} \cdot \hat{k})}{\partial z} = \frac{\partial J_x}{\partial x} + \frac{\partial J_y}{\partial y} + \frac{\partial J_z}{\partial z} \qquad (1.15)$$

This is the rate at which the flux causes the density of the quantity comprising the flux to decrease. *The rate of accumulation of the extensive quantity's density is therefore* minus *the divergence of the flux of that quantity* plus *the rate of production.*

Alternatively, Eq. 1.14 could be derived directly from

$$\begin{aligned}
\dot{M}_i &= -\int_{\mathcal{B}(\Delta V)} \vec{J_i} \cdot d\vec{A} + \int_{\Delta V} \dot{\rho_i}\, dV \\
&= -\int_{\mathcal{B}(\Delta V)} \vec{J_i} \cdot \hat{n}\, dA + \int_{\Delta V} \dot{\rho_i}\, dV \\
&= \int_{\Delta V} (-\nabla \cdot \vec{J_i} + \dot{\rho_i})\, dV
\end{aligned} \qquad (1.16)$$

where $\mathcal{B}(\Delta V)$ is the oriented surface around ΔV and the divergence theorem (Gauss's theorem),

$$\int_{\mathcal{B}(\Delta V)} \vec{J} \cdot \hat{n}\, dA = \int_{\Delta V} \nabla \cdot \vec{J}\, dV \qquad (1.17)$$

has been applied. Note that the divergence theorem has a geometrical interpretation. If the volume is comprised of many neighboring cells, the total accumulation in the volume is the sum of accumulations in all the cells; see the right-hand side of Eq. 1.17. Each cell's accumulation arises from the flux at its surfaces. However, when cells share an interface, they have opposite normal vectors, and the flux terms, $\vec{J} \cdot \hat{n}$, cancel. In a group of abutting cells, the fluxes across the interior interfaces cancel so that the only contribution is due to the exterior surfaces.

1.3.6 Conserved and Nonconserved Quantities

A conserved quantity cannot be created or destroyed and therefore has no sources or sinks; for conserved quantities such as atomic species i or internal energy U,

$$\frac{\partial c_i}{\partial t} = -\nabla \cdot \vec{J_i} \qquad (1.18)$$

$$\frac{\partial u}{\partial t} = -\nabla \cdot \vec{J_u} \qquad (1.19)$$

where u is the internal energy density.[8]

For nonconserved quantities such as entropy, S,

$$\frac{\partial s}{\partial t} = -\nabla \cdot \vec{J_s} + \dot{\sigma} \qquad (1.20)$$

where $\dot{\sigma}$ is the rate of entropy production per unit volume. Entropy flux and entropy production are examined in Chapter 2.

---

[8]Barring processes such as nuclear decay, transmutation, or implantation by ion irradiation.

1.3.7   Matrices, Tensors, and the Eigensystem

In this section we provide a brief review of topics in linear algebra and tensor property relations that are used frequently throughout the book. Nye's book on tensor properties contains a complete overview and is also a valuable resource [6].

A general set of linear equations for the quantities y_i $(i = 1, 2, 3, \ldots, n)$ in terms of variables x_j $(j = 1, 2, 3, \ldots, m)$ can be written as

$$
\begin{aligned}
y_1 &= M_{11}x_1 + M_{12}x_2 + \cdots + M_{1m}x_m \\
y_2 &= M_{21}x_1 + M_{22}x_2 + \cdots + \quad \vdots \\
y_3 &= \ldots \qquad\qquad \ddots \qquad\quad \vdots \\
\vdots & \\
y_n &= M_{n1}x_1 + \ldots \qquad\qquad M_{nm}x_n
\end{aligned} \tag{1.21}
$$

or

$$
y_i = \sum_{j=1}^{m} M_{ij}x_j \quad \text{for } i = 1, 2, \ldots, n \tag{1.22}
$$

The M_{ij} are the elements of a matrix, $\underline{M}$, that multiplies a vector $\vec{x}$ and produces the result, $\vec{y} = \underline{M}\vec{x}$, or in component form,

$$
\begin{bmatrix} y_1 \\ y_2 \\ \vdots \\ \vdots \\ y_n \end{bmatrix} = \begin{bmatrix} M_{11} & M_{12} & \cdots & M_{1m} \\ M_{21} & M_{22} & & \vdots \\ \vdots & & \ddots & \\ \vdots & & & \\ M_{n1} & \cdots & & M_{nn} \end{bmatrix} \begin{bmatrix} x_1 \\ x_2 \\ \vdots \\ \vdots \\ x_n \end{bmatrix} \tag{1.23}
$$

In this book, vector quantities such as $\vec{x}$ and $\vec{y}$ above are normally *column vectors*. When necessary, row vectors are indicated by use of the transpose (e.g., $\vec{b}^T$). If the components of $\vec{x}$ and $\vec{y}$ refer to coordinate axes [e.g., orthogonal coordinate axes (ξ_1, ξ_2, ξ_3) aligned with a particular choice of "right," "forward," and "up" in a laboratory], the square matrix $\underline{M}$ is a rank-two tensor.[9] In this book we denote tensors of rank two and higher using boldface symbols (i.e., $\boldsymbol{M}$). If $\vec{x}$ is an applied force and $\vec{y}$ is the material response to the force (such as a flux), $\boldsymbol{M}$ is a rank-two material-property tensor. For example, the full anisotropic form of Ohm's law gives a charge flux $\vec{J_q}$ in terms of an applied electric field $\vec{E}$ as

$$
\vec{J_q} = \boldsymbol{\chi}\vec{E}
$$
$$
\begin{bmatrix} J_{q1} \\ J_{q2} \\ J_{q3} \end{bmatrix} = \begin{bmatrix} \chi_{11} & \chi_{12} & \chi_{13} \\ \chi_{21} & \chi_{22} & \chi_{23} \\ \chi_{31} & \chi_{32} & \chi_{33} \end{bmatrix} \begin{bmatrix} E_1 \\ E_2 \\ E_3 \end{bmatrix} \tag{1.24}
$$

$\boldsymbol{\chi}$ is the rank-two *conductivity tensor* for a particular material. In Eq. 1.24, $\boldsymbol{\chi}$ is the material property that relates both the magnitude of "effect" $\vec{J_q}$ to the "cause" $\vec{E}$ and their directions—$\vec{J_q}$ is not necessarily parallel to $\vec{E}$.

---

[9] $\underline{M}$ is rank two because it relates two different sets of vector components in a prescribed way: that is, the components of $\vec{x}$ are mapped into components of $\vec{y}$ by the tensor $\boldsymbol{M}$. The vectors $\vec{x}$ and $\vec{y}$ refer to a single coordinate system and are called rank-one tensors.

The physical law in Eq. 1.24 can be expressed as an inverse relationship:

$$\vec{E} = \rho \vec{J}$$

$$\begin{bmatrix} E_x \\ E_y \\ E_z \end{bmatrix} = \begin{bmatrix} \rho_{xx} & \rho_{xy} & \rho_{xz} \\ \rho_{yx} & \rho_{yy} & \rho_{yz} \\ \rho_{zx} & \rho_{zy} & \rho_{zz} \end{bmatrix} \begin{bmatrix} J_{q1} \\ J_{q2} \\ J_{q3} \end{bmatrix} \tag{1.25}$$

where the resistivity tensor, ρ, is the inverse of the conductivity tensor (i.e., $\rho = \chi^{-1}$).[10]

Many materials properties are anisotropic: they vary with direction in the material. When anisotropic materials properties are characterized, the values used to represent the properties must be specified with respect to particular coordinate axes. If the material remains fixed and the properties are specified with respect to some new set of coordinate axes, the properties themselves must remain invariant. The way in which the properties *are described* will change, but the properties themselves (i.e., the material behavior) will not. The *components* of tensor quantities transform in specified ways with changes in coordinate axes; such *transformation laws* distinguish tensors from matrices [6].

For a particular material response or applied field, particular choices of coordinate axis orientations may be especially convenient (e.g., axes aligned with crystal lattice vectors). Linear transformations—such as rotations, reflections, and affine distortions— can be performed on vector forces and responses by matrix multiplication to describe force–response relations in different coordinate systems. For instance, a vector $\vec{E}$ can be transformed between "old" and "new" coordinate systems by a matrix $\underline{A}$:

$$\underline{A}^{\text{old}\rightarrow\text{new}} \vec{E}^{\text{old}} = \vec{E}^{\text{new}}$$
$$\underline{A}^{\text{new}\rightarrow\text{old}} \vec{E}^{\text{new}} = \vec{E}^{\text{old}} \tag{1.26}$$

A simple proof will show that

$$\underline{A}^{\text{old}\rightarrow\text{new}} = \underline{A}^{\text{new}\rightarrow\text{old}^{-1}}$$
$$\underline{A}^{\text{new}\rightarrow\text{old}} = \underline{A}^{\text{old}\rightarrow\text{new}^{-1}} \tag{1.27}$$

i.e., $\underline{A}^{\text{old}\rightarrow\text{new}}$ is the *inverse* of $\underline{A}^{\text{new}\rightarrow\text{old}}$, and vice versa.

It is often convenient to select the coordinate system for which the only nonzero elements of the property tensor lie on its diagonal. This is the *eigensystem*. To find the eigensystem, the general rules for transformation of a tensor must be identified. The transformation of Ohm's law (Eq. 1.24) illustrates the way in which the material properties tensor χ^{old} transforms to χ^{new} and serves to demonstrate the general rule for transforming rank-two tensors:

$$\text{in old coordinate system: } \vec{J}_q^{\text{old}} = \chi^{\text{old}} \vec{E}^{\text{old}}$$
$$\text{in new coordinate system: } \vec{J}_q^{\text{new}} = \chi^{\text{new}} \vec{E}^{\text{new}} \tag{1.28}$$

---

[10]Indices appear as 1, 2, 3 in Eq. 1.24 and as x, y, z in Eq. 1.25. The numerical indices represent any three-dimensional coordinate system (including Cartesian), and the indices in Eq. 1.25 are strictly Cartesian.

The relationship between χ^{old} and χ^{new} can be found by applying the transformations in Eqs. 1.26 to the expressions for Ohm's law in both coordinate systems. For the first equation in Eq. 1.28, using the transformations in Eqs. 1.26,

$$\underline{A}^{\text{new}\rightarrow\text{old}}\vec{J}_q^{\text{new}} = \chi^{\text{old}}\underline{A}^{\text{new}\rightarrow\text{old}}\vec{E}^{\text{new}} \tag{1.29}$$

and for the second equation in Eq. 1.28,

$$\underline{A}^{\text{old}\rightarrow\text{new}}\vec{J}_q^{\text{old}} = \chi^{\text{new}}\underline{A}^{\text{old}\rightarrow\text{new}}\vec{E}^{\text{old}} \tag{1.30}$$

Left-multiplying by the inverse transformations,

$$\underline{A}^{\text{old}\rightarrow\text{new}}\underline{A}^{\text{new}\rightarrow\text{old}}\vec{J}_q^{\text{new}} = \vec{J}_q^{\text{new}} = \underline{A}^{\text{old}\rightarrow\text{new}}\chi^{\text{old}}\underline{A}^{\text{new}\rightarrow\text{old}}\vec{E}^{\text{new}}$$

$$\text{and} \tag{1.31}$$

$$\underline{A}^{\text{new}\rightarrow\text{old}}\underline{A}^{\text{old}\rightarrow\text{new}}\vec{J}_q^{\text{old}} = \vec{J}_q^{\text{old}} = \underline{A}^{\text{new}\rightarrow\text{old}}\chi^{\text{new}}\underline{A}^{\text{old}\rightarrow\text{new}}\vec{E}^{\text{old}}$$

Therefore,

$$\chi^{\text{new}} = \underline{A}^{\text{old}\rightarrow\text{new}}\chi^{\text{old}}\underline{A}^{\text{new}\rightarrow\text{old}}$$

$$\text{and} \tag{1.32}$$

$$\chi^{\text{old}} = \underline{A}^{\text{new}\rightarrow\text{old}}\chi^{\text{new}}\underline{A}^{\text{old}\rightarrow\text{new}}$$

This pattern—a rank-one tensor is transformed by a single matrix multiplication and a rank-two tensor is transformed by two matrix multiplications—holds for tensors of any rank. If $\underline{A}$ is an orthogonal transformation, such as a rigid rotation or a rigid rotation combined with a reflection, its inverse is its transpose. For example, if $\underline{R}$ is a rotation, $R_{ij}R_{ji} = \delta_{ij}$, where δ_{ij} is the Kronecker delta, defined as

$$\delta_{ij} = \begin{cases} 1 & \text{if} \quad i = j \\ 0 & \text{if} \quad i \neq j \end{cases} \tag{1.33}$$

i.e., δ_{ij} is the index form of the identity matrix.

Square matrices and tensors can be characterized by their *eigenvalues* and *eigenvectors*. If $\underline{M}$ is an $n \times n$ square matrix (or tensor), there is a set of n special vectors, $\vec{e}$, each with its own special scalar multiplier λ for which matrix multiplication of a vector is equivalent to scalar multiplication of a vector:

$$\underline{M}\vec{e} = \lambda\vec{e}$$

$$\text{or} \tag{1.34}$$

$$(\underline{M} - \lambda\underline{\mathcal{I}})\vec{e} = \vec{0}$$

where $\vec{0}$ is a vector of zeros that has the same number of entries, n, as $\vec{e}$ and $\underline{\mathcal{I}}$ is the $n \times n$ *identity matrix* (i.e., $\underline{\mathcal{I}}$ has ones along its major diagonal and zeros elsewhere). The solutions λ_i and $\vec{e}_i$ are the eigenvalues and eigenvectors of $\underline{M}$. In general, there are n unique $\lambda_i{:}\vec{e}_i$ pairs for any $\underline{M}$. The eigenvectors of $\underline{M}$ can be interpreted geometrically as the set of vectors that do not change direction when multiplied by $\underline{M}$—instead, they are scaled by a constant λ. The eigenvalues can be determined from the polynomial equation for λ:

$$\det(\underline{M} - \lambda\underline{\mathcal{I}}) = 0 \tag{1.35}$$

which is a requirement that the homogeneous equation, Eq. 1.34, has a nontrivial solution. After the eigenvalues have been determined, the directions of the eigenvectors $\vec{e}$ can be determined by solving Eq. 1.34.

A rank-two property tensor is diagonal in the coordinate system defined by its eigenvectors. Rank-two tensors transform like 3×3 square matrices. The general rule for transformation of a square matrix into its diagonal form is

$$\begin{bmatrix} \text{diagonalized} \\ \text{matrix} \end{bmatrix} = \begin{bmatrix} \text{eigenvector} \\ \text{column} \\ \text{matrix} \end{bmatrix}^{-1} \begin{bmatrix} \text{square} \\ \text{matrix} \end{bmatrix} \begin{bmatrix} \text{eigenvector} \\ \text{column} \\ \text{matrix} \end{bmatrix} \qquad (1.36)$$

where the ith member of the diagonal matrix is the eigenvalue corresponding to the eigenvector used for the ith column vector of the transformation matrix. Nearly all rank-two property tensors can be represented by 3×3 symmetric matrices and necessarily have real eigenvalues.

Bibliography

1. S.M. Allen and E.L. Thomas. *The Structure of Materials*. John Wiley & Sons, New York, 1999.

2. R. Clausius. *The Mechanical Theory of Heat: With Its Applications to the Steam-Engine and to the Physical Properties of Bodies*. Van Voorst, London, 1867.

3. J.W. Gibbs. On the equilibrium of heterogeneous substances (1876). In *Collected Works*, volume 1. Longmans, Green, and Co., New York, 1928.

4. L. Onsager. Reciprocal relations in irreversible processes. II. *Phys. Rev.*, 38(12):2265–2279, 1931.

5. W.C. Carter, J.E. Taylor, and J.W. Cahn. Variational methods for microstructural evolution. *JOM*, 49(12):30–36, 1997.

6. J.F. Nye. *Physical Properties of Crystals*. Oxford University Press, Oxford, 1985.

EXERCISES

1.1 The concentration at any point in space is given by

$$c = A\,(xy + yz + zx) \qquad (1.37)$$

where $A = \text{constant}$.

(a) Find the cosines of the direction in which c changes most rapidly with distance from the point $(1, 1, 1)$.

(b) Determine the maximum rate of change of concentration at that point.

Solution.

(a) The direction of maximum rate of change is along the gradient vector ∇c given by

$$\nabla c = A\,[(y + z)\hat{\imath} + (z + x)\hat{\jmath} + (x + y)\hat{z}] \qquad (1.38)$$

Therefore,

$$\nabla c(1, 1, 1) = 2A\,(\hat{\imath} + \hat{\jmath} + \hat{z}) \qquad (1.39)$$

and the direction cosines are $\left[1/\sqrt{3}, 1/\sqrt{3}, 1/\sqrt{3}\right]$.

(b) The maximum rate of change of c is then

$$|\nabla c(1,1,1)| = 2A\sqrt{3} \tag{1.40}$$

1.2 Consider the radially symmetric flux field

$$\vec{J} = \frac{\vec{r}}{r^3} \tag{1.41}$$

where $\vec{r} = x\hat{i} + y\hat{j} + z\hat{k}$.

(a) Show that the total flux through any closed surface that does not enclose the origin vanishes.

(b) Show that the flux through any sphere centered at the origin is independent of the sphere radius.

Solution.

(a) The problem is most easily solved using the divergence theorem:

$$\int_S \vec{J} \cdot \hat{n} \, dA = \int_V \nabla \cdot \vec{J} \, dV \tag{1.42}$$

Consider first the divergence of radially symmetric vector fields of a general form, including the present field as a special case, i.e.,

$$\vec{F} = \frac{\vec{r}}{r^n} \tag{1.43}$$

For such fields

$$\nabla \cdot \vec{F} = \frac{1}{r^2} \frac{\partial}{\partial r} \left(r^2 F_r \right) = \frac{1}{r^2} \frac{\partial}{\partial r} \left(r^2 r^{1-n} \right) = \frac{3-n}{r^n} \tag{1.44}$$

In this case, $n = 3$ and the divergence of $\vec{J}$ in Eq. 1.42 is zero if the singularity at $r = 0$ is avoided. Therefore, if the closed surface does not include the origin,

$$\int_V \nabla \cdot \vec{J} \, dV = 0 \tag{1.45}$$

and the total flux through the surface, $\int_S \vec{J} \cdot \hat{n} \, dA$, is also zero.

(b) When the closed surface does enclose the origin, the total flux through the surface does not vanish. For a sphere of radius R,

$$\int_S \vec{J} \cdot \hat{n} \, dA = \int_S \frac{\vec{R}}{R^3} \cdot \frac{\vec{R}}{R} \, dA = \frac{1}{R^2} \int_S dA = 4\pi \tag{1.46}$$

Therefore the total flux is independent of R and equal to 4π.

1.3 Suppose that the flux of some substance i is given by the vector field

$$\vec{J_i} = A \left(x\hat{i} + y\hat{j} \right) \tag{1.47}$$

where $A = $ constant. Find the rate, $\dot{M_i}$, at which i flows through the hemispherical surface of the unit sphere

$$x^2 + y^2 + z^2 = 1 \tag{1.48}$$

which lies above the (x, y) plane where $z \geq 0$.

Solution.

$$\dot{M}_i = \int_{\text{hemi}} \vec{J}_i \cdot d\vec{A} = \int_{\text{hemi}} \vec{J}_i \cdot \hat{n} \, dA \tag{1.49}$$

For the hemisphere,

$$\hat{n} = x\hat{i} + y\hat{j} + z\hat{k} \tag{1.50}$$

Also, the integral may be converted to an integral over the projection of the hemisphere on the (x, y) plane (denoted by P) by noting that

$$\hat{k} \cdot \hat{n} \, dA = dx \, dy \tag{1.51}$$

so that

$$\dot{M}_i = \iint_P \vec{J}_i \cdot \hat{n} \frac{dx \, dy}{z} = A \iint_P \frac{x^2 + y^2}{\sqrt{1 - x^2 - y^2}} \, dx \, dy \tag{1.52}$$

Converting to polar coordinates and integrating over P,

$$\dot{M}_i = 4A \int_0^{\pi/2} \int_0^1 \frac{r^2}{\sqrt{1 - r^2}} \, r \, dr \, d\theta = \frac{4\pi A}{3} \tag{1.53}$$

1.4 The matrix $\underline{A}$ is given by

$$\underline{A} = \begin{bmatrix} 8 & -1 & -1 \\ 1 & 6 & 0 \\ -5 & 0 & 2 \end{bmatrix} \tag{1.54}$$

(a) Find the eigenvalues and corresponding eigenvectors of $\underline{A}$.

(b) Find matrices $\underline{P}$ and $\underline{P}^{-1}$ such that $\underline{P}^{-1}\underline{A}\underline{P}$ is a diagonal matrix.

Note: The tedium of completing such exercises, as well as following many derivations in this book, is reduced by the use of symbolic mathematical software. We recommend that students gain a working familiarity with at least one package such as *Mathematica*®, *MATLAB*®, *Mathcad*®, or the public-domain package *MAXIMA*.

Solution.

(a) The characteristic equation of $\underline{A}$ is given by Eq. 1.35 as

$$\lambda^3 - 16\lambda^2 + 72\lambda - 68 = 0 \tag{1.55}$$

The eigenvalues are solutions to the characteristic equation, giving

$$\lambda_1 = 8.36258 \quad \lambda_2 = 6.35861 \quad \lambda_3 = 1.27881 \tag{1.56}$$

The eigenvectors corresponding to the eigenvalues are

$$v_1 = \begin{bmatrix} -1.27252 \\ -0.538613 \\ 1 \end{bmatrix} \quad v_2 = \begin{bmatrix} -0.871722 \\ -2.43084 \\ 1 \end{bmatrix} \quad v_3 = \begin{bmatrix} 0.144238 \\ -0.0305511 \\ 1 \end{bmatrix} \tag{1.57}$$

Note that these eigenvectors are of arbitrary length.

(b) From Eq. 1.36 it is seen that the desired matrix is 3×3 and has the three eigenvectors as its columns:

$$\underline{P} = \begin{bmatrix} -1.27252 & -0.871722 & 0.144238 \\ -0.538613 & -2.43084 & -0.0305511 \\ 1 & 1 & 1 \end{bmatrix} \tag{1.58}$$

The inverse of $\underline{P}$ may be calculated as

$$\underline{P}^{-1} = \begin{bmatrix} -0.832149 & 0.352221 & 0.130788 \\ 0.176139 & -0.491171 & -0.0404117 \\ 0.65601 & 0.13895 & 0.909624 \end{bmatrix} \tag{1.59}$$

By substitution it is readily verified that Eq. 1.36 is obeyed:

$$\underline{P}^{-1}\underline{A}\underline{P} = \begin{bmatrix} \lambda_1 = 8.36258 & 0 & 0 \\ 0 & \lambda_2 = 6.35861 & 0 \\ 0 & 0 & \lambda_3 = 1.27881 \end{bmatrix} \tag{1.60}$$

PART I

MOTION OF ATOMS AND MOLECULES BY DIFFUSION

There are two arenas for describing diffusion in materials, macroscopic and microscopic. Theories of macroscopic diffusion provide a framework to understand particle fluxes and concentration profiles in terms of phenomenological coefficients and driving forces. Microscopic diffusion theories provide a framework to understand the physical basis of the phenomenological coefficients in terms of atomic mechanisms and particle jump frequencies.

We start with the macroscopic aspects of diffusion. The components in a system out of equilibrium will generally experience net forces that can generate corresponding fluxes of the components (diffusion fluxes) as the system tries to reach equilibrium. The first step (Chapter 2) is the derivation of the general coupling between these forces and fluxes using the methods of irreversible thermodynamics. From general results derived from irreversible thermodynamics, specific driving forces and fluxes in various systems of importance in materials science are obtained in Chapter 3. These forces and fluxes are used to derive the differential equations that govern the evolution of the concentration fields produced by these fluxes (Chapter 4). Mathematical methods to solve these equations in various systems under specified boundary and initial conditions are explored in greater depth in Chapter 5. Finally, diffusion in multicomponent systems is treated in Chapter 6.

Microscopic and mechanistic aspects of diffusion are treated in Chapters 7–10. An expression for the basic jump rate of an atom (or molecule) in a condensed system is obtained and various aspects of the displacements of migrating particles are described (Chapter 7). Discussions are then given of atomistic models for diffusivities and diffusion in bulk crystalline materials (Chapter 8), along line and planar imperfections in crystalline materials (Chapter 9), and in bulk noncrystalline materials (Chapter 10).

CHAPTER 2

IRREVERSIBLE THERMODYNAMICS AND COUPLING BETWEEN FORCES AND FLUXES

The foundation of irreversible thermodynamics is the concept of entropy production. The consequences of entropy production in a dynamic system lead to a natural and general coupling of the driving forces and corresponding fluxes that are present in a nonequilibrium system.

2.1 ENTROPY AND ENTROPY PRODUCTION

The existence of a conserved internal energy is a consequence of the first law of thermodynamics. Numerical values of a system's energy are always specified with respect to a reference energy. The existence of the entropy state function is a consequence of the second law of thermodynamics. In classical thermodynamics, the value of a system's entropy is not directly measurable but can be calculated by devising a reversible path from a reference state to the system's state and integrating $dS = \delta q_{rev}/T$ along that path. For a nonequilibrium system, a reversible path is generally unavailable. In statistical mechanics, entropy is related to the number of microscopic states available at a fixed energy. Thus, a state-counting device would be required to compute entropy for a particular system, but no such device is generally available for the irreversible case.

To obtain a local quantification of entropy in a nonequilibrium material, consider a continuous system that has gradients in temperature, chemical potential, and other intensive thermodynamic quantities. Fluxes of heat, mass, and other extensive quantities will develop as the system approaches equilibrium. Assume that

Kinetics of Materials. By Robert W. Balluffi, Samuel M. Allen, and W. Craig Carter. **23**
Copyright © 2005 John Wiley & Sons, Inc.

the system can be divided into small contiguous cells at which the temperature, chemical potential, and other thermodynamic potentials can be approximated by their average values. The *local equilibrium* assumption is that the thermodynamic state of each cell is specified and in equilibrium with the local values of thermo-dynamic potentials. If local equilibrium is assumed for each microscopic cell even though the entire system is out of equilibrium, then Gibbs's fundamental relation, obtained by combining the first and second laws of thermodynamics,

$$d\mathcal{U} = T\, d\mathcal{S} - dW + \sum_{i=1}^{N_c} \mu_i \, d\mathcal{N}_i \tag{2.1}$$

can be used to calculate changes in the local equilibrium states as a result of evo-lution of the spatial distribution of thermodynamic potentials. U and S are the internal energy and entropy of a cell, dW is the work (other than chemical work) done by a cell, N_i is the number of particles of the ith component of the possible N_c components, and μ_i is the chemical potential of the ith component. μ_i depends upon the energetics of the chemical interactions that occur when a particle of i is added to the system and can be expressed as a general function of the atomic fraction X_i:

$$\mu_i = \mu_i^\circ + kT \ln(\gamma_i X_i) \tag{2.2}$$

The activity coefficient γ_i generally depends on X_i but, according to Raoult's law, is approximately unity for $X_i \approx 1$.

Dividing $d\mathcal{U}$ through by a constant reference cell volume, $V_\circ$,

$$T\, ds = du + dw - \sum_{i=1}^{N_c} \mu_i \, dc_i \tag{2.3}$$

where all extensive quantities are now on a per unit volume basis (i.e., densities).[1] For example, $v = \mathcal{V}/V_\circ$ is the cell volume relative to the reference volume, $V_\circ$, and $c_i = \mathcal{N}_i/V_\circ$ is the concentration of component i. The work density, dw, includes all types of (nonchemical) work possible for the system. For instance, the elastic work density introduced by small-strain deformation is $dw = -\frac{1}{2} \sum_i \sum_j \sigma_{ij} \, d\varepsilon_{ij}$ (where σ_{ij} and ε_{ij} are the stress and strain tensors), which can be further separated into hydrostatic and deviatoric terms as $dw = P\, dv - \sum_i \sum_j \tilde{\sigma}_{ij} \, d\tilde{\varepsilon}_{ij}$ (where $\tilde{\sigma}$ and $\tilde{\varepsilon}$ are the deviatoric stress and strain tensors, respectively). The elastic work density therefore includes a work of expansion $P\, dv$. Other work terms can be included in Eq. 2.3, such as electrostatic potential work, $dw = -\phi\, dq$ (where ϕ is the electric potential and q is the charge density); interfacial work, $dw = -\gamma\, dA$, in systems containing extensible interfaces (where γ is the interfacial energy density and A is the interfacial area); magnetization work, $dw = -\vec{H} \cdot d\vec{b}$ (where $\vec{H}$ is the magnetic field and $\vec{b}$ is the total magnetic moment density, including the permeability of vacuum); and electric polarization work, $dw = -\vec{E} \cdot d\vec{p}$ (where $\vec{E}$ is the electric field given by $\vec{E} = -\nabla\phi$ and $\vec{p}$ is the total polarization density, including the contribu-tion from the vacuum). If the system can perform other types of work, there must

---

[1] Use of the reference cell volume, $V_\circ$, is necessary because it establishes a thermodynamic reference state.

be terms in Eq. 2.3 to account for them. To generalize:

$$T ds = du - \sum_j \psi_j \, d\xi_j \qquad (2.4)$$

where ψ_j represents a jth generalized intensive quantity and ξ_j represents its con-jugate extensive quantity density.[2] Therefore,

$$\sum_j \psi_j \, d\xi_j = -P \, dv + \phi \, dq + \tilde{\sigma}_{kl} \, d\tilde{\epsilon}_{kl} + \gamma \, dA + \vec{H} \cdot d\vec{b} + \vec{E} \cdot d\vec{p}$$
$$+ \mu_1 \, dc_1 + \cdots + \mu_{N_c} \, dc_{N_c} + \cdots \qquad (2.5)$$

The ψ_j may be scalar, vector, or, generally, tensor quantities; however, each product in Eq. 2.5 must be a scalar.

Equation 2.4 can be used to define the continuum limit for the change in entropy in terms of measurable quantities. The differential terms are the first-order approx-imations to the increase of the quantities at a point. Such changes may reflect how a quantity changes in time, t, at a fixed point, $\vec{r}$; or at a fixed time for a variable location in a point's neighborhood. The change in the total entropy in the system, $\mathcal{S}$, can be calculated by summing the entropies in each of the cells by integrating over the entire system.[3] Equation 2.4, which is derived by combining the first and second laws, applies to reversible changes. However, because s, u, and the ξ_j are all state variables, the relation holds if all quantities refer to a cell under the local equilibrium assumption. Taking s as the dependent variable, Eq. 2.4 shows how s varies with changes in the independent variables, u and ξ_j.

In equilibrium thermodynamics, entropy maximization for a system with fixed internal energy determines equilibrium. Entropy increase plays a large role in ir-reversible thermodynamics. If each of the reference cells were an isolated system, the right-hand side of Eq. 2.4 could only increase in a kinetic process. However, because energy, heat, and mass may flow between cells during kinetic processes, they cannot be treated as isolated systems, and application of the second law must be generalized to the system of interacting cells.

In a hypothetical system for modeling kinetics, the microscopic cells must be open systems. It is useful to consider entropy as a fluxlike quantity capable of flowing from one part of a system to another, just like energy, mass, and charge. Entropy flux, denoted by $\vec{J}_s$, is related to the heat flux. An expression that relates $\vec{J}_s$ to measurable fluxes is derived below. Mass, charge, and energy *are* conserved quantities and additional restrictions on the flux of conserved quantities apply. However, entropy is not conserved—it can be created or destroyed locally. The consequences of entropy production are developed below.

2.1.1 Entropy Production

The local rate of entropy-density creation is denoted by $\dot{\sigma}$. The total rate of en-tropy creation in a volume $\mathcal{V}$ is $\int_{\mathcal{V}} \dot{\sigma} \, dV$. For an isolated system, $d\mathcal{S}/dt = \int_{\mathcal{V}} \dot{\sigma} \, dV$.

---

[2]The generalized intensive and extensive quantities may be regarded as generalized potentials and displacements, respectively.
[3]Note that S is the entropy of a cell, $\mathcal{S}$ is the entropy of the entire system, and s is the entropy per unit volume of the cell in its reference state.

However, for a more general system, the total entropy increase will depend upon how much entropy is produced within it and upon how much entropy flows through its boundaries.

From Eq. 2.4, the time derivative of entropy density in a cell is

$$\frac{\partial s}{\partial t} = \frac{1}{T}\frac{\partial u}{\partial t} - \frac{1}{T}\sum_j \psi_j \frac{\partial \xi_j}{\partial t} \tag{2.6}$$

Using conservation principles such as Eqs. 1.18 and 1.19 in Eq. 2.6,[4]

$$\frac{\partial s}{\partial t} = -\frac{1}{T}\nabla \cdot \vec{J}_u + \frac{1}{T}\sum_j \psi_j \nabla \cdot \vec{J}_j \tag{2.7}$$

From the chain rule for a scalar field A and a vector $\vec{B}$,

$$A\nabla \cdot \vec{B} = -\vec{B} \cdot \nabla A + \nabla \cdot (A\vec{B}) \tag{2.8}$$

Equation 2.7 can be written

$$\frac{\partial s}{\partial t} = \left(\vec{J}_u \cdot \nabla\frac{1}{T} - \sum_j \vec{J}_j \cdot \nabla\frac{\psi_j}{T} \right) - \nabla \cdot \left(\frac{\vec{J}_u - \sum_j \psi_j \vec{J}_j}{T} \right) \tag{2.9}$$

Comparison with terms in Eq. 1.20 identifies the entropy flux and entropy production:

$$\vec{J}_s = \frac{1}{T}\left(\vec{J}_u - \sum_j \psi_j \vec{J}_j \right) \tag{2.10}$$

$$\dot{\sigma} = \vec{J}_u \cdot \nabla\frac{1}{T} - \sum_j \vec{J}_j \cdot \nabla\frac{\psi_j}{T} \tag{2.11}$$

The terms in Eq. 2.10 for the entropy flux can be interpreted using Eq. 2.4. The entropy flux is related to the sum of all potentials multiplying their conjugate fluxes. Each extensive quantity in Eq. 2.4 is replaced by its flux in Eq. 2.10.

Equation 2.11 can be developed further by introducing the flux of heat, J_Q. Applying the first law of thermodynamics to the cell yields

$$du = \frac{dQ}{V_\circ} + \sum_j \psi_j \, d\xi_j \tag{2.12}$$

where Q is the amount of heat transferred to the cell. By comparison with Eq. 2.4 and with the assumption of local equilibrium, $dQ/V_\circ = T ds$ and therefore

$$\vec{J}_u = \vec{J}_Q + \sum_j \psi_j \vec{J}_j \tag{2.13}$$

Substituting Eq. 2.13 into Eq. 2.11 then yields

$$\dot{\sigma} = \vec{J}_Q \cdot \nabla\frac{1}{T} - \sum_j \frac{\vec{J}_j}{T} \cdot \nabla\psi_j \tag{2.14}$$

---

[4]Here, all the extensive densities are treated as conserved quantities. This is not the general case. For example, polarization and magnetization density are not conserved. It can be shown that for nonconserved quantities, additional terms will appear on the right-hand side of Eq. 2.11.

2.1.2 Conjugate Forces and Fluxes

Multiplying Eq. 2.14 by T gives

$$T\dot{\sigma} = -\frac{\vec{J}_Q}{T} \cdot \nabla T - \sum_j \vec{J}_j \cdot \nabla \psi_j \qquad (2.15)$$

Every term on the right-hand side of Eq. 2.15 is the scalar product of a flux and a gradient. Furthermore, each term has the same units as energy dissipation density, J m^{-3} s^{-1}, and is a flux multiplied by a thermodynamic potential gradient. Each term that multiplies a flux in Eq. 2.15 is therefore a force for that flux. The paired forces and fluxes in the entropy production rate can be identified in Eq. 2.15 and are termed *conjugate* forces and fluxes. These are listed in Table 2.1 for heat, component i, and electric charge. These forces and fluxes have been identified for unconstrained extensive quantities (i.e., the differential extensive quantities in Eq. 2.5 can vary independently). However, many systems have constraints relating changes in their extensive quantities, and these constrained cases are treated in Section 2.2.2. Throughout Chapters 1–3 we assume, for simplicity, that the material is isotropic and that forces and fluxes are parallel. This assumption is removed for anisotropic materials in Chapter 4.

Table 2.1 presents corresponding well-known empirical force–flux laws that apply under certain conditions. These are Fourier's law of heat flow, a modified version of Fick's law for mass diffusion at constant temperature, and Ohm's law for the electric current density at constant temperature.[5] The mobility, M_i, is defined as the velocity of component i induced by a unit force.

Table 2.1: Selected Conjugate Forces, Fluxes, and Empirical Force–Flux Laws for Systems with Unconstrained Components, i.

Extensive Quantity	Flux	Conjugate Force	Empirical Force–Flux Law*	
Heat	$\vec{J}_Q$	$-\frac{1}{T}\nabla T$	Fourier's	$\vec{J}_Q = -K\,\nabla T$
Component i	$\vec{J}_i$	$-\nabla \mu_i = -\nabla \Phi_i$	Modified Fick's	$\vec{J}_i = -M_i c_i\,\nabla \mu_i$
Charge	$\vec{J}_q$	$-\nabla \phi$	Ohm's	$\vec{J}_q = -\rho\,\nabla \phi$

*K = thermal conductivity; M_i = mobility of i; ρ = electrical conductivity

2.1.3 Basic Postulate of Irreversible Thermodynamics

The basic postulate of irreversible thermodynamics is that, near equilibrium, the *local* entropy production is nonnegative:

$$\dot{\sigma} \equiv \frac{\partial s}{\partial t} + \nabla \cdot \vec{J}_s \geq 0 \qquad (2.16)$$

---

[5] Under special circumstances, this form of Fick's law reduces to the classical form $\vec{J}_i = -D_i\,\nabla c_i$, where D_i is the mass diffusivity (see Section 3.1 for further discussion).

Using the empirical laws displayed in Table 2.1, the entropy production can be identified for a few special cases. For instance, if only heat flow is occurring, then, using Eq. 2.15 and Fourier's heat-flux law,

$$\vec{J}_Q = -K\,\nabla T \tag{2.17}$$

results in

$$T\dot{\sigma} = \frac{K|\nabla T|^2}{T} \tag{2.18}$$

which predicts (because of Eq. 2.16) that the thermal conductivity will always be positive.

If diffusion is the only operating process,

$$T\dot{\sigma} = \sum_{i=1}^{N_c} M_i c_i |\nabla \mu_i|^2 \tag{2.19}$$

implying that each mobility is always positive.

2.2 LINEAR IRREVERSIBLE THERMODYNAMICS

In many materials, a gradient in temperature will produce not only a flux of heat but also a gradient in electric potential. This coupled phenomenon is called the *thermoelectric effect*. Coupling from the thermoelectric effect works both ways: if heat can flow, the gradient in electrical potential will result in a heat flux. That a coupling between different kinds of forces and fluxes exists is not surprising; flows of mass (atoms), electricity (electrons), and heat (phonons) all involve particles possessing momentum, and interactions may therefore be expected as momentum is transferred between them. A formulation of these coupling effects can be obtained by generalization of the previous empirical force–flux equations.

2.2.1 General Coupling between Forces and Fluxes

In general, the fluxes may be expected to be a function of all the driving forces acting in the system, F_i; for instance, the heat flux J_Q could be a function of other forces in addition to its conjugate force F_Q; that is,

$$J_Q = J_Q(F_Q, F_q, F_1, \ldots, F_{N_c})$$

Assuming that the system is near equilibrium and the driving forces are small, each of the fluxes can be expanded in a Taylor series near the equilibrium point

$F_Q = F_q = F_1 = \cdots = F_{N_c} = 0$. To first order:

$$J_Q(F_Q, F_q, F_1, F_2, \ldots, F_{N_c}) = \frac{\partial J_Q}{\partial F_Q} F_Q + \frac{\partial J_Q}{\partial F_q} F_q + \frac{\partial J_Q}{\partial F_1} F_1 + \cdots + \frac{\partial J_Q}{\partial F_{N_c}} F_{N_c}$$

$$J_q(F_Q, F_q, F_1, F_2, \ldots, F_{N_c}) = \frac{\partial J_q}{\partial F_Q} F_Q + \frac{\partial J_q}{\partial F_q} F_q + \frac{\partial J_q}{\partial F_1} F_1 + \cdots + \frac{\partial J_q}{\partial F_{N_c}} F_{N_c}$$

$$J_1(F_Q, F_q, F_1, F_2, \ldots, F_{N_c}) = \frac{\partial J_1}{\partial F_Q} F_Q + \frac{\partial J_1}{\partial F_q} F_q + \frac{\partial J_1}{\partial F_1} F_1 + \cdots + \frac{\partial J_1}{\partial F_{N_c}} F_{N_c}$$

$$\vdots \qquad = \qquad \vdots$$

$$J_{N_c}(F_Q, F_q, F_1, F_2, \ldots, F_{N_c}) = \frac{\partial J_{N_c}}{\partial F_Q} F_Q + \frac{\partial J_{N_c}}{\partial F_q} F_q + \frac{\partial J_{N_c}}{\partial F_1} F_1 + \cdots + \frac{\partial J_{N_c}}{\partial F_{N_c}} F_{N_c}$$

$$\text{(2.20)}$$

or in abbreviated form,

$$J_\alpha = \sum_\beta L_{\alpha\beta} F_\beta \qquad (\alpha, \beta = Q, q, 1, 2, \ldots, N_c) \qquad \text{(2.21)}$$

where

$$L_{\alpha\beta} = \frac{\partial J_\alpha}{\partial F_\beta} \qquad \text{(2.22)}$$

is evaluated at equilibrium ($F_\beta = 0$, for all β).[6] In this approximation, the fluxes vary linearly with the forces.

In Eqs. 2.20 and 2.22, the diagonal terms, $L_{\alpha\alpha}$, are called *direct coefficients*; they couple each flux to its conjugate driving force. The off-diagonal terms are called *coupling coefficients* and are responsible for the coupling effects (also called *cross effects*) identified above.

Combining Eqs. 2.15 and 2.21 results in a relation for the entropy production that applies near equilibrium:

$$T\dot{\sigma} = \sum_\beta \sum_\alpha L_{\alpha\beta} F_\alpha F_\beta \qquad \text{(2.23)}$$

The connection between the direct coefficients in Eq. 2.21 and the empirical force–flux laws discussed in Section 2.1.2 can be illustrated for heat flow. If a bar of pure material that is an electrical insulator has a constant thermal gradient imposed along it, and no other fields are present and no fluxes but heat exist, then according to Eq. 2.21 and Table 2.1,

$$\vec{J}_Q = L_{QQ} \left(-\frac{1}{T} \nabla T \right) \qquad \text{(2.24)}$$

Comparison with Eq. 2.17 shows that the thermal conductivity K is related to the direct coefficient L_{QQ} by

$$K = \frac{L_{QQ}}{T} \qquad \text{(2.25)}$$

---

[6]Note that the fluxes and forces are written as scalars, consistent with the assumption that the material is isotropic. Otherwise, terms like $\vec{J}_Q = (\partial \vec{J}_Q / \partial F_Q) F_Q$ must be written as rank-two tensors multiplying vectors, and the equations that result can be written as linear relations (see Section 4.5 for further discussion).

If the material is also electronically conducting, the general force–flux relationships are

$$J_Q = L_{QQ} F_Q + L_{Qq} F_q \tag{2.26}$$

$$J_q = L_{qQ} F_Q + L_{qq} F_q \tag{2.27}$$

If a constant thermal gradient is imposed and no electrically conductive contacts are made at the ends of the specimen, the heat flow is in a steady state and the charge-density current must vanish. Hence $J_q = 0$ and a force

$$F_q = -\frac{L_{qQ}}{L_{qq}} F_Q \tag{2.28}$$

will arise. The existence of the force F_q indicates the presence of a gradient in the electrical potential, $\nabla \phi$, along the bar. Therefore, using Eqs. 2.28 and 2.26,

$$J_Q = \left[L_{QQ} - \frac{L_{Qq} L_{qQ}}{L_{qq}} \right] F_Q = -\left[\frac{L_{QQ}}{T} - \frac{L_{Qq} L_{qQ}}{T L_{qq}} \right] \nabla T = -K \nabla T \tag{2.29}$$

In such a material under these conditions, Fourier's law again pertains, but the thermal conductivity K depends on the direct coefficient L_{QQ}, as in Eq. 2.25, as well as on the direct and coupling coefficients associated with electrical charge flow. In general, the empirical conductivity associated with a particular flux depends on the constraints applied to other possible fluxes.

2.2.2 Force–Flux Relations when Extensive Quantities are Constrained

In many cases, changes in one extensive quantity are coupled to changes in others. This occurs in the important case of substitutional components in a crystal devoid of sources or sinks for atoms, such as dislocations, as explained in Section 11.1. Here the components are constrained to lie on a fixed network of sites (i.e., the crystal structure), where each site is always occupied by one of the components of the system. Whenever one component leaves a site, it must be replaced. This is called a *network constraint* [1]. For example, in the case of substitutional diffusion by a vacancy–atom exchange mechanism (discussed in Section 8.1.2), the vacancies are one of the components of the system; every time a vacancy leaves a site, it is replaced by an atom. As a result of this replacement constraint, the fluxes of components are not independent of one another.

This type of constraint will be absent in amorphous materials because any of the N_c components can be added (or removed) anywhere in the material without exchanging with any other components. The dN_i will also be independent for interstitial solutes in crystalline materials that lie in the interstices between larger substitutional atoms, as, for example, carbon atoms in body-centered cubic (b.c.c.) Fe, as illustrated in Fig. 8.8. In such a system, carbon atoms can be added or removed independently in a dilute solution.

When a network constraint is present,

$$\sum_{i=1}^{N_c} dN_i = 0 \tag{2.30}$$

Solving Eq. 2.30 for dN_{N_c} and putting the result into Eq. 2.3 yields

$$Tds = du + dw - \sum_{i=1}^{N_c-1} (\mu_i - \mu_{N_c}) \, dc_i \qquad (2.31)$$

Starting with Eq. 2.31 instead of Eq. 2.3 and repeating the procedure that led to Eq. 2.15, the conjugate force for the diffusion of component i in a network-constrained crystal takes the new form

$$\vec{F}_i = -\nabla (\mu_i - \mu_{N_c}) \qquad (2.32)$$

The conjugate force for the diffusion of a network-constrained component i therefore depends upon the gradient of the difference between the chemical potential of component i and N_c rather than on the chemical potential gradient of component i alone. If in the case of substitutional diffusion by the vacancy exchange mechanism, the vacancies are taken as the component N_c, the driving force for component i depends upon the gradient of the difference between the chemical potential of component i and that of the vacancies. The difference arises because, during migration, a site's state changes from occupancy by an atom of type i to occupancy by a vacancy. This result has been derived and extended by Larché and Cahn, who investigated coherent thermomechanical equilibrium in multicomponent systems with elastic stress fields [1–4].

In the development above, the choice of the N_cth component in a system under network constraint is arbitrary. However, the flux of each component in Eq. 2.21 must be independent of this choice [3, 4]. This independence imposes conditions on the $L_{\alpha\beta}$ coefficients. To demonstrate, consider a three-component system at constant temperature in the absence of an electric field, where components A, B, and C correspond to $i = 1$, 2, and 3, respectively. If component C is the N_cth component, Eqs. 2.21 and 2.32 yield

$$\begin{aligned}
\vec{J}_A &= -L_{AA}\nabla(\mu_A - \mu_C) - L_{AB}\nabla(\mu_B - \mu_C) \\
\vec{J}_B &= -L_{BA}\nabla(\mu_A - \mu_C) - L_{BB}\nabla(\mu_B - \mu_C) \\
\vec{J}_C &= -L_{CA}\nabla(\mu_A - \mu_C) - L_{CB}\nabla(\mu_B - \mu_C)
\end{aligned} \qquad (2.33)$$

On the other hand, if B is the N_cth component,

$$\begin{aligned}
\vec{J}'_A &= -L_{AA}\nabla(\mu_A - \mu_B) - L_{AC}\nabla(\mu_C - \mu_B) \\
\vec{J}'_B &= -L_{BA}\nabla(\mu_A - \mu_B) - L_{BC}\nabla(\mu_C - \mu_B) \\
\vec{J}'_C &= -L_{CA}\nabla(\mu_A - \mu_B) - L_{CC}\nabla(\mu_C - \mu_B)
\end{aligned} \qquad (2.34)$$

Because $\vec{J}_i$ must be the same as $\vec{J}'_i$ and the gradient terms are not necessarily zero, Eqs. 2.33 and 2.34 imply that

$$\begin{aligned}
L_{AA} + L_{AB} + L_{AC} &= 0 \\
L_{BA} + L_{BB} + L_{BC} &= 0 \\
L_{CA} + L_{CB} + L_{CC} &= 0
\end{aligned} \qquad (2.35)$$

or generally,

$$\sum_{j=1}^{N_c} L_{ij} = 0 \qquad (2.36)$$

If the lattice network defines the coordinate system in which the fluxes are measured, the network constraint requires that

$$\sum_{i=1}^{N_c} \vec{J}_i = 0 \tag{2.37}$$

and this imposes the further condition on the L_{ij} that

$$\sum_{i=1}^{N_c} L_{ij} = 0 \tag{2.38}$$

In other words, the sum of the entries in any row or column of the matrix L_{ij} is zero.

The conjugate forces and fluxes that are obtained when the only constraint is a network constraint are listed in Table 2.2. However, there are many cases where further constraints between the extensive quantities exist. For example, suppose that component 1 is a nonuniformly distributed ionic species that has no network constraint. Each ion will experience an electrostatic force due to the local electric field, as well as a force due to the gradient in its chemical potential. This may be demonstrated in a formal manner with Eq. 2.5, noting that dq in this case is not independent of dc_1 but, instead, $dq = q_1 dc_1$, where q_1 is the electrical charge per ion assuming that all electric current is carried by ions. Thus dq and dc_1 can be combined in Eq. 2.5 into a single term $(\mu_1 + q_1\phi)dc_1$, and when this term is carried through the process leading to Eq. 2.15, the ion flux, $\vec{J}_1$, is found to be conjugate to an ionic force

$$\vec{F}_1 = -\nabla(\mu_1 + q_1\phi) \tag{2.39}$$

The potential that appears in the total force expression is the sum of the chemical potential and the electropotential of the charged ion. This total potential is generally called the *electrochemical potential*.

Additional forces would be added to the chemical potential force if, for example, the particle possessed a magnetic moment and a magnetic field were present. As will be seen, many possibilities for total forces exist depending upon the types of components and fields present.

Table 2.2:   Conjugate Forces and Fluxes for Systems with Network-Constrained Components, i

Quantity	Flux	Conjugate Force
Heat	$\vec{J}_Q$	$-\frac{1}{T}\nabla T$
Component i	$\vec{J}_i$	$-\nabla(\mu_i - \mu_{N_c}) = -\nabla\Phi_i$
Charge	$\vec{J}_q$	$-\nabla\phi$

2.2.3   Introduction of the Diffusion Potential

Any potential that accounts for the storage of energy due to the addition of a component determines the driving force for the diffusion of that component. The sum of all such supplemental potentials, including the chemical potential, gives the total conjugate force for a diffusing component and is called the *diffusion potential* for that component and is represented by the symbol Φ.[7] The conjugate force for the flux of component 1 will always have the form

$$\vec{F}_1 = -\nabla\Phi_1 \qquad (2.40)$$

and thus for the special case leading to Eq. 2.39,

$$\Phi_1 = \mu_1 + q_1\phi \qquad (2.41)$$

2.2.4   Onsager's Symmetry Principle

Three postulates were utilized to derive the relations between forces and fluxes:

- The rate of entropy change and the local rate of entropy production can be inferred by invoking equilibrium thermodynamic variations and the assumption of local equilibrium.

- The entropy production is nonnegative.

- Each flux depends linearly on all the driving forces.

These postulates do not follow from statements of the first and second laws of thermodynamics.

Onsager's principle supplements these postulates and follows from the statistical theory of reversible fluctuations [5]. Onsager's principle states that when the forces and fluxes are chosen so that they are conjugate, the coupling coefficients are *symmetric*:

$$L_{\alpha\beta} = L_{\beta\alpha} \qquad (2.42)$$

which simplifies the coupled force–flux equations and has led to experimentally verifiable predictions [6].

Furthermore, Eq. 2.42 guarantees that all the eigenvalues of Eq. 2.21 will be real numbers. Also, the quadratic form in Eq. 2.23 together with Eq. 2.16 implies that the kinetic matrix ($L_{\alpha\beta}$) will be positive definite; all the eigenvalues are nonnegative.[8]

Equation 2.42 can be rewritten

$$\frac{\partial J_\alpha}{\partial F_\beta} = \frac{\partial J_\beta}{\partial F_\alpha} \qquad (2.43)$$

This equation shows that *the change in flux of some quantity caused by changing the direct driving force for another is equal to the change in flux of the second quantity caused by changing the driving force for the first.* These equations resemble the Maxwell relations from thermodynamics.

---

[7]The potential Φ_i is an aggregate of all reversible work terms that can be transported with the species i. Using Lagrange multipliers, Cahn and Larché derive a potential that is a sum of the diffusant's elastic energy and its chemical potential [4]. Cahn and Larché coined the term diffusion potential to describe this sum. Our use of the term is consistent with theirs.

[8]*Positive definite* means that the matrix when left- and right-multiplied by an arbitrary vector will yield a nonnegative scalar. If the matrix multiplied by a vector composed of forces is proportional to a flux, it implies that the flux always has a positive projection on the force vector. Technically, one should say that $L_{\alpha\beta}$ is nonnegative definite but the meaning is clear.

The statistical-mechanics derivation of Onsager's symmetry principle is based on microscopic reversibility for systems near equilibrium. That is, the time average of a correlation between a driving force of type α and the fluctuations of quantity β is identical with respect to switching α and β [6].

A demonstration of the role of microscopic reversibility in the symmetry of the coupling coefficients can be obtained for a system consisting of three isomers, A, B, and C [7, 8]. Each isomer can be converted into either of the other two, without any change in composition. Assuming a closed system containing these molecules at constant temperature and pressure, the rate of conversion of one type into another is proportional to its number, with the constant of proportionality being a rate constant, K (Fig. 2.1). The rates at which the numbers of A, B, and C change are then

$$\frac{dN_A}{dt} = -(K_{AC} + K_{AB})N_A + K_{BA}N_B + K_{CA}N_C$$

$$\frac{dN_B}{dt} = K_{AB}N_A - (K_{BC} + K_{BA})N_B + K_{CB}N_C \qquad (2.44)$$

$$\frac{dN_C}{dt} = K_{AC}N_A + K_{BC}N_B - (K_{CA} + K_{CB})N_C$$

At equilibrium, the time derivatives in Eq. 2.44 vanish. Solving for equilibrium in a closed system ($N_A + N_B + N_C = N^{\text{tot}}$) yields

$$N_A^{\text{eq}} = \frac{K_\alpha N^{\text{tot}}}{K_\alpha + K_\beta + K_\gamma} \qquad N_B^{\text{eq}} = \frac{K_\beta N^{\text{tot}}}{K_\alpha + K_\beta + K_\gamma} \qquad N_C^{\text{eq}} = \frac{K_\gamma N^{\text{tot}}}{K_\alpha + K_\beta + K_\gamma}$$

$$(2.45)$$

where

$$K_\alpha \equiv K_{BA}K_{CA} + K_{BA}K_{CB} + K_{CA}K_{BC}$$

$$K_\beta \equiv K_{CB}K_{AB} + K_{CB}K_{AC} + K_{AB}K_{CA} \qquad (2.46)$$

$$K_\gamma \equiv K_{AC}K_{BC} + K_{AC}K_{BA} + K_{BC}K_{AB}$$

For the system near equilibrium, let Y_A be the difference between the number of A and its equilibrium value, $Y_A = N_A - N_A^{\text{eq}}$. Introducing this relationship and similar ones for B and C into Eq. 2.44,

$$\frac{dY_A}{dt} = -(K_{AC} + K_{AB})Y_A + K_{BA}Y_B + K_{CA}Y_C \qquad (2.47)$$

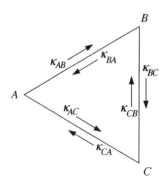

Figure 2.1: Schematic conversion diagram for type A, B, and C molecules.

with similar expressions for B and C.

If Henry's law is obeyed, the activity coefficient is constant and expanding the chemical potential (Eq. 2.2) near equilibrium (small Y_A/N_A^{eq}) yields

$$\Delta\mu_A^{\mathrm{eq}} \equiv \mu_A - \mu_A^{\mathrm{eq}} = kT\ln\left(1 + \frac{Y_A}{N_A^{\mathrm{eq}}}\right) \approx \frac{kTY_A}{N_A^{\mathrm{eq}}} \tag{2.48}$$

Substituting Eq. 2.48 into Eq. 2.47 and carrying out similar procedures for B and C,

$$\frac{dY_A}{dt} = -\frac{(K_{AC}+K_{AB})N_A^{\mathrm{eq}}}{kT}\Delta\mu_A^{\mathrm{eq}} + \frac{K_{BA}N_B^{\mathrm{eq}}}{kT}\Delta\mu_B^{\mathrm{eq}} + \frac{K_{CA}N_C^{\mathrm{eq}}}{kT}\Delta\mu_C^{\mathrm{eq}}$$

$$\frac{dY_B}{dt} = +\frac{K_{AB}N_A^{\mathrm{eq}}}{kT}\Delta\mu_A^{\mathrm{eq}} - \frac{(K_{BC}+K_{BA})N_B^{\mathrm{eq}}}{kT}\Delta\mu_B^{\mathrm{eq}} + \frac{K_{CB}N_C^{\mathrm{eq}}}{kT}\Delta\mu_C^{\mathrm{eq}} \tag{2.49}$$

$$\frac{dY_C}{dt} = +\frac{K_{AC}N_A^{\mathrm{eq}}}{kT}\Delta\mu_A^{\mathrm{eq}} + \frac{K_{BC}N_B^{\mathrm{eq}}}{kT}\Delta\mu_B^{\mathrm{eq}} - \frac{(K_{CA}+K_{CB})N_C^{\mathrm{eq}}}{kT}\Delta\mu_C^{\mathrm{eq}}$$

These constitute a set of linear relationships between the potential differences $\mu_i - \mu_i^{\mathrm{eq}}$, which drive the Y_i toward equilibrium and their corresponding rates, dY_i/dt. In terms of the Onsager coefficients, they have the form

$$\frac{dY_A}{dt} = L_{AA}F_A + L_{AB}F_B + L_{AC}F_C$$

$$\frac{dY_B}{dt} = L_{BA}F_A + L_{BB}F_B + L_{BC}F_C \tag{2.50}$$

$$\frac{dY_C}{dt} = L_{CA}F_A + L_{CB}F_B + L_{CC}F_C$$

When microscopic reversibility is present in a complex system composed of many particles, every elementary process in a forward direction is balanced by one in the reverse direction. The balance of forward and backward rates is characteristic of the equilibrium state, and *detailed balance* exists throughout the system. Microscopic reversibility therefore requires that the forward and backward reaction fluxes in Fig. 2.1 be equal, so that

$$\frac{K_{BA}}{K_{AB}} = \frac{N_A^{\mathrm{eq}}}{N_B^{\mathrm{eq}}} = \frac{K_\alpha}{K_\beta} = \frac{K_{BA}K_{CA} + K_{BA}K_{CB} + K_{CA}K_{BC}}{K_{CB}K_{AB} + K_{CB}K_{AC} + K_{AB}K_{CA}}$$

$$\frac{K_{CB}}{K_{BC}} = \frac{N_B^{\mathrm{eq}}}{N_C^{\mathrm{eq}}} = \frac{K_\beta}{K_\gamma} = \frac{K_{CB}K_{AB} + K_{CB}K_{AC} + K_{AB}K_{CA}}{K_{AC}K_{BC} + K_{AC}K_{BA} + K_{BC}K_{AB}} \tag{2.51}$$

$$\frac{K_{AC}}{K_{CA}} = \frac{N_C^{\mathrm{eq}}}{N_A^{\mathrm{eq}}} = \frac{K_\gamma}{K_\alpha} = \frac{K_{AC}K_{BC} + K_{AC}K_{BA} + K_{BC}K_{AB}}{K_{BA}K_{CA} + K_{BA}K_{CB} + K_{CA}K_{BC}}$$

Comparison of Eq. 2.50 with Eqs. 2.49 and 2.51 shows that $L_{ij} = L_{ji}$ and therefore demonstrates the role of microscopic reversibility in the symmetry of the Onsager coefficients. More demonstrations of the Onsager principle are described in Lifshitz and Pitaerskii [6] and in Yourgrau et al. [8].

Solving Exercise 2.5 shows that the products of the forces and reaction rates in Eq. 2.49 appear in the expression for the entropy production rate for the chemical reactions. The forces and reaction rates are therefore conjugate, as expected.

Bibliography

1. F.C. Larché and J.W. Cahn. A linear theory of thermochemical equilibrium of solids under stress. *Acta Metall.*, 21(8):1051–1063, 1973.

2. F.C. Larché and J.W. Cahn. A nonlinear theory of thermomechanical equilibrium of solids under stress. *Acta Metall.*, 26(1):53–60, 1978.

3. J.W. Cahn and F.C. Larché. An invariant formulation of multicomponent diffusion in crystals. *Scripta Metall.*, 17(7):927–937, 1983.

4. F.C. Larché and J.W. Cahn. The interactions of composition and stress in crystalline solids. *Acta Metall.*, 33(3):331–367, 1984.

5. L. Onsager. Reciprocal relations in irreversible processes. II. *Phys. Rev.*, 38(12):2265–2279, 1931.

6. E.M. Lifshitz and L.P. Pitaevskii. *Statistical Physics, Part 1*. Pergamon Press, New York, 3rd edition, 1980. See pages 365ff.

7. K. Denbigh. *The Principles of Chemical Equilibrium*. Cambridge University Press, New York, 3rd edition, 1971.

8. W. Yourgrau, A. ven der Merwe, and G. Raw. *Treatise on Irreversible and Statistical Thermophysics*. Dover Publications, New York, 1982.

EXERCISES

2.1 Using an argument based on entropy production, what can be concluded about the algebraic sign of the electrical conductivity?

Solution. If electronic conduction is the only operative process in a material at constant T, then Eq. 2.15 reduces to

$$T\dot{\sigma} = -\vec{J}_q \cdot \nabla\phi \tag{2.52}$$

Using Ohm's law, $\vec{J}_q = -\rho\nabla\phi$,

$$T\dot{\sigma} = \rho\,|\nabla\phi|^2 \tag{2.53}$$

Because $\dot{\sigma} \geq 0$ and $|\nabla\phi|^2$ is positive, ρ must be positive.

2.2 An isolated bar of a good electrical insulator contains a rapidly diffusing unconstrained solute (i.e., component 1). Impose a constant thermal gradient along the bar, and find an expression for its thermal conductivity when the system reaches a steady state. Assume that no solute enters or leaves the ends of the bar. Express your result in terms of any of the $L_{\alpha\beta}$ coefficients in Eq. 2.21 that are required.

Solution. Using a similar method as the development that led to Eq. 2.29, the relevant linear force–flux relations are

$$\begin{aligned}J_Q &= L_{QQ}F_Q + L_{Q1}F_1 \\ J_1 &= L_{1Q}F_Q + L_{11}F_1\end{aligned} \tag{2.54}$$

The cross effect between the thermal and diffusion currents causes a redistribution of the unconstrained solute until a steady-state distribution is reached. In this condition $J_1 = 0$ and, therefore, $F_1 = -F_Q(L_{1Q}/L_{11})$. Putting this result into Eq. 2.54 then yields

$$J_Q = -\left(\frac{L_{QQ}}{T} - \frac{L_{Q1}L_{1Q}}{TL_{11}}\right)\nabla T = -K\,\nabla T \tag{2.55}$$

The expression for K is similar to Eq. 2.29, where electrical charge rather than component 1 was forced into a steady-state distribution by the thermal flux.

2.3 A common device used to measure temperature differences is the thermocouple in Fig. 2.2. Wires of metals A and B are connected with their common junction at the temperature $T + \Delta T$ and the opposite ends connected to the terminals of a potentiometer maintained at temperature T. The potentiometer measures a voltage, ϕ_{AB}, across terminals 1 and 2 under conditions where no electric current is flowing. This voltage is then a measure of ΔT. Explain this effect, known as the *Seebeck effect*, in terms of relevant forces and fluxes.

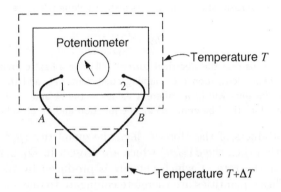

Figure 2.2: Thermocouple composed of metals A and B with a junction at one end.

Solution. The appropriate force–flux equation for this case is

$$J_q = L_{qQ}F_Q + L_{qq}F_q = -L_{qQ}\frac{1}{T}\frac{dT}{dx} - L_{qq}\frac{d\phi}{dx} \qquad (2.56)$$

where, to a good approximation, x measures the distance along the wire. Setting $J_q = 0$,

$$\frac{d\phi}{dT} = -\frac{L_{qQ}}{L_{qq}}\frac{1}{T} \qquad (2.57)$$

The potentials at 1 and 2 relative to the potential at the junction are determined by integrating Eq. 2.57 along each wire. They will differ because the A and B wires possess different values of the coefficients L_{qQ} and L_{qq}. ϕ_{AB} is then the difference between these two potentials.

2.4 Figure 2.3 depicts an apparatus at constant uniform temperature. The battery drives an electrical current around the circuit. Heat is absorbed at one A/B junction and emitted at the other. Explain this phenomenon, known as the *Peltier effect*, in terms of relevant forces and fluxes.

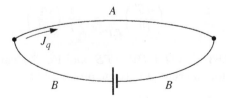

Figure 2.3: Peltier effect apparatus composed of metal wires A and B.

Solution. Both heat and electrical charge currents will be present in the wire, and the generalized linear relationships are therefore

$$J_Q = L_{QQ}F_Q + L_{Qq}F_q = -L_{QQ}\frac{1}{T}\frac{dT}{dx} - L_{Qq}\frac{d\phi}{dx} \tag{2.58}$$

$$J_q = L_{qQ}F_Q + L_{qq}F_q = -L_{qQ}\frac{1}{T}\frac{dT}{dx} - L_{qq}\frac{d\phi}{dx} \tag{2.59}$$

where, to a good approximation, x measures distance along the wires. Because $dT/dx = 0$,

$$J_Q = \frac{L_{Qq}}{L_{qq}}J_q \tag{2.60}$$

Equation 2.60 shows that the electrical current will drive a heat current along each wire by an amount dependent upon the coupling coefficient L_{Qq} and the direct coefficient L_{qq}. These coefficients will have different values in the A and B wires, and therefore heat will accumulate (and be emitted) at one junction and be absorbed at the other.

2.5 Show that products of the "forces" [i.e., the quantities $(\mu_i^{\text{eq}} - \mu_i)$], and the rates of reaction (i.e., the dY_i/dt) which are present in Eq. 2.49 appear in the expression for the rate at which entropy is produced by the corresponding reactions. These quantities are therefore conjugate to one another just as are the conjugate forces and fluxes in Table 2.1.

- Obtain a general expression for the rate at which entropy is produced: the reactions are taking place in a container maintained at constant temperature and pressure.

- The surroundings may be regarded as a reservoir at constant temperature and pressure.

- It may be necessary to transfer a quantity of heat, dQ, from the reservoir into the system in order to maintain constant temperature in the system; the total entropy change of the system plus reservoir, $d\mathcal{S}'$, will then be

$$d\mathcal{S}' = d\mathcal{S} - \frac{dQ}{T} \tag{2.61}$$

where $d\mathcal{S}$ and $-dQ/T$ are the entropy changes of the system and surroundings, respectively. For the system, $d\mathcal{U} = dQ - Pd\mathcal{V}$, and therefore

$$d\mathcal{S}' = \frac{Td\mathcal{S} - d\mathcal{U} - Pd\mathcal{V}}{T} \tag{2.62}$$

- Note that $\mathcal{G} = \mathcal{U} + PV - T\mathcal{S}$, and applying the constant temperature and pressure condition,

$$\left(\frac{\partial \mathcal{S}'}{\partial t}\right)_{T,P} = -\frac{1}{T}\left(\frac{\partial \mathcal{G}}{\partial t}\right)_{T,P} \tag{2.63}$$

- Equation 2.63, $\mathcal{G} = \mathcal{U} + PV - T\mathcal{S}$, and Eq. 2.1 can be combined in the expression for the rate of entropy production:

$$\left(\frac{\partial \mathcal{S}'}{\partial t}\right)_{T,P} = -\frac{1}{T}\sum_i \mu_i \frac{dN_i}{dt} \tag{2.64}$$

Solution. From Eq. 2.64,

$$T\left(\frac{\partial S'}{\partial t}\right)_{T,P} = \mathcal{V}T\dot{\sigma} = -\mu_A\frac{dN_A}{dt} - \mu_B\frac{dN_B}{dt} - \mu_C\frac{dN_C}{dt} \tag{2.65}$$

But $dN_i/dt = dY_i/dt$, and therefore

$$\mathcal{V}T\dot{\sigma} = -\mu_A\frac{dY_A}{dt} - \mu_B\frac{dY_B}{dt} - \mu_C\frac{dY_C}{dt} \tag{2.66}$$

Suppose that the system is at equilibrium with all three species at their equilibrium chemical potentials. We then make small changes in their numbers subject to the conservation condition

$$N_A + N_B + N_C = N^{\text{tot}} \tag{2.67}$$

Since in general for a system at constant T and P,

$$dG = \mu_A\,dN_A + \mu_B\,dN_B + \mu_C\,dN_C \tag{2.68}$$

the change in free energy of the system at equilibrium will be

$$dG = \mu_A^{\text{eq}}\,dN_A + \mu_B^{\text{eq}}\,dN_B + \mu_C^{\text{eq}}\,dN_C = 0 \tag{2.69}$$

Since $Y_i = N_i - N_i^{\text{eq}}$,

$$\mu_A^{\text{eq}}\,dY_A + \mu_B^{\text{eq}}\,dY_B + \mu_C^{\text{eq}}\,dY_C = 0 \tag{2.70}$$

and

$$\mu_A^{\text{eq}}\frac{dY_A}{dt} + \mu_B^{\text{eq}}\frac{dY_B}{dt} + \mu_C^{\text{eq}}\frac{dY_C}{dt} = 0 \tag{2.71}$$

adding Eqs. 2.66 and 2.71 then produces

$$\mathcal{V}T\dot{\sigma} = (\mu_A^{\text{eq}} - \mu_A)\frac{dY_A}{dt} + (\mu_B^{\text{eq}} - \mu_B)\frac{dY_B}{dt} + (\mu_C^{\text{eq}} - \mu_C)\frac{dY_C}{dt} \tag{2.72}$$

CHAPTER 3

DRIVING FORCES AND FLUXES FOR DIFFUSION

Fluxes of chemical components may arise from several different types of driving forces. For example, a charged species tends to flow in response to an applied electrostatic field; a solute atom induces a local volume dilation and tends to flow toward regions of lower hydrostatic compression. Chemical components tend to flow toward regions with lower chemical potential. The last case—flux in response to a chemical potential gradient—leads to Fick's first law, which is an empirical relation between the flux of a chemical species, $\vec{J_i}$, and its concentration gradient, ∇c_i in the form $\vec{J_i} = -D\nabla c_i$, where the quantity D is termed the *mass diffusivity*.

Because different driving forces can arise for a chemical species and because the mechanisms of diffusion comprising the microscopic basis for D are essentially independent of the driving force, all the driving forces can be collected and attributed to the generalized *diffusion potential*, Φ, introduced in Chapter 2.

The flux of a component in a solution can be complicated because components cannot always diffuse independently. This complication necessitates the introduction of different types of diffusion coefficients defined in specified reference frames to distinguish different diffusion systems.

3.1 DIFFUSION IN PRESENCE OF A CONCENTRATION GRADIENT

If a concentration gradient exists in a single phase at uniform temperature that is free of all other fields and any interfaces, that component's diffusion potential is identical to its chemical potential. The gradient in this potential is the driving

Kinetics of Materials. By Robert W. Balluffi, Samuel M. Allen, and W. Craig Carter.   **41**
Copyright © 2005 John Wiley & Sons, Inc.

force for diffusion. A diffusional flux proportional to the diffusion potential gradient will then arise, as discussed in Chapter 2. However, it is much easier to determine a concentration gradient by experiment than a diffusion potential gradient. It is therefore convenient to use a thermodynamic model of the solution to express the chemical-potential gradient in terms of a concentration gradient. The result is a diffusional flux proportional to the concentration gradient. The factor coupling the flux and concentration gradient is termed a *diffusivity* (or *diffusion coefficient*), D, so that Fick's first law in the form

$$\vec{J} = -D\nabla c \tag{3.1}$$

applies. The flux and corresponding diffusivity in this relationship must always be specified relative to a particular *reference frame* (coordinate system) [1, 2].

3.1.1   Self-Diffusion: Diffusion in the Absence of Chemical Effects

During self-diffusion in a pure material, whether a gas, liquid, or solid, the components diffuse in a chemically homogeneous medium. The diffusion can be measured using radioactive tracer isotopes or *marker atoms* that have chemistry identical to that of their stable isotope. The tracer concentration is measured and the tracer diffusivity (self-diffusivity) is inferred from the evolution of the concentration profile.

Figure 3.1 shows a diffusion couple containing a concentration gradient of the tracer atoms that could be used for this purpose. For a crystal where self-diffusion takes place by the vacancy-exchange mechanism, the Fick's law flux equation can be derived.[1] Such a crystal is network-constrained and has three components—inert atoms, radioactive atoms, and vacancies [3]. Planes of atoms provide a local reference frame to quantify the fluxes of these components by allowing the count of the number of atoms that cross a unit area of crystal plane per unit time. The crystal remains rigidly fixed during self-diffusion, and therefore these planes constitute a convenient single reference frame, called the crystal frame, or *C-frame*, for measuring flux. At constant temperature, with vacancies chosen as the N_cth

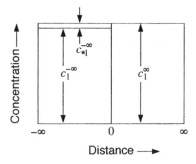

Figure 3.1:   Diffusion couple for measuring self-diffusion in pure material. A small concentration of a radioactive isotope of component 1, c_{*1}^{L}, is present initially on the left side of the couple. During diffusion, the radioactive particles will diffuse in the chemically homogeneous couple and become intermixed with the inert particles.

[1]The vacancy exchange mechanism is described in Section 8.1.2.

component, Eq. 2.21 combined with Eq. 2.32 takes the form (for one-dimensional isotropic diffusion)

$$
\begin{aligned}
J_1^C &= L_{11} F_1 + L_{1*1} F_{*1} \\
&= -L_{11} \frac{d\Phi_1}{dx} - L_{1*1} \frac{d\Phi_{*1}}{dx} \\
&= -L_{11} \frac{\partial(\mu_1 - \mu_V)}{\partial x} - L_{1*1} \frac{\partial(\mu_{*1} - \mu_V)}{\partial x} \\
J_{*1}^C &= L_{*11} F_1 + L_{*1*1} F_{*1} \\
&= -L_{*11} \frac{d\Phi_1}{dx} - L_{*1*1} \frac{d\Phi_{*1}}{dx} \\
&= -L_{*11} \frac{\partial(\mu_1 - \mu_V)}{\partial x} - L_{*1*1} \frac{\partial(\mu_{*1} - \mu_V)}{\partial x} \\
\end{aligned}
\tag{3.2}
$$

$$
\text{and } J_1^C + J_{*1}^C + J_V^C = 0
$$

where the star indicates the radioactive species, and the superscript C indicates that the flux of i is measured in a C-frame.[2] These equations are derived from the vacancy-exchange mechanism; every forward jump of an atom occurs via a backward jump of a vacancy. The vacancies are assumed to be in thermodynamic equilibrium throughout, and therefore $\mu_V = 0$.[3] $c_1 + c_{*1} + c_V = A$, where A is a constant and typical vacancy atomic fractions are less than 10^{-4}. The chemical potentials of species 1 and *1 are given by Eq. 2.2 with Raoultian behavior (i.e., $\gamma_1 = \gamma_{*1} = 1$). Therefore, $\partial \mu_i / \partial x = kT \partial \ln c_i / \partial x$. Because of the assumption of uniform vacancy equilibrium, $c_1 + c_{*1} = A - c_V$ is also uniform. Equations 3.2 become

$$
\begin{aligned}
J_1^C &= -kT \left[\frac{L_{11}}{c_1} - \frac{L_{1*1}}{c_{*1}} \right] \frac{\partial c_1}{\partial x} \\
J_{*1}^C &= -kT \left[\frac{L_{*1*1}}{c_{*1}} - \frac{L_{*11}}{c_1} \right] \frac{\partial c_{*1}}{\partial x}
\end{aligned}
\tag{3.3}
$$

Finally, because $J_V^C = 0$ and $J_1^C + J_{*1}^C = 0$,

$$
J_{*1}^C = -kT \left[\frac{L_{11}}{c_1} - \frac{L_{1*1}}{c_{*1}} \right] \frac{\partial c_{*1}}{\partial x} = -{}^*D \frac{\partial c_{*1}}{\partial x}
\tag{3.4}
$$

Equation 3.4 shows that the self-diffusion of the radioactive tracers obeys Fick's law with a self-diffusivity designated by $^*D$.

---

[2]The vector notation is dropped when there is only one spatial dimension. Unless indicated otherwise, the diffusion is assumed to be isotropic in this chapter. Anisotropic diffusion is treated in Section 4.5.

[3]The vacancy chemical potential is defined by $\mu_V = (\partial G / \partial N_V)_{T,P}$, that is, by the free-energy increase of a large system when a vacancy is created at a source such as a dislocation or a grain boundary. Dislocations act as vacancy sources as they climb in one direction and as sinks (i.e., negative sources), as they climb in the other. Throughout this book we generally refer to vacancy sources with the understanding that source and sink behavior are complementary (see Section 11.4). When this free-energy change is zero (i.e., $\mu_V = 0$), the system will be in a minimum free-energy state with respect to vacancy formation. The vacancy concentration at the source is then in local equilibrium. Assuming that μ_V is uniformly zero in a system requires an adequate density of vacancy sources.

3.1.2   Self-Diffusion of Component i in a Chemically Homogeneous Binary Solution

In Section 3.1.1, the self-diffusivity was obtained for a diffusion couple composed of a chemically pure material but with gradients of an isotope of that material. In this section we discuss self-diffusion of an isotopic species in a chemically homogeneous binary solution consisting of atoms of types 1 and 2 in the presence of a concentration gradient of the isotope.

The self-diffusion of component 1 in such a system is measured by studying the diffusion of a radioactive isotope tracer of component 1 (i.e., ^{*1}) under the condition that while there is a gradient in the tracer's concentration, c_{*1}, the sum $(c_1 + c_{*1})$ and c_2 are both uniform. A possible diffusion couple is shown in Fig. 3.2.

Considering Eq. 2.21 in a case in which diffusion occurs in a crystal by the vacancy exchange mechanism, there are four components, c_1, c_{*1}, c_2, and c_V. Because the crystal remains fixed during the diffusion, the C-frame is again used for measuring the flux. The system is chemically homogeneous, so

$$J_{*1}^C = -kT \left[\frac{L_{11}}{c_1} - \frac{L_{1*1}}{c_{*1}} \right] \frac{\partial c_{*1}}{\partial x} = -^*D_1 \frac{\partial c_{*1}}{\partial x} \tag{3.5}$$

Again, a Fick's-law expression is obtained for the self-diffusion of the radioactive component. The self-diffusivity of component 1 in a binary system of uniform chemical composition is designated by $^*D_1$ to distinguish it from the self-diffusivity of a pure material, $^*D$.

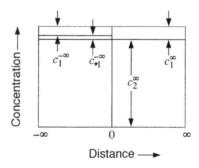

Figure 3.2:   Diffusion couple for measuring solute self-diffusion in a binary system. During diffusion, a c_{*1} gradient develops in the chemically homogeneous material.

3.1.3   Diffusion of Substitutional Particles in a Chemical Concentration Gradient

When a solute of type i diffuses on substitutional sites in an inhomogeneous binary solution, both the solute particles and host particles interdiffuse on the substitutional sites. If one species diffuses more quickly than the other, the region initially richer in that species loses net mass and contracts. On the other hand, the region initially richer in the more slowly diffusing species gains net mass and expands. This process, which establishes a mass flow in the system, is known as the *Kirkendall effect* after E. Kirkendall, who along with A. Smigelskas first observed it in crystalline metals [4].

Description of the Diffusion in a Local C-Frame.   Figure 3.3 illustrates a crystalline binary system in which interdiffusion occurs along x by a vacancy mechanism. The crystal is no longer fixed in structure as during self-diffusion but is flowing due to the Kirkendall effect, and its local planes are moving with a velocity, $v(x)$, with respect to the ends of the sample. Although no unique C-frame exists throughout the diffusing crystal, a local C-frame with a small inert marker particle to indicate its origin can be fixed to any plane as illustrated. The distribution of the faster-diffusing solute ($i = 1$) is also shown in Fig. 3.3a. If the diffusion fluxes are measured at any point in the diffusion zone with respect to its local C-frame, the constraint condition associated with the vacancy mechanism requires that the fluxes satisfy

$$\vec{J}_1^C + \vec{J}_2^C + \vec{J}_V^C = 0 \tag{3.6}$$

In the Kirkendall effect, the difference in the fluxes of the two substitutional species requires a net flux of vacancies. The net vacancy flux requires continuous net vacancy generation on one side of the markers and vacancy destruction on the other side (mechanisms of vacancy generation are discussed in Section 11.4). Vacancy creation and destruction can occur by means of dislocation climb and is illustrated in Fig. 3.3b for edge dislocations. Vacancy destruction occurs when atoms from the extra planes associated with these dislocations fill the incoming vacancies and the extra planes shrink (i.e., the dislocations climb as on the left side in Fig. 3.3b toward which the marker is moving). Creation occurs by the reverse process, where the extra planes expand as atoms are added to them in order to form vacancies, as on the right side of Fig. 3.3b. This contraction and expansion causes a mass flow that is revealed by the motion of embedded inert markers, as indicated in Fig. 3.3 [4].

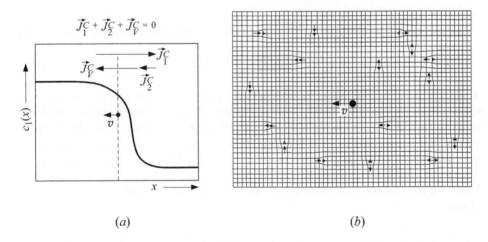

(a)   (b)

Figure 3.3:   Schematic illustration of the Kirkendall effect in a binary crystalline material diffusing by the vacancy mechanism. The sketch illustrates dislocation motion at some time after the diffusion couple's initial condition. **(a)** Concentration vs. distance profile in the diffusion zone. An embedded inert marker is present moving with the velocity, v, and fluxes in the local C-frame associated with the marker are indicated. **(b)** Local arrangement of atom planes across the diffusion zone near the marker. Arrows indicate climb movements of numerous edge dislocations that occur when there is a net diffusion of vacancies from right to left.

In this process, the net flux of substitutional atoms across the interface plane results in local volume changes (i.e., as a crystal plane is removed by climb, the crystal contracts in a direction normal to the plane). However, free expansion in directions parallel to the interface plane is constrained by the specimen ends, where significant diffusion has not occurred, and by the coherence of the interface between the expanding and contracting regions. Therefore, dimensional changes parallel to the interface (i.e., normal to the diffusion direction) are restricted, and in-plane compatibility stresses are generated. No out-of-plane compatibility stresses develop because the diffusion couple can expand freely in the diffusion direction.

In diffusion specimens that have a relatively narrow diffusion zone compared to the extent of the specimen in the diffusion direction, compatibility stresses are pure shear stresses, and if the stresses exceed the crystal's yield stress, the onset of plastic flow enables the cross section of the diffusion specimen to remain constant.[4]

The substitutional binary alloy diffusion illustrated in Fig. 3.3 is discussed in a treatment pioneered by Darken [6] (see also Crank's book [7]). The system has three components, species 1, species 2, and vacancies, and is assumed to be at constant pressure and temperature with sites that can only be created or destroyed at sources (i.e., the system is network constrained except at dislocations or interfaces). The fluxes are obtained from Eqs. 2.21 and 2.32:

$$
\begin{aligned}
J_1^C &= L_{11}F_1 + L_{12}F_2 \\
&= -L_{11}\frac{\partial \Phi_1}{\partial x} - L_{12}\frac{\partial \Phi_2}{\partial x} = -L_{11}\frac{\partial(\mu_1 - \mu_V)}{\partial x} - L_{12}\frac{\partial(\mu_2 - \mu_V)}{\partial x} \\
J_2^C &= L_{21}F_1 + L_{22}F_2 \\
&= -L_{21}\frac{\partial \Phi_1}{\partial x} - L_{22}\frac{\partial \Phi_2}{\partial x} = -L_{21}\frac{\partial(\mu_1 - \mu_V)}{\partial x} - L_{22}\frac{\partial(\mu_2 - \mu_V)}{\partial x}
\end{aligned}
\tag{3.7}
$$

and $J_V^C = -(J_1^C + J_2^C)$ by Eq. 3.6.

Assumption of local equilibrium permits the Gibbs–Duhem relation to be written

$$
c_1\frac{\partial \mu_1}{\partial x} + c_2\frac{\partial \mu_2}{\partial x} + c_V\frac{\partial \mu_V}{\partial x} = 0
\tag{3.8}
$$

A net vacancy flux develops in a direction opposite that of the fastest-diffusing species (species 1 in Fig. 3.3). Nonequilibrium vacancy concentrations would develop in the diffusion zone if they were not eliminated by dislocation climb. However, under usual conditions it is expected that a sufficient density of dislocations will be present to maintain the vacancy concentration near equilibrium [8]. It can therefore be assumed, to a good approximation, that $\mu_V = 0$, and therefore $\nabla\mu_V = 0$ everywhere in the diffusion zone. Using Eqs. 3.7 and 3.8 with $\nabla\mu_V = 0$ yields

$$
\begin{aligned}
J_1^C &= -\left[L_{11} - \frac{c_1}{c_2}L_{12}\right]\frac{\partial \mu_1}{\partial x} \\
J_2^C &= -\left[L_{22} - \frac{c_2}{c_1}L_{21}\right]\frac{\partial \mu_2}{\partial x}
\end{aligned}
\tag{3.9}
$$

---

[4]When the faster-diffusing component is diffused from the vapor phase into a thin sheet, and the diffusion zone is relatively wide compared to the sheet thickness, the constraints on the expansion parallel to the diffusion interface are greatly reduced. Large specimen expansions normal to the diffusion direction have then been observed [5].

The chemical potential gradients can be related to concentration gradients using

$$\mu_i = \mu_i^\circ + kT \ln(\gamma_i \langle \Omega \rangle c_i) \tag{3.10}$$

which is obtained by combining Eq. 2.2 with Eq. A.12, giving the result

$$\frac{\partial \mu_1}{\partial x} = kT \left[1 + \frac{\partial \ln \gamma_1}{\partial \ln c_1} + \frac{\partial \ln \langle \Omega \rangle}{\partial \ln c_1} \right] \frac{1}{c_1} \frac{\partial c_1}{\partial x} \tag{3.11}$$

The local volume expansion arising from the local change of composition contributes to diffusion via the derivative of the average site volume $\langle \Omega \rangle$.[5] The derivative of γ_1 is the contribution associated with the nonideality of the solution. Putting Eq. 3.11 into Eq. 3.9 yields a flux that is proportional to the concentration gradient:

$$J_1^C = -kT \left[\frac{L_{11}}{c_1} - \frac{L_{12}}{c_2} \right] \left[1 + \frac{\partial \ln \gamma_1}{\partial \ln c_1} + \frac{\partial \ln \langle \Omega \rangle}{\partial \ln c_1} \right] \frac{\partial c_1}{\partial x} = -D_1 \frac{\partial c_1}{\partial x} \tag{3.12}$$

This is a form of Fick's law for a chemically inhomogeneous material where the *intrinsic diffusivity*, designated by D_1, measures the flux in the local C-frame. A similar procedure for component 2 yields an analogous Fick's-law expression, $J_2^C = -D_2 \partial c_2 / \partial x$.

The self-diffusivity of species 1 in a chemically homogeneous solution of concentration c_1, corresponding to $^\star D_1$ in Eq. 3.5, can be compared with the intrinsic diffusivity of the same species in a chemically inhomogeneous solution at the same concentration, corresponding to D_1 in Eq. 3.12. Typically, in addition to the approximation of a concentration-independent average site volume $\langle \Omega \rangle$, it is reasonable to assume that the coupling (off-diagonal) terms, L_{12}/c_2 and $L_{\star 1}/c_{\star 1}$ in Eqs. 3.5 and 3.12, are small compared with the direct term L_{11}/c_1. In this approximation,

$$D_1 \cong \left[1 + \frac{\partial \ln \gamma_1}{\partial \ln c_1} \right] {}^\star D_1 \tag{3.13}$$

The primary difference between D_1 and $^\star D_1$ is a thermodynamic factor involving the concentration dependence of the activity coefficient of component 1. The thermodynamic factor arises because mass diffusion has a chemical potential gradient as a driving force, but the diffusivity is measured proportional to a concentration gradient and is thus influenced by the nonideality of the solution. This effect is absent in self-diffusion.

At this point it has been shown that the fluxes of species 1 and 2 can be described by Fick's-law expressions involving two different intrinsic diffusivities, D_1 and D_2, in a local coordinate system (local C-frame) fixed to the lattice plane through which flux is measured. However, because of the Kirkendall effect, these planes (reference frames) move normal to one another at different rates in a nonuniform fashion, and this description is therefore not useful for describing the diffusion throughout the specimen. When there is no change in the total specimen volume, the overall diffusion that occurs during the Kirkendall effect can be described in terms of a single diffusivity (the *interdiffusivity*) measured in a single reference frame (a *volume-fixed*) frame.

---

[5]This expansion can usually be described by *Vegard's law* in the form $\epsilon_{ij} - \epsilon_{ij}^\circ = \Upsilon_{ij}(c - c_\circ)$; ϵ_{ij} is the strain due to a concentration change, and the six Υ_{ij} are the Vegard coefficients.

Diffusion in a Volume-Fixed Frame (V-Frame).   To find the volume-fixed V-frame, assume that a frame, designated an R-frame, exists that relates all local C-frames [9]. If v_i^R is component i's velocity in this R-frame, v_i^C its velocity in a local C-frame, and v_C^R the local C-frame's velocity measured in the R-frame, then

$$c_i \, v_i^R = c_i \left(v_i^C + v_C^R \right) \tag{3.14}$$

Because the flux in a local C-frame is the concentration multiplied by its local velocity, $J_i^C = c_i \, v_i^C$, and using Eq. 3.14,

$$J_i^C = c_i \left(v_i^R - v_C^R \right) = -D_i \frac{\partial c_i}{\partial x} \tag{3.15}$$

Two equations representing the contributions of components 1 and 2 to the volume flux are obtained by multiplying Eq. 3.15 through by Ω_1 and Ω_2. The sum of these two equations, using Eqs. A.8 and A.10, is[6]

$$v_C^R - \left[\Omega_1 c_1 v_1^R + \Omega_2 c_2 v_2^R \right] = (D_1 - D_2) \, \Omega_1 \frac{\partial c_1}{\partial x} \tag{3.16}$$

Using Eqs. 3.15 (for $i = 1$), 3.16, and A.8 yields

$$c_1 \, v_1^R - c_1 \left[\Omega_1 c_1 v_1^R + \Omega_2 c_2 v_2^R \right] = -\left[c_2 \Omega_2 D_1 + c_1 \Omega_1 D_2 \right] \frac{\partial c_1}{\partial x} \tag{3.17}$$

The accumulation equation in the R-frame is

$$\frac{\partial c_i}{\partial t} = -\frac{\partial J_i^R}{\partial x} = -\frac{\partial \left(c_i v_i^R \right)}{\partial x} \tag{3.18}$$

Assuming that Ω_1 and Ω_2 possess constant but generally different values, and thus that there are no changes in overall specimen volume during diffusion, two equations for the component contributions to volume accumulation are obtained by multiplying Eq. 3.18 through by Ω_i and setting $i = 1, 2$.[7] The sum of these two equations is the local volume expansion rate in the R-frame. Using Eq. A.8,

$$\frac{\partial \left(c_1 \Omega_1 + c_2 \Omega_2 \right)}{\partial t} = -\frac{\partial}{\partial x} \left[\Omega_1 c_1 v_1^R + \Omega_2 c_2 v_2^R \right] = 0 \tag{3.19}$$

The quantity $\left[\Omega_1 c_1 v_1^R + \Omega_2 c_2 v_2^R \right]$ can therefore be, at most, a function of t, that is,

$$\Omega_1 c_1 v_1^R + \Omega_2 c_2 v_2^R = \Omega_1 J_1^R + \Omega_2 J_2^R = f(t) \tag{3.20}$$

and is the flux of volume passing through a plane in the R-frame.

There is a particular R-frame, called the V-frame, for which $f(t) = 0$. In this frame, according to Eq. 3.20,

$$\Omega_1 J_1^V + \Omega_2 J_2^V = 0 \tag{3.21}$$

---

[6]For the derivation of Eq. 3.16, c_V has been assumed to be negligible; the exact expression on the right-hand side is $\left[(D_1 - D_2)\Omega_1 \frac{\partial c_1}{\partial x} \right] - \left[D_2 \Omega_V \frac{\partial c_V}{\partial x} \right]$. Since the distance over which the species and the vacancy concentrations vary are similar, the gradient of c_V is negligible compared with the gradient of c_1 as well.

[7]The assumption of constant but different Ω values is generally acceptable. When any small volume changes are taken into account, the analysis of the diffusion becomes more complicated, as discussed elsewhere [2, 7, 10].

and the flux of volume through any plane is zero: hence, the term *volume-fixed frame*. Then, using $J_i^V = c_i v_i^V$ and Eqs. 3.16, 3.17, and 3.21,

$$J_1^V = c_1 v_1^V = -[c_2\Omega_2 D_1 + c_1\Omega_1 D_2]\frac{\partial c_1}{\partial x} \tag{3.22}$$

and

$$v_C^V = (D_1 - D_2)\,\Omega_1 \frac{\partial c_1}{\partial x} \tag{3.23}$$

Equation 3.23 for the velocity of a local C-frame with respect to the V-frame is therefore the velocity of any inert marker with respect to the V-frame. The assumptions that Ω_1 and Ω_2 are each constant throughout the material, and thus that there are no changes in overall specimen volume during diffusion, permit the use of Eq. 3.19 to derive the unique choice of the V-frame.

In Eq. 3.22, the flux of 1 in the V-frame obeys Fick's law and can be written

$$J_1^V = -[c_1\Omega_1 D_2 + c_2\Omega_2 D_1]\frac{\partial c_1}{\partial x} = -\widetilde{D}\frac{\partial c_1}{\partial x} \tag{3.24}$$

where the binary solution *interdiffusivity*, designated by $\widetilde{D}$, is related to the intrinsic diffusivities of components 1 and 2 (measured in a local C-frame) by the relation

$$\widetilde{D} = c_1\Omega_1 D_2 + c_2\Omega_2 D_1 \tag{3.25}$$

which is often approximated through $\Omega_1 = \Omega_2 = \langle\Omega\rangle$ as

$$\widetilde{D} = X_1 D_2 + X_2 D_1 \tag{3.26}$$

Using a similar procedure to find the flux of component 2 in the V-frame yields

$$J_2^V = -\widetilde{D}\frac{\partial c_2}{\partial x} \tag{3.27}$$

The only remaining task is now to relate the V-frame to a laboratory frame suitable for experimental purposes. This is provided by the laboratory frame (L-frame) illustrated in Fig. 3.4. Here, the ends of the specimen are unaffected by the diffusion and are stationary with respect to each other since there is no change in the overall specimen volume. The specimen ends therefore provide the anchoring points for the rigid laboratory frame fixed to the ends as illustrated.

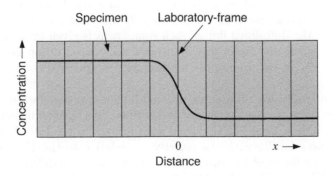

Figure 3.4:   Diffusion couple specimen with superimposed laboratory-frame (L-frame).

We now show that the L-frame in Fig. 3.4 is identical to the V-frame. The fluxes in the two frames are related by

$$J_i^L = J_i^V + c_i v_V^L \tag{3.28}$$

Using Eq. 3.28 with $i = 1$ and $i = 2$ to obtain two equations, multiplying them by Ω_1 and Ω_2, respectively, then adding them, using Eq. A.8, and differentiating the result with respect to x yields

$$\Omega_1 \frac{\partial J_1^L}{\partial x} + \Omega_2 \frac{\partial J_2^L}{\partial x} = \Omega_1 \frac{\partial J_1^V}{\partial x} + \Omega_2 \frac{\partial J_2^V}{\partial x} + \frac{\partial v_V^L}{\partial x} \tag{3.29}$$

However, with the use of Eq. A.10,

$$\Omega_1 \frac{\partial J_1^L}{\partial x} + \Omega_2 \frac{\partial J_2^L}{\partial x} = -\left(\Omega_1 \frac{\partial c_1}{\partial t} + \Omega_2 \frac{\partial c_2}{\partial t} \right) = 0 \tag{3.30}$$

and therefore

$$\frac{\partial v_V^L}{\partial x} = -\left(\Omega_1 \frac{\partial J_1^V}{\partial x} + \Omega_2 \frac{\partial J_2^V}{\partial x} \right) \tag{3.31}$$

which, integrated, gives

$$\int_{x=-L}^{x} dv_V^L = -\left(\Omega_1 \int_{x=-L}^{x} dJ_1^V + \Omega_2 \int_{x=-L}^{x} dJ_2^V \right) \tag{3.32}$$

Because J_i^L, J_i^V, and v_V^L are zero at the specimen ends $x = \pm L$, where L is large compared to the diffusion zone width,

$$\int_0^{v_V^L} dv = -\left(\Omega_1 \int_0^{J_1^V} dJ + \Omega_2 \int_0^{J_2^V} dJ \right) \tag{3.33}$$

Therefore, with Eq. 3.21,

$$v_V^L = -(\Omega_1 J_1^V + \Omega_2 J_2^V) = 0 \tag{3.34}$$

The L-frame and V-frame are thus identical.

Equation 3.23 gives the velocity of the local C-frame with respect to the V-frame (i.e., the velocity of local mass flow measured by the velocity of an embedded inert marker relative to the ends of a diffusion couple such as in Figs. 3.3 and 3.4). The measurement of v_C^V and $\widetilde{D}$ at the same concentration in a diffusion experiment thus produces two relationships involving D_1 and D_2 and allows their determination. In the V-frame, the diffusional flux of each component is given by a simple Fick's-law expression where the factor that multiplies the concentration gradient is the interdiffusivity $\widetilde{D}$. In this frame, the interdiffusion is specified completely by one diffusivity.

Chemical interdiffusion on a substitutional lattice can therefore be considered from two viewpoints. In the V-frame, it is described by a single diffusivity (i.e., the interdiffusivity). In a local C-frame fixed with respect to the local bulk material, the material flows locally with the velocity v_c^V relative to the V-frame and the description of the fluxes of the two components requires two diffusivities (i.e., the intrinsic diffusivities). These three diffusivities are related by Eq. 3.25, and each is generally a function of the local composition.

The Kirkendall effect alters the structure of the diffusion zone in crystalline materials. In many cases, the small supersaturation of vacancies on the side losing mass by fast diffusion causes the excess vacancies to precipitate out in the form of small voids, and the region becomes porous [11]. Also, the plastic flow maintains a constant cross section in the diffusion zone because of compatibility stresses. These stresses induce dislocation multiplication and the formation of cellular dislocation structures in the diffusion zone. Similar dislocation structures are associated with high-temperature plastic deformation in the absence of diffusion [12–14].

In 1946, the Kirkendall effect was observed with inert markers in polymer–solvent systems where the large polymer molecules diffused more slowly than the small solvent molecules [15]. Figure 3.5 shows an analogous phenomenon involving interdiffusion of fluids to help explain marker motion and plastic deformation in solids during the Kirkendall effect. Two fluids comprised of components A and B are encapsulated in a fixed volume and separated in two chambers by a fixed rigid membrane. Initially, there is no pressure difference across the membrane (i.e., $P_A^{left} + P_B^{left} = P_A^{right} + P_B^{right}$), but there is a difference in partial pressures, $P_A^{left} > P_A^{right}$. If the osmotic membrane allows rapid diffusion of A but not of B, the pressures P_A^{left} and P_A^{right} will then relax to equilibrium values until there is no difference in chemical potential across the membrane. This results in a difference in total pressure across the membrane.

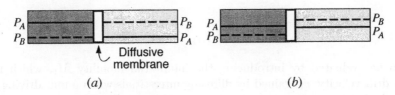

Figure 3.5:   Interface motion and cavitation during interdiffusion in fluids. (**a**) Initial situation: $P_A^{left} > P_A^{right}$, $P_B^{left} < P_B^{right}$, $P_A^{left} + P_B^{left} = P_A^{right} + P_B^{right}$. (**b**) Later situation: Component A diffuses through membrane so that $P_A^{left} = P_A^{right}$. But $P_B^{left} < P_B^{right}$ and $P_A^{left} + P_B^{left} < P_A^{right} + P_B^{right}$.

Consider now the consequences of the pressure difference. If the membrane became free to move, it would move to the left, compressing the left chamber and expanding the right to equilibrate the pressure difference (Fig. 3.6a). However, if the membrane is constrained, the fluid may cavitate in the left chamber to relieve the low pressure, as in Fig. 3.6b. This is analogous to the formation of voids in the Kirkendall effect.

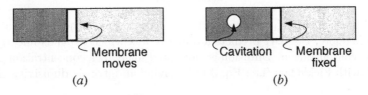

Figure 3.6:   Possible consequences of the situation in Fig. 3.5b. (**a**) If the membrane is allowed to move to the left, the total pressures on the left and right will become equal. (**b**) If the membrane is held fixed, cavitation may occur on the left to relieve the low pressure.

3.1.4   Diffusion of Interstitial Particles in a Chemical Concentration Gradient

Another system obeying Fick's law is one involving the diffusion of small interstitial solute atoms (component 1) among the interstices of a host crystal in the presence of an interstitial-atom concentration gradient. The large solvent atoms (component 2) essentially remain in their substitutional sites and diffuse much more slowly than do the highly mobile solute atoms, which diffuse by the interstitial diffusion mechanism (described in Section 8.1.4). The solvent atoms may therefore be considered to be immobile. The system is isothermal, the diffusion is not network constrained, and a local C-frame coordinate system can be employed as in Section 3.1.3. Equation 2.21 then reduces to

$$\vec{J}_1^C = L_{11}\vec{F}_1 = -L_{11}\nabla\Phi_1 = -L_{11}\nabla\mu_1 \tag{3.35}$$

Assuming a dilute solution, Henry's law [16] applies, and because $c_1 \ll c_2$ and $X_1 \simeq c_1/c_2$, an acceptable approximation, using Eq. 2.2, is

$$\mu_1 = \mu_1^\circ + kT\ln\left(K_1 c_1\right) \tag{3.36}$$

where K_1 is a constant. Therefore,

$$\nabla\mu_1 = \frac{kT}{c_1}\nabla c_1 \tag{3.37}$$

and

$$\vec{J}_1^C = -L_{11}\frac{kT}{c_1}\nabla c_1 \tag{3.38}$$

L_{11} can be evaluated by introducing the interstitial mobility M_1, which is the average drift velocity, $\vec{v}_1$, gained by diffusing interstitials when a unit driving force is applied, that is,

$$\vec{v}_1^C \equiv -M_1\nabla\mu_1 = -\frac{M_1 kT}{c_1}\nabla c_1 \tag{3.39}$$

An interstitial drift velocity causes an interstitial flux,

$$\vec{J}_1^C = \vec{v}_1 c_1 \tag{3.40}$$

so that

$$\vec{J}_1^C = -M_1 kT\,\nabla c_1 \tag{3.41}$$

Comparison of Eqs. 3.38 and 3.41 indicates that

$$L_{11} = M_1 c_1 \tag{3.42}$$

In a dilute solution, M_1 is expected to be independent of the interstitial concentration. Therefore, Eq. 3.41 predicts a diffusive flux which depends linearly on the gradient in concentration. Diffusion is motivated solely by a concentration gradient, consistent with Fick's law (i.e., Eq. 3.1) involving an intrinsic diffusivity given by

$$D_1 = M_1 kT = L_{11}\frac{kT}{c_1} \tag{3.43}$$

Equation 3.43, which expresses the link between the mobility and the diffusivity in this case, is known as the *Nernst–Einstein equation.*

The same diffusion can be described within a V-frame. The analysis is identical to that used to obtain Eqs. 3.23, 3.24, and 3.25 except that in this case, $D_2 = 0$. The marker velocity is therefore

$$v_C^V = \Omega_1 D_1 \frac{\partial c_1}{\partial x} \tag{3.44}$$

and

$$J_1^V = -\widetilde{D} \frac{\partial c_1}{\partial x} \tag{3.45}$$

with the interdiffusivity, $\widetilde{D}$, given by

$$\widetilde{D} = c_2 \Omega_2 D_1 \tag{3.46}$$

In general, for dilute concentrations $c_1 \Omega_1 \ll 1$, and therefore $c_2 \Omega_2 \cong 1$ and $\widetilde{D} \cong D_1$.

3.1.5 On the Algebraic Signs of Diffusivities

If the rate of entropy production, $\dot{\sigma}$, is nonnegative, M_1 is also nonnegative (see Section 2.1.3) and L_{11} must be nonnegative. For a Henrian solution, it can be inferred that the diffusivity D_1 is also nonnegative. However, there are more complicated cases, such as the concentrated solutions discussed in Section 3.1.3. In cases where the derivative of the chemical potential with respect to the concentration is negative, the postulate of positive entropy production implies a negative diffusivity in Fick's law (an experimental observation of a negative diffusivity is presented in Exercise 3.3). A negative diffusivity leads to an ill-posed diffusion equation; so, formulations based on fluxes and their conjugate driving forces (e.g., gradients in chemical potentials) are then preferred to Fick's law and indeed are more physical.[8]

3.1.6 Summary of Diffusivities

Four different types of diffusivities are summarized in Table 3.1. These include the self-diffusivity in a pure material, $^\star D$; the self-diffusivity of solute i in a binary system, $^\star D_i$; the intrinsic diffusivity of component i in a chemically inhomogeneous system, D_i; and the interdiffusivity, $\widetilde{D}$, in a chemically inhomogeneous system. These diffusivities are applicable only in certain reference frames which are also listed in Table 3.1. In the remainder of this book, the type of diffusivity under discussion will be identified by these symbols when this information is relevant. When a diffusivity is identified in this manner, it may be assumed that the diffusion under consideration is being described in the proper corresponding frame.

---

[8] *Ill-posed* means that the solution is unbounded and not robust with respect to initial conditions. In other words, two initial conditions that are very close to each other will give solutions that are very far apart a short time later.

Table 3.1:   Summary of Definitions and Relationships for Diffusivities Described in Chapter 3

Symbol	Name	Relationships	Reference Frame
$^\star D$	Self-diffusivity in a pure material	—	C-frame or V-frame

$^\star D$ is the self-diffusivity in a chemically homogeneous material comprised of only one species. It is usually determined by measuring the diffusion of a radioactive isotope that is chemically indistinguishable from the inert species. Because there is no mass flow, the C-frame is also a V-frame, and either may be used.

Symbol	Name	Relationships	Reference Frame
$^\star D_i$	Self-diffusivity of component i in a binary system	$D_i \cong \left(1 + \frac{\partial \ln \gamma_i}{\partial \ln c_i}\right) {}^\star D_i$	C-frame or V-frame

$^\star D_i$ is the self-diffusivity of component i in a chemically homogeneous binary system. $^\star D_i$ is related to D_i as indicated. $^\star D_i$ is usually determined using an isotope. Because there is no net mass flow, the C-frame is also a V-frame, and either may be used.

Symbol	Name	Relationships	Reference Frame
D_i	Intrinsic diffusivity	$\vec{J}_i^C = -D_i \nabla c_i$	Local C-frame
		$\vec{J}_i^V = -D_i \nabla c_i + \vec{v}_C^V c_i$	V-frame

D_i is the composition-dependent intrinsic diffusivity of component i in a chemically inhomogeneous system. In a binary system, it relates the flux of component i to its corresponding concentration gradient via Fick's law in a local C-frame (which is fixed with respect to the local bulk material of the diffusing system) and is moving with a velocity v with respect to the corresponding V-frame. The D_i are related to $\widetilde{D}$ as indicated.

Symbol	Name	Relationships	Reference Frame
$\widetilde{D}$	Interdiffusivity	$\vec{J}_i^V = -\widetilde{D} \nabla c_i$	V-frame
		$\widetilde{D} = c_1 \Omega_1 D_2 + c_2 \Omega_2 D_1$	V-frame

$\widetilde{D}$ is the composition-dependent interdiffusivity in a chemically inhomogeneous system. In a binary system, it relates the flux of either component 1 or 2 to its corresponding concentration gradient via Fick's law in a V-frame.

3.2   MASS DIFFUSION IN AN ELECTRICAL POTENTIAL GRADIENT

A gradient in electrostatic potential can produce a driving force for the mass diffusion of a species, as discussed in Section 2.2.2. Two examples of this are the potential-gradient-induced diffusional transport of charged ions in ionic conductors such as those used in solid-electrolyte batteries and the electron-current-induced diffusion of interstitial atoms in metals.

3.2.1 Charged Ions in Ionic Conductors

Consider an ionic material that contains a dilute concentration of positively charged ions that diffuse interstitially (interstitial diffusion is described in Section 8.1.4). $\tilde{D}$ is the interdiffusivity of these ions in the absence of any field. As shown in Sections 2.2.2 and 2.2.3, if an electric field, $\vec{E} = -\nabla\phi$, is applied, the diffusion potential will be the electrochemical potential given by Eq. 2.41. According to Eq. 2.21, the flux of charged interstitials is

$$\vec{J_1} = L_{11}\vec{F_1} = -L_{11}\nabla\Phi_1 = -L_{11}\nabla(\mu_1 + q_1\phi) \tag{3.47}$$

Using Eqs. 3.36 and 3.43, Eq. 3.47 takes the form

$$\vec{J_1} = -D_1\nabla c_1 - \frac{D_1 c_1 q_1}{kT}\nabla\phi \tag{3.48}$$

In the absence of a significant concentration gradient, the corresponding flux of charge is then

$$\vec{J_q} = q_1\vec{J_1} = -\frac{D_1 c_1 q_1^2}{kT}\nabla\phi \tag{3.49}$$

By comparison with Ohm's law, $\vec{J_q} = -\rho\nabla\phi$, the electrical conductivity, ρ, is

$$\rho = \frac{D_1 c_1 q_1^2}{kT} \tag{3.50}$$

The conductivity is therefore directly proportional to the diffusivity.

3.2.2 Electromigration in Metals

An applied electrical potential gradient can induce diffusion (electromigration) in metals due to a cross effect between the diffusing species and the flux of conduction electrons that will be present. When an electric field is applied to a dilute solution of interstitial atoms in a metal, there are two fluxes in the system: a flux of conduction electrons, J_q, and a flux of the interstitials, J_1. For a system maintained at constant temperature with $F_q = -\nabla\phi \equiv \vec{E}$, Eq. 2.21 gives

$$\vec{J_1} = -L_{11}\nabla\mu_1 + L_{1q}\vec{E} \tag{3.51}$$

where the interstitial chemical potential is again given by Eq. 3.36.

Evaluating the quantity L_{1q} requires understanding the physical mechanism that couples the mass flux of the interstitials to the electron current. The electron current in a metal produces a force $\vec{F}^e$ on a diffusing particle, such as an interstitial atom, which is proportional to the local current density [17]. The force arises from the change in the self-consistent electronic charge distribution surrounding the interstitial defect. The defect scatters the current-carrying electrons and creates a dipole, which in turn creates a resistance and a voltage drop across the defect. This dipole, known as a *Landauer resistivity dipole*, exerts an electrostatic force on the nucleus of the interstitial. This current-induced force is usually described phenomenologically by ascribing an effective charge to the defect, which couples to the applied electric field to create an effective force. When this force is averaged over all jumps of a diffusing interstitial, an average force $\langle \vec{F}^e \rangle$ is obtained which is proportional to $\vec{E}$, so that

$$\langle \vec{F}^e \rangle = \beta \vec{E} \tag{3.52}$$

where β is a constant. This force, in turn, induces a diffusional drift flux of interstitials given by

$$\vec{J}_1^e = \langle \vec{v}_1 \rangle c_1 = M_1 \langle \vec{F}^e \rangle c_1 = \frac{D_1 c_1 \beta}{kT} \vec{E} \tag{3.53}$$

Therefore, upon comparison with Eq. 3.51,

$$L_{1q} = \frac{D_1 c_1 \beta}{kT} \tag{3.54}$$

Consider now the interstitial flux in a material subjected to both an electrostatic driving force and a concentration gradient. Using Eqs. 3.36, 3.43, 3.51, and 3.54, for a dilute species obeying Henry's law,

$$\vec{J}_1 = -D_1 \left(\nabla c_1 - \frac{c_1 \beta}{kT} \vec{E} \right) \tag{3.55}$$

Data describing this type of electromigration of interstitial atoms in a number of systems are described by Shewmon [18]. β can be measured by passing a fixed current through an isothermal system until a quasi-steady state is achieved where $\vec{J}_1$ in Eq. 3.55 approaches zero. In this case, *uphill diffusion* (flux in the direction of the concentration gradient) takes place until the concentration gradient term (proportional to ∇c_1) cancels the electromigration term. All the quantities can be measured experimentally and β can be inferred from the measurements. Electromigration can also be used to purify a variety of metals by sweeping interstitials to one end of a specimen [18].

Electromigration is also important in the narrow metal conductors that form the current-carrying *vias* in integrated-circuit devices. Here, the electric field induces a directed migration of substitutional atoms by a vacancy mechanism (Exercise 3.10 develops a relation analogous to Eq. 3.55 for this case). With increasing miniaturization, the current densities are pushed higher and higher. In certain cases, electromigration is so severe that mass can be transported away from thin regions of metallic conductors, causing open circuits and the destruction of the electronic device. Additional Joule heating, resulting from increased current density in the thin regions, can exacerbate the damage.

3.3 MASS DIFFUSION IN A THERMAL GRADIENT

Both thermal gradients and electrical-potential gradients can induce mass diffusion. In a system containing a thermal gradient where both heat flow and mass diffusion of a dilute interstitial component 1 can occur, Eq. 2.21 predicts the interstitial flux

$$\vec{J}_1 = -L_{11} \nabla \mu_1 - \frac{L_{1Q}}{T} \nabla T \tag{3.56}$$

The interstitial chemical potential is a function of both concentration and temperature, and therefore

$$d\mu_1(c_1, T) = \left(\frac{\partial \mu_1}{\partial c_1} \right)_T dc_1 + \left(\frac{\partial \mu_1}{\partial T} \right)_{c_1} dT \tag{3.57}$$

The partial atomic entropy, $\bar{s}_1$, is

$$\bar{s}_1 = -\left(\frac{\partial \mu_1}{\partial T}\right)_{c_1} \tag{3.58}$$

and therefore

$$\nabla \mu_1 = \left(\frac{\partial \mu_1}{\partial c_1}\right)_T \nabla c_1 - \bar{s}_1 \nabla T \tag{3.59}$$

Finally, combining Eqs. 3.36, 3.43, 3.56, and 3.59 gives

$$\vec{J}_1 = -D_1 \nabla c_1 - \frac{D_1 c_1 Q_1^{\text{trans}}}{kT^2} \nabla T \tag{3.60}$$

where the parameter Q_1^{trans}, which is seen to have dimensions of energy, is termed the *heat of transport* and is given by

$$Q_1^{\text{trans}} = \frac{L_{1Q}}{L_{11}} - T\bar{s}_1 \tag{3.61}$$

Equation 3.60 indicates that mass diffusion can be induced by gradients in either the composition, or the temperature, or both. The degree of coupling of mass diffusion to the thermal gradient is determined by the heat of transport, Q_1^{trans}. The origin of Q_1^{trans} is in the asymmetry between the energy states before, during, and after a diffusing species jumps to a neighboring site.

Methods of measuring Q_1^{trans} are similar to those for measuring β in an electromigration experiment. A temperature gradient is established, causing a quasi-steady state to exist where $\vec{J}_1 = 0$. The concentration gradient in this state can be determined experimentally and the heat of transport can then be calculated using $Q_1^{\text{trans}} = -kT^2 \nabla c_1 / (c_1 \nabla T)$. In the case of interstitial carbon in b.c.c. α-Fe, the carbon atoms diffuse toward the hot end of a specimen, which indicates that the heat of transport in this system is negative. Various data for Q_1^{trans} are presented and discussed by Shewmon [18].

3.4 MASS DIFFUSION MOTIVATED BY CAPILLARITY

Capillarity is another important motivation for diffusion in many materials systems containing interfaces. The diffusion potentials of the components in the direct vicinity of an interface depend upon the local interface curvature; when interfaces possessing regions of different curvatures are present, differences in diffusion potential will drive diffusional transport between these regions in a direction that reduces the amount of energy in the system.

Figure 3.7 provides a simple example of a pure crystalline material with an undulating surface in which self-diffusion takes place by the vacancy exchange mechanism. In this case, the diffusion potential of the atoms just below the convex surface is higher than in the region where the surface is concave. This tends to establish a diffusion current through the bulk from the convex region to the concave region, as indicated, smoothing the surface and reducing the total interfacial energy. Any creation or destruction of vacancies at dislocations within the bulk

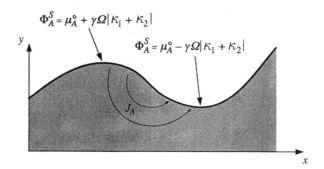

Figure 3.7:   An undulating surface possessing regions of positive and negative curvature. The curvature differences lead to diffusion-potential gradients that result in surface smoothing by diffusional transport.

can be ignored, an approximation that is usually justifiable.[9] The rate of surface smoothing can then be determined by finding expressions for the atom flux and the diffusion equation in the crystal, and then solving the diffusion equation subject to the boundary conditions at the surface. In the following section, the diffusion equation and boundary conditions are established. Exercise 14.1 provides the complete solution to the problem.

3.4.1   The Flux Equation and Diffusion Equation

The system contains two network-constrained components—host atoms and vacancies; the crystal is used as the frame for measuring the diffusional flux, and the vacancies are taken as the N_cth component. Note that there is no mass flow within the crystal, so the crystal C-frame is also a V-frame. With constant temperature and no electric field, Eq. 2.21 then reduces to

$$\vec{J}_A = L_{AA}\vec{F}_A = -L_{AA}\nabla\Phi_A = -L_{AA}\nabla(\mu_A - \mu_V)$$
$$\vec{J}_V = -\vec{J}_A \tag{3.62}$$

An expression for the coefficient L_{AA} may be obtained by considering diffusion in a very large crystal with flat surfaces. The free energy of the system, containing N_A atoms and N_V vacancies (in dilute solution), can be expressed

$$\mathcal{G} = N_A\mu_A^\circ + N_V G_V^f + kT\left[N_A\ln\left(\frac{N_A}{N_A+N_V}\right) + N_V\ln\left(\frac{N_V}{N_A+N_V}\right)\right] \tag{3.63}$$

Here, μ_A° is the free energy per atom in a vacancy-free crystal composed of only A-atoms with a flat (zero curvature) surface, $G_V^f = H_V^f - TS_V^f(\text{vib})$ is the free energy [exclusive of that due to the mixing entropy, $S_V^f(\text{vib})$ is the vibrational entropy] to form a vacancy, and the last term is the free energy of mixing due to the entropy

---

[9]Vacancy creation and destruction is discussed in Sections 11.1 and 11.4.

associated with the random distribution of the vacancies. Therefore,

$$\mu_A = \frac{\partial \mathcal{G}}{\partial N_A} = \mu_A^\circ + kT \ln \left(\frac{N_A}{N_A + N_V} \right) \cong \mu_A^\circ$$

$$\mu_V = \frac{\partial \mathcal{G}}{\partial N_V} = G_V^f + kT \ln \left(\frac{N_V}{N_A + N_V} \right) = G_V^f + kT \ln X_V \tag{3.64}$$

where X_V is the atom fraction of vacancies.[10]

If $\mu_V = 0$ when the vacancies are at their equilibrium fraction, X_V^{eq}, Eq. 3.64 may be written

$$X_V^{\text{eq}} = e^{-G_V^f/(kT)} \tag{3.65}$$

and

$$\mu_V = kT \ln \left(\frac{X_V}{X_V^{\text{eq}}} \right) = kT \ln \left(\frac{c_V}{c_V^{\text{eq}}} \right) \tag{3.66}$$

Putting these expressions into Eq. 3.62 yields

$$\vec{J}_A = -L_{AA} \nabla (\mu_A^\circ - G_V^f - kT \ln X_V) = \frac{L_{AA} kT}{X_V} \nabla X_V = -\vec{J}_V \tag{3.67}$$

Using Eq. A.12, Eq. 3.67 can be written as a Fick's-law expression for the vacancy flux

$$\vec{J}_V = -\frac{L_{AA} kT}{X_V} \nabla (\langle \Omega \rangle c_V) \approx -\frac{L_{AA} kT \langle \Omega \rangle}{X_V} \nabla c_V = -D_V \nabla c_V \tag{3.68}$$

where D_V is the vacancy diffusivity, the volume per site is assumed to be uniform, and the fact that $c_A \gg c_V$ has been incorporated. The diffusion equation for vacancies in the absence of significant dislocation sources or sinks within the crystal is then

$$\frac{\partial c_V}{\partial t} = -\nabla \cdot \vec{J}_V = D_V \nabla^2 c_V \tag{3.69}$$

From Eq. 3.68,

$$L_{AA} = \frac{D_V X_V}{\langle \Omega \rangle kT} \tag{3.70}$$

and an expression for the atom flux can be obtained by substituting Eq. 3.70 into Eq. 3.62 to obtain

$$\vec{J}_A = -\frac{D_V X_V}{\langle \Omega \rangle kT} \nabla (\mu_A - \mu_V) = -\frac{D_V X_V}{\langle \Omega \rangle kT} \nabla \Phi_A \tag{3.71}$$

If the variations in X_V throughout the crystal in Fig. 3.7 are sufficiently small, $D_V X_V / (\langle \Omega \rangle kT)$ can be assumed to be constant, and the conservation equation (see Eq. 1.18) may be written[11]

$$\frac{\partial c_A}{\partial t} = -\nabla \cdot \vec{J}_A = \frac{D_V X_V}{\langle \Omega \rangle kT} \nabla^2 (\mu_A - \mu_V) = \frac{D_V X_V}{\langle \Omega \rangle kT} \nabla^2 \Phi_A \tag{3.72}$$

[10]Note that Eqs. 3.64 for the chemical potentials are of the form given by Eq. 2.2.

[11]Equations 3.71 and 3.72 can be further developed in terms of the self-diffusivity using the atomistic models for diffusion described in Chapters 7 and 8. The resulting formulation allows for simple kinetic models of processes such as dislocation climb, surface smoothing, and diffusional creep that include the operation of vacancy sources and sinks (see Eqs. 13.3, 14.48, and 16.31).

The smoothing of a rough isotropic surface such as illustrated in Fig. 3.7 due to vacancy flow follows from Eq. 3.69 and the boundary conditions imposed on the vacancy concentration at the surface.[12] In general, the surface acts as an efficient source or sink for vacancies and the equilibrium vacancy concentration will be maintained in its vicinity. The boundary condition on c_V at the surface will therefore correspond to the local equilibrium concentration. Alternatively, if c_V, and therefore X_V, do not vary significantly throughout the crystal, smoothing can be modeled using the diffusion potential and Eq. 3.72 subject to the boundary conditions on Φ_A at the surface and in the bulk.[13]

During surface smoothing, differences in the local equilibrium values of X_V maintained in the different regions and differences in vacancy concentration throughout the crystal will be relatively small. Assuming that the crystal has isotropic surface tension, the local equilibrium vacancy concentration at the surface is a function of the local curvature [i.e., $c_V^{\mathrm{eq}} = c_V^{\mathrm{eq}}(\kappa)$], and can be found by minimizing Eq. 3.63 with respect to N_V after adding in the energy required to create the vacancies directly adjacent to the surface. When a vacancy is added to the crystal at a convex region, the crystal expands by the volume $\Delta V = \Omega_V$ and the surface area is increased by ΔA. Work must therefore be done to create the additional area. Because $\Delta A = \kappa \Delta V = \kappa \Omega_V$, the work is

$$\Delta W = \gamma \kappa \Omega_V \tag{3.73}$$

where γ is the isotropic surface tension.[14] When this surface work is added to the free energy in Eq. 3.63 and the sum is minimized,

$$\frac{c_V^{\mathrm{eq}}(\kappa)}{c_V^{\mathrm{eq}}(0)} = e^{-\gamma \kappa \Omega_V / (kT)} \tag{3.74}$$

When typical values are inserted into Eq. 3.74, $c_V^{\mathrm{eq}}(\kappa)/c_V^{\mathrm{eq}}(0)$ does not vary from unity by more than a few percent.

Because only relatively small variations in c_V occur in typical specimens undergoing sintering and diffusional creep (Chapter 16), we prefer to carry out the analyses of surface smoothing, sintering, and diffusional creep in terms of atom diffusion and the diffusion potential using Eq. 3.72. In this approach, the boundary conditions on Φ_A can be expressed quite simply.[15]

To solve the surface smoothing problem in Fig. 3.7, Eq. 3.72 can be simplified further by setting $\partial c_A / \partial t$ equal to zero because the diffusion field is, to a good approximation, in a quasi-steady state, which then reduces the problem to solving the Laplace equation

$$\nabla^2 \Phi_A = 0 \tag{3.75}$$

within the crystal subject to the boundary conditions on Φ_A described below.

[12]Methods for solving diffusion problems by setting up and solving the diffusion equation under specified boundary conditions are discussed in Chapter 5.

[13]The vacancy concentration far from the surface will generally be a function of the total surface curvature. In this case, the crystal can be assumed to be a large block possessing surfaces which on average have zero curvature. The vacancies in the deep interior can then be assumed to be in equilibrium with a flat surface.

[14]See Exercise 3.11 for further explanation.

[15]However, during the annealing of small dislocation loops (treated in Section 11.4.3), larger variations of the vacancy concentration occur and Eq. 3.68 must be employed.

3.4.2 Boundary Conditions

The boundary conditions on the diffusion potential $\Phi_A = \mu_A - \mu_V$ are readily found using results from the preceding section. At the surface where the vacancies are maintained in equilibrium, $\mu_V = 0$. The diffusion potential for the atoms is the surface work term of the form given by Eq. 3.73 plus the usual chemical term, μ_A°:

$$\Phi_A^S = \mu_A^{\circ} + \gamma \kappa \Omega_A \tag{3.76}$$

Deep within the crystal, $\mu_V = 0$ and $\mu_A = \mu_A^{\circ}$, and therefore $\Phi_A^{XL} = \mu_A^{\circ}$. The diffusion potential at the convex region of the surface is greater than that at the concave region, and atoms therefore diffuse to smooth the surface as indicated in Fig. 3.7.

We discuss surface smoothing in greater detail in Chapter 14. Exercise 14.1 uses Eq. 3.75 subject to the boundary condition given by Eq. 3.76 to obtain a quantitative solution for the evolution of the surface profile in Fig. 3.7.

3.5 MASS DIFFUSION IN THE PRESENCE OF STRESS

Because stress affects the mobility, the diffusion potential, and the boundary conditions for diffusion, it both induces and influences diffusion [19]. By examining selected effects of stress in isolation, we can study the main aspects of diffusion in stressed systems.

3.5.1 Effect of Stress on Mobilities

Consider again the diffusion of small interstitial atoms among the interstices between large host atoms in an isothermal unstressed crystal as in Section 3.1.4. According to Eqs. 3.35 and 3.42, the flux is given by

$$\vec{J}_1 = -L_{11}\nabla \mu_1 = -M_1 c_1 \nabla \mu_1 \tag{3.77}$$

The diffusion is isotropic and the mobility, M_1, is a scalar, as assumed previously.

If a general *uniform* stress field is imposed on a material, no force will be exerted on a diffusing interstitial because its energy is independent of position.[16] Assuming no other fields, the flux remains linearly related to the gradient of the chemical potential so that $\vec{J}_1 = -M_1 c_1 \nabla \mu_1$. However, M_1 will be a tensor because the stress will cause differences in the rates of atomic migration in different directions; this general effect occurs in all types of crystals.[17] It may be understood in the following way: there will be a distortion of the host lattice when the jumping atom squeezes its way from one interstitial site to another, and work must be done during the jump against any elements of the stress field that resist this distortion. Jumps in different directions will cause different distortions in the fixed stress field, so different amounts of work, W, must be done against the stress field during these jumps. The rate of a particular jump in the absence of stress is proportional to the exponential factor $\exp[-G^m/(kT)]$, where G^m is the free-energy barrier to the

---

[16]When the stress is nonuniform and stress gradients exist, the stress will exert a force, as discussed in the following section.

[17]The tensor nature of the diffusivity (mobility) in anisotropic materials is discussed in Section 4.5.

jumping process (see Chapter 7). When stress is present, the work, W, must be added to this energy barrier, and the jump rate will therefore be proportional to the factor $\exp[-(G^m + W)/(kT)]$. For almost all cases of practical importance, $W/(kT)$ is sufficiently small so that $\exp[-W/(kT)] \cong 1 - W/(kT)$, and the factor can then be written as $\exp[-G^m/(kT)][1 - W/(kT)]$. The overall interstitial mobility will be the result of the interstitials making numbers of different types of jumps in different directions. As just shown, each type of jump depends linearly on W, which, in turn, is a linear function of the elements of the stress tensor. The latter function depends on the direction of the jump, and it is therefore anticipated that the mobility should vary linearly with stress and be expressible as a tensor in the very general linear form

$$M_{ij} = M_{ij}^\circ + \sum_{kl} M_{ijkl}\sigma_{kl} \tag{3.78}$$

where the stress-dependent terms in the sum are relatively small. Similar considerations hold for the migration of substitutional atoms in a stress field (see Fig. 8.3), and the form of Eq. 3.78 should apply in such cases as well. These and other features of Eq. 3.78 are discussed by Larché and Voorhees [19].

3.5.2 Stress as a Driving Force for Diffusion: Formation of Solute-Atom Atmosphere around Dislocations

In a system containing a nonuniform stress field, a diffusing particle generally experiences a force in a direction that reduces its interaction energy with the stress field. Ignoring any effect of the stress on the mobility and focusing on the force stemming from the nonuniformity of the stress field, the stress-induced diffusion of interstitial solute atoms in the inhomogeneous stress field of an edge dislocation would look like Fig. 3.8. An interstitial in a host crystal is generally oversized for the space available and pushes outward, acting as a positive center of dilation and causing a volume expansion as illustrated in Fig. 3.9. To find the force exerted on an interstitial by a stress field, one must consider the entropy production in a

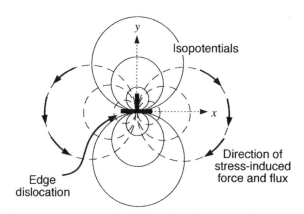

Figure 3.8:   Edge dislocation in an isotropic elastic body. Solid lines indicate isopotential cylinders for the portion of the diffusion potential of any interstitial atom present in the hydrostatic stress field of the dislocation. Dashed cylinders and tangential arrows indicate the direction of the corresponding force exerted on the interstitial atom.

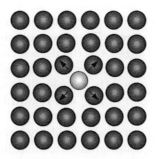

Figure 3.9: Dilation produced by an interstitial atom in a crystal. Arrows indicate outward displacements of the interstitial's nearest neighbors.

small cell embedded in the material as in Section 2.1. Suppose that the interstitial causes a pure dilation $\Delta\Omega_1$ and there are no deviatoric strains associated with the interstitial; then the supplemental work term which must be added to the right side of Eq. 2.4 is

$$dw = -P\Delta\Omega_1 dc_1 \tag{3.79}$$

where P is the hydrostatic pressure. For the case of an edge dislocation in an isotropic elastic material

$$
\begin{aligned}
P &= -\frac{\sigma_{rr} + \sigma_{\theta\theta} + \sigma_{zz}}{3} = -\frac{\sigma_{xx} + \sigma_{yy} + \sigma_{zz}}{3} \\
&= \frac{\mu(1+\nu)b}{3\pi(1-\nu)}\frac{\sin\theta}{r} = \frac{\mu(1+\nu)b}{3\pi(1-\nu)}\frac{y}{x^2 + y^2}
\end{aligned} \tag{3.80}
$$

where μ and ν are the elastic shear modulus and Poisson's ratio, respectively, and b is the magnitude of the Burgers vector [20].

When this work term is added to the chemical potential term, $\mu_1 dc_1$, and the procedure leading to Eq. 2.11 is followed, the force is

$$\vec{F}_1 = -\nabla\left(\mu_1 + \Delta\Omega_1 P\right) \tag{3.81}$$

The diffusion potential is therefore an "elastochemical" type of potential corresponding to[18]

$$\Phi_1 = \mu_1 + \Delta\Omega_1 P \tag{3.82}$$

Therefore, using Eqs. 2.16, 3.37, 3.43, and 3.82,

$$\vec{J}_1 = L_{11}\vec{F}_1 = -L_{11}\nabla\Phi_1 = -L_{11}\nabla\left(\mu_1 + \Delta\Omega_1 P\right) = -D_1\left(\nabla c_1 + \frac{c_1\Delta\Omega_1}{kT}\nabla P\right) \tag{3.83}$$

The flux has two components: the first results from the concentration gradient and the second from the gradient in hydrostatic stress.[19] The solid circles (cylinders

---

[18]The general diffusion potential for stress and chemical effects is $\Phi_1 = \mu_1 + \Delta\epsilon_{ij}\sigma_{ij}c_1$, where $\Delta\epsilon_{ij}$ is the local strain associated with the migrating species.

[19]Several typically negligible effects have been neglected in the derivation of Eq. 3.83, including (1) interactions between the interstitials, (2) effects of the interstitials on the local elastic constants, (3) quadratic terms in the elastic energy, and (4) nonlinear stress-strain behavior. A more complete treatment, applicable to the present problem, takes into account many of these effects and has been presented by Larché and Cahn [21].

in three dimensions) in Fig. 3.8 are isopotential lines for the portion of the diffusion potential due to hydrostatic stress. They were obtained by setting P equal to constant values in Eq. 3.80. Tangents to the dashed circles indicate the directions of the corresponding diffusive force arising from the dislocation stress field (this is treated in Exercise 3.6). Because $\Delta\Omega_1$ is generally positive, this force is directed away from the compressive region ($y > 0$) and toward the tensile region ($y < 0$) of the dislocation, as shown.

In the case where an edge dislocation is suddenly introduced into a region of uniform interstitial concentration, solute atoms will immediately begin diffusing toward the tensile region of the dislocation due to the pressure gradient alone (treated in Exercise 3.7). However, opposing concentration gradients build up, and eventually a steady-state equilibrium solute atmosphere, known as a *Cottrell atmosphere*, is created where the composition-gradient term cancels the stress-gradient term of Eq. 3.83 (this is demonstrated in Exercise 3.8).

From these considerations, Cottrell demonstrated that the rate at which solute atoms diffuse to dislocations and subsequently pin them in place is proportional to time$^{2/3}$ (this time dependence is derived by an approximate method in Exercise 3.9). This provided the first quantifiable theory for the strain aging caused by solute pinning of dislocations [22].

3.5.3 Influence of Stress on the Boundary Conditions for Diffusion: Diffusional Creep

In a process termed *diffusional creep*, the applied stress establishes different diffusion potentials at various sources and sinks for atoms in the material. Diffusion currents between these sources and sinks are then generated which transport atoms between them in a manner that changes the specimen shape in response to the applied stress.

A particularly simple example of this type of stress-induced diffusional transport is illustrated in Fig. 3.10, where a polycrystalline wire specimen possessing a "bamboo" grain structure is subjected to an applied tensile force, $\vec{F}_{\mathrm{app}}$. This force subjects the transverse grain boundaries to a normal tensile stress and therefore reduces the diffusion potential at these boundaries. On the other hand, the applied stress has no normal component acting on the cylindrical specimen surface and, to first order, the diffusion potential maintained there is unaffected by the applied stress. When $\vec{F}_{\mathrm{app}}$ is sufficiently large that the diffusion potential at the transverse boundaries becomes lower than that at the surface, atoms will diffuse from the surface (acting as an atom source) to the transverse boundaries (acting as sinks), thereby causing the specimen to lengthen in response to the applied stress.[20]

A similar phenomenon would occur in a single-crystal wire containing dislocations possessing Burgers vectors inclined at various angles to the stress axis. The diffusion potential at dislocations (each acting as sources or sinks) varies with each dislocation's inclination. Vacancy fluxes develop in response to gradients in diffusion potential and cause the edge dislocations to climb, and as a result, the wire lengthens in the applied tensile stress direction.

The problem of determining the elongation rate in both cases is therefore reduced to a boundary-value diffusion problem where the boundary conditions at the sources

---

[20]Surface sources and grain boundary sinks for atoms are considered in Sections 12.2 and 13.2.

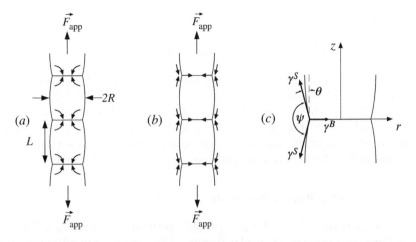

Figure 3.10:   Polycrystalline wire specimen with bamboo grain structure subjected to uniaxial tensile stress, σ_{zz}, arising from the applied force, $\vec{F}_{app}$. The bulk crystal-diffusion fluxes shown in **(a)** and grain-boundary and surface-diffusion fluxes shown in **(b)** cause diffusional elongation. **(c)** Enlarged view at the junction of the grain boundary with the surface.

and sinks are determined by the inclination of the sources and sinks relative to the applied stress and the magnitude of the applied stress. In the following we outline the procedure for obtaining the elongation rate of the polycrystalline wire shown in Fig. 3.10 for the case where the material is a pure cubic metal and the diffusion occurs through the grains as in Fig. 3.10a by a vacancy exchange mechanism. The diffusional creep rate of a single crystal containing various types of dislocations is treated in Chapter 16.

Flux and diffusion equations.   During diffusional creep, the stresses are relatively small, so variations in the vacancy concentration throughout the specimen will generally be small and can be ignored. The flux equation and diffusion equation in the grains are then given by Eqs. 3.71 and 3.75 (with $\Phi_A = \mu_A - \mu_V$), which were derived for diffusion in a crystal during surface smoothing. In both cases, quasi-steady-state diffusion may be assumed, and any creation or destruction of vacancies at dislocations within the grains can be neglected.

Boundary conditions.   The cylindrical wire surface is a source and sink for vacancies, and the condition $\mu_V = 0$ is therefore maintained there. The diffusion potential at the curved surface, Φ_A^S, is given by Eq. 3.76.

At the grain boundaries, the condition $\mu_V = 0$ should also hold. The boundaries will be under a traction, $\sigma_{nn} = \hat{n}^T \cdot \boldsymbol{\sigma} \cdot \hat{n}$, and when an atom is inserted, the tractions will be displaced as the grain expands by the volume Ω_A. For the case in Fig. 3.10, the boundary is oriented so that its normal is parallel to the z-axis and therefore $\sigma_{nn} = \sigma_{zz}$. This displacement contributes work, $\sigma_{nn}\Omega_A = \sigma_{zz}\Omega_A$, and reduces the potential energy of the system by a corresponding amount. This term must be added to the chemical term, μ_A°, and therefore the diffusion potential along the grain boundary is[21]

$$\Phi_A^B = \mu_A^\circ - \sigma_{nn}\Omega_A - \mu_V = \mu_A^\circ - \sigma_{nn}\Omega_A \qquad (3.84)$$

Φ_A^B decreases as the stress increases; an increase in the applied force increases σ_{nn}, and when σ_{nn} is sufficiently large so that $\Phi_A^B < \Phi_A^S$, atoms will diffuse from the surface to the boundaries at a quasi-steady rate. The bamboo wire behaves like a viscous material, due to the quasi-steady-state diffusional transport.[22] Complete solutions for the elongation rates due to the grain boundary and surface diffusion fluxes shown in Fig. 3.10a and b are presented in Sections 16.1.1 and 16.1.3.

3.5.4 Summary of Diffusion Potentials

The diffusion potential Φ_j is the generalized thermodynamic driving force that produces fluxes of atomic or molecular species. The diffusion potential reflects the change in energy that results from the motion of a species; therefore, it includes energy-storage mechanisms and any constraints on motion.

$\Phi_j = \mu_j$: For chemical interactions and entropic effects with no other constraint (e.g., interstitial diffusion). Section 3.1.4.

$\Phi_j = \mu_j - \mu_V$: Reflecting the additional network constraint when sites are conserved (e.g., vacancy substitution). Section 3.1.1.

$\Phi_j = \mu_j + q_j\phi$: When the diffusing species has an associated charge q_j in an electrostatic potential field ϕ (e.g., interstitial Li ions in a separator between an anode and a cathode). Section 3.2.1.

$\Phi_j = \mu_j + \Omega_j P$: Accounting for the work against a hydrostatic pressure, P, to move a species with volume Ω_j (e.g., interstitial diffusion in response to hydrostatic stress gradients). Section 3.5.2.

$\Phi_j = \mu_j + \gamma\kappa\Omega_j$: Accounting for the work against capillary pressure $\gamma\kappa$ to move a species with volume Ω_j to an isotropic surface (e.g., surface diffusion in response to a curvature gradient). Section 3.4.2.

$\Phi_j = \mu_j + \kappa_\gamma\Omega_j$: Accounting for the anisotropic equivalent to capillary pressure. κ_γ, the weighted mean curvature, is the rate of energy increase with volume addition (e.g., surface diffusion on a faceted surface). Section 14.2.2.

$\Phi_j = \mu_j - \sigma_{nn}\Omega_j$: Accounting for the work against an applied normal traction $\sigma_{nn} = \hat{n}^T \cdot (\boldsymbol{\sigma} \cdot \hat{n})$ as an atom with volume Ω_j is added to an interface with normal $\hat{n}$; $\hat{n}^T$ is the transpose of $\hat{n}$ (e.g., diffusion along an incoherent grain boundary in response to gradients in applied stress). Section 3.5.3.

$\Phi_j = \mu_j + \Omega_j\{[(\vec{b}^T \cdot \boldsymbol{\sigma}) \times \hat{\zeta}] \cdot (\hat{\zeta} \times \vec{b})\}/\{[(\hat{\zeta} \times \vec{b}) \times \hat{\zeta}] \cdot \vec{b}\}$: Accounting for the change in energy as a dislocation with Burgers vector $\vec{b}$ and unit tangent $\hat{\zeta}$ climbs

---

[21]Again, as in the derivation of Eq. 3.82, quadratic terms in the elastic energy, which are of lower order in importance, have been neglected (see Larché and Cahn [21]).

[22]For an ideally viscous material, the strain rate $\dot{\epsilon}$ is linearly related to the applied stress σ by the relation $\dot{\epsilon} = (1/\eta)\sigma$, where η is the viscosity.

with stress $\boldsymbol{\sigma}$ due to applied loads and other stress sources (i.e., other defects) for each added volume Ω_j (e.g., diffusion to a climbing dislocation by the substitutional mechanism). Section 13.3.2.[23]

$\Phi_j = \partial^2 f^{\mathrm{hom}}/\partial c_j{}^2 - 2K_c\nabla^2 c_j$: Accounting for the gradient-energy term in the diffuse interface model for conserved order parameters (e.g., "uphill" diffusion during spinodal decomposition). Section 18.3.1.

Bibliography

1. J.S. Kirkaldy and D.J. Young. *Diffusion in the Condensed State*. Institute of Metals, London, 1987.

2. J.G. Kirkwood, R.L. Baldwin, P.J. Dunlap, L.J. Gosting, and G. Kegeles. Flow equations and frames of reference for isothermal diffusion in liquids. *J. Chem. Phys.*, 33(5):1505–1513, 1960.

3. J. Bardeen and C. Herring. Diffusion in alloys and the Kirkendall effect. In J.H. Hollomon, editor, *Atom Movements*, pages 87–111. American Society for Metals, Cleveland, OH, 1951.

4. A.D. Smigelskas and E.O. Kirkendall. Zinc diffusion in alpha brass. *Trans. AIME*, 171:130–142, 1947.

5. R.W. Balluffi and B.H. Alexander. Dimensional changes normal to the direction of diffusion. *J. Appl. Phys.*, 23:953–956, 1952.

6. L.S. Darken. Diffusion, mobility and their interrelation through free energy in binary metallic systems. *Trans. AIME*, 175:184–201, 1948.

7. J. Crank. *The Mathematics of Diffusion*. Oxford University Press, Oxford, 2nd edition, 1975.

8. R.W. Balluffi. The supersaturation and precipitation of vacancies during diffusion. *Acta Metall.*, 2(2):194–202, 1954.

9. R.F. Sekerka, C.L. Jeanfils, and R.W. Heckel. The moving boundary problem. In H.I. Aaronson, editor, *Lectures on the Theory of Phase Transformations*, pages 117–169. AIME, New York, 1975.

10. R.W. Balluffi. On the determination of diffusion coefficients in chemical diffusion. *Acta Metall.*, 8(12):871–873, 1960.

11. R.W. Balluffi and B.H. Alexander. Development of porosity during diffusion in substitutional solid solutions. *J. Appl. Phys.*, 23(11):1237–1244, 1952.

12. R.W. Balluffi. Polygonization during diffusion. *J. Appl. Phys.*, 23(12):1407–1408, 1952.

13. V.Y. Doo and R.W. Balluffi. Structural changes in single crystal copper–alpha-brass diffusion couples. *Acta Metall.*, 6(6):428–438, 1959.

14. R.W. Cahn. Recovery and recrystallization. In R.W. Cahn and P. Haasen, editors, *Physical Metallurgy*, pages 1595–1671. North-Holland, Amsterdam, 1983.

15. C. Robinson. Diffusion and swelling of high polymers. II. The orientation of polymer molecules which accompanies unidirectional diffusion. *Trans. Faraday Soc.*, 42B:12–17, 1946.

16. D.R. Gaskell. *Introduction to Metallurgical Thermodynamics*. McGraw-Hill, New York, 2nd edition, 1981.

---

[23]The expression for this diffusion potential is derived in Exercise 13.3.

17. J. Hoekstra, A.P. Sutton, T.N. Todorov, and A.P. Horsfield. Electromigration of vacancies in copper. *Phys. Rev. B*, 62(13):8568–8571, 2000.

18. P. Shewmon. *Diffusion in Solids*. The Minerals, Metals and Materials Society, Warrendale, PA, 1989.

19. F.C. Larché and P.W. Voorhees. Diffusion and stresses, basic thermodynamics. *Defect and Diffusion Forum*, 129–130:31–36, 1996.

20. J.P. Hirth and J. Lothe. *Theory of Dislocations*. John Wiley & Sons, New York, 2nd edition, 1982.

21. F. Larché and J.W. Cahn. The effect of self-stress on diffusion in solids. *Acta Metall.*, 30(10):1835–1845, 1982.

22. A.H. Cottrell. *Dislocations and Plastic Flow*. Oxford University Press, Oxford, 1953.

23. L.S. Darken. Diffusion of carbon in austenite with a discontinuity in composition. *Trans. AIME*, 180:430–438, 1949.

24. U. Mehmut, D.K. Rehbein, and O.N. Carlson. Thermotransport of carbon in two-phase V–C and Nb–C alloys. *Metall. Trans.*, 17A(11):1955–1966, 1986.

25. A.H. Cottrell and B.A. Bilby. Dislocation theory of yielding and strain ageing of iron. *Proc. Phys. Soc. A*, 49:49–62, 1949.

EXERCISES

3.1 Component 1, which is unconstrained, is diffusing along a long bar while the temperature everywhere is maintained constant. Find an expression for the heat flow that would be expected to accompany this mass diffusion. What role does the heat of transport play in this phenomenon?

Solution. The basic force–flux relations are

$$\vec{J}_1 = -L_{11}\nabla\mu_1 - L_{1Q}\frac{1}{T}\nabla T$$
$$\vec{J}_Q = -L_{Q1}\nabla\mu_1 - L_{QQ}\frac{1}{T}\nabla T \tag{3.85}$$

Under isothermal conditions

$$\vec{J}_1 = -L_{11}\nabla\mu_1$$
$$\vec{J}_Q = -L_{Q1}\nabla\mu_1 \tag{3.86}$$

Therefore, using Eqs. 3.61 and 3.86,

$$\vec{J}_Q = \frac{L_{Q1}}{L_{11}}\vec{J}_1 = T\,\bar{s}_1\vec{J}_1 + Q_1^{\text{trans}}\vec{J}_1 \tag{3.87}$$

The heat flux consists of two parts. The first is the heat flux due to the flux of entropy, which is carried along by the mass flux in the form of the partial atomic entropy, $\bar{s}_1$. Because $\bar{s}_1 = \partial S/\partial N_1$, a flux of atoms will transport a flux of heat given by $\vec{J}_Q = T\vec{J}_S = T\bar{s}_1\vec{J}_1$. The second part is a "cross effect" proportional to the flux of mass, with the proportionality factor being the heat of transport.

3.2 As shown in Section 3.1.4, the diffusion of small interstitial atoms (component 1) among the interstices between large host atoms (component 2) produces an interdiffusivity, $\widetilde{D}$, for the interstitial atoms and host atoms in a V-frame given by Eq. 3.46, that is

$$\widetilde{D} = c_2\Omega_2 D_1 \tag{3.88}$$

and therefore a flux of host atoms given by

$$J_2^V = -\widetilde{D}\frac{\partial c_2}{\partial x} \tag{3.89}$$

This result holds even though the intrinsic diffusivity of the host atoms is taken to be zero and the flux of host atoms across crystal planes in the local C-frame is therefore zero. Give a physical explanation of this behavior.

Solution. When mobile interstitials diffuse across a plane in the V-frame, the material left behind shrinks, due to the loss of the dilational fields of the interstitials. This establishes a bulk flow in the diffusion zone toward the side losing interstitials and causes a compensating flow (influx) of the large host atoms toward that side even though they are not making any diffusional jumps in the crystal.

The rate of loss of volume of the material (per unit area) on one side of a fixed plane in the V-frame due to a loss of interstitials is

$$\frac{dV_1}{dt} = \Omega_1 J_1^V = -\Omega_1 \widetilde{D}\frac{\partial c_1}{\partial x} = -\Omega_1 c_2 \Omega_2 D_1 \frac{\partial c_1}{\partial x} \tag{3.90}$$

In the V-frame this must be compensated for by a gain of volume due to a gain of host atoms so that

$$\frac{dV_1}{dt} + \frac{dV_2}{dt} = 0 \tag{3.91}$$

where dV_2/dt is the rate of volume gain due to the gain of host atoms corresponding to

$$\frac{dV_2}{dt} = \Omega_2 J_2^V \tag{3.92}$$

Substituting Eqs. 3.90 and 3.92 into Eq. 3.91 and using Eq. A.10,

$$J_2^V = c_2 \Omega_1 D_1 \frac{\partial c_1}{\partial x} = -c_2 \Omega_2 D_1 \frac{\partial c_2}{\partial x} = -\widetilde{D}\frac{\partial c_2}{\partial x} \tag{3.93}$$

3.3 In a classic diffusion experiment, Darken welded an Fe–C alloy and an Fe–C–Si alloy together and annealed the resulting diffusion couple for 13 days at 1323 K, producing the concentration profile shown in Fig. 3.11 [23]. Initially, the C concentrations in the two alloys were uniform and essentially equal, whereas the Si concentration in the Fe–C–Si alloy was uniform at about 3.8%. After a diffusion anneal, the C had diffused "uphill" (in the direction of its concentration gradient) out of the Si-containing alloy. Si is a large substitutional atom, so the Fe and Si remained essentially immobile during the

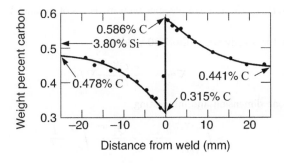

Figure 3.11: Nonuniform concentration of C produced by diffusion from an initially uniform distribution. Carbon migrated from the Fe–Si–C (left) to the Fe–C alloy (right). From Darken [23].

diffusion, whereas the small interstitial C atoms were mobile. Si increases the activity of C in Fe. Explain these results in terms of the basic driving forces for diffusion.

Solution. As the C interstitials are the only mobile species, Eq. 3.35 applies, and therefore

$$\vec{J}_1 = -L_{11}\nabla\mu_1 \tag{3.94}$$

Using the standard expression for the chemical potential,

$$\mu_1 = \mu_1^o + kT\ln a_1 \tag{3.95}$$

where $a_1 = \gamma_1 X_1$ is the activity of the interstitial C,

$$\vec{J}_1 = -\frac{L_{11}kT}{a_1}\nabla a_1 \tag{3.96}$$

The coefficient L_{11} in Eq. 3.96 is positive and the equation therefore shows that the C flux will be in the direction of reduced C activity. Because the C activity is higher in the Si-containing alloy than in the non-Si-containing alloy at the same C concentration, the uphill diffusion into the non-Si-containing alloy occurs as observed. In essence, the C is pushed out of the ternary alloy by the presence of the essentially immobile Si.

3.4 Following Shewmon, consider the metallic couple specimen consisting of two different metals, A and B, shown in Fig. 3.12 [18]. The bonded end is at temperature T_1 and the open end is at T_2. A mobile interstitial solute is present at the same concentration in both metals for which $Q_1^{\text{trans}} = -84$ kJ/mol in one leg and $Q_1^{\text{trans}} = 0$ in the other. Assuming that the interstitial concentration remains the same at the bonded interface at T_1, derive the equation for the steady-state interstitial concentration difference between the two metal legs at T_2. Assume that $T_1 > T_2$.

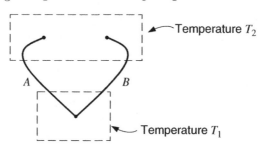
Temperature T_2

A B

Temperature T_1

Figure 3.12: Metallic couple specimen made up of metals A and B.

Solution. In the steady state, Eq. 3.60 yields

$$\nabla c_1 = -\frac{c_1 Q_1^{\text{trans}}}{kT^2}\nabla T \tag{3.97}$$

Reducing to one dimension and integrating,

$$\int_{c_1(T_1)}^{c_1(T_2)} \frac{dc_1}{c_1} = -\frac{Q_1^{\text{trans}}}{k}\int_{T_1}^{T_2} \frac{dT}{T^2} \tag{3.98}$$

Therefore,

$$\ln\left[\frac{c_1(T_2)}{c_1(T_1)}\right] = \frac{Q_1^{\text{trans}}}{k}\frac{T_1 - T_2}{T_1 T_2} \tag{3.99}$$

Therefore, for leg A,

$$c_1^A(T_2) = c_1^A(T_1) \exp\left[\frac{-84000\,(T_1 - T_2)}{N_o k T_1 T_2}\right] \tag{3.100}$$

while for leg B, $c_1^B(T_2) = c_1^B(T_1)$. Finally, because $c_1^A(T_1) = c_1^B(T_1) = c_1^B(T_2) \equiv c_1(T_1)$,

$$\Delta c_1 = c_1(T_1) \left\{\exp\left[\frac{-84000\,(T_1 - T_2)}{N_o k T_1 T_2}\right] - 1\right\} \tag{3.101}$$

3.5 Suppose that a two-phase system consists of a fine dispersion of a carbide phase in a matrix. The carbide particles are in equilibrium with C dissolved interstitially in the matrix phase, with the equilibrium solubility given by

$$c_1 = c_1^\circ e^{-\Delta H/(kT)} \tag{3.102}$$

If a bar-shaped specimen of this material is subjected to a steep thermal gradient along the bar, C atoms move against the thermal gradient (toward the cold end) and carbide particles shrink at the hot end and grow at the cold end, even though the heat of transport is negative! (For an example, see the paper by Mehmut et al. [24].) Explain how this can occur.

- Assume that the concentration of C in the matrix is maintained in local equilibrium with the carbide particles, which act as good sources and sinks for the C atoms. Also, ΔH is positive and larger in magnitude than the heat of transport.

Solution. The local C concentration will be coupled to the local temperature by Eq. 3.102, and therefore

$$\frac{dc_1}{dx} = \frac{dc_1}{dT}\frac{dT}{dx} = c_1 \frac{\Delta H}{kT^2}\frac{dT}{dx} \tag{3.103}$$

Substitution of Eq. 3.103 into Eq. 3.60 then yields

$$J_1 = -\frac{D_1 c_1}{kT^2}\left(\Delta H + Q^{\text{trans}}\right)\frac{dT}{dx} \tag{3.104}$$

Because $\left(\Delta H + Q^{\text{trans}}\right)$ is positive, the C atoms will be swept toward the cold end, as observed.

3.6 Show that the forces exerted on interstitial atoms by the stress field of an edge dislocation are tangent to the dashed circles in the directions of the arrows shown in Fig. 3.8.

Solution. The hydrostatic stress on an interstitial in the stress field is given by Eq. 3.80 and the force is equal to $\vec{F}_1 = -\Omega_1 \nabla P$. Therefore,

$$\vec{F}_1 = -\Omega_1 A \nabla\left[\frac{y'}{x'^2 + y'^2}\right] \tag{3.105}$$

where A is a positive constant. Translating the origin of the (x', y') coordinate system to a new position corresponding to $(x' = R, y' = 0)$, the expression for $\vec{F}_1$ in the new (x, y) coordinate system is

$$\vec{F}_1 = -\Omega_1 A \nabla\left[\frac{y}{x^2 + 2xR + R^2 + y^2}\right] \tag{3.106}$$

Converting to cylindrical coordinates,

$$\vec{F}_1 = -\Omega_1\,A\,\nabla\left[\frac{r\sin\theta}{r^2 + R^2 + 2rR\cos\theta}\right] \tag{3.107}$$

The gradient operator in cylindrical coordinates is

$$\nabla = \hat{u}_r\frac{\partial}{\partial r} + \hat{u}_\theta\frac{1}{r}\frac{\partial}{\partial\theta} \tag{3.108}$$

Therefore, using Eq. 3.107 and Eq. 3.108 yields

$$\vec{F}_1 = -\frac{\Omega_1\,A}{[R^2 + r^2 + 2Rr\cos\theta]^2}\left\{\hat{u}_r(R^2 - r^2)\sin\theta + \hat{u}_\theta\left[(R^2 + r^2)\cos\theta + 2Rr\right]\right\} \tag{3.109}$$

The force on an interstitial lying on a cylinder of radius R centered on the origin where $r = R$ is then

$$\vec{F}_1 = -\frac{\Omega_1\,A}{2R^2(1 + \cos\theta)}\hat{u}_\theta \tag{3.110}$$

The force anywhere on the cylinder therefore lies along $-\hat{u}_\theta$, which is tangential to the cylinder in the direction of decreasing θ.

3.7 Consider the diffusional flux in the vicinity of an edge dislocation after it is suddenly inserted into a material that has an initially uniform concentration of interstitial solute atoms.

(a) Calculate the initial rate at which the solute increases in a cylinder that has an axis coincident with the dislocation and a radius R. Assume that the solute forms a Henrian solution.

(b) Find an expression for the concentration gradient at a long time when mass diffusion has ceased.

Solution.

(a) The diffusion flux is given by Eq. 3.83. Initially, the concentration gradient is zero and the flux is due entirely to the stress gradient. Therefore,

$$\begin{aligned}
\vec{J}_1 &= -\frac{D_1\Omega_1\,c_1}{kT}\left[\frac{\partial P}{\partial r}\hat{u}_r + \frac{1}{r}\frac{\partial P}{\partial\theta}\hat{u}_\theta\right] \\
&= -\frac{D_1\Omega_1\,c_1\mu(1+\nu)b}{3\pi kT(1-\nu)}\left[-\frac{\sin\theta}{r^2}\hat{u}_r + \frac{\cos\theta}{r^2}\hat{u}_\theta\right]
\end{aligned} \tag{3.111}$$

Now, integrate the flux entering the cylinder, noting that the θ component contributes nothing:

$$I_1 = \int_0^{2\pi}\frac{A\sin\theta}{R^2}R\,d\theta = 0 \tag{3.112}$$

where $A = $ constant. Note that this result can be inferred immediately, due to the symmetry of the problem.

(b) When mass flow has ceased, the flux in Eq. 3.83 is zero and therefore

$$\nabla c_1 = -\frac{\Omega_1\,c_1\mu(1+\nu)b}{3\pi kT(1-\nu)}\left[-\frac{\sin\theta}{r^2}\hat{u}_r + \frac{\cos\theta}{r^2}\hat{u}_\theta\right] \tag{3.113}$$

3.8 The diffusion of interstitial atoms in the stress field of a dislocation was considered in Section 3.5.2. Interstitials diffuse about and eventually form an equilibrium distribution around the dislocation (known as a *Cottrell atmosphere*), which is invariant with time. Assume that the system is very large and that the interstitial concentration is therefore maintained at a concentration c_1° far from the dislocation. Use Eq. 3.83 to show that in this equilibrium atmosphere, the interstitial concentration on a site where the hydrostatic pressure, P, due to the dislocation is

$$c_1^{eq} = c_1^\circ e^{-\Omega_1 P/(kT)} \tag{3.114}$$

Solution. According to Eq. 3.83,

$$\vec{J}_1 = -D_1 \left(\nabla c_1 + \frac{\Omega_1 c_1}{kT} \nabla P \right) = -D_1 c_1 \nabla \left(\ln c_1 + \frac{\Omega_1}{kT} P \right) \tag{3.115}$$

At equilibrium, $\vec{J}_1 = 0$ and therefore

$$\ln c_1^{eq} + \frac{\Omega_1 P}{kT} = a_1 = \text{constant} \tag{3.116}$$

Because $c_1^{eq} = c_1^\circ$ at large distances from the dislocation where $P = 0$, $a_1 = \ln c_1^\circ$,

$$c_1^{eq} = c_1^\circ e^{-\Omega_1 P/(kT)} \tag{3.117}$$

3.9 In the *Encyclopedia of Twentieth Century Physics*, R.W. Cahn describes A.H. Cottrell and B.A. Bilby's result that strain aging in an interstitial solid solution increases with time as $t^{2/3}$ as the coming of age of the science of quantitative metallurgy [25]. *Strain aging* is a phenomenon that occurs when interstitial atoms diffuse to dislocations in a material and adhere to their cores and cause them to be immobilized. Especially remarkable is that the $t^{2/3}$ relation was derived even before dislocations had been observed.

Derive this result for an edge dislocation in an isotropic material.

- Assume that the degree of the strain aging is proportional to the number of interstitials that reach the dislocation.

- Assume that the interstitial species is initially uniformly distributed and that an edge dislocation is suddenly introduced into the crystal.

- Assume that the force, $-\Omega_1 \nabla P$, is the dominant driving force for interstitial diffusion. Neglect contributions due to ∇c.

- Find the time dependence of the number of interstitials that reach the dislocation. Take into account the rate at which the interstitials travel along the circular paths in Fig. 3.8 and the number of these paths funneling interstitials into the dislocation core.

Solution. The tangential velocity, v, of an interstitial traveling along a circular path of radius R in Fig. 3.8 will be proportional to the force $\vec{F}_1 = -\Omega_1 \nabla P$ exerted by the dislocation. In cylindrical coordinates, P is proportional to $\sin\theta/r$, so

$$\vec{F}_1 \propto \nabla P \propto \left(\hat{u}_r \frac{\partial P}{\partial r} + \hat{u}_\theta \frac{1}{r} \frac{\partial P}{\partial \theta} \right) \propto \frac{1}{r^2} \left(-\hat{u}_r \sin\theta + \hat{u}_\theta \cos\theta \right) \tag{3.118}$$

Therefore, $v \propto F_1 \propto 1/r^2$. As shown in Fig. 3.8, v at equivalent points on each circle will scale as $1/r^2$, and because r at these points scales as R,

$$v \propto \frac{1}{R^2} \tag{3.119}$$

The average velocity, $\langle v \rangle$, around each circular path will therefore scale as $1/R^2$. Since the distance around a path is $2\pi R$, the time, t_R, required to travel completely around a path of radius R scales as

$$t_R = \frac{2\pi R}{\langle v \rangle} \propto R^3 \tag{3.120}$$

Therefore, at time, t, the circles with radii less than

$$R_{\rm crit} \propto t^{1/3} \tag{3.121}$$

will be depleted of solute. During an increment of time dt, the average distance at which interstitials along the active flux circles approach the dislocation is equal (to a reasonable approximation) to $ds = \langle v \rangle \, dt$. The total volume (per unit length of dislocation) supplying atoms during this period is then

$$dV \propto dt \int_{R_{\rm crit}}^{\infty} \langle v \rangle \, dR \propto \frac{1}{R_{\rm crit}} dt \tag{3.122}$$

where the integral is taken over only the active flux circles. Because the concentration was initially uniform, the number of interstitials reaching the dislocation in time t, designated by N, is therefore proportional to the volume swept out. Therefore, substituting Eq. 3.121 in Eq. 3.122 and integrating,

$$N \propto V \propto \int_0^t \frac{dt}{t^{1/3}} \propto t^{2/3} \tag{3.123}$$

More detailed treatments are given in the original paper by Cottrell and Bilby [25] and in the summary in Cottrell's text on dislocation theory [22].

3.10 Derive the expression

$$\vec{J}_A = -\frac{D_V c_V \beta}{kT} \vec{E}$$

for the electromigration of substitutional atoms in a pure metal, where D_V is the vacancy diffusivity and c_V is the vacancy concentration. Assume that:

- There are two mobile components: atoms and vacancies.
- Diffusion occurs by the exchange of atoms and vacancies.
- There is a sufficient density of sources and sinks for vacancies so that the vacancies are maintained at their local equilibrium concentration everywhere.

Solution. Vacancies are defects that scatter the conduction electrons and are therefore subject to a force which in turn induces a vacancy current. The vacancy current results in an equal and opposite atom current. The components are network constrained so that Eq. 2.21 for the vacancies, which are taken as the N_cth component, is

$$\vec{J}_V = -L_{VA} \nabla \left(\mu_A - \mu_V \right) + L_{Vq} \vec{E}$$

Because $\nabla \mu_A = 0$ (see Eq. 3.64) and $\mu_V = 0$,

$$\vec{J}_V = L_{Vq} \vec{E}$$

The vacancy current is therefore due solely to the cross term arising from the current of conduction electrons (which is proportional to $\vec{E}$). The coupling coefficient for the vacancies is the off-diagonal coefficient L_{Vq} which can be evaluated using the same procedure as that which led to Eq. 3.54 for the electromigration of interstitial atoms in a metal. Therefore, if $\langle \vec{v}_V \rangle$ is the average drift velocity of the vacancies induced by the current and M_V is the vacancy mobility,

$$\vec{J}_V = c_V \langle \vec{v}_V \rangle = c_V M_V \langle \vec{F}_V^e \rangle = \frac{c_V D_V \beta \vec{E}}{kT} = -\vec{J}_A$$

3.11 **(a)** It is claimed in Section C.2.1 that the mean curvature, κ, of a curved interface is the ratio of the increase in its area to the volume swept out when the interface is displaced toward its convex side. Demonstrate this by creating a small localized "bump" on the initially spherical interface illustrated in Fig. 3.13.

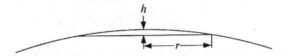

Figure 3.13:   Circular cap (spherical zone) on a spherical interface.

(b) Show that Eq. 3.124 also holds when the volume swept out is in the form of a thin layer of thickness dw, as illustrated in Fig. 3.14.

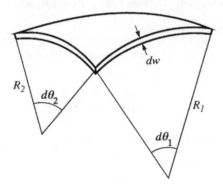

Figure 3.14:   Layer of thickness dw swept out by addition of material at an interface with curvature $\kappa = (1/R_1) + (1/R_2)$.

- Construct the bump in the form of a small circular cap (spherical zone) by increasing h infinitesimally while holding r constant. Then show that

$$\kappa = \frac{dA}{dV} \tag{3.124}$$

where dA and dV are, respectively, the increases in interfacial area and volume swept out due to the construction of the bump.

Solution.

(a) The area of the circular cap in Fig. 3.13 is

$$A = \pi \left(r^2 + h^2 \right)$$

Here r and h are related to the radius of curvature of the spherical surface, R, by the relation

$$R = \frac{r^2}{2h} \left(1 + \frac{h^2}{r^2} \right) \tag{3.125}$$

The volume under the circular cap is given by

$$V = \frac{\pi}{2} h r^2 + \frac{\pi}{6} h^3$$

If the bump is now created by forming a new cap of height $h + dh$ while keeping r constant,

$$dA = 2\pi h \, dh \tag{3.126}$$

$$dV = \frac{\pi r^2}{2} \left(1 + \frac{h^2}{r^2} \right) dh \tag{3.127}$$

Therefore, using Eqs. 3.125, 3.126, and 3.127, and the fact that $h^2/r^2 \ll 1$,

$$\frac{dA}{dV} = \frac{2}{R} = \kappa$$

(b) The increase in area is

$$dA = (R_1 + dw) \, d\theta_1 \, (R_2 + dw) \, d\theta_2 - R_1 \, d\theta_1 \, R_2 \, d\theta_2 = (R_1 + R_2) \, dw \, d\theta_1 \, d\theta_2$$

The volume swept out is

$$dV = R_1 \, d\theta_1 \, R_2 \, d\theta_2 \, dw$$

Therefore,

$$\frac{dA}{dV} = \frac{1}{R_1} + \frac{1}{R_2} = \kappa$$

CHAPTER 4

THE DIFFUSION EQUATION

The diffusion equation is the partial-differential equation that governs the evolution of the concentration field produced by a given flux. With appropriate boundary and initial conditions, the solution to this equation gives the time- and spatial-dependence of the concentration. In this chapter we examine various forms assumed by the diffusion equation when Fick's law is obeyed for the flux. Cases where the diffusivity is constant, a function of concentration, a function of time, or a function of direction are included. In Chapter 5 we discuss mathematical methods of obtaining solutions to the diffusion equation for various boundary-value problems.

4.1 FICK'S SECOND LAW

If the diffusive flux in a system is $\vec{J}$, Section 1.3.5 and Eq. 1.18 are used to write the diffusion equation in the general form

$$\frac{\partial c}{\partial t} = \dot{n} - \nabla \cdot \vec{J} \tag{4.1}$$

where $\dot{n}$ is an added source or sink term corresponding to the rate per unit volume at which diffusing material is created locally, possibly by means of chemical reaction or fast-particle irradiation, and $\vec{J}$ is any flux referred to a V-frame. There frequently are no sources or sinks operating, and $\dot{n} = 0$ in Eq. 4.1. When Fick's law applies (see Section 3.1) and $\dot{n} = 0$, Eq. 4.1 takes the general form

$$\frac{\partial c}{\partial t} = -\nabla \cdot \vec{J} = \nabla \cdot (D\nabla c) \tag{4.2}$$

which is sometimes called *Fick's second law* (note that Fick's second law is simply a consequence of the conservation of the diffusing species).

Accumulation within a volume depends only on the fluxes at its boundary. For example, in one dimension,

$$\frac{\partial N}{\partial t} = \int_{x_1}^{x_2} \frac{\partial c}{\partial t} A \, dx = \int_{x_1}^{x_2} \frac{\partial}{\partial x}\left(D\frac{\partial c}{\partial x}\right) A \, dx = [J(x_1, t) - J(x_2, t)] \, A \tag{4.3}$$

where N is the number of particles and A is the area through which the diffusion occurs. In three dimensions,

$$\frac{\partial N}{\partial t} = \int_V \frac{\partial c}{\partial t} dV = \int_V -\nabla \cdot \vec{J} \, dV = -\int_{\partial V} \vec{J}(\vec{r}, t) \cdot \hat{n} \, dA \tag{4.4}$$

where in the final integral, $\vec{J}(\vec{r}, t)$ is the time-dependent value of flux at the oriented surface ∂V that bounds V. The geometrical interpretation in Fig. 4.1 shows how $c(x, t)$ changes locally; the equations above imply a conservation constraint for the entire concentration field.

Because Eq. 4.2 has one time and two spatial derivatives, its solution requires three independent conditions: an initial condition and two independent boundary conditions. Boundary conditions typically may look like

$$c(\vec{r} = \vec{r}_B) = f(t) = c_B(t) \qquad \text{or} \qquad \vec{J}(\vec{r} = \vec{r}_B) \cdot \hat{n}_B = g(t) = J_B(t) \tag{4.5}$$

where $\hat{n}_B$ is the normal to the boundary and the initial conditions have the form

$$c(x, y, z, t = t_0) = c(\vec{r}, t = t_0) = h(x, y, z) = h(\vec{r}) = c_0(x, y, z) \tag{4.6}$$

In Chapter 3, several different types of diffusivity were introduced for diffusion in a chemically homogeneous system or for interdiffusion in a solution. In each case, Fick's law applies, but the appropriate diffusivity depends on the particular system. The development of the diffusion equation in this chapter depends only on the form of Fick's law, $\vec{J} = -D\nabla c$. D is a placeholder for the appropriate diffusivity, just as $\vec{J}$ and c are placeholders for the type of component that diffuses.

Equation 4.2 can take various forms, depending upon the behavior of D. The simplest case is when D is constant. However, as discussed below, D may be a function of concentration, particularly in highly concentrated solutions where the interactions between solute atoms are significant. Also, D may be a function of time: for example, when the temperature of the diffusing body changes with time. D may also depend upon the direction of the diffusion in anisotropic materials.

4.1.1 Linearization of the Diffusion Equation

Methods to solve the diffusion equation for specific boundary and initial conditions are presented in Chapter 5. Many analytic solutions exist for the special case that D is uniform. This is generally *not* the case for interdiffusivity $\widetilde{D}$ (Eq. 3.25). If $\widetilde{D}$ does not vary rapidly with composition, it can be replaced by successive approximations of a uniform diffusivity and results in a *linearization* of the diffusion equation. The linearized form permits approximate models from known solutions. The diffusivity is expanded about its average value, $\tilde{D}_0$, as follows

$$\tilde{D}(c) = \tilde{D}_0 + \frac{\tilde{D}_1}{\langle c \rangle} \Delta c + \cdots \tag{4.7}$$

where $\Delta c = c - \langle c \rangle$, and

$$\tilde{D}_1 = \left. \frac{\partial \tilde{D}}{\partial c} \right|_{c=\langle c \rangle} \langle c \rangle \tag{4.8}$$

The diffusion equation becomes

$$\frac{\partial c}{\partial t} = \tilde{D}_0 \nabla^2 c + \frac{\tilde{D}_1}{\langle c \rangle} \Delta c \nabla^2 c + \frac{1}{\langle c \rangle} \nabla c \cdot \tilde{D}_1 \nabla c + \cdots \tag{4.9}$$

The lowest-order approximation for small Δc and small $|\nabla c|$ is

$$\frac{\partial c}{\partial t} = \tilde{D}_0 \nabla^2 c \tag{4.10}$$

which is the diffusion equation for constant diffusivity.

4.1.2 Relation of Fick's Second Law to the Heat Equation

For evolution of a temperature field during heat flow, an equation with the same form as Eq. 4.2 arises:

$$\frac{\partial h}{\partial t} = -\nabla \cdot \vec{J}_Q$$
$$c_P \frac{\partial T}{\partial t} = -\nabla \cdot (-K \nabla T) \tag{4.11}$$
$$\frac{\partial T}{\partial t} = \nabla \cdot \left(\frac{K}{c_P} \nabla T \right) = \nabla \cdot (\kappa \nabla T)$$

where h is the enthalpy density and c_P is the heat capacity per unit volume. The ratio K/c_P is called the *thermal diffusivity*, κ. It is assumed that no enthalpy is stored by a phase change and that c_P is constant.

Therefore, any result that follows from considerations of the form of Fick's second law applies to evolution of heat as well as concentration. However, the thermal and mass diffusion equations differ physically. The mass diffusion equation, $\partial c/\partial t = \nabla \cdot D \nabla c$, is a partial-differential equation for the density of an extensive quantity, and in the thermal case, $\partial T/\partial t = \nabla \cdot \kappa \nabla T$ is a partial-differential equation for an intensive quantity. The difference arises because for mass diffusion, the driving force is converted from a gradient in a potential $\nabla \mu$ to a gradient in concentration ∇c, which is easier to measure. For thermal diffusion, the time-dependent temperature arises because the enthalpy density is inferred from a temperature measurement.

4.1.3   Variational Interpretation of the Diffusion Equation

The rate of entropy production, $\dot\sigma$ (Eq. 2.19), for one-dimensional diffusion becomes

$$\dot\sigma = \frac{kD}{c}\left(\frac{dc}{dx}\right)^2 \tag{4.12}$$

when the activity coefficient is independent of concentration. Localized changes in $c(x, t)$ affect the rate of total entropy production. How changes in the evolution of a field affect a functional (such as an integral quantity like total entropy production) is a topic in the calculus of variations [1].

For an adiabatic system, the rate of total entropy production $\dot S_{\text{tot}}$ is a functional of the concentration field $c(x)$,

$$\dot S_{\text{tot}}\left[c(x)\right] = \int_{x_1}^{x_2} \frac{kD}{c}\left(\frac{dc}{dx}\right)^2 dx \tag{4.13}$$

The functional gradient of $\dot S_{\text{tot}}$ indicates the function pointing in the "direction" of fastest increase. Gradients depend on an inner product because it provides a measure of "distance" for functions [2]. One choice of an inner product for functions is the $L2$ inner product, defined by

$$p(x)q(x) = \int pq \, dx \tag{4.14}$$

so the magnitude of a function is related to the integral of its square: $|p(x)| = (p\,p)^{1/2}$. Note that least-squares data fits use this inner product.

The functional gradient of F (or gradient of a vector function) can be defined by G_F, and the inner product with a velocity field v:

$$\left.\frac{d}{dt}F(x + vt)\right|_{t=0} = G_F\, v \tag{4.15}$$

That is, of all possible functions $v(x)$, those that are parallel, subject to choice of norm or inner product, to G_F give the fastest increase in F. For the entropy production with $D = $ constant,

$$\left.\frac{d}{dt}\dot S_{\text{tot}}\left(c + \frac{\partial c}{\partial t}t\right)\right|_{t=0} = \frac{2kD}{c}\int_{x_1}^{x_2}\frac{dc}{dx}\left[\frac{d}{dx}\left(\frac{\partial c}{\partial t}\right)\right]dx = G_{\dot S_{\text{tot}}}\frac{\partial c}{\partial t} \tag{4.16}$$

Integrating by parts,

$$G_{\dot S_{\text{tot}}}\frac{\partial c}{\partial t} = 2kD\left[-\int_{x_1}^{x_2}\frac{d^2c}{dx^2}\frac{\partial c}{\partial t}dx + \left.\left(\frac{\partial c}{\partial x}\frac{\partial c}{\partial t}\right)\right|_{x_1}^{x_2}\right] \tag{4.17}$$

If the boundary conditions are zero flux or fixed composition, the last term vanishes. Comparison with the $L2$ inner product reveals that for evolution according to the diffusion equation, $c(x, t)$ changes so that $\ddot S_{\text{tot}}$ (total entropy "acceleration") is its most negative. Thus, entropy production, which is always positive, decreases in time as rapidly as possible when $dc/dt \propto -G_{\dot S_{\text{tot}}} \propto d^2c/dx^2$.

4.2   CONSTANT DIFFUSIVITY

When D is constant, Eq. 4.2 takes the relatively simple form of the linear second-order partial differential equation

$$\frac{\partial c}{\partial t} = D\nabla^2 c \tag{4.18}$$

Some of the major features of this equation are discussed below, and methods of solving it under a variety of boundary and initial conditions are described at length in Chapter 5.

4.2.1   Geometrical Interpretation of the Diffusion Equation when Diffusivity is Constant

Figure 4.1 illustrates how a one-dimensional concentration field, $c(x,t)$, evolves according to Eq. 4.18. The right-hand side of Eq. 4.18 is proportional to the curvature of the concentration profile. Where the curvature is negative, as on the left-hand side, the concentration must decrease at a rate proportional to the magnitude of the curvature. Conversely, the concentration must increase on the right-hand side, where the curvature is positive.

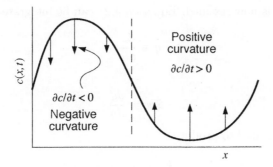

Figure 4.1:   Evolution of concentration field according to Fick's law. $\partial c/\partial t$ is proportional to the curvature of the concentration field.

4.2.2   Scaling of the Diffusion Equation

Under certain conditions, boundary-value diffusion problems can be solved conveniently by scaling. First, introduce the dimensionless variable η,

$$\eta \equiv \frac{x}{\sqrt{4Dt}} \tag{4.19}$$

into the diffusion equation. Using Eq. 4.18 for one-dimensional diffusion and

$$\frac{\partial}{\partial t} = \frac{\partial \eta}{\partial t}\frac{\partial}{\partial \eta} \qquad\qquad \frac{\partial}{\partial x} = \frac{\partial \eta}{\partial x}\frac{\partial}{\partial \eta} \tag{4.20}$$

the diffusion equation becomes

$$-2\eta\frac{\partial c}{\partial \eta} = \frac{\partial^2 c}{\partial \eta^2} \tag{4.21}$$

Next, suppose that for *the particular boundary-value problem under consideration,* the initial and boundary conditions are unchanged by scale change:

$$\bar{x} = \lambda x \quad \bar{t} = \lambda^2 t \tag{4.22}$$

Then η is invariant under the scaling corresponding to Eq. 4.19 and c becomes a function of the single variable, η. The diffusion equation becomes an ordinary differential equation (i.e., $\partial \to d$).

If the boundary-value diffusion problem can be scaled according to Eq. 4.19, it is considerably easier to solve.[1] Consider the one-dimensional step-function diffusion problem shown in Fig. 4.2, where

$$c(x, t = 0) = \left\{ \begin{array}{ll} c^L & -\infty < x < 0 \\ c^R & 0 < x < \infty \end{array} \right. \tag{4.23}$$

$$c(-\infty, t) = c^L; \quad c(\infty, t) = c^R$$

The initial and boundary conditions given by Eq. 4.23 are transformed by scaling into

$$c(-\infty) = c^L \qquad \text{and} \qquad c(\infty) = c^R \tag{4.24}$$

and the diffusion equation has the form in Eq. 4.21. The entire boundary-value diffusion problem is now rescaled. Equation 4.21 can be integrated by letting

$$q \equiv \frac{dc}{d\eta} \tag{4.25}$$

Then

$$-2\eta q = \frac{dq}{d\eta} \tag{4.26}$$

which can be integrated to produce

$$\frac{dc}{d\eta} = a_1 e^{-\eta^2} \tag{4.27}$$

where a_1 is a constant. Integrating again yields

$$c(\eta) - c(\eta = \eta_0) = a_1 \int_{\eta_0}^{\eta} e^{-\zeta^2} d\zeta \tag{4.28}$$

Applying the step-function initial conditions in Eq. 4.24,

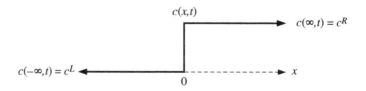

Figure 4.2: One-dimensional step-function initial conditions.

<hr />

[1] The diffusion equation itself can always be rescaled. However, to solve a boundary-value diffusion problem using the scaling method, the initial and boundary conditions must also be scalable.

$$c(\eta) = c\left(\frac{x}{\sqrt{4Dt}}\right) = c^L + a_2\left(\frac{2}{\sqrt{\pi}}\int_{-\infty}^{0} e^{-\zeta^2}\, d\zeta + \frac{2}{\sqrt{\pi}}\int_{0}^{x/\sqrt{4Dt}} e^{-\zeta^2}\, d\zeta\right) \quad (4.29)$$

where a_2 is a constant. The integral with the limit $x/\sqrt{4Dt}$ is known as the *error function*, abbreviated "erf":

$$\mathrm{erf}(z) \equiv \frac{2}{\sqrt{\pi}}\int_0^z e^{-\zeta^2}\, d\zeta \quad (4.30)$$

The error function has the properties $\mathrm{erf}(0) = 0$, $\mathrm{erf}(\infty) = 1$, and $\mathrm{erf}(-z) = -\mathrm{erf}(z)$.

So, after evaluating a_2 by using the boundary conditions, the diffusion problem posed above has the solution

$$c(x,t) = \bar{c} + \frac{\Delta c}{2}\mathrm{erf}\left(\frac{x}{\sqrt{4Dt}}\right) \quad (4.31)$$

where $\bar{c} = (c^R + c^L)/2$ and $\Delta c = c^R - c^L$. When c is assigned units of particles per unit length, Eq. 4.31 describes the one-dimensional diffusion along x from an initial step function on a line in one dimension as in Fig. 4.3a. When c has units of particles per unit area, it describes the one-dimensional diffusion from a step function in a two-dimensional plane as in Fig. 4.3b, and when the units are particles per unit volume, it describes the one-dimensional diffusion from a step function in three dimensions, as in Fig. 4.3c.

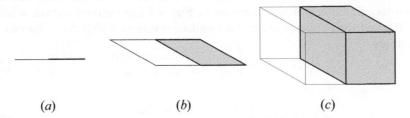

(a)   (b)   (c)

Figure 4.3:   Initial step-function distributions in **(a)** one, **(b)** two, and **(c)** three dimensions.

Scaling as a Means to Compare Similar Systems.   When the diffusion problem is invariant to the scaling parameter $\eta = x/\sqrt{4Dt}$, equal values of η can be used to determine relationships between length, time, and the value of the diffusivity. For example, consider two masses that differ only in their length dimension. Let the first block have length L and the second block have length αL. If at a time, τ, a particular concentration appears at the center of the first block, the same concentration will appear in the second block at time $\alpha^2\tau$.

4.2.3 Superposition

Suppose that $p(x,t)$ is a solution to the one-dimensional diffusion equation

$$\frac{\partial p}{\partial t} = D\frac{\partial^2 p}{\partial x^2} \quad (4.32)$$

with boundary and initial conditions

$$p(x = a, t) = A_p(t) \quad p(x = b, t) = B_p(t) \quad p(x, t = 0) = I_p(x) \tag{4.33}$$

and that $q(x, t)$ is a solution to

$$\frac{\partial q}{\partial t} = D \frac{\partial^2 q}{\partial x^2} \tag{4.34}$$

with boundary and initial conditions

$$q(x = a, t) = A_q(t) \quad q(x = b, t) = B_q(t) \quad q(x, t = 0) = I_q(x) \tag{4.35}$$

Then, because the diffusion equation is a linear second-order differential equation, $r(x, t) = p(x, t) + q(x, t)$ is a solution for the boundary conditions and the initial condition:

$$\begin{aligned} r(x = a, t) &= A_p(t) + A_q(t) \\ r(x = b, t) &= B_p(t) + B_q(t) \\ r(x, t = 0) &= I_p(x) + I_q(x) \end{aligned} \tag{4.36}$$

The superposition of two solutions therefore also solves the diffusion equation with superposed boundary and initial conditions.

Superposition of two displaced step-function initial conditions permits solutions that describe diffusion from an initially localized source into an infinite domain. The two step-function initial conditions in Fig. 4.4 have error-function solutions (Eq. 4.31), and their superposition is a localized source of width Δx. The two step functions are

$$c(x, t = 0) = \begin{cases} 0 & -\infty < x < 0 \\ c_0 & 0 < x < \infty \end{cases} \tag{4.37}$$

and[2]

$$c(x, t = 0) = \begin{cases} 0 & -\infty < x < \Delta x \\ -c_0 & \Delta x < x < \infty \end{cases} \tag{4.38}$$

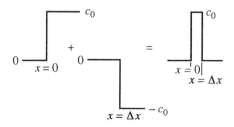

Figure 4.4:   Superposition method for constructing localized source for one-dimensional diffusion along x.

---

[2]Although one initial condition is unphysical (i.e., a negative concentration), the superposition is physical and justifies its use. The negative concentration is similar to the use of a negative electrical image charge to solve the electrostatics problem of the potential field produced by a positive charge in a planar half-space where the plane bounding the half-space is held at zero potential. The negative image charge outside the half-space allows superposition and satisfies the boundary condition at the plane bounding the half-space [3].

Each step function evolves according to an error-function solution of the type given by Eq. 4.31, and their superposition is

$$c(x,t) = \frac{c_0}{2} + \frac{c_0}{2}\frac{2}{\sqrt{\pi}}\int_0^{x/\sqrt{4Dt}} e^{-\zeta^2}\,d\zeta - \frac{c_0}{2} - \frac{c_0}{2}\frac{2}{\sqrt{\pi}}\int_0^{(x-\Delta x)/\sqrt{4Dt}} e^{-\zeta^2}\,d\zeta$$
$$= \frac{c_0}{\sqrt{\pi}}\int_{(x-\Delta x)/\sqrt{4Dt}}^{x/\sqrt{4Dt}} e^{-\zeta^2}\,d\zeta \qquad (4.39)$$

When Δx is small compared to the distance x,

$$c(x,t) = \frac{c_0\Delta x}{\sqrt{4\pi Dt}}e^{-x^2/(4Dt)} = \frac{n_d}{\sqrt{4\pi Dt}}e^{-x^2/(4Dt)} \qquad (4.40)$$

where the *source strength*, n_d, is given by

$$n_d = \int_{-\infty}^{\infty} c(x)\,dx = \int_{-\infty}^{\infty}\frac{n_d}{\sqrt{4\pi Dt}}e^{-x^2/(4Dt)}\,dx \qquad (4.41)$$

When c is assigned units of particles per unit length, n_d corresponds to the total number of particles in the source, and Eq. 4.40 describes the one-dimensional diffusion from a point source as in Fig. 4.5a. Also, when c has units of particles per unit area, n_d has units of particles per unit length and Eq. 4.40 describes the one-dimensional diffusion in a plane in two dimensions from a line source initially containing n_d particles per unit length as in Fig. 4.5b. Finally, when c has units of particles per unit volume, n_d has units of particles per unit area, and Eq. 4.40 describes the one-dimensional diffusion from a planar source in three dimensions initially containing n_d particles per unit area as in Fig. 4.5c. These results are summarized in Table 5.1.

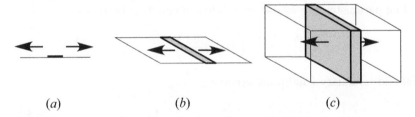

(a)   (b)   (c)

Figure 4.5:   One-dimensional diffusion into an infinite domain. **(a)** Point source diffusing into a line. **(b)** Line source diffusing into a plane. **(c)** Planar source diffusing into a volume.

4.3 DIFFUSIVITY AS A FUNCTION OF CONCENTRATION

When D is a function of concentration [i.e., $D = D(c)$], Eq. 4.2 takes the form

$$\frac{\partial c}{\partial t} = \nabla \cdot [D(c)\nabla c] \qquad (4.42)$$

This differential equation is generally nonlinear [depending upon the form of $D(c)$], and solutions therefore can be obtained analytically only in certain special cases which are not discussed here [4].

When Fick's law applies, the concentration profile generally contains information about the concentration dependence of the diffusivity. For constant D, step-function initial conditions have the error function (Eq. 4.31) as a solution to $\partial c/\partial t = D\partial^2 c/\partial x^2$. When the diffusivity is a function of concentration,

$$\frac{\partial c}{\partial t} = D(c)\frac{\partial^2 c}{\partial x^2} + \frac{dD(c)}{dc}\left(\frac{\partial c}{\partial x}\right)^2 \tag{4.43}$$

For identical initial conditions, the difference between a measured profile and the error-function solution is related to the last (nonlinear) term in Eq. 4.43. When diffusivity is a function of local concentration, the concentration profile tends to be relatively flat at a concentration where $D(c)$ is large and relatively steep where $D(c)$ is small (this is demonstrated in Exercise 4.2). Asymmetry of the diffusion profile in a diffusion couple is an indicator of a concentration-dependent diffusivity.

Matano developed a graphical method which, for certain classes of boundary value problems, relates the form of the diffusion profile with the concentration dependence of the interdiffusivity, $\widetilde{D}(c)$, introduced in Section 3.1.3 [5]. This method can determine $\widetilde{D}(c)$ from the diffusion profile in chemical concentration-gradient diffusion experiments where atomic volumes are sufficiently constant so that changes in overall specimen volume are insignificant and diffusion can be formulated in a V-frame. The method uses scaling, as discussed in Section 4.2.2.

Consider a case where the initial and boundary conditions for a diffusion couple are

$$c(x, t = 0) = \begin{cases} c_1^L & -\infty < x < 0 \\ c_1^R & 0 < x < \infty \end{cases} \tag{4.44}$$

$$c(-\infty, t) = c_1^L \qquad c(\infty, t) = c_1^R$$

Using the scaling parameter $\eta = x/\sqrt{4t}$, the diffusion equation becomes

$$-2\eta\frac{dc_1}{d\eta} = \frac{d}{d\eta}\left(\widetilde{D}(c_1)\frac{dc_1}{d\eta}\right) \tag{4.45}$$

and the initial and boundary conditions become

$$c_1(\eta = \infty) = c_1^R \qquad c_1(\eta = -\infty) = c_1^L \tag{4.46}$$

Equation 4.45 can be integrated as

$$-2\int_\infty^u \eta\frac{dc_1}{d\eta}d\eta = \widetilde{D}(c_1)\frac{dc_1}{d\eta}\bigg|_\infty^u \tag{4.47}$$

If c_1 is a monotonically increasing function, variables can be changed so that

$$-2\int_{c_1^R}^{c_1(u)} \eta\,dc_1 = \widetilde{D}(c_1)\frac{dc_1}{d\eta}\bigg|_{c_1^R}^{c_1(u)} \tag{4.48}$$

If the profile is measured at some particular time $t = \tau$, $x = \eta\sqrt{4\tau}$, so

$$-\int_{c_1^R}^{c_1(x)} x(c_1)\,dc_1 = 2\tau\widetilde{D}(c_1)\frac{dc_1(x)}{dx} \tag{4.49}$$

because $dc_1/dx(c_1 = c_1^R) = 0$. Equation 4.49 is an equation for $\widetilde{D}$ in terms of integrals and derivatives of the function $c_1(x)$ and its inverse $x(c_1)$, which can be determined from a measured profile. The boundary condition

$$\int_{c_1^R}^{c_1^L} x \, dc_1 = 0 \tag{4.50}$$

determines the position of the original interface (commonly termed the *Matano interface*) where $x = 0$ (Exercise 4.1 demonstrates this). The expression for the interdiffusivity is

$$\widetilde{D}(c_1') = -\frac{1}{2\tau} \left. \frac{dx}{dc_1} \right|_{c_1'} \int_{c_1^R}^{c_1'} x(c_1) \, dc_1 \tag{4.51}$$

Equation 4.51 is an integral equation that can be used to determine $\widetilde{D}(c_1)$ by a graphical construction or numerical solution. The derivative required in Eq. 4.51 is provided by the measured concentration profile at time τ and the integration is performed on the inverse of $c_1(x)$ [6]. However, this historically important method is only moderately accurate, and it would be preferable to obtain diffusion profiles for various assumed diffusivities as a function of concentration by computation. $\widetilde{D}(c)$ could be deduced by fitting calculated results for a parametric representation of $\widetilde{D}(c)$ to an experimentally determined diffusion profile.

4.4 DIFFUSIVITY AS A FUNCTION OF TIME

When D is a function of time, but not position, Eq. 4.2 takes the form

$$\frac{\partial c}{\partial t} = \nabla \cdot [D(t)\nabla c] = D(t)\nabla^2 c \tag{4.52}$$

This could be the case for a diffusion specimen that is slowly cooled while a uniform temperature is maintained. Problems of this type can be treated by making the change of variable

$$\tau_D \equiv \int_0^t D(t') \, dt' \tag{4.53}$$

Then $\partial c/\partial t = (\partial c/\partial \tau_D)(\partial \tau_D/\partial t) = (\partial c/\partial \tau_D)D(t)$ and Eq. 4.52 is transformed to

$$\frac{\partial c}{\partial \tau_D} = \nabla^2 c \tag{4.54}$$

with the solution $c = c(x, \tau_D)$ for unit diffusivity. Equation 4.54, with the same form as Eq. 4.18, holds when D is uniform. If the boundary conditions for a time-dependent diffusivity problem are invariant under this change of variable, solutions from known constant-D problems can be applied to the time-dependent D case. Consider, for example, the boundary-value problem in Fig. 4.2, which for the constant-D case was solved by Eq. 4.31. Because $\tau_D = 0$ when $t = 0$, the initial and boundary conditions are invariant under the change of variable, and the solution is

$$c(x, \tau_D) = \bar{c} + \frac{\Delta c}{2} \operatorname{erf}\left(\frac{x}{\sqrt{4\tau_D}}\right)$$

$$c(x, t) = \bar{c} + \frac{\Delta c}{2} \operatorname{erf}\left(\frac{x}{\sqrt{4 \int_0^t D(t')dt'}}\right) \qquad (4.55)$$

4.5 DIFFUSIVITY AS A FUNCTION OF DIRECTION

In the expressions for Fourier's law of heat conductivity and Fick's law for mass flux, it has been assumed that the flux vector is always parallel to the driving force vector. However, these vectors are *not* parallel for general materials. For instance, consider a bar, made of alternating layers of copper and silica glass, which is conducting heat from a reservoir at high temperature to one of lower temperature, as in Fig. 4.6. Because copper's heat conductivity is more than 60 times greater than silica's, the temperature along each inclined copper sheet will be nearly uniform. Furthermore, because the thermal gradient is always normal to lines of uniform temperature, it points in a direction approximately normal to the copper sheets. However, the heat flux is parallel to the bar because the only sources and sinks for heat are the reservoirs at the ends.

This hypothetical example is similar to the case of a graphite single crystal. Graphite has a hexagonal Bravais lattice. Along the basal planes, the carbon bonding is covalent, so the thermal conductivity is $K_\parallel = 355 \ \mathrm{J\,m^{-1}\,s^{-1}\,K^{-1}}$, nearly that of carbon-diamond. Between the graphite layers, where the bonding is very weak, the conductivity is much lower, $K_\perp = 89.3 \ \mathrm{J\,m^{-1}\,s^{-1}\,K^{-1}}$. Figure 4.6 is therefore representative of single-crystal graphite, where the basal plane is parallel to the layers shown.

In general, the properties of crystals and other types of materials, such as composites, vary with direction (i.e., macroscopic materials properties such as mass diffusivity and electrical conductivity will generally be anisotropic). It is possible to generalize the isotropic relations between driving forces and fluxes to account for

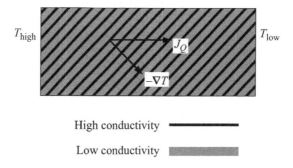

High conductivity ▬▬▬▬▬

Low conductivity ▬▬▬▬▬

Figure 4.6: Thermal conduction in a laminar composite. The macroscopic value of the thermal conductivity depends on the individual values of conductivity for the materials as well as the inclination of the laminates.

anisotropy.[3] The isotropic form for Fick's law is

$$\vec{J} = -D\nabla c \qquad J_i = -D\frac{\partial c}{\partial x_i} \qquad (4.56)$$

where the final expression represents three equations, one for each coordinate axis, written in component form. For the anisotropic case, there is a linear relation between the flux and gradient vectors. As discussed in Section 1.3.7, the matrix of the linear coefficients depends on the particular material and the orientation of the material with respect to the V-frame:

$$\begin{bmatrix} J_1 \\ J_2 \\ J_3 \end{bmatrix} = - \begin{bmatrix} D_{11} & D_{12} & D_{13} \\ D_{12} & D_{22} & D_{23} \\ D_{13} & D_{23} & D_{33} \end{bmatrix} \begin{bmatrix} \frac{\partial c}{\partial x_1} \\ \frac{\partial c}{\partial x_2} \\ \frac{\partial c}{\partial x_3} \end{bmatrix} \qquad (4.57)$$

or in component form,

$$J_i = -\sum_j D_{ij}\frac{\partial c}{\partial x_j} \qquad (4.58)$$

or simply,

$$\vec{J} = -\boldsymbol{D}\nabla c \qquad (4.59)$$

$\boldsymbol{D}$ is called the diffusivity tensor and acts as an object that *connects* one vector to another (e.g., the flux vector with the gradient vector). This connection can be written in matrix form as in Eq. 4.57. The diffusivity tensor $\boldsymbol{D}$ is symmetric (i.e., $D_{ij} = D_{ji}$) for any underlying material symmetry.

The anisotropic form of Fick's law would seem to complicate the diffusion equation greatly. However, in many cases, a simple method for treating anisotropic diffusion allows the diffusion equation to keep its simple form corresponding to isotropic diffusion. Because D_{ij} is symmetric, it is always possible to find a linear coordinate transformation that will make the D_{ij} diagonal with real components (the eigenvalues of $\boldsymbol{D}$). Let elements of such a transformed system be identified by a "hat." Then

$$\hat{\boldsymbol{D}} = \begin{bmatrix} \hat{D}_{11} & 0 & 0 \\ 0 & \hat{D}_{22} & 0 \\ 0 & 0 & \hat{D}_{33} \end{bmatrix} \qquad (4.60)$$

The diagonal elements of $\hat{\boldsymbol{D}}$ are the eigenvalues of $\boldsymbol{D}$, and the coordinate system of $\hat{\boldsymbol{D}}$ defines the *principal axes* $\hat{x}_1, \hat{x}_2, \hat{x}_3$ (the eigensystem). In the principal axes coordinate system, the diffusion equation then has the relatively simple form

$$\frac{\partial c}{\partial t} = -\nabla \cdot \vec{J} = \nabla \cdot \hat{\boldsymbol{D}}\nabla c = \hat{D}_{11}\frac{\partial^2 c}{\partial \hat{x}_1^2} + \hat{D}_{22}\frac{\partial^2 c}{\partial \hat{x}_2^2} + \hat{D}_{33}\frac{\partial^2 c}{\partial \hat{x}_3^2} \qquad (4.61)$$

The eigenvalues of $\boldsymbol{D}$ in Eq. 4.60 are the three roots λ_i of the cubic equation

$$\det \begin{vmatrix} D_{11} - \lambda & D_{12} & D_{13} \\ D_{12} & D_{22} - \lambda & D_{23} \\ D_{13} & D_{23} & D_{33} - \lambda \end{vmatrix} = 0 \qquad (4.62)$$

[3]The quintessential resource for this topic is Nye's book on crystal properties [7].

If $\underline{R}$ is the matrix that rotates the original (x_1, x_2, x_3) coordinate system into the principal $(\hat{x}_1, \hat{x}_2, \hat{x}_3)$ system, then according to Eq. 1.36, $\hat{D}$ must be related to D by

$$\hat{D} = \underline{R}D\underline{R}^{-1} \tag{4.63}$$

To solve the diffusion equation in the principal coordinate system (i.e., Eq. 4.61), the Cartesian space can now be stretched or contracted along the principal axes by scaling:

$$
\begin{aligned}
\hat{x}_1 &= \frac{\hat{D}_{11}^{1/2}}{(\hat{D}_{11}\hat{D}_{22}\hat{D}_{33})^{1/6}}\xi_1 \\
\hat{x}_2 &= \frac{\hat{D}_{22}^{1/2}}{(\hat{D}_{11}\hat{D}_{22}\hat{D}_{33})^{1/6}}\xi_2 \\
\hat{x}_3 &= \frac{\hat{D}_{33}^{1/2}}{(\hat{D}_{11}\hat{D}_{22}\hat{D}_{33})^{1/6}}\xi_3
\end{aligned}
\tag{4.64}
$$

This scaling conserves the volume. Using Eq. 4.64, the diffusion equation can now be written in terms of the ξ_i:

$$\frac{\partial c}{\partial t} = \mathcal{D}\nabla_\xi^2 c = \mathcal{D}\left(\frac{\partial^2 c}{\partial \xi_1^2} + \frac{\partial^2 c}{\partial \xi_2^2} + \frac{\partial^2 c}{\partial \xi_3^2}\right) \tag{4.65}$$

where $\mathcal{D} = (\hat{D}_{11}\hat{D}_{22}\hat{D}_{33})^{1/3}$. Equation 4.65 has the same form as Fick's second law for a material with a constant isotropic diffusivity. Thus, known solutions to the diffusion equation for constant isotropic diffusivity can be used to find solutions for anisotropic constant diffusivities by a simple algorithm. A solution to Eq. 4.65 with $\mathcal{D}$ and coordinates ξ_1, ξ_2, ξ_3 is rescaled back to the the principal axis coordinates $\hat{x}_1, \hat{x}_2, \hat{x}_3$ using Eq. 4.64. If necessary, the system can be transformed back into the original, anisotropic laboratory coordinate system with $D = \underline{R}^{-1}\hat{D}\underline{R}$.

The diffusivity tensor has special forms for particular choices of coordinate axes if the diffusing body itself has special symmetry (e.g., if it is crystalline). *Neumann's principle* states:

> The symmetry elements of any physical property of a material must include the symmetry elements of the point group of the material.[4]

A consequence of Neumann's symmetry principle is that direct tensor Onsager coefficients (such as in the diffusivity tensor) must be symmetric. This is equivalent to the addition of a center of symmetry (an inversion center) to a material's point group. Thus, the direct tensor properties of crystalline materials must have one of the point symmetries of the 11 Laue groups. Neumann's principle can impose additional relationships between the diffusivity tensor coefficients D_{ij} in Eq. 4.57. For a hexagonal crystal, the diffusivity tensor in the principal coordinate system has the form

$$\hat{D} = \begin{bmatrix} \hat{D}_{11} & 0 & 0 \\ 0 & \hat{D}_{11} & 0 \\ 0 & 0 & \hat{D}_{33} \end{bmatrix} \tag{4.66}$$

---

[4]This also applies to the macroscopic properties of composite materials with underlying symmetry—like honeycomb, wood, and woven materials—for which the crystal structure, if any, may play no direct role.

when $\hat{x}_3$ lies along the crystal's c-axis and $\hat{x}_1$ and $\hat{x}_2$ lie anywhere in the basal plane. Exercise 4.6 demonstrates that the diffusivity tensor in a cubic crystal has the form

$$\boldsymbol{D} = \begin{bmatrix} D & 0 & 0 \\ 0 & D & 0 \\ 0 & 0 & D \end{bmatrix} \tag{4.67}$$

and that the diffusion is therefore isotropic. Forms of the diffusivity tensor $\boldsymbol{D}$ for other crystal systems are tabulated in Nye's text [7].

Bibliography

1. I.M. Gelfand and S.V. Fomin. *Calculus of Variations.* Prentice-Hall, Englewood Cliffs, NJ, 1963.

2. W.C. Carter, J.E. Taylor, and J.W. Cahn. Variational methods for microstructural evolution. *JOM,* 49(12):30–36, 1997.

3. P.M. Morse and H. Feshbach. *Methods of Theoretical Physics, Vols. 1 and 2.* McGraw-Hill, New York, 1953.

4. J. Crank. *The Mathematics of Diffusion.* Oxford University Press, Oxford, 2nd edition, 1975.

5. C. Matano. On relation between diffusion coefficients and concentrations of solid metals (the nickel–copper system). *Jpn. J. Phys.,* 8(3):109–113, 1933.

6. P. Shewmon. *Diffusion in Solids.* The Minerals, Metals and Materials Society, Warrendale, PA, 1989.

7. J.F. Nye. *Physical Properties of Crystals.* Oxford University Press, Oxford, 1985.

8. L.C.C. Da Silva and R.F. Mehl. Interface and marker movements in diffusion in solid solutions of metals. *Trans. AIME,* 191(2):155–173, 1951.

EXERCISES

4.1 Consider the Boltzmann–Matano analysis leading to Eq. 4.51. Explain why the condition imposed by Eq. 4.50 determines the location of the $x = 0$ plane (i.e., the position of the original interface).

Solution. The laboratory coordinate system is used and there is no change in the overall specimen volume. The integral in Eq. 4.50 is proportional to the sum area $1 +$ area 2 in Fig. 4.7. Area 1 is positive and area 2 is negative. When $x = 0$ is set at the position of the original interface, area 2 is proportional to the amount of diffusant that has left

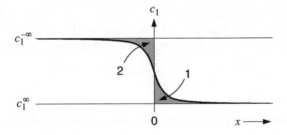

Figure 4.7: Composition profile arising from interdiffusion.

the original block of composition c_1^L, and area 1 is proportional to the amount that has entered the original block of composition c_1^R. Because these quantities must be equal, the condition imposed by Eq. 4.50 determines the $x = 0$ plane.

4.2 The interdiffusivity, $\widetilde{D}$, which measures the interdiffusion between Cu and Zn in the laboratory frame, is a strong function of the concentration of Zn. The curve describing $\widetilde{D}(c_{Zn})$ is concave upward and roughly parabolic in shape, and $\widetilde{D}(c_{Zn})$ increases by a factor of about 20 when the Zn content increases from 0 to 30 at. % [8]. Describe how the shape of the diffusion-penetration curve for a diffusion couple made of Cu/Cu–30 at. % Zn is expected to deviate from the symmetric form of the constant diffusivity error-function solution.

Solution. Base your argument on the Boltzmann–Matano solution (Eq. 4.51), which links the interdiffusivity with the shape of the diffusion curve. Two factors are present: the integral under the diffusion curve from c_{Zn}^R to the concentration in question, c_{Zn}, and the reciprocal of the slope at c_{Zn}. The integral varies from 0 at $x = \infty$ to 0 at $x = -\infty$ and reaches a maximum at $x = 0$ (see Exercise 4.1). The slope varies from 0 at $x = \infty$ to 0 at $x = -\infty$ and reaches a maximum somewhere in between. A little trial and error quickly shows that the only way $\widetilde{D}$ can increase by a factor of 20 with increasing c_{Zn} with a diffusion curve of reasonable shape is to have a curve with a small slope at high c_{Zn} and a large slope at low c_{Zn}. Such a curve is markedly nonsymmetric around its midpoint. Figure 4.8 shows an observed interdiffusion profile for this system.

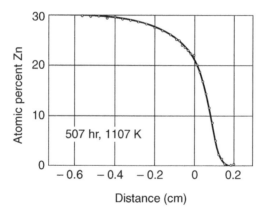

Figure 4.8:   Concentration profile observed in Cu–30 at. % Zn diffusion couple. From Da Silva and Mehl [8].

4.3 The Kirkendall effect can be studied by embedding an inert marker in the original step-function interface ($x = 0$) of the diffusion couple illustrated in Fig. 3.4. Show that this marker will move in the V-frame or, equivalently, with respect to the nondiffused ends of the specimen, according to

$$x_m = \alpha\, t^{1/2}$$

where α is a constant.

Solution. According to Eq. 3.23, the instantaneous velocity of any marker is given by

$$v = \Omega_1 \left[D_1 - D_2 \right]_m \left(\frac{\partial c_1}{\partial x} \right)_m \tag{4.68}$$

Now, in general, $c_1 = c_1(\eta)$, where $\eta = x/\sqrt{t}$. Because $(D_1 - D_2)$ is a function of η, we can represent it as $[D_1 - D_2] = h(\eta)$. Similarly, we can write $\partial c_1/\partial x = (dc_1/d\eta)\,t^{-1/2} = f(\eta)\,t^{-1/2}$. Putting these results into Eq. 4.68 yields

$$v = \frac{dx_m}{dt} = \Omega_1 h(\eta_m) f(\eta_m) t^{-1/2}$$

A solution that satisfies this equation and also the initial condition that $x_m = 0$ when $t = 0$ is

$$x_m = \alpha\, t^{1/2}$$

with the constant α given by $\alpha = 2\,\Omega_1 h(\alpha) f(\alpha)$.

4.4 A diffusion measurement at the temperature T_0 is made by annealing a diffusion couple comprised of two semi-infinite bars. However, there is a complication; after the completion of the isothermal anneal, carried out at T_0 for the time t_0, the specimen must be cooled to room temperature at a finite rate. During this cooling period, a small amount of additional nonisothermal diffusion occurs. If an expression can be found for the amount of time, Δt, required to produce this same additional increment of diffusion at the constant temperature T_0, the specimen could be analyzed very simply at the end of the experiment by assuming that it was annealed at T_0 for the time $t_0 + \Delta t$. Assume that $D = D_0 \exp[-E/(kT)]$ and that the temperature during the cooling period is

$$T = \frac{T_\circ}{1 + [(t - t_\circ)\,/a]} \tag{4.69}$$

where a is constant. When $t = t_0 + a$, $T = T_0/2$ and it may be assumed that beyond that time during the cooling, any further diffusion is negligible. Find an expression for Δt.

Solution. During the cooling the diffusivity is time dependent and the solution is therefore obtained by utilizing the variable τ_D, which is described and defined in Section 4.4. The value of τ_D at the end of the actual experiment is then

$$\tau_D^{\text{end}} = \int_0^{t_\circ} D\,dt + \int_{t_\circ}^{t_\circ + a} D(t)\,dt \tag{4.70}$$

where $D = D_0 \exp[-E/(kT_0)]$ and

$$D(t) = D_\circ\, e^{-[E/(kT_\circ)][1+(t-t_\circ)/a]} \tag{4.71}$$

Putting these expressions into Eq. 4.70 and integrating,

$$\tau_D^{\text{end}} = D_\circ e^{-E/(kT_\circ)} \left[t_\circ + \frac{kT_\circ a}{E}\left(1 - e^{-E/(kT_\circ)}\right) \right] \tag{4.72}$$

On the other hand, the value of τ achieved at the end of an isothermal diffusion anneal of duration $t_0 + \Delta t$ is

$$\tau_D^{\text{end}'} = D_\circ e^{-E/(kT_\circ)} [t_\circ + \Delta t] \tag{4.73}$$

The amounts of diffusion achieved in these two anneals will be equal if $\tau_D^{\text{end}} = \tau_D^{\text{end}'}$. Therefore, setting these two quantities equal,

$$\Delta t = \frac{kT_\circ a}{E}\left[1 - e^{-E/(kT_\circ)}\right] \cong \frac{kT_\circ a}{E} \tag{4.74}$$

4.5 The transport relations for the thermoelectric effect in an isotropic material are

$$\vec{J}_Q = -\frac{L_{QQ}}{T}\nabla T - L_{Qq}\nabla\phi \tag{4.75}$$

$$\vec{J}_q = -\frac{L_{qQ}}{T}\nabla T - L_{qq}\nabla\phi \tag{4.76}$$

(a) Write the corresponding relations that apply for an anisotropic material.

(b) From Onsager's symmetry theorem, what can be said about the symmetry of the rank-two tensors that are involved and any relations between them?

(c) Are the cross-coupling tensors (which couple nonconjugate forces and fluxes) positive definite?

Solution.   For an anisotropic material, the coefficients must be rank-two tensors. Therefore,

$$J_i^Q = \sum_j \left(-\frac{\alpha_{ij}}{T}\frac{\partial T}{\partial x_j} - \beta_{ij}\frac{\partial\phi}{\partial x_j}\right) \tag{4.77}$$

$$J_i^q = \sum_j \left(-\frac{\gamma_{ij}}{T}\frac{\partial T}{\partial x_j} - \eta_{ij}\frac{\partial\phi}{\partial x_j}\right) \tag{4.78}$$

or in a matrix representation involving rank-one tensors (vectors) and rank-two tensors,

$$\begin{bmatrix} \vec{J}^Q \\ \vec{J}^q \end{bmatrix} = -\begin{bmatrix} \boldsymbol{\alpha} & \boldsymbol{\beta} \\ \boldsymbol{\gamma} & \boldsymbol{\eta} \end{bmatrix}\begin{bmatrix} \nabla T/T \\ \nabla\phi \end{bmatrix} \tag{4.79}$$

(a) When written in full, Eq. 4.79 becomes

$$\begin{bmatrix} J_1^Q \\ J_2^Q \\ J_3^Q \\ J_1^q \\ J_2^q \\ J_3^q \end{bmatrix} = -\begin{bmatrix} \alpha_{11} & \alpha_{12} & \alpha_{13} & \beta_{11} & \beta_{12} & \beta_{13} \\ \alpha_{21} & \alpha_{22} & \alpha_{23} & \beta_{21} & \beta_{22} & \beta_{23} \\ \alpha_{31} & \alpha_{32} & \alpha_{33} & \beta_{31} & \beta_{32} & \beta_{33} \\ \gamma_{11} & \gamma_{12} & \gamma_{13} & \eta_{11} & \eta_{12} & \eta_{13} \\ \gamma_{21} & \gamma_{22} & \gamma_{23} & \eta_{21} & \eta_{22} & \eta_{23} \\ \gamma_{31} & \gamma_{32} & \gamma_{33} & \eta_{31} & \eta_{32} & \eta_{33} \end{bmatrix}\begin{bmatrix} (1/T)(\partial T/\partial x_1) \\ (1/T)(\partial T/\partial x_2) \\ (1/T)(\partial T/\partial x_3) \\ \partial\phi/\partial x_1 \\ \partial\phi/\partial x_2 \\ \partial\phi/\partial x_3 \end{bmatrix} \tag{4.80}$$

(b) Onsager's relation applies to the entire 6×6 matrix above, so

$$\begin{aligned} \beta_{ij} &= \gamma_{ji} \quad \text{or} \quad \boldsymbol{\beta} = \boldsymbol{\gamma}^T \\ \alpha_{ij} &= \alpha_{ji} \quad \text{or} \quad \boldsymbol{\alpha} = \boldsymbol{\alpha}^T \\ \eta_{ij} &= \eta_{ji} \quad \text{or} \quad \boldsymbol{\eta} = \boldsymbol{\eta}^T \end{aligned} \tag{4.81}$$

(c) If a flux, $\vec{J}^a$, is coupled to a force, $\vec{X}^b$, by a tensor, $\boldsymbol{\xi}$, so that $\vec{J}^a = \boldsymbol{\xi}\vec{X}^b$, the tensor is positive definite if the projection of $\vec{J}^a$ on $\vec{X}^b$ is always positive; that is,

$$\vec{J}^a \cdot \vec{X}^b = \left(\boldsymbol{\xi}\vec{X}^b\right) \cdot \vec{X}^b > 0 \tag{4.82}$$

The direct-effect $\boldsymbol{\alpha}$ and $\boldsymbol{\eta}$ must be positive definite because of Fourier's law of heat conduction (where the thermal conductivity is always positive, according to

Eq. 2.18) and Ohm's law for electrical conduction (where the electrical conductivity is also always positive, according to Exercise 2.1's solution). However, this is not necessarily the case for the cross-effect tensors β and γ. For example, in mass diffusion in a thermal gradient, the heat of transport can be either positive or negative; the direction of the atom flux in a temperature gradient can then be in either direction. The anisotropic equivalent to the heat of transport relates the direction of the mass diffusion to the direction of the temperature gradient. There is no physical requirement that these quantities could not be in reversed directions, and indeed, sometimes they are.

4.6 Show by the use of symmetry arguments that the diffusivity in a cubic crystal must be isotropic.

Solution. Take the three cubic axes x_1, x_2, and x_3 as the coordinate system and assume initially that the diffusivity has the most general form possible, which, according to Eq. 4.57, is

$$D = \begin{bmatrix} D_{11} & D_{12} & D_{13} \\ D_{12} & D_{22} & D_{23} \\ D_{13} & D_{23} & D_{33} \end{bmatrix} \tag{4.83}$$

Now, establish a concentration gradient of magnitude, g, successively along x_1, x_2, and x_3. The three corresponding flux vectors will then be, respectively,

$$-g \begin{bmatrix} D_{11} \\ D_{12} \\ D_{13} \end{bmatrix} \qquad -g \begin{bmatrix} D_{12} \\ D_{22} \\ D_{23} \end{bmatrix} \qquad -g \begin{bmatrix} D_{13} \\ D_{23} \\ D_{33} \end{bmatrix} \tag{4.84}$$

Because of the cubic crystal symmetry, the fluxes parallel to the directions of the gradients in the three cases (i.e., $-gD_{11}$, $-gD_{22}$, and $-gD_{33}$) must be equal. Therefore,

$$D_{11} = D_{22} = D_{33} \tag{4.85}$$

Preservation of fourfold symmetry along the x_1, x_2, and x_3 axes requires that any flux components in directions perpendicular to the gradients be zero. Therefore,

$$gD_{12} = gD_{13} = gD_{23} = 0 \tag{4.86}$$

and

$$D = \begin{bmatrix} D & 0 & 0 \\ 0 & D & 0 \\ 0 & 0 & D \end{bmatrix} = D \begin{bmatrix} 1 & 0 & 0 \\ 0 & 1 & 0 \\ 0 & 0 & 1 \end{bmatrix} \tag{4.87}$$

According to Eq. 4.63, the diffusivity tensor in any other rotated system (indicated by a prime) will have the form

$$D' = \underline{R}D\underline{R}^{-1} = D\underline{R}\underline{R}^{-1} = D \begin{bmatrix} 1 & 0 & 0 \\ 0 & 1 & 0 \\ 0 & 0 & 1 \end{bmatrix} = D \tag{4.88}$$

Because $D' = D$, the diffusivity is isotropic.

4.7 Consider two-dimensional anisotropic diffusion in an infinite thin film where the initial condition consists of a point source of atoms located at $x_1 = x_2 = 0$. The diffusivity tensor D, in arbitrary units, in the (x_1, x_2) coordinate system is

$$D = \begin{bmatrix} \frac{49}{4} & -\frac{15\sqrt{3}}{4} \\ -\frac{15\sqrt{3}}{4} & \frac{19}{4} \end{bmatrix} \tag{4.89}$$

Determine the fundamental solution, $c = c(x_1, x_2, t)$, of the diffusion equation for the point source in this coordinate system.

- Solve the problem in the principal axes coordinate system and then transform the solution back to the (x_1, x_2) system.
- Note that a rotation matrix, $\underline{R}$, has the properties $\det \underline{R} = 1$ and $\underline{R}^{-1} = \underline{R}^T$, where $\underline{R}^T$ is the transpose of $\underline{R}$.

Solution. First transform the problem to the principal axes system. The eigenvalue equation for $\boldsymbol{D}$ is

$$\det[\boldsymbol{D} - \lambda \boldsymbol{I}] = \begin{bmatrix} \frac{49}{4} - \lambda & -\frac{15\sqrt{3}}{4} \\ -\frac{15\sqrt{3}}{4} & \frac{19}{4} - \lambda \end{bmatrix} = (\lambda - 16)(\lambda - 1) = 0 \qquad (4.90)$$

Therefore, the diffusivity tensor in the principal axes system is

$$\hat{\boldsymbol{D}} = \begin{bmatrix} 16 & 0 \\ 0 & 1 \end{bmatrix} \qquad (4.91)$$

The rotation, $\underline{R}$, required to rotate the original (x_1, x_2) coordinate system into the principal coordinate system must satisfy

$$\hat{\boldsymbol{D}} = \underline{R} \boldsymbol{D} \underline{R}^{-1} \qquad (4.92)$$

yielding a set of linear equations which, in conjunction with the properties of rotation matrices given above produces the result

$$\underline{R} = \begin{bmatrix} \frac{\sqrt{3}}{2} & -\frac{1}{2} \\ \frac{1}{2} & \frac{\sqrt{3}}{2} \end{bmatrix} \qquad (4.93)$$

which corresponds to a rotation of $\pi/6$. Now solve the diffusion problem in the principal coordinate system where the diffusion equation is

$$\frac{\partial c}{\partial t} = \hat{D}_{11} \frac{\partial^2 c}{\partial \hat{x}_1^2} + \hat{D}_{22} \frac{\partial^2 c}{\partial \hat{x}_2^2} \qquad (4.94)$$

Changing variables in Eq. 4.94 using $\hat{x}_1 = (\sqrt{\hat{D}_{11}}/\sqrt{\mathcal{D}}) \xi_1$, $\hat{x}_2 = (\sqrt{\hat{D}_{22}}/\sqrt{\mathcal{D}}) \xi_2$, and $\mathcal{D} = \sqrt{\hat{D}_{11}\hat{D}_{22}}$, a new diffusion equation is obtained with the solution

$$\begin{aligned} c(\xi_1, \xi_2, t) &= \frac{N}{4\pi\mathcal{D}t} e^{-(\xi_1^2 + \xi_2^2)/(4\mathcal{D}t)} = c(\hat{x}_1, \hat{x}_2, t) \\ &= \frac{N}{4\pi\mathcal{D}t} e^{-[\hat{x}_1^2/(4\hat{D}_{11}t) + \hat{x}_2^2/(4\hat{D}_{22}t)]} \end{aligned} \qquad (4.95)$$

The concentration therefore spreads to an ellipse.

We can now find the final solution in the original coordinate system by using

$$\vec{\hat{r}} = \underline{R}\vec{r} \qquad (4.96)$$

where $\vec{\hat{r}} = (\hat{x}_1, \hat{x}_2)$ and $\vec{r} = (x_1, x_2)$ to solve for $\hat{x}_1$ and $\hat{x}_2$ and by putting the results into Eq. 4.95. It is an ellipse of aspect ratio 4 with axes rotated $\pi/6$ with respect to the axes of the original coordinate system.

4.8 To measure self-diffusivities along different directions in a hexagonal crystal, a flat surface is prepared perpendicular to the direction along which the diffusivity is to be measured. A thin instantaneous plane source is deposited on this surface and the specimen is then annealed at a constant diffusion temperature. After the diffusion anneal, the diffusion-penetration profile normal to the surface is measured and D is determined. (The technique is described further in Section 5.2.1.) If the selected crystal direction makes an angle θ with the c-axis of the crystal, show that the measured D [i.e., $D(\theta)$] is given by

$$D(\theta) = D_{33} \cos^2 \theta + D_{11} \sin^2 \theta \qquad (4.97)$$

- Note that the concentration gradient will always be along the selected crystal direction (perpendicular to the planar source), regardless of the direction of the diffusion flux.

Solution. Adopting a principal axis system in which $\hat{k}$ is parallel to the c-axis and $\hat{\imath}$ and $\hat{\jmath}$ lie in the basal plane, the diffusivity tensor will have the form given in Eq. 4.66 and the flux will be given by

$$\vec{J} = -D_{11} \frac{\partial c}{\partial \hat{x}_1} \hat{\imath} - D_{11} \frac{\partial c}{\partial \hat{x}_2} \hat{\jmath} - D_{33} \frac{\partial c}{\partial \hat{x}_3} \hat{k} \qquad (4.98)$$

The normal to the isoconcentration surfaces will be $\hat{n} = \nabla c / |\nabla c|$ and the component of $\vec{J}$ along $-\hat{n}$ [i.e., $J(\theta)$] is

$$J(\theta) = -\vec{J} \cdot \hat{n} = -\vec{J} \cdot \frac{\nabla c}{|\nabla c|} = D(\theta) |\nabla c| \qquad (4.99)$$

Therefore,

$$D(\theta) = -\frac{\vec{J} \cdot \nabla c}{|\nabla c|^2} = \frac{1}{|\nabla c|^2} \left[D_{11} \left(\frac{\partial c}{\partial \hat{x}_1} \right)^2 + D_{11} \left(\frac{\partial c}{\partial \hat{x}_2} \right)^2 + D_{33} \left(\frac{\partial c}{\partial \hat{x}_3} \right)^2 \right] \qquad (4.100)$$

But

$$\cos^2 \theta = \left[\frac{\nabla c \cdot \hat{k}}{|\nabla c|} \right]^2 = \frac{1}{|\nabla c|^2} \left(\frac{\partial c}{\partial \hat{x}_3} \right)^2 \qquad (4.101)$$

$$\sin^2 \theta = \left[\frac{\nabla c \cdot (\hat{\imath} + \hat{\jmath})}{|\nabla c|} \right]^2 = \frac{1}{|\nabla c|^2} \left[\left(\frac{\partial c}{\partial \hat{x}_1} \right)^2 + \left(\frac{\partial c}{\partial \hat{x}_2} \right)^2 \right] \qquad (4.102)$$

Therefore, Eq. 4.100 has the form

$$D(\theta) = D_{11} \sin^2 \theta + D_{33} \cos^2 \theta \qquad (4.103)$$

CHAPTER 5

SOLUTIONS TO THE DIFFUSION EQUATION

In Chapter 4 we described many of the general features of the diffusion equation and several methods of solving it when D varies in different ways. We now address in more detail methods to solve the diffusion equation for a variety of initial and boundary conditions when D is constant and therefore has the relatively simple form of Eq. 4.18; that is,

$$\frac{\partial c}{\partial t} = D\nabla^2 c$$

This equation is a second-order linear partial-differential equation with a rich mathematical literature [1]. For a large class of initial and boundary conditions, the solution has theorems of uniqueness and existence as well as theorems for its maximum and minimum values.[1]

Many texts, such as Crank's treatise on diffusion [2], contain solutions in terms of simple functions for a variety of conditions—indeed, the number of worked problems is enormous. As demonstrated in Section 4.1, the differential equation for the "diffusion" of heat by thermal conduction has the same form as the mass diffusion equation, with the concentration replaced by the temperature and the mass diffusivity replaced by the thermal diffusivity, κ. Solutions to many heat-flow

---

[1]If the diffusivity is imaginary, the diffusion equation has the same form as the time-dependent Schrödinger's equation at zero potential. Also, Eq. 4.18 implies that the velocity of the diffusant can be infinite. Schrödinger's equation violates this relativistic principle.

Kinetics of Materials. By Robert W. Balluffi, Samuel M. Allen, and W. Craig Carter.
Copyright © 2005 John Wiley & Sons, Inc.

boundary-value problems can therefore be adopted as solutions to corresponding mass diffusion problems.[2]

For problems with relatively simple boundary and initial conditions, solutions can probably be found in a library. However, it can be difficult to find a closed-form solution for problems with highly specific and complicated boundary conditions. In such cases, numerical methods could be employed. For simple boundary conditions, solutions to the diffusion equation in the form of Eq. 4.18 have a few standard forms, which may be summarized briefly.

For various instantaneous localized sources diffusing out into an infinite medium, the solution is a spreading Gaussian distribution:

$$c(\vec{r}, t) = \frac{n_d}{2^d (\pi D t)^{d/2}} e^{-\vec{r} \cdot \vec{r}/(4Dt)} \tag{5.1}$$

where d is the dimensionality of the space in which matter is diffusing and n_d is the source strength introduced in Section 4.2.3. When the initial condition can be represented by a distribution of sources, one simply superposes the solutions for the individual sources by integration, as in Section 4.2.3. When the boundaries are planar orthonormal surfaces, solutions to the diffusion equation have the form of trigonometric series. For diffusion in a cylinder, the trigonometric series is replaced by a sum over Bessel functions. For diffusion with spherical symmetry, trigonometric functions apply. All such solutions can be obtained by the *separation-of-variables method*, which is described below.

A third method—solution by Laplace transforms—can be used to derive many of the results already mentioned. It is a powerful method, particularly for complicated problems or those with time-dependent boundary conditions. The difficult part of using the Laplace transform is *back-transforming* to the desired solution, which usually involves integration on the complex domain. Fortunately, Laplace transform tables and tables of integrals can be used for many problems (Table 5.3).

5.1 STEADY-STATE SOLUTIONS

A particularly simple case occurs when the diffusion is in a steady state and the composition profile is therefore *not* a function of time. Steady-state conditions are often achieved for constant boundary conditions in finite samples at very long times.[3] Then $\partial c / \partial t = 0$, all local accumulation (divergence) vanishes, and the diffusion equation reduces to the Laplace equation,

$$\nabla^2 c = 0 \tag{5.2}$$

Solutions to the Laplace equation are called *harmonic functions*. Some harmonic functions are given below for particular boundary conditions.

5.1.1 One Dimension

Consider diffusion through an infinite flat plate of thickness L, with $0 < x < L$, subject to boundary conditions

$$c(0, t) = c^0 \quad c(L, t) = c^L \tag{5.3}$$

---

[2]Carslaw and Jaeger's treatise on heat flow is a primary source [3].
[3]Estimates of times required for "nearly steady-state" conditions are addressed in Section 5.2.6.

Integrating the one-dimensional Laplacian, $d^2c/dx^2 = 0$, twice yields

$$c(x) = a_1 x + a_2 \tag{5.4}$$

where a_1 and a_2 are integration constants. Solving for the integration constants using the boundary conditions, Eq. 5.3, produces the one-dimensional steady-state solution,

$$c(x) = c^0 - (c^0 - c^L)\frac{x}{L} \tag{5.5}$$

i.e., the concentration varies linearly across the plate as illustrated in Fig. 5.1. The flux is constant and proportional to the slope:

$$J = -D\frac{dc}{dx} = D\frac{c^0 - c^L}{L} \tag{5.6}$$

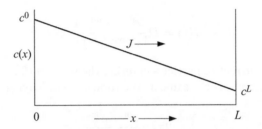

Figure 5.1:   Concentration, $c(x)$, vs. x for steady-state diffusion through a plate.

5.1.2   Cylindrical Shell

Consider steady-state diffusion through a cylindrical shell with inner radius r^{in} and outer radius r^{out} as in Fig. 5.2. The boundary conditions are

$$c(r^{\text{in}}, \theta, z, t) = c^{\text{in}} \quad c(r^{\text{out}}, \theta, z, t) = c^{\text{out}} \tag{5.7}$$

The Laplacian operator operating on $c(r, \theta, z)$ in cylindrical coordinates is

$$\nabla^2 c = \frac{1}{r}\frac{\partial}{\partial r}\left(r\frac{\partial c}{\partial r}\right) + \frac{1}{r^2}\frac{\partial^2 c}{\partial \theta^2} + \frac{\partial^2 c}{\partial z^2} \tag{5.8}$$

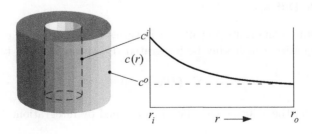

Figure 5.2:   Concentration, $c(r)$, vs. r for steady-state diffusion through a cylindrical shell.

Because the boundary conditions are independent of θ and z, the solution will also be independent of these variables. The solution must therefore satisfy

$$\frac{d}{dr}\left(r\frac{dc}{dr}\right) = 0 \tag{5.9}$$

Integrating twice produces

$$c(r) = a_1 \ln r + a_2 \tag{5.10}$$

and applying the boundary conditions gives

$$c(r) = c^{\text{in}} - \frac{c^{\text{in}} - c^{\text{out}}}{\ln(r^{\text{out}}/r^{\text{in}})} \ln\left(\frac{r}{r^{\text{in}}}\right) \tag{5.11}$$

The flux $J = -D(dc/dr)$ depends inversely on r:

$$J(r) = D\frac{c^{\text{in}} - c^{\text{out}}}{\ln(r^{\text{out}}/r^{\text{in}})} \frac{1}{r} \tag{5.12}$$

Note that the total current of particles entering the inner surface per unit length of cylinder $[I = J(r^{\text{in}})2\pi r^{\text{in}}]$ is the same as the total current leaving the outer surface,

$$I = 2\pi D\frac{c^{\text{in}} - c^{\text{out}}}{\ln(r^{\text{out}}/r^{\text{in}})} \tag{5.13}$$

which is a requirement for the steady state.

5.1.3 Spherical Shell

The Laplacian operator operating on $c(r, \theta, \phi)$ in spherical coordinates is

$$\nabla^2 c = \frac{1}{r^2}\left[\frac{\partial}{\partial r}\left(r^2\frac{\partial c}{\partial r}\right) + \frac{1}{\sin\theta}\frac{\partial}{\partial\theta}\left(\sin\theta\frac{\partial c}{\partial\theta}\right) + \frac{1}{\sin^2\theta}\frac{\partial^2 c}{\partial\phi^2}\right] \tag{5.14}$$

The steady-state solution for diffusion through spherical shells with boundary conditions dependent only on r may be obtained by integrating twice and determining the two constants of integration by fitting the solution to the boundary conditions.

5.1.4 Variable Diffusivity

When steady-state conditions prevail and D varies with position (e.g., $D = D(\vec{r})$), the diffusion equation can readily be integrated. Equation 4.2 then takes the form

$$\nabla \cdot (D\nabla c) = 0$$

In one dimension, the solution can then be obtained by integration,

$$c(x) = c(x_1) + a_1 \int_{x_1}^{x} \frac{d\xi}{D(\xi)} \tag{5.15}$$

5.2   NON-STEADY-STATE (TIME-DEPENDENT) DIFFUSION

When the diffusion profile is time-dependent, the solutions to Eq. 4.18 require considerably more effort and familiarity with applied mathematical methods for solving partial-differential equations. We first discuss some fundamental-source solutions that can be used to build up solutions to more complicated situations by means of superposition.

5.2.1   Instantaneous Localized Sources in Infinite Media

Equation 4.40 gives the solution for one-dimensional diffusion from a point source on an infinite line, an infinite thin line source on an infinite plane, and a thin planar source in an infinite three-dimensional body (summarized in Table 5.1). Corresponding solutions for two- and three-dimensional diffusion can easily be obtained by using products of the one-dimensional solution. For example, a solution for three-dimensional diffusion from a point source is obtained in the form

$$c(x,y,z,t) = \frac{n_{d_x}}{\sqrt{4\pi Dt}}\, e^{-x^2/(4Dt)} \times \frac{n_{d_y}}{\sqrt{4\pi Dt}}\, e^{-y^2/(4Dt)} \times \frac{n_{d_z}}{\sqrt{4\pi Dt}}\, e^{-z^2/(4Dt)} \quad (5.16)$$

where n_{d_x}, n_{d_y}, and n_{d_z} are constants. This may be written simply

$$c(r,t) = \frac{n_d}{(4\pi Dt)^{3/2}}\, e^{-r^2/(4Dt)} \quad (5.17)$$

where $n_d \equiv n_{d_x} \times n_{d_y} \times n_{d_z}$ This result has spherical symmetry and describes the spreading of a point source into an infinite domain. Integration verifies that n_d is equal to the total amount of diffusant in the system. As $t \to 0$, the solution approaches a delta function form, corresponding to the initial localized source [i.e.,

Table 5.1:   **Fundamental Solutions for Instantaneous, Localized Sources in One-, Two-, and Three-Dimensional Infinite Media**

Solution Type	Symmetric Part of ∇^2	Fundamental Solution
One-Dimensional Diffusion		
Point source in 1D Line source in 2D Plane source in 3D	$\frac{d^2}{dx^2}$	$c(x,t) = \frac{n_d}{(4\pi Dt)^{1/2}}\, e^{-x^2/(4Dt)}$
Two-Dimensional Diffusion		
Point source in 2D Line source in 3D	$\frac{1}{r}\frac{d}{dr}r\frac{d}{dr}$	$c(r,t) = \frac{n_d}{(4\pi Dt)}\, e^{-r^2/(4Dt)}$
Three-Dimensional Diffusion		
Point source in 3D	$\frac{1}{r^2}\frac{d}{dr}r^2\frac{d}{dr}$	$c(r,t) = \frac{n_d}{(4\pi Dt)^{3/2}}\, e^{-r^2/(4Dt)}$

$c(r, t = 0) = \delta(r)$].[4] Corresponding results for two-dimensional diffusion are given in Table 5.1.

The form of the solution for one-dimensional diffusion is illustrated in Fig. 5.3. The solution $c(x, t)$ is symmetric about $x = 0$ (i.e., $c(x, t) = c(-x, t)$). Because the flux at this location always vanishes, no material passes from one side of the plane to the other and therefore the two sides of the solution are independent. Thus the general form of the solution for the infinite domain is also valid for the semi-infinite domain ($0 < x < \infty$) with an initial thin source of diffusant at $x = 0$. However, in the semi-infinite case, the initial thin source diffuses into one side rather than two and the concentration is therefore larger by a factor of two, so that

$$c(x, t) = \frac{n_d}{(\pi D t)^{1/2}} e^{-x^2/(4Dt)} \tag{5.18}$$

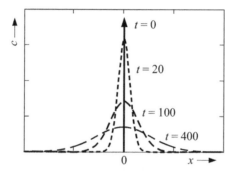

Figure 5.3: Spreading of point, line, and planar diffusion sources with increasing time according to the one-dimensional solution in Table 5.1. Curves were calculated from Eq. 5.18 for times shown and $n_d = 1$, $D = 10^{-4}$, and $-1 < x < 1$ (all units arbitrary).

Equation 5.18 offers a convenient technique for measuring self-diffusion coefficients. A thin layer of radioactive isotope deposited on the surface of a flat specimen serves as an instantaneous planar source. After the specimen is diffusion annealed, the isotope concentration profile is determined. With these data, Eq. 5.18 can be written

$$\ln {}^\star c = \text{constant} - \frac{x^2}{4\,{}^\star D t} \tag{5.19}$$

and ${}^\star D$ can be determined from the slope of a $\ln {}^\star c$ vs. x^2 plot, as shown in Fig. 5.4.[5]

---

[4]A delta function, $\delta(\vec{r})$,is a distribution that equals zero everywhere except where its argument is zero, where it has an infinite singularity. It has the property $\int f(\vec{r})\delta(\vec{r} - \vec{r}_0)d\vec{r} = f(\vec{r}_0)$; so it also follows that $\int \delta(\vec{r} - \vec{r}_0)d\vec{r} = 1$. The singularity of $\delta(\vec{r} - \vec{r}_0)$ is located at $\vec{r}_0$.

[5]This technique can be used to measure the diffusivity in anisotropic materials, as described in Section 4.5. Measurements of the concentration profile in the principal directions can be used to determine the entire diffusion tensor.

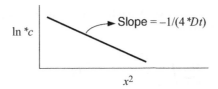

Figure 5.4:   Plot of $\ln {}^*c$ vs. x^2 used to determine self-diffusivity when an instantaneous planar source is used and Eq. 5.18 applies.

5.2.2   Solutions Involving the Error Function

The instantaneous local-source solutions in Table 5.1 can be used to build up solutions for general initial distributions of diffusant by using the method of superposition (see Section 4.2.3).

Section 4.2.2 shows how to use the scaling method to obtain the error function solution for the one-dimensional diffusion of a step function in an infinite medium given by Eq. 4.31. The same solution can be obtained by superposing the one-dimensional diffusion from a distribution of instantaneous local sources arrayed to simulate the initial step function. The boundary and initial conditions are

$$c(x,0) = \begin{cases} c_0 & x > 0 \\ 0 & x < 0 \end{cases} \tag{5.20}$$

$$\frac{dc}{dx}(x = \pm\infty, t) = 0$$

The initial distribution is simulated by a uniform distribution of point, line, or planar sources placed along $x > 0$ as in Fig. 5.5. The *strength*, or the amount of diffusant contributed by each source, must be $c_0\,dx$. The superposition can be achieved by replacing n_d in Table 5.1 with $c(\vec{x})dV$ [$c(x)dx$ in one dimension] and integrating the sources from each point.

Consider the contribution at a general position x from a source at some other position ξ. The distance between the general point x and the source is $\xi - x$, thus

$$c_\xi(x,t) = \frac{c_0\,d\xi}{\sqrt{4\pi Dt}}\, e^{-(\xi-x)^2/(4Dt)} \tag{5.21}$$

So the solution corresponding to the conditions given by Eq. 5.20 must be the integral over all sources,

$$c(x,t) = \frac{c_0}{\sqrt{4\pi Dt}} \int_0^\infty e^{-(\xi-x)^2/(4Dt)}d\xi \tag{5.22}$$

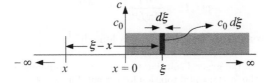

Figure 5.5:   Diagram used to determine the contribution at the general point x of a local source of strength $c_0\,d\xi$ located at ξ.

or by transforming the integration variable by using $u = (\xi - x)/\sqrt{4Dt}$ and the properties of an even integrand,

$$
\begin{aligned}
c(x,t) &= \frac{c_0}{\sqrt{\pi}} \int_{-\infty}^{-x/(2\sqrt{Dt})} e^{-u^2} du \\
&= \frac{c_0}{\sqrt{\pi}} \left(\int_{-\infty}^{0} e^{-u^2} du + \int_{0}^{x/(2\sqrt{Dt})} e^{-u^2} du \right) \\
&= \frac{c_0}{2} + \frac{c_0}{2} \operatorname{erf}\left(\frac{x}{2\sqrt{Dt}} \right)
\end{aligned}
\tag{5.23}
$$

which is consistent with the solution given by Eq. 4.31.

Summations over point-, line-, or planar-source solutions are useful examples of the general method of Green's functions. [6] For instance, the boundary and initial conditions for a *triangular source* are

$$
c(x,t=0) = \begin{cases} 0 & x > a \\ \frac{n_d}{a}\left(1 - \frac{x}{a}\right) & 0 < x < a \\ \frac{n_d}{a}\left(1 + \frac{x}{a}\right) & -a < x < 0 \\ 0 & x < -a \end{cases}
\tag{5.24}
$$

$$
\frac{\partial c}{\partial x}(x = \pm\infty, t) = 0
$$

A solution to this boundary-value problem can be obtained by using Eq. 4.40 with a position-dependent point-source density (this method is useful for solving Exercise 5.7).

As a last example, the solution for two-dimensional diffusion from a line source lying along z in three dimensions can be obtained by integrating over a distribution of point sources lying along the z-axis. If the point sources are distributed so that the source strength along the line is n_d particles per unit length, the contribution of an effective point source of strength $n_d \, d\xi$ at $(0,0,\xi)$ to the point (x,y,z) is

$$
dc = \frac{n_d \, d\xi}{(4\pi Dt)^{3/2}} e^{-[x^2 + y^2 + (\xi - z)^2]/(4Dt)}
\tag{5.25}
$$

so that

$$
c(x,y,z,t) = \frac{n_d}{(4\pi Dt)^{3/2}} e^{-(x^2 + y^2)/(4Dt)} \int_{-\infty}^{\infty} e^{-(\xi - z)^2/(4Dt)} \, d\xi
\tag{5.26}
$$

In cylindrical coordinates, $r^2 = x^2 + y^2$ and, after integration, in agreement with the entry in Table 5.1,

$$
c(r,t) = \frac{n_d}{4\pi Dt} e^{-r^2/(4Dt)}
\tag{5.27}
$$

where the source strength n_d has dimensions length^{-1}.

---

[6]Green's functions arise in the general solution to many partial-differential equations. They are generally obtained from the fundamental solution for a point, line, or planar source. Subsequently, an integral equation for a general solution is obtained by integrating over all the source terms; the fundamental solution becomes the *kernel* to the integral equation, which is the term that multiplies the source density in the integrand.

5.2.3 Method of Superposition

In Section 4.2.3 we described application of the method of superposition to infinite and semi-infinite systems. The method can also be applied, in principle, to finite systems, but it often becomes unwieldy (see Crank's discussion of the reflection method [2]).

5.2.4 Method of Separation of Variables: Diffusion on a Finite Domain

A standard method to solve many partial-differential equations is to assume that the solution can be written as a product of functions, each a function of one of the independent variables. Table 5.2 provides several functional forms of such solutions.

Table 5.2: Product Solutions for the Separation-of-Variables Method in Cartesian and Cylindrical Coordinates

System	Equation	Solution
One dimension, x	$\frac{dc}{dt} = D\frac{d^2c}{dx^2}$	$c(x,t) = X(x)T(t)$
Three dimensions, (x,y,z)	$\frac{dc}{dt} = D\nabla^2 c$	$c(x,y,z,t) = X(x)Y(y)Z(z)T(t)$
Cylindrical, (r,θ,z)	$\frac{dc}{dt} = D\nabla^2 c$	$c(r,\theta,z,t) = R(r)\Theta(\theta)Z(z)T(t)$

The following example illustrates the method. Consider a one-dimensional diffusion problem with the initial and boundary conditions for the domain $0 < x < L$:

$$c(x,0) = c_0 \qquad c(0,t) = 0 \qquad c(L,t) = 0 \tag{5.28}$$

This situation may represent the diffusion of a high-vapor-pressure dopant out of a thin film (thickness L, initial dopant concentration c_0) of silicon when placed in a vacuum. Assume that the variables are separable.[7] Letting $c(x,t) = X(x)T(t)$ and substituting into the diffusion equation gives

$$\frac{1}{DT}\frac{dT}{dt} = \frac{1}{X}\frac{d^2X}{dx^2} \tag{5.29}$$

Because the left side depends only on t and the right side depends only on x, each side must be equal to the same constant. This may be understood by considering Fig. 5.6, in which f is a function of y only and g is a function of x only. Each surface is a "ruled" surface; that is, the surface contains lines of constant value running in one direction. If the two functions are equal as in the separation equation (Eq. 5.29), the surface must be flat in both variables. Thus, if the two functions are equal, they are constant.

Let that constant be $-\lambda$. Then

$$\frac{dT}{dt} = -\lambda DT \tag{5.30}$$

---

[7]If a solution is found for the initial and boundary conditions, there is a uniqueness theorem that justifies the assumption. Whether a solution can be found using separation of variables depends on whether the boundary conditions follow the symmetry of the separation variables.

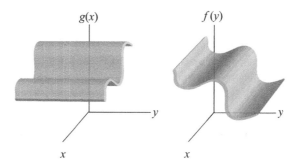

Figure 5.6:   Representation in (x, y) space of $g(x)$ and $f(y)$.

and

$$\frac{d^2 X}{dx^2} = -\lambda X \tag{5.31}$$

Equation 5.31 has solutions of the form

$$X(x) = \begin{cases} A\sin(\sqrt{\lambda}x) + B\cos(\sqrt{\lambda}x) & (\lambda > 0) \\ A'e^{\sqrt{-\lambda}x} + B'e^{-\sqrt{-\lambda}x} & (\lambda < 0) \\ A''x + B'' & (\lambda = 0) \end{cases} \tag{5.32}$$

where A and B are constants that must satisfy the boundary conditions. For the *particular* boundary conditions specified in Eq. 5.28, nontrivial solutions to Eq. 5.31 exist *only* if $\lambda > 0$ and $B = 0$. However, there is no nonzero A that can satisfy the boundary conditions for a general $\lambda > 0$, so λ must take on values appropriate to the boundary conditions. Therefore,

$$\lambda_n = n^2 \frac{\pi^2}{L^2} \tag{5.33}$$

because $\sin\sqrt{\lambda_n}L = \sin n\pi = 0$, where n is an integer. The λ_n are the linear differential equation's eigenvalues for the boundary conditions.

Because use of any λ_n satisfies the boundary conditions in Eq. 5.28, each of the functions

$$X_n(x) = a_n \sin\left(n\pi\frac{x}{L}\right) \tag{5.34}$$

satisfies Eq. 5.31 for the boundary conditions in Eq. 5.28. The X_n are known as the *eigenfunctions* for the boundary conditions.

The general solution to Eq. 5.30 is

$$T(t) = T^\circ e^{-\lambda D t} \tag{5.35}$$

where T° is a constant. But λ must take on the values given in Eq. 5.33, and therefore the time-dependent eigenfunction solutions can be written

$$T_n(t) = T_n^\circ\, e^{-n^2\pi^2 D t / L^2} \tag{5.36}$$

The general solution, satisfying the boundary conditions, is then (by superposition) a sum of the products of the eigenfunction solutions and is of the form

$$c(x, t) = \sum_{n=1}^{\infty} X_n(x)\, T_n(t) = \sum_{n=1}^{\infty} A_n \sin\left(n\pi\frac{x}{L}\right) e^{-n^2\pi^2 D t / L^2} \tag{5.37}$$

where $A_n = a_n T_n^\circ$.

It is now necessary to satisfy the initial condition given in Eq. 5.28. This requires that

$$c_0 = \sum_{n=1}^{\infty} A_n \sin\left(n\pi\frac{x}{L}\right) \tag{5.38}$$

which, as seen in Eq. 5.42, is a Fourier sine series representation of c_0.

Synopsis of Fourier Series

If a function $u(x)$ exists on the interval $-L < x < L$, $u(x)$ can be represented as

$$u(x) = \frac{b_o}{2} + \sum_{n=1}^{\infty} \left[a_n \sin\left(n\pi\frac{x}{L}\right) + b_n \cos\left(n\pi\frac{x}{L}\right)\right] \tag{5.39}$$

where the coefficients are given by

$$a_n = \frac{1}{L} \int_{-L}^{L} u(x) \sin\left(n\pi\frac{x}{L}\right) dx \tag{5.40}$$

$$b_n = \frac{1}{L} \int_{-L}^{L} u(x) \cos\left(n\pi\frac{x}{L}\right) dx \tag{5.41}$$

If $u(x)$ is an odd function $[u(x) = -u(-x)]$ and the sine expansion is applied,

$$u(x) = \sum_{n=1}^{\infty} a_n \sin\left(n\pi\frac{x}{L}\right) \tag{5.42}$$

$$a_n = \frac{2}{L} \int_{0}^{L} u(x) \sin\left(n\pi\frac{x}{L}\right) dx \tag{5.43}$$

Similarly, if $u(x)$ is an even function $[u(x) = u(-x)]$, all a_n will vanish and $u(x)$ can be written as a cosine expansion only:

$$u(x) = \frac{b_o}{2} + \sum_{n=1}^{\infty} b_n \cos\left(n\pi\frac{x}{L}\right) \tag{5.44}$$

$$b_n = \frac{2}{L} \int_{0}^{L} u(x) \cos\left(n\pi\frac{x}{L}\right) dx \tag{5.45}$$

Finally, any function can be written as a sum of an odd and an even function.

Using Eq. 5.43, the coefficients, A_n, are then given by

$$A_n = \frac{2c_0}{L} \int_{0}^{L} \sin\left(n\pi\frac{x}{L}\right) dx = -\frac{2c_0}{n\pi}(\cos n\pi - 1) = \begin{cases} \frac{4c_0}{n\pi} & n \text{ odd} \\ 0 & n \text{ even} \end{cases} \tag{5.46}$$

Therefore, the final solution is given by

$$c(x,t) = \frac{4c_0}{\pi} \sum_{j=0}^{\infty} \left(\frac{1}{2j+1} \sin\left[(2j+1)\pi\frac{x}{L}\right] e^{-[(2j+1)\pi/L]^2 Dt}\right) \tag{5.47}$$

The coefficients of the higher-order (shorter-wavelength) terms in Eq. 5.47 decrease as $1/n$. Not only do the shorter-wavelength terms start out smaller but they also decay exponentially at rates that scale inversely as the *square* of the wavelength. Thus, even at times as short as $L^2/(25D)$, the first term of the series in Eq. 5.47 suffices to a good approximation, such that

$$c(x,t) \approx \frac{4c_0}{\pi} \sin\left(\pi\frac{x}{L}\right) e^{-\pi^2 Dt/L^2} \tag{5.48}$$

with a maximum error of about 1%. The average composition $\bar{c}$ in this "long-time" regime can be obtained by integration of Eq. 5.48:

$$\bar{c}(t) \approx \frac{8c_0}{\pi^2} e^{-\pi^2 Dt/L^2} \tag{5.49}$$

$\bar{c}$ therefore decays exponentially with the characteristic time $\tau = L^2/(\pi^2 D)$. This is reasonably consistent with estimates of diffusion depths and times in Section 5.2.6.

The method of separation of variables can be applied in the same manner to other initial distributions of diffusant. The effort lies only in determining the Fourier coefficients, which, for many cases, can be looked up in tables. If the spatial dimension of the system is higher [e.g., $c(x, y, z, t)$], a separate Fourier series must be obtained for each of the three separate functions in the product $X(x)Y(y)Z(z)$.

Cylindrical Coordinates. The separation-of-variables method also applies when the boundary conditions and initial conditions have cylindrical symmetry (see Eqs. 5.7 and 5.8). If $c(r,t) = R(r)T(t)$, the resulting ordinary differential equation for $R(r)$ is

$$\frac{d^2R}{dr^2} + \frac{1}{r}\frac{dR}{dr} + \alpha^2 R = 0 \tag{5.50}$$

This equation has general solutions

$$R(r) = \sum_{n=1}^{\infty} [a_n J_0(\alpha_n r) + b_n Y_0(\alpha_n r)] \tag{5.51}$$

where J_0 and Y_0 are Bessel functions of order zero of the first and second kind. The α_n are solved by matching boundary conditions, and the coefficients a_n and b_n are determined by matching the initial conditions in a Bessel function series expansion. See Carslaw and Jaeger for examples [3].

5.2.5 Method of Laplace Transforms

The *Laplace transform* method is a powerful technique for solving a variety of partial-differential equations, particularly time-dependent boundary condition problems and problems on the semi-infinite domain. After a Laplace transform is performed on the original boundary-value problem, the transformed equation is often easily solved. The transformed solution is then back-transformed to obtain the desired solution.

The Laplace transform of a function $f(x,t)$ is defined as

$$\mathcal{L}\{f(x,t)\} = \hat{f}(x,p) = \int_0^\infty e^{-pt} f(x,t)\,dt \tag{5.52}$$

The Laplace-transformed f is represented by both the operational form $\mathcal{L}\{f\}$ and the shorthand $\hat{f}$. The variable p is the transformation variable.[8]

The key utility of the Laplace transform involves its operation on time derivatives:

$$\mathcal{L}\left\{\frac{\partial f}{\partial t}\right\} = \int_0^\infty e^{-pt}\frac{\partial f(x,t)}{\partial t}\,dt \tag{5.53}$$

Integrating the right-hand side of Eq. 5.53 by parts,

$$e^{-pt}f(x,t)\bigg|_{t=0}^{t=\infty} + p\int_0^\infty e^{-pt}f(x,t)\,dt = p\mathcal{L}\{f\} - f(x,t=0) \tag{5.54}$$

Therefore,

$$\mathcal{L}\left\{\frac{\partial f}{\partial t}\right\} = p\mathcal{L}\{f\} - f(x,t=0) \tag{5.55}$$

The Laplace transform of a spatial derivative of f is seen from Eq. 5.52 to be equal to the spatial derivative of $\hat{f}$; that is,

$$\mathcal{L}\left\{\frac{\partial^n f}{\partial x^n}\right\} = \frac{\partial^n \hat{f}(x,p)}{\partial x^n} \tag{5.56}$$

The method can be demonstrated by considering diffusion into a semi-infinite body where the surface concentration $c(x=0,t)$ is fixed:

$$c(x=0,t) = c_0$$
$$\frac{\partial c}{\partial x}(x=\infty,t) = 0 \tag{5.57}$$
$$c(x,t=0) = 0 \quad \text{for } 0 \le x < \infty$$

Applying the Laplace transform to both sides of the diffusion equation yields

$$\mathcal{L}\left\{\frac{\partial c}{\partial t}\right\} = D\mathcal{L}\left\{\frac{\partial^2 c}{\partial x^2}\right\}$$
$$\frac{\partial^2 \hat{c}}{\partial x^2} - \frac{p}{D}\hat{c} = -\frac{c(x,t=0)}{D} = 0 \tag{5.58}$$

Thus, the Laplace transform removes the t-dependence and turns the partial-differential equation into an ordinary-differential equation.[9] The solution to Eq. 5.58 is

$$\hat{c}(x,p) = a_1 e^{\sqrt{p/D}\,x} + a_2 e^{-\sqrt{p/D}\,x} \tag{5.59}$$

The boundary conditions must also be transformed:

$$\hat{c}(x=0,p) = c_0\int_0^\infty e^{-pt}\,dt = \frac{c_0}{p}$$
$$\frac{\partial \hat{c}}{\partial x}(x=\infty,p) = 0 \tag{5.60}$$

---

[8]Technical requirements on p are that its real part must be positive and large enough that the integral converges.

[9]Note that it also automatically incorporates the initial condition.

Solving for the coefficients in Eq. 5.59 leads to the solution

$$\hat{c}(x,p) = \frac{c_0}{p} e^{-\sqrt{\frac{p}{D}}\,x} \tag{5.61}$$

This solution is then inversely transformed by use of a table of Laplace transforms (see Table 5.3) to obtain the desired solution:

$$c(x,t) = c_0 \left[1 - \text{erf}\left(\frac{x}{\sqrt{4Dt}}\right) \right] \equiv c_0 \, \text{erfc}\left(\frac{x}{\sqrt{4Dt}}\right) \tag{5.62}$$

where $\text{erfc}(z) \equiv 1 - \text{erf}(z)$ is known as the *complementary error function*. Note that this solution could have been deduced directly from Eq. 5.23 (the solution for the step-function initial conditions for an infinite system) because in that solution, the plane $x = 0$ always maintains a constant composition.

Table 5.3:   Selected Laplace Transform Pairs

$$\hat{c}(x,p) = \int_0^\infty e^{-pt} c(x,t)\, dt$$
$$q \equiv \sqrt{p/D}$$

$\frac{1}{p}$	1
$\frac{1}{p^{\nu+1}}$ $\nu > -1$	$\frac{t^\nu}{\Gamma(\nu+1)}$
$\frac{1}{p+\alpha}$	$e^{-\alpha t}$
$\frac{\omega}{p^2+\omega^2}$	$\sin \omega t$
$\frac{p}{p^2+\omega^2}$	$\cos \omega t$
e^{-qx}	$\frac{x\, e^{-x^2/(4Dt)}}{\sqrt{4\pi Dt^3}}$
$\frac{e^{-qx}}{q}$	$\sqrt{\frac{D}{\pi t}}\, e^{-x^2/(4Dt)}$
$\frac{e^{-qx}}{p}$	$\text{erfc}\left(\frac{x}{\sqrt{4Dt}}\right)$
$\frac{e^{-qx}}{pq}$	$\sqrt{\frac{4Dt}{\pi}}\, e^{-x^2/(4Dt)} - x\, \text{erfc}\left(\frac{x}{\sqrt{4Dt}}\right)$

Example with Time-Dependent Boundary Conditions.   Consider the case where a constant flux, $J_\circ$, is imposed on the surface of a semi-infinite sample:

$$\frac{\partial c}{\partial x}(x = 0, t) = -\frac{J_\circ}{D} = \text{constant}$$
$$c(x = \infty, t) = c_0 \tag{5.63}$$
$$c(x, t = 0) = c_0 \quad \text{for } 0 \le x < \infty$$

This boundary condition might apply for solute absorption with its rate moderated by some thin passive surface layer. Note that the surface concentration at $x = 0$ must be a function of time to maintain the constant-flux condition (see Fig. 5.7).

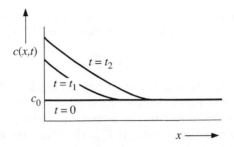

Figure 5.7:   Diffusion profiles necessary to maintain constant flux into a semi-infinite body. Note the fixed value of $dc/dx|_{x=0}$ for $t > 0$.

Using the Laplace transform,

$$\frac{\partial^2 \hat{c}}{\partial x^2} - \frac{p}{D}\hat{c} = -\frac{c(x, t=0)}{D} = -\frac{c_0}{D} \tag{5.64}$$

Equation 5.64 is an inhomogeneous ordinary differential equation and its solution is therefore the sum of the solution of its homogeneous form (i.e., Eq. 5.59) and a particular solution (i.e., $\hat{c} = c_0/p$). Therefore,

$$\hat{c}(x, p) = \frac{c_0}{p} + a_1 e^{\sqrt{p/D}\, x} + a_2 e^{-\sqrt{p/D}\, x} \tag{5.65}$$

The transformed boundary conditions are

$$\frac{\partial \hat{c}(x=0, p)}{\partial x} = -\frac{J_o}{pD}$$
$$\hat{c}(x=\infty, p) = \frac{c_0}{p} \tag{5.66}$$

Solving for the coefficients a_1 and a_2,

$$\hat{c}(x, p) = \frac{c_0}{p} + \frac{J_o}{p^{3/2} D^{1/2}} e^{-\sqrt{p/D}\, x} \tag{5.67}$$

Inversely transforming this solution with the use of Table 5.3 yields

$$c(x, t) = c_0 + \frac{J_o}{D}\left[\sqrt{\frac{4Dt}{\pi}}\, e^{-x^2/(4Dt)} - x\, \operatorname{erfc}\left(\frac{x}{\sqrt{4Dt}}\right)\right] \tag{5.68}$$

The surface concentration must therefore increase as

$$c(0, t) = c_0 + 2J_o\sqrt{\frac{t}{\pi D}} \tag{5.69}$$

5.2.6   Estimating the Diffusion Depth and Time to Approach Steady State

A rough estimate of the diffusion penetration distance from a point source is the location where the concentration has fallen off by $\approx 1/e$ of the concentration at $x = 0$. This occurs when

$$x \approx 2\sqrt{Dt} \tag{5.70}$$

An estimate of the penetration distance for the error-function solution (Eq. 5.23) is the distance where $c(x,t) = c_0/8$, or equivalently, $\text{erf}[x/(2\sqrt{Dt})] = -3/4$, which corresponds to

$$x \approx 1.6\sqrt{Dt} \tag{5.71}$$

A reasonable estimate for the penetration depth is therefore again $2\sqrt{Dt}$.

To estimate the time at which steady-state conditions are expected, the required penetration distance is set equal to the largest characteristic length over which diffusion can take place in the system. If L is the characteristic linear dimension of a body, steady state may be expected to apply at times $\tau \gg L^2/D_{\min}$, where $D_{\min}$ is the smallest value of the diffusivity in the body. Of course, there are many physical situations where steady-state conditions will never arise, such as when the boundary conditions are time dependent or the system is infinite or semi-infinite.

Bibliography

1. P.M. Morse and H. Feshbach. *Methods of Theoretical Physics, Vols. 1 and 2*. McGraw-Hill, New York, 1953.

2. J. Crank. *The Mathematics of Diffusion*. Oxford University Press, Oxford, 2nd edition, 1975.

3. H.S. Carslaw and J.C. Jaeger. *Conduction of Heat in Solids*. Oxford University Press, Oxford, 2nd edition, 1959.

EXERCISES

5.1 A flat bilayer slab is composed of layers of material A and B, each of thickness L. A component is diffusing through the bilayer in the steady state under conditions where its concentration is maintained at $c = c_0 = $ constant at one surface and at $c = 0$ at the other. Its diffusivity is equal to the constants D^A and D^B in the two layers, respectively. No other components in the system diffuse significantly.

Does the flux through the bilayer depend on whether the concentration is maintained at $c = c_0$ at the surface of the A layer or the surface of the B layer? Assume that the concentration of the diffusing component is continuous at the A/B interface.

Solution. Solve for the diffusion in each layer and match the solutions across the A/B interface. Assume that $c = c_0$ at the surface of the A layer and let $c = c^{A/B}$ be the concentration at the A/B interface. Using Eq. 5.5, the concentration in the A layer in the interval $0 < x < L$ is

$$c^A(x) = c_0 - \left(c_0 - c^{A/B}\right)\frac{x}{L} \tag{5.72}$$

$$J^A = -D^A\frac{dc^A}{dx} = \frac{D^A\left(c_0 - c^{A/B}\right)}{L} \tag{5.73}$$

For the B slab in the interval $L \leq x \leq 2L$,

$$c^B(x) = c^{A/B} - c^{A/B}\frac{x - L}{L} \tag{5.74}$$

$$J^B = -D^B\frac{dc^B}{dx} = D^B\frac{c^{A/B}}{L} \tag{5.75}$$

Setting $J^A = J^B$ and solving for $c^{A/B}$,

$$c^{A/B} = c_0 \frac{D^A}{D^A + D^B} \tag{5.76}$$

The steady-state flux through the bilayer is then

$$J = \frac{c_0}{L} \frac{D^A D^B}{D^A + D^B} \tag{5.77}$$

J is invariant with respect to switching the materials in the two slabs, and therefore it does not matter on which surface $c = c_0$.

5.2 Find an expression for the steady-state concentration profile during the radial diffusion of a diffusant through a cylindrical shell of thickness, ΔR, and inner radius, R^{in}, in which the diffusivity is a function of radius $D(r)$. The boundary conditions are $c(r = R^{\text{in}}) = c^{\text{in}}$ and $c(r = R^{\text{in}} + \Delta R) = c^{\text{out}}$.

Solution. The gradient operator in cylindrical coordinates is

$$\nabla = \frac{\partial}{\partial r} \hat{e}_r + \frac{1}{r} \frac{\partial}{\partial \theta} \hat{e}_\theta + \frac{\partial}{\partial z} \hat{e}_z \tag{5.78}$$

The divergence of a flux $\vec{J}$ in cylindrical coordinates is

$$\nabla \cdot \vec{J} = \frac{1}{r} \frac{\partial (r J_r)}{\partial r} + \frac{1}{r} \frac{\partial J_\theta}{\partial \theta} + \frac{\partial J_z}{\partial z} \tag{5.79}$$

Therefore, the steady-state radially-symmetric diffusion equation becomes

$$0 = \frac{\partial}{\partial r} \left(r D(r) \frac{\partial c}{\partial r} \right) \tag{5.80}$$

which can be integrated twice to give

$$c(r) = c^{\text{in}} + a_1 \int_{R^{\text{in}}}^{r} \frac{d\rho}{\rho D(\rho)} \tag{5.81}$$

The integration constant a_1 is determined by the boundary condition at $R^{\text{in}} + \Delta R$:

$$a_1 = \frac{c^{\text{out}} - c^{\text{in}}}{\int_{R^{\text{in}}}^{R^{\text{in}} + \Delta R} d\rho / \rho D(\rho)} \tag{5.82}$$

5.3 Find the steady-state concentration profile during the radial diffusion of a diffusant through a bilayer cylindrical shell of inner radius, R^{in}, where each layer has thickness $\Delta R/2$ and the constant diffusivities in the inner and outer layers are D^{in} and D^{out}. The boundary conditions are $c(r = R^{\text{in}}) = c^{\text{in}}$ and $c(r = R^{\text{in}} + \Delta R) = c^{\text{out}}$. Will the total diffusion current through the cylinder be the same if the materials that make up the inner and outer shells are exchanged? Assume that the concentration of the diffusant is the same in the inner and outer layers at the bilayer interface.

Solution. The concentration profile at the bilayer interface will not have continuous derivatives. Break the problem into separate diffusion problems in each layer and then impose the continuity of flux at the interface. Let the concentration at the bilayer interface be $c^{i/o}$.

Inner region: $R^{\text{in}} \leq r \leq R^{\text{in}} + \frac{\Delta R}{2}$

Using Eq. 5.82,

$$c^{\text{in}}(r) = \frac{c^{\text{i/o}} - c^{\text{in}}}{\ln\left(\frac{R^{\text{in}} + \Delta R/2}{R^{\text{in}}}\right)} \ln\left(\frac{r}{R^{\text{in}}}\right) + c^{\text{in}} \tag{5.83}$$

The flux at the bilayer interface is

$$J^{\text{i/o}} = -D^{\text{in}} \frac{c^{\text{i/o}} - c^{\text{in}}}{\ln\left(\frac{R^{\text{in}} + \Delta R/2}{R^{\text{in}}}\right)} \frac{1}{R^{\text{in}} + \Delta R/2} \tag{5.84}$$

<u>Outer region:</u> $R^{\text{in}} + \frac{\Delta R}{2} \leq r \leq R^{\text{in}} + \Delta R$

$$c^{\text{out}}(r) = \frac{c^{\text{out}} - c^{\text{i/o}}}{\ln\left(\frac{R^{\text{in}} + \Delta R}{R^{\text{in}} + \Delta R/2}\right)} \ln\left(\frac{r}{R^{\text{in}} + \Delta R/2}\right) + c^{\text{i/o}} \tag{5.85}$$

The flux at the bilayer interface is

$$J^{\text{i/o}} = -D^{\text{out}} \frac{c^{\text{out}} - c^{\text{i/o}}}{\ln\left(\frac{R^{\text{in}} + \Delta R}{R^{\text{in}} + \Delta R/2}\right)} \frac{1}{R^{\text{in}} + \Delta R/2} \tag{5.86}$$

Setting the fluxes at the interfaces equal and solving for $c^{\text{i/o}}$ yields

$$c^{\text{i/o}} = \frac{\alpha^{\text{out}} c^{\text{out}} + \alpha^{\text{in}} c^{\text{in}}}{\alpha^{\text{out}} + \alpha^{\text{in}}} \tag{5.87}$$

where

$$\alpha^{\text{out}} \equiv \frac{D^{\text{out}}}{\ln\left(\frac{R^{\text{in}} + \Delta R}{R^{\text{in}} + \Delta R/2}\right)} \qquad \alpha^{\text{in}} \equiv \frac{D^{\text{in}}}{\ln\left(\frac{R^{\text{in}} + \Delta R/2}{R^{\text{in}}}\right)} \tag{5.88}$$

Putting Eq. 5.87 into Eqs. 5.83 and 5.85 yields the concentration profile of the entire cylinder.

The total current diffusing through the cylinder (per unit length) is

$$I = 2\pi\left(R^{\text{in}} + \frac{\Delta R}{2}\right) J^{\text{i/o}} = -\frac{2\pi D^{\text{in}}\left(c^{\text{i/o}} - c^{\text{in}}\right)}{\ln\left(\frac{R^{\text{in}} + \Delta R/2}{R^{\text{in}}}\right)} \tag{5.89}$$

Using Eq. 5.87,

$$c^{\text{i/o}} - c^{\text{in}} = \frac{\alpha^{\text{out}}\left(c^{\text{out}} - c^{\text{in}}\right)}{\alpha^{\text{out}} + \alpha^{\text{in}}} \tag{5.90}$$

If everything is kept constant except D^{in} and D^{out}, use of Eq. 5.90 in Eq. 5.89 shows that

$$I \propto \frac{D^{\text{in}} D^{\text{out}}}{\alpha_1 D^{\text{out}} + \alpha_2 D^{\text{in}}} \tag{5.91}$$

where α_1 and α_2 are constants. Clearly, I will be different if the materials making up the inner and outer shells are exchanged and the values of D^{out} and D^{in} are therefore exchanged. This contrasts with the result for the two adjoining flat slabs in Exercise 5.1.

5.4 Suppose that a very thin planar layer of radioactive Au tracer atoms is placed between two bars of Au to produce a thin source of diffusant as illustrated in Fig. 5.8. A diffusion anneal will cause the tracer atoms to spread by self-diffusion as illustrated in Fig. 5.3. (A mathematical treatment of this spreading out is presented in Section 4.2.3.) Suppose that the diffusion ex-

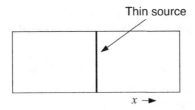

Figure 5.8: Thin planar tracer-atom source between two long bars.

periment is now carried out with a constant electric current passing through the bars along x.

(a) Using the statement of Exercise 3.10, describe the difference between the way in which the tracer atoms spread out when the current is present and when it is absent.

(b) Assuming that $D_V c_V$ is known, how could you use this experiment to determine the electromigration parameter β for Au?

Solution.

(a) The electric current produces a flux of vacancies in one direction and an equal flux of atoms in the reverse direction, so that

$$\vec{J}_A = -\vec{J}_V \tag{5.92}$$

Using the statement of Exercise 3.10, this will result in an average drift velocity for each atom, given by

$$\langle \vec{v}_A \rangle = \frac{\vec{J}_A}{c_A} = -\frac{1}{c_A} \frac{D_V c_V \beta \vec{E}}{kT} \tag{5.93}$$

The tracer atoms will spread out as they would in the absence of current; however, they will also be translated bodily by the distance $\Delta x = \langle v_A \rangle t$ relative to an embedded inert marker as illustrated in Fig. 5.9.

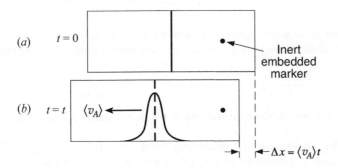

Figure 5.9: (a) The initially thin distribution of tracer atoms that, subsequently, will spread due to diffusion and drift due to electromigration. (b) The electromigration has caused the distribution to spread out and to be translated bodily by $\Delta x = \langle v_A \rangle t$ relative to the fixed marker.

This may be shown by choosing an origin at the initial position of the source in a coordinate system fixed with respect to the marker. The diffusion equation is then

$$\frac{\partial \,^\star c}{\partial t} = -\nabla \cdot \vec{J} = \frac{\partial}{\partial x}\left(^\star D\frac{\partial \,^\star c}{\partial x} - \langle v_A\rangle c\right) \tag{5.94}$$

where $\langle v_A\rangle c$ is the flux due to the drift. Defining a moving (primed) coordinate system with its origin at $x = \langle v_A\rangle t$,

$$x' = x - \langle v_A\rangle t \tag{5.95}$$

Using $[\partial(\ )/\partial x]_t = [\partial(\ )/\partial x']_t$, the drift velocity does not appear in the resulting diffusion equation in the primed coordinate system, which is

$$\frac{\partial \,^\star c}{\partial t} = \,^\star D\frac{\partial^2 \,^\star c}{\partial x'^2} \tag{5.96}$$

The solution in this coordinate system can be obtained from Table 5.1;

$$^\star c(x', t) = \frac{n_d}{\sqrt{4\pi^\star Dt}}\, e^{-x'^2/(4^\star Dt)} \tag{5.97}$$

The distribution therefore *spreads* independently of $\langle v_A\rangle$, but is translated with velocity $\langle v_A\rangle$ with respect to the marker.

(b) The velocity $\langle v\rangle_A$ can be measured experimentally and then β can be obtained through use of Eq. 5.93 if $D_V c_V$ is known. It will be seen in Chapter 8 that $D_V c_V$ can be determined by use of Eq. 8.17 if $^\star D$ is known. $^\star D$ can be determined from the measured distribution illustrated in Fig. 5.9*b* using Eq. 5.97 and the method outlined in Section 5.2.1.

5.5 Obtain the instantaneous plane-source solution in Table 5.1 by representing the plane source as an array of instantaneous point sources in a plane and integrating the contributions of all the point sources.

Solution. Assume an infinite plane containing m point sources per unit area each of strength n_d. The plane is located in the (y,z) plane at $x = 0$. All the point sources in the plane lying within a thin annular ring of radius r and thickness dr centered on the x-axis will contribute a concentration at the point P located along the x-axis at a distance, x, given by

$$dc = m2\pi r\, dr \frac{n_d}{(4\pi Dt)^{3/2}}\, e^{-(x^2+r^2)/(4Dt)} \tag{5.98}$$

where the point-source solution in Table 5.1 has been used. The total concentration is then obtained by integrating over all the point sources in the plane, so that

$$c = \frac{2\pi mn_d\, e^{-x^2/(4Dt)}}{(4\pi Dt)^{3/2}}\int_0^\infty e^{-r^2/(4Dt)} r\, dr = \frac{M}{\sqrt{4\pi Dt}}\, e^{-x^2/(4Dt)} \tag{5.99}$$

where $M = mn_d$ is the total strength of the planar source per unit area.

5.6 Consider an infinite bar extending from $-\infty$ to $+\infty$ along x. Starting at $t = 0$, heat is generated at a constant rate in the $x = 0$ plane. Show that the temperature distribution along the bar is

$$T(x,t) = \int_0^t \frac{P}{2c_P}\sqrt{\frac{1}{\pi\kappa t}}\, e^{-x^2/(4\kappa t)}\, dt \tag{5.100}$$

where P = power input at $x = 0$ (per unit area) and c_p = specific heat per unit volume. Next, show that

$$T(x,t) = \frac{P}{c_P}\sqrt{\frac{t}{\pi\kappa}}\, e^{-x^2/(4\kappa t)} - \frac{Px}{2c_P\kappa}\,\mathrm{erfc}\left(\frac{x}{2\sqrt{\kappa t}}\right) \tag{5.101}$$

Finally, verify that this solution satisfies the conservation condition

$$\int_{-\infty}^{\infty} T(x,t)\,c_P\,dx = Pt$$

Solution. The amount of heat added (per unit area) at $x = 0$ in time dt is $P\,dt$. Using the analogy between problems of mass diffusion and heat flow (Section 4.1), each added amount of heat, $P\,dt$, spreads according to the one-dimensional solution for mass diffusion from a planar source in Table 5.1:

$$dT = \frac{1}{c_P}\left[\frac{P\,dt}{2(\pi\kappa t)^{1/2}}\right]e^{-x^2/(4\kappa t)} \tag{5.102}$$

Because the term in brackets represents an incremental energy input per unit volume, the factor $(c_P)^{-1}$ must be included to obtain an expression for the corresponding incremental temperature rise, dT. Let

$$a_1 = \frac{P}{2c_P\sqrt{\pi\kappa}} \quad a_2 = \frac{x^2}{4\kappa} \quad t = y^{-2} \tag{5.103}$$

Then

$$T(x,t) = -2a_1\int_{\infty}^{1/\sqrt{t}}\frac{\exp\left(-a_2 y^2\right)}{y^2}\,dy \tag{5.104}$$

Integrating by parts yields

$$T(x,t) = 2a_1\, e^{-a_2/t}\sqrt{t} + 4a_1\sqrt{a_2}\int_{\infty}^{\sqrt{a_2/t}} e^{-\zeta^2}\,d\zeta \tag{5.105}$$

Substituting for a_1 and a_2, we finally obtain

$$T(x,t) = \frac{P}{c_P}\sqrt{\frac{t}{\pi\kappa}}\,\exp\left(-\frac{x^2}{4\kappa t}\right) - \frac{Px}{2c_P\kappa}\left[1 - \mathrm{erf}\frac{x}{2\sqrt{\kappa t}}\right] \tag{5.106}$$

Note that the solution given by Eq. 5.106 holds for $x \geq 0$ because the positive root of $\sqrt{a_2}$ was used. The symmetric solution for $x \leq 0$ is easily obtained by changing the sign of x. All the heat stored in the specimen at the time t is represented by the integral

$$\begin{aligned}
Q &= 2\int_0^{\infty} T(x,t)\,c_P\,dx \\
&= \left[2P\sqrt{\frac{t}{\pi\kappa}}\int_0^{\infty}\exp\left(-\frac{x^2}{4\kappa t}\right)dx\right] - \left[\frac{P}{\kappa}\int_0^{\infty} x\left(1 - \mathrm{erf}\frac{x}{2\sqrt{\kappa t}}\right)dx\right]
\end{aligned} \tag{5.107}$$

The first bracketed term in Eq. 5.107 has the value $2Pt$. The second term can be integrated by parts and has the value Pt. Therefore,

$$Q = 2Pt - Pt = Pt$$

and the stored heat is equal to the heat generated during the time t, given by Pt.

5.7 Consider the following boundary-value problem:

$$c(x, t = 0) = \begin{cases} 0 & x > a \\ \frac{n_d}{a}\left(1 - \frac{x}{a}\right) & 0 < x < a \\ \frac{n_d}{a}\left(1 + \frac{x}{a}\right) & -a < x < 0 \\ 0 & x < -a \end{cases}$$ (5.108)

$$\frac{\partial c}{\partial x}(x = \pm\infty, t) = 0$$

(a) Use the superposition method to find the time-dependent solution.

(b) Show that when $2\sqrt{Dt} \gg a$, the solution in (a) reduces to a standard instantaneous planar-source solution in which the initial distribution given by Eq. 5.108 serves as the source. Use the following expansions for small ϵ:

$$\text{erf}(z + \epsilon) = \text{erf}(z) + \frac{2\epsilon}{\sqrt{\pi}}e^{-z^2} + \cdots \qquad e^{\epsilon} = 1 + \epsilon + \cdots$$ (5.109)

Solution.

(a) The concentration of diffusant located between ζ and $\zeta + d\zeta$ in the initial distribution acts as a planar source of thickness, $d\zeta$, and produces a concentration increment at a distance, x, given by

$$dc = \frac{c(\zeta)d\zeta}{\sqrt{4\pi Dt}}e^{-(x-\zeta)^2/(4Dt)}$$ (5.110)

The total concentration produced at x is then obtained by integrating over the distribution. Therefore,

$$c(x, t) = \frac{n_d}{a\sqrt{\pi}}\left[\int_{-a}^{0}\left(1 + \frac{\zeta}{a}\right)e^{-(x-\zeta)^2/\lambda^2}\frac{d\zeta}{\lambda} + \int_{0}^{a}\left(1 - \frac{\zeta}{a}\right)e^{-(x-\zeta)^2/\lambda^2}\frac{d\zeta}{\lambda}\right]$$ (5.111)

where $\lambda \equiv \sqrt{4Dt}$. Let $x - \zeta = \lambda u$; then

$$c(x, t) = -\frac{n_d}{a\sqrt{\pi}}\left[\left(1 + \frac{x}{a}\right)\int_{(x+a)/\lambda}^{x/\lambda}e^{-u^2}du - \frac{\lambda}{a}\int_{(x+a)/\lambda}^{x/\lambda}ue^{-u^2}du\right.$$
$$\left. - \left(1 - \frac{x}{a}\right)\int_{(x-a)/\lambda}^{x/\lambda}e^{-u^2}du - \frac{\lambda}{a}\int_{(x-a)/\lambda}^{x/\lambda}ue^{-u^2}du\right]$$

Using the relations

$$\int_{\alpha}^{\beta}e^{-u^2}du = \frac{\sqrt{\pi}}{2}[\text{erf}(\beta) - \text{erf}(\alpha)]$$ (5.112)

$$\int_{\alpha}^{\beta}ue^{-u^2}du = \frac{1}{2}\left[e^{-\alpha^2} - e^{-\beta^2}\right]$$ (5.113)

The solution is

$$c(x, t) = \frac{n_d}{2a}\left\{\text{erf}\left(\frac{x+a}{\lambda}\right) - \text{erf}\left(\frac{x-a}{\lambda}\right)\right.$$
$$+ \frac{x}{a}\left[\text{erf}\left(\frac{x+a}{\lambda}\right) - 2\,\text{erf}\left(\frac{x}{\lambda}\right) + \text{erf}\left(\frac{x-a}{\lambda}\right)\right]$$
$$\left. + \frac{\lambda}{\sqrt{\pi}a}\left[e^{-[(x+a)/\lambda]^2} - 2e^{-(x/\lambda)^2} + e^{-[(x-a)/\lambda]^2}\right]\right\}$$ (5.114)

(b) Expanding Eq. 5.114 for small values of $a/\lambda = a/\sqrt{4Dt}$ produces the result

$$c(x,t) = \frac{n_d}{\sqrt{4\pi Dt}} e^{-x^2/(4Dt)} \qquad (5.115)$$

This is just the solution for a planar source of strength n_d corresponding to the content per unit area of the original distribution given by Eq. 5.108.

5.8 (a) Find the solution $c(x, y, z, t)$ of the constant-D diffusion problem where the initial concentration is uniform at c_0, inside a cube of volume a^3 centered at the origin. The concentration is initially zero outside the cube. Therefore,

$$c(x, y, z, t = 0) = \begin{cases} c_0 & \text{if } |x| \leq \frac{a}{2} \text{ and } |y| \leq \frac{a}{2} \text{ and } |z| \leq \frac{a}{2} \\ 0 & \text{otherwise} \end{cases}$$

and

$$c(x = \pm\infty, y = \pm\infty, z = \pm\infty, t) = 0$$

(b) Show that when $2\sqrt{Dt} \gg a$, the solution reduces to a standard instantaneous point-source solution in which the contents of the cube serve as the point source. Use the $\text{erf}(z + \epsilon)$ expansion in Eq. 5.109.

Solution.

(a) The method of superposition of point-source solutions can be applied to this problem. Taking the number of particles in a volume $dV = d\chi\, d\eta\, d\zeta$ equal to $dN = c_0\, d\chi\, d\eta\, d\zeta$ as a point source and integrating over all point sources in the cube using the point-source solution in Table 5.1, the concentration at x, y, z is

$$c(x, y, z, t)$$
$$= \int_{-a/2}^{a/2} \int_{-a/2}^{a/2} \int_{-a/2}^{a/2} \frac{c_0\, d\chi\, d\eta\, d\zeta}{(4\pi Dt)^{3/2}} e^{-[(x-\chi)^2 + (y-\eta)^2 + (z-\zeta)^2]/(4Dt)} \qquad (5.116)$$

The integral can be factored

$$c(x, y, z, t) = \frac{c_0}{(4\pi Dt)^{3/2}} \int_{-a/2}^{a/2} e^{-(x-\chi)^2/(4Dt)}\, d\chi$$
$$\times \int_{-a/2}^{a/2} e^{-(y-\eta)^2/(4Dt)}\, d\eta \times \int_{-a/2}^{a/2} e^{-(z-\zeta)^2/(4Dt)}\, d\zeta \qquad (5.117)$$

The integrals all have similar forms. Consider the first one. Let $u \equiv (x-\chi)/\sqrt{4Dt}$; then

$$\int_{-a/2}^{a/2} e^{-(x-\chi)^2/(4Dt)}\, d\chi = \sqrt{4Dt} \int_{(x-a/2)/\sqrt{4Dt}}^{(x+a/2)/\sqrt{4Dt}} e^{-u^2}\, du$$
$$= \sqrt{4Dt} \left(\int_0^{(x+a/2)/\sqrt{4Dt}} e^{-u^2}\, du - \int_0^{(x-a/2)/\sqrt{4Dt}} e^{-u^2}\, du \right) \qquad (5.118)$$
$$= \sqrt{\pi Dt} \left[\text{erf}\left(\frac{x + a/2}{\sqrt{4Dt}} \right) - \text{erf}\left(\frac{x - a/2}{\sqrt{4Dt}} \right) \right]$$

Therefore, the solution can be written

$$c(x, y, z, t) = \frac{c_o}{8} \times \left[\text{erf}\left(\frac{x + a/2}{\sqrt{4Dt}} \right) - \text{erf}\left(\frac{x - a/2}{\sqrt{4Dt}} \right) \right]$$
$$\times \left[\text{erf}\left(\frac{y + a/2}{\sqrt{4Dt}} \right) - \text{erf}\left(\frac{y - a/2}{\sqrt{4Dt}} \right) \right]$$
$$\times \left[\text{erf}\left(\frac{z + a/2}{\sqrt{4Dt}} \right) - \text{erf}\left(\frac{z - a/2}{\sqrt{4Dt}} \right) \right]$$

(b) Expansion of Eq. 5.119 using Eq. 5.109 produces the result

$$c = \frac{c_0 a^3}{(4\pi Dt)^{3/2}} e^{-r^2/(4Dt)} \tag{5.119}$$

which is just the solution for a point source containing the contents of the cube corresponding to $c_0 a^3$ particles.

5.9 Determine the temperature distribution $T = T(x, y, z, t)$ produced by an initial point source of heat in an infinite graphite crystal. Plot isothermal curves for a fixed temperature as a function of time in:

(a) The basal plane containing the point source

(b) A plane containing the point source with a normal that makes a $60°$ angle with the c-axis

(c) A plane containing the c-axis and the point source

The thermal diffusivity in the basal plane is isotropic and the diffusivity along the c-axis is smaller than in the basal plane by a factor of 4.

Solution. Using Eq. 4.61 and the analogy between mass diffusion and thermal diffusion, the basic differential equation for the temperature distribution in graphite can be written

$$\frac{\partial T}{\partial t} = \hat{\kappa}_\parallel \left(\frac{\partial^2 T}{\partial \hat{x}_1^2} + \frac{\partial^2 T}{\partial \hat{x}_2^2} \right) + \hat{\kappa}_\perp \frac{\partial^2 T}{\partial \hat{x}_3^2} \tag{5.120}$$

where $\hat{x}_1$ and $\hat{x}_2$ are the two principal coordinate axes in the basal plane and $\hat{x}_3$ is the principal coordinate along the c-axis.

In order to make use of the point-source solution for an isotropic medium as in Section 4.5, rescale the axes

$$\hat{x}_1 = \frac{\sqrt{\hat{\kappa}_\parallel}}{\left(\hat{\kappa}_\parallel^2 \hat{\kappa}_\perp \right)^{1/6}} \xi_1 \qquad \hat{x}_2 = \frac{\sqrt{\hat{\kappa}_\parallel}}{\left(\hat{\kappa}_\parallel^2 \hat{\kappa}_\perp \right)^{1/6}} \xi_2 \qquad \hat{x}_3 = \frac{\sqrt{\hat{\kappa}_\perp}}{\left(\hat{\kappa}_\parallel^2 \hat{\kappa}_\perp \right)^{1/6}} \xi_3 \tag{5.121}$$

Then Eq. 5.120 becomes

$$\frac{\partial T}{\partial t} = \left(\hat{\kappa}_\parallel^2 \hat{\kappa}_\perp \right)^{1/3} \left[\frac{\partial^2 T}{\partial \xi_1^2} + \frac{\partial^2 T}{\partial \xi_2^2} + \frac{\partial^2 T}{\partial \xi_3^2} \right] \tag{5.122}$$

The solution of Eq. 5.122 for the point source in these coordinates then has the form (Table 5.1)

$$T = \frac{\alpha}{t^{3/2}} e^{-(\xi_1^2 + \xi_2^2 + \xi_3^2)/[4(\hat{\kappa}_\parallel^2 \kappa_\perp)^{1/3} t]} \tag{5.123}$$

where α is a constant. Converting back to the principal axis coordinates yields

$$T(\hat{x}_1, \hat{x}_2, \hat{x}_3, t) = \frac{\alpha}{t^{3/2}} \exp \left[-\left(\frac{\hat{x}_1^2 + \hat{x}_2^2}{4\hat{\kappa}_\parallel t} + \frac{\hat{x}_3^2}{4\hat{\kappa}_\perp t} \right) \right] \tag{5.124}$$

(a) Isotherms in the basal plane: In the basal plane passing through the origin, $\hat{x}_3 = 0$ and

$$T(\hat{x}_1, \hat{x}_2, \hat{x}_3 = 0, t) = \frac{\alpha}{t^{3/2}} e^{-(\hat{x}_1^2 + \hat{x}_2^2)/(4\hat{\kappa}_\| t)} \tag{5.125}$$

Isotherms for a fixed temperature at increasing times are shown in Fig. 5.10. They are circles, as expected, because the thermal conductivity is isotropic in the basal plane. Initially, the isotherms spread out and expand because of the heat conduction but they will eventually reverse themselves and contract toward the origin, due to the finite nature of the initial point source of heat.

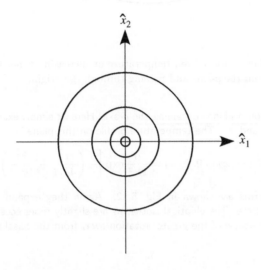

Figure 5.10:   Isotherms for a fixed temperature at increasing times in a basal plane that passes through the origin.

(b) Isotherms in a 60° inclined plane: The isotherms on a plane with a normal inclined 60° with respect to the c-axis can be determined by expressing the solution (Eq. 5.124) in a new coordinate system rotated 60° about the $\hat{x}_1$ axis. The new (primed) coordinates are

$$\begin{pmatrix} x_1' \\ x_2' \\ x_3' \end{pmatrix} = \begin{pmatrix} 1 & 0 & 0 \\ 0 & \cos 60° & \sin 60° \\ 0 & -\sin 60° & \cos 60° \end{pmatrix} \begin{pmatrix} \hat{x}_1 \\ \hat{x}_2 \\ \hat{x}_3 \end{pmatrix}$$

In the new coordinates, with $x_3' = 0$, the temperature profile in the inclined plane passing through the origin is

$$T(x_1', x_2', x_3' = 0, t) = \frac{\alpha}{t^{3/2}} \exp\left\{ -\left[\frac{1}{4\hat{\kappa}_\| t} {x_1'}^2 + \left(\frac{1}{16\hat{\kappa}_\| t} + \frac{3}{16\hat{\kappa}_\perp t} \right) {x_2'}^2 \right] \right\} \tag{5.126}$$

Figure 5.11 shows the isotherms as a function of time. Again the curves expand and contract with increasing time. However, the isotherms are elliptical because the thermal conductivity coefficient is different along the c-axis and in the basal plane.

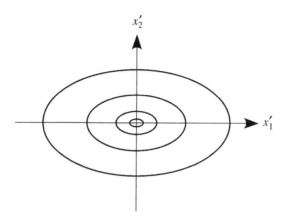

Figure 5.11:   Isotherms for a fixed temperature at increasing times in a plane with its normal inclined 60° from the c-axis and passing through the origin.

(c) Isotherms on a plane containing the c-axis: Here we simply examine the plane containing $\hat{x}_1$ and $\hat{x}_3$. The temperature profile on this plane is

$$T\left(\hat{x}_1, \hat{x}_2 = 0, \hat{x}_3, t\right) = \frac{\alpha}{t^{3/2}} \exp\left[-\left(\frac{\hat{x}_1^2}{4\hat{\kappa}_\parallel t} + \frac{\hat{x}_3^2}{4\hat{\kappa}_\perp t}\right)\right] \qquad (5.127)$$

The isotherms are shown in Fig. 5.12. Again they expand and contract with increasing time. The elliptical isotherms are slightly more eccentric than those in Fig. 5.11 because of the greater rotation away from the basal plane.

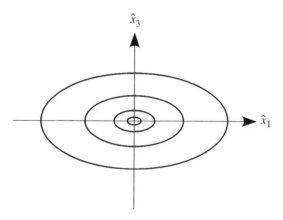

Figure 5.12:   Isotherms for a fixed temperature at increasing times in a plane containing $\hat{x}_3$ (the c-axis) and $\hat{x}_1$.

5.10 Consider one-dimensional diffusion in an infinite medium with a periodic "square wave" initial condition given by

$$c(x, t = 0) = \begin{cases} c_0 & \text{if } 0 \leq x + n\lambda \leq \frac{\lambda}{2} \\ 0 & \text{otherwise} \end{cases} \qquad (5.128)$$

where n takes on all (positive and negative) integer values.

(a) Obtain a solution involving an infinite sine series.

(b) Investigate the accuracy of truncating the full series solution. How many sine terms must be retained in order for the concentration at $x = \lambda/4$ to agree with the full solution to within 1% when $Dt/\lambda^2 = 0.002$?

Solution.

(a) Use the method of separation of variables. Let $c(x,t) = Y(x)T(t)$. Substituting this into the diffusion equation yields

$$\frac{1}{Y}\frac{d^2Y}{dx^2} = \frac{1}{D}\frac{1}{T}\frac{dT}{dt} \equiv -q \qquad (5.129)$$

where q is a positive constant. The solutions to these two ordinary differential equations are

$$T = a_1 \exp(-qDt) \qquad Y(x) = b_1 \sin(\sqrt{q}\,x) + b_2 \cos(\sqrt{q}\,x) \qquad (5.130)$$

where a_1, b_1, and b_2 are constants. The constant b_2 must be zero because the initial concentration profile is an odd function if the origin is shifted upward by $-c_0/2$. Further, the periodicity requires that

$$Y(\sqrt{q}\,(x+\lambda)) = Y(\sqrt{q}\,x) \qquad (5.131)$$

This condition will be satisfied if $\sqrt{q}\,\lambda = 2\pi m$, where m is an integer. Therefore, solving for q and assigning it an index,

$$\sqrt{q_m} = \frac{2\pi m}{\lambda} \qquad (5.132)$$

The general solution is then the sum of all the terms with different indices, plus a constant, A_0. Thus,

$$c(x,t) = A_0 + \sum_{m=1}^{\infty} A_m \sin\left(\frac{2\pi m}{\lambda}x\right) e^{-4\pi^2 m^2 Dt/\lambda^2} \qquad (5.133)$$

The coefficients can be determined by using the initial condition given by Eq. 5.128. When $t = 0$, Eq. 5.133 is a standard Fourier series with coefficients given by

$$A_0 = \frac{1}{\lambda}\int_0^{\lambda} c(x,0)\,dx \qquad A_m = \frac{2}{\lambda}\int_0^{\lambda} c(x,0)\sin\left(\frac{2m\pi x}{\lambda}\right)dx \qquad (5.134)$$

Inserting Eq. 5.128 and integrating, $A_0 = c_0/2$ and $A_m = 2c_0/(m\pi)$ for m odd and $A_m = 0$ for m even. Therefore,

$$c(x,t) = \frac{c_0}{2} + \frac{2c_0}{\pi}\sum_{m=\text{odd}}^{\infty}\frac{1}{m}\sin\left(\frac{2\pi m}{\lambda}x\right)e^{-4\pi^2 m^2 Dt/\lambda^2} \qquad (5.135)$$

(b) Symbolic algebra software can efficiently calculate the partial sums in Eq. 5.135 for the specified values of x and Dt/λ^2. Setting $c_0 = 1$, partial sums for 1 to 10 sine terms give the values: 1.08829, 0.984021, 1.00171, 0.999809, 0.999927, 0.999923, 0.999923, 0.999923, 0.999923, 0.999923, respectively. The series converges fairly rapidly to a value of approximately 0.9999. From the partial sums calculated, three sine terms are required to give a concentration value that is within 1% of that given by the complete series. Note that successive terms in the sum are of opposite sign, causing the partial sums to oscillate about the exact value of the complete sum.

5.11 Consider a plate of thickness L $(0 < x < L)$ with the following boundary and initial conditions:

$$T(x = 0, t) = 0 \qquad T(x = L, t) = 0 \qquad T(x, t = 0) = T_\circ \sin\left(\pi\frac{x}{L}\right)$$

Assume that the thermal diffusivity, κ, is constant.

(a) Find an exact expression for the temperature as a function of time.

(b) Find an exact relation for $t_{1/2}$, the time when the temperature at the center of the plate drops to $T_\circ/2$ (half its initial value).

(c) If the heat flux at the surface of the plate is set to zero, would the time calculated in part (b) be longer, shorter, or the same?

Solution.

(a) Use the separation-of-variables method as in Exercise 5.10. Assume a solution of the form $T(x, t) = Y(x)T(t)$. Putting this into the thermal diffusion equation, two ordinary differential equations are obtained whose solutions are

$$\begin{aligned} T(t) &= a_1 \exp(-q\kappa t) \\ Y(x) &= b_1 \sin\left(\sqrt{q}x\right) + b_2 \cos\left(\sqrt{q}x\right) \end{aligned} \tag{5.136}$$

where a_1, b_1, b_2, and q are constants. The resulting product solution can be fitted to the initial and boundary conditions by setting $a_1 = 1$, $b_1 = T_\circ$, $b_2 = 0$, and $\sqrt{q} = \pi/L$, so that

$$T = T_\circ \sin\left(\pi\frac{x}{L}\right) e^{-\pi^2 \kappa t / L^2} \tag{5.137}$$

Equation 5.137 is a sine-series solution to the diffusion equation, but because of the sinusoidal initial condition, it consists of only a single term.

(b) Setting $T = T_\circ/2$, $x = L/2$, and $t = t_{1/2}$ in Eq. 5.137, the time for the temperature in the center of the plate to drop to half its initial value is

$$t_{1/2} = -\frac{L^2}{\pi^2\kappa}\ln(0.5) \tag{5.138}$$

(c) Much longer! In fact, if no heat is allowed to leave the plate, the fixed amount of heat in the plate will spread until the temperature everywhere is uniform at the value T_∞ given by

$$T_\infty = \frac{1}{L}\int_0^L T_\circ \sin\left(\pi\frac{x}{L}\right) dx = \frac{2}{\pi}T_\circ \tag{5.139}$$

The temperature in the center will therefore never drop to the level $T = T_\circ/2$!

5.12 It is desired to de-gas a thick plate of material containing a uniform concentration of dissolved gas by annealing it in a vacuum. The rate at which the gas leaves the plate surface is proportional to its concentration at the surface; that is,

$$J_{\text{surf}} = -\alpha\, c_{\text{surf}} \tag{5.140}$$

where $\alpha = $ constant.

(a) Solve this diffusion problem during the early de-gassing period before the outward diffusion of gas has any significant effect at the center of the plate. The initial and boundary conditions are therefore

$$c(x,0) = c_0 \quad c(\infty,t) = c_0 \quad J = -D\left(\frac{\partial c}{\partial x}\right)_{x=0} = -\alpha\, c(0,t) \quad (5.141)$$

Incorporate the parameter $h = \alpha/D$ into the solution.

(b) Show that when the dimensionless parameter $h\sqrt{Dt}$ is large, $c(0,t) \approx 0$, and that when $h\sqrt{Dt}$ is small, $c(0,t) \approx c_0$. You will need the following series expansions of $\operatorname{erf}(z)$:

For small z,

$$\operatorname{erfc}(z) = 1 - \operatorname{erf}(z) = 1 - \frac{2}{\sqrt{\pi}}z + \frac{2}{\sqrt{\pi}}\frac{z^3}{3} - \cdots \quad (5.142)$$

and for large z,

$$\operatorname{erfc}(z) = 1 - \operatorname{erf}(z) = \frac{e^{-z^2}}{\sqrt{\pi}}\left(\frac{1}{z} - \frac{1}{2z^3} + \frac{1\times 3}{2^2 z^5} - \frac{1\times 3\times 5}{2^3 z^7} + \cdots\right)$$
$$(5.143)$$

(c) Give a physical interpretation of the results in (b).

Solution.

(a) Use the Laplace transform method. Transforming the diffusion equation along with the initial condition given by Eq. 5.141 yields the same result as Eq. 5.64:

$$\frac{\partial^2 \hat{c}}{\partial x^2} - \frac{p}{D}\hat{c} = -\frac{c_0}{D} \quad (5.144)$$

The solution is therefore Eq. 5.65 with $a_1 = 0$:

$$\hat{c}(x,p) = \frac{c_0}{p} + a_2\, e^{-\sqrt{p/D}\,x} \quad (5.145)$$

To determine a_2, transform the boundary condition given by Eq. 5.141 to obtain

$$-D\left(\frac{\partial \hat{c}}{\partial x}\right)_{x=0} = -\alpha\, \hat{c}(0,p) \quad (5.146)$$

Therefore, putting Eq. 5.145 into Eq. 5.146 yields

$$a_2 = -\frac{\alpha\, c_0}{p\sqrt{D}\left(\sqrt{p} + \frac{\alpha}{\sqrt{D}}\right)} \quad (5.147)$$

and

$$\hat{c} = -\frac{h\, c_0}{p\left(h + \sqrt{p/D}\right)}e^{-\sqrt{p/D}\,x} + \frac{c_0}{p} \quad (5.148)$$

where $h = \alpha/D$. Find the desired solution by taking the inverse transform using a table of transforms to obtain

$$c(x,t) = c_0\left[\operatorname{erf}\left(\frac{x}{2\sqrt{Dt}}\right) + e^{hx + h^2 Dt}\operatorname{erfc}\left(\frac{x}{2\sqrt{Dt}} + h\sqrt{Dt}\right)\right] \quad (5.149)$$

(b) From Eq. 5.149 the concentration at the surface is

$$c(0,t) = c_0\, e^{-h^2/(Dt)}\operatorname{erfc}\left(h\sqrt{Dt}\right) \quad (5.150)$$

For large $h\sqrt{Dt}$, use the series expansion for a large argument, to obtain

$$c(0,t) = \frac{c_0}{h\sqrt{\pi Dt}} \tag{5.151}$$

Therefore, $c(0,t)$ approaches zero for large $h\sqrt{Dt}$. For small $h\sqrt{Dt}$, use the small-argument expansion to obtain

$$c(0,t) = c_0 \left(1 + h^2 Dt\right)\left(1 - \frac{2}{\sqrt{\pi}}h\sqrt{Dt}\right)\cdots \tag{5.152}$$

Therefore, $c(0,t)$ approaches c_0 for small $h\sqrt{Dt}$.

(c) We can rewrite $h\sqrt{Dt}$ as $\alpha\sqrt{t/D}$. Therefore, as α becomes small, or at short times t, or as D increases, $c(0,t)$ approaches c_0. For small α, surface desorption is compensated by diffusion from the bulk, so that $c(0,t)$ decreases slowly. However, at short times, the concentration gradients near the surface will be large, so $c(0,t)$ will initially change rapidly. With large D, bulk diffusion to the surface compensates the surface desorption. The reverse applies for large values of $h\sqrt{Dt}$.

5.13 Solve the following boundary value problem on the semi-infinite domain with discontinuous initial conditions,

$$c(x, t = 0) = \begin{cases} c_0 & 0 < x < L \\ 0 & L \leq x < \infty \end{cases}$$

with zero flux conditions at $x = 0$ and $x = \infty$.

Suggestion: Use superposition of known solutions, or split the problem into two parts and use continuity to match Laplace-transformed solutions.

Solution. Designate the region $x < L$ as region I and the region $x > L$ as region II. The Laplace transform method will be used to solve the problem in each region and the solutions will then be matched across the interface at $x = L$. In region I the diffusion equation and initial condition are the same as in the problem leading to Eq. 5.64, and therefore the general solution after Laplace transforming corresponds to Eq. 5.65. Similarly, in region II, the initial condition is the same as in the problem leading to Eq. 5.58 and the general solution therefore corresponds to Eq. 5.59. The four constants of integration can be determined from the boundary conditions imposed at $x = 0$, $x = L$, and $x = \infty$. After Laplace transforming, these become

$$\left[\frac{\partial \hat{c}^I}{\partial x}\right]_{x=0} = \left[\frac{\partial \hat{c}^{II}}{\partial x}\right]_{x=\infty} = 0 \tag{5.153}$$

$$\left[\frac{\partial \hat{c}^I}{\partial x}\right]_{x=L} = \left[\frac{\partial \hat{c}^{II}}{\partial x}\right]_{x=L} \tag{5.154}$$

$$\hat{c}^I(L,p) = \hat{c}^{II}(L,p) \tag{5.155}$$

After determining the four constants of integration and putting them into Eqs. 5.59 and 5.65, the solutions in regions I and II are

$$\hat{c}^I = \frac{c_0}{p} - \frac{c_0}{2p}\exp(-qL)\left[\exp(qx) + \exp(-qx)\right]$$

$$\hat{c}^{II} = -\frac{c_0}{2p}\exp(-qx)\left[\exp(-qL) - \exp(qL)\right] \tag{5.156}$$

where $q = \sqrt{p/D}$. Using standard tables of transforms to transform back to (x, t) coordinates, the final solutions are

$$c^I(x, t) = \frac{c_0}{2} \left[\mathrm{erf}\left(\frac{L+x}{2\sqrt{Dt}}\right) + \mathrm{erf}\left(\frac{L-x}{2\sqrt{Dt}}\right) \right]$$

$$c^{II}(x, t) = \frac{c_0}{2} \left[\mathrm{erf}\left(\frac{x+L}{2\sqrt{Dt}}\right) - \mathrm{erf}\left(\frac{x-L}{2\sqrt{Dt}}\right) \right] \tag{5.157}$$

Because $\mathrm{erf}(z) = -\mathrm{erf}(-z)$, the solutions are identical in the two regions and thus

$$c(x, t) = \frac{c_0}{2} \left[\mathrm{erf}\left(\frac{x+L}{2\sqrt{Dt}}\right) - \mathrm{erf}\left(\frac{x-L}{2\sqrt{Dt}}\right) \right] \tag{5.158}$$

CHAPTER 6

DIFFUSION IN MULTICOMPONENT SYSTEMS

In earlier chapters we examined systems with one or two types of diffusing chemical species. For binary solutions, a single interdiffusivity, $\tilde{D}$, suffices to describe composition evolution. In this chapter we treat diffusion in ternary and larger multicomponent systems that have two or more independent composition variables. Analysis of such diffusion is complex because multiple cross terms and particle–particle chemical interaction terms appear. The cross terms result in N^2 independent interdiffusivities for a solution with N independent components. The increased complexity of multicomponent diffusion produces a wide variety of diffusional phenomena.

The general treatment for multicomponent diffusion results in linear systems of diffusion equations. A linear transformation of the concentrations produces a simplified system of uncoupled linear diffusion equations for which general solutions can be obtained by methods presented in Chapter 5.

6.1 GENERAL FORMULATION

In Chapter 2 we considered diffusion in a closed system containing N components, exclusive of any mediating point defects.[1] If only chemical potential gradients are present and all other driving forces—such as thermal gradients or electric fields—

[1]Such defects, if present, will be assumed to be in local thermal equilibrium at very small concentrations.

are absent, the general formulation presented in Eq. 2.21 and developed further in Chapter 3 applies, and for one-dimensional diffusion,[2]

$$
\begin{aligned}
\vec{J}_1 &= -L_{11}\frac{\partial \mu_1}{\partial x} - L_{12}\frac{\partial \mu_2}{\partial x} + \cdots - L_{1N}\frac{\partial \mu_N}{\partial x} \\
\vec{J}_2 &= -L_{21}\frac{\partial \mu_1}{\partial x} - L_{22}\frac{\partial \mu_2}{\partial x} + \cdots - L_{2N}\frac{\partial \mu_N}{\partial x} \\
&\;\;\vdots \\
\vec{J}_N &= -L_{N1}\frac{\partial \mu_1}{\partial x} - L_{N2}\frac{\partial \mu_2}{\partial x} + \cdots - L_{NN}\frac{\partial \mu_N}{\partial x}
\end{aligned}
\tag{6.1}
$$

Equation 2.15 for the rate of entropy production is then

$$
T\dot{\sigma} = -\sum_{i=1}^{N} J_i \frac{\partial \mu_i}{\partial x}
\tag{6.2}
$$

Assuming that the atomic volumes of the components are constants and the fluxes are measured in a V-frame, as defined in Section 3.1.3, Eq. 3.22 holds for all N components,

$$
\sum_{i=1}^{N} \Omega_i J_i = 0
\tag{6.3}
$$

The Nth flux can now be eliminated in Eq. 6.2 by using Eq. 6.3 and putting the result into Eq. 6.2, so that

$$
J_N = -\frac{1}{\Omega_N} \sum_{i=1}^{N-1} \Omega_i J_i
\tag{6.4}
$$

and

$$
T\dot{\sigma} = -\sum_{i=1}^{N-1} J_i \frac{\partial}{\partial x}\left(\mu_i - \frac{\Omega_i}{\Omega_N}\mu_N \right)
\tag{6.5}
$$

The force, F_i, conjugate to the flux, J_i, is

$$
F_i = -\left(\frac{\partial \mu_i}{\partial x} - \frac{\Omega_i}{\Omega_N}\frac{\partial \mu_N}{\partial x} \right)
\tag{6.6}
$$

and therefore the general linear relation between the independent fluxes and the $N-1$ independent driving forces is

$$
J_i = \sum_{j=1}^{N-1} L_{ij} F_j = -\sum_{j=1}^{N-1} L_{ij}\left(\frac{\partial \mu_j}{\partial x} - \frac{\Omega_j}{\Omega_N}\frac{\partial \mu_N}{\partial x} \right)
\tag{6.7}
$$

The chemical potential gradients and Onsager coefficients in Eq. 6.7 can be converted to concentration gradients and interdiffusivities (Table 3.1). Each chemical potential in Eq. 6.7 is a function of the local concentration:

$$
\mu_i = \mu_i(c_1, c_2, \ldots, c_{N-1}) \quad \text{where } i = 1, 2, \ldots, N-1
\tag{6.8}
$$

[2]This treatment is similar to that of Kirkaldy and Young [1].

There are $N - 1$ independent concentrations because Eq. A.10 provides a single relation between concentrations and their atomic volumes.

Under the assumption of local equilibrium, the Gibbs–Duhem relation applies, which places an additional constraint on chemical potential changes in Eq. 6.7 and implies that only $N - 1$ of the μ_i can vary independently:

$$\frac{\partial \mu_i}{\partial x} = \frac{\partial \mu_i}{\partial c_1} \frac{\partial c_1}{\partial x} + \frac{\partial \mu_i}{\partial c_2} \frac{\partial c_2}{\partial x} + \cdots + \frac{\partial \mu_i}{\partial c_{N-1}} \frac{\partial c_{N-1}}{\partial x} = \sum_{j=1}^{N-1} \frac{\partial \mu_i}{\partial c_j} \frac{\partial c_j}{\partial x} \tag{6.9}$$

and

$$J_i = \sum_{j=1}^{N-1} L_{ij} F_j = -\sum_{j=1}^{N-1} L_{ij} \sum_{k=1}^{N-1} \left(\frac{\partial \mu_j}{\partial c_k} - \frac{\Omega_j}{\Omega_N} \frac{\partial \mu_N}{\partial c_k} \right) \frac{\partial c_k}{\partial x} \tag{6.10}$$

Interdiffusivities, $\widetilde{D}_{ij}$, are defined by Eq. 6.10,

$$J_i = -\sum_{j=1}^{N-1} \widetilde{D}_{ij} \frac{\partial c_j}{\partial x} \tag{6.11}$$

with

$$\widetilde{D}_{ij} = \sum_{k=1}^{N-1} L_{ik} \left(\frac{\partial \mu_k}{\partial c_j} - \frac{\Omega_k}{\Omega_N} \frac{\partial \mu_N}{\partial c_j} \right) \tag{6.12}$$

$\widetilde{D}_{ij}$ is the product of two matrices,

$$\widetilde{D}_{ij} = \sum_{k=1}^{N-1} L_{ik} \Upsilon_{kj} \tag{6.13}$$

where the L_{ik} are Onsager coefficients and Υ_{kj} are thermodynamic factors that couple chemical potentials to concentrations,

$$\Upsilon_{kj} = \frac{\partial \mu_k}{\partial c_j} - \frac{\Omega_k}{\Omega_N} \frac{\partial \mu_N}{\partial c_j} \tag{6.14}$$

The analysis of the diffusion for N components requires $N - 1$ independent concentrations and $(N - 1)^2$ interdiffusivities. For the ternary case,

$$\widetilde{D}_{ij} = \begin{bmatrix} \widetilde{D}_{11} & \widetilde{D}_{12} \\ \widetilde{D}_{21} & \widetilde{D}_{22} \end{bmatrix} \tag{6.15}$$

The eigenvalues, $\lambda_\pm$, of the interdiffusivity matrix (see Eq. 4.62) must be real and positive [1]. For the ternary case,

$$\lambda_\pm = \frac{\mathrm{Tr}(\underline{\widetilde{D}}) \pm \sqrt{(\mathrm{Tr}(\underline{\widetilde{D}}))^2 - 4 \det(\underline{\widetilde{D}})}}{2} = \frac{\mathrm{Tr}(\underline{\widetilde{D}}) \pm \Delta}{2} \tag{6.16}$$

where

$$\mathrm{Tr}(\underline{\widetilde{D}}) \equiv \widetilde{D}_{11} + \widetilde{D}_{22} \quad \text{and} \quad \det(\underline{\widetilde{D}}) \equiv \widetilde{D}_{11}\widetilde{D}_{22} - \widetilde{D}_{12}\widetilde{D}_{21}$$
$$\Delta \equiv \sqrt{(\mathrm{Tr}(\underline{\widetilde{D}}))^2 - 4 \det(\underline{\widetilde{D}})} = \sqrt{(\widetilde{D}_{11} - \widetilde{D}_{22})^2 + 4\widetilde{D}_{12}\widetilde{D}_{21}} \tag{6.17}$$

The real and positive condition on the eigenvalues places physical limits on the interdiffusivities. For the ternary case,

$$\tilde{D}_{11} + \tilde{D}_{22} > 0$$
$$\left(\tilde{D}_{11} + \tilde{D}_{22}\right)^2 \geq 4(\tilde{D}_{11}\tilde{D}_{22} - \tilde{D}_{12}\tilde{D}_{21}) \qquad (6.18)$$
$$(\tilde{D}_{11}\tilde{D}_{22} - \tilde{D}_{12}\tilde{D}_{21}) \geq 0$$

The sum $(\tilde{D}_{11} + \tilde{D}_{22})$ must be positive, but a direct interdiffusivity, $\tilde{D}_{ii}$, could be negative and still satisfy the conditions in Eq. 6.18. The off-diagonal terms, $\tilde{D}_{ij}$, need not be symmetric with respect to the exchange of i and j.

In the steps leading to Eq. 6.5, the choice of the Nth component is arbitrary, and a different set of values for the four interdiffusivities will be obtained for each choice. However, each set leads to the same physical behavior predicted for the system: the diffusion profiles of the three components predicted by the equations are independent of the choice for N.[3] For some choices, the interpretation of interdiffusivities in terms of kinetic and thermodynamic data may be more straightforward [1].

6.2 SOLUTIONS OF MULTICOMPONENT DIFFUSION EQUATIONS

Generally, a set of coupled diffusion equations arises for multiple-component diffusion when $N \geq 3$. The least complicated case is for ternary ($N = 3$) systems that have two independent concentrations (or fluxes) and a 2×2 matrix of interdiffusivities. A matrix and vector notation simplifies the general case. Below, the equations are developed for the ternary case along with a parallel development using compact notation for the more extended general case. Many characteristic features of general multicomponent diffusion can be illustrated through specific solutions of the ternary case.

The coupled ternary diffusion equations in one dimension are obtained from the accumulation fluxes in Eq. 6.11:

$$\frac{\partial c_1}{\partial t} = -\nabla \cdot \vec{J}_1 = \frac{\partial}{\partial x}\left(\tilde{D}_{11}\frac{\partial c_1}{\partial x}\right) + \frac{\partial}{\partial x}\left(\tilde{D}_{12}\frac{\partial c_2}{\partial x}\right)$$
$$\frac{\partial c_2}{\partial t} = -\nabla \cdot \vec{J}_2 = \frac{\partial}{\partial x}\left(\tilde{D}_{21}\frac{\partial c_1}{\partial x}\right) + \frac{\partial}{\partial x}\left(\tilde{D}_{22}\frac{\partial c_2}{\partial x}\right) \qquad (6.19)$$

or

$$\frac{\partial}{\partial t}\left[\begin{array}{c} c_1 \\ c_2 \end{array}\right] = \frac{\partial}{\partial x}\left\{\left[\begin{array}{cc} \tilde{D}_{11} & \tilde{D}_{12} \\ \tilde{D}_{21} & \tilde{D}_{22} \end{array}\right]\frac{\partial}{\partial x}\left[\begin{array}{c} c_1 \\ c_2 \end{array}\right]\right\} \qquad (6.20)$$

and generally,

$$\frac{\partial \vec{c}}{\partial t} = \frac{\partial}{\partial x}\left[\tilde{\underline{D}}\frac{\partial \vec{c}}{\partial x}\right] \qquad (6.21)$$

In general, the $\tilde{D}_{ij}$ are functions of the concentrations, so these equations are nonlinear. Numerical methods must then be employed. However, solutions can be obtained for a variety of special cases, several of which are described below.

----

[3]This independence is similar to a constrained system's insensitivity to the choice of N_c in Section 2.2.2.

6.2.1 Constant Diffusivities

If the interdiffusivities are each constant and uniform, the coupled ternary diffusion equations, Eq. 6.20, are a linear system,

$$\frac{\partial}{\partial t}\begin{bmatrix} c_1 \\ c_2 \end{bmatrix} = \begin{bmatrix} \tilde{D}_{11} & \tilde{D}_{12} \\ \tilde{D}_{21} & \tilde{D}_{22} \end{bmatrix} \nabla^2 \begin{bmatrix} c_1 \\ c_2 \end{bmatrix} \tag{6.22}$$

and in general,

$$\frac{\partial \vec{c}}{\partial t} = \tilde{\underline{D}} \nabla^2 \vec{c} \tag{6.23}$$

It is possible to uncouple the expressions for the fluxes by diagonalizing the diffusivity matrix through a coordinate transformation. The transformed interdiffusivity matrix will have eigenvalues λ_i as its diagonal entries. For the ternary system, the eigenvalues are the $\lambda_{\pm}$ from Eq. 6.16. There will be N positive eigenvalues λ_N in the general case, where N is the number of independent components.

According to Eq. 1.36, the eigenvectors $\vec{\chi}_i$ of $\tilde{\underline{D}}$ form the columns of the matrix that diagonalizes $\tilde{\underline{D}}$ by the coordinate transformation. For the ternary system, let the eigenvector for the (fast) λ_+ eigenvalue be $\vec{f}$ and for the (slow) λ_- be $\vec{s}$:

$$\text{for } \lambda_-\colon \vec{s} = \begin{bmatrix} \frac{\tilde{D}_{11}-\tilde{D}_{22}-\Delta}{2\tilde{D}_{21}} \\ 1 \end{bmatrix} \quad \text{for } \lambda_+\colon \vec{f} = \begin{bmatrix} \frac{\tilde{D}_{11}-\tilde{D}_{22}+\Delta}{2\tilde{D}_{21}} \\ 1 \end{bmatrix} \tag{6.24}$$

The slow and fast eigendirections are related by an angle θ,

$$\cos\theta = \frac{\vec{s}\cdot\vec{f}}{|\vec{s}||\vec{f}|} = \frac{\tilde{D}_{21}-\tilde{D}_{12}}{\sqrt{(\tilde{D}_{11}-\tilde{D}_{22})^2+(\tilde{D}_{12}+\tilde{D}_{21})^2}} \tag{6.25}$$

The transformation matrix, $\underline{A}$, that diagonalizes $\tilde{\underline{D}}$ has columns formed by the eigenvectors $\vec{f}$ and $\vec{s}$. For the ternary case,

$$\underline{A} = \begin{bmatrix} \frac{\tilde{D}_{11}-\tilde{D}_{22}-\Delta}{2\tilde{D}_{21}} & \frac{\tilde{D}_{11}-\tilde{D}_{22}+\Delta}{2\tilde{D}_{21}} \\ 1 & 1 \end{bmatrix} \text{ and } \underline{A}^{-1} = \begin{bmatrix} \frac{-\tilde{D}_{21}}{\Delta} & \frac{\tilde{D}_{11}-\tilde{D}_{22}+\Delta}{2\Delta} \\ \frac{\tilde{D}_{21}}{\Delta} & -(\frac{\tilde{D}_{11}-\tilde{D}_{22}-\Delta}{2\Delta}) \end{bmatrix} \tag{6.26}$$

and for the general case,

$$\underline{A} = \begin{bmatrix} \vec{\chi}_1 & \vec{\chi}_2 & \cdots & \vec{\chi}_{N-1} \end{bmatrix} \tag{6.27}$$

and therefore

$$\underline{A}^{-1}\tilde{\underline{D}}\,\underline{A} = \begin{bmatrix} \lambda_- & 0 \\ 0 & \lambda_+ \end{bmatrix} \tag{6.28}$$

Transforming Eq. 6.43 yields

$$\underline{A}^{-1}\begin{bmatrix} J_1 \\ J_2 \end{bmatrix} = -\underline{A}^{-1}\tilde{\underline{D}}\,\underline{A}\,\underline{A}^{-1}\nabla\begin{bmatrix} c_1 \\ c_2 \end{bmatrix} \tag{6.29}$$

or

$$\begin{bmatrix} J_\alpha \\ J_\beta \end{bmatrix} = -\begin{bmatrix} \lambda_- & 0 \\ 0 & \lambda_+ \end{bmatrix}\nabla\begin{bmatrix} c_\alpha \\ c_\beta \end{bmatrix} \tag{6.30}$$

or

$$J_\alpha = -\lambda_- \nabla c_\alpha$$
$$J_\beta = -\lambda_+ \nabla c_\beta \tag{6.31}$$

and the fluxes are seen to be uncoupled in the diagonalized system. J_α and J_β are the two fluxes in the diagonalized system given by

$$\begin{bmatrix} J_\alpha \\ J_\beta \end{bmatrix} = \underline{A}^{-1} \begin{bmatrix} J_1 \\ J_2 \end{bmatrix} \tag{6.32}$$

and c_α and c_β are the two concentrations given by

$$\begin{bmatrix} c_\alpha \\ c_\beta \end{bmatrix} = \underline{A}^{-1} \begin{bmatrix} c_1 \\ c_2 \end{bmatrix} \tag{6.33}$$

These quantities therefore have the forms

$$J_\alpha = \frac{-D_{21}}{\Delta} J_1 + \frac{D_{11} - D_{22} + \Delta}{2\Delta} J_2$$
$$J_\beta = \frac{D_{21}}{\Delta} J_1 - \frac{D_{11} - D_{22} - \Delta}{2\Delta} J_2 \tag{6.34}$$

and

$$c_\alpha = \frac{-D_{21}}{\Delta} c_1 + \frac{D_{11} - D_{22} + \Delta}{2\Delta} c_2$$
$$c_\beta = \frac{D_{21}}{\Delta} c_1 - \frac{D_{11} - D_{22} - \Delta}{2\Delta} c_2 \tag{6.35}$$

The flux problem can now be easily solved in the diagonalized system using Eq. 6.31; the solution can then be transformed back to the original concentration coordinates by using the inverse relationships

$$\begin{bmatrix} J_1 \\ J_2 \end{bmatrix} = \underline{A} \begin{bmatrix} J_\alpha \\ J_\beta \end{bmatrix} \tag{6.36}$$

and

$$\begin{bmatrix} c_1 \\ c_2 \end{bmatrix} = \underline{A} \begin{bmatrix} c_\alpha \\ c_\beta \end{bmatrix} \tag{6.37}$$

Steady-State Solutions. For the steady-state case, Eq. 6.22 becomes

$$\begin{bmatrix} 0 \\ 0 \end{bmatrix} = \begin{bmatrix} \widetilde{D}_{11} & \widetilde{D}_{12} \\ \widetilde{D}_{21} & \widetilde{D}_{22} \end{bmatrix} \nabla^2 \begin{bmatrix} c_1 \\ c_2 \end{bmatrix} \tag{6.38}$$

and in general,

$$\vec{0} = \underline{\widetilde{D}} \nabla^2 \vec{c} \tag{6.39}$$

If $\underline{\widetilde{D}}$ has an inverse,

$$\underline{\widetilde{D}}^{-1} \vec{0} = \underline{\widetilde{D}}^{-1} \underline{\widetilde{D}} \nabla^2 \vec{c} \tag{6.40}$$

or

$$\begin{bmatrix} 0 \\ 0 \end{bmatrix} = \nabla^2 \begin{bmatrix} c_1 \\ c_2 \end{bmatrix} \tag{6.41}$$

that is,

$$\nabla^2 c_i = 0 \tag{6.42}$$

Therefore, Laplace's equation holds for each component separately. However, the steady-state fluxes are interdependent, as may be seen from Eq. 6.11 for the ternary case,

$$\left[\begin{array}{c} J_1 \\ J_2 \end{array} \right] = -\underline{\widetilde{D}} \nabla \left[\begin{array}{c} c_1 \\ c_2 \end{array} \right] \tag{6.43}$$

Time-Dependent Solutions. In the time-dependent case, the diffusion equations given by Eq. 6.23 are coupled. However, they can be uncoupled by again using the diagonalization method.

Using the transformation matrix $\underline{A}$ on Eq. 6.23 gives for the ternary case,

$$\frac{\partial}{\partial t} \left(\underline{A}^{-1} \vec{c} \right) = \underline{A}^{-1} \underline{\widetilde{D}} \nabla^2 \vec{c}$$

$$\frac{\partial}{\partial t} \left(\underline{A}^{-1} \vec{c} \right) = \underline{A}^{-1} \underline{\widetilde{D}} \, \underline{A} \left(\nabla^2 \underline{A}^{-1} \vec{c} \right) \tag{6.44}$$

$$\frac{\partial}{\partial t} \left[\begin{array}{c} c_\alpha \\ c_\beta \end{array} \right] = \left[\begin{array}{cc} \lambda_- & 0 \\ 0 & \lambda_+ \end{array} \right] \nabla^2 \left[\begin{array}{c} c_\alpha \\ c_\beta \end{array} \right]$$

and for the general case,

$$\frac{\partial}{\partial t} \left[\begin{array}{c} \check{c}_1 \\ \check{c}_2 \\ \vdots \\ \check{c}_{N-1} \end{array} \right] = \left[\begin{array}{ccccc} \lambda_1 & 0 \cdots & 0 & 0 \\ 0 & \lambda_2 & 0 & \vdots \\ \vdots & 0 & \ddots & 0 \\ 0 & \cdots & 0 & \lambda_{N-1} \end{array} \right] \nabla^2 \left[\begin{array}{c} \check{c}_1 \\ \check{c}_2 \\ \vdots \\ \check{c}_{N-1} \end{array} \right] \tag{6.45}$$

where the concentrations $\check{c}_i$ are linear combinations given by the eigensystem transformation of the actual components c_i.

Each partial-differential equation in the diagonal frame is independent:

$$\frac{\partial c_\alpha}{\partial t} = \left(\frac{\mathrm{Tr}(\underline{\widetilde{D}}) - \Delta}{2} \right) \nabla^2 c_\alpha$$

$$\frac{\partial c_\beta}{\partial t} = \left(\frac{\mathrm{Tr}(\underline{\widetilde{D}}) + \Delta}{2} \right) \nabla^2 c_\beta \tag{6.46}$$

In general,

$$\frac{\partial \check{c}_1}{\partial t} = \lambda_1 \nabla^2 \check{c}_1$$

$$\frac{\partial \check{c}_2}{\partial t} = \lambda_2 \nabla^2 \check{c}_2$$

$$\vdots \quad = \quad \vdots \tag{6.47}$$

$$\frac{\partial \check{c}_{N-1}}{\partial t} = \lambda_{N-1} \nabla^2 \check{c}_{N-1}$$

For one-dimensional ternary diffusion, the boundary and initial conditions—$c_1(x = L, t)$, $c_1(x = R, t)$, $c_1(x, t = 0)$, $c_2(x = L, t)$, $c_2(x = R, t)$, and $c_2(x, t = 0)$—become

$$c_\alpha(x = L, t) = \frac{-\widetilde{D}_{21}}{\Delta} c_1(x = L, t) + \frac{\widetilde{D}_{11} - \widetilde{D}_{22} + \Delta}{2\Delta} c_2(x = L, t)$$

$$c_\alpha(x = R, t) = \frac{-\widetilde{D}_{21}}{\Delta} c_1(x = R, t) + \frac{\widetilde{D}_{11} - \widetilde{D}_{22} + \Delta}{2\Delta} c_2(x = R, t)$$

$$c_\alpha(x, t = 0) = \frac{-\widetilde{D}_{21}}{\Delta} c_1(x, t = 0) + \frac{\widetilde{D}_{11} - \widetilde{D}_{22} + \Delta}{2\Delta} c_2(x, t = 0)$$

$$c_\beta(x = L, t) = \frac{\widetilde{D}_{21}}{\Delta} c_1(x = L, t) - \frac{\widetilde{D}_{11} - \widetilde{D}_{22} - \Delta}{2\Delta} c_2(x = L, t) \qquad (6.48)$$

$$c_\beta(x = R, t) = \frac{\widetilde{D}_{21}}{\Delta} c_1(x = R, t) - \frac{\widetilde{D}_{11} - \widetilde{D}_{22} - \Delta}{2\Delta} c_2(x = R, t)$$

$$c_\beta(x, t = 0) = \frac{\widetilde{D}_{21}}{\Delta} c_1(x, t = 0) - \frac{\widetilde{D}_{11} - \widetilde{D}_{22} - \Delta}{2\Delta} c_2(x, t = 0)$$

In general,

$$\vec{\tilde{c}}(x = L, t) = \underline{A}^{-1}\vec{c}(x = L, t)$$

$$\vec{\tilde{c}}(x = R, t) = \underline{A}^{-1}\vec{c}(x = R, t) \qquad (6.49)$$

$$\vec{\tilde{c}}(x, t = 0) = \underline{A}^{-1}\vec{c}(x, t = 0)$$

The system is reduced to a set of uncoupled diffusion equations with diffusivities constructed from the component interdiffusivities by a prescribed algorithm. Each equation can be solved by methods described in Chapter 5.

The transient behavior at the interface of two ternary alloy compositions in a system with complete solid solubility will lead to a path in composition "space" as shown in Fig. 6.1. Evolution is initially parallel to the fast eigendirection $\vec{f}$ and, after its gradients become small, finally proceeds parallel to the slow direction $\vec{s}$.

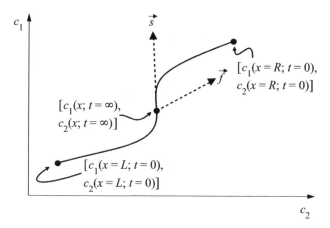

Figure 6.1:   General evolution of a ternary diffusion couple with initial conditions specified by a concentration pair on the left ($x = L$) and on the right ($x = R$).

The solution for a diffusion couple in which two semi-infinite ternary alloys are bonded initially at a planar interface is worked out in Exercise 6.1 by the same basic method. Because each component has step-function initial conditions, the solution is a sum of error-function solutions (see Section 4.2.2). Such diffusion couples are used widely in experimental studies of ternary diffusion. In Fig. 6.2 the diffusion profiles of Ni and Co are shown for a ternary diffusion couple fabricated by bonding together two Fe–Ni–Co alloys of differing compositions. The Ni, which was initially uniform throughout the couple, develops transient concentration gradients. This example of uphill diffusion results from interactions with the other components in the alloy. Coupling of the concentration profiles during diffusion in this ternary case illustrates the complexities that are present in multicomponent diffusion but absent from the binary case.

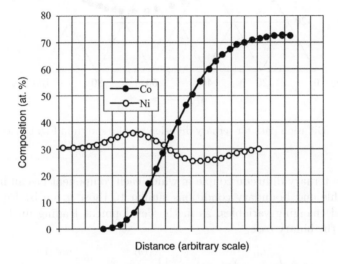

Figure 6.2:   Concentration profiles for Ni and Co in ternary diffusion couple fabricated with Fe–Ni–Co alloys. From Kirkaldy and Young [1], and Vignes and Sabatier [2].

The results of ternary diffusion experiments are often presented in the form of *diffusion paths*, which are plots of the concentrations measured across the diffusion zone. The diffusion path corresponding to the measurements in Fig. 6.2 is shown in Fig. 6.3; note the characteristic S-shape, due to the inflections in the c_{Ni} profile.

6.2.2 Concentration-Dependent Diffusivities

If the diffusivities are functions of concentration, the Boltzmann–Matano method, described in Section 4.3 for the binary case, can be employed if the initial and boundary conditions are appropriate. The diffusion equations are

$$\frac{\partial c_i}{\partial t} = -\nabla \cdot \vec{J}_i = \frac{\partial}{\partial x}\left(\tilde{D}_{i1}\frac{\partial c_1}{\partial x}\right) + \frac{\partial}{\partial x}\left(\tilde{D}_{i2}\frac{\partial c_2}{\partial x}\right) \tag{6.50}$$

and when the scaling parameter $\eta = x/\sqrt{t}$ is employed, these equations become

$$-\frac{\eta}{2}\frac{dc_i}{d\eta} = \frac{d}{d\eta}\left(\tilde{D}_{i1}\frac{dc_1}{d\eta} + \tilde{D}_{i2}\frac{dc_2}{d\eta}\right) \tag{6.51}$$

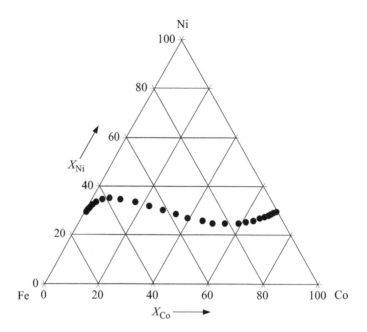

Figure 6.3:   Diffusion path for ternary diffusion profiles shown in Fig. 6.2. From Vignes and Sabatier [2].

Consider a ternary diffusion couple in which each component has an initial step-function profile and boundary conditions similar to those given by Eq. 4.44. Integrating and changing variables, as in the development leading up to Eqs. 4.48 and 4.49 for the binary case,

$$-2 \int_{c_i^R}^{c_i(\eta)} \eta \, dc_i = \widetilde{D}_{i1} \left. \frac{dc_1}{d\eta} \right|_{c_1^R}^{c_1(\eta)} + \widetilde{D}_{i2} \left. \frac{dc_2}{d\eta} \right|_{c_2^R}^{c_2(\eta)} \tag{6.52}$$

and

$$-\frac{1}{2\tau} \int_{c_1^R}^{c_i(x)} x \, dc_i = \widetilde{D}_{i1} \frac{dc_1}{dx} + \widetilde{D}_{i2} \frac{dc_2}{dx} = -J_i \tag{6.53}$$

Because the diffusion profiles $c_1(x)$ and $c_2(x)$ are known, the fluxes J_1 and J_2 can be determined at any x by inverting $c_1(x)$ and $c_2(x)$ and evaluating the integrals in Eqs. 6.53 for $i = 1, 2$. Because dc_1/dx and dc_2/dx are known at the selected x, two equations relating the four diffusivities are obtained. Therefore, if two ternary diffusion experiments are analyzed at a point of common concentration, four equations relating the four diffusivities at those common concentrations will be obtained and all four diffusivities can be determined. However, use of this method to derive diffusivities from experimental results is highly labor intensive and subject to significant error [3].

Alternatively, coupled diffusion equations with concentration-dependent diffusivities and comparison with experimental results can be solved with numerical methods (see Kirkaldy and Young [1] and Glicksman [4]).

In many systems, complex particle–particle interactions can produce a plane inside the diffusion zone where the flux of one of the components is zero. Such planes can be found by evaluating the integral in Eq. 6.53 along the diffusion profile to find points where $J_i = 0$. This approach can be used to obtain information regarding diffusion in the system [5]. Exercise 6.3 provides an illustration.

6.3   MEASUREMENTS AND INTERPRETATIONS OF DIFFUSIVITIES

There are extensive reviews of the many measurements of the $\widetilde{D}_{ij}$, particularly in ternary systems [1]. Numerous systems exhibit uphill diffusion, due to strong particle–particle interactions, and efforts have been made to interpret the diffusivity behavior in terms of thermodynamic activity data and particle–particle interaction models. In many cases the diffusion behavior has been explained, and more details and discussion are found in Kirkaldy and Young's text [1].

Bibliography

1. J.S. Kirkaldy and D.J. Young. *Diffusion in the Condensed State*. Institute of Metals, London, 1987.

2. A. Vignes and J.P. Sabatier. Ternary diffusion in Fe–Co–Ni alloys. *Trans. TMS-AIME*, 245:1795–1802, 1969.

3. C. Cserhati, U. Ugaste, M.J.H. van Dal, N.J.H.G.M. Lousberg, A.A. Kodentsov, and F.J.J. van Loo. On the relationship between interdiffusion and tracer diffusion coefficients in ternary solid solutions. *Defect and Diffusion Forum*, 194–199:189–194, 2001.

4. M.E. Glicksman. *Diffusion in Solids*. John Wiley & Sons, New York, 2000.

5. M.A. Dayananda. Zero-flux planes, flux reversals and diffusion paths in ternary and quaternary diffusion. In *Diffusion in Solids: Recent Developments*, pages 195–230, Warrendale, PA, 1985. The Metallurgical Society.

EXERCISES

6.1 Find the diffusion profile solutions for ternary diffusion in a diffusion couple fabricated by bonding two semi-infinite alloy blocks face to face along a planar interface. Assume constant diffusivities.

Solution. The method of diagonalization described in Section 6.2.1 is employed. The initial conditions for all components are step-function concentration profiles and hence error-function profiles having the form

$$c_\alpha = a_1 + a_2 \operatorname{erf}\left(\frac{x}{\sqrt{4\lambda_- Dt}}\right)$$
$$c_\beta = a_3 + a_4 \operatorname{erf}\left(\frac{x}{\sqrt{4\lambda_+ Dt}}\right) \tag{6.54}$$

are tried as solutions to the uncoupled diffusion equations given by Eq. 6.44. Using Eqs. 6.26 and 6.37, the relationships between the concentrations c_1 and c_2 and the concentrations c_α and c_β in the diagonalized system are

$$c_1 = \frac{D_{11} - D_{22} - \Delta}{2D_{21}} c_\alpha + \frac{D_{11} - D_{22} + \Delta}{2D_{21}} c_\beta$$
$$c_2 = c_\alpha + c_\beta \tag{6.55}$$

By combining the equations above, the solutions take the forms

$$c_1 = \frac{D_{11} - D_{22} - \Delta}{2D_{21}} \left[a_1 + a_2 \operatorname{erf}\left(\frac{x}{\sqrt{4\lambda_- Dt}} \right) \right]$$
$$+ \frac{D_{11} - D_{22} + \Delta}{2D_{21}} \left[a_3 + a_4 \operatorname{erf}\left(\frac{x}{\sqrt{4\lambda_+ Dt}} \right) \right] \tag{6.56}$$
$$c_2 = a_1 + a_2 \operatorname{erf}\left(\frac{x}{\sqrt{4\lambda_- Dt}} \right) + a_3 + a_4 \operatorname{erf}\left(\frac{x}{\sqrt{4\lambda_+ Dt}} \right)$$

The boundary conditions on c_1 and c_2 are

$$\begin{array}{ll} c_1(\infty, t) = c_1(x, 0) = c_1^R & c_2(\infty, t) = c_2(x, 0) = c_2^R \\ c_1(-\infty, t) = c_1(-x, 0) = c_1^L & c_2(-\infty, t) = c_2(-x, 0) = c_2^L \end{array} \tag{6.57}$$

and therefore

$$c_1^R = \frac{D_{11} - D_{22} - \Delta}{2D_{21}} (a_1 + a_2) + \frac{D_{11} - D_{22} + \Delta}{2D_{21}} (a_3 + a_4)$$
$$c_1^L = \frac{D_{11} - D_{22} - \Delta}{2D_{21}} (a_1 - a_2) + \frac{D_{11} - D_{22} + \Delta}{2D_{21}} (a_3 - a_4) \tag{6.58}$$
$$c_2^R = a_1 + a_2 + a_3 + a_4$$
$$c_2^L = a_1 - a_2 + a_3 - a_4$$

Solving the four equations above for the four coefficients,

$$a_1 = \frac{(D_{11} - D_{22} + \Delta)(c_2^R + c_2^L) - 2D_{21}(c_1^R + c_1^L)}{4\Delta}$$
$$a_2 = \frac{(D_{11} - D_{22} + \Delta)(c_2^R - c_2^L) - 2D_{21}(c_1^R - c_1^L)}{4\Delta}$$
$$a_3 = \frac{(D_{22} - D_{11} + \Delta)(c_2^R + c_2^L) + 2D_{21}(c_1^R + c_1^L)}{4\Delta} \tag{6.59}$$
$$a_4 = \frac{(D_{22} - D_{11} + \Delta)(c_2^R - c_2^L) + 2D_{21}(c_1^R - c_1^L)}{4\Delta}$$

Substituting Eq. 6.59 into Eq. 6.56, the final solution for c_2 is

$$c_2 = \frac{c_2^R + c_2^L}{2}$$
$$+ \frac{(D_{11} - D_{22} + \Delta)(c_2^R - c_2^L) - 2D_{21}(c_1^R - c_1^L)}{4\Delta} \operatorname{erf}\left(\frac{x}{\sqrt{4\lambda_- t}} \right) \tag{6.60}$$
$$+ \frac{(D_{22} - D_{11} + \Delta)(c_2^R - c_2^L) + 2D_{21}(c_1^R - c_1^L)}{4\Delta} \operatorname{erf}\left(\frac{x}{\sqrt{4\lambda_+ t}} \right)$$

A similar expression holds for c_1.

6.2 Point sources of components 1 and 2, containing N_1 and N_2 atoms, respectively, are located at the origin $r = 0$ in a large piece of pure component 3. Solve for the resulting three-dimensional diffusion field, assuming that the diffusivities D_{11}, D_{21}, and D_{22}, are independent of concentration.

Solution. Assume solutions in the diagonalized system having the standard point-source form

$$c_\alpha = \frac{a_\alpha}{(4\pi\lambda_- t)^{3/2}} \exp\left(-\frac{r^2}{4\lambda_- t}\right)$$

$$c_\beta = \frac{a_\beta}{(4\pi\lambda_+ t)^{3/2}} \exp\left(-\frac{r^2}{4\lambda_+ t}\right) \tag{6.61}$$

Using Eqs. 6.26 and 6.37,

$$c_1 = \frac{D_{11} - D_{22} - \Delta}{2D_{21}} c_\alpha + \frac{D_{11} - D_{22} + \Delta}{2D_{21}} c_\beta \tag{6.62}$$

$$c_2 = c_\alpha + c_\beta$$

Conservation of components 1 and 2 requires that

$$4\pi \int_0^\infty c_1 r^2 dr = N_1 = \text{constant}$$

$$4\pi \int_0^\infty c_2 r^2 dr = N_2 = \text{constant} \tag{6.63}$$

By substituting Eq. 6.61 into Eq. 6.62 and requiring that Eq. 6.63 be satisfied,

$$\frac{D_{11} - D_{22} - \Delta}{2D_{21}} a_\alpha + \frac{D_{11} - D_{22} + \Delta}{2D_{21}} a_\beta = N_1 \tag{6.64}$$

$$a_\alpha + a_\beta = N_2$$

By solving Eq. 6.64 for a_α and a_β and putting the results into Eq. 6.62 with the help of Eq. 6.61, the final solution for c_2 is

$$c_2 = \frac{2D_{21}N_1 + (\Delta + D_{22} - D_{11})N_2}{16\Delta(\pi\lambda_+ t)^{3/2}} \exp\left(-\frac{r^2}{4\lambda_+ t}\right)$$

$$+ \frac{(D_{11} - D_{22} + \Delta)N_2 - 2D_{21}N_1}{16\Delta(\pi\lambda_- t)^{3/2}} \exp\left(-\frac{r^2}{4\lambda_- t}\right) \tag{6.65}$$

A similar expression may be found for c_1.

6.3 Show that the simple relation

$$\frac{X_2}{X_1} \frac{L_{11}}{L_{12}} = -\frac{d(X_2/X_3)}{d(X_1/X_3)} \tag{6.66}$$

holds for a plane of zero flux of component 1 in the diffusion zone of a ternary system when component 3 is chosen as the Nth component, the approximation $\Omega_1 = \Omega_2 = \Omega_3$ is used, and the activity coefficients are constants.

Solution. From Eq. 6.7,

$$J_1 = -L_{11} \frac{\partial}{\partial x}(\mu_1 - \mu_3) - L_{12} \frac{\partial}{\partial x}(\mu_2 - \mu_3) = 0 \tag{6.67}$$

Therefore,

$$\frac{L_{11}}{L_{12}} = -\frac{d(\mu_2 - \mu_3)}{d(\mu_1 - \mu_3)} \tag{6.68}$$

Then, substitution of Eq. 2.2 into Eq. 6.68 produces the desired result.

6.4 Construct a diffusion path for the Cu/Zn binary diffusion zone in Fig. 4.8.

Solution. For the binary Cu–Zn system,

$$X_{\text{Cu}} + X_{\text{Zn}} = 1 \tag{6.69}$$

Therefore, the path appears as in Fig. 6.4.

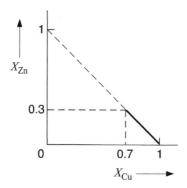

Figure 6.4:   Diffusion path for the Cu/Zn binary diffusion zone in Fig. 4.8.

CHAPTER 7

ATOMIC MODELS FOR DIFFUSION

Macroscopic treatments of diffusion result in continuum equations for the fluxes of particles and the evolution of their concentration fields. The continuum models involve the diffusivity, D, which is a kinetic factor related to the diffusive motion of the particles. In this chapter, the microscopic physics of this motion is treated and atomistic models are developed. The displacement of a particular particle can be modeled as the result of a series of thermally activated discrete movements (or *jumps*) between neighboring positions of local minimum energy. The rate at which each jump occurs depends on the vibration rate of the particle in its minimum-energy position and the excitation energy required for the jump. The average of such displacements over many particles over a period of time is related to the macroscopic diffusivity. Analyses of *random walks* produce relationships between individual atomic displacements and macroscopic diffusivity.

7.1 THERMALLY ACTIVATED ATOMIC JUMPING

The fundamental process in atomistic diffusion models is the thermally activated jump between neighboring sites of local minimum energy. The duration of any jump is typically very short compared to the particle's residence time in a minimum-energy site. Therefore, the average jump rate—the basis for any model of atomistic diffusive motion—is essentially inversely proportional to the average residence time.

The residence time depends upon the probability that the local potential energy will undergo a fluctuation large enough to enable the particle to surmount the potential-energy barrier that it will encounter while making a jump. This barrier can readily be visualized by considering, as an example, the diffusion of interstitial atoms among the interstices of large substitutional host atoms as described in Section 3.1.4. In this case, a jumping interstitial atom must squeeze its way past the large intervening substitutional atoms to make a successful jump between interstices. This squeezing increases the potential energy in the local region, and a potential-energy barrier to the jump therefore exists. Similar barriers exist for the jumps of particles in other systems. The height of the barrier will depend upon the interaction between the jumping particle and its surroundings and can vary depending upon the path of the jumping particle and the positions of its neighbors. For example, neighboring atoms may cooperatively enlarge the gap through which the jumping particle passes. The complexity of any analysis is increased by this multiplicity of possible activated configurations. However, useful approximations of varying accuracy can be obtained.

There are numerous approaches to modeling the jump rate.[1] Below, three progressively more realistic models are presented. All three approaches produce the same basic result—the jump rate is a product of the vibration frequency in the initial stable site and a Boltzmann probability of a sufficient energy fluctuation for the jump.

7.1.1   One-Particle Model with Square Potential-Energy Wells

The simplest model is composed of identical noninteracting particles sitting in rectangular potential-energy wells separated by flat potential-energy barriers. The barriers have widths L^A, as illustrated in Fig. 7.1. The rate at which the particle traverses a barrier is calculated as a *one-particle event* that occurs in one dimension. The many-bodied aspects are ignored and it is assumed that the migrating particle's surroundings—and therefore the potential-energy landscape—is static. Furthermore, the system is assumed to be in thermal equilibrium, so that the local temperature provides a statistical probability of a particle's kinetic energy fluctuations. Under this condition, a given particle spends most of its time in the energy

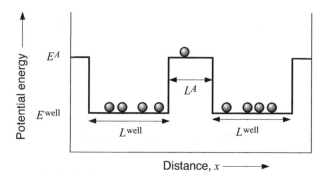

Figure 7.1:   Square potential-energy wells and an energy barrier for a particle jumping in one dimension. The number of particles is proportional to the occupation probability of well states along L^{well} and activated states along L^A.

[1]See Glasstone et al., Wert and Zener, Vineyard, Rice, Flynn, Girifalco, Christian, and Franklin for examples [1–8].

wells that correspond to low-energy (high-probability) *well states*. However, fluctuations produce brief intervals during which a particle is situated atop the energy barrier along L^A in high-energy (low-probability) *activated states*. The jump rate is the inverse of the average migration time, the average amount of time between an atom's arrival at one site and its arrival at a neighboring site. The average migration time is the sum of two distinct terms: the time an atom waits to reach an activated state and the migration time in the activated state. It is assumed that the system behaves classically and that any contribution from quantum-mechanical tunneling between the energy wells is negligible. Quantum tunneling can become important for light particles at low temperatures and is discussed elsewhere [5, 8].

The average time for a particle in an activated state to cross the activation barrier is

$$\tau^{\text{cross}} = \frac{L^A}{\langle v \rangle} = L^A \sqrt{\frac{2\pi m}{kT}} \tag{7.1}$$

where $\langle v \rangle$ is the particle's average velocity along L^A and m is its mass. The expression for $\langle v \rangle$ is found by determining its average momentum $\langle p \rangle$ and then using $\langle v \rangle = \langle p \rangle / m$. In this classical system, the probability that the particle has momentum in the forward direction (i.e., with positive values of p) between p and $p + dp$ is proportional to $\exp[-p^2/(2mkT)]dp$ [1]. Therefore,

$$\langle p \rangle = \frac{\int_0^\infty p \, e^{-p^2/(2mkT)} \, dp}{\int_{-\infty}^\infty e^{-p^2/(2mkT)} \, dp} = \sqrt{\frac{mkT}{2\pi}} \tag{7.2}$$

In a system of N particles, the total rate at which particles cross the barrier, $\dot{N}^{\text{cross}}$, is

$$\dot{N}^{\text{cross}} = \frac{\text{number of particles in an activated state}}{\tau^{\text{cross}}} \tag{7.3}$$

Consider a total time $\tau \gg \tau^A$, where τ^A is the time that a particle spends in an activated state. Then,

$$\dot{N}^{\text{cross}} = \frac{N \left(\tau^A / \tau^{\text{well}} \right)}{\tau^{\text{cross}}} \tag{7.4}$$

where $\tau^{\text{well}} \simeq \tau$ is the average duration in a well state. Therefore, for one particle, the crossing frequency (i.e., the *jump frequency*, Γ'), is

$$\Gamma' = \frac{\tau^A}{\tau^{\text{well}}} \frac{1}{\tau^{\text{cross}}} \tag{7.5}$$

or

$$\Gamma' = \frac{\langle v \rangle}{L^A} \frac{\tau^A}{\tau^{\text{well}}} = \sqrt{\frac{kT}{2\pi m}} \frac{1}{L^A} \frac{\tau^A}{\tau^{\text{well}}} \tag{7.6}$$

The ratio of times spent in a well state and in an activated state is

$$\frac{\tau^A}{\tau^{\text{well}}} = \frac{Z^A}{Z^{\text{well}}} = \frac{\sum_A e^{-E^A/(kT)}}{\sum_{\text{well}} e^{-E^{\text{well}}/(kT)}} \tag{7.7}$$

where Z^A and Z^{well} are the partition functions for the activated and well states.[2]

---

[2]The partition function plays a central role in statistical mechanics [9]. If the probability of finding a system in a state i with energy E_i is proportional to $\exp[-E_i/(kT)]$, the partition

If $\phi(x)$ is the potential energy function illustrated in Fig. 7.1, the classical limit is

$$\frac{\tau^A}{\tau^{\text{well}}} = \frac{\int_{L^A} e^{-\phi(x)/(kT)}dx}{\int_{L^{\text{well}}} e^{-\phi(x)/(kT)}dx} = \frac{L^A}{L^{\text{well}}}e^{-(E^A - E^{\text{well}})/(kT)} \tag{7.8}$$

Therefore,

$$\Gamma' = \sqrt{\frac{kT}{2\pi m}}\frac{1}{L^{\text{well}}}e^{-(E^A - E^{\text{well}})/(kT)} = \left[\sqrt{\frac{kT}{2\pi m}}\frac{1}{L^{\text{well}}}\right]e^{-E^m/(kT)} \tag{7.9}$$

where $E^m = E^A - E^{\text{well}}$ (i.e., the height of the barrier) is termed the *activation energy for migration* of the particle. The bracketed term that multiplies the Boltzmann–Arrhenius term $\exp[-E^m/(kT)]$ has dimensions $(\text{time})^{-1}$ and represents the number of attempts at the barrier per unit time—the average attempt frequency. The Boltzmann–Arrhenius exponential term is the activation success probability for each attempt. As demonstrated in less simple models below, this simple result—that there is a characteristic attempt frequency multiplied by a Boltzmann–Arrhenius factor containing the activation energy—is quite robust.

7.1.2 One-Particle Model with Parabolic Potential-Energy Wells

An improved approximation to the potential-energy landscape can be obtained by introducing parabolic wells and a smooth barrier as in Fig. 7.2.[3] This model is more realistic, as particles that are displaced small distances from their average positions of minimum energy in a solid will generally experience restoring forces that increase linearly with the displacements. This leads to a potential energy that increases as the square of the particle displacement, which corresponds to a static (i.e., non-many-bodied) harmonic model for a solid [9]. The energies of the states of the particles are approximately

$$E(x) = E^{\text{min}} + \frac{\beta}{2}\left(x - x^{\text{min}}\right)^2 \tag{7.10}$$

In Fig. 7.2, the well states are located in the region denoted by L^{well} near the minima. The particles spend the remainder of their time at the approximately flat region denoted by L^A, where the changes in average particle velocity are small. Particles at other positions experience significant forces from $-\nabla\phi(x)$ and therefore tend to accelerate, resulting in low occupation probabilities for those positions. The analysis method is the same as that for the rectangular-well model, and Eqs. 7.6 and 7.7 again hold. Using the harmonic potential, the ratio of partition functions

---

function is related to the normalization factor for the probabilities $Z = \sum_{\text{states } i}\exp[-E_i/(kT)]$ $= \sum_{\text{energies } j}\Omega(E_j)\exp[-E_j/(kT)] = \sum_{\text{energies } j}\exp(S_j/k)\exp[-E_j/(kT)]$, where $\Omega(E_j)$ is the number of states of identical energy E_j. Because Z_A is proportional to a sum of probabilities, it is proportional to the total probability of finding a particle in the activated state and therefore the average time in that state.

[3]By illustrating the potential-energy landscape in one dimension in Fig. 7.2, it appears that the activated state is one of maximum energy. The single dimension represents the most likely trajectory between the minimum states which requires the least energy, so the activation energy is that of the trajectory's saddle point—the minimum of all the maximum energies of the trajectories between two minima (see Fig. 7.3).

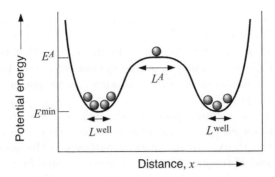

Figure 7.2:   Parabolic potential-energy well for one-dimensional particle jumps. Unlike the square potential energy function in Fig. 7.1, the energy barrier is no longer perfectly flat.

(Eqs. 7.6 and 7.7) becomes

$$\frac{Z^A}{Z^{\text{well}}} = \frac{\int_{L^A} e^{-E(x)/(kT)}dx}{e^{-E^{\min}/(kT)}} \int_{-L^{\text{well}}/2}^{L^{\text{well}}/2} e^{-[\beta(x-x^{\min})^2]/(2kT)}dx \qquad (7.11)$$

and because states far from the well minimum do not contribute significantly, the limits of integration can be approximated with

$$\int_{-L^{\text{well}}/2}^{L^{\text{well}}/2} e^{-[\beta(x-x^{\min})^2]/(2kT)}dx \approx \int_{-\infty}^{\infty} e^{-[\beta(x-x^{\min})^2]/(2kT)}dx \qquad (7.12)$$

and integrated to give

$$\Gamma' = \frac{1}{2\pi}\sqrt{\frac{\beta}{m}}\, e^{-(E^A - E^{\min})/(kT)} = \nu e^{-E^m/(kT)} \qquad (7.13)$$

where

$$\nu = \frac{1}{2\pi}\sqrt{\frac{\beta}{m}} \qquad (7.14)$$

is the characteristic attempt frequency. Again, the activation energy is the height of the energy barrier and the jump rate is given by an attempt frequency multiplied by a Boltzmann–Arrhenius factor of the form $\exp[-E^m/(kT)]$. The frequency, ν, is that of a simple harmonic oscillator of mass, m, with a restoring-force constant given by β (demonstrated in Exercise 7.1).

7.1.3   Many-Body Model

The two simple single-body models can be improved by including many-body aspects and by allowing the jumping particle to differ from the remaining particles. A treatment similar to Vineyard's is developed [3]. The N-body system consists of $N-1$ identical particles of mass m and a single migrating particle of mass m_J. The state of such a system of N interacting particles can be defined by $3N$ spatial coordinates, q_i, and $3N$ momenta, p_i, and can then be represented by a point in a $6N$-dimensional phase space with coordinates $(q_1, q_2, \ldots, q_{3N}, p_1, p_2, \ldots, p_{3N})$ [9]. Furthermore, the total energy of such a system can be expressed as the sum of its kinetic energy (a function of the $3N$ momenta) and its potential energy (a function of the $3N$ spatial coordinates).

Assuming that there are numerous sites in the system that the jumping particle can occupy while maintaining a stable system structure, the rate at which this particle jumps from one stable site to another can be determined. Figure 7.3a depicts how the total potential energy of the system, $\phi = \phi(q_1, q_2, \ldots, q_{3N})$, varies as the jumping particle occupies positions throughout the system, including the sites of local minimum energy. Because it is impossible to make such a plot in three dimensions, the many displacements of particles in the system that accompany the displacement of the jumping particle are suggested by the added multiple axes. Point P represents the situation of the jumping particle in a stable site while point Q represents the corresponding situation when the jumping particle is in a neighboring stable site. In both cases, the system is stable because it is at a local potential-energy minimum, as indicated by the two minima in the hypersurface shown in the $3N$-dimensional space of Fig. 7.3. In Fig. 7.3b, hypersurfaces of constant potential energy are a function of the $3N$ coordinates indicated in Fig. 7.3a. [These hypersurfaces are of dimensionality $3N - 1$ because they are defined by the level sets of $\phi = \phi(q_1, q_2, \ldots, q_{3N})$, so that $3N - 1$ coordinates are independent.] There are many choices for a particle trajectory between P and Q, but the trajectories that cross the saddle point (located at the point P^A) require the smallest energy fluctuation and are the most probable. Therefore, the saddle point energies in the potential-energy landscape determine the transition probabilities. The saddles are present because each minimum is surrounded by neighboring maxima. Neighboring minima pairs have at least one connecting path that has an associated saddle energy; the path between P and Q in Fig. 7.3 passes through the saddle point P^A. The force on a particle exactly at P^A is zero, but the configuration is an unstable equilibrium. A unique hypersurface, S^A, passes through P^A and is

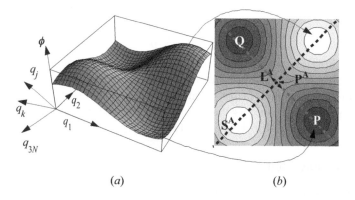

(a)   (b)

Figure 7.3: In a system with $3N$ spatial coordinates, the potential-energy landscape consists of minima for each of the stable atomic sites, such as P and Q. The potential-energy landscape surface (**a** is the surface and **b** shows its isopotentials) represents such a landscape, but only for two spatial coordinates. The surface is impossible to illustrate as a function of all coordinates $q_1, q_2, \ldots, q_{3N}$. The migrating atom traverses the region between P and Q via the saddle point P^A. During migration the landscape changes in response to the geometrical configuration of the activated state.

perpendicular to the contours of constant ϕ. S^A constitutes an energy ridge and is analogous to a "continental divide" separating the region associated with P from the region associated with Q.

In an equilibrium system, a migrating particle spends most of its time vibrating with small amplitude about low-energy states such as P and Q. The crossing time is very short compared to the equilibrium duration, but as the migrating atom crosses, it slows near the saddle point. For the crossing interval depicted in Fig. 7.3, most of the active migration time is spent near saddle points such as P^A in a volume of the hyperspace centered on the saddle point of width L^A. Considering the states in the activated volume and the states near the minima, Eqs. 7.6 and 7.7 can be applied to model the jump rate, Γ', but the analysis must be modified if the system is at constant pressure instead of constant volume. The jump rate for a system at constant temperature and pressure is

$$
\begin{aligned}
\Gamma' &= \sqrt{\frac{kT}{2\pi m_J}} \frac{1}{L^A} \frac{Z_P^A}{Z_P^{\text{well}}} \\
&= \sqrt{\frac{kT}{2\pi m_J}} \frac{1}{L^A} \frac{e^{-G^A/(kT)}}{e^{-G^{\text{well}}/(kT)}} \\
&= \sqrt{\frac{kT}{2\pi m_J}} \frac{1}{L^A} \frac{e^{-F^A/(kT)}}{e^{-F^{\text{well}}/(kT)}} \frac{e^{-PV^A/(kT)}}{e^{-PV^{\text{well}}/(kT)}} \\
&= \sqrt{\frac{kT}{2\pi m_J}} \frac{e^{-P(V^A-V^{\text{well}})/(kT)}}{L^A} \frac{Z^A}{Z^{\text{well}}} \\
&= \sqrt{\frac{kT}{2\pi m_J}} \frac{e^{-P(V^A-V^{\text{well}})/(kT)}}{L^A} \frac{\sum_{i=1}^{N} e^{-U_i^A/(kT)}}{\sum_{i=1}^{N} e^{-U_i^{\text{well}}/(kT)}}
\end{aligned}
\tag{7.15}
$$

where Z_P is the fixed-pressure partition function, Z is the fixed-volume partition function, and V^A and V^{well} are the system volumes in the activated and well states.

Using $U_i = p_i^2/2m_i + \phi_i(x)$ in the classical limit [the number of energetically degenerate states $= (dp\,dq)/h$] and integrating the momentum terms for Z yields

$$
Z = \left(\frac{2\pi m kT}{h^2}\right)^{\frac{3(N-1)}{2}} \left(\frac{2\pi m_J kT}{h^2}\right)^{\frac{3}{2}} \int e^{-\phi(q_1,q_2,\dots,q_{3N})/(kT)}\,dq_1\,dq_2\cdots dq_{3N}
\tag{7.16}
$$

To find an expression for the potential energy of the vibrating system in the well state at P, the harmonic approximation is used and $\phi(\vec{q})$ is expanded about P, so the potential-energy surface near P has the form

$$
\phi^{\text{well}}(q_1,q_2,\dots,q_{3N}) = N\phi^\circ + \frac{1}{2}\sum_{i=1}^{3N}\sum_{j=1}^{3N} \left(\frac{\partial^2 \phi}{\partial q_i \partial q_j}\right)\bigg|_{q_i=q_i^\circ} (q_i - q_i^\circ)(q_j - q_j^\circ)
\tag{7.17}
$$

where i and j sum over the $3N$ displacements of the J atom and the $N-1$ other atoms, q_i° is the average coordinate of the ith vibrating atom, and ϕ° is the potential energy per atom when all atoms are located in their average positions. The elements of the matrix of second derivatives are the linearized spring constants for each atom site pair and correspond to the quantity β in Eq. 7.10, which was employed in the static harmonic model. Typically, only the near neighbors have nonnegligible entries, the number of which depends on the interatomic potential length scale. The matrix of second derivatives is real and symmetric and therefore has real eigenvalues, all of which are positive in the stable state. If the δq_i are transformed to its diagonalized coordinate eigensystem $\delta \eta_i$,

$$\phi^{\text{well}}(\eta_1, \eta_2, \ldots, \eta_{3N}) = N\phi^\circ + \frac{1}{2}\sum_{i=1}^{3N} m_i \omega_i^2 (\delta \eta_i)^2 \tag{7.18}$$

where m_i and ω_i are the effective masses and characteristic angular frequencies of the three vibrational modes of each harmonic mode and i now runs over the $3N$ modes of the J atom and the $N-1$ other atoms. The modes are decoupled in the eigensystem and this transformation exists for any interatomic potential. If the interactions are short-range, the matrix will be sparse and the effective masses and characteristic frequencies will be nearly the same in the eigensystem (η_i) and the lattice system (q_i).

In Eq. 7.16, ϕ can be approximated using Eq. 7.18; the integral's value is dominated by $\delta \eta_i \approx 0$, and therefore its limits can be taken to be $\pm\infty$, even though the parabolic approximation is valid only near equilibrium,

$$\int e^{-\phi(q_1, q_2, \ldots, q_{3N})/(kT)} \, dq_1 \, dq_2 \cdots dq_{3N}$$

$$\approx e^{-N\phi^\circ/(kT)} \left(\int_{-\infty}^{\infty} e^{-m_1 \omega_1^2 \delta \eta_1^2/(2kT)} d\delta \, \eta_1 \right) \left(\int_{-\infty}^{\infty} e^{-m_2 \omega_2^2 \delta \eta_2^2/(2kT)} d\delta \, \eta_2 \right)$$

$$\cdots \left(\int_{-\infty}^{\infty} e^{-m_{3N} \omega_{3N}^2 \delta \eta_{3N}^2/(2kT)} d\delta \, \eta_{3N} \right)$$

$$= e^{-N\phi^\circ/(kT)} \sqrt{\frac{2\pi kT}{m_1 \omega_1^2}} \sqrt{\frac{2\pi kT}{m_2 \omega_2^2}} \cdots \sqrt{\frac{2\pi kT}{m_{3N} \omega_{3N}^2}}$$

$$= e^{-N\phi^\circ/(kT)} \left(\frac{2\pi kT}{m_J \omega_J^2} \right)^{\frac{3}{2}} \left(\frac{2\pi kT}{m\omega^2} \right)^{\frac{3(N-1)}{2}}$$

$$\tag{7.19}$$

where ω is the same for all masses except the migrating atom. If the masses and characteristic frequencies differ, a product would appear above for each unique effective mass and characteristic frequency.

When the jumping atom is located along L^A, where the potential-energy hypersurface has principal curvatures of opposite sign, there is one negative eigenvalue corresponding to the unstable direction along the path between P and Q and

$$\phi^A = \sum_i^N \phi_i^A + \frac{1}{2} \left(\frac{\partial^2 \phi^A}{\partial \eta_{J1}^{A^2}} \right) \Bigg|_{\eta_{J1}^A = 0}$$

$$+ \frac{1}{2} m_J (\omega_{J2}^A)^2 (\delta \eta_{J2}^A)^2 + \frac{1}{2} m_J (\omega_{J3}^A)^2 (\delta \eta_{J3}^A)^2 + \frac{1}{2} \sum_{j=1}^{3N-3} m_j (\omega_j^A)^2 (\delta \eta_j^A)^2 \tag{7.20}$$

where i iterates over all N atoms and j iterates over all but the J atom.

It is reasonable to assume that only a relatively small number of atoms surrounding the jumping atom are affected when the system goes from the well state to the activated state. Let this number be N^A. Also, approximate the potential energy of the jumping atom along L^A at the saddle point by a parabola in a single variable, η_{J1}. Let the approximating parabola have its maximum ϕ_J^A at $\delta\eta_{J1} = 0$ and decrease by a factor $1 - \epsilon$ at $\delta\eta_{J1} = \pm L^A/2$:

$$
\phi^A = \left[\phi_J^A + \sum_{i=1}^{N^A} \phi_i^A + (N - N^A - 1)\phi^\circ\right] - \frac{4\varepsilon}{(L^A)^2}\phi_J^A\left(\delta\eta_{J1}^A\right)^2
$$
$$
+ \frac{1}{2}m_J(\omega_{J2}^A)^2(\delta\eta_{J2}^A)^2 + \frac{1}{2}m_J(\omega_{J3}^A)^2(\delta\eta_{J3}^A)^2 \tag{7.21}
$$
$$
+ \frac{1}{2}\sum_{j=1}^{3N^A} m(\omega_j^A)^2(\delta\eta_j^A)^2 + \frac{1}{2}m\omega^2 \sum_{k=1}^{3(N-N^A-1)} (\delta\eta_k)^2
$$

where i iterates over all N^A affected atoms, j iterates over the $3N^A$ modes of the affected atoms, and k iterates over all modes of the $(N - N^A - 1)$ nonaffected atoms. Equation 7.21 can be integrated with the same approximation employed to obtain Eq. 7.19. To lowest order in L^A and ϵ,

$$
\int e^{-\phi^A/(kT)} d\eta_{J1}^A\, d\eta_{J2}^A\, d\eta_{J3}^A\, d\eta_1^A\, d\eta_2^A \cdots d\eta_{3N^A}^A d\eta_1 \cdots d\eta_{3(N-N^A-1)}
$$
$$
= L^A e^{-[\phi_J^A + \sum_{i=1}^{N^A}\phi_i^A + (N-N^A-1)\phi^\circ]/(kT)}
$$
$$
\times \left[\frac{2\pi kT}{m_J(\omega_J^A)^2}\right]\left[\frac{2\pi kT}{m\omega^2}\right]^{\frac{3(N-N^A-1)}{2}} \left(\prod_{\ell=1}^{3N^A} \sqrt{\frac{2\pi kT}{m(\omega_\ell^A)^2}}\right) \tag{7.22}
$$

where i iterates over all N^A affected atoms and ℓ iterates over the modes of the affected atoms and it is assumed that $\omega_{J2}^A = \omega_{J3}^A$. Therefore,

$$
\frac{Z^A}{Z^{\text{well}}} = L^A e^{-[\phi_J^A + \sum_{i=1}^{N^A}\phi_i^A + (N-N^A-1)\phi^\circ]/(kT)} e^{-N\phi^\circ/(kT)} \frac{\frac{2\pi kT}{m_J(\omega_J^A)^2}}{\left[\frac{2\pi kT}{m_J(\omega_J)^2}\right]^{3/2}} \left(\prod_{\ell=1}^{3N^A} \frac{\omega}{\omega_\ell^A}\right) \tag{7.23}
$$

and

$$
\Gamma' = \frac{\omega_J}{2\pi} e^{-[\phi_J^A + \sum_{i=1}^{N^A}\phi_i^A - (N^A+1)\phi^\circ + P(V^A - V^{\text{well}})]/(kT)} \left(\frac{\omega_J}{\omega_J^A}\right)^2 \left(\prod_{\ell=1}^{3N^A} \frac{\omega}{\omega_\ell^A}\right) \tag{7.24}
$$

The final expression for the jump rate is then

$$
\Gamma' = \nu e^{-G^m/(kT)} = \nu e^{S^m/k} e^{-H^m/(kT)} = \nu e^{-PV^m/(kT)} e^{S^m/k} e^{-U^m/(kT)} \tag{7.25}
$$

where

$$
\nu = \frac{\omega_J}{2\pi}
$$
$$
U^m = \phi_J^A - \phi^\circ + \sum_{i=1}^{N^A}(\phi_i^A - \phi^\circ)
$$
$$
V^m = V^A - V \tag{7.26}
$$
$$
S^m = k\left[2\ln\frac{\omega_J}{\omega_J^A} + \sum_{\ell=1}^{3N^A}\ln\left(\frac{\omega}{\omega_\ell^A}\right)\right] \approx \frac{2k(\omega_J^A - \omega_J)}{\omega_J} + \frac{3kN^A\langle\Delta\omega\rangle}{\omega}
$$

and $\langle\Delta\omega\rangle$ is the average difference between the activated and stable frequencies, $\omega_\ell^A - \omega$.

The interpretation of the jump rate for this multibody harmonic model is the same as for the simpler models. Equation 7.25 is the product of an attempt frequency and a Boltzmann–Arrhenius exponential factor containing a migration activation energy. This result will be used throughout the remainder of this book. If *none* of the migrating atom's neighbors are affected in Eq. 7.25, the activation energy is simply the difference between the migrating atom's energy in the activated and well states, and the entropy is proportional to the difference between the migrating atom's frequencies in the activated and well states.

7.2 DIFFUSION AS A SERIES OF DISCRETE JUMPS

In general, a particle migrates in a material by a series of thermally activated jumps between positions of local energy minima. Macroscopic diffusion is the result of all the migrations executed by a large ensemble of particles. The spread of the ensemble due to these migrations connects the macroscopic diffusivity to the microscopic particle jumping.

If a particle jumps with average frequency Γ in a sequence of displacements $\vec{r}_i$ ($\vec{r}_i$ gives the magnitude and direction of the ith jump), then after a period of time, τ, the particle will execute $N_\tau = \Gamma\tau$ individual jumps.[4] The position relative to its starting point is the sum of the individual displacements,

$$\vec{R}(N_\tau) = \sum_{i=1}^{N_\tau} \vec{r}_i = \sum_{i=1}^{\Gamma\tau} \vec{r}_i \tag{7.27}$$

The random walk process can be characterized by the distribution of total displacements for either a large set of noninteracting walkers or for repeated trials of an isolated walker. The average displacement is a vector $\langle\vec{R}(N_\tau)\rangle$ and the *mean-square displacement* $\langle\vec{R}(N_\tau)\cdot\vec{R}(N_\tau)\rangle = \langle R^2(N_\tau)\rangle$ is a scalar that characterizes the spread or *diffuseness* of the distribution of total displacements about its average.

The square displacement for a given sequence of random steps is

$$R^2(N_\tau) = \vec{R}(N_\tau)\cdot\vec{R}(N_\tau) = \begin{array}{l} \vec{r}_1\cdot\vec{r}_1 + \vec{r}_1\cdot\vec{r}_2 + \vec{r}_1\cdot\vec{r}_3 + \cdots + \vec{r}_1\cdot\vec{r}_{N_\tau} \\ + \quad \vec{r}_2\cdot\vec{r}_1 + \vec{r}_2\cdot\vec{r}_2 + \vec{r}_2\cdot\vec{r}_3 + \cdots + \vec{r}_2\cdot\vec{r}_{N_\tau} \\ + \quad \vdots \qquad\qquad \cdots \qquad\qquad \ddots \\ + \quad \vec{r}_{N_\tau}\cdot\vec{r}_1 + \vec{r}_{N_\tau}\cdot\vec{r}_2 + \vec{r}_{N_\tau}\cdot\vec{r}_3 \cdots + \vec{r}_{N_\tau}\cdot\vec{r}_{N_\tau} \end{array} \tag{7.28}$$

---

[4] Γ represents the average total jump rate during the sequence, which generally consists of jumps of different lengths and directions. This contrasts with Γ', which represents the jump rate between two specified sites.

This sum consists of $N_\tau \times N_\tau$ products that can be collected into diagonal $\vec{r}_i \cdot \vec{r}_i$ and off-diagonal $(\vec{r}_i \cdot \vec{r}_j = \vec{r}_j \cdot \vec{r}_i)$ parts,

$$
\begin{aligned}
R^2(N_\tau) &= \sum_{i=1}^{N_\tau} \vec{r}_i \cdot \vec{r}_i + 2 \sum_{j=1}^{N_\tau - 1} \sum_{i=j+1}^{N_\tau} \vec{r}_i \cdot \vec{r}_j \\
&= \sum_{i=1}^{N_\tau} |\vec{r}_i|^2 + 2 \sum_{j=1}^{N_\tau - 1} \sum_{i=1}^{N_\tau - j} \vec{r}_i \cdot \vec{r}_{i+j}
\end{aligned}
\tag{7.29}
$$

The $\vec{r}_i \cdot \vec{r}_{i+j}$ can be expressed as products of the jump distances and the cosine of the angle between the jump-vector directions, $\theta_{i,i+j}$ (i.e., $\vec{r}_i \cdot \vec{r}_{i+j} = |\vec{r}_i||\vec{r}_{i+j}| \cos \theta_{i,i+j}$). The mean-square jump distance is

$$
\langle r^2 \rangle = \frac{1}{N_\tau} \sum_{i=1}^{N_\tau} |\vec{r}_i|^2
\tag{7.30}
$$

Averaging over a large number of independent walkers or trials for a single walker,

$$
\langle R^2(N_\tau) \rangle = N_\tau \langle r^2 \rangle + 2 \left\langle \sum_{j=1}^{N_\tau - 1} \sum_{i=1}^{N_\tau - j} |\vec{r}_i||\vec{r}_{i+j}| \cos \theta_{i,i+j} \right\rangle
\tag{7.31}
$$

Equation 7.31 is general. No assumptions have been made about the randomness of the displacements, the lengths of the various displacements, the allowed values of $\theta_{i,i+j}$, or the number of dimensions in which the random walk is occurring.

7.2.1 Relation of Diffusivity to the Mean-Square Particle Displacement

A relationship between the macroscopic diffusivity, D, of a component i and the mean-square displacement, $\langle R^2(N_\tau) \rangle$, can be obtained from the behavior of $c_i(x,t)$ as it evolves from an initial point source at the origin. Using the solution for diffusion from an instantaneous point source in three dimensions in Table 5.1, the distribution of particles after a time τ will be given by

$$
c(r, \tau) = \frac{n_d}{(4\pi D\tau)^{3/2}} e^{-r^2/(4D\tau)}
\tag{7.32}
$$

The second moment of the distribution in Eq. 7.32,

$$
\langle R^2(\tau) \rangle = \frac{\int_0^\infty r^2 c(r,\tau)\, 4\pi r^2\, dr}{\int_0^\infty c(r,\tau)\, 4\pi r^2\, dr}
\tag{7.33}
$$

gives the mean-square displacement away from the original point source. Using Eq. 7.32 and the relationship

$$
\int_0^\infty x^4 e^{-ax^2}\, dx = \frac{3}{8a^2} \sqrt{\frac{\pi}{a}}
\tag{7.34}
$$

in Eq. 7.33, the mean-square displacement for isotropic three-dimensional diffusion is related to the diffusivity by

$$
\langle R^2(\tau) \rangle = 6D\tau
\tag{7.35}
$$

For diffusion in one and two dimensions, similar calculations show that $\langle R^2(\tau)\rangle = 2D\tau$ and $4D\tau$, respectively. An analogous expression for $\langle R^2(\tau)\rangle$ when the diffusivity is anisotropic is explored in Exercise 7.4.

Equation 7.35 is a fundamental relationship between the diffusivity and the mean-square displacement of a particle diffusing for a time τ. Because diffusion processes in condensed matter are comprised of a sequence of jumps, the mean-square displacement in Eq. 7.31 should be equivalent to Eq. 7.35. This equivalence, as demonstrated below, results in relations between macroscopic and microscopic diffusion parameters.

7.2.2   Diffusion and Random Walks

If a particle moves by a series of displacements, each of which is independent of the one preceding it, the particle moves by a *random walk*. Random walks can involve displacements of fixed or varying length and direction. The theory of random walks provides distributions of the positions assumed by particles; such distributions can be compared directly to those predicted to result from macroscopic diffusion. Furthermore, the results from random walks provide a basis for understanding non-random diffusive processes.

The distribution of positions is easily formulated for a random walker on a one-dimensional lattice and illustrates important aspects of all random walks. Such a distribution can be compared to the solution for macroscopic diffusion in Table 5.1. Extensions to two and three dimensions are not difficult. The particles are assumed to migrate independently. A given particle starts at the origin and jumps either forward (along $+x$) with probability p_R or backward (along $-x$) with probability p_L, where $0 < p_R < 1$ and $p_L + p_R = 1$. Suppose that each displacement is of length one, then after $N_\tau \gg 1$ displacements it is possible that the particle will end up at $-N_\tau, -N_\tau + 1, \ldots - 1, 0, 1, \ldots N_\tau - 1, N_\tau$. It is highly unlikely that the particle will make all positive jumps to reach site N_τ or all negative jumps to reach $-N_\tau$. If $p_R = p_L = 1/2$, then, on average, the particle will be located at the origin because the equally probable positive and negative displacements negate each other. To find the probability that a particle occupies a position n after N_τ jumps, let N_R be the number of positive displacements and N_L be the number of negative displacements,

$$N_R - N_L = n \tag{7.36}$$

$$N_R + N_L = N_\tau \tag{7.37}$$

The number, $\Omega(n, N_\tau)$, of different ways (trajectories, sequences, etc.) the walker can get to site n from the origin is given by the binomial coefficient

$$\Omega(n, N_\tau) = \frac{N_\tau!}{N_R! \, N_L!} = \frac{N_\tau!}{[(N_\tau + n)/2]! \, [(N_\tau - n)/2]!} \tag{7.38}$$

Therefore, the probability of getting to site n after N_τ jumps is

$$p(n, N_\tau) = \frac{N_\tau!}{[(N_\tau + n)/2]! \, [(N_\tau - n)/2]!} \, p_R^{N_R} p_L^{N_L} \tag{7.39}$$

If the probability of jumping right, p_R, is equal to the probability of jumping left, p_L,

$$p(n, N_\tau) = \frac{N_\tau!}{[(N_\tau + n)/2]! \, [(N_\tau - n)/2]!} \left(\frac{1}{2}\right)^{N_\tau} \tag{7.40}$$

Using Stirling's formula,

$$Q! = \sqrt{2\pi Q}\, Q^Q e^{-Q} \tag{7.41}$$

and taking the limit $n/N_\tau \ll 1$ yields

$$p(n, N_\tau) \propto e^{-n^2/(2N_\tau)} \tag{7.42}$$

Equation 7.42 shows that the distribution of a point source in one dimension spreads as a Gaussian.

Letting $R = n\langle r^2\rangle^{1/2}$, the probability distribution for the displacement R is

$$p(R, N_\tau) \propto e^{-R^2/(2\langle r^2\rangle \Gamma\tau)} \tag{7.43}$$

The probability distribution must be normalized, so that

$$\int_{-\infty}^{\infty} p(R, \tau)\, dR = 1 \tag{7.44}$$

Therefore, the probability distribution becomes

$$p(R, N_\tau) = \frac{1}{\sqrt{2\pi N_\tau \langle r^2\rangle}} e^{-R^2/(2\langle r^2\rangle \Gamma\tau)} \tag{7.45}$$

which is of the same form as the macroscopic solution for one-dimensional diffusion from a point source in Table 5.1.

The first and second moments of Eq. 7.45 are readily evaluated:

$$\langle R\rangle = \int_{-\infty}^{\infty} p(R, N_\tau) R\, dR = 0 \tag{7.46}$$

and

$$\langle R^2\rangle = \int_{-\infty}^{\infty} p(R, N_\tau) R^2\, dR = N_\tau \langle r^2\rangle \tag{7.47}$$

Equation 7.46 demonstrates that if each jump of a walk occurs randomly (i.e., is uncorrelated), the average displacement is zero and the center of mass of a large number of individual random jumpers is not displaced. Equation 7.47 gives the mean-square displacement of a random walk, $N_\tau \langle r^2\rangle$. Although Eqs. 7.46 and 7.47 were derived here for one-dimensional random walks, both are valid for two- and three-dimensional random walks.

The probability distribution of a random walk shows that the mean-square displacement after N_τ jumps is $\langle R^2\rangle = N_\tau \langle r^2\rangle = \Gamma\tau\langle r^2\rangle$ (Eq. 7.47). Comparison of the probability distribution (Eq. 7.45) to the point-source solution for one-dimensional diffusion from a point source (Table 5.1) indicates that

$$D = \frac{\Gamma\langle r^2\rangle}{2} \tag{7.48}$$

Equation 7.48 relates the macroscopic diffusivity and microscopic jump parameters for uncorrelated diffusion in one dimension.

7.2.3   Diffusion with Correlated Jumps

The calculated root-mean-square displacement for a general sequence of jumps has two terms in Eq. 7.31. The first term, $N_\tau \langle r^2 \rangle$, corresponds to an ideal random walk (see Eq. 7.47) and the second term arises from possible correlation effects when successive jumps do not occur completely at random.

For walks with correlations,[5] a *correlation factor*, $\mathbf{f}$, can be defined

$$\mathbf{f} = 1 + \frac{2}{N_\tau \langle r^2 \rangle} \left\langle \sum_{j=1}^{N_\tau - 1} \sum_{i=1}^{N_\tau - j} |\vec{r}_i| |\vec{r}_{i+j}| \cos \theta_{i,i+j} \right\rangle \tag{7.49}$$

so that Eq. 7.31 becomes

$$\langle R^2 \rangle = N_\tau \langle r^2 \rangle \mathbf{f} \tag{7.50}$$

For a random walk, $\mathbf{f} = 1$ because the double sum in Eq. 7.49 is zero and Eq. 7.50 reduces to the form of Eq. 7.47. In principle, $\mathbf{f}$ can have a wide range of values corresponding to physical processes relating to specific diffusion mechanisms. This is readily apparent in extreme cases of perfectly correlated one-dimensional diffusion on a lattice via nearest-neighbor jumps. When each jump is identical to its predecessor, Eq. 7.49 shows that the correlation factor $\mathbf{f}$ equals N_τ.[6] Another extreme is the case of $\mathbf{f} = 0$, which occurs if each individual jump is exactly opposite the previous jump. However, there are many real diffusion processes that are nearly ideal random walks and have values of $\mathbf{f} \approx 1$, which are described in more detail in Chapter 8.

The relationship between the macroscopic isotropic diffusivity, D, and microscopic jump processes can be evaluated in three dimensions. The equivalence of Eqs. 7.31 and 7.35 means that

$$\langle R^2(N_\tau) \rangle = N_\tau \langle r^2 \rangle + 2 \left\langle \sum_{j=1}^{N_\tau - 1} \sum_{i=1}^{N_\tau - j} |\vec{r}_i| |\vec{r}_{i+j}| \cos \theta_{i,i+j} \right\rangle = 6 D \tau \tag{7.51}$$

Substitution of Eq. 7.49 into Eq. 7.51 yields the relation between the macroscopic isotropic diffusivity and microscopic parameters

$$D = \frac{\langle r^2 \rangle N_\tau}{6\tau} \mathbf{f} = \frac{\Gamma \tau \langle r^2 \rangle}{6\tau} \mathbf{f} = \frac{\Gamma \langle r^2 \rangle}{6} \mathbf{f} \tag{7.52}$$

Equation 7.52 is of central importance for atomistic models for the macroscopic diffusivity in three dimensions (see Chapter 8). For isotropic diffusion in a system of dimensionality, d, the generalized form of Eq. 7.52 is

$$D = \frac{\Gamma \langle r^2 \rangle}{2d} \mathbf{f} \tag{7.53}$$

Equations 7.52 and 7.53 reduce to D for random-walking particles (i.e., Eq. 7.48) where there are no correlations and $\mathbf{f} = 1$.

Values of $\mathbf{f}$ for several diffusion mechanisms are discussed in Section 8.2.1.

---

[5] Correlated jumps are discussed in Chapter 8.
[6] There are $(N_\tau^2 - N_\tau)/2$ cosine terms in the double sum and all are equal to unity.

Bibliography

1. S. Glasstone, K.J. Laidler, and H. Eyring. *The Theory of Rate Processes*. McGraw-Hill, New York, 1941.

2. C.A. Wert and C. Zener. Interstitial atom diffusion coefficients. *Phys. Rev.*, 76(8):1169–1175, 1949.

3. G.H. Vineyard. Frequency factors and isotope effects in solid state rate processes. *J. Chem. Phys. Solids*, 3(1–2):121–127, 1957.

4. S.A. Rice. Dynamical theory of diffusion in crystals. *Phys. Rev.*, 112(3):804–811, 1958.

5. C.P. Flynn. *Point Defects and Diffusion*. Oxford University Press, Oxford, 1972.

6. L.A. Girifalco. *Statistical Physics of Materials*. John Wiley & Sons, New York, 1973.

7. J.W. Christian. *The Theory of Transformations in Metals and Alloys*. Pergamon Press, Oxford, 1975.

8. W.M. Franklin. Classical and quantum theory of diffusion in solids. In A.S. Nowick and J.J. Burton, editors, *Diffusion in Solids, Recent Developments*, pages 1–72. Academic Press, New York, 1975.

9. D.A. McQuarrie. *Statistical Mechanics*. HarperCollins, New York, 1976.

EXERCISES

7.1 Prove that the pre-exponential frequency factor given by Eq. 7.14 is indeed the frequency of a linear oscillator of mass, m, and force constant, β.

Solution. The equation of motion of a linear oscillator is

$$F(x) = -\beta x(t) = m\frac{d^2x}{dt^2} \tag{7.54}$$

where $x(t)$ is the displacement of the mass from the position where the restoring force, F, is zero. The solution of Eq. 7.54 is of the form $x(t) = A\sin(\omega t)$ where $A = $ constant. Substitution of $x(t)$ in Eq. 7.54 shows that

$$\omega = 2\pi\nu = \sqrt{\frac{\beta}{m}} \tag{7.55}$$

7.2 The quantity V^m, given by Eq. 7.26, is the difference between the volume of the system in an activated state and a well state. This volume difference is generally termed the *activation volume* for migration and is a positive quantity because of the atomic squeezing and resulting expansion of the system that occurs in the activated state. The activation volume can be measured experimentally by measuring the pressure dependence of the jump frequency, Γ'. Find an expression for the pressure dependence of Γ' and describe how it can be used to determine V^m.

Solution. Use Eq. 7.25 for Γ' and differentiate Γ' with respect to pressure, so that

$$\left[\frac{\partial \ln \Gamma'}{\partial P}\right]_T = \left[\frac{\partial \ln \nu}{\partial P}\right]_T - \frac{1}{kT}\left[\frac{\partial G^m}{\partial P}\right]_T \tag{7.56}$$

Using the standard thermodynamic relation $[\partial G/\partial P]_T = V$ and realizing that the pressure dependence of $\ln \nu$ will be relatively very small, we may write to a good approximation

$$\left[\frac{\partial \ln \Gamma'}{\partial P}\right]_T \cong -\frac{1}{kT}V^m \tag{7.57}$$

If a plot of $\ln \Gamma'$ vs. P is now constructed using the experimental data, V^m can be determined from its slope.

7.3 Consider small interstitial atoms jumping by the interstitial mechanism in b.c.c. Fe with the diffusivity D for a time τ.

  (a) What is the most likely expected total displacement after a large number of diffusional jumps?

  (b) What is the standard deviation of the total displacement?

Solution.

  (a) The expected total displacement will be zero because there is no correlation between successive jumps—after a jump the interstitial loses its memory of its jump and makes its next jump randomly into any one of its nearest-neighbor sites.

  (b) The distribution of displacements will be Gaussian (Eq. 7.32) and the standard deviation will be the root-mean-square displacement given by Eq. 7.35 as $\sqrt{6D\tau}$.

7.4 Suppose the random walking of a diffusant in a primitive orthorhombic crystal where the particle makes N_1 jumps of length a_1 along the x_1 axis, N_2 jumps of length a_2 along the x_2 axis, and N_3 jumps of length a_3 along the x_3 axis. The three axes are orthogonal and aligned along the crystal axes of the orthorhombic unit cell and the diffusivity tensor in this axis system is

$$\boldsymbol{D} = \begin{bmatrix} D_{11} & 0 & 0 \\ 0 & D_{22} & 0 \\ 0 & 0 & D_{33} \end{bmatrix} \tag{7.58}$$

  (a) Find an expression for the mean-square displacement in terms of the numbers of jumps and jump distances.

  (b) Find another expression for the mean-square displacement in terms of the three diffusivities in the diffusivity tensor and the diffusion time. Your answer should be analogous to Eq. 7.35, which holds for the isotropic case.

Solution.

  (a) Using Eqs. 7.30 and 7.31,

$$\langle R^2 \rangle = \sum_{i=1}^{N} \vec{r}_i \cdot \vec{r}_i = N_1 a_1^2 + N_2 a_2^2 + N_3 a_3^2 \tag{7.59}$$

  (b) The diffusion equation will have the form of Eq. 4.61. By using the method of scaling described in Section 4.5 (based on the scaling relationships in Eq. 4.64), the solution can be written

$$c(x_1, x_2, x_3, t) = \frac{A}{t^{3/2}} \exp\left[-\left(\frac{x_1^2}{4D_{11}t} + \frac{x_2^2}{4D_{22}t} + \frac{x_3^2}{4D_{33}t}\right)\right] \tag{7.60}$$

where $A = $ constant. The mean-square displacement is then

$$\langle R^2 \rangle = \frac{\int_0^\infty \int_0^\infty \int_0^\infty c(x_1, x_2, x_3, t)\, r^2\, dx_1\, dx_2\, dx_3}{\int_0^\infty \int_0^\infty \int_0^\infty c(x_1, x_2, x_3, t)\, dx_1\, dx_2\, dx_3}$$

$$= \frac{\int_0^\infty \int_0^\infty \int_0^\infty e^{-\frac{x_1^2}{4D_{11}t}}\, e^{-\frac{x_2^2}{4D_{22}t}}\, e^{-\frac{x_3^2}{4D_{33}t}}\, \left(x_1^2 + x_2^2 + x_3^2\right)\, dx_1\, dx_2\, dx_3}{\int_0^\infty \int_0^\infty \int_0^\infty c(x_1, x_2, x_3, t)\, dx_1\, dx_2\, dx_3}$$

$$\tag{7.61}$$

Equation 7.61 can be factored into standard definite integrals and the result is

$$\langle R^2 \rangle = 2D_{11}t + 2D_{22}t + 2D_{33}t \tag{7.62}$$

Comparison of Eqs. 7.59 and 7.62 shows that the mean-square displacement consists of three terms, each of which is the mean-square displacement that would be achieved in one dimension along one of the three coordinate directions.

7.5 Suppose a random walk occurs on a primitive cubic lattice and successive jumps are uncorrelated. Show explicitly that $\mathbf{f} = 1$ in Eq. 7.49. Base your argument on a detailed consideration of the values that the $\cos\theta_{i,i+j}$ terms assume.

Solution. Because all jumps are of the same length,

$$\mathbf{f} = 1 + \frac{2}{N_\tau} \sum_{j=1}^{N_\tau} \sum_{i=1}^{N_\tau - j} \langle \cos\theta_{i,i+j} \rangle$$

$$= 1 + \frac{2}{N_\tau} \langle \cos\theta_{1,2} + \cos\theta_{1,3} \cdots + \cos\theta_{2,3} + \cos\theta_{2,4} \cdots + \cos\theta_{N_\tau - 1, N_\tau} \rangle$$

$$\tag{7.63}$$

and thus,

$$\mathbf{f} = 1 + \frac{2}{N_\tau} \left[\langle \cos\theta_{1,2} \rangle + \langle \cos\theta_{1,3} \rangle + \cdots + \langle \cos\theta_{2,3} \rangle \cdots + \langle \cos\theta_{N_\tau - 1, N_\tau} \rangle \right] \tag{7.64}$$

Any jump can be one of the six vectors: $[a00]$, $[\bar{a}00]$, $[0a0]$, $[0\bar{a}0]$, $[00a]$, and $[00\bar{a}]$. Each occurs with equal probability. For each pair of jump vectors, i and $i+j$, the six possible values of $\cos\theta_{i,i+j}$ are $1, -1, 0, 0, 0,$ and 0, and these occur with equal probability. For a large number of trajectories, each mean value in Eq. 7.64 is zero and therefore $\mathbf{f} = 1$.

7.6 For the diffusion of vacancies on a face-centered cubic (f.c.c.) lattice with lattice constant a, let the probability of first- and second-nearest-neighbor jumps be p and $1 - p$, respectively. At what value of p will the contributions to diffusion of first- and second-nearest-neighbor jumps be the same?

Solution. There is no correlation and, using Eq. 7.29,

$$\langle R^2 \rangle = N_\tau \langle r^2 \rangle = \sum_{i=1}^{N_\tau} \vec{r_i} \cdot \vec{r_i} \tag{7.65}$$

The number of first nearest-neighbor jumps is $N_\tau p$ and the number of second nearest-neighbor jumps is $N_\tau(1 - p)$. Therefore,

$$\langle R^2 \rangle = N_\tau p \frac{a^2}{2} + N_\tau(1 - p)a^2 \tag{7.66}$$

They make equal contributions when $N_\tau p a^2 / 2 = N_\tau(1 - p)a^2$ or $p = 2/3$.

CHAPTER 8

DIFFUSION IN CRYSTALS

The driving forces necessary to induce macroscopic fluxes were introduced in Chapter 3 and their connection to microscopic random walks and activated processes was discussed in Chapter 7. However, for diffusion to occur, it is necessary that *kinetic mechanisms* be available to permit atomic transitions between adjacent locations. These mechanisms are material-dependent. In this chapter, diffusion mechanisms in metallic and ionic crystals are addressed. In crystals that are free of line and planar defects, diffusion mechanisms often involve a point defect, which may be charged in the case of ionic crystals and will interact with electric fields. Additional diffusion mechanisms that occur in crystals with dislocations, free surfaces, and grain boundaries are treated in Chapter 9.

8.1 ATOMIC MECHANISMS

Atom jumping in a crystal can occur by several basic mechanisms. The dominant mechanism depends on a number of factors, including the crystal structure, the nature of the bonding in the host crystal, relative differences of size and electrical charge between the host and the diffusing species, and the type of crystal site preferred by the diffusing species (e.g., anion or cation, substitutional or interstitial).

Kinetics of Materials. By Robert W. Balluffi, Samuel M. Allen, and W. Craig Carter.   **163**
   Copyright © 2005 John Wiley & Sons, Inc.

8.1.1 Ring Mechanism

A substitutional atom (indicated by shading in Fig. 8.1) may jump and replace an adjacent nearest-neighbor substitutional atom. In the *ring mechanism*, the substitutional atom exchanges places with a neighboring atom by a cooperative ringlike rotational movement.

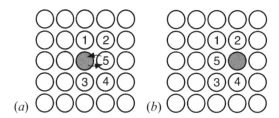

Figure 8.1: Ring mechanism for diffusion of substitutional atoms.

8.1.2 Vacancy Mechanism

A substitutional atom can migrate to a neighboring substitutional site without cooperative motion and with a relatively small activation energy if the neighboring substitutional site is unoccupied. This is equivalent to exchange with a neighboring vacancy.[1] In Fig. 8.2a, the vacancy is initially separated from a particular substitutional atom (again indicated by shading). In Fig. 8.2b, it has migrated by exchanging places with host atoms to a nearest-neighbor substitutional site of the shaded atom. In Fig. 8.2c the vacancy has exchanged sites with the substitutional atom, and in Fig. 8.2d the vacancy has migrated some distance away. As a result, the particular substitutional atom is displaced by one nearest-neighbor distance while the vacancy has undergone at least five individual displacements.

The atomic environment during a vacancy-exchange mechanism can be illustrated in a three-dimensional cubic lattice. Figure 8.3 shows an atom-vacancy exchange between two face-centered sites in an f.c.c. crystal. The migrating atom (A in Fig. 8.3) moves in a $\langle 110 \rangle$-direction through a rectangular "window" framed by two cube corner atoms and two opposing face-centered atoms. The f.c.c. crystal is close-packed and each site has 12 equivalent nearest-neighbor sites [1]. In

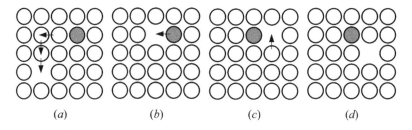

Figure 8.2: Vacancy mechanism for diffusion of substitutional atoms.

[1]Vacancies will always exist in equilibrium in a crystal because their enthalpy of formation can always be compensated by a configurational entropy increase at finite temperatures (see the derivation of Eq. 3.65). Therefore, vacancies function as a component that occupies substitutional sites.

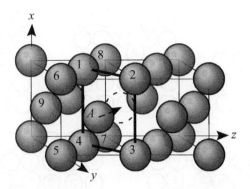

Figure 8.3: Atom-vacancy exchange in f.c.c. crystal. Atom initially at A jumps into a nearest-neighbor vacancy (dashed circle). The four nearest-neighbor atoms common to A and the vacant site (joined by the bold rectangle) form a "window" 1234 through which the A atom must pass. The A atom is centered in unit-cell face 2356. The vacancy is centered in unit-cell face 2378.

a hard-sphere model, in which nearest-neighbor atoms are in contact, the atom must "squeeze" through a window that is about 27% smaller than its diameter. The potential-energy increases required for such distortions create the energetic migration barriers discussed in Section 7.1.3.

8.1.3 Interstitialcy Mechanism

A substitutional atom can migrate to a neighboring substitutional site by the two-step process illustrated in Fig. 8.4. The first step is an exchange with an interstitial defect in which the migrating substitutional atom becomes the interstitial atom.[2] The second step is to exchange the migrating atom with a neighboring substitutional atom. This mechanism is only possible when substitutional atoms can occupy interstitial sites. This cooperative and serpentine motion constitutes the interstitialcy mechanism, and when large normally substitutional atoms are involved, can occur with a much lower migration energy than the interstitial mechanism (see below).

Interstitialcy migration depends on the geometry of the interstitial defect. However, an a priori prediction of interstitial defect geometry is not straightforward in real materials. For an f.c.c. crystal, a variety of conceivable interstitial defect candidates are illustrated in Fig. 8.5. The lowest-energy defect will be stable and predominant. For example, in the f.c.c. metal Cu, the stable configuration is the $\langle 100 \rangle$ split-dumbbell configuration in Fig. 8.5d [3].

The $\langle 100 \rangle$ split-dumbbell defect in Fig. 8.5d, while having the lowest energy of all interstitial defects, still has a large formation energy ($E^f = 2.2$ eV) because of the large amount of distortion and ion-core repulsion required for its insertion into the close-packed Cu crystal. However, once the interstitial defect is present, it persists until it migrates to an interface or dislocation or annihilates with a vacancy. The

---

[2]Interstitial point defects involving normally substitutional atoms will always exist (although typically at very low concentration) at equilibrium in a crystal at finite temperatures because, as in the case of vacancies described above, their enthalpy of formation can always be compensated by a configurational entropy increase.

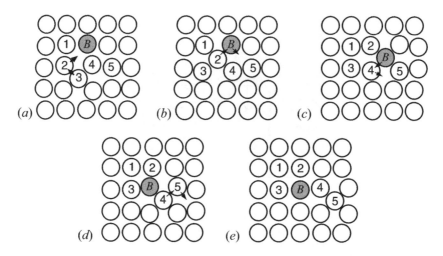

Figure 8.4: Substitutional diffusion by the interstitialcy mechanism. **(a)** The interstitial defect corresponding to the interstitial atom (3) is separated from a particular substitutional atom B (shaded). **(b)** The interstitial defect moved adjacent to B when the previously interstitial atom (3) replaced the substitutional atom (2). (2) then became the interstitial atom. **(c)** Atom (2) has replaced B, and B has become the interstitial atom. **(d)** B has replaced atom (4), which has become the interstitial atom. **(e)** The interstitial defect has migrated away from B. As a result, B has completed one nearest-neighbor jump and the interstitial defect has moved at least four times.

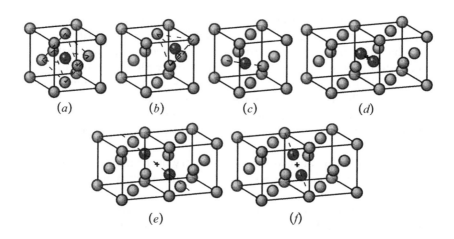

Figure 8.5: Geometric configurations for a self-interstitial defect atom in an f.c.c. crystal: **(a)** octahedral site, **(b)** tetrahedral site, **(c)** $\langle 110 \rangle$ crowdion, **(d)** $\langle 100 \rangle$ split dumbbell, **(e)** $\langle 111 \rangle$ split, **(f)** $\langle 110 \rangle$ split crowdion [2].

activation energy for migration ($E^m = 0.1$ eV) is small compared to E^f because little additional distortion is required for its serpentine motion, which is illustrated in Fig. 8.6. It therefore migrates relatively rapidly.

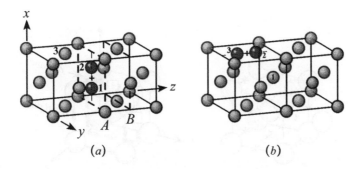

Figure 8.6: Diffusional migration of a [100] split-dumbbell self-interstitial in an f.c.c. crystal. The dumbbell in (**a**) can jump into four nearest-neighbor sites, creating [001] dumbbells [one of which is shown in (**b**)] and into four others, creating [010] dumbbells.

8.1.4 Interstitial Mechanism

An interstitial atom can simply migrate between interstitial sites as in Fig. 8.7. The interstitial atom must attain enough energy to distort the host crystal as it migrates between substitutional sites. This mechanism is expected for small solute atoms that normally occupy interstitial sites in a host crystal of larger atoms.

Diffusion by the interstitial mechanism and by the interstitialcy mechanism are quite different processes and should not be confused. Diffusion by the vacancy and interstitialcy mechanisms requires the presence of point defects in the system, whereas diffusion by the ring and interstitial mechanisms does not.

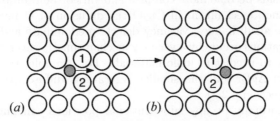

Figure 8.7: Interstitial mechanism for diffusion of interstitial atoms. The smaller shaded interstitial atom migrates through the opening between host atoms (1) and (2) to a neighboring interstitial site.

8.1.5 Diffusion Mechanisms in Various Materials

Diffusion of relatively small atoms that normally occupy interstitial sites in the solvent crystal generally occurs by the interstitial mechanism. For example, hydrogen atoms are small and migrate interstitially through most crystalline materials. Carbon is small compared to Fe and occupies the interstitial sites in b.c.c. Fe illustrated in Fig. 8.8 and migrates between neighboring interstitial sites.

Migration of atoms that occupy substitutional sites may occur through a range of mechanisms involving either vacancy- or interstitial-type defects. In f.c.c., b.c.c., and hexagonal close-packed (h.c.p.) metals, *self-diffusion* occurs predominantly by the vacancy mechanism [4, 5]. However, in some cases self-diffusion by the

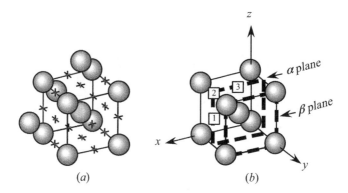

Figure 8.8:   Interstitial sites for C atoms in b.c.c. Fe. **(a)** The interstitial sites have point-group symmetry $4/mmm$, and the orientations of the fourfold axes are indicated by the shorter, grey spokes on the symbols. **(b)** Nomenclature used in the model for diffusion of interstitial atoms in b.c.c. Fe discussed in Section 8.2.1 Three different types of sites are present: sites 1, 2, and 3 have nearest-neighbor Fe atoms lying along x, y, and z, respectively.

interstitialcy mechanism contributes a small amount to the overall diffusion (see Section 8.2.1). In Ge, which has the less closely packed diamond-cubic structure, self-diffusion occurs by a vacancy mechanism. In Si (which like Ge has covalent bonding), self-diffusion occurs by the vacancy mechanism at low temperatures and by an interstitialcy mechanism at elevated temperatures [6–8]. In ionic materials, diffusion mechanisms become more complex and varied. Self-diffusion of Ni in NiO occurs by a vacancy mechanism; in Cu_2O the diffusion of O involves interstitial defects [9]. In the alkali halides, vacancy defects predominate and the diffusion of both anions and cations occurs by a vacancy mechanism. However, the predominant defect is not easy to predict in ionic materials. For example, vacancy–interstitial pairs dominate in AgBr and the smaller Ag cations diffuse by an interstitialcy mechanism (see Section 8.2.2).

Solutes that normally occupy substitutional positions can migrate by a variety of mechanisms. In many systems they migrate by the same mechanism as for self-diffusion of the host atoms. However, the details of migration become more complex if there is an interaction or binding energy between the solute atoms and point defects—this is described in Section 8.2.1 for vacancy–solute-atom binding. Certain solute atoms can migrate by more than one mechanism. For example, while Au solute atoms in Si are mainly substitutional, under equilibrium conditions, a relatively small number of Au atoms occupy interstitial sites. The rate of migration of the interstitial Au atoms is orders of magnitude faster than the rate of the substitutional Au atoms, and the small population of interstitial Au atoms therefore makes an important contribution to the overall solute-atom diffusion rate [6, 8]. The solute atoms transfer from substitutional sites to interstitial sites by either kick-out or dissociative mechanisms (Fig. 8.9). In the *kick-out mechanism*, an interstitial host atom, H_I, pushes the substitutional solute atom, S_S, into an interstitial position and simultaneously takes up a substitutional position according to the reaction

$$H_I + S_S = S_I + H_S \qquad (8.1)$$

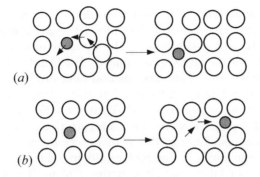

Figure 8.9:   Transfer of a solute atom (filled atom) from a substitutional site to an interstitial site by **(a)** the kick-out mechanism and **(b)** the dissociative mechanism.

In the *dissociative mechanism*, a substitutional solute atom enters an interstitial site, leaving a vacancy, V, behind according to the reaction

$$S_S = V_S + S_I \qquad (8.2)$$

These reactions are reversible. This dual-site occupancy leads to complicated solute diffusion behavior and has been described for several solute species in Si [4, 6, 8]. There is no compelling evidence that the ring mechanism in Fig. 8.1 contributes significantly to diffusion in any material.

8.2   ATOMIC MODELS FOR DIFFUSIVITIES

Atomic models for the diffusivity can be constructed when the diffusion occurs by a specified mechanism in various crystalline materials. A number of cases are considered below.

8.2.1   Metals

Diffusion of Solute Atoms by the Interstitial Mechanism in the B.C.C. Structure.   The general expression that connects the jump rate, Γ, the intersite jump distance, r, and the correlation factor, Eq. 7.52, then takes the form

$$D_I = \frac{\Gamma r^2}{6} \qquad (8.3)$$

Because each interstitial site has four nearest-neighbors, the jump rate, Γ, is given by $4\Gamma'$, where Γ' has the form of Eq. 7.25.[3] If a is the lattice constant for the b.c.c. unit cell in b.c.c. Fe, then $r = a/2$ and Eq. 8.3 yields

$$D_I = \frac{a^2}{6}\nu\, e^{S^m/k} e^{-H^m/(kT)}$$
$$= D_I^\circ e^{-E/(kT)} \qquad (8.4)$$

[3]The quantity Γ', introduced in Section 7.1.1, is the jump rate of an atom from one specified site to a specified neighboring site. Γ is the total jump rate of the atom in the material. If the atom is diffusing among equivalent sites in a crystal where each site has z equivalent nearest-neighbors, then $\Gamma = z\Gamma'$.

where the weakly temperature-dependent terms have been collected into D_I°. Because D_I° is relatively temperature independent, the Arrhenius form of Eq. 8.4 indicates a thermally activated process. The enthalpy of migration, H^m, is the activation energy, E, for the interstitial diffusion. For C in Fe, $D_I^\circ = 0.004 \text{ cm}^2\,\text{s}^{-1}$ and $H^m = 80.1 \text{ kJ mol}^{-1}$ [10]. This experimental value of D_I° is consistent with the value predicted by Eq. 8.4 for $a = 2.9 \times 10^{-10}$ m, $\nu = 10^{13}$ s^{-1}, and $S^m = 1k/$atom.

The relationship between jump rate and diffusivity in Eq. 8.3 can be obtained by an alternate method that considers the local concentration gradient and the number of site-pairs that can contribute to flux across a crystal plane. A concentration gradient of C along the y-axis in Fig. 8.8b results in a flux of C atoms from three distinguishable types of interstitial sites in the α plane (labeled 1, 2, and 3 in Fig. 8.8). The sites are assumed to be occupied at random with small relative populations of C atoms that can migrate between nearest-neighbor interstitial sites. If c' is the number of C atoms in the α plane per unit area, the carbon concentration on each type of site is $c'/3$. Carbon atoms on the types 1 and 3 sites jump from plane α to plane β at the rate $(c'/3)\Gamma'$. The jump rate from type-2 sites in plane α to plane β is zero. The contribution to the flux from all three site types is

$$J^{\alpha \to \beta} = \frac{2\Gamma' c'}{3} \tag{8.5}$$

If c is the number of C atoms per unit volume, $c = 2c'/a$, and therefore

$$J^{\alpha \to \beta} = \frac{a\Gamma' c}{3} \tag{8.6}$$

The reverse flux can be obtained by using a first-order expansion of the concentration in the β plane, so that

$$J^{\beta \to \alpha} = \frac{a\Gamma'}{3}\left(c + \frac{a}{2}\frac{\partial c}{\partial y}\right) \tag{8.7}$$

Therefore, the net flux is

$$J^{\text{net}} = J^{\alpha \to \beta} - J^{\beta \to \alpha} = -\frac{a^2\Gamma'}{6}\frac{\partial c}{\partial y} \tag{8.8}$$

Comparison of Eq. 8.8 with the Fick's law expression,

$$J^{\text{net}} = -D_I\frac{\partial c}{\partial y} \tag{8.9}$$

produces

$$D_I = \frac{a^2\Gamma'}{6} \tag{8.10}$$

The total jump frequency for a given C atom is $\Gamma = 4\Gamma'$, and therefore

$$D_I = \frac{a^2\Gamma}{24} = \frac{r^2\Gamma}{6} \tag{8.11}$$

which is identical to Eq. 8.3. The same result would have been obtained with the α and β planes chosen at any arbitrary inclination in the Fe crystal because D_I is isotropic in all cubic crystals (see Exercise 4.6).

Self-Diffusion by the Vacancy Mechanism in the F.C.C. Structure. Each site on an f.c.c. lattice has 12 nearest-neighbors, and if vacancies occupy sites randomly and have a jump frequency Γ_V,

$$\Gamma_V = 12\Gamma'_V \tag{8.12}$$

where Γ'_V is given by Eq. 7.25,

$$\Gamma'_V = \nu\, e^{S^m_V/k} e^{-H^m_V/(kT)} \tag{8.13}$$

If the fraction of sites randomly occupied by the vacancies is X_V, the jump rate of the host atoms must be

$$\Gamma_A = X_V\Gamma_V \tag{8.14}$$

Using Eq. 7.52, the self-diffusivity is

$$^\star D = \frac{\Gamma_A r^2 \mathbf{f}}{6} = X_V \Gamma'_V a^2 \mathbf{f} \tag{8.15}$$

where $r = a/\sqrt{2}$ is the nearest-neighbor jump distance in an f.c.c. crystal. The diffusion of the vacancies is uncorrelated[4] and the vacancy diffusivity is

$$D_V = \frac{\Gamma_V r^2}{6} = \Gamma'_V a^2 \tag{8.16}$$

which is related to the self-diffusivity by

$$^\star D = X_V D_V \mathbf{f} \tag{8.17}$$

If the vacancies are in thermal equilibrium, $X_V = X^{\mathrm{eq}}_V$, where, according to Eq. 3.65,

$$X^{\mathrm{eq}}_V = e^{-G^f_V/(kT)} = e^{S^f_V/k} e^{-H^f_V/(kT)} \tag{8.18}$$

where S^f_V and H^f_V are the vacancy vibrational entropy and enthalpy of formation, respectively.[5] Using Eqs. 8.13 and 8.15

$$\begin{aligned} ^\star D &= \mathbf{f} a^2 \nu\, e^{(S^m_V + S^f_V)/k} e^{-(H^m_V + H^f_V)/(kT)} \\ &= {}^\star D^\circ e^{-E/(kT)} \quad \text{where } {}^\star D^\circ \equiv \mathbf{f} a^2 \nu\, e^{(S^m_V + S^f_V)/k} \end{aligned} \tag{8.19}$$

and the activation energy is given by

$$E = H^m_V + H^f_V \tag{8.20}$$

Equation 8.19 contains the correlation factor, $\mathbf{f}$, which in this case is not unity since the self-diffusion of tracer atoms by the vacancy mechanism involves correlation. Correlation is present because the jumping sequence of each tracer atom produced by atom-vacancy exchanges is not a random walk. This may be seen by

---

[4]The diffusion of vacancies is uncorrelated for the same reasons given above for diffusion of the interstitial atoms. After each jump, a vacancy will have the possibility of jumping into any one of its 12 nearest-neighbor sites with equal probability.

[5]Because G^f_V is the free energy to form a vacancy exclusive of the configurational mixing entropy (see Section 3.4.1), the only entropy included in S^f_V in the relation $G^f_V = H^f_V - TS^f_V$ is the thermal vibrational entropy.

considering a tracer atom immediately after a jump. The vacancy with which it just exchanged will be one of its 12 nearest-neighbors. For its next jump, the tracer atom can either jump back into the vacancy with which it has just exchanged, jump into another vacancy which happens to be present in another nearest-neighbor site, or wait for another vacancy to arrive at a nearest-neighbor site into which it jumps. Because the first possibility is most probable, the atom jumping is nonrandom. The second jump is correlated with respect to the first because the first jump creates a situation (i.e., the existence of a nearest-neighbor vacancy) that biases the second jump.

A rough estimate for $\mathbf{f}$ can be obtained based on the number of nearest-neighbors and the probability that a tracer atom which has just jumped and vacated a site will return to the vacant site on the vacancy's next jump. A vacancy jumps randomly into its nearest-neighbor sites, and the probability that the return will occur is $1/z$. This event will then occur on average once during every z jumps of an atom. For each return jump, two atom jumps are effectively eliminated by cancellation, and the overall number of tracer-atom jumps that contribute to diffusion is reduced by the fraction $2/z$. According to Eq. 8.3, $^*D$ is proportional to the product $\Gamma\mathbf{f}$, and since the number of effective jumps is reduced by $2/z$, $\mathbf{f}$ can be assigned the value $\mathbf{f} \approx 1 - 2/z = 0.83$ for f.c.c. crystals. More accurate calculations (see below) show that $\mathbf{f} = 0.78$.

To find a more accurate value of $\mathbf{f}$, Eq. 7.49 is applied [4, 11–13]. If all displacements are of equal length,

$$\mathbf{f} = 1 + \frac{2}{N_\tau} \sum_{j=1}^{N_\tau - 1} \sum_{i=1}^{N_\tau - j} \langle \cos \theta_{i,i+j} \rangle \tag{8.21}$$

The quantity $\langle \cos \theta_{i,i+j} \rangle$ can be evaluated with the aid of the law of cosines from spherical trigonometry:

$$\langle \cos \theta_{i,i+2} \rangle = \langle \cos \theta_{i,i+1} \cos \theta_{i+1,i+2} \rangle + \langle \sin \theta_{i,i+1} \sin \theta_{i+1,i+2} \cos \alpha \rangle \tag{8.22}$$

where α is the angle between the two planes defined by the successive jump vectors $\vec{r}_i$ and $\vec{r}_{i+1}$ and the successive jump vectors $\vec{r}_{i+1}$ and $\vec{r}_{i+2}$. For cubic crystals, contributions from angle α will be canceled by those from angle $(180° - \alpha)$ on the average. Therefore, the last term in Eq. 8.22 containing $\cos \alpha$ will average to zero:

$$\langle \cos \theta_{i,i+2} \rangle = \langle \cos \theta_{i,i+1} \cos \theta_{i+1,i+2} \rangle \tag{8.23}$$

The average cosine of the angle between successive jumps must be the same for all pairs of successive jumps.[6] Therefore,

$$\langle \cos \theta_{i,i+2} \rangle = \langle \cos \theta_{i,i+1} \rangle \langle \cos \theta_{i+1,i+2} \rangle = \langle \cos \theta \rangle^2 \tag{8.24}$$

where $\cos \theta$ denotes the angle between successive jumps. By induction,

$$\langle \cos \theta_{i,i+3} \rangle = \langle \cos \theta_{i,i+2} \rangle \langle \cos \theta_{i+2,i+3} \rangle = \langle \cos \theta \rangle^3$$
$$\langle \cos \theta_{i,i+j} \rangle = \langle \cos \theta \rangle^j \tag{8.25}$$

---

[6]The immediate surroundings after each jump must be the same (excepting a change of orientation of the vacancy–atom pair), consisting of the atom with a nearest-neighbor vacancy next to it. The average of what happens to produce the next jump must be the same for all pairs of jumps (in their respective orientations).

Putting Eq. 8.25 into Eq. 8.21 yields

$$\mathbf{f} = 1 + \frac{2}{N_\tau} \sum_{j=1}^{N_\tau - 1} (N_\tau - j) \langle \cos\theta \rangle^j$$
$$= 1 + 2 \sum_{j=1}^{N_\tau - 1} \left(1 - \frac{j}{N_\tau} \right) \langle \cos\theta \rangle^j \tag{8.26}$$

As N_τ becomes large, the term j/N_τ in Eq. 8.26 can be neglected and the finite sum becomes an infinite sum,

$$\lim_{N_\tau \to \infty} = 1 + 2 \sum_{j=1}^{\infty} \langle \cos\theta \rangle^j \tag{8.27}$$

This infinite sum is known as

$$\sum_{j=1}^{\infty} x^j = \frac{x}{1-x} \tag{8.28}$$

and therefore

$$\mathbf{f} = \frac{1 + \langle \cos\theta \rangle}{1 - \langle \cos\theta \rangle} \tag{8.29}$$

Equation 8.29 is exact, and the accuracy of the determination of $\mathbf{f}$ then depends on the accuracy with which $\langle \cos\theta \rangle$ can be determined.

Equation 8.29 can be used to obtain another approximation for $\mathbf{f}$ by employing an estimated value of $\langle \cos\theta \rangle$. To estimate this quantity, consider an atom that has just exchanged with a vacancy; there is a probability $1/z$ that the vacancy's next exchange will be with the same atom and, therefore, the probability that $\cos\theta = -1$ is $1/z$ or $\langle \cos\theta \rangle = -1/z$. If the vacancy separates from the particular atom, that particular atom cannot migrate until it obtains another (or the same) vacancy as a neighbor; the contribution to $\langle \cos\theta \rangle$ from these displacements will be small compared to $-1/z$, and therefore Eq. 8.29 can be written

$$\mathbf{f} \approx \frac{1 - 1/z}{1 + 1/z} = \frac{z-1}{z+1} \tag{8.30}$$

and $\mathbf{f} = 0.85$ for f.c.c. crystals, which is close to the previous estimate.

An accurate determination of $\mathbf{f}$ can be obtained by considering all contributing vacancy trajectories to determine $\langle \cos\theta \rangle$ by use of Eq. 8.29 [13]. For f.c.c., the accurate value of $\mathbf{f}$ is found to be 0.78; thus, correlations affect the diffusivity value by about 22% in Eq. 7.52.[7] Correlations can have a considerably larger effect on the diffusivity for substitutional solute atoms by the vacancy mechanism.

For the vacancy self-diffusion mechanism in many metals, experimental values of $^\star D^\circ$ are approximately 0.1–1.0 cm^2 s^{-1}, which correspond to physically reasonable values of the quantities in $^\star D^\circ$ according to Eq. 8.19: $\mathbf{f} \approx 1$, $a \approx 3.5 \times 10^{-10}$ m, $\nu \approx 10^{13}$ s^{-1}, and $(S_V^f + S_V^m) \approx 2k$. In metals, as in many classes of materials, the

---

[7]A calculation of $\mathbf{f}$ in a two-dimensional lattice that takes into account multiple return vacancy trajectories appears in Exercise 8.8.

activation energy for vacancy self-diffusion, E in Eq. 8.20, scales with the melting temperature because the crystal binding energy correlates with both melting temperature and vacancy formation energy. Activation energies are typically 0.5–6 eV (1 eV $= 96.46$ kJ mol^{-1}) and $^{*}D^{\circ}$ is often nearly the same for materials with the same crystal structure and bonding [4].

Vacancy formation and migration energies, such as H_V^f and H_V^m, have been obtained by independent experiments. For example, to obtain H_V^f, equilibrium vacancy concentrations can be measured from simultaneous thermal expansion and lattice expansion during quasi-equilibrium heating [14] and by positron annihilation [15]. The vacancy migration rate can be determined by measuring the decay of a supersaturated population of quenched-in vacancies to their equilibrium population in order to measure H_V^m [16]. The results of these independent determinations are generally consistent with the measured values of the substitutional diffusivity activation energies inferred from Eq. 8.19 [4].

In Section 3.1.1, self-diffusion was analyzed by studying the diffusion of radioactive tracer atoms, which were isotopes of the inert host atoms, thereby eliminating any chemical differences. Possible effects of a small difference between the masses of the two species were not considered. However, this difference has been found to have a small effect, which is known as the *isotope effect*. Differences in atomic masses result in differences of atomic vibrational frequencies, and as a result, the heavier isotope generally diffuses more slowly than the lighter. This effect can—if migration is approximated as a single-particle process—be predicted from the mass differences and Eq. 7.14. If m_1 and m_2 are the atomic masses of two isotopes of the same component, Eqs. 7.13 and 7.52 predict the jump-rate ratio,

$$\frac{^{*}D(\text{mass 1})}{^{*}D(\text{mass 2})} = \frac{\Gamma_1}{\Gamma_2} = \frac{\Gamma_1'}{\Gamma_2'} = \frac{\nu_1\, e^{-G^m/(kT)}}{\nu_2\, e^{-G^m/(kT)}} = \frac{\nu_1}{\nu_2} = \frac{\sqrt{\beta/m_1}}{\sqrt{\beta/m_2}} = \sqrt{\frac{m_2}{m_1}} \qquad (8.31)$$

Jump rate and diffusivity scale inversely with the square root of atomic mass. However, if migration involves many-body effects and collective motion, the assumptions leading to Eq. 8.31 are no longer valid and this model must be discarded.

Diffusion of Solute Atoms by Vacancy Mechanism in Close-Packed Structure. Diffusion of substitutional solutes in dilute solution by the vacancy mechanism is more complex than self-diffusion because the vacancies may interact with the solute atoms and no longer be randomly distributed. If the vacancies are attracted to the solute atoms, any resulting association will strongly affect the solute-atom diffusivity.

The effect is demonstrated in a simple manner in two dimensions in Fig. 8.10, which shows an isolated solute atom with a vacancy occupying a nearest-neighbor site [4]. Three jump frequencies are considered: the intrinsic host–vacancy jump rate, $\Gamma'_{H \rightleftharpoons V}$; the solute–vacancy jump rate, $\Gamma'_{S \rightleftharpoons V}$; and the jump rate for a vacancy and a host atom that also has a neighboring solute, $\Gamma'_{V \circlearrowleft S}$. When there is an attractive interaction between a solute atom and the vacancy (a negative binding energy), $\Gamma'_{H \rightleftharpoons V}$ is decreased because of the increase in activation energy of the jump due to the binding energy between the vacancy and the solute atom. The activation energies for the remaining two types of exchange are not influenced by the binding energy, and two extremes can be considered:

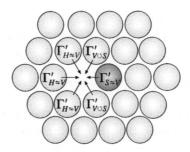

Figure 8.10:   A solute atom (darker shading) with a nearest-neighbor vacancy in a close-packed atomic plane. The vacancy and its three different nearest-neighbor types exchange places with differing jump frequencies.

- **Case A:** $\Gamma'_{S\leftrightarrows V} \gg \Gamma'_{V \circlearrowright S} \gg \Gamma'_{H\leftrightarrows V}$

- **Case B:** $\Gamma'_{V \circlearrowright S} \gg \Gamma'_{S\leftrightarrows V} \gg \Gamma'_{H\leftrightarrows V}$

Case A is an example of strong correlation. Since the jump rate $\Gamma'_{H\leftrightarrows V}$ is relatively small, the vacancy remains bound to the solute atom for relatively long periods and the solute atom and bound vacancy exchange positions repeatedly at the rate $\Gamma'_{S\leftrightarrows V}$, which is relatively high in comparison to $\Gamma'_{V \circlearrowright S}$. However, eventually the vacancy will exchange with a host atom that is a nearest-neighbor of the solute atom (at the rate $\Gamma'_{V \circlearrowright S}$) and a new mode of oscillation of the solute atom is established with the bound vacancy in a new nearest-neighbor site. This allows the solute atom to occupy a new site outside the first oscillating mode. If this occurs repeatedly, the solute atom can occupy new sites and execute long-range migration by a sort of tumbling motion of the oscillating mode's axis. The effective jump frequency of the solute atom during the period when the vacancy is bound to the solute atom is then $\Gamma'_{V \circlearrowright S}$ and the self-diffusivity of the solute atom during the time that the vacancy is bound to it can be written as

$$^\star D_S^b = \alpha_A \Gamma'_{V \circlearrowright S} \tag{8.32}$$

where α_A is a constant that includes various geometrical factors. An approximate expression for the diffusivity over a much longer period, including many $H\leftrightarrows V$ jumps, may now be obtained by using a simple nearest-neighbor model for the binding of a vacancy to a solute atom. According to Boltzmann statistics, the probability of finding a bound vacancy in a nearest-neighbor site to a solute atom at equilibrium is

$$p_V^{\text{eq}}(\text{bound}) = e^{-G_V^b/(kT)} X_V^{\text{eq}}(\text{free}) \tag{8.33}$$

where G_V^b is the binding energy (negative when attractive) of the vacancy to the solute atom and $X_V^{\text{eq}}(\text{free})$ is the fraction of free vacancies in the bulk crystal. The number of bound vacancies (per unit volume) in a system where the solute concentration is c_S is therefore $12 c_S p_V^{\text{eq}}(\text{bound})$, and since $12 p_V^{\text{eq}}(\text{bound}) \ll 1$, the probability of finding more than one vacancy bound to a solute atom at any time is very small. The fraction of solute atoms with a bound vacancy is then approximately $12 p_V^{\text{eq}}(\text{bound})$. Over a long period of time, the fraction of time that any solute atom has a vacancy bound to it is then also given by $12 p_V^{\text{eq}}(\text{bound})$. The effective jump rate of the solute during this long time period is therefore lower than the jump rate when a bound vacancy is present by the factor $12p_V^{\text{eq}}(\text{bound})$, and putting this result into Eq. 8.32 and using Eqs. 3.63 and 8.33, the solute diffusivity over a long period of time, including many $H \leftrightarrows V$ jumps, is

$$^\star D_S = 12\alpha_A \Gamma'_{V \circlearrowleft S} \, e^{-(G_V^f + G_V^b)/(kT)} \tag{8.34}$$

The self-diffusivity of solute atoms is then proportional to the rate at which bound vacancies circulate around them rather than the rate at which they exchange with vacancies.

In Case B, vacancies are again bound to solute atoms for relatively long periods of time. However, a bound vacancy will spend most of its time circling around a stationary solute atom by making a large number of $V \circlearrowleft S$ jumps, although it will occasionally make an $S \leftrightarrows V$ jump, allowing the solute atom to occupy a new site. Repetition of this process leads to long-range migration of the solute atom. Using an analysis similar to the above, the solute self-diffusivity is then

$$^\star D_S = 12\alpha_B \Gamma'_{S \leftrightarrows V} \, e^{-(G_V^f + G_V^b)/(kT)} \tag{8.35}$$

In contrast to Case A, the solute self-diffusivity is now proportional to the rate at which solute atoms exchange with vacancies.

If the binding energy was negligible and all the frequencies in Fig. 8.10 were equal, the solute atom self-diffusivity would be the same as that of the host atoms.

Additional features of solute-atom diffusion can be studied using this three-frequency model. These, as well as models in three dimensions, are described by Shewmon [4].

Diffusion of Self-Interstitial Imperfections by the Interstitialcy Mechanism in the F.C.C. Structure.   For f.c.c. copper, self-interstitials have the $\langle 100 \rangle$ split-dumbbell configuration shown in Fig. 8.5d and migrate by the interstitialcy mechanism illustrated in Fig. 8.6. The jumping is uncorrelated,[8] ($\mathbf{f} = 1$), and $a/\sqrt{2}$ is the nearest-neighbor distance, so

$$
\begin{aligned}
D_I &= \frac{\Gamma r^2}{6} = \frac{\Gamma a^2}{12} \\
\Gamma &= 8\Gamma' = 8\nu \, e^{S_I^m/k} e^{-H_I^m/(kT)} \\
D_I &= \frac{2}{3}a^2\nu \, e^{S_I^m/k} e^{-H_I^m/(kT)}
\end{aligned}
\tag{8.36}
$$

These defects will always be present at thermal equilibrium, but their concentrations will be very small because of their high energy of formation. They can also be created by nonequilibrium processes such as irradiation [3].

Self-Diffusion by the Interstitialcy Mechanism.   If their formation energy is not too large, the equilibrium population of self-interstitials may be large enough to contribute to the self-diffusivity. In this case, the self-diffusivity is similar to that for self-diffusion via the vacancy mechanism (Eq. 8.19) with the vacancy formation and migration energies replaced by corresponding self-interstitial quantities. The

---

[8]After each jump the $\langle 100 \rangle$ dumbbell has an equal probability of making any of eight different jumps. Its next jump is therefore made at random.

correlation factor **f**—similar to self-diffusion by the vacancy mechanism—is less than unity because the atom jumps produced by the interstitialcy mechanism are correlated.

For example, the $\langle 100 \rangle$ split-dumbbell configuration of the self-interstitial defect in Cu has a formation energy that is considerably larger than the vacancy formation energy [3]. However, the relatively small population of equilibrium self-interstitials may contribute significantly to the self-diffusivity because the activation energy for interstitial migration is considerably lower than that for vacancy migration (as described in Section 8.1.3).

8.2.2   Ionic Solids

Diffusion in ionically bonded solids is more complicated than in metals because site defects are generally electrically charged. Electric neutrality requires that point defects form as neutral complexes of charged site defects. Therefore, diffusion always involves more than one charged species.[9] The point-defect population depends sensitively on stoichiometry; for example, the high-temperature oxide semiconductors have diffusivities and conductivities that are strongly regulated by the stoichiometry. The introduction of extrinsic aliovalent solute atoms can be used to fix the low-temperature population of point defects.

Intrinsic Crystal Self-Diffusion.   A simple example of intrinsic self-diffusion in an ionic material is pure stoichiometric KCl, illustrated in Fig. 8.11a. As in many alkali halides, the predominant point defects are cation and anion vacancy complexes (Schottky defects), and therefore self-diffusion takes place by a vacancy mechanism. For stoichiometric KCl, the anion and cation vacancies are created in equal numbers because of the electroneutrality condition. These vacancies can be created

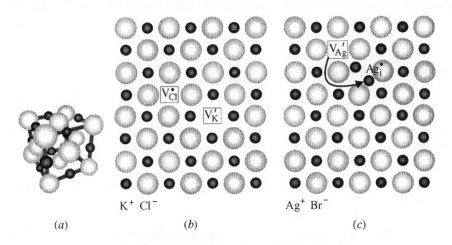

K$^+$ Cl$^-$ Ag$^+$ Br$^-$

(a)   (b)   (c)

Figure 8.11:   (a) Rocksalt structure of KCl and AgBr with (100) planes delineated. (b) Schottky defect on a (100) plane in KCl composed of anion vacancy and cation vacancy. (c) Frenkel defect on a (100) plane in AgBr composed of cation self-interstitial and cation vacancy.

[9]For general discussions, see Kingery et al. [17] or Chiang et al. [18].

by removing K^+ and Cl^- ions from the bulk and placing them at an interface or dislocation, or on a surface ledge as illustrated in Fig. 8.12. A vacancy on a K^+ cation site will have an effective negative electronic charge and a vacancy on a Cl^- site will have a corresponding effective positive charge. This defect creation is a reaction written in Kröger–Vink notation as

$$K_K^\times + Cl_{Cl}^\times = V_K' + V_{Cl}^\bullet + K_K^\times + Cl_{Cl}^\times$$
$$\text{or,} \tag{8.37}$$
$$\text{null} = V_K' + V_{Cl}^\bullet$$

In the Kröger–Vink notation used here, the subscript indicates the type of site the species occupies and the superscript indicates the excess effective charge associated with the species in that site [17]. A positive unit of charge (equal in magnitude to the electron charge) is indicated by a dot ($\bullet$) superscript, a corresponding negative charge by a prime ($'$) superscript, and zero charge (a neutral situation) by a times ($\times$) superscript.

The equilibrium constant, K^{eq}, for Eq. 8.37 is related to the free energy of formation, G_S^f, of the Schottky pair

$$G_S^f = -kT \ln K^{eq} \tag{8.38}$$

or

$$K^{eq} = e^{-G_S^f/(kT)} = a_{AV}\, a_{CV} \tag{8.39}$$

where the a's are the activities of the anion and cation vacancies. For dilute concentrations of the vacancies, activities are equal to their site fractions by Raoult's law,

$$[V_K']\, [V_{Cl}^\bullet] = K^{eq} = e^{-G_S^f/(kT)} \tag{8.40}$$

where the square brackets indicate a site fraction. Equation 8.40 is a general mass-action law for the combined anion and cation vacancy site fractions. Furthermore, electrical neutrality requires that

$$[V_K'] = [V_{Cl}^\bullet] \tag{8.41}$$

Combining Eqs. 8.40 and 8.41,

$$[V_K'] = [V_{Cl}^\bullet] = e^{-G_S^f/(2kT)} \tag{8.42}$$

The vacancy populations enter the expressions for the self-diffusivity of the K^+ cations and Cl^- anions. Starting with Eq. 7.52 and using the method that led to Eq. 8.19 for vacancy self-diffusion in a metal,

$$^\star D^K = ga^2 \mathbf{f} \nu\, e^{(S_{CV}^m + S_S^f/2)/k}\, e^{-(H_{CV}^m + H_S^f/2)/(kT)} \tag{8.43}$$

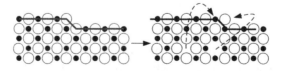

Figure 8.12:   Creation of Schottky defect by transfer of anion and cation from regular crystal sites to ledges at the surface.

where g is a geometrical factor and the correlation factor, $\mathbf{f}$, has a value slightly less than unity. The activation energy for the self-diffusion is therefore

$$E = H_{CV}^m + H_S^f/2 \qquad (8.44)$$

A similar expression applies to Cl self-diffusion on the anion sublattice.

Self-diffusion of Ag cations in the silver halides involves Frenkel defects (equal numbers of vacancies and interstitials as seen in Fig. 8.11b). In a manner similar to the Schottky defects, their equilibrium population density appears in the diffusivity. Both types of sites in the Frenkel complex—vacancy and interstitial—may contribute to the diffusion. However, for AgBr, experimental data indicate that cation diffusion by the interstitialcy mechanism is dominant [4]. The cation Frenkel pair formation reaction is

$$Ag_{Ag}^\times = Ag_i^\bullet + V_{Ag}' \qquad (8.45)$$

The activity of $Ag_{Ag}^\times$ is unity and, therefore,

$$K^{eq} = [Ag_i^\bullet]\,[V_{Ag}'] = e^{-G_F^f/(kT)} \qquad (8.46)$$

The electrical neutrality condition constrains the two site fractions:

$$[Ag_i^\bullet] = [V_{Ag}'] = e^{-G_F^f/(2kT)} \qquad (8.47)$$

The activation energy for self-diffusivity of the Ag cations by the interstitialcy mechanisms is the sum of one-half the Frenkel defect formation enthalpy and the activation enthalpy for migration,

$$E = H_I^m + \frac{H_F^f}{2} \qquad (8.48)$$

Extrinsic Crystal Self-Diffusion. Charged point defects can be induced to form in an ionic solid by the addition of substitutional cations or anions with charges that differ from those in the host crystal. Electrical neutrality demands that each addition results in the formation of defects of opposite charge that can contribute to the diffusivity or electronic conductivity. The addition of aliovalent solute (impurity) atoms to an initially pure ionic solid therefore creates *extrinsic* defects.[10]

For example, the self-diffusivity of K in KCl depends on the population of both extrinsic and intrinsic cation-site vacancies. Extrinsic cation-site vacancies can be created by incorporation of Ca^{++} by doping KCl with $CaCl_2$ and can be considered a two-step process. First, two cation vacancies and two anion vacancies form as illustrated in Fig. 8.12.[11] Second, the single Ca^{++} cation and two Cl anions from $CaCl_2$ are inserted into the cation and anion vacancies, respectively; electric neutrality requires that each substitutional divalent cation impurity in KCl be balanced

---

[10] *Extrinsic* has the same meaning as in doped semiconductors.

[11] This process involves *creation* of additional sites in the crystal. Cation and anion sites must be created in the same proportion as the ratio of cation to anion sites in the host crystal—in this case, 1:1. These defects can also be formed at point-defect sources such as dislocations and grain boundaries (see Sections 11.4 and 13.4).

by the formation of a cation vacancy. The cationic and anionic vacancy populations are related to the site fraction of the extrinsic Ca^{++} impurity,

$$[Ca_K^\bullet] + [V_{Cl}^\bullet] = [V_K'] \tag{8.49}$$

The mass-action relationship in Eq. 8.40 for the product of the cation and anion vacancy site fractions combined with Eq. 8.49 yields

$$[V_K'] \left([V_K'] - [Ca_K^\bullet]\right) = e^{-G_S^f/(kT)} = [V_K']_{pure}^2 \tag{8.50}$$

The last term on the right-hand side of Eq. 8.50 is the square of the cation vacancy site fraction in pure (intrinsic) KCl. Solving the quadratic equation for the cation vacancy site fraction yields

$$[V_K'] = \frac{[Ca_K^\bullet]}{2} \left(1 + \sqrt{1 + \frac{4\,[V_K']_{pure}^2}{[Ca_K^\bullet]^2}}\right) \tag{8.51}$$

There are two limiting cases for the behavior of $[V_K']$ according to Eq. 8.51:

- **Intrinsic:** $[V_K']_{pure} \gg [Ca_K^\bullet]$, then $[V_K'] = [V_K']_{pure}$

- **Extrinsic:** $[V_K']_{pure} \ll [Ca_K^\bullet]$, then $[V_K'] = [Ca_K^\bullet]$

The intrinsic case applies at small doping levels or at high temperatures where the thermal equilibrium site fraction of the intrinsic cation vacancy population exceeds that due to the aliovalent solute atoms. In this case, the effect of the added solute atoms is negligible. The activation energy for cation self-diffusion is therefore the same as in the pure material and is given by Eq. 8.44.

The extrinsic case applies at low temperatures or large doping levels. The site fraction of cation vacancies is equal to the solute-atom site-fraction and is therefore temperature independent. In the extrinsic regime, no thermal defect formation is necessary for cation self-diffusion and the activation energy consists only of the activation energy for cation vacancy migration.

The expected Arrhenius plot for cation self-diffusion in KCl doped with Ca^{++} is shown in Fig. 8.13. The two-part curve reflects the intrinsic behavior at high temperatures and extrinsic behavior at low temperatures.

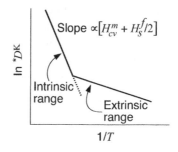

Figure 8.13:   Arrhenius plot for self-diffusivity on the cation sublattice, $^*D^K$, in KCl doped with Ca^{++}. The intrinsic and extrinsic ranges have different activation energies.

Crystal Self-Diffusion in Nonstoichiometric Materials. Nonstoichiometry of semiconductor oxides can be induced by the material's environment. For example, materials such as FeO (illustrated in Fig. 8.14), NiO, and CoO can be made metal-deficient (or O-rich) in oxidizing environments and TiO_2 and ZrO_2 can be made O-deficient under reducing conditions. These induced stoichiometric variations cause large changes in point-defect concentrations and therefore affect diffusivities and electrical conductivities.

In pure FeO, the point defects are primarily Schottky defects that satisfy mass-action and equilibrium relationships similar to those given in Eqs. 8.39 and 8.42. When FeO is oxidized through the reaction

$$FeO + \frac{x}{2}O_2 = FeO_{1+x} \tag{8.52}$$

each O atom takes two electrons from two Fe^{++} ions, as illustrated in Fig. 8.14a, according to the reactions

$$2Fe^{++} = 2Fe^{+++} + 2e^-$$
$$\frac{1}{2}O_2 + 2e^- = O^{--} \tag{8.53}$$

corresponding to the combined reaction,

$$2Fe^{++} + \frac{1}{2}O_2 = 2Fe^{+++} + O^{--} \tag{8.54}$$

Electrical neutrality requires that a cation vacancy be created for every O atom added, as in Fig. 8.14b; this, combined with site conservation, becomes

$$2Fe_{Fe}^{\times} + \frac{1}{2}O_2 = 2Fe_{Fe}^{\bullet} + O_O^{\times} + V_{Fe}'' \tag{8.55}$$

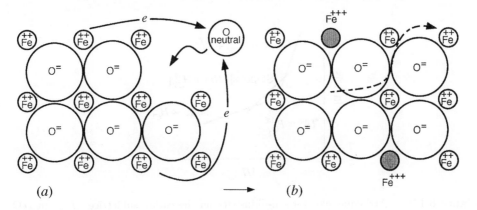

Figure 8.14: Addition of a neutral O atom to FeO to produce O-rich (metal-deficient) oxide. **(a)** An O atom receives two electrons from Fe^{++} ions in the bulk material. **(b)** The final structure contains defects in the form of two Fe^{+++} ions and a cation (Fe^{++}) vacancy.

which can be written in terms of holes, h, in the valence band created by the loss of an electron from an Fe^{++} ion producing an Fe^{+++} ion,

$$\frac{1}{2}O_2 = O_O^\times + V_{Fe}'' + 2h_{Fe}^\bullet$$

$$h_{Fe}^\bullet \equiv Fe_{Fe}^\bullet - Fe_{Fe}^\times \tag{8.56}$$

Equation 8.56 predicts a relationship between the cation vacancy site fraction and the oxygen gas pressure. The equilibrium constant for this reaction is important for oxygen-sensing materials:

$$K^{eq} = \frac{[V_{Fe}''][h_{Fe}^\bullet]^2}{P_{O_2}^{1/2}} = e^{-\Delta G/(kT)} \tag{8.57}$$

For the regime in which the dominant charged defects are the oxidation-induced cation vacancies and their associated holes, the electrical neutrality condition is

$$[h_{Fe}^\bullet] = 2[V_{Fe}''] \tag{8.58}$$

Therefore, inserting Eq. 8.58 into Eq. 8.57 and solving for $[V_{Fe}'']$ yields

$$[V_{Fe}''] = \left(\frac{1}{4}\right)^{1/3} e^{-\Delta G/(3kT)} (P_{O_2})^{1/6} \tag{8.59}$$

The cation self-diffusivity due to the vacancy mechanism varies as the one-sixth power of the oxygen pressure at constant temperature and the activation energy is

$$E = \frac{\Delta H}{3} + H_{CV}^m \tag{8.60}$$

The dominance of oxidation-induced vacancies creates an additional behavior regime. The effect of this additional regime on diffusivity behavior is illustrated in Fig. 8.15. Other types of environmental effects create defects through other mechanisms and may lead to other behavior regimes.

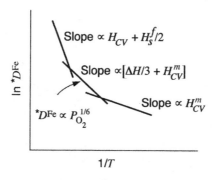

Figure 8.15: Arrhenius plot for self-diffusivity on the cation sublattice, $^*D^{Fe}$, in FeO made O-rich by exposure to oxygen gas at a pressure P_{O_2} or doped with an aliovalent impurity. Three regimes of behavior are possible, each with a different activation energy.

8.3 DIFFUSIONAL ANELASTICITY (INTERNAL FRICTION)

In this section, pedagogical models for the time dependence of mechanical response are developed. Elastic stress and strain are rank-two tensors, and the compliance (or stiffness) are rank-four material property tensors that connect them. In this section, a simple spring and dashpot analog is used to model the mechanical response of anelastic materials. Scalar forces in the spring and dashpot model become analogs for a more complex stress tensor in materials. To enforce this analogy, we use the terms *stress* and *strain* below, but we do not treat them as tensors.

For an ideally elastic material, the stress is linearly related to the strain by

$$\sigma = C\varepsilon \tag{8.61}$$

(where the constant C represents the elastic stiffness), and conversely, the strain is linearly related to the stress by

$$\varepsilon = S\sigma \tag{8.62}$$

(where the constant $S = 1/C$ represents the compliance). For each level of stress, such a material responds immediately with a unique value of the strain. However, in many real materials, stress-induced diffusional processes cause additional time-dependent *anelastic strains* and nonlinear behavior. This anelastic behavior degrades the mechanical work performed by the stresses into heat so that the material exhibits internal friction, which can damp out mechanical oscillations in a material.

Anelasticity therefore affects the mechanical properties of materials. As seen below, its study yields unique information about a number of kinetic processes in materials, such as diffusion coefficients, especially at relatively low temperatures.

8.3.1 Anelasticity due to Reorientation of Anisotropic Point Defects

Anelastic behavior can be produced by the stress-induced diffusional jumping of anisotropic point defects. An example of such a process is described in Exercise 8.5, in which an f.c.c. metal contains a concentration of self-interstitial point defects having the $\langle 100 \rangle$ split-dumbbell configuration (see Fig. 8.5d). Each defect produces a tetragonal distortion of the crystal, elongating it preferentially along its dumbbell axis. The three types of sites in the crystal in which the interstitials can lie with their axes along [100], [010], or [001] exist in equal numbers and will be occupied equally in the absence of any stress. However, if the crystal is suddenly stressed uniaxially along [100], an excess of dumbbells will jump to sites where they are aligned along [100], because the crystal is elongated along the direction of the applied stress and the applied stress performs work. This principle applies to loading along the other cube directions as well. (Note that this is a good example of LeChatelier's principle.) When the stress is released suddenly, the defects repopulate the sites in equal numbers and the crystal regains its original shape. The relaxation time for this re-population is

$$\tau = \frac{2}{3\Gamma} \tag{8.63}$$

where Γ is the total jump frequency of a dumbbell (see Exercise 8.5). This process therefore causes the crystal to elongate or to contract in response to the applied stress at a rate dependent upon the rate at which the dumbbells jump between the different types of sites.

The overall response of the crystal to such a stress cycle is shown in Fig. 8.16. When the stress $\sigma_\circ$ is applied suddenly, the crystal instantaneously undergoes an ideally elastic strain following Eq. 8.62. As the stress is maintained, the crystal undergoes further time-dependent strain due to the re-population of the interstitials. When the stress is released, the ideally elastic strain is recovered instantaneously and the remaining anelastic strain will be recovered in a time-dependent fashion as the interstitials regain their random distribution.

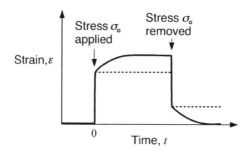

Figure 8.16:   Strain vs. time for an anelastic solid during a stress cycle in which stress is applied suddenly at $t = 0$, held constant for a period of time, and then suddenly removed.

General Formulation of Anelastic Behavior.   Anelastic behavior where the strain is a function of both stress and time may be described by generalizing Eq. 8.62 and expressing the compliance in the more general form

$$S(t) = \frac{\varepsilon(t)}{\sigma_\circ} \quad \text{for } t \geq 0 \tag{8.64}$$

The initial value of the compliance, corresponding to

$$S(0) = S_U \tag{8.65}$$

is the *unrelaxed compliance*, which corresponds to ideal elastic behavior because there is no time for point-defect re-population.   The value of $S(t)$ at long times, corresponding to

$$S(\infty) = S_R \tag{8.66}$$

is the *relaxed compliance*, since it includes the maximum possible additional strain due to the stress-induced re-population of the defects. Clearly, $S_R > S_U$.

Suppose now that the crystal is subjected to a periodic applied stress of amplitude $\sigma_\circ$ corresponding to

$$\sigma = \sigma_\circ e^{i\omega t} \tag{8.67}$$

The resulting strain is also periodic with the same angular frequency but generally lags behind the stress because time is required for the growth (or decay) of the anelastic strain contributed by the point-defect re-population during each cycle. The strain may therefore be written

$$\varepsilon = \varepsilon_\circ e^{i(wt-\phi)} \tag{8.68}$$

where ϕ is the phase angle by which the strain lags behind the stress. Note that $\phi = 0$ at both very high and very low frequencies. At very high frequencies, the cycling is so rapid that the point defects have insufficient time to repopulate and therefore make no contribution to the strain. At very low frequencies, there is sufficient time for the defects to re-populate (relax) at every value of the stress, and the stress and strain are therefore again in phase. To proceed with the more general intermediate case, it is convenient to write the expression for the strain,

$$\varepsilon = \varepsilon_1 e^{i\omega t} - i\varepsilon_2 e^{i\omega t} \tag{8.69}$$

In this formulation, the first term on the right-hand side is the component of ε that is in phase with the stress, and the second term is the component that lags behind the stress by 90°. Also,

$$\frac{\varepsilon_2}{\varepsilon_1} = \tan \phi \tag{8.70}$$

The compliance (again the ratio of strain over stress) is then

$$S(\omega) = \frac{(\varepsilon_1 - i\varepsilon_2) e^{i\omega t}}{\sigma_\circ e^{i\omega t}} = \frac{\varepsilon_1}{\sigma_\circ} - i\frac{\varepsilon_2}{\sigma_\circ} \tag{8.71}$$

Because the strain lags behind the stress, the stress-strain curve for each cycle consists of a hysteresis loop, as in Fig. 8.17, and an amount of mechanical work, given by the area enclosed by the hysteresis loop,

$$\Delta W = \oint \sigma d\varepsilon \tag{8.72}$$

will be dissipated (converted to heat) during each cycle. To determine ΔW, only the part of the strain that is out of phase with the stress must be considered. The stress and strain in Eq. 8.72 can then be represented by

$$\sigma = \sigma_\circ \cos \omega t \quad \text{and} \quad \varepsilon = \varepsilon_2 \cos(\omega t - \pi/2) \tag{8.73}$$

and

$$\Delta W = -\sigma_\circ \varepsilon_2 \int_0^{2\omega/\pi} \cos \omega t \, \sin\left(\omega t - \frac{\pi}{2}\right) \omega \, dt = \pi \sigma_\circ \varepsilon_2 \tag{8.74}$$

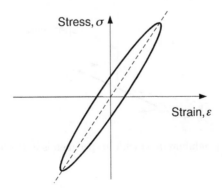

Figure 8.17: Hysteresis loop shown by the stress-strain curve of an anelastic solid subjected to an oscillating stress.

The energy dissipated can be compared with the maximum elastic strain energy, W, which is stored in the material during the stress cycle. Because the elastic strain is proportional to the applied stress, W is equal to just half of the product of the maximum stress and strain (i.e., $W = \sigma_o \varepsilon_1 / 2$), and therefore

$$\frac{\Delta W}{W} = 2\pi \frac{\varepsilon_2}{\varepsilon_1} \tag{8.75}$$

$\Delta W / W$ can be measured with a torsion pendulum, in which a specimen in the form of a wire containing the point defects is made the active element and strained periodically in torsion as in Fig. 8.18. If the pendulum is put into free torsional oscillation, its amplitude will slowly decay (damp out), due to the dissipation of energy. As shown in Exercise 8.20, the maximum potential energy (the elastic energy, W) stored in the pendulum is proportional to the square of the amplitude of its oscillation, A. The amplitude of the oscillations therefore decreases according to

$$\ln\left(\frac{A_N}{A_{N+1}}\right) = \ln\left(\frac{W_N}{W_{N+1}}\right)^{1/2} = \frac{1}{2}\ln\left(\frac{W_N}{W_{N+1}}\right) = \frac{1}{2}\ln\left(W_N W_N - \Delta W\right) = \frac{1}{2}\frac{\Delta W}{W} \tag{8.76}$$

where N is the number of the oscillation and it is realistically assumed that $\Delta W \ll W$. The logarithmic decay of the amplitude is the *logarithmic decrement*, designated by δ. Therefore,

$$\delta = \ln\frac{A_N}{A_{N+1}} = \frac{1}{2}\frac{\Delta W}{W} = \pi\frac{\varepsilon_2}{\varepsilon_1} = \pi\tan\phi \tag{8.77}$$

Measurements of δ yield direct information about the magnitude of the energy dissipation and the phase angle. ϕ measures the fractional energy loss per cycle due to the anelasticity and is often termed the *internal friction*. According to the discussion above, δ will be a function of the frequency, ω; should approach zero at both low and high frequencies; and will have a maximum at some intermediate frequency. The maximum occurs at a frequency that is the reciprocal of the relaxation time for the re-population of the point defects.

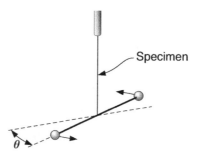

Figure 8.18:   Torsion pendulum in which the specimen is in the form of a wire subjected to an oscillating stress.

Analog Model for Standard Anelastic Solid.   To find the dependence of δ on frequency, a model that relates the stress and strain and their time derivatives must be constructed. Figure 8.19's analog model for a standard anelastic solid serves this purpose; it consists of two linear springs, $S1$ and $S2$, and a dashpot, D, which is a plunger immersed in a viscous fluid. The dashpot changes length at a rate proportional to the force exerted on it. This model gives a good account of the anelastic behavior illustrated in Fig. 8.16. When a force, F, is first applied, $S2$ elongates instantaneously. At the same time, in the upper section of the model, F is fully supported by D and the force on $S1$ is zero. However, with increasing time, D extends and the force is gradually transferred to $S1$, which extends under its influence. Eventually, the force is fully transferred, as both $S1$ and $S2$ experience the full force while D experiences nothing. At this point, the model reaches its fullest extension. The extension remains constant until F is suddenly removed. $S2$ then contracts instantaneously and $S1$ gradually relaxes by forcing D back to its original extension and the model recovers its original state. $S2$ therefore accounts for the ideal elasticity of the solid, and the combination of $S1$ and D accounts for the anelasticity.

The linear spring element $S1$ will undergo an extension Δx_{S1} according to

$$\Delta x_{S1} = a_{S1} F_{S1} \tag{8.78}$$

where F_{S1} is the force on $S1$ and a_{S1} is a constant. Similarly for $S2$,

$$\Delta x_{S2} = a_{S2} F_{S2} \tag{8.79}$$

Also, for the dashpot,

$$\frac{d(\Delta x_D)}{dt} = a_D F_D \tag{8.80}$$

where Δx_D is the extension of D, F_D is the force on D, and a_D is a constant. In addition,

$$\Delta x_{S1} = \Delta x_D \tag{8.81}$$

$$F = F_{S2} \tag{8.82}$$

and

$$F_{S2} = F_{S1} + F_D \tag{8.83}$$

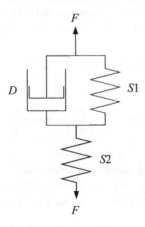

Figure 8.19: Analog model for a standard anelastic solid.

Finally, the stress, σ, and strain, ε, may be expressed

$$\varepsilon = a_\varepsilon(\Delta x_{S1} + \Delta x_{S2}) \tag{8.84}$$

and

$$\sigma = \frac{1}{a_\sigma}F \tag{8.85}$$

where a_ε and a_σ are constants. By combining Eqs. 8.78–8.85, the following equation, which contains three independent constants (bracketed) corresponding to the three elements in the model, can be obtained:

$$[a_\varepsilon a_\sigma a_{S1} + a_\varepsilon a_\sigma a_{S2}]\sigma + \left[\frac{a_{S1}a_{S2}a_\sigma a_\varepsilon}{a_D}\right]\frac{d\sigma}{dt} = \varepsilon + \left[\frac{a_{S1}}{a_D}\right]\frac{d\varepsilon}{dt} \tag{8.86}$$

Equation 8.86 may be solved for the time period in Fig. 8.16 during which the stress is held constant at $\sigma_\circ$. Under this condition, it reduces to

$$[a_\varepsilon a_\sigma a_{S1} + a_\varepsilon a_\sigma a_{S2}]\sigma_\circ = \varepsilon + \left[\frac{a_{S1}}{a_D}\right]\frac{d\varepsilon}{dt} \tag{8.87}$$

Equation 8.87's general solution can be written

$$\varepsilon(t) = [a_\varepsilon a_\sigma a_{S1} + a_\varepsilon a_\sigma a_{S2}]\sigma_\circ - A\,e^{-a_D t/a_{S1}} \tag{8.88}$$

The constant of integration, A, can be evaluated by recalling that at $t = 0$ only $S2$ is extended. The strain is then

$$\varepsilon(0) = a_\varepsilon \Delta x_{S2} = a_\varepsilon a_{S2}F_{S2} = a_\varepsilon a_{S2}F = a_\varepsilon a_{S2}a_\sigma \sigma_\circ \tag{8.89}$$

and therefore from Eq. 8.88, $A = a_\varepsilon a_\sigma a_{S1}$ and

$$\varepsilon(t) = a_\varepsilon a_\sigma a_{S2}\sigma_\circ + a_\varepsilon a_\sigma a_{S1}\sigma_\circ \left(1 - e^{-a_D t/a_{S1}}\right) \tag{8.90}$$

Examining the forms of $\varepsilon(0)$ and $\varepsilon(\infty)$ and comparing the results with Eqs. 8.64–8.66 shows that $a_\varepsilon a_\sigma a_{S2} = S_U$ and $a_\varepsilon a_\sigma a_{S1} = S_R - S_U$. Also, the anelastic relaxation occurs exponentially, in agreement with the results in Exercise 8.5, and the relaxation time corresponds to $\tau = a_{S1}/a_D$. Equation 8.90 then takes the simpler form

$$\varepsilon(t) = S_U\sigma_\circ + (S_R - S_U)\sigma_\circ(1 - e^{-t/\tau}) \tag{8.91}$$

and Eq. 8.86 takes the form

$$S_R\sigma + \tau S_U\frac{d\sigma}{dt} = \varepsilon + \tau\frac{d\varepsilon}{dt} \tag{8.92}$$

Frequency Dependence of the Logarithmic Decrement.   The frequency dependence of δ can now be found. Putting Eqs. 8.67 and 8.69 into Eq. 8.92 and equating the real and imaginary parts yields two equations which can be solved for ε_1 and ε_2 in the forms

$$\varepsilon_1 = \sigma_\circ \left(S_U + \frac{S_R - S_U}{1 + \omega^2\tau^2}\right) \tag{8.93}$$

$$\varepsilon_2 = \sigma_\circ \left[\frac{(S_R - S_U)\omega\tau}{1 + \omega^2\tau^2} \right] \qquad (8.94)$$

Therefore,

$$\delta(\omega) = \pi\frac{\varepsilon_2}{\varepsilon_1} = \frac{\pi(S_R - S_U)\omega\tau}{S_R + S_U\omega^2\tau^2} \qquad (8.95)$$

Because $(S_R - S_U) \ll S_U$ in the majority of cases,

$$\delta(\omega) = \pi\frac{S_R - S_U}{S_U}\frac{\omega\tau}{1 + \omega^2\tau^2} \qquad (8.96)$$

The decrement $\delta(\omega)$ forms a *Debye peak*, as shown in Fig. 8.20.

The maximum damping (anelasticity) occurs when the applied angular frequency is tuned to the relaxation time of the anelastic process so that $\omega\tau = 1$. Also, $\delta(\omega)$ approaches zero at both high and low frequencies, as anticipated.

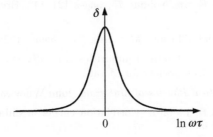

Figure 8.20:   Curve of the decrement, $\delta(\omega)$, according to Eq. 8.96, vs. $\ln \omega\tau$. The curve exhibits a Debye peak at $\ln \omega\tau = 0$ (or $\omega = 1/\tau$).

8.3.2   Determination of Diffusivities

The preceding analysis provides a powerful method for determining the diffusivities of species that produce an anelastic relaxation, such as the split-dumbbell interstitial point defects. A torsional pendulum can be used to find the frequency, ω_p, corresponding to the Debye peak. The relaxation time is then calculated using the relation $\tau = 1/\omega_p$, and the diffusivity is obtained from the known relationships among the relaxation time, the jump frequency, and the diffusivity. For the split-dumbbell interstitials, the relaxation time is related to the jump frequency by Eq. 8.63, and the expression for the diffusivity (i.e., $D = \Gamma a^2/12$), is derived in Exercise 8.6. Therefore, $D = a^2/18\tau$. This method has been used to determine the diffusivities of a wide variety of interstitial species, particularly at low temperatures, where the jump frequency is low but still measurable through use of a torsion pendulum. A particularly important example is the determination of the diffusivity of C in b.c.c. Fe, which is taken up in Exercise 8.22.

Bibliography

1. S.M. Allen and E.L. Thomas. *The Structure of Materials*. John Wiley & Sons, New York, 1999.

2. R.A. Johnson. Empirical potentials and their use in calculation of energies of point-defects in metals. *J. Phys. F*, 3(2):295–321, 1973.

3. W. Schilling. Self-interstitial atoms in metals. *J. Nucl. Mats.*, 69–70(1–2):465–489, 1978.

4. P. Shewmon. *Diffusion in Solids*. The Minerals, Metals and Materials Society, Warrendale, PA, 1989.

5. G. Neumann. Diffusion mechanisms in metals. In G.E. Murch and D.J. Fischer, editors, *Defect and Diffusion Forum*, volume 66–69, pages 43–64, Brookfield, VT, 1990. Sci-Tech Publications.

6. W. Frank, U. Gösele, H. Mehrer, and A. Seeger. Diffusion in silicon and germanium. In G.E. Murch and A.S. Nowick, editors, *Diffusion in Crystalline Solids*, pages 63–142, Orlando, Florida, 1984. Academic Press.

7. T.Y. Tan and U. Gösele. Point-defects, diffusion processes, and swirl defect formation in silicon. *Appl. Phys. A*, 37(1):1–17, 1985.

8. W. Frank. The interplay of solute and self-diffusion—A key for revealing diffusion mechanisms in silicon and germanium. In D. Gupta, H. Jain, and R.W. Siegel, editors, *Defect and Diffusion Forum*, volume 75, pages 121–148, Brookfield, VT, 1991. Sci-Tech Publications.

9. A. Atkinson. Interfacial diffusion. *Mat. Res. Soc. Symp.*, 122:183–192, 1988.

10. D. Beshers. Diffusion of interstitial impurities. In *Diffusion*, pages 209–240, Metals Park, OH, 1973. American Society for Metals.

11. L.A. Girifalco. *Statistical Physics of Materials*. John Wiley & Sons, New York, 1973.

12. A.D. LeClaire and A.B. Lidiard. Correlation effects in diffusion in crystals. *Phil. Mag.*, 1(6):518–527, 1956.

13. K. Compaan and Y. Haven. Correlation factors for diffusion in solids. *Trans. Faraday Soc.*, 52:786–801, 1956.

14. R.O. Simmons and R.W. Balluffi. Measurements of equilibrium vacancy concentrations in aluminum. *Phys. Rev.*, 117:52–61, 1960.

15. A. Seeger. The study of point defects in metals in thermal equilibrium. I. The equilibrium concentration of point defects. *Cryst. Lattice Defects*, 4:221–253, 1973.

16. R.W. Balluffi. Vacancy defect mobilities and binding energies obtained from annealing studies. *J. Nucl. Mats.*, 69–70:240–263, 1978.

17. W.D. Kingery, H.K. Bowen, and D.R. Uhlmann. *Introduction to Ceramics*. John Wiley & Sons, New York, 1976.

18. Y.-M. Chiang, D. Birnie, and W.D. Kingery. *Physical Ceramics*. John Wiley & Sons, New York, 1996.

19. D. Halliday and R. Resnick. *Fundamentals of Physics*. John Wiley & Sons, New York, 1974.

EXERCISES

8.1 It has sometimes been claimed that the observation of a Kirkendall effect implies that the diffusion occurred by a vacancy mechanism. However, a Kirkendall effect can be produced just as well by the interstitialcy mechanism. Explain why this is so.

Solution. Substitutional atoms of type 1 may diffuse more rapidly than atoms of type 2 if they diffuse independently by the interstitialcy mechanism in Fig. 8.4. To sustain the unequal fluxes, interstitial-atom defects can be created at climbing dislocations acting as interstitial sources in the region richer in 1 and destroyed at dislocations acting as interstitial sinks in the region poorer in 1. This will cause the region richer in 1 to contract and the other region to expand, thereby producing a Kirkendall effect.

8.2 For copper self-diffusion by the vacancy mechanism, demonstrate that Eq. 7.14 predicts that the pre-exponential "attempt" frequency factor is on the order of 10^{13} s^{-1}. Use a harmonic one-particle model for the configuration illustrated in Fig. 8.3. For Cu, Young's modulus is $E = 12 \times 10^{10}$ MPa, the lattice constant is $a = 0.36$ nm, the atomic weight is 63.5 g, and the structure is f.c.c. (12 nearest-neighbors).

- Assume a simple ball-and-spring model in which the atoms are replaced by balls of mass m, which are coupled by nearest-neighbor bonds represented by linear springs having a restoring force spring constant, S. Make reasonable approximations to estimate the restoring force experienced by the atom as it vibrates along its jump path. Remember that in the one-particle model the environment of the jumping particle remains fixed.

Solution. A value of the spring constant, S, can be obtained by applying a tensile stress, σ, to the ball and spring model along [100], finding the elastic strain, ε, resulting from the stretching of the springs, and then using the relation

$$\sigma = E\varepsilon \tag{8.97}$$

Each atom in a (200) plane has four nearest-neighbors lying in an adjacent (200) plane. The springs connecting it to these nearest-neighbors lie at $45°$ with respect to [100]. Because there are $2/a^2$ atoms per unit area in a (200) plane, the force stretching each spring along the spring axis due to the applied stress is

$$F_S = \frac{a^2 \sigma}{4\sqrt{2}} \tag{8.98}$$

The extension of each spring is then

$$\Delta L_S = \frac{F_S}{S} \tag{8.99}$$

and the strain along [100] is

$$\varepsilon = \frac{2\sqrt{2}\,\Delta L_S}{a} \tag{8.100}$$

Therefore, using Eqs. 8.97, 8.98, 8.100, and 8.99,

$$E = \frac{\sigma}{\varepsilon} = \frac{2}{a}\frac{F_S}{\Delta L_S} = \frac{2}{a}S \tag{8.101}$$

and $S = aE/2 = 2.2 \times 10^3$ MPa. The restoring force experienced by an atom vibrating in the direction of a nearest-neighbor vacancy (e.g., atom A in Fig. 8.3) in the one-particle model can be estimated. Atom A is in a cage of 11 nearest-neighbors. These include atoms 1, 2, 3, and 4 in the window in the $(1\bar{1}0)$ plane on one side, and four atoms (including atoms 5 and 6) in a similar window configuration in the $(1\bar{1}0)$ plane on the back side of atom A, atom 9 in the same $(1\bar{1}0)$ plane as atom A along with another atom symmetrically disposed on the other side of A in the direction [011], and a final atom behind A along $[01\bar{1}]$. Making the one-particle assumption that the environment of the jumping particle is fixed, simple geometry shows that if the A atom moves toward the vacancy by the distance ΔL, eight springs will change their lengths by $\Delta L/2$ to first order when $\delta L \ll a$, and one spring will stretch by ΔL. When the forces induced by these changes in spring length are resolved along $[0\,\bar{1}\,1]$, the total restoring force on A is found to be $3S\,\Delta L$: the total effective linear restoring-force constant is then $\beta = 3S$. Putting this value into Eq. 7.14 yields $\nu \approx 0.4 \times 10^{13}$ s^{-1}.

8.3 The self-diffusivity in an f.c.c. crystal for diffusion by a vacancy mechanism can be written

$$^{\star}D = g\mathbf{f}a^2\nu\, e^{(S_V^m + S_V^f)/k} e^{-(H_V^m + H_V^f)/(kT)}$$

where $g = 1$. Find the value of g for self-diffusion in a b.c.c. crystal.

Solution. The diffusion of the atoms will be correlated because of the vacancy exchange mechanism and, therefore, using Eq. 7.52,

$$^{\star}D = \frac{\Gamma r^2}{6}\mathbf{f} \tag{8.102}$$

But

$$\Gamma = X_V^{\mathrm{eq}}\Gamma_V \tag{8.103}$$

where X_V^{eq} and Γ_V are the equilibrium atom fraction of vacancies and the vacancy jump rate, respectively. Also, $r^2 = (3/4)a^2$ and $\Gamma_V = 8\Gamma_V'$, so that

$$^{\star}D = X_V^{\mathrm{eq}}\Gamma_V' a^2\mathbf{f} \tag{8.104}$$

Using Eqs. 8.13 and 8.18,

$$^{\star}D = \mathbf{f}a^2\nu\, e^{(S_V^m + S_V^f)/k} e^{-(H_V^m + H_V^f)/(kT)} \tag{8.105}$$

and therefore $g = 1$.

8.4 An interstitial C atom will generally diffuse in b.c.c. Fe by jumping almost exclusively between nearest-neighbor interstitial sites such as sites 1 and 2 in Fig. 8.8b. However, very occasionally it may jump between next-nearest-neighbor sites such as 1 and 3. Find an expression for the overall diffusivity of the C atoms, D_I, as a result of both nearest-neighbor and next-nearest-neighbor jumps.

Solution. The diffusion is uncorrelated and therefore

$$D_I = \frac{\Gamma \langle r^2 \rangle}{6} \tag{8.106}$$

Let Γ_A' and Γ_B' be the frequencies for type $1 \to 2$ (A-type) and type $1 \to 3$ (B-type) jumps, respectively, in Fig. 8.8b. Then, because there are four nearest-neighbors for A-type jumps and eight next-nearest-neighbors for B-type jumps, the frequencies for A-type and B-type jumps are $\Gamma_A = 4\Gamma_A'$ and $\Gamma_B = 8\Gamma_B'$, respectively. The mean-square displacement during time τ is then

$$\langle r^2 \rangle = \frac{(a^2/4)\Gamma_A + (a^2/2)\Gamma_B}{\Gamma_A + \Gamma_B} \tag{8.107}$$

and

$$\Gamma = \Gamma_A + \Gamma_B \tag{8.108}$$

Therefore,

$$D_I = \frac{a^2\Gamma_A}{24} + \frac{a^2\Gamma_B}{12} \tag{8.109}$$

The quantities $a^2\Gamma_A/24$ and $a^2\Gamma_B/12$ may be regarded as the hypothetical diffusivities of the C atoms if they are allowed to make only A- type and B-type jumps, respectively, and therefore Eq. 8.109 may be written

$$D_I = D_{IA} + D_{IB} \tag{8.110}$$

where D_{IA} and D_{IB} are the two hypothetical diffusivities. In general, $D_{IB} \ll D_{IA}$.

8.5 As discussed in Section 8.3.1, the $\langle 100 \rangle$ split-dumbbell self-interstitial in the f.c.c. structure can exist with its axis along [100], [010], or [001]. Under stress, certain of these orientation states are preferentially populated due to the tetragonality of the defect as a center of dilation. When the stress is suddenly released, the defects repopulate the available states until the populations in the three states become equal. Show that the relaxation time for this re-population is

$$\tau = \frac{2}{3\Gamma} \tag{8.111}$$

where Γ is the total jump frequency of a dumbbell.

- Derive the differential equation that describes the rate at which [100] dumbbells convert to [010] and [001] dumbbells, and then solve the equation.

Solution. According to Fig. 8.6, a [100] dumbbell can jump into a neighboring site in eight different ways, four with [010] orientations and four with [001] orientations. Therefore,

$$\frac{dc^{[100]}}{dt} = -8\Gamma' c^{[100]} + 4\Gamma' c^{[010]} + 4\Gamma' c^{[001]} \tag{8.112}$$

where Γ' is the jump rate into a specific adjacent site, and the c's are the concentrations in the three orientations. However, the total concentration, c^{tot}, is constant, and therefore

$$c^{\text{tot}} = c^{[100]} + c^{[010]} + c^{[001]} \tag{8.113}$$

Combining Eqs. 8.112 and 8.113 yields

$$\frac{dc^{[100]}}{dt} = -12\Gamma' \left(c^{[100]} - \frac{c^{\text{tot}}}{3} \right) \tag{8.114}$$

Integrating and applying the condition that $c^{[100]}(t = \infty) = c^{\text{tot}}/3$,

$$c^{[100]}(t) - \frac{c^{\text{tot}}}{3} = \left[c^{[100]}(o) - \frac{c^{\text{tot}}}{3} \right] e^{-12\Gamma' t} \tag{8.115}$$

Because the total jump rate is $\Gamma = 8\Gamma'$, the relaxation time is

$$\tau = \frac{1}{12\Gamma'} = \frac{2}{3\Gamma} \tag{8.116}$$

8.6 It is possible to express the diffusivity of the split-dumbbell self-interstitial in an f.c.c. crystal (illustrated in Fig. 8.6) in terms of its total jump frequency, Γ, and the lattice constant of the crystal, a. Show that the following approaches lead to the same result.

Approach 1: Start with Eq. 8.3.

Approach 2: Start by determining the net flux between two adjacent (002) planes when the gradient of the interstitial concentration is normal to these planes.

Solution.

Approach 1: As seen in Fig. 8.6, the jump distance for the dumbbell is equal to the displacement of its center of mass, $a/\sqrt{2}$. Every neighboring site to the dumbbell is equally probable for the next jump, so $\mathbf{f} = 1$. Thus,

$$D_I = \frac{\Gamma r^2}{6}\mathbf{f} = \frac{\Gamma(a/\sqrt{2})^2}{6} = \frac{\Gamma a^2}{12}$$

Approach 2: Alternatively, we can analyze diffusion arising from a gradient of interstitial concentration along [002] in Fig. 8.6. Consider the jumping of interstitials between two adjacent (002) planes. If there are c' interstitials per unit area with centers of mass on plane A, one-third will have their axes along [100], one-third along [010], and one-third along [001]. Each [001] interstitial on plane A has four sites on an adjacent (002) plane (i.e., plane B) in which to jump. Each [100] interstitial and [010] interstitial has two sites in which to jump. The total concentration of interstitials per unit volume associated with plane A is $c = c'/(a/2) = 2c'/a$ and the flux from plane A to plane B is

$$J^{A \to B} = \frac{4c'\Gamma'}{3} + \frac{2c'\Gamma'}{3} + \frac{2c'\Gamma'}{3} = \frac{8c'\Gamma'}{3} = \frac{4ac\Gamma'}{3}$$

Expanding c to first order, the flux from plane B to plane A is

$$J^{B \to A} = \frac{4}{3}\Gamma'a\left(c + \frac{a}{2}\frac{\partial c}{\partial z}\right)$$

and the net flux is

$$J^{\text{net}} = J^{A \to B} - J^{B \to A} = -\frac{2}{3}a^2\Gamma'\frac{\partial c}{\partial z}$$

Therefore, $D_I = (2/3)a^2\Gamma'$. However, the total jump rate is $\Gamma = 8\Gamma'$ and

$$D_I = \frac{2}{3}a^2\Gamma' = \frac{\Gamma a^2}{12}$$

in agreement with the results of Approach 1.

8.7 Consider the diffusion of particles along x in a dilute system where no fields are present and there is only a concentration gradient. Under these conditions, the potential energy of the system will vary as shown in Fig. 8.21a when a diffusing particle jumps from a site in a plane at $x = x_{\text{o}}$ into an equivalent site in an adjacent plane at $x = x_{\text{o}} + a$. Suppose now that a conservative field is imposed that interacts with the diffusing particles so that the potential energy varies with the position of the jumping particle as shown in Fig. 8.21b. ΔU is the increase in the potential energy when a particle advances by one planar spacing and is given by

$$\Delta U = a\frac{d\psi}{dx} \tag{8.117}$$

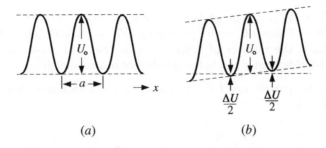

Figure 8.21:   Barrier to atom jumping. **(a)** No field present. **(b)** After imposition of a field that interacts with jumping particles.

where ψ is the potential associated with the imposed field. Obtain an expression for the net flux of particles between planes and show that it will have the form of Eq. 3.48 with the electrical potential, ϕ, replaced by ψ.

- Assume that the barrier to jumping is modified by the field as indicated by Fig. 8.21*b* and that the quantity $\Delta U/2 \ll kT$.

Solution. The net forward flux along x between planes can be written as

$$J = J(\rightarrow) - J(\leftarrow) = cA\,e^{-(U_\circ + \Delta U/2)/(kT)} - \left(c + \frac{dc}{dx}a\right)A\,e^{-(U_\circ - \Delta U/2)/(kT)} \quad (8.118)$$

where A is a constant. Expanding exponentials which involve the powers $\pm\Delta U/(2kT)$ to first order and neglecting higher-order terms yields

$$J = -A\,e^{-U_\circ/(kT)}\left[a\frac{dc}{dx} + \frac{c\Delta U}{kT}\right] \quad (8.119)$$

Finally, identifying $aA\exp[-U_\circ/(kT)]$ with D and using Eq. 8.117,

$$J = -D\frac{dc}{dx} - \frac{Dc}{kT}\frac{d\psi}{dx} \quad (8.120)$$

8.8 Calculate the correlation factor for tracer self-diffusion by the vacancy mechanism in the two-dimensional close-packed lattice illustrated in Fig. 8.22. The tracer atom at site 7 has just exchanged with the vacancy, which is now at site 6. Following Shewmon [4], let p_k be the probability that the tracer will make its next jump to its kth nearest-neighbor (i.e., a $7 \rightarrow k$ jump). θ_k is the angle between the initial $6 \rightarrow 7$ jump and the $7 \rightarrow k$ jump. The average of the cosines of the angles between successive tracer jumps is then

$$\langle\cos\theta\rangle = \sum_{k=1}^{z} p_k \cos\theta_k \quad (8.121)$$

and **f** is given by Eq. 8.29. The quantity p_k can be expressed in the form

$$p_k = \sum_{i=1}^{m} n_{ik}P_i = \sum_{i=1}^{m} n_{ik}\left(\frac{1}{z}\right)^i \quad (8.122)$$

where $P_i = (1/z)^i$ is the probability that the vacancy on its ith jump will make a $k \rightarrow 7$ jump (thereby producing a $7 \rightarrow k$ tracer jump) for the first time. n_{ik} is the number of different paths that will allow the vacancy to accomplish this, and $z = 6$ is the number of nearest-neighbors.

Calculate p_k, $\langle \cos \theta \rangle$, and $\mathbf{f}$ if all vacancy trajectories longer than four jumps are neglected.

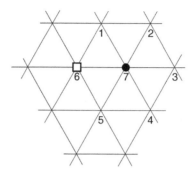

Figure 8.22:   Two-dimensional close-packed lattice. The tracer atom at 7 has just exchanged with the vacancy at 6.

Solution. First evaluate the n_{ik} in Eq. 8.122. Consider $k = 6$ first. For $i = 1$, the only possibility is a direct $6 \rightarrow 7$ jump. Therefore, $n_{16} = 1$. For $i = 2$, no possible paths exist, so $n_{26} = 0$. For $i = 3$, there are five paths, so $n_{36} = 5$. For $i = 4$, there are eight paths, so $n_{46} = 8$.

Similar inspections produce the results shown in Table 8.1 for $k = 5, 4,$ and 3. Note that by symmetry the results will be the same for $k = 1$ and $k = 5$ and for $k = 2$ and $k = 4$, respectively. Putting these results into Eq. 8.122,

$$p_6 = 1\left(\frac{1}{6}\right) + 0\left(\frac{1}{6}\right)^2 + 5\left(\frac{1}{6}\right)^3 + 8\left(\frac{1}{6}\right)^4 = 0.1960$$

$$p_5 = p_1 = 0\left(\frac{1}{6}\right) + 1\left(\frac{1}{6}\right)^2 + 1\left(\frac{1}{6}\right)^3 + 11\left(\frac{1}{6}\right)^4 = 0.0409$$

$$p_4 = p_2 = 0\left(\frac{1}{6}\right) + 0\left(\frac{1}{6}\right)^2 + 1\left(\frac{1}{6}\right)^3 + 2\left(\frac{1}{6}\right)^4 = 0.0061$$

$$p_3 = 0\left(\frac{1}{6}\right) + 0\left(\frac{1}{6}\right)^2 + 0\left(\frac{1}{6}\right)^3 + 2\left(\frac{1}{6}\right)^4 = 0.0015$$

(8.123)

Table 8.1:   Values of n_{ik} in Eq. 8.122

k	$i = 1$	$i = 2$	$i = 3$	$i = 5$
6	1	0	5	8
5	0	1	1	11
4	0	0	1	2
3	0	0	0	2

Substituting these values into Eq. 8.121 yields

$$\langle \cos \theta \rangle = 0.1960\,(-1) + 0.0409\left(-\frac{1}{2}\right)2 + 0.0061\left(\frac{1}{2}\right)2 + 0.0015\,(1)$$

$$= -0.2293$$

Finally,

$$\mathbf{f} = \frac{1 + \langle \cos \theta \rangle}{1 - \langle \cos \theta \rangle} = \frac{0.7707}{1.2293} = 0.627 \tag{8.124}$$

When all relevant trajectories including those beyond $i = 4$ are taken into account, the true value of $\mathbf{f}$ is 0.560 [13]. The truncation at $i = 4$ therefore causes $\mathbf{f}$ to be overestimated by about 12%.

8.9 Consider the diffusion of a randomly walking diffusant in the h.c.p. structure, which is composed of close-packed basal planes stacked in the sequence $ABABA\ldots.$ The lattice constants are a and c. The probability of a first-nearest-neighbor jump within a basal plane (jump distance $= a$) is p, and the probability of a jump between basal planes (jump distance $= \sqrt{a^2/3 + c^2/4}$) is $1 - p$. If axes x_1 and x_2 are located in a basal plane, derive the following expressions for the diffusivities D_{11} and D_{33}:

$$D_{11} = \frac{a^2 N_\tau p}{4\tau} + \frac{a^2 N_\tau (1-p)}{12\tau} \tag{8.125}$$

$$D_{33} = \frac{c^2 N_\tau (1-p)}{8\tau} \tag{8.126}$$

where N_τ is the total number of jumps in time, τ. Note that we have employed a principal coordinate system in which the diffusivity tensor is given by Eq. 4.66.

Solution. We will determine the D_{ii} by the general method used to obtain Eq. 8.11. According to Eq. 4.66, the diffusion is isotropic in directions perpendicular to x_3. We shall therefore determine the net flux, J^{net}, parallel to x_1 across the CD plane illustrated in Fig. 8.23.

$$J(\rightarrow) = J^{2\rightarrow3} + J^{1\rightarrow3} + J^{2\rightarrow4} = \frac{3n_2}{2}a\Gamma'_o + n_2 a\Gamma'_x + \frac{n_1}{2}a\Gamma'_o \tag{8.127}$$

$$J(\leftarrow) = \frac{3n_3}{2}a\Gamma'_o + n_3 a\Gamma'_x + \frac{n_4}{2}a\Gamma'_o \tag{8.128}$$

Here, n_i is the concentration on plane i, Γ'_o is the jump frequency from one site to a single neighboring site in the basal plane, and Γ'_x is the jump frequency from a site in

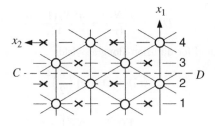

Figure 8.23: View of h.c.p. structure looking along $-x_3$. Open circles lie in the A plane; X's lie in the B plane.

a basal plane to a single neighboring site in an adjacent basal plane. Therefore,

$$J^{\text{net}} = J(\rightarrow) - J(\leftarrow) = \frac{3}{2}a\Gamma'_\circ(n_2 - n_3) + a\Gamma'_x(n_2 - n_3) + \frac{a}{2}\Gamma'_\circ(n_1 - n_4) \quad (8.129)$$

Making the usual Taylor expansions for the concentration differences yields

$$-J^{\text{net}} = D_{11}\frac{dn}{dx} = \left(\frac{3a^2}{4}\Gamma'_\circ + \frac{3a^2}{4}\Gamma'_\circ + \frac{a^2}{2}\Gamma'_x\right)\frac{dn}{dx} \quad (8.130)$$

Now

$$\Gamma'_\circ = \tfrac{1}{6}\Gamma_\circ = pN_\tau/(6\tau)$$
$$\Gamma'_x = \tfrac{1}{6}\Gamma_x = (1-p)N_\tau/(6\tau) \quad (8.131)$$

where $\Gamma_\circ$ is the total jump frequency for jumping in the basal plane and Γ_x is the total jump frequency for jumping between basal planes. Using Eqs. 8.130 and 8.131,

$$D_{11} = \frac{a^2 N_\tau p}{4\tau} + \frac{a^2 N_\tau(1-p)}{12\tau} \quad (8.132)$$

The diffusivity D_{33} is obtained by analyzing the net flux parallel to x_3 passing between two adjacent basal planes, A and B. In this case,

$$J^{\text{net}} = J(\rightarrow) - J(\leftarrow) = \frac{3}{2}n_A c\Gamma'_x - \frac{3}{2}n_B c\Gamma'_x = \frac{3}{2}c\Gamma'_x(n_A - n_B) \quad (8.133)$$

Using a Taylor expansion to evaluate $(n_A - n_B)$ and employing Eq. 8.131,

$$D_{33} = \frac{c^2 N_\tau(1-p)}{8\tau} \quad (8.134)$$

8.10 Show that the results obtained in Exercise 8.9 (i.e., Eqs. 8.125 and 8.126), can be obtained in a simpler way by using Eq. 7.53 in one dimension if $\langle R^2 \rangle$ is taken as the mean value of the squares of the jump vector components along the chosen direction.

Solution. For diffusion along axis x_1 in Fig. 8.23, Eq. 7.53 is written

$$D_{11} = \frac{\Gamma\langle r_1^2 \rangle}{2} \quad (8.135)$$

where $\langle r_1^2 \rangle$ is the mean square of the jump vector components along axis 1:

$$\langle r_1^2 \rangle = \frac{N_\tau p\left[\frac{a^2+(a/2)^2+(a/2)^2}{3}\right] + N_\tau(1-p)\left[\frac{(a/2)^2+(a/2)^2+0}{3}\right]}{N_\tau} \quad (8.136)$$

Putting Eq. 8.136 into Eq. 8.135 and using the relation $\Gamma = N_\tau/\tau$,

$$D_{11} = \frac{a^2 N_\tau p}{4\tau} + \frac{a^2 N_\tau(1-p)}{12\tau} \quad (8.137)$$

Using the same method for diffusion along x_3 yields

$$\langle r_3^2 \rangle = (1-p)\left(\frac{c}{2}\right)^2 \quad (8.138)$$

and

$$D_{33} = \frac{\Gamma\langle r_3^2 \rangle}{2} = \frac{N_\tau(1-p)c^2}{8\tau} \quad (8.139)$$

8.11 Exercise 7.4 demonstrated that the mean-square displacement during random three-dimensional diffusion in a primitive orthorhombic crystal is equal to the sum of the mean-square displacements achieved during one-dimensional diffusion along each of the three coordinate axes.

Demonstrate this result for the diffusion of a randomly walking diffusant in an h.c.p. crystal using the information and results in Exercises 8.9 and 8.10.

Solution. Using the same procedure as in Exercise 7.4,

$$c(x_1, x_2, x_3, t) = \frac{A}{\sqrt{t}} \exp\left[-\left(\frac{x_1^2}{4D_{11}t} + \frac{x_2^2}{4D_{11}t} + \frac{x_3^2}{4D_{33}t}\right)\right] \quad (8.140)$$

and the mean-square displacement is then

$$\langle R^2 \rangle = \frac{\int_0^\infty \int_0^\infty \int_0^\infty c(x_1, x_2, x_3, t)\, r^2\, dx_1\, dx_2\, dx_3}{\int_0^\infty \int_0^\infty \int_0^\infty c(x_1, x_2, x_3, t)\, dx_1\, dx_2\, dx_3} \quad (8.141)$$

$$= 2D_{11}\tau + 2D_{11}\tau + 2D_{33}\tau$$

But according to Eq. 7.31,

$$\langle R^2 \rangle = N\langle r^2 \rangle \quad (8.142)$$

Because $\langle r^2 \rangle = a^2 p + (1-p)(a^2/3) + (1-p)c^2/4$,

$$\langle R^2 \rangle = \left[\frac{Na^2 p}{2} + \frac{N(1-p)a^2}{6}\right] + \left[\frac{Na^2 p}{2} + \frac{N(1-p)a^2}{6}\right] + \frac{N(1-p)c^2}{4} \quad (8.143)$$

Substituting Eqs. 8.125 and 8.126 into Eq. 8.143 yields the relation

$$\langle R^2 \rangle = 2D_{11}\tau + 2D_{11}\tau + 2D_{33}\tau \quad (8.144)$$

which is consistent with Eq. 8.141.

8.12 Exercise 7.5 shows explicitly for a random walker on a primitive-cubic lattice that the mean values of the $\cos\theta_{i,i+j}$ terms in Eq. 7.49 sum to zero and, therefore, that $\mathbf{f} = 1$. Use Eq. 8.29 to demonstrate the same result.

Solution. First evaluate $\langle\cos\theta\rangle$. Possible values of $\cos\theta$ are $1, -1, 0, 0, 0$, and 0, all of which occur with equal probability. Therefore, $\langle\cos\theta\rangle = 0$ and

$$\mathbf{f} = \frac{1 + \langle\cos\theta\rangle}{1 - \langle\cos\theta\rangle} = 1 \quad (8.145)$$

8.13 Using Eq. 7.52, calculate an expression for self-diffusivity by the vacancy mechanism in a primitive cubic lattice. Suppose that the *back-jump probability* (i.e., an atom returns to the site from which it jumped previously) is p. Consider first-neighbor jumps only.

Evaluate the case $p = 0$ and compare it to an uncorrelated random walk.

Solution. There are six first-neighbor sites in the primitive cubic lattice, and the first-neighbor jump distance, r, is equal to the lattice constant, a. Once an atom has jumped into a given site, the probability that it will next jump into any of its nearest-neighbor sites (with the exception of the site from which it just jumped) is $(1-p)/5$. Therefore,

$$\langle\cos\theta\rangle = (-1)p + (0+0+0+0)\frac{1-p}{5} + (1)\frac{1-p}{5} = \frac{1}{5}(1-6p) \quad (8.146)$$

and using Eq. 8.29,

$$f = \frac{1 + \langle \cos \theta \rangle}{1 - \langle \cos \theta \rangle} = \frac{3(1-p)}{2+3p} \tag{8.147}$$

The self-diffusivity is then

$$^{\star}D = \frac{\Gamma \langle r^2 \rangle}{6} f = \frac{\Gamma a^2}{2} \frac{1-p}{2+3p} \tag{8.148}$$

In the random case when $p = 1/6$, $f = 1$. In the most correlated case when $p = 1$, $f = 0$. When $p = 0$ and the atom cannot jump backward to erase its previous jump, $f = 3/2$ and diffusion is enhanced relative to the random, uncorrelated case.

8.14 A computer simulation of diffusion via the vacancy mechanism is performed on a square lattice of screen pixels with a spacing of $a = 0.5$ mm. The computer performs the calculations so that the vacancy jumps at a constant rate of $\Gamma = 1000\,\mathrm{s}^{-1}$. The simulation cell is a square of edge length 5 cm containing 10,000 pixels. There is just one vacancy in the simulation cell, and as it moves by nearest-neighbor jumps, it remains within the cell (by using periodic boundary conditions or reflection at the borders).

(a) Estimate the *vacancy diffusion coefficient* in this simulation if the vacancy moves by a random walk.

(b) One *tracer atom*, represented by a specially marked pixel, is initially located at the center of the simulation cell. The vacancy is introduced in the cell at a random location and then moves by a random walk. Estimate the value of the *tracer diffusion coefficient* in this simulation.

(c) Estimate the average *time* for the tracer atom to move from the center of the cell to the cell border.

Solution.

(a) Diffusion of a vacancy in a lattice is uncorrelated, so $f = 1$. The vacancy diffusivity D_V for this two-dimensional diffusion is

$$D_V = \frac{\Gamma r^2}{4} f = \frac{\Gamma r^2}{4} = \frac{1000 \text{ s}^{-1}\, 0.5^2 \text{ mm}^2}{4} = 6.25 \times 10^{-5} \text{ m}^2\,\text{s}^{-1}$$

(b) Self-diffusion of a tracer by vacancy exchange is correlated, so in this square lattice we have $f \approx (z-1)/(z+1) \cong 0.6$. The tracer self-diffusivity $^{\star}D$ is

$$^{\star}D = \frac{\Gamma r^2}{4} X_V f = D_V X_V f$$
$$= 6.25 \times 10^{-5} \text{ m}^2\,\text{s}^{-1} \times 10^{-4} \times 0.6 = 3.8 \times 10^{-9} \text{ m}^2\,\text{s}^{-1}$$

(c) A very simple estimate can be made by using the relation $\langle R^2 \rangle = 4\,^{\star}Dt$ and taking $R \approx 2.5$ cm. This gives

$$t \approx \frac{\langle R^2 \rangle}{4\,^{\star}D} = \frac{(2.5 \text{ cm})^2}{4 \times 3.8 \times 10^{-9} \text{ m}^2\,\text{s}^{-1}} = 4.1 \times 10^4 \text{ s}$$

which is probably an *overestimate*. The time required is the average time for the tracer atom to *first* hit the wall. Also, depending on where along the wall the tracer first hits, the path will be somewhat longer because of the square shape of the simulation cell. Nevertheless, this method gives an estimate. A more accurate value could be determined most easily by doing a computer simulation and keeping statistics on the times for the tracer to "hit the wall."

8.15 Schottky defects form at equilibrium in stoichiometric ZrO_2. Show that the equilibrium site fraction of anion vacancies is given by

$$[V_O^{\bullet\bullet}] = 2^{1/3} e^{-G_S^f/(3kT)}$$

Solution. First write the Schottky reaction:

$$\text{null} = V_{Zr}^{''''} + 2V_O^{\bullet\bullet} \tag{8.149}$$

The corresponding mass-action equilibrium equation for this reaction is

$$[V_{Zr}^{''''}][V_O^{\bullet\bullet}]^2 = K^{\text{eq}} = e^{-G_S^f/(kT)} \tag{8.150}$$

where G_S^f is the free energy of formation of a Schottky defect, which in this case consists of a cation vacancy and two anion vacancies. Then, charge neutrality requires that

$$2[V_{Zr}^{''''}] = [V_O^{\bullet\bullet}] \tag{8.151}$$

Substituting Eq. 8.151 into Eq. 8.150 yields

$$[V_O^{\bullet\bullet}] = 2^{1/3} e^{-G_S^f/(3kT)} \tag{8.152}$$

8.16 Schottky defects are the predominant equilibrium point defects in stoichiometric zirconia ZrO_2 (see Exercise 8.15). Suppose that the soluble oxide Ta_2O_5 is added to ZrO_2. Assume that cation vacancies form without the formation of any interstitial defects.

(a) Find an expression for the equilibrium cation vacancy site fraction that will form.

(b) Discuss how the self-diffusion of the cations will be affected by the addition of Ta_2O_5.

Solution.

(a) When two units of Ta_2O_5 are added to ZrO_2, four Ta ions will be put into four existing Zr sites. The four displaced Zr ions will be put into four new normal Zr cation sites. The ten incoming O ions will be put into new normal O sites, and one Zr ion will be removed from its existing site and placed in a new normal Zr site, thereby creating a cation vacancy. This process preserves electrical neutrality and may be expressed

$$2Ta_2O_5 \xrightarrow{ZrO_2} 10O_O^{\times} + 4Ta_{Zr}^{\bullet} + V_{Zr}^{''''} \tag{8.153}$$

In addition to this reaction, $V_{Zr}^{''''}$ and $V_O^{\bullet\bullet}$ defects will be produced by the Schottky reaction (Eq. 8.149), and the mass-action equilibrium in Eq. 8.150 will hold. The condition for electrical neutrality may be obtained by realizing that the introduction of 1 unit of $Ta_{Zr}^{\bullet}$ produces 1/4 unit of $V_{Zr}^{''''}$. Also, for every unit of $V_O^{\bullet\bullet}$ formed,

1/2 unit of V_{Zr}'''' is produced. Therefore, the charge-balanced site fraction of V_{Zr}'''' must be

$$[V_{Zr}''''] = \frac{1}{4}[Ta_{Zr}^{\bullet}] + \frac{1}{2}[V_O^{\bullet\bullet}] \tag{8.154}$$

Combining Eq. 8.154 and Eq. 8.150 yields

$$[V_{Zr}''''] \left(2[V_{Zr}''''] - \frac{1}{2}[Ta_{Zr}^{\bullet}]\right)^2 = e^{-G_S^f/(kT)} \tag{8.155}$$

(b) The self-diffusivity on the cation sublattice will be proportional to the cation vacancy site fraction, $[V_{Zr}'''']$. At high temperatures, the numbers of anion and cation vacancies produced by the Schottky-pair reaction will be much larger than the number of $Ta_{Zr}^{\bullet}$ defects, so that

$$[V_{Zr}''''] = \frac{1}{2}[V_O^{\bullet\bullet}] \gg [Ta_{Zr}^{\bullet}]$$

Therefore, from Eq. 8.155,

$$[V_{Zr}''''] = \left(\frac{1}{4}\right)^{1/3} e^{-G_S^f/(3kT)} \tag{8.156}$$

At low temperatures, the site fraction of cation vacancies due to Schottky-pair formation will be negligible and their site fraction will therefore be fixed at the level

$$[V_{Zr}''''] = \frac{1}{4}[Ta_{Zr}^{\bullet}]$$

An Arrhenius plot of the cation self-diffusivity will then possess two linear regions. In the high-temperature intrinsic regime, the slope will be $-(H_S^f/3 + H^m)/k$; in the low-temperature extrinsic regime, the slope will be simply H^m/k, where H^m is the migration enthalpy of a cation vacancy.

8.17 ZrO_2 can be made O deficient in a sufficiently reducing atmosphere.

(a) Show that the oxygen anion vacancy site fraction increases with a decrease in the oxygen pressure in the atmosphere according to

$$[V_O^{\bullet\bullet}] \propto \frac{1}{P_{O_2}^{1/6}}$$

(b) Show that the self-diffusivity on the anion (oxygen) sublattice, $^\star D^O$, increases with decreasing oxygen pressure (at constant temperature) according to

$$\frac{1}{P_{O_2}^{1/6}}$$

and varies with temperature (at constant oxygen pressure) according to

$$e^{-(\Delta G/3 + G_V^m)/(kT)}$$

where ΔG is the free-energy change due to the reduction reaction and G_V^m is the free energy of migration of an oxygen anion vacancy.

Solution.

(a) The reduction reaction involves removing an O^{2-} anion from the structure and transferring two electrons from it to two Zr^{4+} cations. This makes two Zr^{3+} cations, a neutral O atom, and a cation vacancy $V_O^{\bullet\bullet}$. Therefore, the reaction can be written

$$2Zr_{Zr}^{\times} + O_O^{\times} = 2Zr_{Zr}' + V_O^{\bullet\bullet} + \frac{1}{2}O_2$$

and for this reaction,

$$\left[Zr_{Zr}'\right]^2 [V_O^{\bullet\bullet}] P_{O_2}^{1/2} = K^{\mathrm{eq}} = e^{-\Delta G/(kT)}$$

The neutrality condition is

$$\left[Zr_{Zr}'\right] = 2[V_O^{\bullet\bullet}]$$

Therefore, combining these equations,

$$[V_O^{\bullet\bullet}] = \frac{1}{4^{1/3}} \frac{1}{P_{O_2}^{1/6}} e^{-\Delta G/(3kT)}$$

(b) Because $^{\star}D^O$ is proportional to $[V_O^{\bullet\bullet}]$,

$$^{\star}D^O \propto e^{-\Delta G/(3kT)}$$

at constant T. Because $^{\star}D^O$ is proportional to $[V_O^{\bullet\bullet}]$ and also to the Boltzmann factor $\exp[-G_V^m/(kT)]$,

$$^{\star}D^O \propto \frac{1}{P_{O_2}^{1/6}} e^{-(\Delta G/3 + G_V^m)/(kT)}$$

8.18 Consider an oxygen-deficient oxide MO_{2-x} containing a low concentration of solute A_M'', due to the addition of the soluble oxide AO. Oxygen diffusion occurs by a vacancy mechanism. Assume that all oxygen vacancies are doubly ionized.

(a) Write the reduction reaction for reducing MO_{2-x} and a corresponding equation for its equilibrium constant, K^{eq}.

(b) Write a defect reaction for the incorporation of the solute A into MO_2.

(c) Write the charge neutrality condition for the impure, nonstoichiometric oxide.

(d) How would P_{O_2} qualitatively affect the self-diffusivity of oxygen on the anion sublattice, $^{\star}D^O$, in the intrinsic and extrinsic regimes?

Solution.

(a) The reduction occurs by removing one O ion and transferring two electrons from it to two M ions, creating two $[M_M']$ defects and one $[V_O^{\bullet\bullet}]$ defect and one free O atom. This reaction may be written

$$2M_M^{\times} + O_O^{\times} = 2M_M' + V_O^{\bullet\bullet} + \frac{1}{2}O_2 \tag{8.157}$$

At equilibrium,

$$[M_M']^2[V_O^{\bullet\bullet}]P_{O_2}^{1/2} = K^{\mathrm{eq}} = e^{-\Delta G/(kT)} \tag{8.158}$$

(b) When AO is added, A is added at a new M site and O goes to a new O site. This creates one A_M'' defect. To maintain electrical neutrality, an O ion is removed from an O site and placed in a new O site, thereby creating a $V_O^{\bullet\bullet}$ defect. The reaction may be written as

$$AO = A_M'' + V_O^{\bullet\bullet} + O_O^{\times} \tag{8.159}$$

(c) Overall charge neutrality requires that

$$[V_O^{\bullet\bullet}] = \frac{1}{2}[M_M'] + [A_M''] \tag{8.160}$$

(d) The mass-action law given by Eq. 8.158 will hold and, therefore, putting the neutrality condition given by Eq. 8.160 into Eq. 8.158,

$$4[V_O^{\bullet\bullet}]\left([V_O^{\bullet\bullet}] - [A_M'']\right)^2 P_{O_2}^{1/2} = e^{-\Delta G/(kT)} \tag{8.161}$$

At high temperatures, $[V_O^{\bullet\bullet}]$ is entirely due to the reduction process and $[V_O^{\bullet\bullet}] \gg [A_M'']$. Therefore, in this intrinsic regime,

$$[V_O^{\bullet\bullet}] = \frac{1}{4^{1/3}}\frac{1}{P_{O_2}^{1/6}}\, e^{-\Delta G/(3kT)} \tag{8.162}$$

Because $^\star D^O$ is proportional to $[V_O^{\bullet\bullet}]$, $^\star D^O \propto P_{O_2}^{-1/6}$. At low temperatures, the contribution of the reduction process to $[V_O^{\bullet\bullet}]$ is essentially negligible and $[V_O^{\bullet\bullet}]$ becomes constant at the value

$$[V_O^{\bullet\bullet}] = [A_M'']$$

which is determined by the amount of solute AO that has been added. In this extrinsic regime, $^\star D^O$ is therefore independent of P_{O_2}.

8.19 The relationship between the intrinsic diffusivity, D_1, of charged interstitial ions in an ionic solid and the ionic electrical conductivity, ρ, due to the motion of these ions in the absence of a significant concentration gradient is given by Eq. 3.50; that is,

$$\rho = \frac{D_1 c_1 q_1^2}{kT}$$

Suppose that an ionic solid contains charged cation vacancies such as NaCl containing Na^+ vacancies. Find a relationship, comparable to Eq. 3.50, between the cation tracer self-diffusion coefficient, $^\star D^{cation}$, and the electrical conductivity, ρ, due to voltage-induced motion of the cations.

Solution. In this case, the charged cation vacancies, possessing a diffusivity D_V^{cation}, will respond to the voltage just as the charged interstitials did in Section 3.2.1. The relationship between D_V^{cation} and ρ will then be given by the same type of relation as Eq. 3.50; that is,

$$\rho = \frac{D_V^{cation} c_V q_V^2}{kT} \tag{8.163}$$

where c_V is the cation vacancy concentration and q_V the charge carried by the vacancy. On the other hand, the cation tracer self-diffusivity, $^\star D^{cation}$, will be related to the cation vacancy concentration by a relationship similar to Eq. 8.17:

$$^\star D^{cation} = X_V D_V^{cation}\mathbf{f} \tag{8.164}$$

where X_V is the fraction of cation sites occupied by the vacancies and $\mathbf{f}$ is the correlation factor for the operative vacancy exchange mechanism. Combining Eqs. 8.163 and 8.164 and setting c_{cation} equal to the number of cation sites per unit volume,

$$\frac{\rho}{^*D^{cation}} = \frac{c_{cation}q_V^2}{kT} \frac{1}{\mathbf{f}} \tag{8.165}$$

Note that independent measurements of ρ and $^*D^{cation}$ will yield information about $\mathbf{f}$, because the other factors in Eq. 8.165 are known.

8.20 Show that the maximum potential energy stored in a torsion pendulum is proportional to the square of the amplitude of its oscillation.

Solution. The restoring torque for a torsion pendulum is $-k\theta$, where θ is the angle of rotation (see Fig. 8.18) and k is the torsion constant. The equation of motion [19] is then

$$\frac{d^2\theta}{dt^2} = -\frac{\kappa}{I}\theta \tag{8.166}$$

where I is the moment of inertia. Equation 8.166's solution is

$$\theta = \theta_{max}\cos(\omega t + \phi) \tag{8.167}$$

The stored energy is a maximum when $\theta = \theta_{max}$ and is therefore

$$W = \int_0^{\theta_{max}} \kappa\theta\,d\theta = \frac{\kappa}{2}\theta_{max}^2 \tag{8.168}$$

8.21 Figure 8.17 shows a hysteresis loop for an anelastic solid subjected to an oscillating stress. If the amplitude of the stress is σ_o, find the shape of the hysteresis loop:

(a) When $\omega\tau \ll 1$

(b) When $\omega\tau \gg 1$

(c) When $\omega\tau = 1$

Specify the direction in which the loop is traversed with increasing time, the width of the loop at $\sigma = 0$, and the slope of the dashed line in Fig. 8.17. Express your answer in terms of σ_o, S_R, and S_U.

Solution. By using Eqs. 8.67–8.69 and constructing the diagram for ε in the complex plane,

$$\tan\phi = \frac{\varepsilon_2}{\varepsilon_1} \qquad \sin\phi = \frac{\varepsilon_2}{\varepsilon_o} \qquad \cos\phi = \frac{\varepsilon_1}{\varepsilon_o} \tag{8.169}$$

Also, using the real parts of σ and ϵ yields

$$\sigma = \sigma_o\cos\omega t$$
$$\varepsilon = \varepsilon_o\cos(\omega t - \phi) = \varepsilon_o(\cos\omega t\cos\phi + \sin\omega t\sin\phi) \tag{8.170}$$

When $\sigma = \sigma_o$, $\cos\omega t = 1$ and $\sin\omega t = 0$. Therefore, $\varepsilon = \varepsilon_o\cos\phi = \varepsilon_1$ and the slope of the dashed line is σ_o/ε_1. When $\sigma = 0$, $\cos\omega t = 0$, and $\sin\omega t = 1$, $\varepsilon = \varepsilon_o\sin\phi = \varepsilon_2$. Also, when $\sigma = 0$, $\cos\omega t = 0$, and $\sin\omega t = -1$, $\varepsilon = -\varepsilon_o\sin\phi = -\varepsilon_2$.

(a) When $\omega t \ll 1$, use of Eqs. 8.93 and 8.94 shows that

$$\varepsilon_1 = S_R\sigma_o \quad \text{and} \quad \varepsilon_2 = 0 \tag{8.171}$$

The hysteresis loop will therefore appear as a line of negligible width and slope $1/S_R$ as in Fig. 8.24a. Negligible internal friction therefore occurs.

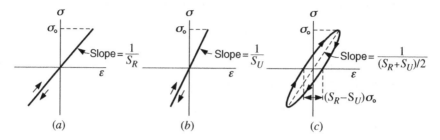

Figure 8.24:   Frequency dependence of anelastic behavior. **(a)** $\omega\tau \ll 1$. **(b)** $\omega\tau \gg 1$. **(c)** $\omega\tau = 1$.

(b) When $\omega t \gg 1$,
$$\varepsilon_1 = S_U\sigma_\circ \quad \text{and} \quad \varepsilon_2 = 0 \tag{8.172}$$
The hysteresis loop will therefore appear again as a line of negligible width but with a larger slope, as in Fig. 8.24b. Negligible internal friction occurs.

(c) When $\omega t = 1$,
$$\varepsilon_1 = \frac{S_R + S_U}{2}\sigma_\circ \quad \text{and} \quad \varepsilon_2 = \frac{S_R - S_U}{2}\sigma_\circ \tag{8.173}$$
The hysteresis loop will therefore appear as in Fig. 8.24c. The slope of the dashed line is
$$\frac{\sigma_\circ}{\varepsilon_1} = \frac{1}{(S_R + S_U)/2} \tag{8.174}$$
and the width of the loop at $\sigma = 0$ is
$$2\varepsilon_2 = (S_R - S_U)\sigma_\circ \tag{8.175}$$

Also, because the strain lags behind the stress, the direction of traversal of the loop must be as indicated. In this situation, maximum internal friction occurs.

8.22 Describe in detail how to determine the diffusivity of C in b.c.c. Fe using a torsion pendulum. Include all of the necessary equations.

- See Section 8.3.1 and Fig. 8.8, where C atoms in sites 1, 2, and 3 expand the crystal preferentially along x, y, and z, respectively.

Solution. Using a torsion pendulum, find the anelastic relaxation time, τ, by measuring the frequency of the Debye peak, ω_p, and applying the relation $\omega_p\tau = 1$. Having τ, the relationship between τ and the C atom jump frequency Γ is found by using the procedure to find this relationship for the split-dumbbell interstitial point defects in Exercise 8.5. Assume the stress cycle shown in Fig. 8.16 and consider the anelastic relaxation that occurs just after the stress is removed. A C atom in a type 1 site can jump into two possible nearest-neighbor type 2 sites or two possible type 3 sites. Therefore,
$$\frac{dc_1}{dt} = -4\Gamma'c_1 + 2\Gamma'c_2 + 2\Gamma'c_3 \tag{8.176}$$
Because $c_1 + c_2 + c_3 = c^{\text{tot}} = $ constant,
$$\frac{dc_1}{dt} = -6\Gamma'\left(c_1 - \frac{c^{\text{tot}}}{3}\right) \tag{8.177}$$

which may be integrated to obtain

$$\left[c_1(t) - \frac{c^{\text{tot}}}{3}\right] = \left[c_1(0) - \frac{c^{\text{tot}}}{3}\right] = e^{-6\Gamma' t} \tag{8.178}$$

The relaxation time is then $\tau = 1/(6\Gamma')$, and because the total jump frequency is $\Gamma = 4\Gamma'$, $\tau = 2/(3\Gamma)$. According to Eq. 7.52, $D_I = \Gamma r^2/6$ because $f = 1$, and because $r = a/2$, $D_I = \Gamma a^2/24$. Substituting for Γ,

$$D_I = \frac{a^2}{36\tau} \tag{8.179}$$

Finally, insert the experimentally determined value of τ into Eq. 8.179 to obtain D_I.

8.23 Under equilibrium conditions in a stressed b.c.c. Fe crystal, interstitial C atoms are generally unequally distributed among the three types of sites identified in Fig. 8.8b. This occurs because the C atoms in sites 1, 2, and 3 in Fig. 8.8b expand the crystal preferentially along the x, y, and z directions, respectively. These directions are oriented differently in the stress field, and the C atoms in the various types of sites therefore have different interaction energies with the stress field. In the absence of applied stress, this effect does not exist and all sites are populated equally. In Exercise 8.22 it was shown that when the stress on an equilibrated specimen is suddenly released, the relaxation time for the nonuniformly distributed C atoms to achieve a random distribution, τ, is $\tau = 2/(3\Gamma)$, where Γ is the total jump frequency of a C atom in the unstressed crystal.

Show that when stress is suddenly applied to an unstressed crystal, the relaxation time for the randomly distributed C atoms to assume the nonrandom distribution characteristic of the stressed state is again $\tau = 2/(3\Gamma)$.

- Assume the energy-level system for the specimen shown in Fig. 8.25. Write the kinetic equations for the rates of change of the concentrations of the interstitials in the various types of sites and solve them subject to the appropriate initial and final conditions. Assume that the barriers to the jumping interstitials shown in Fig. 8.25 are distorted by the differences in the site energies (indicated in Fig. 8.21).

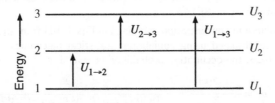

Figure 8.25:   Energy-level diagram for a stressed b.c.c. specimen containing an interstitial atom in sites 1, 2, or 3 illustrated in Fig. 8.8.

Solution. Let c_1, c_2, and c_3 be the concentrations of interstitials occupying sites of types 1, 2, and 3, respectively. Also, $c_1 + c_2 + c_3 = c^{\text{tot}} = $ constant. Since an interstitial in a given type of site can jump into two sites of each other type,

$$\frac{dc_1}{dt} = -2\left(\Gamma'_{1\to2} + \Gamma'_{1\to3} + \Gamma'_{3\to1}\right)c_1 + 2\left(\Gamma'_{2\to1} - \Gamma'_{3\to1}\right)c_2 + 2\Gamma'_{3\to1}c^{\text{tot}}$$
$$\frac{dc_2}{dt} = -2\left(\Gamma'_{2\to1} + \Gamma'_{2\to3} + \Gamma'_{3\to2}\right)c_2 + 2\left(\Gamma'_{1\to2} - \Gamma'_{3\to2}\right)c_1 + 2\Gamma'_{3\to2}c^{\text{tot}}$$

(8.180)

If the barrier to the jump of an interstitial between two sites of differing energy is deformed as indicated in Fig. 8.21, the information given in Fig. 8.25 may be used to derive expressions for the various jump rates that appear in the coefficients of Eq. 8.180. Neglecting small differences in the entropies of activation in the presence and absence of stress, and expanding Boltzmann factors of the form $\exp[-U_{i\to j}/(kT)]$ to first order so that $\exp[-U_{i\to j}/(kT)] = 1 + U_{i\to j}/(kT)$,

$$\Gamma'_{1\to2} = \Gamma'\left(1 - \frac{U_{1\to2}}{2kT}\right) \qquad \Gamma'_{2\to1} = \Gamma'\left(1 + \frac{U_{1\to2}}{2kT}\right)$$
$$\Gamma'_{2\to3} = \Gamma'\left(1 - \frac{U_{2\to3}}{2kT}\right) = \Gamma'\left(1 + \frac{U_{1\to2}}{2kT} - \frac{U_{1\to3}}{2kT}\right) \quad \Gamma'_{3\to2} = \Gamma'\left(1 - \frac{U_{1\to2}}{2kT} + \frac{U_{1\to3}}{2kT}\right)$$
$$\Gamma'_{1\to3} = \Gamma'\left(1 - \frac{U_{1\to3}}{2kT}\right) \qquad \Gamma'_{3\to1} = \Gamma'\left(1 + \frac{U_{1\to3}}{2kT}\right)$$

(8.181)

where Γ' is the jump rate between any two adjacent sites in the absence of stress. Equation 8.180 is a pair of simultaneous linear first-order equations with constant coefficients. The initial and final conditions are

$$c_1(0) = c_2(0) = \frac{c^{\text{tot}}}{3} \qquad c_1(\infty) = c_1^{\text{eq}} \qquad c_2(\infty) = c_2^{\text{eq}} \qquad (8.182)$$

where $c_1(\infty)$ and $c_2(\infty)$ are the final equilibrium concentrations reached at long times in the presence of the applied stress. In view of the symmetry of Eqs. 8.180, we try

$$c_1(t) = \left(\frac{c^{\text{tot}}}{3} - c_1^{\text{eq}}\right)e^{-k't} + c_1^{\text{eq}} \qquad c_2(t) = \left(\frac{c^{\text{tot}}}{3} - c_2^{\text{eq}}\right)e^{-k't} + c_1^{\text{eq}} \quad (8.183)$$

which satisfy the conditions in Eq. 8.182. Direct substitution shows that Eqs. 8.183 indeed satisfy Eqs. 8.180 when higher-order terms involving products of the small quantities $U_{i\to j}/(kT)$ are neglected and

$$k' = 6\Gamma'$$
$$c_1^{\text{eq}} = \frac{c^{\text{tot}}}{3}\left(1 + \frac{U_{1\to2}}{3kT} + \frac{U_{1\to3}}{3kT}\right)$$
$$c_2^{\text{eq}} = \frac{c^{\text{tot}}}{3}\left(1 - \frac{2U_{1\to2}}{3kT} + \frac{U_{1\to3}}{3kT}\right)$$

(8.184)

This shows that relaxation to the equilibrium distribution occurs exponentially with a relaxation time $\tau = 1/(6\Gamma')$. Since $\Gamma = 4\Gamma'$, where Γ is the total jump frequency in the unstressed crystal, $\tau = 2/(3\Gamma)$.

Finally, the equilibrium concentrations obtained in Eqs. 8.184 from the kinetic equations agree with those obtained using equilibrium statistical mechanics. In the three-level system in Fig. 8.25, the occupation probability for level 1 is

$$p_1 = \frac{e^{-U_1/(kT)}}{Z} = \frac{e^{-U_1/(kT)}}{e^{-U_1/(kT)} + e^{-U_2/(kT)} + e^{-U_3/(kT)}}$$
$$= \frac{1}{3\left(1 - \frac{U_{1\to2}}{3kT} - \frac{U_{1\to3}}{3kT}\right)} = \frac{1}{3}\left(1 + \frac{U_{1\to2}}{3kT} + \frac{U_{1\to3}}{3kT}\right)$$

(8.185)

Since $c_1 = c^{\text{tot}}p_1$, the result for c_1 is the same as that given by Eq. 8.184. Similar agreement is obtained for c_2.

CHAPTER 9

DIFFUSION ALONG CRYSTAL IMPERFECTIONS

Experiments demonstrate that along crystal imperfections such as dislocations, internal interfaces, and free surfaces, diffusion rates can be orders of magnitude faster than in crystals containing only point defects. These line and planar defects provide *short-circuit diffusion paths*, analogous to high-conductivity paths in electrical systems. Short-circuit diffusion paths can provide the dominant contribution to diffusion in a crystalline material under conditions described in this chapter.

9.1 THE DIFFUSION SPECTRUM IN IMPERFECT CRYSTALS

Rapid diffusion along line and planar crystal imperfections occurs in a thin region centered on the defect core. For a dislocation, the region is cylindrical, roughly two interatomic distances in diameter, and includes the "bad material" in the dislocation core.[1] For a grain boundary, the region is a thin slab, roughly two interatomic distances thick, including the bad material in the grain boundary core. For a free surface, this region is the first few atomic layers of the material at the surface. These regions are very thin in comparison to the usual diffusional transport distances. To model the diffusion due to these imperfections, we replace them by thin slabs or cylinders of effective thickness, δ, possessing effective diffusivities which are much larger than the diffusivity in the adjoining crystalline material. Table 9.1 lists the

---

[1] *Bad material* is disordered material in which the regular atomic structure characteristic of the crystalline state no longer exists. *Good bulk material* is free of line or planar imperfections.

Kinetics of Materials. By Robert W. Balluffi, Samuel M. Allen, and W. Craig Carter.   **209**
Copyright © 2005 John Wiley & Sons, Inc.

Table 9.1:   Notation for Short-Circuit Diffusivities

D^D(undissoc)	diffusivity along an undissociated dislocation core (i.e., a cylinder, or a "pipe" of diameter, δ)
D^D(dissoc)	diffusivity along a dissociated dislocation core (i.e., a cylinder, or a "pipe" of diameter, δ)
D^B	diffusivity along a grain boundary (i.e., a slab of thickness, δ)
D^S	diffusivity along a free surface (i.e., a slab of thickness, δ)
D^{XL}	diffusivity in a bulk crystal free of line or planar imperfections
D^L	diffusivity in a liquid

notation to be used to describe the diffusivities in various regions of crystalline materials containing line and planar imperfections.

Figure 9.1 presents self-diffusivity data for $^\star D^D$(dissoc), $^\star D^D$(undissoc), $^\star D^B$, $^\star D^S$, $^\star D^{XL}$, and $^\star D^L$, for f.c.c. metals on a single Arrhenius plot. With the exception of the surface diffusion data, the data are represented by ideal straight-line Arrhenius plots, which would be realistic if the various activation energies were constants (independent of temperature). However, the data are not sufficiently accurate or extensive to rule out some possible curvature, at least for the grain boundary and dislocation curves, as discussed in Section 9.2.3.

Dislocations, grain boundaries, and surfaces can possess widely differing structures, and these structural variations affect their diffusivities to significant degrees. If the defective core region is less dense or "looser" than defect-free material, or if a defect possesses structurally "open" channels in its core structure, transport will generally be more rapid along the defect, particularly in the open directions. Some grain boundary structures can be represented by dislocation arrays, and their boundary diffusivity can be modeled in terms of transport along the grain-boundary

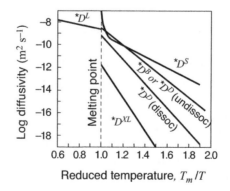

Figure 9.1:   Master Arrhenius plot of $^\star D^{XL}$, $^\star D^D$(dissoc), $^\star D^D$(undissoc), $^\star D^B$, $^\star D^S$, and $^\star D^L$ characteristic of f.c.c. metals. Data for various f.c.c. metals have been normalized by using a reduced reciprocal temperature scale, $(1/T)/(1/T_m) = T_m/T$. All diffusivities were derived from experimental data by assuming that all $\delta = 0.5$ nm. From Gjostein [1].

dislocation cores. General grain boundary structures cannot support discrete local-ized dislocations but, nevertheless, still act as short-circuit diffusion paths.

Short-circuit diffusion along grain boundaries has been studied extensively via experiments and modeling. Because diffusion along dislocations and crystal sur-faces is comparatively less well characterized, particular attention is paid to grain-boundary transport in this chapter. However, briefer discussions of diffusion along dislocations and free surfaces are also presented.

To describe the effects of grain-boundary structure on boundary diffusion, it is necessary to review briefly some important aspects of boundary structure. Addi-tional details appear in Appendix B. It takes a minimum of five geometric pa-rameters to define a crystalline interface. Three describe the crystal/crystal *mis-orientation*: e.g., two to specify the axis about which one crystal is rotated with respect to the other, and one for the rotation angle. The remaining two parameters define the *inclination* of the plane along which the crystals abut at the interface.[2] If the interface is a free surface, just two parameters are required to specify the surface's inclination (unit normal). Crystal symmetries determine special values of the parameters at which the interfacial energies take on extreme values. Depending on the specific nature of a system with interfaces, some of the parameters may be constrained and others free to vary as the system seeks a lower-energy state.

Small-angle grain boundaries have crystal misorientations less than about 15° and consist of regular arrays of discrete dislocations (i.e., where the cores are sep-arated by regions of defect-free material). As the crystal misorientation across the boundary increases beyond about 15°, the dislocation spacing becomes so small that the cores overlap and the boundary becomes a continuous slab of bad mate-rial; these are called *large-angle boundaries*. Large-angle boundaries can be further classified into singular boundaries, vicinal boundaries, and general boundaries.[3]

An interface is regarded as *singular* with respect to a degree of freedom if it is at a local minimum in energy with respect to changes in that degree of freedom. It is therefore stable against changes in that degree of freedom.

A *vicinal interface* is an interface that deviates from being singular by a rela-tively small variation of one or more of its geometric parameters from their singular-interface values. A vicinal interface can therefore minimize its energy by adopting a fit–misfit structure consisting of patches of the nearby minimum-energy singular interface delineated by arrays of discrete interfacial dislocations or steps as illus-trated in Figs. B.4 and B.9. These line defects serve to accommodate the relatively small deviations of the vicinal interfaces from the singular interfaces.

A *general interface* is not energy-minimized with respect to any of its degrees of freedom, and is far from any singular-interface values of the parameters that set its degrees of freedom. Such an interface cannot reduce its energy by adopting a fit–misfit structure (as in the vicinal case) and therefore cannot support localized dislocations or steps. Two examples serve to clarify these distinctions:

Example 1 The tilt grain boundary in Fig. B.4a is singular with respect to its tilt angle.[4] The boundary in Fig. B.4c is vicinal to the singular boundary

---

[2]Additional variables may be required, such as three that specify a relative translation of one crystal with respect to the other.

[3]Similar terminology is used for classification of free-surface structure.

[4]See Appendix B for descriptions of tilt, twist, and mixed grain boundaries.

with respect to its tilt angle. It consists of patches of the singular boundary delineated by dislocations that accommodate the change in tilt angle.

Example 2 A surface corresponding to the patch of light-colored atoms in Fig. B.1 is singular with respect to its inclination about an axis parallel to the surface steps in the figure. The stepped surface in Fig. B.1 is vicinal to such a flat surface and consists of patches of the flat surface delineated by steps that accommodate the change in surface inclination.

Because the structure of general large-angle grain boundaries is usually less regular and rigid than that of singular or vicinal boundaries, its activation energies for diffusion are typically lower and the diffusivities correspondingly higher. The diffusion rate along small-angle grain boundaries is generally lower than along large-angle grain boundaries and, indeed, approaches D^{XL} as the crystal misorientation approaches zero. This is due to two factors: first, the diffusion rate along the bad material in dislocation cores is about the same as, or lower than, that along large-angle grain boundary cores (see Fig. 9.1); second, because small-angle grain boundaries consist of periodic arrays of lattice dislocations at discrete spacings that approach infinity as the crystal misorientation approaches zero, the density of fast-diffusion paths is smaller in small-angle boundaries than in large-angle boundaries.

Figure 9.2 presents diffusivity data for a series of tilt boundaries as a function of the misorientation tilt angle.

The structures of these boundaries vary considerably as the misorientation changes. In the central part of the plot, the minima occur at crystal misorientations (values of θ) corresponding to singular and vicinal boundaries. The ends of the plot (where the crystal misorientation approaches zero) correspond to small-angle boundaries, and the diffusivities are correspondingly low. The regions centered around the maxima in Fig. 9.2 correspond to general grain boundaries. Polycrystalline materials not subjected to special processing conditions possess mainly general boundaries; the grain-boundary data in Fig. 9.1 are for general boundaries that have fairly similar diffusivities and can therefore be described reasonably well by average normalized values.

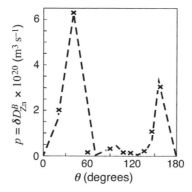

Figure 9.2:   Grain-boundary diffusivity of Zn along the tilt axes of [110] symmetric tilt grain boundaries in Al as a function of tilt angle, θ. From *Interfaces in Crystalline Materials* by A.P. Sutton and R.W. Balluffi (1995). Reprinted by permission of Oxford University Press. Data from I. Herbeuval and M. Biscondi [2].

The wide range of diffusivity magnitudes evident in the diffusivity spectrum in Fig. 9.1 may be expected intuitively; as the atomic environment for jumping becomes progressively less free, the jump rates, Γ, decrease accordingly in the sequence $\Gamma^S > \Gamma^B \approx \Gamma^D(\text{undissoc}) > \Gamma^D(\text{dissoc}) > \Gamma^{XL}$. The activation energies for these diffusion processes consistently follow the reverse behavior,

$$E^S < E^B \approx E^D(\text{undissoc}) < E^D(\text{dissoc}) < E^{XL} \tag{9.1}$$

The diffusivity in free surfaces is larger than that in general grain boundaries, which is about the same as that in undissociated dislocations. Furthermore, the diffusivity in undissociated dislocations is greater than that in dissociated dislocations, which is greater than that in the crystal:[5]

$$^*\!D^S > {}^*\!D^B \approx {}^*\!D^D(\text{undissoc}) > {}^*\!D^D(\text{dissoc}) > {}^*\!D^{XL} \tag{9.2}$$

Free-surface and grain-boundary diffusivities in metals at $0.5T_m$ are seven to eight orders of magnitude larger than crystal diffusivities. Provided that defects are present at sufficiently high densities, significant amounts of mass transport can occur in crystals at $0.5T_m$ via surface and grain-boundary diffusion even though the cross-sectional area through which the diffusional flux occurs is relatively very small. As the temperature is lowered further, the ratio of diffusivities becomes larger and short-circuit diffusion assumes even greater importance. Generally, similar behavior is found in ionically bonded crystals, as shown in Fig. 9.3.

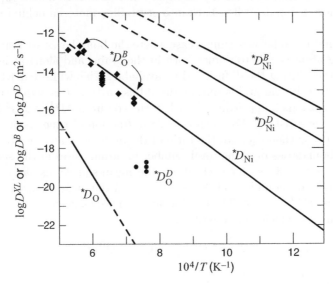

Figure 9.3:   Self-diffusivities of O and Ni on their respective sublattices in a NiO single crystal free of significant line imperfections and along grain boundaries in a polycrystal. The grain-boundary diffusivities of both Ni and O in the oxide semiconductor NiO are very much greater than corresponding crystal diffusivities. From Atkinson [3].

There are many situations, particularly at low temperatures, where short-circuit diffusion along grain boundaries and free surfaces is the dominant mode of diffusional transport and therefore controls important kinetic phenomena in materials;

---

[5]We discuss diffusion along dislocations and free surfaces in Sections 9.3 and 9.4.

several examples are discussed in Sections 9.2 and 9.4. Similar conclusions hold for dislocation diffusional short-circuiting, although to a lesser degree because of the relatively small cross sections of the high-diffusivity pipes.

9.2 DIFFUSION ALONG GRAIN BOUNDARIES

9.2.1 Regimes of Grain-Boundary Short-Circuit Diffusion in a Polycrystal

In a polycrystal containing a network of grain boundaries, atoms may migrate in both the grain interiors and the grain boundary slabs [4]. They may jump into or out of boundaries during the time available, and spend various lengths of time jumping in the grains and along the boundaries. Widely different situations may occur, depending upon such variables as the grain size, the temperature, the diffusion time, and whether the boundary network is stationary or moving. For example, as the grain size is reduced and more boundaries become available, the overall diffusion will be enhanced due to the relatively fast diffusion along the boundaries. At elevated temperatures where the ratio of the boundary diffusivity to the crystal diffusivity is lower than at low temperatures (Fig. 9.1), the importance of the boundary diffusion will be diminished. At very long diffusion times, the distance each atom diffuses will be relatively large, and each atom will be able to sample a number of grains and grain boundaries. If the boundaries are moving, an atom in a grain may be overrun by a moving boundary and be able to diffuse rapidly in the boundary before being deposited back into crystalline material behind the moving boundary.

Consider first the relatively simple case where the boundaries are stationary and each diffusing atom is able to diffuse both in the grains and along at least several grain boundaries during the diffusion time available. This will occur whenever the diffusion distance in the grains during the diffusion time t is significantly larger than the grain size [i.e., approximately when the condition $^\star D^{XL} t > s^2$ (where s is the grain size) is satisfied]. For each atom, the fraction of time spent diffusing in grain boundaries is then equal to the ratio of the number of atomic sites that exist in the grain boundaries over the total number of atomic sites in the specimen [5]. This fraction is $\eta \approx 3\delta/s$: for each atom, the mean-square displacement due to diffusion along grain boundaries is then $^\star D^B \eta t$, and the mean-square displacement in the grains is $^\star D^{XL}(1 - \eta)t$. The total mean-square displacement is then the sum of these quantities, which can be written

$$\langle ^\star D \rangle t =^\star D^{XL}(1 - \eta)t +^\star D^B \eta t \tag{9.3}$$

and because $\eta \ll 1$,

$$\langle ^\star D \rangle = \,^\star D^{XL} + (3\delta/s)\,^\star D^B \qquad ^\star D^{XL} t > s^2 \tag{9.4}$$

The quantity $\langle ^\star D \rangle$ is the average effective diffusivity, which describes the overall diffusion in the system. The diffusion in the system therefore behaves macroscopically as if bulk diffusion were occurring in a homogeneous material possessing a uniform diffusivity given by Eq. 9.4. The situation is illustrated schematically in Fig. 9.4a, and experimental data for diffusion of this type are shown in Fig. 9.5. This diffusion regime is called the *multiple-boundary diffusion regime* since the diffusion field

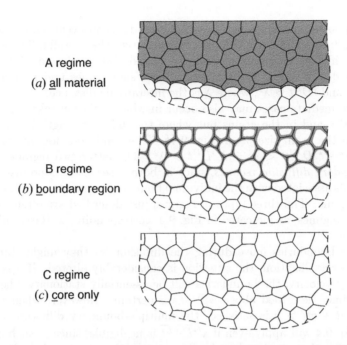

A regime

(*a*) a̱ll material

B regime

(*b*) ḇoundary region

C regime

(*c*) c̱ore only

Figure 9.4: The A, B, and C regimes for self-diffusion in polycrystal with stationary grain boundaries according to Harrison [6]. The tracer atoms are diffusing into the semi-infinite specimen from the surface located along the top of each figure. Regions of relatively high tracer concentration are shaded. **(a)** Regime A: the diffusion length in the grains is considerably longer than the average grain size. **(b)** Regime B: the diffusion length in the grains is significant but smaller than the grain size. **(c)** Regime C: the diffusion length in the grains is negligible, but significant diffusion occurs along the grain boundaries. In all figures, preferential penetration within the grain boundaries is too narrow to be depicted.

overlaps multiple boundaries. Note that in Fig. 9.4*a*, fast grain-boundary diffusion will cause preferential diffusion to occur along the narrow grain-boundary cores beyond the main diffusion front, but the number of atoms will be relatively small and this effect cannot be depicted.

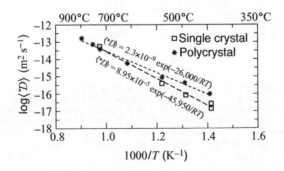

Figure 9.5: Values of the average self-diffusivity, $\langle {}^*D \rangle$, in single- and polycrystalline silver. At lower temperatures, grain-boundary diffusion makes significant contributions to the overall measured average diffusivity in the polycrystal. From Turnbull [7].

At the opposite extreme when essentially no diffusion occurs in the grains but significant diffusion still occurs along the boundaries, the overall diffusion will consist of only diffusion penetration along the boundaries, as illustrated in Fig. 9.4c. This will tend to occur at low temperatures or short times under the conditions $^\star D^{XL}t < \lambda^2$ and $^\star D^B t > \lambda^2$, where λ is the interatomic distance.

Many intermediate cases may also occur in which diffusion takes place in both the boundaries and in the grains but where the diffusion length in the grains is smaller than the grain size, as in Fig. 9.4b. The conditions for this type of diffusion are $\lambda^2 <^\star D^{XL}t < s^2$ and $^\star D^B t > \lambda^2$. The latter two regimes are called *isolated-boundary diffusion regimes*, since in both cases there is no overlap of the diffusion fields associated with the individual boundary segments, as in the multiple-boundary regime. The three types of regimes just described are often termed the A, B, and C regimes, as indicated in Fig. 9.4, corresponding to Harrison's original designation [6].

When the boundaries move during the diffusion, as they might during grain growth or recrystallization, the situation is considerably altered. If v is the average boundary velocity, the boundaries will be essentially stationary when $vt < \lambda$, and the regimes described above will again pertain. However, when the condition $[\sqrt{^\star D^{XL}t} + vt] > s$ is satisfied, the multiple-boundary diffusion regime will hold, and Eq. 9.4 will apply even if $\sqrt{^\star D^{XL}t}$ is negligible, since in such a case the boundaries visit the atoms rather than vice versa. Conversely, when the condition $[\sqrt{^\star D^{XL}t} + vt] < s$ is satisfied, the isolated-boundary diffusion regime will exist.

The various regimes of possible diffusion behavior can be represented graphically in an approximate manner, as shown in Fig. 9.6 [8]. The axes are taken to be $\log(^\star D^{XL}t)$ and $\log(vt)$: logarithmic scales have been used to show the details near the origin because s/λ is typically 10^3 or more. The stationary-boundary regimes

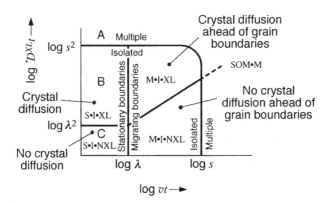

Figure 9.6: The regimes of diffusion behavior in a polycrystal in which diffusion may occur both in the grains and along the grain boundaries and the boundaries may be stationary or moving. λ is the interatomic spacing and s is the grain size. On the left side, where the boundaries are essentially stationary, Harrison's A, B, and C regimes are shown. **S·I·XL** ≡ Stationary boundaries, Isolated boundary diffusion, and crystal (**XL**) diffusion penetration into adjacent grains. **S·I·NXL** ≡ Stationary boundaries, Isolated boundary diffusion, and **N**o crystal (**XL**) diffusion penetration into adjacent grains. **M·I·XL** ≡ Moving boundaries, Isolated boundary diffusion, crystal (**XL**) diffusion ahead of boundaries. **M·I·NXL** ≡ Moving boundaries, Isolated boundary diffusion, **N**o crystal (**XL**) diffusion ahead of boundaries. **SOM·M** ≡ Stationary **O**r **M**oving boundaries, **M**ultiple boundary diffusion. From Cahn and Balluffi [8].

($vt < \lambda$) are shown on the left and include Harrison's A, B, and C regimes. The isolated-boundary regimes are enclosed in a region that includes the origin and extends out along the vertical and horizontal axes to distances where $^{\star}D^{XL}t = s^2$ and $vt = s$, respectively. Beyond the isolated-boundary regimes the multiple-boundary regime holds sway in all locations.

The isolated-boundary regime for moving boundaries in Fig. 9.6 is subdivided into two regimes, depending on whether the crystal diffusion is fast enough so that the atoms are able to diffuse out into the grains ahead of the advancing boundaries. To analyze this, consider a boundary segment between two grains moving with velocity v as in Fig. 9.7a.

Atoms are diffusing into the boundary laterally from its edges and can diffuse out through its front face into the forward grain. At the same time, atoms will be deposited in the backward grain in the wake of the boundary. In the quasi-steady state in a coordinate system fixed to the moving boundary, the diffusion flux in the forward grain is $J = -^{\star}D^{XL}(dc/dx) - vc$ and the diffusion equation is

$$-\nabla \cdot \vec{J} = -\frac{d}{dx}\left[-^{\star}D^{XL}\frac{dc}{dx} - vc\right] = 0 \qquad (9.5)$$

with the solution

$$^{\star}D^{XL}\frac{dc}{dx} + vc = A \qquad (9.6)$$

where A is a constant. At a large distance in front of the boundary, $dc/dx \to 0$ and $c \to 0$ and therefore $A = 0$. Finally, upon integration,

$$c = c^{\circ}\,e^{-vx/^{\star}D^{XL}} \qquad (9.7)$$

where c° is the concentration maintained at the boundary. The resulting concentration profile is shown in Fig. 9.7b. According to Eq. 9.7, the concentration in front of the boundary will be negligible when $^{\star}D^{XL}/v < \lambda$. Therefore, the curve separating the regimes indicated by **M·I·XL** and **M·I·NXL** in Fig. 9.6 should follow the straight-line relationship $^{\star}D^{XL}t = \lambda vt$, as indicated.

The diagram in Fig. 9.6 is highly approximate, but it is useful for visualizing the various regimes that might be expected during diffusion in a polycrystal. With increasing time, the point representing the system will start at the origin and move

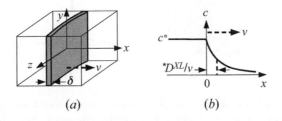

(a) $\qquad\qquad\qquad\qquad\qquad$ (b)

Figure 9.7:   (a) Grain boundary moving with constant velocity v. Tracer atoms are diffusing rapidly transversely into the boundary slab from its edges, are also diffusing normally out of the boundary into the grain ahead of the boundary, and are being deposited in the grain behind the boundary. Diffusion is in a steady state in a coordinate system moving with the boundary. (b) Tracer concentration in the vicinity of the boundary according to Eq. 9.7.

progressively away from it. If diffused long enough it will inevitably reach the multiple-boundary regime, regardless of whether the boundaries are stationary or moving.

9.2.2   Analysis of Diffusion in the A, B, and C Regimes

A Regime: Since diffusion in this regime is macroscopically similar to diffusion in a homogeneous material possessing an effective bulk diffusivity (Eq. 9.4), it may be analyzed by the methods described in Chapters 4 and 5.

B Regime: In this regime, the diffusant diffuses along a boundary while simultaneously leaking out by diffusing into the adjoining grains. Analysis of this type of diffusion is therefore considerably more complex than for diffusion in the A and C regimes since it involves solving for the coupled diffusion fields in the grain boundary and in the adjoining grains. This problem has been solved to different degrees of accuracy for several boundary conditions [9]. Solutions are generally obtained that contain the lumped grain-boundary diffusion parameter $p = \delta^{\star}D^B$ and the crystal diffusivity $^{\star}D^{XL}$. The analyses can then be applied to experimental results to obtain values of p when $^{\star}D^{XL}$ is known.

Fisher has produced a relatively simple solution for a specimen geometry that is convenient for experimentalists and which has been widely used in the study of boundary self-diffusion by making several approximations which are justified over a range of conditions [9, 10]. The geometry is shown in Fig. 9.8; it is assumed that the specimen is semi-infinite in the y direction and that the boundary is stationary. The boundary condition at the surface corresponds to constant unit tracer concentration, and the initial condition specifies zero tracer concentration within the specimen. Rapid diffusion then occurs down the boundary slab along y, while tracer atoms simultaneously leak into the grains transversely along x by means of crystal diffusion. The diffusion equation in the boundary slab then has the form

$$\frac{\partial c^B}{\partial t_1} = \frac{\partial^2 c^B}{\partial y_1^2} + 2\left(\frac{\partial c^{XL}}{\partial x_1}\right)_{x_1=0} \tag{9.8}$$

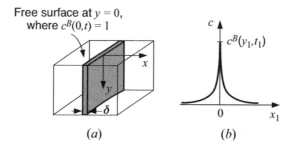

(a) (b)

Figure 9.8:   Isolated-boundary (Type-B) self-diffusion associated with a stationary grain boundary. **(a)** Grain boundary of width δ extending downward from the free surface at $y = 0$. The surface feeds tracer atoms into the grain boundary and maintains the diffusant concentration at the grain boundary's intersection with the surface at the value $c^B(y = 0, t) = 1$. Diffusant penetrates the boundary along y and simultaneously diffuses transversely into the grain interiors along x. **(b)** Diffusant distribution as a function of scaled transverse distance, x_1, from the boundary at scaled depth, y_1, from the surface. Penetration distance in grains is assumed large relative to δ.

where x_1, y_1, and t_1 are reduced dimensionless variables defined by $x_1 = x/\delta$, $y_1 = (y/\delta)\sqrt{{}^\star D^{XL}/{}^\star D^B}$, and $t_1 = t^\star D^{XL}/\delta^2$. In the process of solving this equation, Fisher found that the combination of relatively fast diffusion along the grain boundary and slower leakage into the grains causes the concentration in the boundary slab to quickly "saturate" so that the concentration in the adjoining grains at the time t_1 is essentially the same as the concentration that would have been there if the concentration distribution which existed in the boundary at t_1 had been maintained constant there since the start of the diffusion at $t_1 = 0$. Also, since the diffusion along the boundary is rapid compared to the transverse diffusion in the grains, gradients along y in the grains are much smaller than gradients along x in the grains and can therefore be neglected. The transverse concentration profile along x_1 in the grains at constant y_1 is therefore, to a good approximation, an error-function type of solution (Eq. 5.23) of the form

$$c^{XL}(x_1, y_1, t_1) = c^B(y_1, t_1)\left[1 - \text{erf}\left(\frac{x_1}{2t_1^{1/2}}\right)\right] \tag{9.9}$$

Also, the rapid saturation found in the boundary slab produces a quasi-steady-state condition along the boundary: $\partial c^B/\partial t_1$ in Eq. 9.8 can then be set equal to zero so that

$$0 = \frac{\partial^2 c^B(y_1, t_1)}{\partial y_1^2} - 2\,c^B(y_1, t_1)\left[\frac{\partial}{\partial x_1}\text{erf}\left(\frac{x_1}{2t_1^{1/2}}\right)\right]_{x_1=0} \tag{9.10}$$

The solution to Eq. 9.10 must satisfy the boundary condition $c^B(0, t_1) = 1$ and is therefore

$$c^B(y_1, t_1) = \exp\left[-\left(\frac{4}{\pi t_1}\right)^{1/4}y_1\right] \tag{9.11}$$

Putting this result into Eq. 9.9 in order to find $c^{XL}(x_1, y_1, t_1)$,

$$c^{XL}(x_1, y_1, t_1) = \exp\left[-\left(\frac{4}{\pi t_1}\right)^{1/4}y_1\right]\left[1 - \text{erf}\left(\frac{x_1}{2t_1^{1/2}}\right)\right] \tag{9.12}$$

The results above and the approximations made to obtain them have been shown to be valid when the dimensionless parameter $\beta = \delta^\star D^B({}^\star D^{XL})^{-3/2}t^{-1/2} \geq 20$ [9]. When this condition is not satisfied, more rigorous but complex analysis is required.

Experimentalists have frequently used Eq. 9.12 to determine values of the lumped grain boundary diffusion parameter $p = \delta^\star D^B$. The specimen is diffused for the time t and is sectioned by removing thin slices parallel to the surface of thickness Δy. The tracer content of each slice, ΔN, is then measured and plotted logarithmically against y. From Eq. 9.12 the resulting curve should have the slope

$$\frac{d\ln\Delta N}{dy} = -\left[\frac{4\,{}^\star D^{XL}}{\pi t}\right]^{1/4}\left[\frac{1}{\delta\,{}^\star D^B}\right]^{1/2} \tag{9.13}$$

and $p = \delta^\star D^B$ can then be determined when ${}^\star D^{XL}$ is known.

Further analyses of B-regime diffusion, including diffusion under different boundary conditions, are described by Kaur and Gust [9].

When solute atoms rather than tracer isotope atoms diffuse in the B regime, further analysis is necessary. Solute atoms may be expected to segregate in the grain boundary, and the concentration in the boundary slab $c_2^B(y_1, t_1)$ will then differ from the concentration in the grains in the direct vicinity of the boundary slab $c_2^{XL}(0, y_1, t_1)$. Assuming local equilibrium between these concentrations and that a simple McLean-type segregation isotherm typical of a dilute solution applies [4], the two concentrations will be related by

$$\frac{c_2^B(y_1, t_1)}{c_2^{XL}(0, y_1, t_1)} = k \tag{9.14}$$

where k is a constant equilibrium segregation ratio. When Eq. 9.14 is substituted into Eq. 9.9,

$$c_2^{XL}(x_1, y_1, t_1) = \frac{c_2^B(y_1, t_1)}{k} \left[1 - \mathrm{erf}\left(\frac{x_1}{2t_1^{1/2}} \right) \right] \tag{9.15}$$

When Eq. 9.15 is used instead of Eq. 9.9 and the same procedure is used that produced Eq. 9.13 for self-diffusion, it is found (see Exercise 9.3) that

$$\frac{d\ln \Delta N}{dy} = -\left(\frac{4D_2^{XL}}{\pi t} \right)^{1/4} \left(\frac{1}{k\delta D_2^B} \right)^{1/2} \tag{9.16}$$

holds for the solute diffusion, where D_2^{XL} and D_2^B are the solute diffusivities in the crystal and the boundary, respectively. The lumped grain-boundary diffusion parameter $k\delta D_2^B$ can be determined experimentally as before from a plot of $\ln \Delta N$ vs. y, but it now contains the segregation ratio k. Values of δD_2^B can therefore be obtained only when independent information about k is available. Further analysis is required if the simple McLean isotherm does not apply and k is concentration-dependent.

C Regime: In this regime, diffusion occurs only in the thin grain-boundary slabs. Since the number of diffusing atoms within the slabs is exceedingly small, the experimental measurement of boundary concentration profiles is difficult. Recourse has therefore been made to accumulation methods where the number of atoms which have diffused along a grain boundary are collected in a form that can readily be measured. For example, solute atoms have been deposited on one surface of a thin-film specimen possessing a columnar grain structure and then diffused through the film along the grain boundaries so that they accumulated on the reverse surface [11, 12]. The diffusion was carried out at low temperatures where no crystal diffusion occurred, and where, according to Fig. 9.1, the diffusion along the surfaces was much more rapid than the diffusion along the grain boundaries. Diffusion through the film specimen was therefore controlled by the rate of grain-boundary diffusion. Measurement of the rate of accumulation of the solute on the reverse surface then allowed the measurement of the lumped parameter δD_2^B as detailed in Exercise 9.4.

9.2.3 Mechanism of Fast Grain-Boundary Diffusion

The mechanisms by which fast grain-boundary diffusion occurs are not well established at present. There is extensive evidence that a net diffusional transport of atoms can be induced along grain boundaries, ruling out the ring mechanism and implicating defect-mediated mechanisms as responsible for grain-boundary diffusion [13]. Due to the small amount of material present in the grain boundary, it has not been possible, so far, to gain critical information about defect-mediated processes using experimental techniques. Recourse has been made to computer simulations which indicate that vacancy and interstitial point defects can exist in the boundary core as localized bona fide point defects (see the review by Sutton and Balluffi [4]). Calculations also show that their formation and migration energies are often lower than in the bulk crystal. Figure 9.9 shows the calculated trajectory of a vacancy in the core of a large-angle tilt grain boundary in b.c.c. Fe. Calculations showed that vacancies were more numerous and jump faster in the grain boundary than in the crystal, indicating a vacancy mechanism for diffusion in this particular boundary. However, there is an infinite number of different types of boundaries, and computer simulations for other types of boundaries indicate that the dominant mechanism in some cases may involve interstitial defects [4, 12].

During defect-mediated grain-boundary diffusion, an atom diffusing in the core will move between the various types of sites in the core. Because various types of jumps have different activation energies, the overall diffusion rate is not controlled by a single activation energy. Arrhenius plots for grain-boundary diffusion therefore should exhibit at least some curvature. However, when the available data are of only moderate accuracy and exist over only limited temperature ranges, such curvature may be difficult to detect. This has been the case so far with grain-boundary diffusion data, and the straight-line representation of the data in the Arrhenius

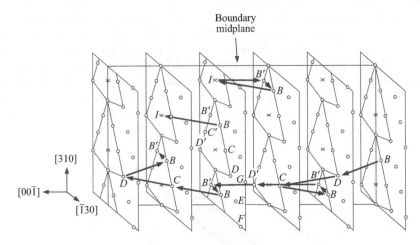

Figure 9.9: Calculated atom jumps in the core of a $\Sigma 5$ symmetric $\langle 001 \rangle$ tilt boundary in b.c.c. Fe. A pair-potential–molecular-dynamics model was employed. For purposes of clarity, the scales used in the figure are $[\bar{1}30] : [310] : [00\bar{1}] = 1 : 1 : 5$. All jumps occurred in the fast-diffusing core region. Along the bottom, a vacancy was inserted at B, and subsequently executed the series of jumps shown. The trajectory was essentially parallel to the tilt axis. Near the center of the figure, an atom in a B site jumped into an interstitial site at I. At the top an atom jumped between B, I and B' sites. From Balluffi et al. [14].

plot in Fig. 9.3 must be regarded as an approximation that yields an effective activation energy, E^B, for the temperature range of the data. Some evidence for curvature of Arrhenius plots for grain-boundary diffusion has been reviewed [4].

9.3   DIFFUSION ALONG DISLOCATIONS

As with grain boundaries, dislocation-diffusion rates vary with dislocation structure, and there is some evidence that the rate is larger along a dislocation in the edge orientation than in the screw orientation [15]. In general, dislocations in close-packed metals relax by dissociating into partial dislocations connected by ribbons of stacking fault as in Fig. 9.10 [16]. The degree of dissociation is controlled by the stacking fault energy. Dislocations in Al are essentially nondissociated because of its high stacking fault energy, whereas dislocations in Ag are highly dissociated because of its low stacking energy. The data in Fig. 9.1 (averaged over the available dislocation orientations) indicate that the diffusion rate along dislocations in f.c.c. metals decreases as the degree of dislocation dissociation into partial dislocations increases. This effect of dissociation on the diffusion rate may be expected because the core material in the more relaxed partial dislocations is not as strongly perturbed and "loosened up" for fast diffusion, as in perfect dislocations.

In Fig. 9.1, $^\star D^D$ for nondissociated dislocations is practically equal to $^\star D^B$, which indicates that the diffusion processes in nondissociated dislocation cores and large-angle grain boundaries are probably quite similar. Evidence for this conclusion also comes from the observation that dislocations can support a net diffusional transport of atoms due to self-diffusion [15]. As with grain boundaries, this supports a defect-mediated mechanism.

The overall self-diffusion in a dislocated crystal containing dislocations throughout its volume can be classified into the same general types of regimes as for a polycrystal containing grain boundaries (see Section 9.2.1). Again, the diffusion may be multiple or isolated, with or without diffusion in the lattice, and the dislocations may be stationary or moving. However, the critical parameters include $^\star D^D$ rather than $^\star D^B$ and the dislocation density rather than the grain size. The multiple-diffusion regime for a dislocated crystal is analyzed in Exercise 9.1.

Figure 9.11 shows a typical diffusion penetration curve for tracer self-diffusion into a dislocated single crystal from an instantaneous plane source at the surface [17]. In the region near the surface, diffusion through the crystal directly from the surface source is dominant. However, at depths beyond the range at

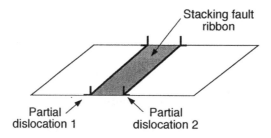

Figure 9.10:   Dissociated lattice dislocation in f.c.c. metal. The structure consists of two partial dislocations separated by a ribbon of stacking fault.

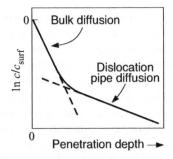

Figure 9.11:   Typical penetration curve for tracer self-diffusion from a free surface at tracer concentration c_{surf} into a single crystal containing dislocations. Transport near the surface is dominated by diffusion in the bulk; at greater depths, dislocation pipe diffusion is the major transport path.

which atoms can be delivered by crystal diffusion alone, long penetrating "tails" are present, due to fast diffusion down dislocations with some concurrent spreading into the adjacent lattice and no overlap of the diffusion fields of adjacent dislocations. This behavior corresponds to the dislocation version of the B regime in Fig. 9.4.

9.4 DIFFUSION ALONG FREE SURFACES

The general macroscopic features of fast diffusion along free surfaces have many of the same features as diffusion along grain boundaries because the fast-diffusion path is again a thin slab of high diffusivity, and a diffusing species can diffuse in both the surface slab and the crystal and enter or leave either region. For example, if a given species is diffusing rapidly along the surface, it may leak into the adjoining crystal just as during type-B kinetics for diffusion along grain boundaries. In fact, the mathematical treatments of this phenomenon in the two cases are similar.

The structure of crystalline surfaces is described briefly in Sections 9.1 and 12.2.1 and in Appendix B. All surfaces have a tendency to undergo a "roughening" transition at elevated temperatures and so become general. Even though a considerable effort has been made, many aspects of the atomistic details of surface diffusion are still unknown.[6]

For singular and vicinal surfaces at relatively low temperatures, surface-defect-mediated mechanisms involving single jumps of adatoms and surface vacancies are predominant.[7] Calculations indicate that the formation energies of these defects are of roughly comparable magnitude and depend upon the surface inclination [i.e., (hkl)]. Energies of migration on the surface have also been calculated, and in most cases, the adatom moves with more difficulty. Also, as might be expected, the diffusion on most surfaces is anisotropic because of their low two-dimensional symmetry. When the surface structure consists of parallel rows of closely spaced atoms, separated by somewhat larger inter-row distances, diffusion is usually easier parallel to the dense rows than across them. In some cases, it appears that the

---

[6]Our discussion follows reviews by of Shewmon [18] and Bocquet et al. [19].
[7]Adatoms, surface vacancies, and other features of surface structure are depicted in Fig. 12.1.

transverse diffusion occurs by a replacement mechanism in which an atom lying between dense rows diffuses across a row by replacing an atom in the row and pushing the displaced atom into the next valley between dense rows. Repetition of this process results in a mechanism that resembles the bulk interstitialcy mechanism described in Section 8.1.3. In addition, for vicinal surfaces, diffusion rates along and over ledges differs from those in the nearby singular regions.

At more elevated temperatures, the diffusion mechanisms become more complex and jumps to more distant sites occur, as do collective jumps via multiple defects. At still higher temperatures, adatoms apparently become delocalized and spend significant fractions of their time in "flight" rather than in normal localized states. In many cases, the Arrhenius plot becomes curved at these temperatures (as in Fig. 9.1), due to the onset of these new mechanisms. Also, the diffusion becomes more isotropic and less dependent on the surface orientation.

The mechanisms above allow rapid diffusional transport of atoms along the surface. We discuss the role of surface diffusion in the morphological evolution of surfaces and pores during sintering in Chapters 14 and 16, respectively.

Bibliography

1. N.A. Gjostein. Short circuit diffusion. In *Diffusion*, pages 241–274. American Society for Metals, Metals Park, OH, 1973.

2. I. Herbeuval and M. Biscondi. Diffusion of zinc in grains of symmetric flexion of aluminum. *Can. Metall. Quart.*, 13(1):171–175, 1974.

3. A. Atkinson. Diffusion in ceramics. In R.W. Cahn, P. Haasen, and E. Kramer, editors, *Materials Science and Technology–A Comprehensive Treatment*, volume 11, pages 295–337, Wienheim, Germany, 1994. VCH Publishers.

4. A.P. Sutton and R.W. Balluffi. *Interfaces in Crystalline Materials*. Oxford University Press, Oxford, 1996.

5. E.W. Hart. On the role of dislocations in bulk diffusion. *Acta Metall.*, 5(10):597, 1957.

6. L.G. Harrison. Influence of dislocations on diffusion kinetics in solids with particular reference to the alkali halides. *Trans. Faraday Soc.*, 57(7):1191–1199, 1961.

7. D. Turnbull. Grain boundary and surface diffusion. In J.H. Holloman, editor, *Atom Movements*, pages 129–151, Cleveland, OH, 1951. American Society for Metals. Special Volume of ASM.

8. J.W. Cahn and R.W. Balluffi. Diffusional mass-transport in polycrystals containing stationary or migrating grain boundaries. *Scripta Metall. Mater.*, 13(6):499–502, 1979.

9. I. Kaur and W. Gust. *Fundamentals of Grain and Interphase Boundary Diffusion*. Ziegler Press, Stuttgart, 1989.

10. J.C. Fisher. Calculation of diffusion penetration curves for surface and grain boundary diffusion. *J. Appl. Phys.*, 22(1):74–77, 1951.

11. J.C.M. Hwang and R.W. Balluffi. Measurement of grain-boundary diffusion at low-temperatures by the surface accumulation method 1. Method and analysis. *J. Appl. Phys.*, 50(3):1339–1348, 1979.

12. Q. Ma and R.W. Balluffi. Diffusion along [001] tilt boundaries in the Au/Ag system 1. Experimental results. *Acta Metall.*, 41(1):133–141, 1993.

13. R.W. Balluffi. Grain boundary diffusion mechanisms in metals. In G.E. Murch and A.S. Nowick, editors, *Diffusion in Crystalline Solids*, pages 319–377, Orlando, FL, 1984. Academic Press.

14. R.W. Balluffi, T. Kwok, P.D. Bristowe, A. Brokman, P.S. Ho, and S. Yip. Determination of the vacancy mechanism for grain-boundary self-diffusion by computer simulation. *Scripta Metall. Mater.*, 15(8):951–956, 1981.

15. R.W. Balluffi. On measurements of self diffusion rates along dislocations in f.c.c. metals. *Phys. Status Solidi*, 42(1):11–34, 1970.

16. R.E. Reed-Hill and R. Abbaschian. *Physical Metallurgy Principles*. PWS-Kent, Boston, 1992.

17. Y.K. Ho and P.L. Pratt. Dislocation pipe diffusion in sodium chloride crystals. *Radiat. Eff.*, 75:183–192, 1983.

18. P. Shewmon. *Diffusion in Solids*. The Minerals, Metals and Materials Society, Warrendale, PA, 1989.

19. J.L. Bocquet, G. Brebec, and Y. Limoge. Diffusion in metals and alloys. In R.W. Cahn and P. Haasen, editors, *Physical Metallurgy*, pages 535–668. North-Holland, Amsterdam, 2nd edition, 1996.

EXERCISES

9.1 In a Type-A regime, short-circuit grain-boundary self-diffusion can enhance the effective bulk self-diffusivity according to Eq. 9.4. A density of lattice dislocations distributed throughout a bulk single crystal can have a similar effect if the crystal diffusion distance for the diffusing atoms is large compared with the dislocation spacing.

Derive an equation similar to Eq. 9.4 for the effective bulk self-diffusivity, $\langle {}^{*}D \rangle$, in the presence of fast dislocation diffusion. Assume that the dislocations are present at a density, ρ, corresponding to the dislocation line length in a unit volume of material.

Solution. During self-diffusion, the fraction of the time that a diffusing atom spends in dislocation cores is equal to the fraction of all available sites that are located in the dislocation cores. This fraction will be $\eta = \rho \pi \delta^2 / 4$. The mean-square displacement due to self-diffusion along the dislocations is then ${}^{*}D^D \eta t$, while the corresponding displacement in the crystal is ${}^{*}D^{XL}(1 - \eta)t$. Therefore,

$$\langle {}^{*}D \rangle t = {}^{*}D^{XL}(1 - \eta)t + {}^{*}D^D \eta t \tag{9.17}$$

and because $\eta \ll 1$,

$$\langle {}^{*}D \rangle = {}^{*}D^{XL} + \frac{\rho \pi \delta^2}{4}\,{}^{*}D^D \tag{9.18}$$

9.2 Exercise 9.1 yielded an expression, Eq. 9.18, for the enhancement of the effective bulk self-diffusivity due to fast self-diffusion along dislocations present in the material at the density, ρ. Find a corresponding expression for the enhancement of the effective bulk self-diffusivity of solute atoms due to fast solute self-diffusion along dislocations. Assume that the solute atoms segregate to the dislocations according to simple McLean-type segregation where $c_2^D / c_2^{XL} = k = \text{constant}$, where c_2^D is the solute concentration in the dislocation cores and c_2^{XL} is the solute concentration in the crystal.

Solution. Because the fraction of solute sites in the dislocations is small, the number of occupied solute-atom sites (per unit volume) in the crystal is c_2^{XL}, and the number of occupied sites in the dislocations is $\rho\pi\delta^2 kc_2^{XL}/4$. The fraction of time that a diffusing solute atom spends in dislocation cores is then $\eta = \rho\pi\delta^2 k/4$. Therefore, following the same argument as in Exercise 9.1,

$$\langle {}^\star D_2 \rangle t = {}^\star D_2^{XL}(1 - \eta)t + {}^\star D_2^D \eta t \tag{9.19}$$

and thus

$$\langle {}^\star D_2 \rangle = {}^\star D_2^{XL} + \frac{\rho\pi\delta^2 k}{4} {}^\star D_2^D \tag{9.20}$$

9.3 For Type-B diffusion along a grain boundary, Eq. 9.9, which holds for self-diffusion, takes the form of Eq. 9.15 for solute diffusion when simple McLean-type segregation occurs with $c_2^B/c_2^{XL} = k$. Show that this causes Eq. 9.13, which holds for self-diffusion, to take the form

$$\frac{d\ln\Delta N}{dy} = -\left(\frac{4D_2^{XL}}{\pi t}\right)^{1/4}\left(\frac{1}{k\delta D_2^B}\right)^{1/2} \tag{9.21}$$

for solute diffusion.

Solution. As indicated in the text, Eq. 9.9 must have the form of Eq. 9.15 in order to satisfy the segregation condition $k = c_2^B/c_2^{XL}$ at the boundary slab. Equation 9.10 then becomes

$$0 = \frac{\partial^2 c_2^B(y_1, t_1)}{\partial y_1^2} - \frac{2c_2^B(y_1, t_1)}{k}\left[\frac{\partial}{\partial x_1}\text{ erf }\frac{x_1}{2t_1^{1/2}}\right]_{x_1=0} \tag{9.22}$$

Equation 9.11 becomes

$$c_2^B(y_1, t_1) = \exp\left[-\left(\frac{4}{\pi k^2 t_1}\right)^{1/4} y_1\right] \tag{9.23}$$

Equation 9.12 becomes

$$c_2^{XL}(x_1, y_1, t_1) = \frac{1}{k}\exp\left[-\left(\frac{4}{\pi k^2 t_1}\right)^{1/4} y_1\right]\left[1 - \text{erf }\frac{x_1}{2t_1^{1/2}}\right] \tag{9.24}$$

and, finally, Eq. 9.13 becomes

$$\frac{d\ln\Delta N}{dy} = -\left[\frac{4D_2^{XL}}{\pi t}\right]^{1/4}\left[\frac{1}{k\delta D_2^B}\right]^{1/2} \tag{9.25}$$

9.4 As described in Section 9.2.2, grain-boundary diffusion rates in the Type-C diffusion regime can be measured by the surface-accumulation method illustrated in Fig. 9.12. Assume that the surface diffusion is much faster than the grain-boundary diffusion and that the rate at which atoms diffuse from the "source" surface to the "accumulation" surface is controlled by the diffusion rate along the transverse boundaries. If the diffusant, designated component 2, is initially present on the source surface and absent on the accumulation surface and the specimen is isothermally diffused, a quasi-steady rate of accumulation of the diffusant is observed on the accumulation surface after a short initial transient. Derive a relationship between the rate of accumulation and the parameter δD_2^B that can be used to determine δD_2^B experimentally. Assume that each grain is a square of side d in the plane of the surface.

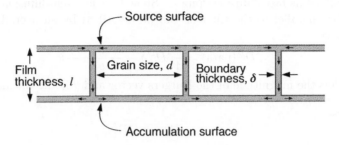

Figure 9.12:   Transport of diffusant through a thin polycrystalline film by grain-boundary diffusion.

Solution. Because of the fast surface diffusion, the concentrations of the diffusant on both surfaces are essentially uniform over their areas. After the initial transient, the quasi-steady rate (per unit area of surface) at which the diffusant diffuses along the transverse boundaries between the two surfaces is

$$\frac{dN}{dt} = -\frac{2\delta J^B}{d} = \frac{2\delta D_2^B (\partial c^B/\partial x)}{d} = \frac{2\delta D_2^B}{ld} \left[c^B(0) - c^B(l) \right] \qquad (9.26)$$

Here, d is the average grain size of the columnar grains, J^B is the diffusional flux along the grain boundaries, $\partial c^B/\partial x = \left[c^B(0) - c^B(l) \right]/l$, where $c^B(0)$ and $c^B(l)$ are the diffusant concentrations in the boundaries at the source surface and accumulation surface, respectively, and l is the specimen thickness. In the early stages, $c^B(l) \approx 0$ and, therefore, to a good approximation,

$$\delta D_2^B = \frac{ld}{2c^B(0)} \frac{dN}{dt} \qquad (9.27)$$

All quantities on the right-hand side of Eq. 9.27 are measurable, which allows the determination of δD_2^B [12].

9.5 Using the result of Exercise 9.1 and data in Fig. 9.1, estimate the density of dissociated dislocations necessary to enhance the average bulk self-diffusivity by a factor of 2 at $T_m/2$, where T_m is the absolute melting temperature of the material. *Note:* typical dislocation densities in annealed f.c.c. metal crystals are in the range 10^6–10^8 cm^{-2}.

Solution. Equation 9.18 may be solved for ρ in the form

$$\rho = \frac{4}{\pi \delta^2} \frac{^*D^{XL}}{^*D^D} \left[\frac{\langle ^*D \rangle}{^*D^{XL}} - 1 \right] \qquad (9.28)$$

It is estimated from Fig. 9.1 that $^*D^D(\text{dissoc})/^*D^{XL} = 3 \times 10^6$ at $T_m/T = 2.0$. Also, $\delta \approx 6 \times 10^{-8}$ cm. Using these values and $\langle ^*D \rangle/^*D^{XL} = 2$ in Eq. 9.28,

$$\rho \approx 10^8 \text{ cm}^{-2}$$

Therefore, it appears that the dislocations could make a significant contribution to diffusion under many common conditions.

9.6 The asymmetric small-angle tilt boundary in Fig. B.5a consists of an array of parallel edge dislocations running parallel to the tilt axis. During diffusion they will act as fast diffusion "pipes." Show that fast self-diffusion along this boundary parallel to the tilt axis can be described by an overall boundary diffusivity,

$$^\star D^B(\text{para}) = \frac{\pi}{4}\, ^\star D^D \delta\, \frac{\sin\phi + \cos\phi}{b}\, \theta \tag{9.29}$$

where b is the magnitude of the Burgers vector and θ is the tilt angle.

- Use

$$^\star D^D \gg\, ^\star D^L \tag{9.30}$$

Solution. As usual, take the boundary as a slab that is δ thick. In considering diffusion along the tilt axis, any contribution of the crystal regions in the slab can be neglected and only the contributions of the dislocation pipes are included because $^\star D^D \gg\, ^\star D^{XL}$. The flux through a unit cross-sectional area of the boundary slab is then

$$J = -\left[\frac{\pi\delta^2}{4}\, ^\star D^D\, \frac{\partial c}{\partial x}\right]\left[\frac{1}{\delta}\frac{\theta}{b}(\sin\phi + \cos\phi)\right] \tag{9.31}$$

where the first bracketed term is the flux along a single pipe and the second is the number of pipes per unit area of the boundary slab. The desired expression is obtained by equating this result with $J = -\, ^\star D^B(\text{para})\, \partial c/\partial x$ and solving for $^\star D^B$.

9.7 Self-diffusion along the boundary in Exercise 9.6 is highly anisotropic because diffusion along the tilt axis (parallel to the dislocations) is much greater than diffusion transverse to it (i.e., perpendicular to the dislocations but still in the boundary plane). Find an expression for the anisotropy factor,

$$\frac{^\star D^B(\text{para})}{^\star D^B(\text{transv})} \tag{9.32}$$

where $^\star D^B(\text{transv})$ is the boundary diffusivity in the transverse direction.

Solution. The transverse diffusion rate is controlled by the relatively slow crystal diffusion rate because the diffusing atoms must traverse the patches of perfect crystal between the dislocation pipes. Therefore, when the dislocations are discretely spaced, a good approximation is the simple result

$$\frac{^\star D^B(\text{para})}{^\star D^B(\text{transv})} = \frac{^\star D^B(\text{para})}{^\star D^{XL}} \tag{9.33}$$

CHAPTER 10

DIFFUSION IN NONCRYSTALLINE MATERIALS

Noncrystalline materials exist in many different forms. A huge variety of atomic and molecular structures, ranging from liquids to simple monatomic amorphous structures to network glasses to dense long-chain polymers, are often complex and difficult to describe. Diffusion in such materials occurs by a correspondingly wide variety of mechanisms, and is, in general, considerably more difficult to analyze quantitatively than is diffusion in crystals.

The understanding of diffusion in many noncrystalline materials has lagged behind the understanding of diffusion in crystalline material, and a unified treatment of diffusion in noncrystalline materials is impossible because of its wide range of mechanisms and phenomena. In many cases, basic mechanisms are still controversial or even unknown. We therefore focus on selected cases, although some of the models discussed are still under development and not yet firmly established.

10.1 FREE-VOLUME MODEL FOR SELF-DIFFUSION IN LIQUIDS

Self-diffusion in simple monatomic liquids at temperatures well above their glass-transition temperatures may be interpreted in a simple manner.[1] Within such liquids, regions with *free volume* appear due to displacement fluctuations. Occasionally, the fluctuations are large enough to permit diffusive displacements.

---

[1]This section closely follows Cohen and Turnbull's original derivation [1]. The original paper should be consulted for further details.

Kinetics of Materials. By Robert W. Balluffi, Samuel M. Allen, and W. Craig Carter.   **229**
Copyright © 2005 John Wiley & Sons, Inc.

The *hard-sphere model* for the liquid serves as a reasonably good approximation for the atomic interactions [2]. Here, the potential energy between any pair of approaching particles is assumed to be constant until they touch, at which point it becomes infinite. On average, the particles in the liquid maintain a volume larger than that which they would have if they all touched; the resulting volume difference is the free volume. Each particle effectively traverses a small confined volume within which the interatomic potentials are essentially flat [3]. The average velocity of a particle in the region of flat potential inside the confining volume is the same as the velocity of a gas particle. Most of the time a particular particle is confined to a particular region. However, there will occasionally be a fluctuation in local density that opens a space large enough to permit a considerable displacement of the particle. If another particle jumps into that space before the displaced first particle returns, a diffusive-type jump will have occurred. Diffusion therefore occurs as a result of the redistribution of the free volume that occurs at essentially constant energy because of the flatness of the interatomic potentials.

According to the kinetic theory of gases, the self-diffusivity of a hard-sphere gas is given by $^\star D^G = (2/5)\langle u \rangle L$, where $\langle u \rangle$ is the average velocity and L is the mean free path [4]. Because the mean free path of a confined particle in the liquid is about equal to the diameter of its confining volume, the contribution of the confined particle to the self-diffusivity of the liquid may be written

$$^\star D(V) = C_{\text{geom}}\, a(V)\, \langle u \rangle \tag{10.1}$$

where $a(V)$ is the diameter of the confining volume, V is the free volume associated with the particle, $\langle u \rangle$ is the average velocity of the particle, and C_{geom} is a geometrical constant. It is reasonable to assume that the diffusivity is very small, $^\star D(V) = 0$, unless the local free volume V exceeds a critical volume, V^{crit}. Therefore, the overall diffusivity may be expressed

$$^\star D^L = \int_{V^{\text{crit}}}^{\infty} {}^\star D(V)\, p(V)\, dV \tag{10.2}$$

where $p(V)\, dV$ is the free volume's probability that it lies between V and $V + dV$. To determine this probability distribution, consider a system containing $\mathcal{N}$ particles and divide the total range of possible free volumes for a particle into bins indexed by i. Let $N_i(V_i)$ be the number of particles with free volume V_i. If $\mathcal{V}^{\text{free}}$ is the total free volume, the condition

$$\mathcal{V}^{\text{free}} = \gamma \sum_i N_i V_i \tag{10.3}$$

must hold. The factor γ accounts for all free-volume overlap between adjacent particles. γ lies between zero and one because of the physical limits of complete and no overlap; its value is probably closer to one. The total number of particles, $\mathcal{N}$, is

$$\mathcal{N} = \sum_i N_i \tag{10.4}$$

The entropy associated with the number of ways that the free volume can be distributed at constant energy is

$$\mathcal{S}^{\text{conf}} = k \ln \left(\frac{\mathcal{N}!}{\prod_i N_i!} \right) \tag{10.5}$$

for bin populations given by N_i. The equilibrium probability distribution in Eq. 10.2 is the continuum limit of the bin populations N_i that maximize $\mathcal{S}^{\text{conf}}$ subject to constraints, Eqs. 10.3 and 10.4. Introducing Lagrange multipliers β and λ for the total free volume and fixed-number constraints, the extremal conditions are

$$\frac{\partial \mathcal{S}^{\text{conf}}|_{V_{\text{const}} N_{\text{const}}}}{\partial N_i} = \frac{\partial \left(\mathcal{S}^{\text{conf}} + \lambda \mathcal{V}^{\text{free}} + \beta \mathcal{N} \right)}{\partial N_i} = 0 \qquad (10.6)$$

which, using Stirling's formula $N_i! \approx N_i \ln N_i$ and the limit of $N_i \gg 1$, reduces to

$$N_i = e^{-\lambda/k} e^{-\beta \gamma V_i / k} \qquad (10.7)$$

The constraints Eqs. 10.3 and 10.4 determine the Lagrange multipliers. With Eq. 10.7,

$$\mathcal{N} = e^{-\lambda/k} \sum_i e^{-\beta \gamma V_i / k}$$

$$N_i = \frac{\mathcal{N} e^{-\beta V_i / k}}{\sum_i e^{-\beta V_i / k}} \qquad (10.8)$$

and

$$\mathcal{V}^{\text{free}} = \gamma \sum_i V_i N_i = \gamma \mathcal{N} \frac{\sum_i V_i e^{-\beta \gamma V_i / k}}{\sum_i e^{-\beta \gamma V_i / k}} \approx \gamma \mathcal{N} \frac{\int_0^\infty V e^{-\beta \gamma V / k} dV}{\int_0^\infty e^{-\beta \gamma V / k} dV} = \frac{\mathcal{N} k}{\beta} \qquad (10.9)$$

The average free volume per particle is $\langle \Omega^{\text{free}} \rangle = \mathcal{V}^{\text{free}} / \mathcal{N} = k/\beta$. Therefore, comparison of the average free volume to its definition from a probability distribution,

$$\frac{\mathcal{V}^{\text{free}}}{\mathcal{N}} = \frac{\sum_i \gamma V_i e^{-\beta \gamma V_i / k}}{\sum_i e^{-\beta \gamma V_i / k}} = \frac{\int_0^\infty V p(V) dV}{\int_0^\infty p(V) dV} \equiv \langle \Omega^{\text{free}} \rangle \qquad (10.10)$$

shows that the probability distribution $p(V)$ must be proportional to $\gamma \exp[-\gamma \beta V / k]$ $= \gamma \exp[-\gamma V / \langle \Omega^{\text{free}} \rangle]$. The proportionality factor can be determined by setting the sum of probabilities equal to one, and

$$p(V) = \frac{\gamma e^{-\gamma V / \langle \Omega^{\text{free}} \rangle}}{\langle \Omega^{\text{free}} \rangle} \qquad (10.11)$$

The probability distribution, Eq. 10.11, can be used in Eq. 10.2 as an estimate for evaluating $^\star D^L$. Above the critical free volume V^{crit}, $^\star D(V)$ is probably nearly constant; therefore,

$$
\begin{aligned}
^\star D^L &= \int_{V^{\text{crit}}}^\infty \frac{^\star D(V) \gamma e^{-\gamma V / \langle \Omega^{\text{free}} \rangle}}{\langle \Omega^{\text{free}} \rangle} dV \\
&\approx {}^\star D(V^{\text{crit}}) e^{-\gamma V^{\text{crit}} / \langle \Omega^{\text{free}} \rangle} = C_{\text{geom}} \, a(V^{\text{crit}}) \langle u \rangle \, e^{-\gamma V^{\text{crit}} / \langle \Omega^{\text{free}} \rangle}
\end{aligned} \qquad (10.12)
$$

Equation 10.12 matches diffusivities measured in simple liquids if the characteristic "cage" diameter, $a(V^{\text{crit}})$, is approximately the particle diameter and γV^{crit} is approximated by the particle volume [1]. $^\star D^L$ is not thermally activated—it does not exhibit Arrhenius behavior as does, for example, the diffusivity in crystals, because $\langle u \rangle \propto T^{1/2}$ and $\langle \Omega^{\text{free}} \rangle$ increases approximately linearly with T [4]. Less approximate models for diffusion in liquids have been reviewed by Frohberg [5].

10.2   DIFFUSION IN AMORPHOUS METALS

Amorphous metallic alloys (*metallic glasses*) can be produced by rapid cooling (quenching) from the liquid phase. If the initially stable liquid avoids solidification by crystallization by being quenched rapidly below its ordinary melting temperature, T_m, it first becomes a supercooled liquid, and then, at a still lower temperature, it undergoes a *glass transition* to an amorphous glassy state as in Fig. 10.1. Occurring over a range of temperatures that is dependent upon the cooling rate, the glass transition is characterized by an abrupt change in the rate at which the volume and other physical properties change with decreasing temperature. The glass transition temperature, T_g, which occurs at a given cooling rate, is obtained from the intersection of the extrapolated cooling curves from well above and well below the transition. Because the glass transition occurs at a higher temperature during rapid cooling than during slow cooling, less free volume remains in glasses formed at low temperatures. Below the glass transition temperature, the combined effects of the low temperature and the loss of free volume cause the initially liquid material to lose its characteristic fluidity and become relatively rigid and unable to reorganize itself quickly as the temperature is decreased further (i.e., it becomes a frozen-in glass).

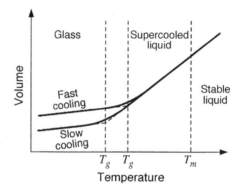

Figure 10.1:   Volume of metallic glass during fast and relatively slow cooling from the liquid phase. T_m is the melting temperature; T_g is the glass-transition temperature (shown for both fast and relatively slow cooling).

10.2.1   Self-Diffusion

If a rapidly cooled metallic glass is reheated and annealed isothermally at a temperature below T_g, the excess free volume that is frozen-in will anneal out as the system attempts to relax and equilibrate without crystallizing. The free volume is mobile and is presumably annihilated when it encounters regions of higher than average atomic density [6, 7]. The self-diffusivity that is measured during such annealing decreases initially. However, it eventually reaches an asymptotic value and becomes time independent, as in Fig. 10.2. The asymptotic value of the diffusivity is then that of the relaxed glassy state in which the supersaturated excess volume has annealed out. This dense structure is randomly packed, and the atoms are arranged with the highest density compatible with their hard-sphere radii and

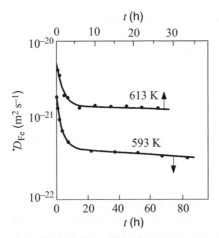

Figure 10.2:   Self-diffusion coefficient of ^{59}Fe in amorphous $Fe_{40}Ni_{40}B_{20}$ during isothermal annealing below T_g after rapid quenching from liquid state as in Fig. 10.1. Arrows indicate different time scales used at each temperature. Reprinted from "Tracer Diffusion of Fe-59 in Amorphous $Fe_{40}Ni_{40}B_{20}$," J. Horvath and H. Mehrer, 1986, *Crystal Lattice Defects and Amorphous Materials*, Taylor and Francis, http://www.tandf.co.uk/journals/ [8].

lack of translational symmetry. Locally, the atoms form various polyhedral units in definite ratios with neither microcrystallites nor large holes present. Even though relaxed, this structure is still metastable with respect to the crystalline state.

Extensive measurements show that self-diffusivities in the relaxed glassy state are time independent and closely exhibit Arrhenius behavior (i.e., $\ln{}^*D_i$ vs. $1/T$ plots appear as essentially straight lines) [8–11]. The diffusion therefore is thermally activated (in contrast to self-diffusion in the liquid above T_g as described in Section 10.1).

The mechanism by which the self-diffusion in the relaxed state occurs is not firmly established at present. However, there are reasons to believe that for certain atoms in glassy systems, self-diffusion occurs by a direct collective mechanism and is not aided by point defects in thermal equilibrium as in the vacancy mechanism for self-diffusion in crystals (Section 8.2.1).[2] These reasons include:

- Sudden changes in temperature during diffusion cause instantaneous changes in the diffusivity [9, 12]. This result is unexpected if diffusion occurs by a point-defect mechanism because significant time is required to obtain the new equilibrium defect concentrations corresponding to the temperature changes.

- The activation volume for diffusion, as measured by the pressure dependence of the diffusivity, is zero to within experimental accuracy [13, 14]. This is unexpected for defect-mediated diffusion, as in such cases, the activation volume for diffusion should consist of the sum of the volume of formation of the defect and the activation volume for the defect migration, and this is usually measurable.

- Computer simulations of the diffusion process in relaxed Fe–Zr glasses reveal diffusion which takes place directly via thermally activated displacement

[2]The ring mechanism in Section 8.1.1 is an example of a direct mechanism.

chains like that in Fig. 10.3 [7, 9, 15]. These chains do not start at localized

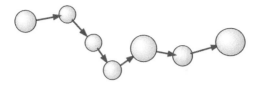

Figure 10.3: Mechanism of diffusion in amorphous glasses by thermally activated displacement chains.

point defects but in regions where the initial density deviations are small. Furthermore, when the displacement sequences are completed, any large-density deviations disperse gradually and do not leave behind localized point defects. The entire displacement process, from beginning to end, involves a relatively large number of atoms and, therefore, is of a collective nature. Such a direct collective diffusion process, which is spread over a considerable volume and involves relatively little ion-core overlap and repulsion, presumably occurs with relatively little volume change, in agreement with the small activation volume cited above.

- The observation that the self-diffusion exhibits Arrhenius behavior is consistent with a direct collective mechanism because the thermally activated displacement chains are spread over a considerable number of atomic distances. Irregularities in the disordered glassy structure are therefore averaged in the activated state, and all activation energies for displacements are then closely the same.

- No isotope effect is observed (see Eq. 8.31) during self-diffusion in relaxed glasses [16, 17]. In tracer self-diffusion studies of crystalline materials, where the atomic displacements that lead to vacancy migration and diffusion are highly localized, the harmonic model for the isotope effect is justified. However, if the migration process involves a relatively large number of atoms and is highly collective, this estimate of the effective attempt frequency is no longer valid. Instead, it is expected that two isotopes diffuse at close to the same rates because the mass difference of the two isotopes hardly affects their jump frequencies when relatively large numbers of atoms are strongly involved in the activated state.

Further discussion of self-diffusion in relaxed metallic glasses and other disordered systems may be found in key articles [7, 10, 14, 18, 19].

10.2.2 Diffusion of Small Interstitial Solute Atoms

Small solute atoms in the interstices between the larger host atoms in a relaxed metallic glass diffuse by the direct interstitial mechanism (see Section 8.1.4). The host atoms can be regarded as immobile. A classic example is the diffusion of H solute atoms in glassy $Pd_{80}Si_{20}$. For this system, a simplified model that retains the essential physics of a thermally activated diffusion process in disordered systems is used to interpret experimental measurements [20–22].

Because many different types of interstitial sites exist in the disordered glassy structure, the energy of the system varies as an interstitial atom jumps between the sites. The trace of the energy during successive jumps has the general form illustrated in Fig. 10.4a, where, for simplicity, the energy at each saddle point is assumed to be the same [20, 22, 23]. This approximation has the realistic feature that a diffusing interstitial encounters sites of varying energy and jump barriers of various heights.

The following quantities will be of use in describing the interstitial self-diffusion and intrinsic chemical diffusion:

N = total number of interstitial sites

p = fraction of all interstitial sites that are occupied

$^{\star}p$ = fraction of all sites that are tracer-interstitial occupied

p_k° = fraction of all sites that are type k sites (Fig. 10.4a)

$p(k)$ = fraction of type k sites that are occupied

p_k = fraction of all sites that are occupied type k sites

$^{\star}p_k$ = fraction of all sites that are tracer-occupied type k sites

The occupation probability at the various sites should follow Fermi–Dirac statistics because each site can accommodate only one interstitial. Therefore,

$$p(k) = \frac{1}{1 + e^{(G_k - \mu)/(kT)}} \tag{10.13}$$

where G_k is the energy corresponding to occupation of the type k site and μ is the chemical potential of the interstitials [24]. The fraction of all interstitial sites that are occupied is then

$$p = \frac{\sum_k p(k) N p_k^{\circ}}{N} = \sum_k \left[\frac{p_k^{\circ}}{1 + e^{(G_k - \mu)/(kT)}} \right] \tag{10.14}$$

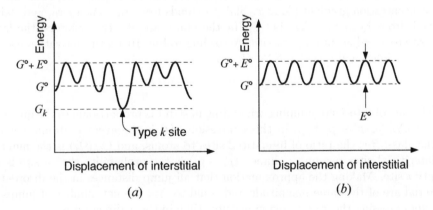

(a) (b)

Figure 10.4:   (a) The energy variation of an amorphous glass with the displacement of a diffusing interstitial atom as it jumps between successive interstitial sites. (b) A plot similar to (a) for interstitial jumping in a hypothetical material containing only sites of the reference state and having activation energies corresponding to E°.

and the partial concentration, p_k, can be written

$$p_k = \frac{p_k^\circ N p(k)}{N} = \frac{p_k^\circ}{1 + e^{(G_k - \mu)/(kT)}} \tag{10.15}$$

Also,

$$\sum_k p_k = p \tag{10.16}$$

A model for the tracer self-diffusivity of the interstitials is now developed for a system in which the total concentration of inert interstitials and chemically similar radioactive-tracer interstitials is constant throughout the specimen but there is a gradient in both concentrations. Since the inert and tracer interstitials are randomly intermixed in each local region,

$$\frac{{}^\star p_k}{p_k} = \frac{{}^\star p}{p} \tag{10.17}$$

Therefore, with the use of Eq. 10.15,

$$^\star p_k = \frac{{}^\star p}{p} \frac{p_k^\circ}{1 + e^{(G_k - \mu)/(kT)}} \tag{10.18}$$

In a typical tracer self-diffusion experiment, the tracer concentration probability, ${}^\star p$, depends upon position, whereas the total interstitial concentration probability, p, does not.

An expression for the tracer self-diffusivity, of the interstitials, ${}^\star D$, can be derived by employing the same basic method applied to a crystalline material to obtain the self-diffusivity given by Eq. 8.19. This involves finding the net flux of tracer interstitials jumping through a unit cross-sectional area in the diffusion zone perpendicular to the concentration gradient. For a crystalline material, this flux is found by considering the jumping of atoms between well-defined adjacent atomic planes lying parallel to the unit cross section. This approach, however, cannot be applied to a glassy material because of the disorder that is present, and therefore the flux must be determined by a slightly modified method. Consider two thin slabs in the material, each of thickness Δx and having unit area, lying perpendicular to the concentration gradient along x. Slab 1 extends from $x_0 - \Delta x$ to x_0, and slab 2 extends from x_0 to $x_0 + \Delta x$. Let Γ'_{ki} be the jump rate of a tracer interstitial from an i site to an adjacent empty i site. According to Fig. 10.4a, the activation energy for such a jump will be $G^\circ + E^\circ - G_k$, so

$$\Gamma'_{ki} = \nu\, e^{-(G^\circ + E^\circ - G_k)/(kT)} \tag{10.19}$$

The rate of k-to-i site jumping originating in slab 1 is proportional to the quantity $(\beta/\langle\Omega\rangle)\Delta x\, {}^\star p_k (p_i^\circ - p_i)\Gamma'_{ki}$. In this expression, $\langle\Omega\rangle$ is the average atomic volume in the glass, β is the ratio of interstitial sites to atoms, and $(\beta/\langle\Omega\rangle)$ is the number of interstitial sites per unit volume. $(p_i^\circ - p_i)$ is the probability that a site is an empty i site. Making the approximation that all jump distances in the disordered material are of the same magnitude and equal to Δx, the net number of jumps of all types crossing the $x = x_\circ$ plane per unit time in the x direction is

$$^\star J_I = g\frac{\beta}{\langle\Omega\rangle}\Delta x \sum_i \sum_k \left\{ [{}^\star p_k(p_i^\circ - p_i)\Gamma'_{ki}]_{x_\circ - \Delta x/2} - [{}^\star p_k(p_i^\circ - p_i)\Gamma'_{ki}]_{x_\circ + \Delta x/2} \right\} \tag{10.20}$$

where g is a purely geometrical constant and the double summation ensures that all types of different jumps between the various sites are included. The first term represents the jumps that originate in slab 1 and cross $x = x_o$ in the x direction, while the second term represents the jumps that originate in slab 2 and cross $x = x_o$ in the $-x$ direction. During tracer self-diffusion, the total concentration of inert and tracer interstitial atoms is constant, so both p and $(p_i^\circ - p_i)$ are independent of x. Making the usual Taylor expansion to evaluate the small difference between the terms and using Eqs. 10.18 and 10.19,

$$^\star J_I = \frac{-g\beta\,(\Delta x)^2}{\langle\Omega\rangle}\nu\,e^{-E^\circ/(kT)}\sum_i (p_i^\circ - p_i)\sum_k \left[\frac{p_k^\circ\,e^{(G_k-G^\circ)/(kT)}}{1 + e^{(G_k-\mu)/(kT)}}\right]\frac{1}{p}\frac{\partial^\star p}{\partial x} \quad (10.21)$$

Using Eq. 10.16 and the fact that $\sum_i p_i^\circ = 1$, $\sum_i (p_i^\circ - p_i) = 1 - p$. Also, using this result and Eq. 10.15,

$$\sum_k \left[\frac{p_k^\circ\,e^{(G_k-G^\circ)/(kT)}}{1 + e^{(G_k-\mu)/(kT)}}\right] = e^{(\mu-G^\circ)/(kT)}\sum_k (p_k^\circ - p_k) = e^{(\mu-G^\circ)/(kT)}(1 - p) \quad (10.22)$$

Putting these results into Eq. 10.21 then yields

$$^\star J_I = \frac{-g\beta\,(\Delta x)^2}{\langle\Omega\rangle}\nu\,e^{-E^\circ/(kT)}\frac{(1 - p)^2}{p}\,e^{(\mu-G^\circ)/(kT)}\frac{\partial^\star p}{\partial x} \quad (10.23)$$

Equation 10.23 can be put into the simpler form

$$^\star J_I = -{}^\star D_I^\circ\frac{(1 - p)^2}{p}\,e^{(\mu-G^\circ)/(kT)}\frac{\beta}{\langle\Omega\rangle}\frac{\partial^\star p}{\partial x} \quad (10.24)$$

where $^\star D_I^\circ \equiv g\,(\Delta x)^2\,\nu\exp[-E^\circ/(kT)]$ is the self-diffusivity in a hypothetical material that contains only sites of the reference state with the energy G°, and in which jumps may occur between them with the activation energy E°, as illustrated in Fig. 10.4b. Equation 10.24 is a Fick's-law equation with a tracer interstitial self-diffusivity corresponding to

$$^\star D_I = {}^\star D_I^\circ\frac{(1 - p)^2}{p}\,e^{(\mu-G^\circ)/(kT)} \quad (10.25)$$

Having this result, an expression can be obtained for the "intrinsic" chemical diffusivity, D_I, which describes the diffusion arising from an inert-interstitial concentration gradient. According to Eqs. 3.35 and 3.42, the flux in such a system is

$$-J_I = -M_I\frac{\beta}{\langle\Omega\rangle}p\frac{\partial\mu}{\partial x} \quad (10.26)$$

Also,

$$\mu = \mu^\circ + kT\ln\gamma p \quad (10.27)$$

where the activity coefficient γ is generally a function of concentration and therefore of position. Putting Eq. 10.27 into Eq. 10.26 leads to the Fick's-law-type expression

$$J_I = -M_I kT\left(1 + \frac{\partial\ln\gamma}{\partial\ln p}\right)\frac{\beta}{\langle\Omega\rangle}\frac{\partial p}{\partial x} \quad (10.28)$$

and, therefore,

$$D_I = M_I kT \left(1 + \frac{\partial \ln \gamma}{\partial \ln p} \right) \tag{10.29}$$

For tracer self-diffusion, a similar initial equation for the flux is

$${}^{\star}J_I = -{}^{\star}M_I \frac{\beta^{\star}}{\Omega} p \frac{\partial {}^{\star}\mu}{\partial x} \tag{10.30}$$

However, in this system, the ideal free energy of mixing of the inert and tracer interstitials is the only component that varies with x. By taking the derivative of the free energy to obtain the chemical potential, the x-dependent component of the chemical potential of the tracer interstitials is simply $kT \ln({}^{\star}p/p)$, and therefore, because p is constant, $(\partial {}^{\star}\mu / \partial x) = (kT/{}^{\star}p)(\partial {}^{\star}p / \partial x)$. Putting this result into Eq. 10.30,

$${}^{\star}J_I = -{}^{\star}M_I kT \frac{\beta}{\langle \Omega \rangle} \frac{\partial {}^{\star}p}{\partial x} \tag{10.31}$$

which is a Fick's-law-type expression with an interstitial tracer self-diffusivity given by

$${}^{\star}D_I = {}^{\star}MkT \tag{10.32}$$

Neglecting any small isotope effect, $M_I = {}^{\star}M_I$, and comparing Eqs. 10.29 and 10.32,

$$D_I = \left(1 + \frac{\partial \ln \gamma}{\partial \ln p} \right) {}^{\star}D_I \tag{10.33}$$

which is of the same form as Eq. 3.13.

The model above has been compared to experimental results for the diffusion of H in glassy $Pd_{80}Si_{20}$ by Kirchheim and coworkers [21, 22]. D_I increases strongly with increasing H concentration as seen in Fig. 10.5. By assuming that the energies of the interstitial sites follow a Gaussian distribution around a mean value, good agreement was obtained between the model and experiment. The increase of D_I

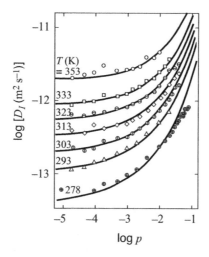

Figure 10.5:   Logarithm of the diffusivity of H in amorphous $Pd_{80}Si_{20}$ as a function of the H concentration probability at different temperatures. Points are experimental data. The curves are the predictions of the model leading to Eq. 10.25. From Kirchheim [22].

with p arises from the successive saturation of the lower-energy sites as the concentration is increased. This causes a progressive decrease of the activation energy and a corresponding increase in the diffusivity. For example, at very low concentrations, essentially all of the interstitials become trapped at the lowest-energy sites and they engage in long-range diffusion only with difficulty. Further aspects are discussed elsewhere [22].

Figure 10.6 plots the tracer diffusivity data for a number of solute species in glassy $Ni_{80}Zr_{50}$ as a function of their metallic radius. The diffusivity increases rapidly as the metallic radius decreases. The relatively rapid diffusion of the small atoms in this case may result from the fact that they diffuse by the interstitial mechanism [10, 18].

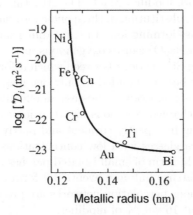

Figure 10.6:   Tracer diffusivities in glassy $Ni_{80}Zr_{50}$ of various solute atoms as a function of their size (as measured by their metallic radii) [25]. Reprinted, by permission, from H. Hahn and R.S. Averback, "Dependence of tracer diffusion on atomic size in amorphous Ni–Zr," *Phys. Rev. B*, Vol. 37, p. 6534. Copyright ©1988 by the American Physical Society.

10.3   SMALL ATOMS (OR MOLECULES) IN GLASSY POLYMERS

Some small atoms and molecules, such as He, Ar, CO_2, and N_2, dissolve in glassy polymers from the gas phase. These particles then diffuse in the bulk polymer presumably by occupying interstices in the glassy structure and jumping between them by the direct interstitial mechanism. The solubilities increase with increasing partial pressure, and the behavior observed can be well explained on the basis of a model in which the dissolved species occupy interstitial sites, the site occupancy obeys Fermi–Dirac statistics, and the site energies are distributed about a mean value in the form of a Gaussian distribution [26, 27]. The corresponding diffusivities of these species increase with increasing concentration, in a manner similar to the diffusion of small solute atoms in amorphous metals. This behavior can be explained by the same interstitial diffusion model. Here, the diffusing particles must again occupy progressively higher-energy sites as their concentration increases, causing the average activation energy for diffusion to decrease and the diffusivity to increase. The diffusion of small particles in glassy polymers therefore appears to be quite similar to that in glassy metals.

10.4 DIFFUSION OF ALKALI IONS IN NETWORK OXIDE GLASSES

The structure of a pure oxide network glass having stoichiometry G_2O_3, free of any alkali ions, is illustrated in Fig. 10.7a [28]. In this structure, cations are three-coordinated and the oxygen anions are two-coordinated. In three-dimensional silica glass, each glass-forming Si^{4+} cation is enclosed in a polyhedron of oxygen anions, and these polyhedra are arranged in a network lacking special symmetry and periodicity. The oxygen polyhedra share corners, not edges or faces, and each oxygen ion is covalently bonded to no more than two cations.

The oxide glass structure changes significantly when *modifying* alkali ions are added, as in Fig. 10.7b, where the G_2O_3 glass has been altered by adding a significant amount of the network modifier M_2O. The structure accommodates the network modifier M^+ ions by substitution of three one-coordinated modifier cations for one three-coordinated glass-forming ion. In three-dimensional silica glass, the addition of Na ions (e.g., via Na_2O) causes oxygen ions, previously covalently bonded to two of the glass-forming Si^{4+} cations between which it formed a *bridge*, to reduce this bonding so that they become bonded to only one glass-forming cation. These oxygen ions, called *nonbridging oxygens*, possess an effective negative charge. The corresponding positively charged Na^+ ions are then ionically bonded to the nonbridging oxygens, resulting in a partly covalent and partly ionic overall structure. Studies show that in silica glasses with low concentrations of Na_2O, the ionically bound material exists in the form of small isolated patches or *lakes*. As the concentration increases, these patches link and eventually form a network of continuous channels [29–32]. Continuous percolation networks are present at and above a percolation threshold of about 16 vol. % of modifier.

Na^+ ions are highly mobile compared to the glass-forming components and possess a diffusivity which follows Arrhenius behavior [21, 26, 29, 31, 33]. Furthermore, the activation energy for diffusion decreases markedly (and the diffusivity increases correspondingly) as the modifier concentration is increased, as in Fig. 10.8. The

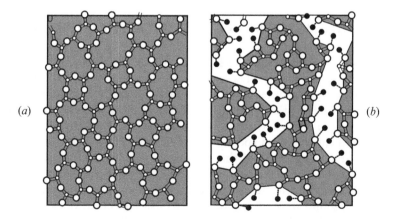

(a) (b)

Figure 10.7: **(a)** Two-dimensional schematic of pure, oxide network glass of composition G_2O_3. Small open circles are glass-forming cations G^{3+}. Large open circles represent oxygen anions. From Kingery et al. [28]. **(b)** Schematic of glass as modified by the addition of alkali M^+ cations (filled small circles). At high modifier-ion content, the modifier ions aggregate and form high-diffusivity "lakes" or channels in the glass. Adapted from Greaves [29].

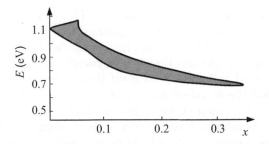

Figure 10.8:   Activation energy for diffusion of Na^+ ions in sodium silicate glass of composition $(Na_2O)_x(SiO_2)_{1-x}$ as a function of Na^+ ion concentration as measured by the fraction of Na_2O, x. From Frischat [33].

mechanism for the diffusion of the Na^+ ions is not thoroughly understood, but the results above can be explained with a model in which the Na^+ ions diffuse in the modified random-network structure by a direct interstitial mechanism.

To engage in long-range diffusion at low concentrations, the Na^+ ions must dissociate themselves from their ionic bonding with the nonbridging oxygens and diffuse through the interstices in the covalently bonded network glass regions. This generally requires a relatively large activation energy. At higher concentrations above the percolation threshold, the Na^+ ions diffuse relatively easily along the interstices in the ionically bonded percolation channels with a low activation energy. As the modifier concentration increases, the activation energy decreases progressively. Also, the correlation coefficient decreases as the modifier concentration increases, due to the increased degree of correlation arising from the restriction of the diffusion to the narrow channels [31].

10.5 DIFFUSION OF POLYMER CHAINS

Polymer structure is characterized by long chains of molecules arranged in a wide variety of ways. Diffusion of these long chains occurs in two important and characteristic situations.[3] In the first, each chain is essentially isolated and embedded in a solvent melt made up of much smaller molecules, and the long chains diffuse by a type of Brownian motion. In the second situation, the long chains are in a dense entangled arrangement, much like cooked spaghetti, and a given chain can then only diffuse in the entangled structure by a snakelike process called *reptation*.

10.5.1 Structure of Polymer Chains

A polymer chain is typically composed of a large number of units (i.e., monomers) arranged in a chainlike configuration held together by covalent bonds. When the monomers are identical, it is termed a *homopolymer chain*. The chain possesses considerable flexibility since the covalently bonded monomers can change their bonding angles with one another, allowing the chain to act as if were almost "freely jointed." This flexibility allows a chain to adopt a huge number of possible configurations. An important parameter for describing the chain configuration is the mean-square

[3]For detailed treatments of this topic, see de Gennes [34, 35] and Lodge et al. [36].

distance between the two ends of the chain $\langle h^2 \rangle$. This can be approximated using the *freely jointed chain model*, in which it is assumed that the direction of each bond between monomers is random. This model ignores the fact that covalent bonds tend to assume specific angles and also that it is physically impossible for the chain to bend sharply backward and overlap on itself (i.e., it ignores the existence of an *excluded volume* that the flexible chain cannot enter). Nevertheless, the model is a reasonable first approximation under many conditions and retains much of the essential physics of the problem. If each monomer is of length b, a chain consisting of N monomers can be constructed by joining them together sequentially end to end. If the angles at which they are added are at random, the problem of determining the distance between the ends, $\langle h^2 \rangle$, is identical to that of finding the mean-square displacement resulting from a random walk given by Eq. 7.47. Therefore,

$$\langle h^2 \rangle = Nb^2 \tag{10.34}$$

A typical freely jointed chain will therefore be quite compact since the root-mean-square value of its end-to-end length, $\sqrt{\langle h^2 \rangle} = \sqrt{N}\, b$, will be small compared with its length if it were stretched out (i.e., Nb, when N is large). Figure 10.9 shows a simulated molecule of polyethylene, $(-CH_2-CH_2-)_N$, which approximates a freely jointed configuration.

The relationship $\sqrt{\langle h^2 \rangle} \propto N^{1/2}$ derived on the basis of the freely jointed model is often quite satisfactory despite the approximate nature of the model. Even though the excluded volume clearly exists, its effect can essentially be canceled out under many circumstances. When the chains are in dilute solution in a *theta solvent*, interactions between the solvent and the chain monomers favor compression of the chain so that the relation $\sqrt{\langle h^2 \rangle} \propto N^{1/2}$ is nearly obeyed. On the other hand, in a "good solvent" where monomer–monomer and monomer–solvent interactions are closely the same, any interactions favoring compression are absent and the excluded volume then acts to reduce the degree of compactness of the chain and produce swelling. In these cases, the relation $\sqrt{\langle h^2 \rangle} \propto N^{3/5}$ holds to a good approximation. In the case of homopolar polymer melts where the polymer chains are densely entangled, the excluded volume does not play an important role in determining the degree of swelling since each monomer is surrounded by similar monomers, and is unable to distinguish whether they belong to its own chain or another nearby chain. Even though the relationship $\sqrt{\langle h^2 \rangle} \propto N^{1/2}$ will hold closely for a given

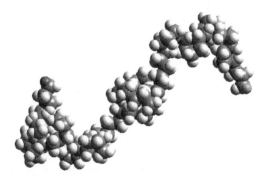

Figure 10.9:   Conformation of polyethylene, $(-CH_2-CH_2-)_N$. Degree of polymerization $N = 50$.

type of chain of fixed N in different theta solvents and in its melt, the magnitudes of $\sqrt{\langle h^2 \rangle}$ for the chain will differ.

10.5.2   Diffusion of Isolated Polymer Chains in Dilute Solutions

From a hydrodynamical standpoint, a single isolated chain immersed in a liquid solvent consisting of relatively small molecules may be regarded as a porous sponge having a uniform density [35]. Vorticity cannot penetrate this sponge except over a certain screening length which is negligible for a long chain. As far as global properties (such as viscosity and sedimentation) are concerned, the coil possesses an effective radius, R_h, which is proportional to $\sqrt{\langle h^2 \rangle}$. Using previous results, R_h is therefore

$$R_h \propto N^{3/5}b \qquad (10.35)$$

for a chain in a good solvent, and for a chain in a theta solvent is

$$R_h \propto N^{1/2}b \qquad (10.36)$$

Using this approximation, the diffusivity of the chain is the diffusivity of the effective sponge of radius, R_h, due to its classical *Brownian motion* in the solvent. The Brownian motion of the sponge is its irregular motion due to random collisions with surrounding solvent molecules that induce the sponge to follow a random walk. The diffusivity of a small particle due to Brownian motion was determined in the early part of the twentieth century by Einstein, Smoluchowski, and Langevin. We follow here a more recent description which imagines that the particle (i.e., the sponge) is embedded in an effectively uniform medium having a viscosity η [37]. At the same time, the particle is subjected to random collisions with molecules in the surrounding medium. The model therefore has a continuum-molecular duality. The Newtonian force equation for all forces in the x direction is

$$m\frac{d^2X}{dt^2} = -\mathcal{F}\frac{dX}{dt} + F_x \qquad (10.37)$$

where X is the instantaneous x-coordinate of the sponge's position. The first term in Eq. 10.37 is the usual inertial term (m is the effective mass of the chain). The second term is the frictional force exerted on the sponge by the viscous medium and is proportional to the velocity via the friction factor, $\mathcal{F}$, as would be expected on the basis of Stokes's law. The third term includes all forces associated with collisions with the surrounding molecules. Multiplying Eq. 10.37 by X yields

$$mX\frac{d^2X}{dt^2} = -\mathcal{F}X\frac{dX}{dt} + XF_x \qquad (10.38)$$

But

$$X\frac{d^2X}{dt^2} = \frac{1}{2}\frac{d}{dt}\left[\frac{d(X^2)}{dt}\right] - \left(\frac{dX}{dt}\right)^2 \qquad (10.39)$$

and

$$X\frac{dX}{dt} = \frac{1}{2}\frac{d(X^2)}{dt} \qquad (10.40)$$

Putting these expressions into Eq. 10.38 then yields

$$\frac{m}{2}\frac{d}{dt}\left[\frac{d\left(X^2\right)}{dt}\right] - m\left(\frac{dX}{dt}\right)^2 = -\frac{\mathcal{F}}{2}\frac{d\left(X^2\right)}{dt} + XF_x \qquad (10.41)$$

Next, the mean values of these terms over a long period are introduced so that

$$\frac{m}{2}\left\langle \frac{d}{dt}\left[\frac{d\left(X^2\right)}{dt}\right]\right\rangle - \left\langle m\left(\frac{dX}{dt}\right)^2\right\rangle = -\frac{\mathcal{F}}{2}\left\langle \frac{d\left(X^2\right)}{dt}\right\rangle + \langle XF_x\rangle \qquad (10.42)$$

Now, according to equipartition,

$$\frac{1}{2}\left\langle m\left(\frac{dX}{dt}\right)^2\right\rangle = \frac{kT}{2} \qquad (10.43)$$

Also,

$$\left\langle \frac{d}{dt}\left(\frac{d\left(X^2\right)}{dt}\right)\right\rangle = \frac{d}{dt}\left\langle \frac{d\left(X^2\right)}{dt}\right\rangle \qquad (10.44)$$

and

$$\langle XF_x\rangle = 0 \qquad (10.45)$$

because the F_x forces are exerted randomly. Therefore,

$$\frac{m}{2}\frac{du}{dt} + \frac{\mathcal{F}}{2}u = kT \qquad (10.46)$$

where

$$u = \left\langle \frac{d(X^2)}{dt}\right\rangle \qquad (10.47)$$

Equation 10.47 has the solution

$$u = A\,e^{-\mathcal{F}t/m} + \frac{2kT}{\mathcal{F}} \qquad (10.48)$$

where A is constant. The exponential term in Eq. 10.48 is negligible for all times of interest, and therefore

$$\left\langle \frac{d(X^2)}{dt}\right\rangle = \frac{d}{dt}\left\langle X^2\right\rangle = \frac{2kT}{\mathcal{F}} \qquad (10.49)$$

By integrating Eq. 10.49,

$$\left\langle X^2\right\rangle = \frac{2kT}{\mathcal{F}}\,t \qquad (10.50)$$

Results for the mean-square displacement along y and z will be the same and, therefore,

$$\left\langle R^2\right\rangle = \left\langle X^2\right\rangle + \left\langle Y^2\right\rangle + \left\langle Z^2\right\rangle = \frac{6kT}{\mathcal{F}}\,t \qquad (10.51)$$

Comparing this result with Eq. 7.52, the diffusivity of the isolated chain is

$$^\star D_1 = \frac{kT}{\mathcal{F}} \qquad (10.52)$$

An expression for the friction factor, $\mathcal{F}$, can be obtained from Stokes's law, which gives the force, F, exerted by a viscous medium on a sphere of radius, R, moving through it with a velocity, v, in the form [38]

$$F = 6\pi\eta Rv \qquad (10.53)$$

where η is the viscosity. Therefore,

$$\mathcal{F} = \frac{F}{v} = 6\pi\eta R \tag{10.54}$$

Putting this result into Eq. 10.52 and setting $R = R_h$ for the chain (sponge),

$$^\star D_1 \propto \frac{kT}{6\pi\eta R_h} \tag{10.55}$$

where R_h is given by Eq. 10.35 or 10.36. Therefore, $^\star D_1$ for the chain varies inversely with the viscosity and decreases as the chain length increases since it scales approximately as $N^{-1/2}$ or $N^{-3/5}$. Because the viscosity, η, is generally thermally activated, the diffusivity is similarly thermally activated. The determination of the proportionality constant implicit in Eq. 10.55 requires more detailed calculations.

10.5.3 Diffusion of Densely Entangled Polymer Chains by Reptation

In a polymer melt or a concentrated polymer solution, chains are densely packed and highly entangled. An unattached chain in such an entangled structure is able to diffuse through a process called *reptation*, first proposed by de Gennes [39].[4] In the densely packed and entangled environment of a polymer melt, a given chain will be unable to move bodily in directions perpendicular to itself because of the resistance provided by its closely packed environment. On the other hand, it will be able to move in directions tangential to itself by a sliding type of motion much like the movement of a snake—hence, the term reptation. The freely jointed "head" at the leading end of the chain can always find an optimum region of low density in the material in front of it to advance into, and the remainder of the chain can then follow along by an appropriate sliding motion. This type of motion is similar in many respects to the motion of a train along a curved track. The chain can move equally well in the reverse direction in the same manner since its "head" and "tail" are interchangeable. For conceptual purposes, the highly constricting and anisotropic environment just described can be represented by a fictitious rigid tube within which the chain can slide backward and forward. The sliding can be accomplished by the propagation along the chain length of localized dispiration defects [41], as in Fig. 10.10. The motion of the chain is then visualized essentially as a quasi one-dimensional Brownian motion in which the chain randomly walks forward and backward along its tube.

As a result of this motion, the tube that initially surrounds the chain (i.e., the initial *primitive tube*), will be replaced progressively by a new tube, as in Fig. 10.10. By executing excursions in random directions, the ends of the chain iteratively change their surroundings. After a large number of excursions, the chain's conformation eventually loses contact with its original primitive tube. At this loss of contact, the chain's trailing end ceases to touch the original primitive tube (Fig. 10.10d) and a new primitive tube is defined. Thus, each successive primitive tube is connected to its predecessor as in Fig. 10.11.

An approximate expression for the chain's self-diffusivity can be obtained using the theory of random walks. Let $\tau_{\rm rep}$ be the average time required for the chain to move from one primitive tube to its successor. During an interval $\tau_{\rm rep}$, the chain's

---

[4]This discussion of reptation is a simplified version of other rigorous treatments [36, 40].

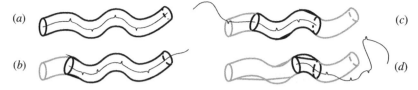

Figure 10.10: Stages in the elimination of the initial primitive tube associated with a reptating polymer chain. **(a)** Configuration of the initial chain (thin line) and its associated primitive tube (cylinder). The chain can diffuse along the tube by the propagation along its length of the small defects shown. **(b)** The chain has diffused toward the right by emerging from the end of the primitive tube on the right and creating a newly configured segment outside the primitive tube. **(c)** The chain has diffused toward the left by emerging from the end of the portion of the primitive tube that it occupied in (b) and creating a newly configured segment outside the primitive tube. The portion of the initial primitive tube that it now occupies is indicated by the heavy lines. **(d)** The chain has diffused toward the right by emerging from the portion of the primitive tube that it occupied in (c) and creating a newly configured segment outside the primitive tube. The chain has almost completely escaped from the initial primitive tube and is almost completely enclosed in a new successor primitive tube (not shown). From Lodge et al. [36].

center of mass will undergo a displacement that is approximately the mean length of a primitive tube, $\langle h^2 \rangle = Nb^2$ (Eq. 10.34). The relationship between $\tau_{\rm rep}$ and the mean-square displacement for random walks (Eq. 7.53) provides a model for the chain's self-diffusivity:

$$^{\star}D = \frac{\Gamma r^2}{6} = \frac{r^2}{6\tau_{\rm rep}} \approx \frac{Nb^2}{6\tau_{\rm rep}} \tag{10.56}$$

An estimate for $\tau_{\rm rep}$ may be found by considering the Brownian motion of the chain. At the transition from one isolated primitive tube to its successor, the chain must have executed enough excursions to travel the chain length Nb. If the distance associated with each excursion is u and the number of excursions is n, Eq. 7.47 also establishes the estimate $u\sqrt{N} \approx Nb$. Therefore, the number of excursions, n, is approximately $n = N^2b^2/u^2$. If ν is the "excursion" frequency,

$$\tau_{\rm rep} \approx \frac{n}{\nu} \approx \frac{N^2b^2}{\nu u^2} \tag{10.57}$$

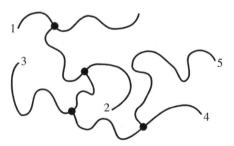

Figure 10.11: Successive configurations of noncorrelated primitive tubes for a reptating polymer chain, formed in the sequence $1 \to 5$. One end of each tube configuration must be in contact with its predecessor at points indicated by dots. From Lodge et al. [36].

It is reasonable that ν will be smaller for long chains. Therefore, as an approximation, let $\nu = \nu_o/N$, where $\nu_o = $ constant. Therefore,

$$\tau_{\text{rep}} \approx \frac{N^2 b^2}{\nu u^2} = \frac{N^3 b^2}{\nu_o u^2} \tag{10.58}$$

and, putting Eq. 10.58 into Eq. 10.56,

$$^\star D \approx \frac{\nu_o u^2}{6N^2} \tag{10.59}$$

The diffusivity scales as $1/N^2$, decreasing rapidly with increasing chain length.

A comparison of Eqs. 10.55 and 10.59 shows that the diffusivity of an isolated chain due to Brownian motion falls off more slowly with increasing chain size than the diffusivity of an entangled chain diffusing by the reptation mechanism.

Bibliography

1. M.H. Cohen and D. Turnbull. Molecular transport in liquids and glasses. *J. Chem. Phys.*, 31(5):1164–1169, 1959.

2. G.H. Vineyard. The theory and structure of liquids. In *Liquid Metals and Solidification*, pages 1–48, Cleveland, OH, 1958. American Society for Metals.

3. D. Turnbull and M.H. Cohen. Free-volume model of the amorphous phase glass transition. *J. Chem. Phys.*, 34(1):120–125, 1961.

4. E.H. Kennard. *Kinetic Theory of Gases*. McGraw-Hill, New York, 1938.

5. G. Frohberg. Diffusion in liquid metals between the glass transition and the evaporation temperature. *Defect and Diffusion Forum*, 143–147:869–874, 1997.

6. M.H. Cohen and G.S. Grest. The nature of the glass transition. *J. Non-Cryst. Solids*, 61–62:749–759, 1984.

7. W. Frank, U. Hamlescher, H. Kronmuller, P. Scharwaechter, and T. Schuler. Diffusion in amorphous metallic alloys—Experiments, molecular-dynamics simulations, interpretation. *Phys. Scripta*, T66:201–206, 1996.

8. J. Horvath and H. Mehrer. Tracer diffusion of Fe-59 in amorphous $Fe_{40}Ni_{40}B_{20}$. *Lattice Defects Amorphous Mater.*, 13(1):1–14, 1986.

9. W. Frank, J. Horvath, and H. Kronmuller. Diffusion mechanisms in amorphous alloys. *Mater. Sci. Eng.*, 97:415–418, 1988.

10. H. Mehrer and W. Dorner. Diffusion in amorphous alloys. *Defect and Diffusion Forum*, 66–69:189–206, 1989.

11. W. Frank, A. Horner, P. Scharwaechter, and H. Kronmuller. Diffusion in amorphous metallic alloys. *Mater. Sci. Eng. A*, 179:36–40, 1994.

12. C.J. Rank. *Evaluation of diffusion and relaxation in metallic glasses after short-term annealing in a mirror furnace*. PhD thesis, Stuttgart University, 1992. In German.

13. K. Ratzke and F. Faupel. Pressure dependence of cobalt diffusion in amorphous $Fe_{39}Ni_{40}B_{21}$. *J. Non-Cryst. Solids*, 181(3):261–265, 1995.

14. H. Mehrer and G. Rummel. Amorphous metallic alloys—diffusional aspects. In H. Jain and D. Gupta, editors, *Diffusion in Amorphous Materials*, pages 163–176, Warrendale, PA, 1994. The Minerals, Metals and Materials Society.

15. A. Horner. *Self-diffusion in metallic glasses: Approximation of the effective medium and molecular simulation*. PhD thesis, Stuttgart University, 1993. In German.

16. K. Ratzke, P.W. Huppe, and F. Faupel. Transition from single-jump type to highly cooperative diffusion during structural relaxation of a metallic-glass. *Phys. Rev. Lett.*, 68(15):2347–2349, 1992.

17. K. Ratzke, A. Heesemann, and F. Faupel. The vanishing isotope effect of cobalt diffusion in $Fe_{39}Ni_{40}B_{21}$ glass. *J. Phys. Condens. Matter*, 7(39):7663–7668, 1995.

18. R.S. Averback. Defects and diffusion in amorphous alloys. *Mater. Res. Soc. Bull.*, 16(11):47–52, 1991.

19. F. Faupel. Diffusion in noncrystalline metallic and organic media. *Phys. Status Solidi. A*, 134(1):9–59, 1992.

20. U. Stolz, R. Kirchheim, J.E. Sadoc, and M. Laridjani. Hydrogen in liquid-quenched and vapor-quenched amorphous $Pd_{80}Si_{20}$. *J. Less-Common Mater.*, 103(1):81–90, 1984.

21. R. Kirchheim and U. Stolz. Modeling tracer diffusion and mobility of interstitials in disordered materials. *J. Non-Cryst. Solids*, 70(3):323–341, 1985.

22. R. Kirchheim. Hydrogen solubility and diffusivity in defective and amorphous metals. *Prog. Mater. Sci.*, 32(4):261–325, 1988.

23. R. Kirchheim. Solubility, diffusivity, and the trapping of hydrogen in dilute alloys, deformed and amorphous metals. 2. *Acta Metall.*, 30(6):1069–1078, 1982.

24. D.A. McQuarrie. *Statistical Mechanics*. HarperCollins, New York, 1976.

25. H. Hahn and R.S. Averback. Dependence of tracer diffusion on atomic size in amorphous Ni–Zr. *Phys. Rev. B*, 37(11):6533–6535, 1988.

26. R. Kirchheim. Interstitial diffusion in glass. *Defect and Diffusion Forum*, 95–98:1159–1164, 1993.

27. R. Kirchheim. Interstitial diffusion in glasses and the mixed alkali effect. In H. Jain and D. Gupta, editors, *Diffusion in Amorphous Materials*, pages 43–54, Warrendale, PA, 1994. The Minerals, Metals and Materials Society.

28. W.D. Kingery, H.K. Bowen, and D.R. Uhlmann. *Introduction to Ceramics*. John Wiley & Sons, New York, 1976.

29. G.N. Greaves. EXAFS and the structure of glass. *J. Non-Cryst. Solids*, 71(1–3):203–217, 1985.

30. C. Huang and A.N. Cormack. The structure of sodium silicate glass. *J. Chem. Phys.*, 93(11):8180–8186, 1990.

31. G.N. Greaves, S.J. Gurman, C.R.A. Catlow, A.V. Chadwick, S. Houdewalter, C.M.B. Henderson, and B.R. Dobson. A structural basis for ionic diffusion in oxide glasses. *Phil. Mag. A*, 64(5):1059–1072, 1991.

32. Y. Cao and A.N. Cormack. A structural model for interpretation of an anomaly in alkali aluminosilicate glasses at Al/alkali = 0.2–0.4. In H. Jain and D. Gupta, editors, *Diffusion in Amorphous Materials*, pages 137–151, Warrendale, PA, 1994. The Minerals, Metals and Materials Society.

33. G.H. Frischat. *Ionic Diffusion in Oxide Glasses*. Trans Tech Publications, Bay Village, OH, 1975.

34. P.G. de Gennes. *Scaling Concepts in Polymer Physics*. Cornell University Press, Ithaca, NY, 1979.

35. P.G. de Gennes. *Introduction to Polymer Dynamics*. Cambridge University Press, Cambridge, 1990.

36. T.P. Lodge, N.A. Rotstein, and S. Prager. Dynamics of entagled polymer liquids: Do linear chains reptate? In I. Prigogine and S.A. Rice, editors, *Advances in Chemical Physics*, volume 79, pages 1–132, New York, 1990. John Wiley & Sons.

37. J.F. Lee, F.W. Sears, and D.L. Turcotte. *Statistical Thermodynamics*. Addison-Wesley, Reading, MA, 1963.

38. K. Huang. *Statistical Mechanics*. John Wiley & Sons, New York, 1963.

39. P.G. de Gennes. Reptation of a polymer chain in the presence of fixed obstacles. *J. Chem. Phys.*, 55(2):572–579, 1971.

40. M. Doi and S.F. Edwards. *The Theory of Polymer Dynamics*. Oxford University Press, Oxford, 1986.

41. S.M. Allen and E.L. Thomas. *The Structure of Materials*. John Wiley & Sons, New York, 1999.

PART II

MOTION OF DISLOCATIONS AND INTERFACES

A host of important kinetic processes in materials depends upon the motion of dislocations and various types of interfaces, and we now describe the basic features of the different types of motion that can occur. Different atomic mechanisms are involved—including jumping processes of the diffusive type described in Chapter 7, long-range diffusion of components in the system, and atomic shuffles (short-range displacements that produce local structural changes as the dislocation, or interface, moves). The situation is compounded further by the fact that dislocations and interfaces are themselves generally complex entities that can have a variety of structures, and therefore varying responses, to the forces that drive their motion.

We start with dislocations and describe both glissile (conservative) and climb (nonconservative) motion in Chapter 11. The motion of vapor/crystal interfaces and liquid/crystal interfaces is taken up in Chapter 12. Finally, the complex subject of the motion of crystal/crystal interfaces is treated in Chapter 13, including both glissile and nonconservative motion.

CHAPTER 11

MOTION OF DISLOCATIONS

The motion of dislocations by glide and climb is fundamental to many important kinetic processes in materials. Gliding dislocations are responsible for plastic deformation of crystalline materials at relatively low temperatures, where any dislocation climb is negligible. They also play important roles in the motion of glissile interfaces during twinning and diffusionless martensitic phase transformations. Both gliding and climbing dislocations cause much of the deformation that occurs at higher temperatures where self-diffusion rates become significant, and significant climb is then possible. Climbing dislocations act as sources and sinks for point defects. This chapter establishes some of the basic kinetic features of both dislocation glide and climb.

11.1 GLIDE AND CLIMB

The general motion of a dislocation can always be broken down into two components: glide motion and climb motion. *Glide* is movement of the dislocation along its glide (slip) plane, which is defined as the plane that contains the dislocation line and its Burgers vector. *Climb* is motion normal to the glide plane. Glide motion is a conservative process in the sense that there is no need to deliver or remove atoms at the dislocation core during its motion. In contrast, the delivery or removal of atoms at the core is necessary for climb. This is illustrated for the simple case of the glide and climb of an edge dislocation in Fig. 11.1. The glide along x in Fig. 11.1a and b is accomplished by the local conservative shuffling of atoms at the disloca-

Kinetics of Materials. By Robert W. Balluffi, Samuel M. Allen, and W. Craig Carter. **253**
Copyright © 2005 John Wiley & Sons, Inc.

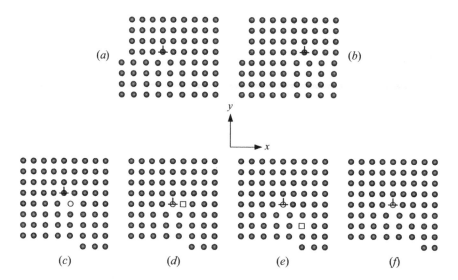

Figure 11.1:   Glide and climb of edge dislocation in primitive cubic crystal ($\vec{b} = [b00]$, $\hat{\zeta} = [001]$) [1]. **(a)** and **(b)** Glide from left to right. **(c)–(f)** Downward climb along $-y$. In (d), the lighter-shaded substitutional atom shown adjacent to the dislocation core in (c) has joined the extra half plane and created a vacancy. In (e), the vacancy has migrated away from the dislocation core by diffusion. In (f), the vacancy has been annihilated at the surface step. This overall process is equivalent to removing an atom from the surface and transporting it to the dislocation at its core. A new site was created at the dislocation, which acted as a vacancy source. This site was subsequently annihilated at the surface, which acted as an atom source.

tion core as it moves. The climb along $-y$, however, requires that the extra plane associated with the edge dislocation be extended in the $-y$ direction. This requires a diffusive flux of atoms to the dislocation core, and when self-diffusion occurs by a vacancy mechanism, the corresponding creation of an equivalent number of new lattice sites in the form of vacancies. In this case, the dislocation acts as a sink for atoms and, equivalently, as a source for vacancies. Glide can therefore occur at any temperature, whereas significant climb is possible only at elevated temperatures where the required diffusion can occur.[1]

Defects such as dislocations can be sources or sinks for atoms or for vacancies. Whether such point entities are created or destroyed depends on the type of defect, its orientation, and the stresses acting on it. It is convenient to adopt a single term *source*, which describes a defect's capability for creation and destruction of crystal sites and vacancies in the crystal. "Source" will generically indicate creation of point entities (i.e., "positive" source action) as well as destruction of point entities (i.e., "negative" source action). Thus, a climbing edge dislocation that destroys vacancies will be, equivalently, both a (positive) source of atoms and a (negative) source of vacancies. If the sense of climb is reversed, the dislocation would be a (negative) source of atoms.

[1]Provided that the Peierls force is not too large (see Section 11.3.1).

11.2 DRIVING FORCES ON DISLOCATIONS

Dislocations in crystals tend to move in response to forces exerted on them. In general, an effective driving force is exerted on a dislocation whenever a displacement of the dislocation causes a reduction in the energy of the system. Forces may arise in a variety of ways.

11.2.1 Mechanical Force

In general, a segment of dislocation in a crystal in which there is a stress field is subjected to an effective force because the stress does an increment of work (per unit length), δW, when the dislocation is moved in a direction perpendicular to itself by the vector, $\delta \vec{r}$. In this process, the material on one side of the area swept out by the dislocation during its motion is displaced relative to the material on the opposite side by the Burgers vector, $\vec{b}$, of the dislocation. Work δW is generally done by the stress during this displacement. This results in a corresponding reduction in the potential energy of the system. The magnitude of the effective force on the dislocation (often termed the "mechanical" force) is then just $f = \delta W / \delta r$. A detailed analysis of this force yields the Peach–Koehler equation:

$$\vec{f}_{\sigma} = \left(\vec{b}^{T} \cdot \boldsymbol{\sigma} \right) \times \hat{\zeta} \equiv \vec{d} \times \hat{\zeta} \tag{11.1}$$

where $\vec{f}_{\sigma}$ is the mechanical force exerted on the dislocation (per unit length), $\boldsymbol{\sigma}$ the stress tensor in the material at the dislocation, and $\hat{\zeta}$ the unit vector tangent to the dislocation along its positive direction [2]. Equation 11.1 is consistent with the convention that the Burgers vector of the dislocation is the closure failure (from start to finish) of a Burgers circuit taken in a crystal in a clockwise direction around the dislocation while looking along the dislocation in the positive direction.[2] When written in full, Eq. 11.1 has the form

$$\vec{f}_{\sigma} = \hat{\imath}[d_y \zeta_z - d_z \zeta_y] + \hat{\jmath}[d_z \zeta_x - d_x \zeta_z] + \hat{k}[d_x \zeta_y - d_y \zeta_x] \tag{11.2}$$

where

$$\vec{d} = \begin{bmatrix} d_x \\ d_y \\ d_z \end{bmatrix} = \begin{bmatrix} b_x \sigma_{xx} + b_y \sigma_{xy} + b_z \sigma_{xz} \\ b_x \sigma_{xy} + b_y \sigma_{yy} + b_z \sigma_{yz} \\ b_x \sigma_{xz} + b_y \sigma_{yz} + b_z \sigma_{zz} \end{bmatrix} \tag{11.3}$$

With this result, the mechanical force exerted on any straight dislocation by any stress field can be calculated. For example, if the edge dislocation ($\vec{b} = [b00]$, $\hat{\zeta} = [001]$) in Figure 11.1 is subjected to a shearing stress σ_{xy}, it experiences a force urging it to glide on its slip plane in the x direction. However, if the dislocation is subjected to the tensile stress, σ_{xx}, Eq. 11.1 shows that it will experience the force $\vec{f}_{\sigma} = -\hat{\jmath}\, b\, \sigma_{xx}$ (i.e., a force urging it to climb in the $-y$ direction).

In a more general stress field, the force (which is always perpendicular to the dislocation line) can have a component in the glide plane of the dislocation as well as a component normal to the glide plane. In such a case, the overall force will tend to produce both glide and climb. However, if the temperature is low enough that no significant diffusion is possible, only glide will occur.

---

[2]The Burgers circuit is constructed so that it will close if mapped step by step into a perfect reference crystal. See Hirth and Lothe [2].

11.2.2 Osmotic Force

A dislocation is generally subjected to another type of force if nonequilibrium point defects are present (see Fig. 11.2). If the point defects are supersaturated vacancies, they can diffuse to the dislocation and be destroyed there by dislocation climb. A diffusion flux of excess vacancies to the dislocation is equivalent to an opposite flux of atoms taken from the extra plane associated with the edge dislocation. This causes the extra plane to shrink, the dislocation to climb in the $+y$ direction, and the dislocation to act as a vacancy sink. In this situation, an effective "osmotic" force is exerted on the dislocation in the $+y$ direction, since the destruction of the excess vacancies which occurs when the dislocation climbs a distance δy causes the free energy of the system to decrease by $\delta \mathcal{G}$. The osmotic force is then given by $\vec{f}_\mu = -\hat{\jmath}\,\delta\mathcal{G}/\delta y$.

By evaluating $\delta\mathcal{G}$ and δy when δN_V vacancies are destroyed, an expression for $\vec{f}_\mu$ can be obtained. The quantity $\delta\mathcal{G}$ is just $-\mu_V \delta N_V$, where the chemical potential of the vacancies, μ_V, is given by Eq. 3.66. If a climbing edge dislocation destroys δN_V vacancies per unit length, the climb distance will be $\delta y = (\Omega/b)\,\delta N_V$. The osmotic force is therefore

$$\vec{f}_\mu = -\hat{\jmath}\frac{\delta\mathcal{G}}{\delta y} = \hat{\jmath}\,\frac{\mu_V\,\delta N_V}{(\Omega/b)\,\delta N_V} = \hat{\jmath}\,\frac{b}{\Omega}kT\ln\left(\frac{X_V}{X_V^{\mathrm{eq}}}\right) \tag{11.4}$$

This result is easily generalized for mixed dislocations which are partly screw-type and partly edge-type, and also for cases having subsaturated vacancies. For a mixed dislocation, b must be replaced by the edge component of its Burgers vector

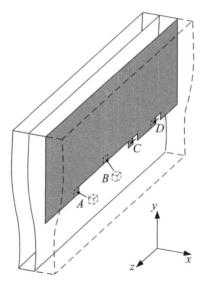

Figure 11.2:   Oblique view of edge dislocation climb due to destruction of excess vacancies. The extra plane associated with the edge dislocation is shaded. At A, a vacancy from the crystal is destroyed directly at a jog. At B, a vacancy from the crystal jumps into the core. At C, an attached vacancy is destroyed at a jog. At D, an attached vacancy diffuses along the core.

and the result (see Exercise 11.1) is

$$\vec{f}_{\mu} = \hat{\zeta} \times \vec{B} \qquad (11.5)$$

where

$$\vec{B} = \vec{b}\,\frac{kT}{\Omega}\ln\left(\frac{X_V}{X_V^{\mathrm{eq}}}\right) \qquad (11.6)$$

If the vacancies are subsaturated, the dislocation tends to produce vacancies and therefore acts as a vacancy source. In that case, Eq. 11.5 will still hold, but μ_V will be negative and the climb force and climb direction will be reversed. Equation 11.5 also holds for interstitial point defects, but the sign of $\vec{f}_{\mu}$ will be reversed.

11.2.3 Curvature Force

Still another force will be present if a dislocation is curved. In such cases, the dislocation can reduce the energy of the system by moving to decrease its length. An effective force therefore tends to induce this type of motion. Consider, for example, the simple case of a circular prismatic dislocation loop of radius, R. The energy of such a loop is

$$W = R\frac{\mu b^2}{2(1-\nu)}\left[\ln\left(\frac{4R}{R_\circ}\right) - 1\right] \qquad (11.7)$$

where $R_\circ$ is the usual cutoff radius (introduced to avoid any elastic singularity at the origin) [2]. The energy of such a loop can be reduced by reducing its radius and therefore its length. Thus, a climb force, $\vec{f}_\kappa$, exists which is radial and in the direction to shrink the loop. A calculation of the reduction in the loop energy achieved when its radius shrinks by δR shows that $(\partial W/\partial R)\,\delta R = 2\pi R f_\kappa\,\delta R$. The force is therefore

$$|\vec{f}_\kappa| = \frac{1}{2\pi R}\frac{\partial W}{\partial R} = \frac{1}{R}\frac{\mu b^2}{4\pi(1-\nu)}\ln\left(\frac{4R}{R_\circ}\right) \qquad (11.8)$$

This result may be generalized. Any segment of an arbitrarily curved dislocation line will be subjected to a curvature force of similar magnitude because the stress fields of other segments of the dislocation line at some distance from the segment under consideration exert only minimal forces on it. For most curved dislocation geometries, the magnitude of the right-hand side of Eq. 11.8 is approximately equal to $\mu b^2\,(1/R)$. Therefore, for a general dislocation with radius of curvature, R,

$$\left|\vec{f}_\kappa\right| \approx \mu b^2(1/R) \qquad (11.9)$$

The quantity μb^2 has the dimensions of a force (or, equivalently, energy per unit length) and is known as the *line tension* of the dislocation. Equation 11.9 can also be obtained by taking the line tension to be a force acting along the dislocation in a manner tending to decrease its length.[3] This approximation is supported by detailed calculations for other forms of curved dislocations [2].

---

[3]This is explored further in Exercise 11.2.

The vector form of Eq. 11.9 is readily obtained. If $\vec{r}$ is the position vector tracing out the dislocation line in space and ds is the increment of arc length traversed along the dislocation when $\vec{r}$ increased by $d\vec{r}$,[4]

$$\frac{d\hat{\zeta}}{ds} = \frac{d^2\vec{r}}{ds^2} = \kappa\,\hat{n} = \frac{1}{R}\,\hat{n} \tag{11.10}$$

where at the point $\vec{r}$ on the line, $\hat{n}$ is the *principal normal*, which is a unit vector perpendicular to $\hat{\zeta}$ and directed toward the concave side of the curved line, κ is the curvature, and R is the radius of curvature. Therefore,

$$\vec{f}_\kappa \approx \frac{\mu b^2}{R}\,\hat{n} = \mu b^2 \frac{d\hat{\zeta}}{ds} \tag{11.11}$$

11.2.4 Total Driving Force on a Dislocation

The total driving force on a dislocation, $\vec{f}$, is the sum of the forces previously considered and, therefore,

$$\vec{f} = \vec{f}_\sigma + \vec{f}_\mu + \vec{f}_\kappa = \left(\vec{d} \times \hat{\zeta}\right) + \left(\hat{\zeta} \times \vec{B}\right) + \mu b^2 \frac{d\vec{\zeta}}{ds} = \hat{\zeta} \times (\vec{B} - \vec{d}) + \mu b^2 \frac{d\hat{\zeta}}{ds} \tag{11.12}$$

11.3 DISLOCATION GLIDE

Of central interest is the rate at which a dislocation is able to glide through a crystal under a given driving force. Many factors play potential roles in determining this rate. In perfect crystals, relativistic effects can come into play as dislocation velocities approach the speed of sound in the medium. At elevated temperatures, dissipative phonon effects can produce frictional drag forces opposing the motion. Also, the atom shuffling at the core, which is necessary for the motion, may be difficult in certain types of crystals and thus inhibit glide. In imperfect crystals, any point, line, and planar defects and inclusions can serve as additional obstacles hindering dislocation glide. We begin by discussing glide in a perfect single crystal, which for the present is taken to be a linear elastic continuum.

11.3.1 Glide in Perfect Single Crystals

Relativistic Effects. Consider the relatively simple case of a screw dislocation moving along x at the constant velocity $\vec{v}$ (see Fig. 11.3). The elastic displacements, u_1, u_2, and u_3, around such a dislocation may be determined by solving the Navier equations of isotropic linear elasticity [3].[5] For this screw dislocation, the only nonzero displacements are along z, and for the moving dislocation the Navier equations therefore reduce to

$$\rho\,\frac{\partial^2 u_3}{\partial t^2} = \mu\left(\frac{\partial^2 u_3}{\partial x^2} + \frac{\partial^2 u_3}{\partial y^2}\right) \tag{11.13}$$

where ρ is the density of the medium, μ is the shear modulus, and on the left is the inertial term due to the acceleration of mass caused by the moving dislocation.

---

[4]See Appendix C for a brief survey of mathematical relations for curves and surfaces.
[5]See standard references on dislocation mechanics [2, 4, 5].

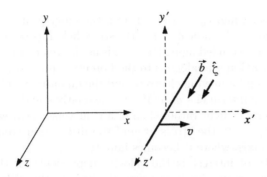

Figure 11.3:   Screw dislocation with $\vec{b} = [00b]$, $\hat{\zeta} = [001]$ moving in the $+x$ direction at a constant velocity $\vec{v}$. The origin of the primed (x', y', z') coordinate system is fixed to the moving dislocation.

Equation 11.13 is readily solved after making the changes of variable

$$
\begin{aligned}
x' &= \frac{x - vt}{\gamma_L} \\
\gamma_L &\equiv \sqrt{1 - \left(\frac{v}{c}\right)^2} \\
y' &= y \\
z' &= z \\
t' &= \frac{t - vx}{c^2 \gamma_L}
\end{aligned}
\tag{11.14}
$$

where $c = \sqrt{\mu/\rho}$ is the velocity of a transverse shear sound wave in the elastic medium. The origin of the (x', y', z') coordinate system is fixed on the moving dislocation as in Fig. 11.3. These changes of variable transform Eq. 11.13 into

$$
\frac{\partial^2 u_3}{\partial x'^2} + \frac{\partial^2 u_3}{\partial y'^2} = 0
\tag{11.15}
$$

because u_3 is a not a function of t' in the moving coordinate system and $\partial u_3/\partial t' = 0$. Equation 11.15 has the form of the Navier equation for a static screw dislocation and its solution[6] has the form

$$
u_3(x', y') = \frac{b}{2\pi} \tan^{-1}\left(\frac{y'}{x'}\right)
\tag{11.16}
$$

Transforming this solution back to x, y, t space,

$$
u_z(x, y, t) = \frac{b}{2\pi} \tan^{-1}\left(\frac{\gamma_L y}{x - vt}\right)
\tag{11.17}
$$

The shear stress of the dislocation in cylindrical coordinates, $\sigma_{\theta z}$, may now be found by using the standard relations $\sigma_{xz} = \mu(\partial u_z/\partial x)$, $\sigma_{yz} = \mu(\partial u_z/\partial y)$, and $\sigma_{\theta z} = \sigma_{yz} \cos\theta - \sigma_{xz} \sin\theta$. The result is

$$
\sigma_{\theta z} = \frac{\mu b}{2\pi} \frac{\gamma_L \left(x_0^2 + y_0^2\right)^{1/2}}{x_0^2 + \gamma_L^2 y_0^2}
\tag{11.18}
$$

[6]Further discussion of this can be found in Hirth and Lothe [2].

where the distances x_0 and y_0 (measured from the moving dislocation) have been introduced. Equation 11.18 indicates that the stress field is progressively contracted along the x_0 axis and extended along the y_0 axis as the velocity of the dislocation is increased. This distortion is analogous to the Lorentz contraction and expansion of the electric field around a moving electron, and the quantity γ_L plays a role similar to the Lorentz–Einstein term $(1 - v^2/c^2)^{1/2}$ in the relativistic theory of the electron, where c is the velocity of light rather than of a transverse shear wave. In the limit when $v \to c$ and $\gamma_L \to 0$, the stress around the dislocation vanishes everywhere except along the y'-axis, where it becomes infinite.

Another quantity of interest is the velocity dependence of the energy of the dislocation. The energy density in the material around the dislocation, w, is the sum of the elastic strain-energy density and the kinetic-energy density,

$$w = 2\mu\varepsilon_{xz}^2 + 2\mu\varepsilon_{yz}^2 + \frac{1}{2}\rho\left(\frac{\partial u_z}{\partial t}\right)^2 = \frac{\mu}{2}\left[\left(\frac{\partial u_z}{\partial x}\right)^2 + \left(\frac{\partial u_z}{\partial y}\right)^2 + \frac{1}{2}\left(\frac{\partial u_z}{\partial t}\right)^2\right]$$

(11.19)

where the first two terms in each expression make up the elastic strain-energy density and the third term is the kinetic-energy density [3]. The total energy may then be found by integrating the energy density over the volume surrounding the dislocation, yielding

$$W = \frac{W^\circ}{\gamma_L} = \frac{W^\circ}{(1 - v^2/c^2)^{1/2}}$$

(11.20)

where W° is the elastic energy of the dislocation per unit length at rest [2, 4, 5],

$$W^\circ = \frac{\mu b^2}{4\pi}\ln\left(\frac{R}{R_\circ}\right)$$

(11.21)

Here, $R_\circ$ is again the usual cutoff radius at the core and R is the dimension of the crystal containing the dislocation. According to Eq. 11.20, the energy of the moving dislocation will approach infinity as its velocity approaches the speed of sound. Again, the relationship for the moving dislocation is similar to that for a relativistic particle as it approaches the speed of light.

These results indicate that in the present linear elastic model, the limiting velocity for the screw dislocation will be the speed of sound as propagated by a shear wave. Even though the linear model will break down as the speed of sound is approached, it is customary to consider c as the limiting velocity and to take the relativistic behavior as a useful indication of the behavior of the dislocation as $v \to c$. It is noted that according to Eq. 11.20, relativistic effects become important only when v approaches c rather closely.

The behavior of an edge dislocation is more complicated since its displacement field produces both shear and normal stresses. The solution consists of the superposition of two terms, each of which behave relativistically with limiting velocities corresponding to the speed of transverse shear waves and longitudinal waves, respectively [2, 4, 5]. The relative magnitudes of these terms depend upon v.

Drag Effects. Dislocations gliding in real crystals encounter dissipative frictional forces which oppose their motion. These frictional forces generally limit the dislocation velocity to values well below the relativistic range. Such drag forces originate from a variety of sources and are difficult to analyze quantitatively.

Drag by Emission of Sound Waves. When a straight dislocation segment glides in a crystal, its core structure varies periodically with the periodicity of the crystal along the glide direction. The potential energy of the system, a function of the core structure, will therefore vary with this same periodicity as the dislocation glides. Because of this position dependence, there is a spatially periodic *Peierls force* that must be overcome to move a dislocation. Therefore, the force required to displace a dislocation continuously must exceed the Peierls force, indicated by the positions where the derivative of potential energy in Fig. 11.4 is maximal [2].[7] As the dislocation traverses the potential-energy maxima and minima, it alternately decelerates and accelerates and changes its structure periodically in a "pulsing" manner. These structural changes radiate energy in the form of sound waves (phonons). The energy required to produce this radiation must come from the work done by the applied force driving the dislocation. The net effect is the conversion of work into heat, and a frictional drag force is therefore exerted on the dislocation.

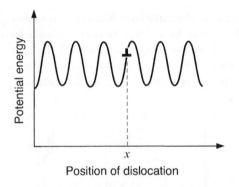

Position of dislocation

Figure 11.4: Variation of potential energy of crystal plus dislocation as a function of dislocation position. Periodicity of potential energy corresponds to periodicity of crystal structure.

In a crystal, sound waves of a given polarization and direction of propagation are dispersive—their velocity is a decreasing function of their wavenumber, which produces a further drag force on a dislocation. The dispersion relation is

$$\omega = \frac{2c}{d} \sin\left(\frac{kd}{2}\right) \tag{11.22}$$

where ω is the angular frequency, d is the distance between successive atomic planes in the direction of propagation, $k = 2\pi/\lambda$ is the wavenumber, and λ is the wavelength.[8] In the long-wavelength limit ($\lambda \gg d$) corresponding to an elastic wave in a homogeneous continuum, the phase velocity is c (as expected). However, at the shortest wavelength that the crystal can transmit ($\lambda = 2d$), the phase velocity is lower and, according to Eq. 11.22, is given by $2c/\pi$. The displacement field of the dislocation can now be broken down into Fourier components of different wavelengths. If the dislocation as a whole is forced to travel at a velocity lower than c but higher than $2c/\pi$, the short-wavelength components will be compelled to travel faster than their phase velocity and will behave as components of a su-

---

[7] However, dislocations will still move by thermally activated processes below the Peierls force.
[8] For more about the dispersion relation, see a reference on solid-state physics, such as Kittel [6].

personic dislocation. These components will radiate energy and therefore impose a viscous drag force on the dislocation (see Section 11.3.4).

Drag by Scattering of Phonons and Electrons.   A dislocation scatters phonons by two basic mechanisms. First, there are density changes in its displacement field which produce scattering. Second, the dislocation moves under the influence of an impinging sound wave and, as it oscillates, re-radiates a cylindrical wave. If the dislocation undergoes no net motion and is exposed to an isotropic flux of phonons it will experience no net force. However, if it is moving, the asymmetric phonon scattering will exert a net retarding force, since, in general, any entity that scatters plane waves experiences a force in the direction of propagation of the waves. If, in addition, free electrons are present, they will be scattered by an effective scattering potential produced by the displacement field of the dislocation. This produces a further retarding force on a moving dislocation.

Peierls Force: Continuous vs. Discontinuous Motion.   In some crystals (e.g., covalent crystals) the Peierls force may be so large that the driving force due to the applied stress will not be able to drive the dislocation forward. In such a case the dislocation will be rendered immobile. However, at elevated temperatures, the dislocation may be able to surmount the Peierls energy barrier by means of stress-aided thermal activation, as in Fig. 11.5.

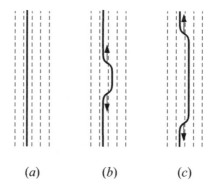

(a) (b) (c)

Figure 11.5:   Movement of dislocation across a Peierls energy barrier by thermally activated generation of double kinks. Dashed lines represent positions of energy minima shown in Fig. 11.4.

In Fig. 11.5*a*, the dislocation is forced up against the side of a Peierls "hill" by an applied stress as in Fig. 11.4. With the aid of thermal activation, it then generates a *double kink* in which a short length of the dislocation moves over the Peierls hill into the next valley (Fig. 11.5*b*).[9] The two kinks then glide apart transversely under the influence of the driving force (Fig. 11.5*c*), and eventually, the entire dislocation advances one periodic spacing. By repeating this process, the dislocation will advance in a discontinuous manner with a waiting period between each advance, and the overall forward rate will be thermally activated. This is an

[9]A *kink* is an offset of the dislocation in its glide plane; it differs fundamentally from a *jog*, an offset normal to the glide plane.

example of discontinuous motion, which results when the driving force is not large enough to drive the dislocations forward continuously in purely mechanical fashion.

Figure 11.6 illustrates the energy that must be supplied by thermal activation. The curve of σb vs. A shows the force that must be applied to the dislocation (per unit length) if it were forced to surmount the Peierls barrier in the manner just described in the absence of thermal activation. The quantity A is the area swept out by the double kink as it surmounts the barrier and is a measure of the forward motion of the double kink. $A = 0$ corresponds to the dislocation lying along an energy trough (minimum) as in Fig. 11.5a. A_2 is the area swept out when maximum force must be supplied to drive the double kink. A_4 is the area swept out when the saddle point has been reached and the barrier has been effectively surmounted. The area under the curve is then the total work that must be done by the applied stress to surmount the barrier in the absence of thermal activation. When the applied stress is σ_A (and too small to force the barrier), the swept-out area is A_1, and the energy that must be supplied by thermal activation is then the shaded area shown in Fig. 11.6. The activation energy is then

$$E = b \int_{A_1}^{A_3} (\sigma - \sigma_A) \, dA \tag{11.23}$$

and the overall dislocation velocity will be of the form

$$v = v_o e^{-E/(kT)} \tag{11.24}$$

where v_o is proportional to an attempt frequency. The area $A_3 - A_1$ swept out during the activation event is termed the *activation area*. Of particular interest from a kinetics standpoint is the result (Eq. 11.23) that the activation energy decreases as the applied stress increases: hence, the term *stress-aided thermal activation*.

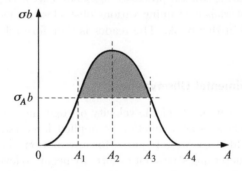

Figure 11.6:   Curve of applied force, σb, vs. area swept out, A, when dislocation surmounts an obstacle to glide.

11.3.2   Glide in Imperfect Crystals Containing Various Obstacles

Real crystals can contain a large variety of different types of point, line, and planar crystal defects and other entities, such as embedded particles, which interact with dislocations and can act as obstacles to glide. Solute atoms are good examples of point defects that hinder dislocation glide by acting as centers of dilation (see Fig. 3.9) and therefore possess stress fields that interact with dislocation stress fields, causing localized dislocation–solute-atom attraction or repulsion. If a dispersion of solute atoms is present in solution, a dislocation will not move through it as a rigid line but will consist of segments that bulge in and out as the dislocation experiences close encounters with nearby solute atoms. The overall dislocation motion therefore consists of a uniform motion with superimposed rapid forward or backward localized bulging. This type of rapid bulging motion dissipates extra energy by a number of the mechanisms already discussed and therefore exerts a drag force. At sufficiently high temperatures, solute atoms may migrate in the stress field of dislocations (Section 3.5.2), and such induced diffusion can dissipate energy and produce a drag force, particularly for slowly moving dislocations. In addition, solute atoms with anisotropic displacement fields can change orientations under the influence of the stress field of a moving dislocation, thereby producing an increment of macroscopic strain (see Section 8.3.1). This can also lead to a dissipative drag force. Solute atoms can also segregate to the cores of dislocations and form atmospheres around dislocations and thus hinder, or even pin, their motion.

Dislocations attract and repel other dislocations. Perhaps the most important example is the work hardening that occurs during the plastic deformation of crystals. Here, large numbers of dislocations are generated during the deformation; many remain in the crystal, where they act as obstacles to the passage of further dislocations, causing the material to strengthen and harden. At elevated temperatures during creep, gliding dislocations, which are held up at obstacles in their slip planes, can climb around them with the help of thermal activation (see the following section) and thus continue their glide.

Grain boundaries act as barriers to slip, since, in general, a gliding dislocation will encounter a discontinuity in its slip plane and Burgers vector when it impinges on a boundary and attempts to pass through it into the adjoining grain.

The host of interesting kinetic processes associated with the movement of dislocations through materials containing various obstacles to their motion is far too large to be described in this book. The reader is therefore referred to specialized texts [2, 7–9].

11.3.3 Some Experimental Observations

Figure 11.7 shows measurements of the velocity of edge and screw dislocation segments in LiF single crystals as a function of applied force (stress) [10]. Stresses above a yield threshold stress were required for any motion. The velocity then increased rapidly with increasing stress but eventually began to level off as the velocity of sound was approached. Results within the significantly relativistic range were not achieved in these experiments, since for all measurements $\gamma_L \approx 1$. It is likely that at the lower stresses (where the results are impurity sensitive), the velocities were limited by impediments arising from dislocation–dislocation and dislocation–defect interactions [2]. This regime holds for the plastic deformation of essentially all crystalline materials deformed at normal strain rates. At the higher stresses in Fig. 11.7 (where the smaller slope is impurity insensitive and decreases with increasing temperature), the higher velocities were limited by phonon-viscosity drag. High dislocation velocities may be achieved at the start of even low-strain-rate deformation if the initial concentration of mobile dislocations is unusually low [11]. In such cases, a small number of dislocations must move very rapidly to accom-

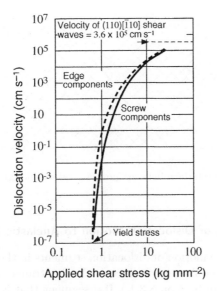

Figure 11.7:   Velocity vs. resolved shear stress for dislocation motion in LiF single crystals. From Johnston and Gilman [10].

plish the strain required. Further experimental evidence has been presented for the strong frictional drag forces that come into play at high velocities approaching the relativistic range [11]. Finally, it is noted that the viscous damping of dislocation motion converts mechanical energy to heat. This produces internal friction when a crystal containing a dislocation network is subjected to an oscillating applied stress (see Section 11.3.5).

11.3.4 Supersonic Glide Motion

If a dislocation is injected into a crystal at a speed greater than the speed of sound in the crystal, it will radiate energy in the form of sound waves similar to the way that a charged particle emits electromagnetic Cherenkov radiation when it is injected into a material at a velocity greater than the speed of light in that medium [5]. This causes rapid deceleration of the dislocation. However, steady-state supersonic motion of dislocations is possible in special cases where the motion of the dislocation in its glide plane causes a sufficiently large reduction in the energy of the system [2, 5]. In such a case, this reduction of energy provides the energy that must be radiated during the supersonic motion. Conceivable examples include motion of a partial dislocation that removes its associated fault (see Fig. 9.10) or dislocation motion in a glissile martensitic interface (Section 24.3), which converts the higher free-energy parent phase to the lower-energy martensitic phase. Models for the motion of such dislocations are entirely different from those discussed in Section 11.3.1 and are described by Nabarro [5] and Hirth and Lothe [2]. So far, there is no clear evidence for the supersonic motion of martensitic interfaces, probably due to the influence of frictional drag forces. However, there is some evidence that supersonic dislocations are present in shock-wave fronts, as in Fig. 11.8 [12]. Models for the motion of such fronts have been described [11, 13], and some evidence for the existence of dislocations in them has been obtained by computer simulation [14].

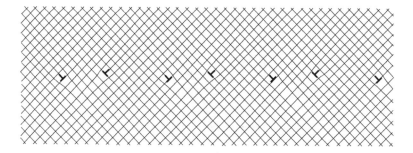

Figure 11.8:   Possible interface between normal (lower) and compressed (upper) lattice in a shock-wave front.

11.3.5 Contributions of Dislocation Motion to Anelastic Behavior

The stress-induced glide motion of dislocation segments in the dislocation networks usually present in materials can produce anelastic strains. (See the general discussion of anelasticity in Section 8.3.1.) If a segment that is pinned at its ends is subjected to an oscillating stress, it will periodically bow in and out in a manner similar to a vibrating string, as illustrated in Fig. 11.9. This will produce a small oscillating strain in the material. Because dissipative drag forces will be in play and the dislocation velocity is not infinite, this strain will lag the stress, causing internal friction [15]. Further aspects are considered in Exercise 11.12.

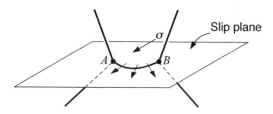

Figure 11.9:   Dislocation segment pinned at A and B bowing out in the slip plane due to the applied shear stress, σ.

11.4 DISLOCATION CLIMB

Figure 11.2 presents a simplified three-dimensional representation of the climb of an edge dislocation arising from the destruction of excess vacancies in the crystal. The jogs (steps in the edge of the extra plane) in the dislocation core are the sites where vacancies are created or permanently destroyed. Vacancies can reach a jog by either jumping directly into it or else by first jumping into the dislocation core and then diffusing along it to a jog, where they are destroyed. The elementary processes involved include:

- The jumping of a vacancy directly into a jog and its simultaneous destruction, as at A

- The jumping of a vacancy into the core, where it becomes attached as at B

- The destruction of an attached vacancy at a jog, as at C

- The diffusion of an attached vacancy along the core, as at D

In many cases, vacancies are bound to the dislocation core by an attractive binding energy and diffuse along the dislocation more rapidly than in the crystal. Many more vacancies may therefore reach jogs by fast diffusion along the dislocation core than by diffusion directly to them through the crystal.

The jogs required for the climb process can be generated by the nucleation and growth of strings of attached excess vacancies along the core. When a string becomes long enough, it will collapse to produce a fully formed jog pair, as, for example, in the region along the core bounded by A and C in Fig. 11.2. The spacing of the jog pair then increases due to the continued destruction of excess vacancies at the jogs until a complete row of atoms has been stripped from the edge of the extra plane. During steady-state climb, this process then repeats itself.

11.4.1 Diffusion-Limited vs. Source-Limited Climb Kinetics

A detailed kinetic model for the overall climb rate based on the above mechanisms has been developed [2, 16–20]. In this model it is assumed that because the vacancies are easily destroyed at jogs, they are maintained at their equilibrium concentration in the immediate vicinity of the jogs. If the vacancies experience an attractive binding energy to the core and also diffuse relatively rapidly along it, a typical attached vacancy will diffuse a significantly large mean distance, $\langle Z \rangle$, along the core before it jumps back off into the crystal. The magnitude of $\langle Z \rangle$ increases with the binding energy of the vacancy to the dislocation and the relative rate of diffusion of the vacancy along the dislocation core. Each jog is therefore capable of maintaining the vacancy concentration essentially at equilibrium over a distance along the dislocation on either side of it equal approximately to the distance $\langle Z \rangle$. Each jog, with the assistance of the two adjoining segments of high-diffusivity core, therefore acts effectively as an ellipsoidal sink of semiaxes b and $\langle Z \rangle$ having a surface on which the vacancy concentration is maintained in local equilibrium with the jog. The overall effectiveness of the dislocation as a sink then depends upon the magnitude of $\langle Z \rangle$ and the mean spacing of the jogs along the dislocation, $\langle S \rangle$. When the vacancy supersaturation is small and the system is near equilibrium, the jog spacing will be given approximately by the usual Boltzmann equilibrium expression $\langle S \rangle \cong b \exp[-E_j/(kT)]$, where E_j is the energy of formation of a jog. However, at high supersaturations when excess vacancies can aggregate quickly along the dislocation and nucleate jog-pairs rapidly, the number of jogs will be increased above the equilibrium value and their spacing will be reduced correspondingly [17, 18]. A wide range of dislocation sink efficiencies is then possible. When $2\langle Z \rangle/\langle S \rangle \geq 1$, the effective jog sinks overlap along the dislocation line, which then acts as a highly efficient line sink capable of maintaining local vacancy equilibrium everywhere along its length. The rate of vacancy destruction is limited only by the rate at which the vacancies can diffuse to the dislocation, and the rate of destruction will then be the maximum possible. The kinetics are therefore *diffusion-limited*, and the dislocation is considered an "ideal" sink. Conditions that promote this situation are a high binding energy for attached vacancies, a relatively fast diffusion rate along the core, a small jog formation energy, and a large vacancy supersaturation.

On the other hand, when the fast diffusion of attached vacancies to the jogs is impeded and $\langle Z \rangle$ is therefore small (i.e., $\langle Z \rangle \cong b$), each jog acts as a small isolated spherical sink of radius b. If, at the same time, $\langle S \rangle$ is large, the jog sinks are far apart and the overall dislocation sink efficiency is relatively small. Under these conditions the rate of vacancy destruction will be limited by the rate at which the vacancies can be destroyed along the dislocation line, and the overall rate of vacancy destruction will be reduced. In the limit where the rate of destruction is slow enough so that it becomes essentially independent of the rate at which vacancies can be transported to the dislocation line over relatively long distances by diffusion, the kinetics are *sink-limited*.

When the dislocation acts as a sink for a flux of diffusing vacancies (or alternatively, as a source of atoms) or as a source for a flux of vacancies, it is useful to introduce a source or sink efficiency, η, defined by

$$\eta = \frac{\text{flux of atoms created at actual source}}{\text{flux of atoms created at corresponding "ideal" source}} \tag{11.25}$$

A dislocation source that climbs rapidly enough so that ideal diffusion-limited conditions are achieved therefore operates with an efficiency of unity. On the other hand, slowly acting sources can have efficiencies approaching zero. Applications of these concepts to the source action of interfaces are discussed in Section 13.4.2.

The climb of mixed dislocations possessing some screw character can proceed by basically the same jog-diffusion mechanism as that for the pure edge dislocation.[10] On the other hand, a pure screw dislocation can climb if the excess vacancies convert it into a helix, as in Fig. 11.10. Here the turns of the helical dislocation possess

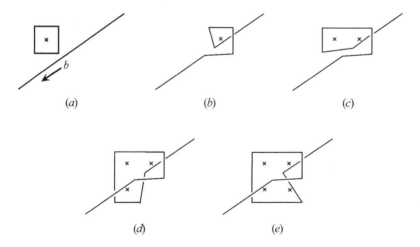

Figure 11.10:   Formation of a helical segment on an initially straight screw dislocation lying along [100] in a primitive-cubic crystal by progressive addition of vacancies to the core. For graphic purposes, each vacancy is represented by a vacancy-type prismatic loop of atomic size. **(a)** Vacancy in a crystal with an initial straight screw dislocation nearby. **(b)** Configuration after a vacancy has joined the dislocation. **(c)**–**(e)** Configurations after two, three, and four vacancies have been added.

[10]Details are discussed by Balluffi and Granato [19].

strong edge components and, once formed, continue to climb as mixed dislocation segments.

Additional factors may play a role during dislocation climb in many systems. These include the possibility that jogs may be able to nucleate heterogeneously at nodes or regions of sharp curvature. Also, in low stacking-fault-energy materials, the dislocation may be dissociated into two partial dislocations bounding a ribbon of stacking fault as shown in Fig. 9.10. In such cases, the jogs may also be dissociated and possess a relatively high formation energy, causing the climb to be more difficult [2, 19].

11.4.2 Experimental Observations

Reviews of experimental observations of the efficiency with which dislocations climb under different driving forces have been published [18–22]. A wide range of semi-quantitative results is available only for metals, including:

- Vacancy quenching experiments where the destruction rate at climbing dislocations of supersaturated vacancies obtained by quenching the metal from an elevated temperature is measured (see the analysis of this phenomenon in the following section)

- Dislocation loop annealing where the rate at which dislocation loops shrink by means of climb is measured (see analysis in following section)

- Sintering experiments where the rate at which vacancies leave voids and are then destroyed at climbing dislocations is measured

Of main interest is the efficiency of climb and its dependence on the magnitude of the force driving the climb process. In general, the efficiency of climbing dislocations as sources increases as the driving force increases, since more energy is then available to drive the climb. A convenient measure of the relative magnitude of this force is the energy change, g_s, which is achieved per crystal site created as a result of the climb.

All dislocations, including dissociated dislocations in lower-stacking-fault-energy metals and relatively nondissociated dislocations in high-stacking-fault-energy metals, operate as highly efficient sources when $|g_s|$ is large, as in rapidly quenched metals [20]. However, when $|g_s|$ is reduced, lower efficiencies, which may become very small, are found for the lower-stacking-fault-energy metals. The efficiencies for the higher-stacking-fault-energy metals appear to fall off less rapidly with $|g_s|$. This may be understood on the basis of the tendency of the dislocations to contain more jogs as $|g_s|$ increases and the greater difficulty in forming jogs on dissociated dislocations than on undissociated dislocations because of the larger jog energies of the former.

11.4.3 Analyses of Two Climb Problems

Climbing Dislocations as Sinks for Excess Quenched-in Vacancies. Dislocations are generally the most important vacancy sources that act to maintain the vacancy concentration in thermal equilibrium as the temperature of a crystal changes. In the following, we analyze the rate at which the usual dislocation network in a crystal destroys excess supersaturated vacancies produced by rapid quenching from an elevated temperature during isothermal annealing at a lower temperature. If the dislocations in the network are present at a density ρ_d (dislocation line length per unit volume), a reasonable approximation is that each dislocation segment acts as the dominant vacancy sink in a cylindrical volume centered on it and of radius $R = (\pi \rho_d)^{-1/2}$. The problem is then reduced to the determination of the rate at which excess vacancies in the cylinder diffuse to the dislocation line as illustrated in Fig. 11.11. The diffusion system is assumed to contain two components (A-type atoms and vacancies) and is network constrained. Equation 3.68 for the diffusion of vacancies is applicable in this case, and therefore

$$\vec{J}_V = -D_V \nabla c_V \tag{11.26}$$

According to the results in Section 11.4.2, the dislocations should act as highly effective sinks for the highly supersaturated vacancies. We therefore assume diffusion-limited kinetics in which each dislocation segment is capable of maintaining the vacancies in local thermal equilibrium at its core, represented as a cylinder of effective radius R_o, where R_o is of atomic dimensions. Also, in this type of problem, the effect of the dislocation climb motion on the diffusion of the vacancies to the dislocation can be neglected to a good approximation [2, 23]. Using the separation-of-variables method (Section 5.2.4), the diffusion equation corresponding to Eq. 3.69,

$$\frac{\partial c_V}{\partial t} = D_V \frac{1}{R} \frac{\partial}{\partial r} \left(r \frac{\partial c_V}{\partial r} \right) \tag{11.27}$$

may be solved subject to the conditions

$$
\begin{aligned}
c_V &= c_V^{\text{eq}} & &\text{for } r = R_o \text{ and } t \geq 0 \\
c_V &= c_V^{\circ} & &\text{for } R_o < r \leq R \text{ and } t = 0 \\
\frac{\partial c_V}{\partial r} &= 0 & &\text{for } r = R \text{ and } t \geq 0
\end{aligned}
\tag{11.28}
$$

where c_V° is the quenched-in vacancy concentration and c_V^{eq} is the equilibrium vacancy concentration maintained at the "surface" of the dislocation core at the an-

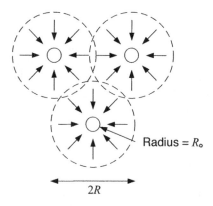

Figure 11.11:   Vacancy diffusion fields in cylindrical cells (of radius R) around dislocations acting as line sinks (of radius R_o).

nealing temperature. The solution shows that the fraction of the excess vacancies remaining in the system decays with time according to

$$f(t) = \frac{4}{R^2 - R_\circ^2} \sum_{n=1}^{\infty} \frac{J_1^2(R\alpha_n)}{\alpha_n^2 \left[J_\circ^2(R_\circ \alpha_n) - J_1^2(R\alpha_n) \right]} e^{-\alpha_n^2 D_V t} \tag{11.29}$$

where the α_n are the roots of

$$Y_\circ(R_\circ \alpha_n) J_1(R\alpha_n) - J_\circ(R_\circ \alpha_n) Y_1(R\alpha_n) = 0 \tag{11.30}$$

and J_n and Y_n are Bessel functions of the first and second kind of order n [24]. For typical values of $R_\circ$ and R, the first term in Eq. 11.29 will be dominant except at very early times when the fraction decayed is small [24]. The major portion of the excess vacancy decay will therefore be essentially exponential [i.e., $f(t) \cong \exp(-\alpha_1^2 D_V t)$]. Finally, it is noted that the above treatment does not take account of the effect of the dislocation stress field on the diffusivity of the vacancies, as discussed in Section 3.5.2. In general, this stress field is of importance only within a relatively small distance from the dislocation. Under these circumstances, its effect during the major portion of the decay can be approximated in a simple manner by making a relatively small change in the value of the effective dislocation core radius, $R_\circ$ [25]. Since the roots of Eq. 11.30 are fairly insensitive to the value of $R_\circ$, the decay rate is also rather insensitive to this choice of $R_\circ$. The effect of the stress field will therefore be relatively small.

Shrinkage of Dislocation Loops by Climb. Prismatic dislocation loops are often formed in crystals by the precipitation of excess vacancies produced by quenching or by fast-particle irradiation (see Exercise 11.7). Once formed, these loops tend to shrink and be eliminated by means of climb during subsequent thermal annealing. A number of measurements of loop shrinkage rates have been made, and analysis of this phenomenon is therefore of interest [2]. In this section we calculate the isothermal annealing rate of such a loop located near the center of a thin film in a high-stacking-fault-energy material (such as Al) where the climb efficiency will be high, and the shrinkage rate is therefore diffusion-limited.

The situation is illustrated in Fig. 11.12a. The loop is taken as an effective torus of large radius, R_L, with much smaller core radius, $R_\circ$, and the film thickness is $2d$ with $d \gg R_L$. The vacancy concentration maintained in equilibrium with the loop

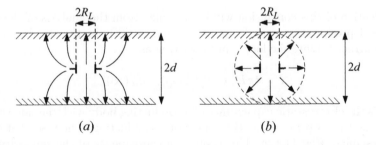

(a)   (b)

Figure 11.12:   (a) Vacancy diffusion fluxes around a dislocation loop (of radius R_L) shrinking by climb in a thin film of thickness $2d$. (b) Spherical approximation of a diffusion field in (a).

at the surface of the torus, $c_V^{\text{eq}}(\text{loop})$, is larger than the equilibrium concentration, $c_V^{\text{eq}}(\infty)$, maintained at the flat film surfaces. These concentrations can differ considerably for small loops, and the approximation leading to Eq. 3.72, which ignored variations in c_V throughout the system, cannot be employed. Equation 3.69 can be used to describe the vacancy diffusion. Vacancies therefore diffuse away from the "surface" of the loop to the relatively distant film surfaces, and the loop shrinks as it generates vacancies by means of climb.

The concentration, $c_V^{\text{eq}}(\text{loop})$, can be found by realizing that the formation energy of a vacancy at the climbing loop is lower than at the flat surface because the loop shrinks when a vacancy is formed, and this allows the force shrinking the loop (see Section 11.2.3) to perform work. In general, $N_V^{\text{eq}} = \exp[-G_V^f/(kT)]$ according to Eq. 3.65, and therefore

$$\frac{c_V^{\text{eq}}(\text{loop})}{c_V^{\text{eq}}(\infty)} = e^{[G_V^f(\infty) - G_V^f(\text{loop})]/(kT)} \tag{11.31}$$

where $G_V^f(\infty) - G_V^f(\text{loop})$ is the work performed by the force on the shrinking loop during the formation of a vacancy. The number of vacancies stored in the loop is $N_V = \pi R_L^2 b/\Omega$. The reduction in the radius of the loop due to formation of a vacancy by climb is then $\delta R_L/\delta N_V = -\Omega/(2\pi b R_L)$. Therefore,

$$\begin{aligned} G_V^f(\infty) - G_V^f(\text{loop}) &= \frac{\Omega}{2\pi b R_L} 2\pi R_L |\vec{f}_\kappa| \\ &= \frac{\mu b \Omega}{4\pi(1-\nu)} \frac{1}{R_L} \ln\left(\frac{4R_L}{R_\circ}\right) = kT \ln\left[\frac{c_V^{\text{eq}}(\text{loop})}{c_V^{\text{eq}}(\infty)}\right] \end{aligned} \tag{11.32}$$

where the force has been evaluated with Eq. 11.8.

The vacancy diffusion field around the toroidal loop will be quite complex, but at distances from it greater than about $2R_L$, it will appear approximately as shown in Fig. 11.12a. A reasonably accurate solution to this complex diffusion problem may be obtained by noting that the total flux to the two flat surfaces in Fig. 11.12a will not differ greatly from the total flux that would diffuse to a spherical surface of radius d centered on the loop as illustrated in Fig. 11.12b. Furthermore, when $d \gg R_L$, the diffusion field around such a source will quickly reach a quasi-steady state [20, 26], and therefore

$$\nabla^2 c_V = 0 \tag{11.33}$$

(A justification of this conclusion will be obtained from the analysis of the growth of spherical precipitates carried out in Section 13.4.2.) In the steady state, the vacancy current leaving the loop can be written as

$$I = 4\pi D_V C \left[c_V^{\text{eq}}(\text{loop}) - c_V^{\text{eq}}(\infty)\right] \tag{11.34}$$

where C is the electrostatic capacitance of a conducting body with the same toroidal geometry as the loop placed at the center of a conducting sphere so that the geometry resembles Fig. 11.12b. This result is a consequence of the similarity of the concentration fields, $c(x, y, z)$, and electrostatic-potential fields, $\phi(x, y, z)$, which are obtained by solving Laplace's equation in steady-state diffusion ($\nabla^2 c = 0$) and electrostatic potential ($\nabla^2 \phi = 0$) problems, respectively [20, 26]. The shrinking rate of the loop is then

$$\frac{\delta R_L}{\delta t} = -\frac{\Omega}{2\pi b R_L}\frac{\delta N_V}{\delta t} = -\frac{\Omega I}{2\pi b R_L}$$

$$= -\frac{2\pi\,^*D}{\mathbf{f}b\ln(8R_L/R_\circ)}\,e^{\mu b\Omega/[4\pi(1-\nu)kTR_L]\ln(4R_L/R_\circ)} - 1 \tag{11.35}$$

This final result is obtained by using Eqs. 8.17, 11.31, 11.32, 11.34, and the relation $C = \pi R_L/\ln(8R_L/R_\circ)$ for the capacitance of a torus in a large space when $R_L \gg R_\circ$ [27].

Analyses of the climbing rates of many other dislocation configurations are of interest, and Hirth and Lothe point out that these problems can often be solved by using the method of superposition (Section 4.2.3) [2]. In such cases the dislocation line source or sink is replaced by a linear array of point sources for which the diffusion solutions are known, and the final solution is then found by integrating over the array. This method can be used to find the same solution of the loop-annealing problem as obtained above.

As in Fig. 11.13, the loop can be represented by an array of point sources each of length $R_\circ$. Using again the spherical-sink approximation of Fig. 11.12b and recalling that $d \gg R_L \gg R_\circ$, the quasi-steady-state solution of the diffusion equation in spherical coordinates for a point source at the origin shows that the vacancy diffusion field around each point source must be of the form

$$c'_V(r') - c_V^{eq}(\infty) = \frac{a_1}{r'} \tag{11.36}$$

where a_1 is a constant to be determined. The value of a_1 is found by requiring that the concentration everywhere along the loop be equal to $c_V^{eq}(\text{loop})$. This concentration is due to the contributions of the diffusion fields of all the point sources around the loop, and therefore, from Fig. 11.13 and using $R_\circ \ll R_L$,

$$c_V^{eq}(\text{loop}) - c_V^{eq}(\infty) = 2a_1\int_{R_\circ/2}^{\pi R_L}\frac{dl}{R'R_\circ} = \frac{2a_1}{R_\circ}\int_{R_\circ/(2R_L)}^{\pi}\frac{d\theta/2}{\sin(\theta/2)}$$

$$= \frac{2a_1}{R_\circ}\ln\left(\frac{8R_L}{R_\circ}\right) \tag{11.37}$$

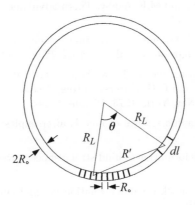

Figure 11.13:   Annealing prismatic dislocation loop taken as a circular array of vacancy point sources.

Note that the integral is terminated at the cutoff distance $R_\circ/2$ in order to avoid a singularity. The vacancy concentration at a distance from the loop appreciably greater than R_L can now be found by treating the loop itself as an effective point source made up of all the point sources on its circumference. The number of these sources is $2\pi R_L/R_\circ$, and therefore

$$c_V(r) - c_V^{\text{eq}}(\infty) = \frac{2\pi R_L a_1}{R_\circ}\frac{1}{r} = \frac{\pi R_L\,[c_V^{\text{eq}}(\text{loop}) - c_V^{\text{eq}}(\infty)]}{\ln(8R_L/R_\circ)}\frac{1}{r} \qquad (11.38)$$

The vacancy current leaving the loop is then

$$I = 4\pi r^2 D_V \frac{\partial}{\partial r}\left[c_V(r)\right] = 4\pi D_V \frac{\pi R_L}{\ln(8R_L/R_\circ)}\left[c_V^{\text{eq}}(\text{loop}) - c_V^{\text{eq}}(\infty)\right] \qquad (11.39)$$

in agreement with the results of the previous analysis.

Applications of Eq. 11.35 and closely related equations to the observed annealing rate of loops have been described [19, 28].

Bibliography

1. S.M. Allen and E.L. Thomas. *The Structure of Materials.* John Wiley & Sons, New York, 1999.

2. J.P. Hirth and J. Lothe. *Theory of Dislocations.* John Wiley & Sons, New York, 2nd edition, 1982.

3. I.S. Sokolnikoff. *Mathematical Theory of Elasticity.* McGraw-Hill, New York, 1956.

4. J. Weertman. High velocity dislocations. In P.G. Shewmon and V.F. Zackay, editors, *Response of Metals to High Velocity Deformation*, pages 205–247, New York, 1961. Interscience.

5. F.R.N. Nabarro. *Theory of Crystal Dislocations.* Clarendon Press, Oxford, 1967.

6. C. Kittel. *Introduction to Solid State Physics.* John Wiley & Sons, New York, 3rd edition, 1967.

7. F.R.N. Nabarro, editor. *Dislocations in Solids (Series)*, volume 1–12. Elsevier North-Holland, New York, 1979–2004.

8. J. Friedel. *Dislocations.* Pergamon Press, Oxford, 1964.

9. U.F. Kocks, A.S. Argon, and M.F. Ashby. Thermodynamics and kinetics of slip. *Prog. Mater. Sci.*, 19:1–288, 1975.

10. W.G. Johnston and J.J. Gilman. Dislocation velocities, dislocation densities, and plastic flow in lithium fluoride crystals. *J. Appl. Phys.*, 30(2):129–144, 1959.

11. J. Weertman. Dislocation mechanics at high strain rates. In R.W. Rohde, B.M. Butcher, J.R. Holland, and C.H. Karnes, editors, *Metallurgical Effects at High Strain Rates*, pages 319–332, New York, 1973. Plenum Press.

12. C.S. Smith. Metallographic studies of metals after explosive shock. *Trans. AIME*, 212(10):574–589, 1958.

13. J. Weertman. Plastic deformation behind strong shock waves. *Mech. Mater.*, 5(1):13–28, 1986.

14. B.L. Holian. Modeling shock-wave deformation via molecular-dynamics. *Phys. Rev. A*, 37(7):2562–2568, 1988.

15. A.S. Nowick and B.S. Berry. *Anelastic Relaxation in Crystalline Solids.* Academic Press, New York, 1972.

16. J. Lothe. Theory of dislocation climb in metals. *J. Appl. Phys.*, 31(6):1077–1087, 1960.

17. R.M. Thomson and R.W. Balluffi. Kinetic theory of dislocation climb I. General models for edge and screw dislocations. *J. Appl. Phys.*, 33(3):803–817, 1962.

18. R.W. Balluffi. Mechanisms of dislocation climb. *Phys. Status Solidi*, 31(2):443–463, 1969.

19. R.W. Balluffi and A. V. Granato. Dislocations, vacancies and interstitials. In F.R.N. Nabarro, editor, *Dislocations in Solids*, volume 4, pages 1–133, Amsterdam, 1979. North-Holland.

20. A.P. Sutton and R.W. Balluffi. *Interfaces in Crystalline Materials.* Oxford University Press, Oxford, 1996.

21. D.N. Seidman and R.W. Balluffi. Dislocation as sources and sinks for point defects in metals. In R.R. Hasiguti, editor, *Lattice Defects and Their Interactions*, pages 911–960, New York, 1967. Gordon and Breach.

22. R.W. Balluffi. Voids, dislocation loops and grain boundaries as sinks for point defects. In M. T. Robinson and F.W. Young, editors, *Proceedings of the Conference on Fundamental Aspects of Radiation Damage in Metals*, volume 2, pages 852–874, Springfield, VA, 1975. National Technical Information Service, U.S. Department of Commerce.

23. R.W. Balluffi and D.N. Seidman. Diffusion-limited climb rate of a dislocation: Effect of climb motion on climb rate. *J. Appl. Phys.*, 36(7):2708–2711, 1965.

24. D.N. Seidman and R.W. Balluffi. Sources of thermally generated vacancies in single crystal and polycrystalline gold. *Phys. Rev.*, 139(6A):1824–1840, 1965.

25. F.S. Ham. Stress assisted precipitation on dislocations. *J. Appl. Phys.*, 30(6):915–926, 1959.

26. C.P. Flynn. Monodefect annealing kinetics. *Phys. Rev.*, 133(2A):A587, 1964.

27. H. Buchholz. *Electrische und Magnetische Potentialfelder.* Springer-Verlag, Berlin, 1957.

28. D.N. Seidman and R.W. Balluffi. On the annealing of dislocation loops by climb. *Phil. Mag.*, 13:649–654, 1966.

29. J. Bardeen and C. Herring. Diffusion in alloys and the Kirkendall effect. In J.H. Hollomon, editor, *Atom Movements*, pages 87–111. American Society for Metals, Cleveland, OH, 1951.

30. W.T. Read. *Dislocations in Crystals.* McGraw-Hill, New York, 1953.

EXERCISES

11.1 Show that Eq. 11.4 for the osmotic force on an edge dislocation may be generalized for a mixed dislocation in the form

$$\vec{f}_\mu = \hat{\zeta} \times \vec{B} \qquad (11.40)$$

where

$$\vec{B} = \vec{b}\,\frac{kT}{\Omega}\ln\left(\frac{X_V}{X_V^{\mathrm{eq}}}\right) \qquad (11.41)$$

Solution. The climb force is normal to the glide plane, which contains both the Burgers vector and the tangent vector. The unit normal vector to the glide plane is therefore

$$\hat{n} = \frac{\hat{\zeta} \times \vec{b}}{\left| \hat{\zeta} \times \vec{b} \right|}$$

so

$$\vec{f}_\mu = -\frac{\hat{\zeta} \times \vec{b}}{\left| \hat{\zeta} \times \vec{b} \right|} \frac{\partial G}{\partial y} \tag{11.42}$$

Since the climb distance that results from the destruction of δN_V vacancies is now $\delta y = (\Omega / \left| \hat{\zeta} \times \vec{b} \right|) \delta N_V$, using Eqs. 3.64 and 3.66,

$$\vec{f}_\mu = -\frac{\hat{\zeta} \times \vec{b}}{\Omega} \frac{\partial G}{\partial N_V} = \frac{\hat{\zeta} \times \vec{b}}{\Omega} \mu_V = \frac{\hat{\zeta} \times \vec{b}}{\Omega} kT \ln \left(\frac{X_V}{X_V^{eq}} \right) \tag{11.43}$$

11.2 Interpret the line tension, μb^2, of a dislocation to be a force that acts along its length in a direction to decrease its length (see Eq. 11.9). Using a simple geometrical argument, show that the curvature force per unit length of dislocation acting locally on a curved segment of dislocation is then just $\left| \vec{f}_\kappa \right| = \mu b^2 / R$, where R is the radius of curvature.

Solution. The line-tension forces acting on a curved differential segment of dislocation having a radius of curvature R due to its line tension will be as shown in Fig. 11.14. The net force exerted on the segment toward the concave side is then

$$df = 2\mu b^2 \sin \left(\frac{d\theta}{2} \right) \approx \mu b^2 \, d\theta = \frac{\mu b^2 \, ds}{R} \tag{11.44}$$

and therefore

$$\left| \vec{f}_\kappa \right| = \frac{df}{ds} = \frac{\mu b^2}{R} \tag{11.45}$$

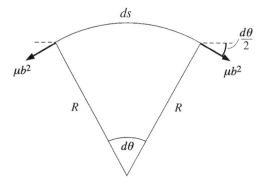

Figure 11.14:   Line tension forces on a curved dislocation segment.

11.3 Use Eq. 11.12 to show that a dislocation in a crystal possessing a uniform nonequilibrium concentration of point defects and a uniform stress field will tend to adopt a helical form. Note that both a circle and a straight line are special forms of a helix.

Solution. The dislocation will tend to adopt a form for which the net force on it given by Eq. 11.12 is everywhere zero. We therefore want to show that a helical dislocation will possess a tangent vector $\hat{\zeta}$ that satisfies

$$\mu b^2 \frac{d\hat{\zeta}}{ds} + \hat{\zeta} \times (\vec{B} - \vec{d}) = 0 \tag{11.46}$$

To evaluate $\hat{\zeta}$, we recall that the equation for a helix with its axis along z is

$$\vec{r} = \hat{i}a\cos\theta + \hat{j}a\sin\theta + \hat{k}p\theta \tag{11.47}$$

where θ is the polar angle in the xy-plane, a is the radius of the circular projection on the xy-plane, and $2\pi p$ is the distance between successive turns along z. Therefore,

$$\hat{\zeta} = \frac{d\vec{r}}{ds} = [-\hat{i}a\sin\theta + \hat{j}a\cos\theta + \hat{k}p]\frac{1}{\sqrt{a^2 + p^2}} \tag{11.48}$$

and

$$\frac{d\hat{\zeta}}{ds} = \frac{d^2\vec{r}}{ds^2} = -[\hat{i}\cos\theta + \hat{j}\sin\theta]\frac{a}{a^2 + p^2} \tag{11.49}$$

Comparing the last result with Eq. 11.10, we see that the curvature given by $\kappa = a/(a^2 + p^2)$ is constant everywhere and that the principal normal vector at any point on the helix is pointed toward the helix axis and is perpendicular to it. Also, the vectors $\vec{B}$ and $\vec{d}$ are constant vectors independent of $\hat{\zeta}$. If we now take the axis of the helix to be along $(\vec{B} - \vec{d})$ so that $(\vec{B} - \vec{d}) = \hat{k}|\vec{B} - \vec{d}|$ and put the results above into Eq. 11.46, we find that it is satisfied if

$$\frac{\mu b^2}{\sqrt{a^2 + p^2}} = |\vec{B} - \vec{d}| \tag{11.50}$$

Note that the solution has the form of a circle when $p = 0$ and a straight line along the axis when $p \to \infty$.

11.4 When a metal crystal free of applied stress and containing screw dislocation segments is quenched so that supersaturated vacancies are produced, the screw segments are converted into helices by climb. Show that the converted helices can be at equilibrium with a certain concentration of supersaturated vacancies and find an expression for this critical concentration in terms of appropriate parameters of the system. Use the simple line-tension approximation leading to Eq. 11.12. We note that the helix will grow by climb if the vacancy concentration in the crystal exceeds this critical concentration and will contract if it falls below it.

Solution. In this case, $\vec{d} = 0$, and therefore we must have

$$\vec{f} = \mu b^2 \frac{d\hat{\zeta}}{ds} + \hat{\zeta} \times \vec{B} = 0 \tag{11.51}$$

The axis of the helix is parallel to $\vec{B}$, which, in turn, is parallel to $\vec{b}$, and therefore Eq. 11.50 of Exercise 11.3 applies in the form

$$\frac{\mu b^2}{\sqrt{a^2 + p^2}} = |\vec{B}| = \frac{bkT}{\Omega}\ln\left(\frac{X_V}{X_V^{\mathrm{eq}}}\right) \tag{11.52}$$

The concentration is therefore

$$X_V = X_V^{\text{eq}} \, e^{\mu b \Omega / \left(kT \sqrt{a^2 + p^2} \right)} \tag{11.53}$$

11.5 Show that regardless of the orientation of a straight dislocation line and its Burgers vector, there will exist a stress system that will convert the dislocation line into a helix whose axis is along the position of the original dislocation when the point-defect concentration is at the equilibrium value characteristic of the stress-free crystal. Use the simple line-tension approximation leading to Eq. 11.12.

Solution. In this case, $\vec{B} = 0$ and the equilibrium shape of the dislocation must satisfy the equation

$$\vec{f} = \mu b^2 \frac{d\hat{\zeta}}{ds} - \hat{\zeta} \times \vec{d} = 0 \tag{11.54}$$

The vector $\vec{d}$ is given by Eq. 11.3, and because the six stresses are independent of each other, a value of the stress tensor may be found so that $\vec{d}$ is parallel to the dislocation line. As shown in Exercise 11.3, a solution of Eq. 11.54 corresponding to a helix with its axis along $\vec{d}$ can then be found.

11.6 Show that the effective mass per unit length of a moving screw dislocation is $m^* = W^\circ / c^2$ when its velocity is relatively small compared with c.

- Expand Eq. 11.20 to first order in $(v/c)^2$ and examine the form of the result.

Solution. Expanding Eq. 11.20 to first order in $(v/c)^2$ produces the result

$$W \cong W^\circ + \frac{1}{2} \left(\frac{W^\circ}{c^2} \right) v^2 \tag{11.55}$$

which is seen to have the form of a rest potential energy plus a kinetic energy involving an effective mass given by W° / c^2. Therefore, at relatively small velocities, the inertia of the regions that must be displaced as the dislocation moves produces the same effects as if the dislocation possessed the mass m^*.

11.7 (a) Describe how excess supersaturated vacancies in a crystal can precipitate and collapse to form a prismatic dislocation loop. (A prismatic dislocation loop is a loop whose Burgers vector possesses a component normal to the plane of the loop. It therefore can glide out of the loop plane.)

(b) Find an expression for the number of vacancies required to form a circular loop of radius R.

Solution.

(a) The vacancies can precipitate in the form of a planar layer as shown in Fig. 11.15b. If the vacancy precipitate then collapses as shown in Fig. 11.15c, a prismatic dislocation loop will be produced. In essence, this process removes a circular patch of one of the (hkl) planes as shown. The Burgers vector of the loop will have a component normal to the loop plane equal to the interplanar spacing d_{hkl}. If there is a shear displacement, S, parallel to the loop plane during the collapse, the Burgers vector will also have a component in the loop plane equal to S. The free energy of the supersaturated vacancies is generally greater than the free energy of the resulting loop and therefore there is an excess of free energy available to drive the loop formation.

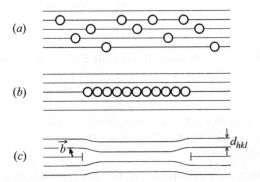

Figure 11.15: Formation of prismatic dislocation by vacancy precipitation and collapse. (a) Excess vacancies dispersed in crystal. (b) Precipitation of excess vacancies. (c) Collapse of vacancy precipitate to form dislocation loop.

(b) The loop may be constructed formally by making a cut parallel to the loop plane and removing a circular disc of material of area πR^2 and thickness d_{hkl} and then displacing the two circular faces of the cut by $\vec{b}$ so that they are joined together. The volume of material that must be removed in this operation is then $V = \pi R^2 d_{hkl} = \pi R^2 \vec{b} \cdot \hat{n}$, where $\hat{n}$ is a unit vector normal to the loop plane. The number of vacancies required to remove this volume of material is then

$$N = \frac{\pi R^2 \vec{b} \cdot \hat{n}}{\Omega} \tag{11.56}$$

11.8 Find the number of supersaturated vacancies that must diffuse to a unit length of climbing screw dislocation to convert it into a helix of radius a, turn spacing $2\pi p$, and Burgers vector length b.

Solution. The helical dislocation may be regarded as equivalent to a stack of circular prismatic edge dislocation loops of radius a as illustrated in Fig. 11.16. This may be confirmed by realizing that the stack of loops can be converted into a helix in a conservative fashion by cutting each loop at its intersection with AB and then sliding

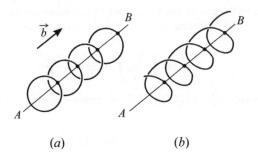

(a) (b)

Figure 11.16: (a) A stack of four prismatic edge dislocation loops perpendicular to AB and also tangent to AB at the points indicated. (b) Helix formed from loops in (a).

various segments of each loop along the cylindrical glide surface in an appropriate way to form a continuous helix after joining the free ends. Each prismatic loop contains $N = \pi a^2 b/\Omega$ collapsed vacancies. The total number of vacancies, N^{tot}, required per unit length of helix is then

$$N^{\text{tot}} = \frac{N}{2\pi p} = \frac{a^2 b}{2p\Omega} \tag{11.57}$$

11.9 Bardeen and Herring first pointed out that if the pinned dislocation segment in Fig. 11.17 has its Burgers vector normal to the plane of the paper, it can act as an infinite source or sink for vacancies (and also as a source for dislocation loops) [29]. As the segment climbs by creating (destroying) vacancies, it bulges as shown in Fig. 11.17b. Upon further climb it creates a surrounding closed loop as seen in Fig. 11.17c and d. This process can then repeat itself, producing further loops.

Suppose now that such a source is present in a crystal that is rapidly quenched from a temperature T_q to a temperature T_a to produce supersaturated vacancies. Find an expression for the critical value of the quenching temperature, T_q, which must be used to produce sufficient supersaturation to activate the source so that it will be able to create dislocations loops capable of destroying the supersaturated vacancies by climb. The vacancy formation energy is E_V^f and the segment length is L.

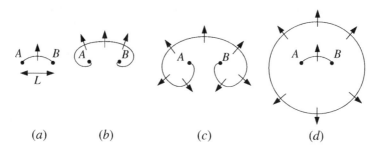

(a)   (b)   (c)   (d)

Figure 11.17:   Generation of a dislocation loop by expansion of a dislocation segment with ends pinned at A and B.

Solution. The source will be able to become active if the driving osmotic climb force is large enough to overcome the restraining curvature force that reaches a maximum when the dislocation segment has bowed out to the minimum radius of curvature corresponding to $R = L/2$. Setting $f_\mu = f_\kappa$, we then have the critical condition

$$\frac{bkT_a}{\Omega} \ln\left(\frac{X_V}{X_V^{\text{eq}}}\right) = \frac{\mu b^2}{R} = \frac{2\mu b^2}{L} \tag{11.58}$$

Since $X_V = A\exp[-E_V^f/(kT_q)]$ and $X_V^{\text{eq}} = A\exp[-E_V^f/(kT_a)]$, we obtain T_q in the form

$$T_q = \frac{T_a}{1 - 2\mu b\Omega/(LE_V^f)} \tag{11.59}$$

11.10 Imagine that the Burgers vector of the dislocation segment in Fig. 11.17a lies in the plane of the figure rather than normal to it as in Exercise 11.9. Frank and Read first pointed out that the configuration in Fig. 11.17a could then act as a source for a large number of dislocation loops which would expand by glide rather than climb (as in Exercise 11.9) in the presence of a sufficiently large shear stress, σ, parallel to the plane of the paper [30]. Find an expression for the critical stress, σ_c, required to activate such a source.

Solution. The source will become active when the forward driving force due to the stress is large enough to overcome the maximum restoring force due to curvature which occurs when the dislocation segment has bowed out to the minimum radius of curvature corresponding to $R = L/2$. Setting $f_\sigma = f_\kappa$, we then have the critical condition

$$\sigma_c b = \frac{\mu b^2}{R} = \frac{2\mu b^2}{L} \tag{11.60}$$

so that

$$\sigma_c = \frac{2\mu b}{L} \tag{11.61}$$

11.11 Two parallel edge dislocations of opposite sign, spaced a distance l apart, lie along the center of a cylindrical body of radius, R, as shown in Fig. 11.18, with $R \gg l \gg R_\circ$, where $R_\circ$ is the dislocation core diameter. The two dislocations attract one another with a force (per unit length) given by

$$f = \frac{\mu b^2}{2\pi(1-\nu)l} \tag{11.62}$$

At an elevated temperature, they will therefore climb toward one another and eventually anneal out. Obtain an expression for the rate at which they approach each other. Assume diffusion-limited climb, that the diffusion is by a vacancy mechanism, and that no other sources or sinks are present except for the surface.

- Use the method of superposition described at the end of Section 11.4.3.

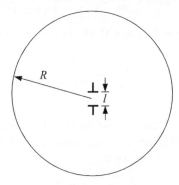

Figure 11.18: Two closely spaced edge dislocations of opposite sign at the center of a large cylindrical body.

Solution. Because of the interaction between the two dislocations, the work required to produce a vacancy at either dislocation and cause it to climb in the direction of the attractive force is reduced by the force by the amount $f\Omega/b$. The equilibrium vacancy concentration at each dislocation, $c_V^{eq}(\text{disl})$, is therefore higher than the concentration in equilibrium with the surface of the cylinder, $c_V^{eq}(\infty)$. Vacancies will therefore diffuse from each climbing dislocation to the surface of the cylinder and the dislocations will climb toward one another. Using the same procedures used to treat the problem of the shrinkage of dislocation loops in Section 11.4.3, and described for this particular problem by Hirth and Lothe [2], we therefore have

$$c_V^{eq}(\text{disl}) - c_V^{eq}(\infty) = c_V^{eq}(\infty)\left[e^{f\Omega/(bkT)} - 1\right] \tag{11.63}$$

The solution of the diffusion equation for the quasi-steady state in cylindrical coordinates shows that each dislocation line source will have a vacancy concentration diffusion field around it of the form

$$c_V(r') - c_V^{eq}(\infty) = a_1 \ln\left(\frac{R}{r'}\right) + a_2 \tag{11.64}$$

where a_1 and a_2 are constants. The constants can be obtained from the conditions that the concentration at each dislocation produced by the superimposed diffusion fields of the two dislocations must correspond to the equilibrium concentration, $c_V^{eq}(\text{disl})$, and that the concentration at the surface must be $c_V^{eq}(\infty)$. Taking an origin at one dislocation, these conditions are therefore

$$a_1 \ln\left(\frac{R}{R_\circ}\right) + a_1 \ln\left(\frac{R}{l}\right) + 2a_2 = c_V^{eq}(\text{disl})$$
$$a_2 = \frac{c_V^{eq}(\infty)}{2} \tag{11.65}$$

At a distance from the two closely spaced dislocations appreciably greater than l, the concentration due to their superimposed diffusion fields will then be

$$c_V(r) - c_V^{eq}(\infty) = \frac{c_V^{eq}(\text{disl}) - c_V^{eq}(\infty)}{\ln(R/\sqrt{R_\circ l})} \ln\left(\frac{R}{r}\right) \tag{11.66}$$

The vacancy current diffusing away from the two dislocations is then

$$I = -2\pi r D_V \frac{\partial c_V}{\partial r} = \frac{2\pi D_V\left[c_V^{eq}(\text{disl}) - c_V^{eq}(\infty)\right]}{\ln(R/\sqrt{R_\circ l})} \tag{11.67}$$

Finally, making use of Eqs. 8.17 and 11.63, the velocity of approach is

$$v = \frac{\Omega}{b}I = \frac{2\pi\,{}^\star D\left[e^{f\Omega/(kTb)} - 1\right]}{fb\ln(R/\sqrt{R_\circ l})} \tag{11.68}$$

11.12 Consider a segment of dislocation in the dislocation network of a crystal pinned at its ends and lying in its slip plane as in Fig. 11.9. Suppose that an oscillating shear stress of the form

$$\sigma = \sigma_\circ e^{-i\omega t} \tag{11.69}$$

is imposed along the slip plane, causing it to bow out.

(a) Describe qualitatively the stress-induced motion of the segment.

(b) Write the equation of motion of the segment taking into account all of the forces acting on it, including the influence of its effective mass (Exercise 11.6), the frictional drag force (Section 11.3.1), and the curvature force (Section 11.2.3).

- Assume that the stress is relatively small.
- Any lateral displacement of the segment will be small compared to its length; and the drag force can be taken as viscous (i.e., proportional to the dislocation velocity).

Solution.

(a) The applied stress exerts a force on each element of the segment, bowing it out laterally in the slip plane. As the stress oscillates, the segment is forced to bow backward and forward much in the manner of a vibrating string.

(b) The equation of motion will be of the form

$$m^\star \frac{\partial^2 u}{\partial t^2} + B\frac{\partial u}{\partial t} - \mu b^2 \frac{\partial^2 u}{\partial y^2} - b\sigma_\circ e^{-i\omega t} = 0 \tag{11.70}$$

where $u(y,t)$ is the displacement in the x direction. The first term is the inertial term proportional to the acceleration $\partial^2 u/\partial t^2$ and the effective mass per unit length $m^\star$. The second is the viscous damping term that is proportional to the velocity $\partial u/\partial t$ (B is a damping constant). The third term is the restoring force proportional to the curvature $\partial^2 u/\partial y^2$, and the fourth term is the oscillating driving force $b\sigma$. Solutions to Eq. 11.70 have been found, and further development of the model is given by Nowick and Berry [15].

CHAPTER 12

MOTION OF CRYSTAL/VAPOR AND CRYSTAL/LIQUID INTERFACES

A vast number of engineering materials are used in solid form, but during processing may be found in vapor or liquid phases. The vapor→solid (condensation) and liquid→solid (solidification) transformations take place at a distinct interface whose motion determines the rate of formation of the solid. In this chapter we consider some of the factors that influence the kinetics of vapor/solid and liquid/solid interface motion. Because vapor and liquid phases lack long-range structural order, the primary structural features that may influence the motion of these interfaces are those at the solid surface.

12.1 THERMODYNAMIC SOURCES OF INTERFACE MOTION

An interface has an *effective driving pressure* if its displacement decreases the total system's free energy. This effective pressure can derive from any mechanism by which a material stores energy, but for many cases it arises from only two sources: the volumetric free-energy differences between the interface's adjacent phases, and mechanical pressure differences due to reduction of the interfacial energy.

The conditions and kinetic equations for phase transformations are treated in Chapters 17 and 20 and involve local changes in free-energy density. The quantification of thermodynamic sources for kinetically active interface motion is approximate for at least two reasons. First, the system is out of equilibrium (the transformations are not reversible). Second, because differences in normal component of mechanical stresses (pressures, in the hydrostatic case) can exist and because the thermal conditions of interfacial motion are system-specific (i.e., interfacial motion can involve the generation of heat), identification of the appropriate minimizing system free energy is troublesome. However, if the interface's adjacent phases are treated as near-equilibrium and if the total free energy, $\mathcal{G}$, can be used as a placeholder for the appropriate free energy, the effective driving pressure due to phase transformations can be modeled roughly with

$$\Pi^{\text{phase trans}} = -\frac{1}{A}\frac{\delta \mathcal{G}}{\delta x} = -\frac{\delta \mathcal{G}}{\delta V} \approx -\frac{\Delta G}{\Omega} \tag{12.1}$$

where $\delta \mathcal{G}$ is the change in total free energy when interface with area A moves normal to itself by a distance δx, ΔG is the difference in free energy per atom, and Ω is related to the atomic volumes, which may differ, in each phase.

For a liquid/crystal interface that moves as the undercooled liquid crystallizes at temperature T, a model for $\Pi^{\text{phase trans}}$ can be developed. Using $\Delta G(T) = \Delta H(T) - T\,\Delta S(T)$ and treating the volumetric heat capacities as temperature independent, $\Delta S(T_m) = \Delta H(T_m)/T_m$, leads to the approximation

$$\Delta G(T) = \Delta H(T_m) - T\frac{\Delta H(T_m)}{T_m} = \frac{\Delta H(T_m)}{T_m}\Delta T \tag{12.2}$$

where the undercooling is $\Delta T = T_m - T$. The approximation from Eq. 12.1 shows that

$$\Pi^{\text{phase trans}} = -\frac{\Delta G}{\Omega} = -\frac{\Delta U(T_m)}{T_m\Omega}\Delta T \tag{12.3}$$

is proportional to the undercooling.

The surface energy per area, γ, has the same units as a force per length and for some interfacial geometries can lead to an interfacial net force that is balanced by a difference in pressure between the two adjacent phases. If γ is isotropic, this pressure difference is directly proportional to the interfacial curvature through the the Gibbs–Thomson equation (see Sections C.2.1 and C.4.1),

$$\Delta P = \gamma\left(\frac{1}{R_1} + \frac{1}{R_2}\right) = \gamma\kappa \tag{12.4}$$

where the R_i are the principal radii of curvature and κ is the mean curvature. The pressure is higher on the side that contains the center of curvature (e.g., the pressure inside a soap bubble is larger than the ambient pressure). $\Delta P = \gamma\kappa$ is the thermodynamic potential to move a small volume Ω across an isotropic interface, absent any chemical or other effects.

12.2   MOTION OF CRYSTAL/VAPOR INTERFACES

Crystals grow from their supersaturated vapor by the addition of vapor atoms at their free surfaces. In this process, the surface is subjected to an effective pressure due to the difference in free energy between the solid and vapor. The interface moves outward toward the vapor as it acts as a sink for the incoming flux of atoms. The mechanism by which atoms leave the vapor phase and eventually become permanently incorporated in the crystal is often relatively complex, and the kinetics of the process depends upon the type of surface involved (i.e., singular, vicinal, or general). Also, crystal growth from the vapor is generally anisotropic because surfaces of different inclination (indices) usually possess different structures.

Another type of motion of crystal/vapor interfaces occurs when a supersaturation of vacancies anneals out by diffusing to the surface where they are destroyed. In this process, the surface acts as a sink for the incoming vacancy flux and the surface moves inward toward the crystal as the vacancies are destroyed. This may be regarded as a form of crystal dissolution, and the kinetics again depend upon the type of surface that is involved.

12.2.1 Structure of Crystal/Vapor Surfaces

Just as for grain boundaries, crystal surfaces at equilibrium with their vapor can be singular, vicinal, or general.[1] As explained in Section 9.1 and Appendix B, an interface is regarded as *singular* with respect to a degree of freedom if it is at a local minimum of energy with respect to changes in that degree of freedom. Such surfaces correspond to the bases of cusps on the surface of the Wulff plot. These surfaces are generally of low index and high atomic density and exist at low temperatures relative to their melting points. Surfaces with inclinations close to those of singular surfaces and having energies lying on the sides of cusps are composed of discrete steps and terraces corresponding to the structures of their nearby singular surfaces and are classified as *vicinal* (see Fig. B.1 and Exercise 12.9). Surfaces far from singular and vicinal inclinations have highly stepped and irregular structures of higher energies and are therefore *general*.

As the temperature is raised, singular and vicinal surfaces tend to undergo a *roughening* transition in which they develop highly stepped and irregular structures. During this transition, the ledges roughen as a result of the thermal generation of surface and ledge vacancies, adatoms, and vacancy and adatom clusters (Fig. 12.1). Ledge vacancy clusters and adatom clusters (which form in strings along the ledge) are topologically equivalent to pairs of kinks on the ledges. The roughening occurs because of the configurational entropy that is contributed by the structural disorder (see Exercise 12.7). This process is therefore thermodynamically similar to the thermal generation of equilibrium vacancies in the bulk. In general, the more singular a surface, the higher the temperature at which it will roughen. The roughening

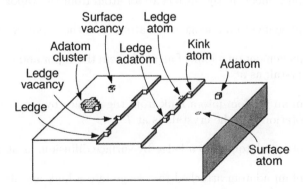

Figure 12.1: Various point and line defects on a vicinal crystal/vapor surface.

[1]See Appendix B for further discussion of the structure of surfaces.

of a (40 0 1) vicinal surface during heating, as determined by computer simulation, is shown in Fig. 12.2. Surfaces that have become roughened in this manner must now be classified as general.

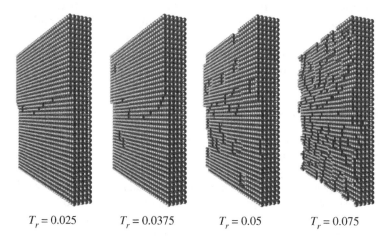

$T_r = 0.025$ $T_r = 0.0375$ $T_r = 0.05$ $T_r = 0.075$

Figure 12.2:   A Monte-Carlo simulation of roughening of a [40 0 1] vicinal surface in a primitive cubic crystal with first- and second-neighbor pair interactions, E_1 and E_2. The binding enthalpy $\Delta H_{\text{bind}} = 6E_1 + 12E_2$ and the calculations were performed at the indicated values of the reduced temperature defined by $T_r = kT/\Delta H_{\text{bind}}$, with $E_2 = 0.66E_1$.

12.2.2   Crystal Growth from a Supersaturated Vapor

Singular and Vicinal Crystal Surfaces.   Kinks on ledges are the only places on a vicinal surface where an incoming atom from the vapor can become fully incorporated in the crystal with its full binding energy. These kinks therefore play the same role during crystal growth as jogs on dislocations during dislocation climb.[2] If ledges with kinks are present, the incoming atoms can be incorporated at the kinks by several mechanisms. The mechanisms are illustrated in Fig. 12.3 and include:

- Direct impingement on the surface of an atom from the vapor, as at A

- Direct impingement on a ledge of an atom from the vapor, as at B

- Direct impingement on a kink of an atom from the vapor and its incorporation into the crystal, as at C

- Jumping of an adatom into a kink from the surface by surface diffusion and its incorporation into the crystal, as at D

- Jumping of an adatom onto a ledge by surface diffusion, as at E

- Jumping of an adatom attached to a ledge into a kink and its incorporation into the crystal, as at F

---

[2]This is demonstrated in Exercise 12.1.

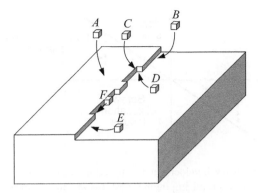

Figure 12.3:   Elementary processes at vicinal surface during crystal growth from a supersaturated vapor.

Because the adatoms diffuse relatively rapidly along the surface and the ledges to the kinks, many more atoms reach the kinks by these routes than by direct impingement from the vapor. Note the close similarity between this crystal growth process on a vicinal surface and the climb of dislocations depicted in Fig. 11.2.

When the surface is initially vicinal, it contains a population of ledges, and growth can then occur by the incorporation of adatoms at the ledges. The efficiency of the surface as a sink for the incoming atoms varies widely depending upon a number of factors. It is usually assumed that sufficient kinks are present at normal growth temperatures so that the ledges act as highly efficient line sinks for the incorporation of adatoms (see Exercise 12.1). If most adatoms jump back into the vapor phase before reaching a ledge, the sink efficiency will be low. This is a result of a small energy of adatom adsorption, a low rate of diffusion on the surface, and a large spacing between ledges. On the other hand, a high surface sink efficiency, corresponding to $\eta = 1$ (see Eq. 11.25), can be achieved if the relative magnitudes of the above parameters are reversed and essentially all adatoms that arrive from the vapor become incorporated at ledges before rejoining the vapor. These variables have been considered in analytical treatments of crystal growth [1] and in Exercise 12.1.

In Exercise 12.2 it is shown that the growth velocity can be expressed in the form

$$v = K(P - P^{\mathrm{eq}}) \qquad (12.5)$$

where P is the pressure of the vapor in the system, P^{eq} is the vapor pressure that would be in equilibrium with the surface, and K is a rate constant that is proportional to the efficiency, η, of the surface as a sink for the impinging vapor atoms. If the surface acts as an ideal sink, $\eta = 1$ and v will be proportional to $(P - P^{\mathrm{eq}})$ as indicated by the ideal sink curve in Fig. 12.4. Otherwise, the curve of v vs. $(P - P^{\mathrm{eq}})$ will fall below the ideal sink curve.

An initially singular and atomically flat surface advances only if ledges, possessing kinks, are nucleated and formed on the surface. This can occur if adatoms cluster together on the surface and nucleate new ledges at small pillbox-shaped clusters of adatoms as indicated in Fig. 12.1. The free energy to form such a cluster

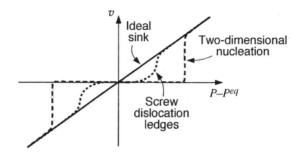

Figure 12.4: Crystal growth velocity v as a function of excess vapor pressure $(P - P^{eq})$. Ideal sink curve pertains when all impinging adatoms are incorporated. Screw dislocation ledges curve pertains when ledges associated with screw dislocations limit the kinetics. Two-dimensional nucleation curve pertains when two-dimensional nucleation of ledges is rate-limiting.

is

$$\Delta G^{\mathrm{clust}} = 2\pi R g^L - \frac{\pi R^2 h}{\Omega} kT \ln \left(\frac{P}{P^{eq}} \right) \qquad (12.6)$$

where R is the cluster radius, g^L is the ledge energy per unit ledge length, h is the ledge height, P is the actual vapor pressure, and P^{eq} is the equilibrium vapor pressure of the bulk crystal. The radius of the critical cluster for nucleation (see Section 19.1) obtained from the condition $\partial \Delta G^{\mathrm{clust}} / \partial R = 0$ is

$$R_c = \frac{g^L \Omega}{hkT \ln(P/P^{eq})} \qquad (12.7)$$

The role of ledge nucleation on the kinetics of crystal growth can now be explored. If the surface is initially singular, no growth occurs at very low supersaturations where, according to Eq. 12.7, R_c is relatively large. There are no ledges under these conditions because the critical free energy for nucleation is so large that the nucleation rate is negligible.[3] However, as the supersaturation is increased and R_c is decreased, the nucleation rate of ledges will at some point increase rapidly, and a significant density of ledges and kinks will appear on the surface. For the initially singular surface, the curve of v vs. $P - P^{eq}$ as a function of increasing $P - P^{eq}$ (Fig. 12.4) therefore exhibits a negligible growth rate at small values of $P - P^{eq}$ but increases extremely rapidly when the supersaturation reaches a value where copious ledge nucleation occurs, and from there it follows closely the ideal sink curve. In this situation, the initially singular surface has become effectively roughened in the presence of the large driving force for its motion.

Ledge nucleation for crystal growth can be bypassed in many cases if the crystal contains dislocations. If a dislocation impinges on a singular surface and has a component of its Burgers vector normal to the surface, it is topologically necessary that a ledge be present on the surface as illustrated in Fig. 12.5a. Adatoms can then become incorporated in the crystal at the ledge, causing it to move normal to itself in the directions of the arrows in Fig. 12.5. The ledge therefore wraps itself up into a spiral configuration centered on the dislocation and the total length

---

[3]See Chapter 19 for a treatment of the kinetics of nucleation.

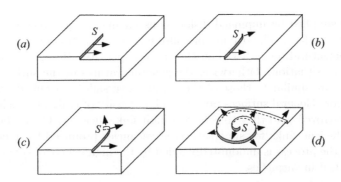

Figure 12.5:   Formation of a spiral ledge on the surface of a crystal growing from the vapor phase at the intersection of a screw dislocation with the surface at the point S.

of ledge available to support crystal growth increases continuously. Dislocations of this type provide ready-made sources of ledges and crystal growth occurs in the absence of any ledge nucleation. When such sources of ledges are present, significant crystal growth occurs at lower vapor supersaturations than those required for ledge nucleation and analysis of the growth kinetics shows that the v vs. $P - P^{eq}$ curve appears as in Fig. 12.4 [1].

General Crystal Surfaces. General surfaces possess a high density of sites where atoms from the vapor can be incorporated in the crystal and are therefore expected to operate as essentially ideal sinks.

Crystal growth from the vapor is often anisotropic, with singular and vicinal surfaces having lower mobilities than general surfaces. In such cases, any fast-moving surface inclinations bounding a growing crystal grow out of existence, leaving behind only slow-growing inclinations. The result is a faceted crystal bounded by slow-moving inclinations. Anisotropic growth when the growth velocity is a known function of surface inclination is analyzed in Section 14.2.2.

There is a body of experimental evidence generally supporting the models above for growth at the different types of surfaces [1].

12.2.3 Surfaces as Sinks for Supersaturated Lattice Vacancies

When a crystal surface acts as a sink for supersaturated vacancies, atoms are supplied from the surface to replace the incoming flux of vacancies. The process of crystal dissolution that occurs is, in many respects, the opposite of the crystal growth process described above. Vacancies diffusing to the surface become surface vacancies and may then either leave the surface by jumping back into the crystal bulk or, in vicinal surfaces, diffuse in the surface to ledges, where they become ledge vacancies, which are eventually destroyed at kinks by being replaced by lattice atoms. The sink efficiency of the surface is high if most of the surface vacancies are destroyed at the kinks before re-entering the lattice. The analysis of the sink efficiency is then analogous to the preceding analysis of crystal growth at vicinal surfaces.

When the surface is initially singular, the required ledges can be nucleated by the clustering of supersaturated surface vacancies in monolayer surface cavities or they can be supplied by impinged dislocations having Burgers vector components normal to the surface. The driving pressure for the nucleation of the ledges is the vacancy supersaturation rather than the vapor supersaturation as in the previous crystal growth situation. Otherwise, the general features of the surface as a sink for vacancies are similar to those of the surface as a sink for atoms during growth from the vapor. General surfaces possess a high density of kinks at which vacancies are easily destroyed, and they thus have a high sink efficiency. Gold (100) surfaces are highly efficient sinks for supersaturated vacancies produced by quenching [2]. In this case, the process is undoubtedly aided by the high degree of supersaturation of the quenched-in vacancies.

12.3   MOTION OF INTERFACES DURING SOLIDIFICATION

12.3.1   Structure of Crystal/Liquid Interfaces

Because experimental study of the structure of crystal/liquid interfaces has been difficult due to the buried nature of the interface and rapid structural fluctuations in the liquid, it has been investigated by computer simulation and theory. Figure B.3 provides several views of crystal/liquid (or amorphous phase) interfaces, which must be classified as diffuse interfaces because the phases adjoining the interface are perturbed significantly over distances of several atomic layers.

A critical question has been whether crystal/liquid interfaces are usually rough (and therefore general) or singular or vicinal. Jackson developed a simple statistical mechanical model in which adatoms were added to a crystal/liquid interface possessing an initially smooth crystal face [3]. The model shows that the interface remains smooth, and hence singular, whenever the dimensionless parameter $\alpha = (\Delta S_m/k)\,(\eta/z)$ is of magnitude $\alpha > 2$ and is rough (general) otherwise. Here, ΔS_m is the entropy of melting, η is the number of nearest-neighbor adatom sites around an adatom on the crystal face, and z is the number of nearest-neighbors in the bulk. The model therefore predicts that the equilibrium interface is smooth when the entropy of melting is large and the interface is low-index (close-packed). This result is reasonable because a smooth interface (one without any intermixing of the solid and liquid due to the presence of adatoms or adatom clusters) is expected when the liquid and solid are strongly dissimilar with respect to their vibrational and configurational character and when the interface plane is relatively close-packed and therefore strongly bonded.

In other work it has been argued that crystal/liquid interfaces roughen under the nonequilibrium conditions that exist during solidification when the driving pressure becomes sufficiently high [4, 5]. Under such conditions, the barrier to the nucleation of new growth layers is effectively eliminated and the interface becomes rough. This behavior is similar to a dislocation becoming highly jogged in the presence of a large vacancy supersaturation (Section 11.4.2) and of a crystal/vapor interface developing a high ledge density at a large vapor supersaturation.

12.3.2   Crystal Growth from an Undercooled Liquid

Singular and Vicinal Interfaces.   The crystal/liquid interface during crystal growth from an undercooled liquid can be singular, vicinal, or general, depending upon the type of material and the driving force [1]. Many types of crystals require a relatively large amount of undercooling (corresponding to 1° to 2°) to obtain significant growth rates and solidify in faceted forms. This behavior is expected when singular (or vicinal) interfaces are dominant and the growth occurs by the movement of discrete ledges across the interface.[4] Crystals exhibiting this behavior have highly directional bonding or strongly bonded complex structures and possess values of the Jackson α parameter greater than two. On the other hand, most metals and some organic materials, possessing α values less than two, solidify with nonfaceted interfaces at exceedingly small undercoolings. This behavior is expected when the interface is rough (general) and there is a high density of sites everywhere in the interface where atoms can easily be incorporated into the solid.

The manner in which a singular (or vicinal) crystal/liquid interface moves during solidification into an undercooled liquid by the lateral movement of ledges is similar to the way a singular (or vicinal) crystal/vapor interface advances in a supersaturated vapor. Data showing the velocity of (001) and (111) interfaces in Ga as a function of the degree of undercooling resulting from the two-dimensional nucleation and growth of atomic layers are in Fig. 12.6. The velocity increases rapidly with increased undercooling in a manner similar to the rapid increase of growth rate predicted schematically by the two-dimensional nucleation and growth curve in Fig. 12.4 for the crystal/vapor case.

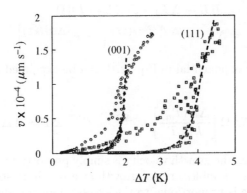

Figure 12.6:   Velocity v of singular (or vicinal) (001) and (111) surfaces as a function of undercooling during solidification. The growth rate is limited by the two-dimensional nucleation of ledges. The dashed lines are model predictions. From Howe [1]. Original figure from Peteves and Abbaschian [6].

General Interfaces.   When the crystal/liquid interface is rough (general), atoms from the liquid can easily be incorporated into the solid at a high density of sites in the interface. A primitive expression for the rate of solidification is easily obtained by using Fig. 12.7, which shows the change in energy when an atom is transferred across the crystal/liquid interface. For an atom to leave the liquid, it must surmount the energy barrier $\Delta G^{L/XL}$, and to leave the crystal, it must surmount the barrier $(\Delta G + \Delta G^{L/XL})$, where ΔG is given by Eq. 12.2. If N is the number of atoms (per unit area) in a position to jump from liquid to solid (or vice versa), the liquid-to-crystal flux is

$$J(L \to XL) = N\nu\, e^{-\Delta G^{L/XL}/(kT)} \tag{12.8}$$

[4]This is demonstrated in Exercise 12.1.

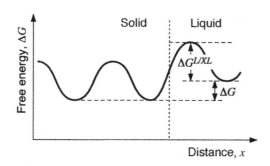

Figure 12.7:   Changes in energy during solidification of an undercooled liquid when an atom is transferred across a crystal/liquid interface.

where ν is a vibrational frequency. The reverse flux is

$$J(XL \to L) = N\nu\, e^{-(\Delta G^{L/XL} + \Delta G)/(kT)} \qquad (12.9)$$

and the rate of growth is

$$\begin{aligned}
v &= \left[J(L \to XL) - J(XL \to L) \right] \Omega \\
&= N\nu\,\Omega\, e^{-\Delta G^{L/XL}/(kT)} \left[1 - e^{-\Delta G/(kT)} \right]
\end{aligned} \qquad (12.10)$$

When $\Delta G \ll kT$, the exponential in Eq. 12.10 can be expanded to first order and, using Eq. 12.2,

$$v = N\nu\,\Omega \left[\frac{\Delta H(T_m)}{kT_m T} \right] e^{-\Delta G^{L/XL}/(kT)}\, \Delta T = A\,\Delta T \qquad (12.11)$$

Under these conditions, the growth velocity is simply proportional to the undercooling, and the situation resembles crystal growth at a rough crystal/vapor interface where the growth rate is proportional to the excess vapor pressure (Fig. 12.4).[5]

Bibliography

1. J.M. Howe. *Interfaces in Materials*. John Wiley & Sons, New York, 1997.

2. K.C. Jain and R.W. Siegel. Temperature dependence of the vacancy sink efficiency of stacking-fault tetrahedra in quenched gold. *Phil. Mag.*, 25(1):105–115, 1972.

3. K.A. Jackson. Mechanism of growth. In *Liquid Metals and Solidification*, pages 174–186, Cleveland Park, OH, 1958. American Society for Metals.

4. J.W. Cahn. Theory of crystal interface motion in crystalline materials. *Acta Metall.*, 8(8):554–562, 1960.

5. J.W. Cahn, W.B. Hillig, and G.W. Sears. The molecular mechanism of solidification. *Acta Metall.*, 12(12):1421–1439, 1964.

6. S.D. Peteves and R. Abbaschian. Growth-kinetics of solid-liquid Ga interfaces. 1. Experimental. *Metall. Trans. A*, 22(6):1259–1270, 1991.

[5]Note that this primitive single-jump model neglects the diffuse nature of the crystal/liquid interface.

EXERCISES

12.1 Construct a simple steady-state model for the growth from the vapor of a crystal with a vicinal surface and investigate the efficiency of the surface as a sink for the incoming vapor atoms. For simplicity assume that (i) the surface contains an array of evenly spaced straight parallel ledges at the spacing λ, and (ii) the ledges contain enough kinks to act as ideal line sinks for adatoms.

Consider first the surface free of any ledges (i.e., the ledge spacing is infinite). Atoms from the vapor then land on the surface as adatoms and spend a mean time, τ, on the surface before jumping back into the vapor. During the time τ they are able to migrate in the surface layer, which is of thickness δ, a mean distance

$$\langle x^2 \rangle^{1/2} = 2\sqrt{D_A^S \tau}$$

where D_A^S is their diffusivity in the surface layer. The steady-state concentration of adsorbed atoms in the surface layer is then $c^\infty = \tau \phi_i = \tau \phi_e$, where ϕ_i and ϕ_e are, respectively, the rates of impingement and evaporation (per unit volume of surface slab), which in this case are equal.

Consider now the vicinal surface where the impingement rate is the same as above. Adsorbed atoms will now either evaporate or diffuse to a ledge and become incorporated into the growing crystal.

(a) Show that the diffusion equation for the adsorbed atoms in the surface layer can be written as

$$\frac{d^2 c}{dx^2} = \frac{4(c - c^\infty)}{\langle x^2 \rangle} \tag{12.12}$$

where x is the distance coordinate running perpendicular to the ledges.

(b) Show that the solution to the diffusion equation above, subject to the applicable boundary conditions, is

$$\frac{c - c^\infty}{c^{\text{eq}} - c^\infty} = \cosh\left(\frac{2x}{\sqrt{\langle x^2 \rangle}}\right) + \sinh\left(\frac{2x}{\sqrt{\langle x^2 \rangle}}\right) \tag{12.13}$$

where c^{eq} is the equilibrium concentration maintained at the perfect ledge sinks. Ignore any effects of the motion of the ledges (see Exercise 12.3).

(c) Use the results to show that the efficiency (see Eq. 11.25) of the surface as a sink for the incoming atoms from the vapor is

$$\eta = \frac{1}{\lambda / \sqrt{\langle x^2 \rangle}} \tanh\left(\frac{\lambda}{\sqrt{\langle x^2 \rangle}}\right) \tag{12.14}$$

Therefore, when $\lambda \gg \sqrt{\langle x^2 \rangle}$, $\eta \approx \sqrt{\langle x^2 \rangle}/\lambda \ll 1$ and the efficiency is very low. This is the case where the ledge spacing is much greater than the adatom diffusion distance on the surface and very few adatoms reach the ledge sinks before jumping back into the vapor phase. On the other hand, when $\lambda << \sqrt{\langle x^2 \rangle}$, $\eta \approx 1$. This is the case where the ledge spacing is relatively small and essentially all adatoms reach the ledge sinks before they jump back into the vapor. The surface therefore acts essentially as a perfect sink.

Solution.

(a) The rate of accumulation of adatoms in unit volume of the surface slab is

$$\frac{\partial c}{\partial t} = -\nabla \cdot J + \phi_i - \phi_e = D_A^S \frac{\partial^2 c}{\partial x^2} + \phi_i - \phi_e \tag{12.15}$$

where $\phi_i = c^\infty / \tau$ and $\phi_e = c/\tau$ (because the evaporation rate will be proportional to the concentration c). Because the diffusion is in the steady state,

$$\frac{d^2 c}{dx^2} = \frac{c - c^\infty}{D_A^S \tau} = \frac{4(c - c^\infty)}{\langle x^2 \rangle} \tag{12.16}$$

(b) The general solution of Eq. 12.16 is

$$c - c^\infty = a_1 \sinh\left(\frac{2x}{\sqrt{\langle x^2 \rangle}}\right) + a_2 \cosh\left(\frac{2x}{\sqrt{\langle x^2 \rangle}}\right) \tag{12.17}$$

where a_1 and a_2 are constants. Fitting Eq. 12.17 to the boundary conditions

$$c(x = 0) = c^{\text{eq}} \qquad \left.\frac{\partial c}{\partial x}\right|_{x=\lambda/2} = 0 \tag{12.18}$$

then yields the desired result.

(c) The rate at which adatoms reach a unit length of ledge on the vicinal surface is

$$\phi = 2\delta D_A^S \left.\frac{\partial c}{\partial x}\right|_{x=0} \tag{12.19}$$

The corresponding rate at which adatoms would reach the ledge if the surface acted as a perfect sink would be

$$\phi_{\text{ideal}} = \left(\frac{c^\infty}{\tau} - \frac{c^{\text{eq}}}{\tau}\right) \lambda \delta = \frac{4(c^\infty - c^{\text{eq}})\lambda \delta D_A^S}{\langle x^2 \rangle} \tag{12.20}$$

Therefore,

$$\eta = \frac{\phi}{\phi_{\text{ideal}}} = \frac{\langle x^2 \rangle}{2\lambda(c^\infty - c^{\text{eq}})} \left.\frac{\partial c}{\partial x}\right|_{x=0} = \frac{1}{\lambda/\sqrt{\langle x^2 \rangle}} \tanh\left(\frac{\lambda}{\sqrt{\langle x^2 \rangle}}\right) \tag{12.21}$$

12.2 In Fig. 12.4 the velocity of crystal growth from the vapor is plotted as a function of the excess vapor pressure $(P - P^{\text{eq}})$. When the surface acts as an ideal sink for incoming vapor atoms, the plot indicates that the velocity of growth should vary linearly with $(P - P^{\text{eq}})$. When the sink efficiency is lower, the curve of v vs. $(P - P^{\text{eq}})$ falls below the ideal curve. Use the results of Exercise 12.1 to demonstrate that the velocity of growth for the model employed there can be expressed in the form

$$v = \Omega J_{\text{net}} = K(P - P^{\text{eq}}) \tag{12.22}$$

where K is a rate constant that is proportional to the sink efficiency, η, given by Eq. 12.14. The behavior predicted by this expression is seen to be consistent with the curves in Fig. 12.4.

Solution. The rate at which adatoms diffuse into a unit length of ledge is

$$\phi = 2\delta D_A^S \left(\frac{dc}{dx}\right)_{x=0} \tag{12.23}$$

Using Eq. 12.13 yields

$$\phi = \frac{4\delta D_A^S (c^\infty - c^{eq})}{\sqrt{\langle x \rangle^2}} \tanh\left(\frac{\lambda}{\sqrt{\langle x \rangle^2}}\right) \tag{12.24}$$

The rate of impingement of atoms from the vapor (per unit area) is $J_i = \delta\phi_i = \delta c^\infty/\tau$, while the rate of evaporation from the surface—if the surface were in equilibrium (i.e., in detailed balance) with the vapor—would be $J_e^{eq} = \delta\phi_e^{eq} = \delta c^{eq}/\tau$. Since, according to the kinetic theory of gases, the rate of impingement from the vapor phase is proportional to the vapor pressure, P,

$$J_i = \frac{\delta c^\infty}{\tau} = K'P \tag{12.25}$$

In detailed balance, the rate of evaporation is equal to the rate of impingement from the vapor at its equilibrium pressure, P^{eq}, and therefore

$$J_e^{eq} = \frac{\delta c^{eq}}{\tau} = K'P^{eq} \tag{12.26}$$

Solving these equations for c^∞ and c^{eq} and using Eq. 12.24, the net flux of atoms incorporated into the crystal from the vapor is

$$\begin{aligned} J_{net} &= \frac{\phi}{\lambda} = \frac{1}{\lambda/\sqrt{\langle x^2 \rangle}} \tanh\left(\frac{\lambda}{\sqrt{\langle x^2 \rangle}}\right) K'(P - P^{eq}) \\ &= \eta K'(P - P^{eq}) = K(P - P^{eq}) \end{aligned} \tag{12.27}$$

12.3 One aspect of crystal growth from the vapor is the diffusive transport of adatoms to ledges (see Exercise 12.1). A suggestion is that the motion of ledges can be ignored in the analysis of crystal growth if

$$\frac{\langle x^2 \rangle^{1/2} v_{max}}{4D_A^S} \ll 1 \tag{12.28}$$

where $\langle x^2 \rangle^{1/2}$ is the root-mean-square distance that an adatom diffuses in the surface during a time τ while the ledge moves with a maximum velocity v_{max} corresponding to a large separation between ledges. Explain and derive this criterion.

Solution. The effect of the ledge motion on the diffusion of adatoms to it will be small if the mean distance that adatoms can diffuse in the surface during a time τ is large compared to the distance that the ledge moves in the time τ. Under these circumstances, the surface adatoms diffuse quickly enough in the surface so that their concentration in the vicinity of a moving ledge is only negligibly affected by the ledge motion. The mean diffusion distance is $\langle x^2 \rangle^{1/2} = 2(D_A^S \tau)^{1/2}$ and the distance traveled by the ledge is $v_{max}\tau$. The criterion is therefore $\langle x^2 \rangle^{1/2} \gg v_{max}\tau$. Substituting for τ, we then obtain $\langle x^2 \rangle^{1/2} v_{max}/(4D_A^S) \ll 1$.

12.4 During the evaporation of crystals with vicinal (or singular) faces, surface ledges of height, h, nucleate preferentially at the edges of the crystals where the faces meet. Develop a model for this type of heterogeneous nucleation and show that the critical free energy for the nucleation is much smaller than the critical free energy for the corresponding homogeneous nucleation of ledges by the formation of pillbox-shaped cavities in the crystal surfaces.

- An overview of heterogeneous nucleation is provided in Section 19.2.

Solution. The homogeneous nucleation geometry is shown in Fig. 12.8a. Using the approach that developed Eqs. 12.6 and 12.7 for homogeneous ledge nucleation during crystal growth, the free energy of the corresponding evaporation nucleus is

$$\Delta G = 2\pi R g^L - \frac{\pi R^2 h}{\Omega} kT \ln\left(\frac{P^{\text{eq}}}{P}\right) \tag{12.29}$$

R_c is found by setting $\partial \Delta G/\partial R = 0$. Putting the result into Eq. 12.29 then yields the critical free energy of nucleation

$$\Delta G_c(\text{homo}) = \frac{\pi \Omega \left(g^L\right)^2}{hkT \ln\left(P^{\text{eq}}/P\right)} \tag{12.30}$$

A model for the heterogeneous nucleation of a ledge at a crystal edge is shown in Fig. 12.8b. The free energy for its formation is

$$\Delta G = \pi R g^L - 2Rh\gamma^S - \frac{\pi R^2 h}{2\Omega} kT \ln\left(\frac{P^{\text{eq}}}{P}\right) \tag{12.31}$$

Using the same procedure as above,

$$\Delta G_c(\text{edge}) = \frac{\pi \Omega [g^L - (2h\gamma^S/\pi)]^2}{2hkT \ln(P^{\text{eq}}/P)} \tag{12.32}$$

Therefore,

$$\frac{\Delta G_c(\text{edge})}{\Delta G_c(\text{homo})} = \frac{1}{2}\left(1 - \frac{2h\gamma^S}{\pi g^L}\right) \tag{12.33}$$

Because $g^L \approx h\gamma^S$, the critical free energy for heterogeneous edge nucleation is indeed considerably smaller than the energy for homogeneous nucleation.

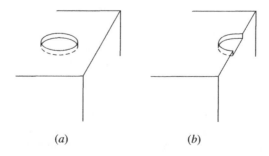

(a)   (b)

Figure 12.8:   (a) Homogeneous nucleus for surface evaporation. (b) Corresponding heterogeneous nucleus at crystal edge. Cavity depth and radius of curvature are h and R, respectively.

12.5 A crystal growing from the vapor phase possesses a singular surface with two screw dislocations intersecting it. The dislocations are very close to one another (relative to the dimensions of the surface) and have opposite Burgers vectors. Describe the form of the step structure that is produced because of the presence of the two dislocations.

Solution. Before any growth, the two dislocations are associated with steps that may be as indicated in Fig. 12.9a. During growth, each dislocation rotates about its point of intersection to produce a spiral step, as in Fig. 12.5. However, the spirals will rotate in opposite directions, and sections will annihilate one another when they meet as in (a) and (b). The process will then continue as in (c)–(f) generating a potentially unlimited series of concentric steps.

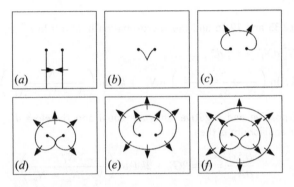

Figure 12.9: Step structure generated on growing crystal surface at the intersections of two screw dislocations with opposite Burgers vectors.

12.6 Suppose that the Burgers vectors of the two screw dislocations in Exercise 12.5 are now the same. Describe the ledge structure that is produced.

Solution. Start with the ledge structure in Fig. 12.10a. The growth spirals of the two dislocations now rotate in the same direction as indicated. This will generate an interleaved double growth spiral of potentially unlimited extent as illustrated in Fig. 12.10b–d.

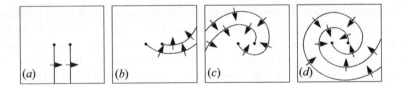

Figure 12.10: Step structures generated on growing crystal surface at the intersections of two screw dislocations with the same Burgers vectors.

12.7 Consider a ledge on a surface which, on average, lies parallel to a closely spaced row of atoms. The ledge contains N sites, with spacing a, where either a positive or a negative kink can exist. (When traveling along a ledge, positive and negative kinks displace the ledge in opposite directions.) Show that the total number of kinks present in thermal equilibrium, N_k^{eq}, is

$$\frac{N_k^{\text{eq}}}{N} = 2\, e^{-G_k^f/(kT)} \tag{12.34}$$

where G_k^f is the kink free energy of formation (excluding any configurational entropy).

Solution. The increment of free energy due to the presence of N_k^+ positive kinks and N_k^- kinks is

$$\Delta \mathcal{G} = N_k^+ G_k^f + N_k^- G_k^f - kT \ln P \tag{12.35}$$

where the last term is the configurational entropy. P is the number of distinguishable ways of arranging the positive and negative kinks on the N sites and is given by

$$P = \frac{N!}{(N_k^+)! \, (N_k^-)! \, (N - N_k^+ - N_k^-)!} \tag{12.36}$$

Using Eqs. 12.35 and 12.36 and the approximation $\ln x! = x \ln x - x$,

$$
\begin{aligned}
d\,\Delta \mathcal{G} = {}& G_k^f \, dN_k^+ + G_k^f \, dN_k^- \\
& + kT \ln \left(\frac{N_k^+}{N - N_k^+ - N_k^-} \right) dN_k^+ + kT \ln \left(\frac{N_k^-}{N - N_k^+ - N_k^-} \right) dN_k^-
\end{aligned}
\tag{12.37}
$$

Numbers of positive and negative kinks are equal and therefore $dN_k^+ = dN_k^-$. The condition for equilibrium is then

$$\frac{\partial \Delta \mathcal{G}}{\partial N_k^+} = 0 = 2G_k^f + 2kT \ln \left(\frac{N_k^+}{N - N_k^+ - N_k^-} \right) \tag{12.38}$$

Therefore, because $N \gg N_k^+ + N_k^-$,

$$\left[\frac{N_k^+}{N} \right]^{\text{eq}} = e^{-G_k^f/(kT)} \tag{12.39}$$

and

$$\frac{N_k^{\text{eq}}}{N} = \left[\frac{N_k^+}{N} \right]^{\text{eq}} + \left[\frac{N_k^-}{N} \right]^{\text{eq}} = 2\,e^{-G_k^f/(kT)} \tag{12.40}$$

12.8 Exercise 12.7 showed that the total equilibrium concentration of positive and negative kinks on a ledge running, on average, along a relatively close-packed direction is given by Eq. 12.34. Find the concentrations of positive and negative kinks on a ledge lying at an angle θ with respect to the close-packed direction. Assume that the direction of θ requires the building-in of positive kinks.

Solution. The population of kinks consists of $\theta a N/d$ built-in positive kinks, where d is the kink height, along with equal numbers of thermally generated positive and negative kinks, represented by N_k^+ and N_k^-, respectively. Following the same procedure as in Exercise 12.7, the derivative of the total kink free energy, $\Delta \mathcal{G}$, is

$$
\begin{aligned}
d\,\Delta \mathcal{G} = {}& G_k^f dN_k^+ + G_k^f dN_k^- \\
& + kT \ln \left[\frac{(\theta a N/d) + N_k^+}{N - (\theta a N/d) - N_k^+ - N_k^-} \right] dN_k^+ \\
& + kT \ln \left[\frac{N_k^-}{N - (\theta a N/d) - N_k^+ - N_k^-} \right] dN_k^-
\end{aligned}
$$

The numbers of thermally generated positive and negative kinks are equal and therefore $dN_k^+ = dN_k^-$ and the condition of equilibrium is

$$\frac{d\,\Delta\mathcal{G}}{dN_k^+} = 0 = 2G_k^f + kT \ln\left\{\frac{[(\theta aN/d) + N_k^+]N_k^-}{[N - (\theta aN/d) - N_k^+ - N_k^-]^2}\right\} \tag{12.41}$$

Assuming that θ is small enough so that $N \gg (\theta aN/d) + N_k^+ + N_k^-$,

$$\left[\frac{N_k^+}{N} + \frac{\theta a}{d}\right]^{eq}\left[\frac{N_k^-}{N}\right]^{eq} = e^{-2G_k^f/(kT)} \tag{12.42}$$

12.9 Consider the energy of a vicinal surface at a low temperature that consists of an array of straight parallel ledges separated by patches of singular terraces as in Fig. B.1. Express the form of the energy cusp in which the surface lies (i.e., express the surface energy as a function of θ, the angle by which the inclination of the vicinal surface deviates from that of the singular terraces). Note that this model will break down at higher temperatures where the system's entropy increases and free energy decreases as the ledges wander and become nonparallel and other roughening processes occur.

Solution. The structure of the vicinal surface is shown in Fig. 12.11. The energy of unit area of surface can be expressed as the sum of two parts: the first is that of the singular terraces, which is $\gamma^S \cos\theta$, where γ^S is the energy per unit area of the terraces; the second is that of the ledges, which is $(g^L \sin\theta)/h$, where h is the ledge height and g^L is the energy per unit length of a ledge. Therefore,

$$\gamma = \gamma^S \cos\theta + \frac{g^L \sin\theta}{h} \tag{12.43}$$

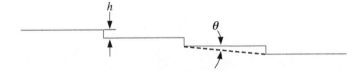

Figure 12.11: Structure of vicinal surface.

CHAPTER 13

MOTION OF CRYSTAL/CRYSTAL INTERFACES

Crystal/crystal interfaces possess more degrees of freedom than vapor/crystal or liquid/crystal interfaces. They may also contain line defects in the form of interfacial dislocations, dislocation-ledges, and pure ledges. Therefore, the structures and motions of crystal/crystal interfaces are potentially more complex than those of vapor/crystal and liquid/crystal interfaces. Crystal/crystal interfaces experience many different types of pressures and move by a wide variety of atomic mechanisms, ranging from rapid glissile motion to slower thermally activated motion. An overview of crystal/crystal interface structure is given in Appendix B.

13.1 THERMODYNAMICS OF CRYSTALLINE INTERFACE MOTION

In Section 12.1, common sources of driving pressures for the motion of vapor/crystal and liquid/crystal interfaces were described. These and additional sources of pressure exist for crystal/crystal interfaces. For example, during recrystallization, the interfaces between the growing recrystallized grains and the deformed matrix are subjected to a pressure that is due to the bulk free-energy difference (per atom) ΔG between the free energy of the deformed matrix and that of the recrystallized grains.[1] Also, compatibility stresses are often generated in stressed polycrystals so

---

[1]Recrystallization occurs when a crystalline material is plastically deformed at a relatively low temperature and then heated [1]. The as-deformed material possesses excess bulk free energy resulting from a high density of dislocations and point-defect debris produced by the plastic

Kinetics of Materials. By Robert W. Balluffi, Samuel M. Allen, and W. Craig Carter.   **303**
Copyright © 2005 John Wiley & Sons, Inc.

that two crystals adjacent to a grain boundary possess markedly different stresses. Equation 12.1 then holds, with ΔF equal to the difference between the elastic strain energy per atom in the two adjoining crystals.

Pressures on crystal/crystal interfaces can also arise when the motion of the interface causes a change in the shape of the body in which it moves. Figure 13.1 shows the motion of a small-angle symmetric tilt grain boundary under applied shear stress. The motion occurs by the forward glide motion of the edge dislocations which comprise the boundary along x, causing the specimen to shear in the y direction. When this occurs, the applied stresses perform work, and the potential energy of the system is reduced. A displacement of the interface by δx allows the applied stress, σ_{xy}, to perform the work, $\sigma_{xy}\theta\,\delta x$. The pressure on the interface is

$$P = \frac{\sigma_{xy}\theta\,\delta x}{\delta x} = \sigma_{xy}\theta \tag{13.1}$$

There is a close similarity between this type of pressure and the mechanical force exerted on a dislocation by a stress (see Section 11.2.1 and Exercise 13.2).

Estimated values of the magnitudes of the pressures commonly applied to crystal/crystal interfaces extend over a wide range of values, spanning about six orders of magnitude [2]. Generally, the pressures generated by phase transformations are the highest (10^7–10^9 Pa), whereas those generated by interface curvature are relatively small (10^3–10^5 Pa).

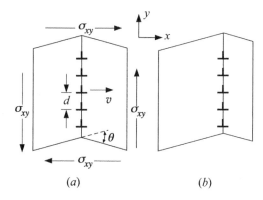

(a) $\qquad\qquad\qquad\qquad\qquad$ (b)

Figure 13.1:   Motion of small-angle symmetric tilt boundary by means of glissile motion of its edge dislocations. Dislocation spacing, d, is equal to b/θ, where b is the magnitude of the Burgers vector and θ is the misorientation between the two grains.

13.2   CONSERVATIVE AND NONCONSERVATIVE INTERFACE MOTION

The motion of a crystal/crystal interface is either conservative or nonconservative. As in the case of conservative dislocation glide, *conservative interface motion* occurs in the absence of a diffusion flux of any component of the system to or from the

deformation. Upon heating, nuclei of relatively perfect recrystallized material form and then grow into the deformed material, eliminating most of the crystal defects at the moving interfaces between the recrystallized and deformed material.

interface. On the other hand, nonconservative motion occurs when the motion of the interface is coupled to long-range diffusional fluxes of one or more of the components of the system.

Conservative motion can be achieved under steady-state conditions only when the atomic fraction of each component is the same in the adjoining crystals (see Exercise 13.1). For sharp interfaces, atoms are simply transferred locally across the interface from one adjoining crystal to the other and there is no need for the long-range diffusion of any species to the boundary. This local transfer can occur by the simple shuffling of atoms across the interface and/or by the creation of crystal defects (vacancies or interstitials) in one grain which then diffuse across the boundary and are destroyed in the adjoining grain, thus transferring atoms across the interface.[2] Examples of conservative motion are the glissile motion of martensitic interfaces (see Chapter 24) and the thermally activated motion of grain boundaries during grain growth in a polycrystalline material.

During *nonconservative interface motion*, the boundary must act as a source for the fluxes. To accomplish this for sharp interfaces, atoms must be added to, or removed from, one or both of the the crystals adjoining the interface. This generally causes crystal growth or shrinkage of one or both of the adjoining crystals and hence interface motion with respect to one or both of the crystals. This can occur by the creation at the interface of the point defects necessary to support the long-range diffusional fluxes of substitutional atoms or by atom shuffling to accommodate the addition or removal of interstitial atoms. Nonconservative interface motion and the role of interfaces as sources or sinks for diffusional fluxes are of central importance in a wide range of phenomena in materials. For example, during diffusional creep and sintering of polycrystalline materials (Chapter 16), and the thermal equilibration of point defects, atoms diffuse to grain boundaries acting as point-defect sources. In these cases, the fluxes require the creation or destruction of lattice sites at the boundaries. In multicomponent-multiphase materials, the growth or shrinkage of the phases adjoining heterophase interfaces often occurs via the long-range diffusion of components in the system. In such cases, heterophase interfaces again act as sources for the diffusing components.

Further aspects of the conservative and nonconservative motion of sharp interfaces are presented below. The mechanism for the motion of a diffuse interface is discussed in Section 13.3.4.

13.3 CONSERVATIVE MOTION

13.3.1 Glissile Motion of Sharp Interfaces by Interfacial Dislocation Glide

Sharp boundaries of several different types can move conservatively by the glide of interfacial dislocations. In many cases, this type of motion occurs over wide ranges of temperature, including low temperatures where little thermal activation is available.

Small-Angle Grain Boundaries. As described in Appendix B, these semicoherent boundaries are composed of arrays of discretely spaced lattice dislocations. For

---

[2]Shuffles are small displacements of atoms (usually smaller than an atomic spacing) in a local region, such as the displacements that occur in the core of a gliding dislocation.

certain small-angle boundaries, these dislocations can glide forward simultaneously, allowing the boundary to move without changing its structure. The simplest example is the motion of a symmetric tilt boundary by the simultaneous glide of its edge dislocations as in Fig. 13.1. An important aspect of this type of motion is the change in the macroscopic shape of the bicrystal specimen which occurs because the transfer of atoms across the boundary from grain 2 to grain 1 by shuffling is a highly correlated process. Each atom in the shrinking grain is moved to a predetermined position in the growing grain as it is overrun by the displacement field of the moving dislocation array and shuffled across the boundary. The positions of all the atoms in the bicrystal are therefore correlated with the position of the interface and there is a change in the corresponding macroscopic shape of the specimen as the boundary moves. This type of interface motion has been termed *military* to distinguish it from the disorganized *civilian* type of interface motion that occurs when an incoherent general interface moves as described in Section 13.3.3 [3]. In the latter case, there is no change in specimen shape.

Numerous experimental observations of the glissile motion of small-angle boundaries have been made [2]. Most general small-angle boundaries possess more than one family of dislocations having different Burgers vectors. Glissile motion of such boundaries without change of structure is possible only when the glide planes of all the dislocation segments in the array lie on a common zone with its axis out of the boundary plane. When this is not the case, the boundary can move conservatively only by the combined glide and climb of the dislocations as described in Section 13.3.2.

Large-Angle Grain Boundaries.   Semicoherent large-angle grain boundaries containing localized line defects with both dislocation and ledge character can often move forward by means of the lateral glissile motion of their line defects. A classic example is the motion of the interface bounding a (111) mechanical twin in the f.c.c. structure illustrated in Fig. 13.2. This boundary can be regarded alternatively as a large-angle grain boundary having a misorientation corresponding to a 60° rotation around a [111] axis. The twin plane is parallel to the (111) matrix plane, and the twin (i.e., island grain) adopts a lenticular shape in order to reduce its elastic energy (discussed in Section 19.1.3). The macroscopically curved upper and lower sections of the interface contain arrays of line defects that have both dislocation and ledge character, as seen in the enlarged view in Fig. 13.2b. Note that the interface is semicoherent with respect to a reference structure (see Section B.6) taken to be a bicrystal containing a flat twin boundary parallel to (111). The line defects are glissile in the (111) plane and their lateral glissile motion across the interface in the directions of the arrows causes the upper and lower sections of the interface to move normal to themselves in directions that expand the thickness of the lenticular twin. In essence, the gliding line defects provide special sites where atoms can be transferred locally across the interface relatively easily by a military shuffling process, making the entire boundary glissile. This type of glissile interface motion produces a macroscopic shape change of the specimen for the same geometric reasons that led to the shape changes illustrated in Fig. 13.1. When a line defect with Burgers vector $\vec{b}$ passes a point on the interface, the material is sheared parallel to the interface by the amount b. At the same time, the interface advances by h, the height of the ledge associated with the line defect. These effects, in combination, produce the shape change. A pressure urging the interface sections to move to

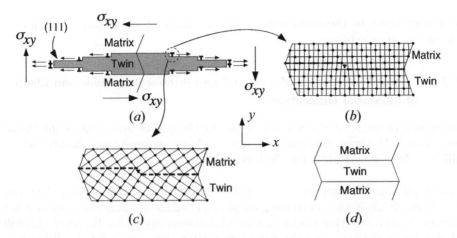

Figure 13.2:   **(a)** A lenticular twin in an f.c.c. structure bounded by glissile interfaces containing dislocations possessing ledge character viewed along $[\bar{1}10]$. **(b)** An enlarged view of the dislocation-step region. The interface is semicoherent with respect to a reference structure, corresponding to the bicrystal formed by a 60° rotation around [111]. The Burgers vector of the dislocation is a translation vector of the DSC-lattice of the reference bicrystal, which is the fine grid shown in the figure (see Section B.6). **(c)** The same atomic structure as in (b). The interface now is considered to be coherent with respect to a reference structure, corresponding to the f.c.c. matrix crystal. In this framework, the dislocation is regarded as a coherency dislocation (see Section B.6). **(d)** The shape change produced by formation of a twin across the entire specimen cross section.

expand the twin and produce this shape change can be generated by applying the shear stress, σ_{xy}, shown in Fig. 13.2a. The magnitude of this pressure is readily found through use of Eq. 12.1. The force (per unit length) tending to glide the line defects laterally is given by Eq. 11.1, $f = b\,\sigma_{xy}$. The work done by the applied force in moving a unit area of the boundary a distance δx is then $(\delta x/h)\,b\sigma_{xy}$, and the pressure is therefore

$$P = -\frac{\delta G}{\delta x} = \frac{(\delta x/h)\,b\sigma_{xy}}{\delta x} = \frac{b\sigma_{xy}}{h} \tag{13.2}$$

This type of glissile boundary motion occurs during mechanical twinning when twins form in matrix grains under the influence of applied shear stresses [4]. The glissile lateral motion of the line defects can be very rapid, approaching the speed of sound (see Section 11.3.1), and the large number of line defects that must be generated on successive (111) planes can be obtained in a number of ways, including a dislocation "pole" mechanism. Glissile motion of other types of large-angle grain boundaries by the same basic mechanism have been observed [2].

Heterophase Interfaces.   In certain cases, sharp heterophase interfaces are able to move in military fashion by the glissile motion of line defects possessing dislocation character. Interfaces of this type occur in martensitic displacive transformations, which are described in Chapter 24. The interface between the parent phase and the newly formed martensitic phase is a semicoherent interface that has no long-range stress field. The array of interfacial dislocations can move in glissile fashion and shuffle atoms across the interface. This advancing interface will transform the parent phase to the martensite phase in military fashion and so produce a macroscopic shape change.

13.3.2   Thermally Activated Motion of Sharp Interfaces by Glide and Climb of Interfacial Dislocations

The motion of many interfaces requires the combined glide and climb of interfacial dislocations. However, this can take place only at elevated temperatures where sufficient thermal activation for climb is available.

Small-Angle Grain Boundaries.   As mentioned, a small-angle grain boundary can move in purely glissile fashion if the glide planes of all the segments in its dislocation structure lie on a zone that has its axis out of the boundary plane. However, this will not usually be the case, and the boundary motion then requires both dislocation glide and climb. Figure 13.3 illustrates such an interface, consisting of an array of two types of edge dislocations with their Burgers vectors lying at 45° to the boundary plane, subjected to the shear stress σ_{xy}.

Equation 11.1 shows that the shear stress exerts a pure climb force $f = b\sigma_{xy}$ on each dislocation, which therefore tends to climb in response to this force. However, mutual forces between the dislocations in the array will tend to keep them at the regular spacing corresponding to the boundary structure of minimum energy. All dislocations will then move steadily along $+x$ by means of combined glide and climb. The boundary as a whole will therefore move without changing its structure, and its motion will produce a specimen shape change, the same as that produced by the glissile motion of the boundary in Fig. 13.1. Successive dislocations in the array must execute alternating positive and negative climb, which can be accomplished by establishing the diffusion currents of atoms between them as shown in Fig. 13.3. Each current may be regarded as crossing the boundary from the shrinking crystal to the growing crystal.

An approximate model for the rate of boundary motion can be developed if it is assumed that the rate of dislocation climb is diffusion limited [2]. Neglecting any effects of the dislocation motion and the local stress fields of the dislocations on

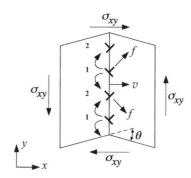

Figure 13.3:   Thermally activated conservative motion of a small-angle symmetric tilt boundary containing two arrays of edge dislocations with orthogonal Burgers vectors. $\vec{f}$ is the force exerted on each dislocation, by the applied stress. Arrows indicate atom fluxes between dislocations.

the diffusion, a flux equation for the atoms can be obtained by combining Eqs. 3.71 and 8.17:[3]

$$\vec{J}_A = -\frac{{}^{\star}D}{\Omega kT \mathbf{f}} \nabla \Phi_A = -\frac{{}^{\star}D}{\Omega kT \mathbf{f}} \nabla(\mu_A - \mu_V) \tag{13.3}$$

Under diffusion-limited conditions, the vacancies can be assumed to be maintained at equilibrium at the dislocations. The dislocations act as ideal sources (Section 11.4.1) and, therefore, at the dislocations $\mu_V = 0$. When an atom is inserted at a dislocation of type 2 acting as a sink (Fig. 13.3), the dislocation will move forward along x by the distance $\sqrt{2}\,\Omega/b$. The force on it acting in that direction is $\sigma_{xy} b/\sqrt{2}$, and the work performed by the stress is therefore $(\sqrt{2}\,\Omega/b)(\sigma_{xy}b/\sqrt{2}) = \sigma_{xy}\,\Omega$. The boundary value for the diffusion potential Φ_A at the cores of these dislocations is, therefore,

$$\Phi_A^D(\text{sink}) = \mu_A^{\circ} - \sigma_{xy}\,\Omega \tag{13.4}$$

where μ_A° is the chemical potential of atoms in stress-free material. Similarly, at dislocations of type 1, acting as sources, $\Phi_A^D(\text{source}) = \mu_A^{\circ} + \sigma_{xy}\,\Omega$.

The average potential gradient in the region between adjacent dislocations is then $\langle \nabla \Phi_A \rangle = 2\Omega\,\sigma_{xy}/d$, where d is the dislocation spacing. The approximate area per unit length, A, through which the diffusion flux passes is of order $A \approx d$. Using these quantities and Eq. 13.3, and assuming that the variations in ${}^{\star}D$ due to local variations in the vacancy concentration are small enough to be neglected, the total atom current per unit length entering a dislocation of type 2 is given by

$$I_A = 2A J_A = \frac{4\,{}^{\star}D \sigma_{xy}}{\mathbf{f} kT} \tag{13.5}$$

where ${}^{\star}D$ is the self-diffusivity as measured under equilibrium conditions. The volume of atoms causing climb (per unit length per unit time) is then $I_A\Omega$, and the corresponding climb rate is therefore $v_c = I_A\Omega/b$. Each dislocation moves along x by combined climb and glide at a rate that exceeds its climb rate by $\sqrt{2}$, and the boundary velocity is then $v = \sqrt{2}\,v_c$, or

$$v = \frac{4\sqrt{2}\,\Omega\,{}^{\star}D}{b\mathbf{f} kT}\,\sigma_{xy} \tag{13.6}$$

Since $d = b/(\theta\sqrt{2})$ [7] and the pressure on the boundary is $P = \theta\sigma_{xy}$, Eq. 13.6 may be expressed

$$v = \frac{8\Omega\,{}^{\star}D d}{\mathbf{f} kT b^2} P = M^B P \tag{13.7}$$

Equation 13.7 shows that the velocity is proportional to the pressure through a boundary mobility, M^B, itself proportional to the self-diffusivity, ${}^{\star}D$. The activation energy for boundary motion will therefore be that for crystal self-diffusion as expected for a crystal diffusion-limited process.

Large-Angle Grain Boundaries. Semicoherent large-angle boundaries may move conservatively through the lateral motion of their dislocations (which also generally possess ledge character) by means of combined glide and climb. In these boundaries, the coherent patches of the boundary between the dislocations are relatively

---

[3]Equation 13.3 was first obtained by Herring and is useful in modeling the kinetics of diffusional creep [5] and sintering [6] in pure metals.

stable and therefore resistant to any type of motion. The dislocations, however, are special places in the boundary that support the transfer of atoms across the interface from the shrinking to the growing crystal relatively easily as the dislocations glide and climb.

The example in Fig. 13.4 is an extension of the model for the motion of a small-angle boundary by the glide and climb of interfacial dislocations (Fig. 13.3). Figure 13.4 presents an expanded view of the internal "surfaces" of the two crystals that face each other across a large-angle grain boundary. Crystal dislocations have

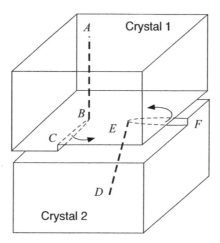

Figure 13.4:   Expanded view of the "internal surfaces" of two crystals facing each other across a grain boundary. Lattice dislocations AB and DE have impinged upon the boundary, creating line defects with both ledge and dislocation character which may glide and climb in the boundary in the directions of the arrows creating growth or dissolution spirals.

impinged upon the boundary from crystals 1 and 2, causing the formation of extrinsic dislocation segments in the boundary along CB and EF, respectively.[4] These extrinsic segments have Burgers vector components perpendicular to the boundary plane and possess ledge character. Crystal 1 can grow and crystal 2 can shrink if the segments CB and EF climb and glide in the directions of the arrows under the influence of the pressure driving the boundary. This can be achieved by the diffusion of atoms across the boundary from segment EF to segment CB, thus allowing the boundary to move conservatively. The continued motion of the segments in these directions will cause them to wrap themselves up into spirals around their pole dislocations in the grains (i.e., AB and ED). The dislocations will therefore form crystal growth or dissolution spirals in the boundary similar to the growth spirals that form on crystal free surfaces at points where lattice dislocations impinge on the surface (see Fig. 12.5). There is therefore a close similarity between this mode of dislocation-induced boundary motion and the motion of free surfaces due to the action of growth or dissolution ledge spirals as discussed in Section 12.2.2. Probable observed examples of such dislocation growth or dissolution spirals on grain boundaries are shown in Fig. 13.5.

The rates of boundary motion will depend strongly upon the available densities of boundary dislocations with ledge character. The formation of such dislocations by

[4]See Section B.7 for a discussion of extrinsic vs. intrinsic interfacial dislocations.

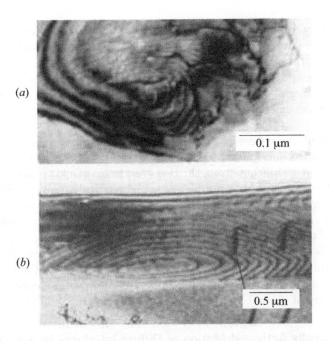

Figure 13.5: Observed examples of apparent dislocation growth or dissolution spirals on grain boundaries. From Gleiter [8] and Dingley and Pond [9].

the homogeneous nucleation of dislocation loops in the boundary is highly unlikely at the pressures that are usually exerted on boundaries [2]. An important source may then be impinged lattice dislocations, as described above. However, under many conditions, the rate of this type of boundary motion may be very slow.

13.3.3 Thermally Activated Motion of Sharp Interfaces by Atom Shuffling

Shuffling at Pure Ledges. Interfaces capable of supporting pure ledges (see Section B.7) may migrate by the transverse motion of the ledges across their faces much like the motion of free surfaces described in Section 12.2.2. However, the ledges in interfaces can move conservatively by the shuffling of single atoms or small groups of atoms from the shrinking crystal to the growing crystal at kinks in the interface ledges. This type of motion does not produce a specimen shape change. Its conservative nature is in contrast to the nonconservative nature of free-surface motion via surface-ledge migration. The shuffles will be thermally activated, and a simple analysis shows that the interface velocity can then be written

$$v = \left[\frac{N_k N_s^2 \Omega^2 \nu_\circ}{kT} e^{S^s/k} \, e^{-E^s/(kT)} \right] P = M^B P \qquad (13.8)$$

where N_k is the number of kink sites per unit interface area, N_s is the average number of atoms transferred per shuffle, $\nu_\circ$ is a frequency, and S^s and E^s are the activation entropy and energy for the shuffling. As in Eq. 13.7, the velocity is proportional to the driving pressure, P, through a boundary mobility, M^B. This mobility is critically dependent upon the density of kink sites, which may vary widely for different interfaces. Ledges will be present initially in vicinal interfaces, but these will tend to be grown off during the interface motion and can therefore support only a limited amount of motion. Ledges cannot be nucleated homogeneously in the form of small pillboxes at significant rates at the driving pressures usually encountered. However, heterogeneous nucleation could be of assistance in certain cases. In general, widely different boundary mobilities may be expected under different circumstances [2].

Uncorrelated Shuffling at General Interfaces. Interfaces that are general with respect to all degrees of freedom possess irregular structures and cannot support localized line defects of any significant strength. However, in many places along an irregular general interface, the structure can be perturbed relatively easily to allow atoms to be shuffled from the shrinking crystal to the growing crystal by means of thermal activation. In this case, a simple analysis of the interface velocity leads again to a relationship of the form of Eq. 13.8 [2]. However, the quantity N_k appearing in the mobility M^B is now the density of sites in the interface at which successful shuffles can occur. Under most circumstances, the intrinsic density of these sites will be considerably larger than the density of kink sites on vicinal stepped boundaries, and the mobility of general interfaces will be correspondingly larger.

13.3.4 Thermally Activated Motion of Diffuse Interfaces by Self-Diffusion

Diffuse interfaces of certain types can move by means of self-diffusion. One example is the motion of diffuse antiphase boundaries which separate two ordered regions arranged on different sublattices (see Fig. 18.7). Self-diffusion in ordered alloys allows the different types of atoms in the system to jump from one sublattice to the other in order to change the degree of local order as the interface advances. This mechanism is presented in Chapter 18.

13.3.5 Impediments to Conservative Interface Motion

The conservative motion of interfaces can be severely impeded by a variety of mechanisms, including solute-atom drag, pinning by embedded particles, and pinning at grooves that form at the intersections of the interfaces with free surfaces. We take up the first two of these mechanisms below and defer discussion of surface grooving and pinning at surface grooves to Section 14.1.2 and Exercise 14.3.

Solute-Atom Drag. Solute atoms, which are present either by design or as unwanted impurities, often segregate to interfaces where they build up "atmospheres" or segregates. This effect is similar in many respects to the buildup of solute-atom atmospheres at dislocations (discussed in Section 3.5.2). For the interface to move, it must either drag the solute atmosphere along with it or tear itself away. The dragging process requires that the solute atoms diffuse along with the moving interface under the influence of the attractive interaction forces exerted on them by the interface. In many cases, the forced diffusive motion of the solute atmosphere will be slow compared to the rate at which the interface would move in the absence of the solute atoms. The solute atoms then exert a *solute-atom drag force* on the moving interface and impede its motion. In cases where the applied pressure moving the interface is sufficiently large, the interface will be torn away from the solute atmosphere. A number of models for solute-atom drag, involving various simplifications, have been developed [2]. Figure 13.6 shows some of the main behavior predicted by Cahn's model [10].

When the driving pressure, P, is zero, the steady-state interface velocity, v, is also zero and the distribution of solute atoms around the interface, shown in Fig. 13.6a, is symmetric. No net drag force is therefore exerted on the interface by the solute atoms in the atmosphere. However, as v increases, the atmosphere becomes increasingly asymmetric and increasing numbers of atoms cannot keep up the pace and are lost from the atmosphere. Figure 13.6b shows the steady-state velocity as a function of P. For the pure material ($c^{XL} = 0$), the velocity is simply proportional to the pressure. This is known as *intrinsic behavior.* When solute atoms are added to the system, the velocity is reduced by the drag effect and the system now exhibits *extrinsic behavior.* At low pressures, the extrinsic velocity increases monotonically with increasing pressure, but at high pressures the interface eventually leaves behind its atmosphere and the velocity approaches the intrinsic velocity. When the solute concentration is sufficiently high, a region of instability appears in which the interface suddenly breaks free of its atmosphere as the pressure is increased. Figure 13.6c shows that essentially intrinsic behavior is obtained at elevated temperatures at all solute concentrations because of thermal desorption of the atmospheres. However, extrinsic behavior appears at the lower temperatures in a manner that is stronger the higher the solute concentration. Finally, Fig. 13.6d shows that essentially intrinsic behavior can be obtained over a range of solute concentrations as long as the driving pressure is sufficiently high. To summarize, the drag effect becomes more important as the solute concentration increases and the driving pressure and temperature decrease.

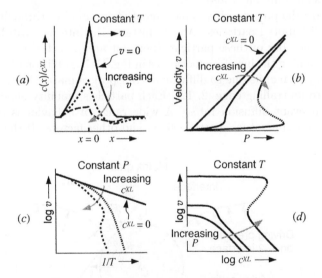

Figure 13.6: Grain-boundary solute-drag phenomena predicted by Cahn's model. **(a)** Segregated solute concentration profile $c(x)$ across boundary as a function of increasing boundary velocity v (the x axis is perpendicular to the boundary). c^{XL} is the solute concentration in the adjoining crystals. **(b)** Boundary velocity vs. pressure, P, on boundary as a function of increasing c^{XL}. **(c)** $\ln v$ vs. $1/T$ as a function of increasing c^{XL}. **(d)** $\ln v$ vs. $\ln c^{XL}$ as a function of increasing P. From Cahn [10].

Pinning Due to Embedded Second-Phase Particles. A single embedded second-phase particle can pin a patch of interface as illustrated in Fig. 13.7. Here, an interface between matrix grains 1 and 2, in contact with a spherical particle, is subjected to a driving pressure tending to move the interface forward along y past the particle. Interfacial energy considerations cause the interface to be held up at the particle, as analyzed below, and therefore to bulge around it. Inspection of the figure shows that static equilibrium of the tangential capillary forces exerted by the particle/grain 1 interface, the particle/grain 2 interface, and the grain 1/grain 2 interface requires that the angle α satisfy the relation

$$\cos\alpha = \frac{\gamma^{P2} - \gamma^{P1}}{\gamma^{12}} \tag{13.9}$$

The net restraining force along y exerted on the interface by the particle (i.e., the negative of the force exerted by the interface on the particle) is

$$F = 2\pi R\cos\phi\,\gamma^{12}\cos(\alpha - \phi) \tag{13.10}$$

The maximum force, $F_{\max}$, occurs when $\partial F/\partial\phi = 0$, corresponding to $\phi = \alpha/2$. Applying this condition to Eq. 13.10, the maximum force is

$$F_{\max} = \pi R\gamma^{12}(1 + \cos\alpha) \tag{13.11}$$

For the simple case where $\gamma^{P1} \cong \gamma^{P2}$, $\cos\alpha \cong 0$ and $\phi \cong \pi/4$ and thus

$$F_{\max} \cong \pi R\gamma^{12} \tag{13.12}$$

and $F_{\max}$ depends only on R and γ^{12}.

Consider now the pressure-driven movement of an interface through a dispersion of randomly distributed particles. At any instant, the interface will be in contact with a certain number of these particles (per unit area), each acting as a pinning point and restraining the interface motion as in Fig. 13.7. Additionally, the particles themselves may be mobile due to diffusional transport of matter from the particle's leading edge to its trailing edge [2, 12]. Each particle's mobility depends upon its size and the relevant diffusion rates. A wide range of behavior is then possible depending upon temperature, particle sizes, and other factors. If the particles are

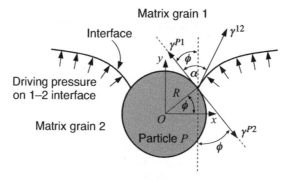

Figure 13.7: Spherical particle pinning an interface between grains 1 and 2. The interface is subjected to a driving pressure that tends to move it in the y direction. From Nes et al. [11].

immobile and the driving pressure is low, the particles may be able to pin the interface and hold it stationary. At higher pressures, the interface may be able to break free of any stationary pinning particles and thereby move freely through the distribution. The breaking-free process may also be aided by thermal activation (thermally activated unpinning, as analyzed in Exercise 13.5) if the temperature is sufficiently high or the particles sufficiently small. Also, if the particles are mobile, the interface and its attached particles may move forward together.[5]

13.3.6   Observations of Thermally Activated Grain-Boundary Motion

The motion of large-angle grain boundaries has been studied more thoroughly than that of any other type of interface. Many measurements of thermally activated motion have been made as a function of temperature, the geometric degrees of freedom, driving pressure, specimen purity, etc. The results have been reviewed [2, 13]. According to the models described earlier, intrinsic motion is expected in materials of extremely high purity at elevated temperatures under large driving pressures.[6] In addition, intrinsic mobilities of general boundaries should commonly be higher than those of singular or vicinal boundaries because of insufficient densities of kinks at dislocation-ledges on boundaries of the latter types. However, in almost all cases, observed interface motion has been influenced to at least some extent by solute-atom drag effects, so that the motion has been extrinsic and not intrinsic. For example, general grain boundaries moved as much as three orders of magnitude faster in Al that had been zone-refined (see Section 22.1.2) using twelve rather than four passes [14]. Also, as in Fig. 13.8, activation energies for the motion of a number of [100] tilt grain boundaries in 99.99995% pure Al were about half as large as the energies for corresponding boundaries in 99.9992% pure Al. Such results show that grain boundary mobilities are extremely sensitive to solute-atom drag effects, and can be strongly affected by them even at exceedingly small solute-atom concentrations.

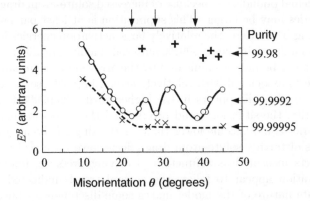

Figure 13.8:   Activation energy, E^B, for the motion of {100} tilt boundaries in Al as a function of tilt angle. The arrows at the top indicate misorientations of singular boundaries. Data for Al of 99.99995%, 99.9992%, and 99.98% purity. From Fridman et al. [15].

---

[5] Detailed analyses of these processes are given by Sutton and Balluffi [2].
[6] If motion is unaffected by drag effects due to impurity atoms, it is called *intrinsic*.

The degree of solute segregation and drag is a function of the intrinsic grain-boundary structure as well as the type and concentration of the solute atoms. When solute drag is rate controlling, the intrinsic boundary structure is only one of several factors that influences the drag and therefore the boundary mobility. The interpretation of boundary-motion experiments solely in terms of the nature of the intrinsic boundary structure then becomes rather indirect and exceedingly treacherous.

For general boundaries, essentially all measurements are consistent with the linear relationship between velocity and pressure given by Eq. 13.7 (i.e., $v = M^B P$), as might be expected on the basis of the preceding shuffling model. Available mobility data have been collected for the motion of general grain boundaries in exceptionally high-purity Al and the activation energy of 55 kJ mol^{-1} is significantly lower than that of boundary self-diffusion, which is expected to be about 69 kJ mol^{-1} [2]. Also, the data can be fit to the uncorrelated atom-shuffling model for intrinsic motion in Section 13.3.3 using reasonable values of the parameters. These results are at least consistent with the shuffling mechanism.

Regarding the motion of singular or vicinal grain boundaries, Fig. 13.5 shows direct electron microscopy images of dislocation-ledge spirals on such boundaries. The importance of line defects with ledge-dislocation character to the mobility of singular boundaries has been demonstrated in a particularly clear manner for highly singular {111} twin boundaries in Cu [16]. These boundaries were essentially immobile in the annealed state but became mobile after picking up dislocation line defects which impinged upon them during plastic deformation. In other work, electron-microscope observations of the motion of vicinal boundaries by atom shuffling at pure ledges have been made [17–19]. The motion of boundaries by shuffling at pure ledges has also been studied by computer simulation [20].

Evidence from measurements of the generally faster intrinsic motion of general boundaries relative to that of singular or vicinal boundaries has been collected [2]. The situation is complicated because the degree of solute segregation at singular or vicinal boundaries is often expected to be lower than that at general boundaries. The extrinsic mobilities of general boundaries may therefore be smaller than those of singular or vicinal boundaries (because of increased solute-atom drag), while their intrinsic mobilities may be larger. This supposition is at least partially supported by the data in Fig. 13.8. Here, the relatively large activation energies for the motion of general tilt boundaries in the 99.9992% material (having misorientations between those of the singular boundaries indicated by the arrows) most probably arose from strong drag effects associated with relatively strong impurity segregation at these boundaries. This effect disappears in the higher-purity, 99.99995%, material. At even higher purity, the situation could reverse and the activation energies for the general boundaries could become lower than for the singular boundaries [13].

Observations of further solute-atom drag effects have been reviewed [2, 13]. A number of effects measured as a function of driving pressure, temperature, and solute concentration appear to follow the general trends indicated in Fig. 13.6. The approximate nature of the model makes some discrepancies unsurprising. In Fig. 13.9, the discontinuous increases in boundary mobility as the temperature is increased are presumably caused by successive detachments of portions of a solute-atom atmosphere that exerted a drag on the boundaries.

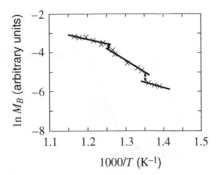

Figure 13.9:   Experimentally determined plot of $\ln M^B$ vs. $1/T$ for a (111) tilt boundary with a $46.5°$ tilt angle in Al containing Fe solute atoms. M^B = boundary mobility. From Molodov et al. [21].

13.4 NONCONSERVATIVE MOTION: INTERFACES AS SOURCES AND SINKS FOR ATOMIC FLUXES

The basic mechanisms by which various types of interfaces are able to move non-conservatively are now considered, followed by discussion of whether an interface that is moving nonconservatively is able to operate rapidly enough as a source to maintain all species essentially in local equilibrium at the interface. When local equilibrium is achieved, the kinetics of the interface motion is determined by the rate at which the atoms diffuse to or from the interface and not by the rate at which the flux is accommodated at the interface. The kinetics is then *diffusion-limited*. When the rate is limited by the rate of interface accommodation, it is *source-limited*. Note that the same concepts were applied in Section 11.4.1 to the ability of dislocations to act as sources during climb.

13.4.1 Source Action of Sharp Interfaces

By the Climb of Dislocations in Vicinal Interfaces.   The climb of the discrete lattice dislocations that comprise small-angle grain boundaries allows them to act as sources for fluxes of point defects (e.g., vacancies). In such cases, the various dislocation segments in the array making up the boundary will attempt to climb in the manner described for individual dislocations in Section 11.4. However, they will be constrained by the tendency to maintain the basic equilibrium structure of the boundary array. In the simplest case of the symmetric tilt boundary illustrated in Fig. 13.1, the edge dislocations will all be able to climb in unison relatively easily.[7] However, a pure twist boundary will act as a source only if the screw dislocations are able to climb into helices as illustrated in Fig. 11.10. This climb process will seriously perturb the structure of the boundary and will be possible only at large driving forces (i.e., large super- or subsaturations of the point defects).

Figure 13.10 shows evidence for small-angle boundaries in Au acting as efficient sinks for supersaturated vacancies under a large driving force. Here, the supersaturated vacancies have collapsed in the form of vacancy precipitates in the region of

---

[7]Note that this process will cause the boundary to move relative to inert markers embedded in either crystal adjoining the boundary.

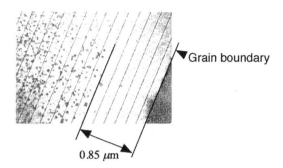

Figure 13.10:   Denudation of vacancy precipitation in a zone lying alongside a small-angle grain boundary in quenched and subsequently annealed Au. The boundary (lower right) acted as a sink for the supersaturated vacancies. Vacancy precipitates are small dislocation configurations resulting from the collapse of vacancy aggregates (as illustrated schematically in Fig. 11.15). From Siegel et al. [22].

the bulk away from the boundary. However, a precipitate-denuded zone is present adjacent to the boundary due to the annihilation of supersaturated vacancies in that region by the sink action of the boundary.

Large-angle singular or vicinal grain boundaries containing localized line defects with dislocation-ledge character can also act as sources for point defects by means of the climb (and possibly accompanying glide) of these defects across their faces. The patches of coherent interface between the line defects remain inactive since they are relatively stable and difficult to perturb. The source efficiency then depends upon the ability of the climbing dislocations to collect or disperse the point defects by diffusion along their lengths as well as in the grain-boundary core. (Note the similarity of this situation to the growth of a crystal at a vicinal surface in a supersaturated vapor as in Fig. 12.3.)

A vicinal grain boundary acting as a sink for supersaturated self-interstitial defects is shown in Fig. 13.11. The interfacial line defects needed to support the source action may often be produced by impinged lattice dislocations as in Fig. 13.4. However, at sufficiently high driving forces, the necessary line defects with dislocation character may be nucleated homogeneously in the boundary in the form of small loops possessing ledge character.[8] In the case of supersaturated point defects, the free energy to nucleate such a loop may be written approximately as

$$\Delta F = \frac{\mu b^2 R}{2(1-\nu)} \left[\ln\left(\frac{4R}{R_\circ}\right) - 1 \right] + 2\pi R f^L - \frac{\pi R^2 (\vec{b} \cdot \hat{n})}{\Omega} E^{\text{defect}} \qquad (13.13)$$

where R is the loop radius, f^L is the energy per unit length of ledge, and E^{defect} is the energy supplied per precipitated defect. The first term is the elastic energy of the loop, the second the core-ledge energy, and the third the energy supplied by the precipitated point defects. For a material with a high vacancy supersaturation, such as one subjected to high-temperature annealing and rapid quenching, Eq. 13.13 may be used to evaluate ΔF, and it may then be shown that loops may be nucleated at significant rates. During high-energy irradiation, boundaries can act as sinks for highly super-saturated self-interstitials by the nucleation (and subsequent growth)

[8]Nucleation theory is presented in Chapter 19.

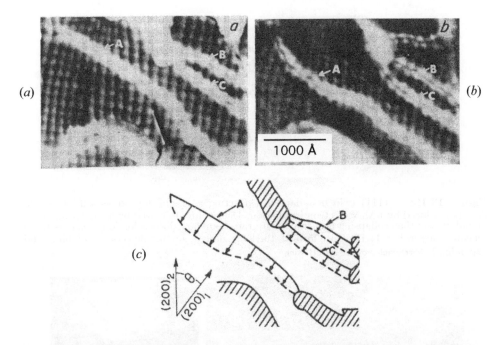

Figure 13.11:   Experimentally observed climb of extrinsic grain-boundary dislocations A, B, and C in vicinal ⟨001⟩ twist grain boundary in Au. Static array of screw dislocations in background accommodates the twist deviation of the vicinal boundary shown from the crystal misorientation of the nearby singular twist boundary to which it is vicinal. Excess self-interstitial defects were produced in the specimen by fast-ion irradiation and were destroyed at the grain-boundary dislocations by climb, causing the boundary to act as a defect sink. **(a)** Prior to irradiation. **(b)** Same area as in (a) after irradiation. **(c)** Diagram showing the extent of the climb. From Komen et al. [24].

of boundary dislocation loops [23]. Triangular dislocation loops formed on twin boundaries in irradiated Cu are shown in Fig. 13.12.

Experimental evidence shows generally that vicinal grain boundaries can act as efficient sinks for point defects under high driving forces where grain boundary dislocation climb is possible [2]. In the case of large-angle boundaries, line defects with dislocation character may be generated as the boundary absorbs vacancies from the bulk. However, at low driving forces the efficiency is often relatively low.

Vicinal heterophase interfaces can act as overall sources (or sinks) for fluxes of solute atoms by the motion across their faces of line defects possessing both dislocation and ledge character of the general type illustrated in Fig. B.6. The line defects act as line sources; during their lateral motion, lattice sites are shuffled from one adjoining crystal to the other, and the interface moves with respect to both phases. If the two phases adjoining the interface have different compositions, solute atoms must be either supplied or removed at the ledge by long-range diffusion. The motion of the ledge is therefore essentially a shuffling process coupled to the long-range diffusional transport of solute atoms.

Figure 13.13 illustrates how platelet precipitates grow and thicken by the movement of line defects of the type just described. The efficiency of the growing precipitate platelet as a sink for the flux of incoming solute atoms then depends upon the density of ledges and their ability to move while incorporating the solute atoms.

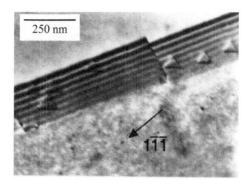

Figure 13.12: (111) twin boundary in Cu acting as a sink for excess self-interstitial defects produced by 1 MeV electron irradiation. Defects are destroyed by aggregating in the boundary and then collapsing into triangular grain-boundary dislocation loops as illustrated schematically in Fig. 11.15. Once formed, the loops destroy further defects by climbing (and expanding). Micrograph provided by A.H. King.

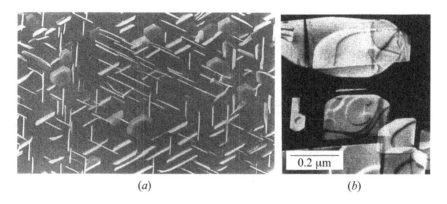

(a)   (b)

Figure 13.13: (a) Ag_2Al precipitates in the form of thin platelets in Al–Ag alloy. The broad faces of the platelets are parallel to {111} planes of the matrix, which lie at different angles to the viewer. (b) Precipitate platelets in Cu–Al alloy. Line defects that possess both dislocation and ledge character are present on the broad faces. Platelets grow in thickness by climb of these line defects across their faces. From Rajab and Doherty [25] and Weatherly [26].

Available experimental information about the source (or sink) efficiency of heterophase interfaces for fluxes of solute atoms indicates that low efficiencies are often associated with a lack of appropriate ledge defects [2].

By the Uncorrelated Shuffling of Atoms in General Interfaces. Homophase and heterophase interfaces that are general with respect to all degrees of freedom are incoherent interfaces unable to sustain localized line defects. However, in many cases, such interfaces are able to act as highly efficient sources for fluxes of point defects or solute atoms by means of atom shuffling in the interface core. The process may be modeled by assuming that the core is a slab of bad material containing a density of favorable sites where point defects can be created and destroyed, or where atomic sites containing solute atoms can be transferred across the interface, by the uncorrelated local shuffling of atoms. It has been shown that high source efficiencies can be obtained in many cases for reasonably low densities of the favorable sites [2]. From experimental results, this appears to be the case for homophase (grain) boundaries as sources for point defects. However, it may not be the case for heterophase boundaries when one of the adjoining phases has a relatively high binding energy and a correspondingly high melting temperature and thermodynamic stability.

13.4.2 Diffusion-Limited Vs. Source-Limited Kinetics

The efficiency of an interface as a source or sink can be specified by using the same parameter, η, which defined the source or sink efficiency of climbing dislocations (see Eq. 11.25).[9] To illustrate this explicitly under diffusion- and source-limited conditions, consider the rate at which a dilute concentration of supersaturated B atoms, which are interstitially dissolved in an A-rich α solution, diffuse to a distribution of growing spherical B-rich β-phase precipitate particles during a precipitation process. The rate depends upon the efficiency of the α/β interfaces as sinks for the incoming B atoms. Consider first the case where the interfaces perform as ideal sinks and the solute concentration at each α/β interface is therefore maintained at the equilibrium solubility limit of the B atoms in the α phase, $c_{eq}^{\alpha\beta}$. If the particles are initially randomly distributed, an approximately spherically symmetric diffusion field will be established around each particle in a spherical cell as in Fig. 13.14. The problem of determining the rate at which the B atoms precipitate is then reduced to solving the appropriate boundary-value diffusion problem within a given cell. The particle radius is R and the cell radius is given, to a good approximation, by $R_c = [3/(4\pi n)]^{1/3}$, where n is the density of precipitate particles. We assume that the particle radius is always considerably smaller than the cell radius and first find a solution for the case where the particle radius is assumed (artificially) to be constant during the precipitation. The more realistic case where it increases due to the incoming flux of B atoms is then considered.

The diffusion equation is $\partial c/\partial t = D_B \nabla^2 c$, where D_B and c are, respectively, the diffusivity and concentration of the B atoms in the α phase. The initial condition

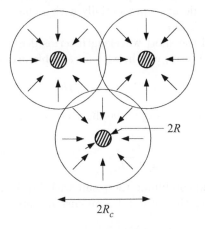

Figure 13.14: Spherical diffusion cells surrounding particles during precipitation.

---

[9]Much of this section closely follows *Interfaces in Crystalline Materials*, by A.P. Sutton and R.W. Balluffi [2].

in the cell is

$$c(r, 0) = c_\circ \tag{13.14}$$

and the boundary conditions are

$$c(R, t) = c_{eq}^{\alpha\beta} \tag{13.15}$$

$$\left[\frac{\partial c(r, t)}{\partial t} \right]_{r=R_c} = 0$$

The separation-of-variables method (Section 5.2.4) then gives the series solution [27]

$$c(r, t) = c_{eq}^{\alpha\beta} + \sum_{i=0}^{\infty} \frac{a_i}{r} e^{-\lambda_i^2 D_B t} \sin[\lambda_i(r - R)] \tag{13.16}$$

where the eigenvalues, λ_i, are the roots of

$$\tan[\lambda_i(R_c - R)] = \lambda_i R_c \tag{13.17}$$

and the coefficients, a_i, are given by

$$a_i = \frac{2(c_\circ - c_{eq}^{\alpha\beta})R(\lambda_i^2 R_c^2 + 1)}{\lambda_i[\lambda_i R_c^2(R_c - R) - R]} \tag{13.18}$$

The diffusion current into the particle carried by the ith term in the series given in Eq. 13.16 is

$$I_i = 4\pi R^2 D_B [\partial c/\partial r]_{r=R}$$

$$= 4\pi D_B(c_\circ - c_{eq}^{\alpha\beta})R \frac{2R(\lambda_i^2 R_c^2 + 1)}{\lambda_i^2 R_c^2(R_c - R) - R} e^{-\lambda_i^2 D_B t} \tag{13.19}$$

The total diffusion current into the particle is therefore a sum of the terms given by Eq. 13.19. Each term decays exponentially with a characteristic relaxation time corresponding to $\tau_i = 1/\lambda_i^2 D_B$.

When $R \ll R_c$, all of the short-wavelength terms decay very rapidly compared to the lowest-order $i = 0$ term, which, with acceptable accuracy, is

$$I_\circ = 4\pi D_B(c_\circ - c_{eq}^{\alpha\beta})R e^{-t/\tau_\circ} \tag{13.20}$$

where $\tau_\circ = R_c^3/3D_B R$ [27]. Letting $\langle c \rangle$ be the average concentration in the spherical cell,

$$\frac{d\langle c \rangle}{dt} = -\frac{I_\circ}{(4/3)\pi R_c^3} = -\frac{c_\circ - c_{eq}^{\alpha\beta}}{\tau_\circ} e^{-t/\tau_\circ} \tag{13.21}$$

Integrating Eq. 13.21 and combining the result with Eq. 13.20 leads to the remarkably simple expression for the total diffusion current entering the particle:

$$I = 4\pi D_B R \left(\langle c \rangle - c_{eq}^{\alpha\beta} \right) \tag{13.22}$$

The analysis shows that the diffusion current quickly settles down to the value given by Eq. 13.22 during all but very early times and that the transients which occur at the early times due to the higher-order eigenfunctions can be neglected whenever the degree of precipitation is significant. Because the effects of any transients are small, Eq. 13.22 also describes, with acceptable accuracy, the instantaneous quasi-steady current of atoms to the particle when it is growing due to the incoming diffusion. Section 20.2.1 (see Eq. 20.47) shows that Eq. 13.22 also holds with acceptable accuracy for an isolated sphere which is growing in an infinite matrix. It also holds for an isolated sphere of constant radius in an infinite matrix (see Exercise 13.6). The result given by Eq. 13.22 is therefore insensitive to the effects due to sphere growth or to the volume in which it is growing as long as $R \ll R_c$. In Exercise 13.8 this result is used to determine the growth of the precipitates in Fig. 13.14 as a function of time.

The situation becomes quite different when the α/β interface is no longer capable of maintaining the concentration of B atoms in its vicinity at the equilibrium value $c_{\mathrm{eq}}^{\alpha\beta}$. If the concentration there rises to the value $c^{\alpha\beta}$, the instantaneous quasi-steady-state current of atoms delivered to the particle by the diffusion field (obtained from Eq. 13.22) will be given by

$$I = 4\pi D_B R \left(\langle c \rangle - c^{\alpha\beta} \right) \tag{13.23}$$

This must be equal to the rate at which these atoms are incorporated into the particle locally at the interface. The rate at which B atoms in the matrix transfer to the particle across the α/β interface will be proportional to the local matrix concentration. The reverse rate of transfer from the particle to the matrix will be the same as the rate of transfer from the matrix to the particle that would occur under equilibrium conditions when detailed balance prevails. The net rate of transfer will then be

$$I' = 4\pi R^2 K \left(c^{\alpha\beta} - c_{\mathrm{eq}}^{\alpha\beta} \right) \tag{13.24}$$

where K is a rate constant. This rate-constant model should apply over a range of situations and has been widely used in the literature. The rate at which incoming B atoms are permanently incorporated in the particle depends upon the product of the impingement rate of B atoms on the particle (which is proportional to the B atom concentration in the matrix at the interface) and the fraction of the impinging B atoms that is permanently incorporated (this fraction depends upon the efficiency with which the particle collects and incorporates these atoms). This efficiency depends upon sink characteristics of the interface, such as the density of incorporation sites, the binding energy of a B atom to the interface, and its rate of diffusion along the interface. These factors can be lumped together in the form of a rate constant, K, so that the rate of permanent incorporation (per unit area of interface) is expressed as the product $Kc^{\alpha\beta}$. The rate at which B atoms are permanently removed from the particle at the interface can be determined by a stratagem in which the particle (with the identical interfacial sink structure) is imagined to be in detailed balance with the equilibrium concentration of B atoms in the matrix. The rate of permanent removal is then equal to the rate of permanent incorporation. However, the rate of incorporation depends upon the same rate constant as in the nonequilibrium case and is therefore given by $Kc_{\mathrm{eq}}^{\alpha\beta}$. This must also be equal to the rate of permanent removal because of the detailed balance. However, it will also be equal to the rate of removal under nonequilibrium conditions since the sink structure is assumed to be unchanged. The net rate of incorporation (per unit area) in the nonequilibrium situation is then $K(c^{\alpha\beta} - c_{\mathrm{eq}}^{\alpha\beta})$. The magnitude of the rate constant, K, can vary widely depending upon the sink efficiency of the particle and it can evolve with time if the structure of the interface (sink) changes. A detailed analysis of the analogous problem of crystal growth due to the impingement of atoms from the vapor phase in Exercises 12.1 and 12.2 shows that the growth rate can be represented by a rate-constant expression of similar form. Setting $I = I'$, solving for $c^{\alpha\beta}$, and putting the result into Eq. 13.23 yields the rate of precipitation

$$I = \frac{4\pi D_B R \left(\langle c \rangle - c_{eq}^{\alpha\beta} \right)}{1 + D_B/(KR)} \qquad (13.25)$$

which is smaller than the rate of precipitation under diffusion-limited conditions (Eq. 13.22) by the factor $1 + D_B/(KR)$. In fact, the efficiency of the particle as a sink is just

$$\eta = \frac{1}{1 + D_B/(KR)} \qquad (13.26)$$

if the previous definition of the efficiency given by Eq. 11.25 is employed. When the transfer rate is very high (or the diffusivity is very small) so that $D_B/(KR) \ll 1$, $\eta \cong 1$, $c^{\alpha\beta} \cong c_{eq}^{\alpha\beta}$, and the kinetics is diffusion-limited. At the other extreme, when $D_B/(KR) \gg 1$, $\eta \cong 0$, $c^{\alpha\beta} \cong \langle c \rangle$, and the kinetics is source-limited. When the kinetics is between these limits, it is regarded as *mixed*.[10]

Bibliography

1. R.W. Cahn. Recovery and recrystallization. In R.W. Cahn and P. Haasen, editors, *Physical Metallurgy*, pages 1595–1671. North-Holland, Amsterdam, 1983.

2. A.P. Sutton and R.W. Balluffi. *Interfaces in Crystalline Materials*. Oxford University Press, Oxford, 1996.

3. J.W. Christian. *The Theory of Transformations in Metals and Alloys*. Pergamon Press, Oxford, 1975.

4. J.P. Hirth and J. Lothe. *Theory of Dislocations*. John Wiley & Sons, New York, 2nd edition, 1982.

5. C. Herring. Diffusional viscosity of a polycrystalline solid. *J. Appl. Phys.*, 21:437–445, 1950.

6. C. Herring. Surface tension as a motivation for sintering. In W.E. Kingston, editor, *The Physics of Powder Metallurgy*, pages 143–179, New York, 1951. McGraw-Hill.

7. W.T. Read. *Dislocations in Crystals*. McGraw-Hill, New York, 1953.

8. H. Gleiter. The mechanism of grain boundary migration. *Acta Metall.*, 17(5):565–573, 1969.

9. D.J. Dingley and R.C. Pond. On the interaction of crystal dislocations with grain boundaries. *Acta Metall.*, 27(4):667–682, 1979.

10. J.W. Cahn. The impurity-drag effect in grain boundary motion. *Acta Metall.*, 10(9):789–798, 1962.

11. E. Nes, N. Ryum, and O. Hunderi. On the Zener drag. *Acta Metall.*, 33:11–22, 1985.

12. M.F. Ashby. The influence of particles on boundary mobility. In N. Hansen, A.R. Jones, and T. Leffers, editors, *Recrystallization and Grain Growth of Multi-Phase and Particle Containing Materials*, pages 325–336, Roskilde, Denmark, 1980. Riso National Laboratory.

13. G. Gottstein and L.S. Shvindlerman. *Grain Boundary Migration in Metals: Thermodynamics, Kinetics, Applications*. CRC Press, London, 1999.

---

[10]See Exercise 13.7 for further results.

14. K.T. Aust and J.W. Rutter. Effect of grain boundary mobility and energy on preferred orientation in annealed high purity lead. *Trans. AIME*, 224:111–115, 1962.

15. E.M. Fridman, C.V. Kopezky, and L.S. Shvindlerman. Effects of orientation and concentration factors on migration of individual grain-boundaries in aluminum. *Z. Metallkd.*, 66(9):533–539, 1975.

16. P.R. Howell, J.O. Nilsson, and G.L. Dunlap. The effect of creep deformation on the structure of twin boundaries. *Phil. Mag. A*, 38(1):39–47, 1978.

17. D.A. Smith, C.M.F. Rae, and C.R.M. Grovenor. Grain boundary migration. In R.W. Balluffi, editor, *Grain Boundary Structure and Kinetics*, pages 337–571, Metals Park, OH, 1980. American Society for Metals.

18. H. Ichinose and Y. Ishida. In situ observation of grain boundary migration of silicon $\Sigma 3$ boundary and its structural transformation at 1000 K. *J. Phys. Colloq. (Paris)*, 51(suppl. no. 1):C1:185–190, 1990.

19. T. Kizuka, M. Iijima, and N. Tanaka. Grain boundary migration at atomic scale in MgO. *Mater. Sci. Forum*, 233–234:405–412, 1997.

20. R.J. Jahn and P.D. Bristowe. A molecular dynamic study of grain boundary migration without the participation of secondary grain boundary dislocations. *Scripta Metall.*, 24(7):1313–1318, 1990.

21. D.A. Molodov, C.V. Kopetskii, and L.S. Shvindlerman. Detachment of a special $(\Sigma = 19, \langle 111 \rangle)$ tilt boundary from an impurity in iron-doped aluminum bicrystals. *Sov. Phys. Solid State*, 23(10):1718–1721, 1981.

22. R.W. Siegel, S.M. Chang, and R.W. Balluffi. Vacancy loss at grain-boundaries in quenched polycrystalline gold. *Acta Metall. Mater.*, 28(3):249–257, 1980.

23. A.H. King and D.A. Smith. On the mechanisms of point-defect absorption by grain and twin boundaries. *Phil. Mag. A*, 42(4):495–512, 1980.

24. Y. Komem, P. Petroff, and R.W. Balluffi. Direct observation of grain boundary dislocation climb in ion-irradiated gold bicrystals. *Phil. Mag.*, 26:239–252, 1972.

25. K.E. Rajab and R.D. Doherty. Kinetics of growth and coarsening of faceted hexagonal precipitates in an fcc matrix. 1. Experimental-observations. *Acta Metall. Mater.*, 37(10):2709–2722, 1989.

26. G.C. Weatherly. The structure of ledges at plate-shaped precipitates. *Acta Metall.*, 19(3):181–192, 1971.

27. F.S. Ham. Theory of diffusion limited precipitation. *J. Phys. Chem. Solids*, 6(4):335–351, 1958.

EXERCISES

13.1 Consider the conservative motion of a heterophase α/β interface in an A–B binary system. c_A^α, c_B^α, c_A^β, and c_B^β are the concentrations of A and B in the two phases facing each other across the interface. Show that the conservative motion requires that $X_A^\alpha = X_A^\beta$ and that it is generally expected that $c_A^\alpha \neq c_A^\beta$.

Solution. If the interface moves conservatively into the α phase, it will convert a slab of the α phase of thickness δ^α into a slab of β phase of thickness δ^β. Each slab must contain the same number of A and B atoms (i.e., $N_A^\alpha = N_A^\beta = N_A$ and $N_B^\alpha = N_B^\beta = N_B$). Each slab will then have the same mass, but because the slab densities will generally differ, the slab thicknesses will generally differ. The condition $X_A^\alpha = X_A^\beta$ must then be satisfied since

$$X_A^\alpha = \frac{N_A}{N_A + NB} = X_A^\beta \tag{13.27}$$

However, since (for unit area of boundary)

$$c_A^\alpha = \frac{N_A}{\delta^\alpha} \quad \text{and} \quad c_A^\beta = \frac{N_A}{\delta^\beta} \tag{13.28}$$

the concentrations will be expected to differ since in general $\delta^\alpha \neq \delta^\beta$.

13.2 Equation 13.1 for the pressure exerted on the small-angle tilt boundary by a shear stress was derived by considering the work done by the shear stress during the change in macroscopic shape of the bicrystal which occurred when the boundary moved (see Fig. 13.1). Obtain the same result by considering the force exerted on each moving edge dislocation by the applied stress and summing the forces on all dislocations.

Solution. Using Eq. 11.2, the force per unit length on each dislocation is $f_\sigma = \sigma_{xy} b$. The spacing, d, of the dislocations in a symmetric tilt boundary is $d = b/\theta$ [7]. The pressure on the boundary due to all dislocations is then

$$P = \sigma_{xy} b \frac{1}{d} = \sigma_{xy} b \frac{\theta}{b} = \sigma_{xy} \theta \tag{13.29}$$

Note that this result is expected since the force exerted on an individual dislocation by a stress is the result of the change in crystal shape that occurs when the dislocation moves. Since the change in shape of the bicrystal due to the boundary motion is just the sum of the changes due to the motion of each of its individual dislocations, the total force on the boundary must be just the sum of the forces on the individual dislocations.

13.3 An expression for the diffusion potential at an edge dislocation in a small-angle tilt boundary subjected to a pure shear stress has been derived in Section 13.3.2. Derive a general expression for the diffusion potential at an isolated general (mixed) straight dislocation acted on by a general stress field. Express your answer in terms of the stress tensor, $\boldsymbol{\sigma}$, the Burgers vector, $\vec{b}$, and the unit tangent vector, $\hat{\zeta}$. Remember that the potential is related to the work performed by the stress field when an atom is inserted at the dislocation (per unit dislocation length) during the dislocation climb. The force exerted on the dislocation is given by the Peach–Koehler equation (Eq. 11.1). Also, only the edge component of the dislocation plays a role in the climb.

Solution. The dislocation will climb in a direction perpendicular to its glide plane (i.e., the plane containing both $\vec{b}$ and $\hat{\zeta}$). The unit normal, $\hat{n}$, to the glide plane is

$$\hat{n} = \frac{\hat{\zeta} \times \vec{b}}{|\hat{\zeta} \times \vec{b}|} \tag{13.30}$$

The force per unit length, f, exerted on the dislocation by the stress field in the climb direction normal to the glide plane is then

$$f = \vec{f_\sigma} \cdot \hat{n} \tag{13.31}$$

The climb distance per unit length, d_c, due to the insertion of an atom is

$$d_c = -\frac{\Omega}{b_e} = -\frac{\Omega}{(\hat{n} \times \hat{\zeta}) \cdot \vec{b}}$$ (13.32)

where $b_e = (\hat{n} \times \hat{\zeta}) \cdot \vec{b}$ is the magnitude of the edge component of the Burgers vector. The potential relative to the reference potential, μ_A°, is then the negative of the work done by the stress during the climb and is therefore given by

$$\Phi_A^D = \mu_A^\circ + \frac{(\vec{f}_\sigma \cdot \hat{n})\Omega}{b_e} = \mu_A^\circ + \frac{[(\vec{b}^T \cdot \sigma) \times \hat{\zeta}] \cdot (\hat{\zeta} \times \vec{b})\Omega}{[(\hat{\zeta} \times \vec{b}) \times \hat{\zeta}] \cdot \vec{b}}$$ (13.33)

Note that the sign of the final expression in Eq. 13.33 must be consistent with the convention for determining the Burgers vector (see the text following Eq. 11.1).

13.4 A single-phase bicrystal sheet of thickness, h, is produced in a laboratory and cut into a symmetrical wedge as in Fig. 13.15. Upon heating, the boundary

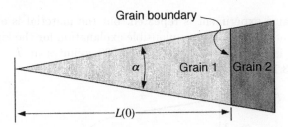

Figure 13.15:   Bicrystal specimen with a planar grain boundary initially located a distance $L(0)$ from the apex.

is found to migrate toward the apex and data for $L(t)$ as a function of time, t, are shown in Fig. 13.16.

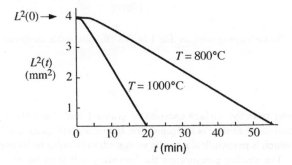

Figure 13.16:   Migration data for boundary in Fig. 13.15.

(a) Develop a plausible model for the observed data in Fig. 13.16. State two assumptions of the model.

(b) Give a plausible explanation for the observed initial transient behavior in Fig. 13.16.

(c) Using only the information in Fig. 13.16, estimate the activation energy for boundary motion.

(d) Closer examination of the data in Fig. 13.16 shows that growth is not uniform but oscillates between slow and fast growth, as shown in Fig. 13.17. Give a plausible explanation for this behavior.

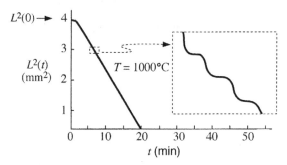

Figure 13.17:   Magnified view of data from Fig. 13.16 at 1000°C.

(e) The same experiment is repeated, but the material is obtained from a different supplier. Give a plausible explanation for the kinetic transition which is now observed after about 30 minutes at $T = 800°C$, as in Fig. 13.18.

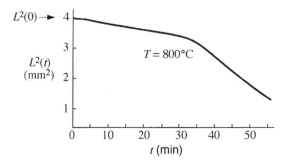

Figure 13.18:   Same experiment as for Fig. 13.16, but with material from a different supplier.

Solution.

(a) Assuming that the surface energies of grains 1 and 2 are the same and that the grain-boundary energy is isotropic, the boundary will quickly adopt a circular-arc shape which is perpendicular to the wedge edges in order to balance surface-tension forces. The driving pressure on the boundary will then be $P = \gamma/R$, where the radius of curvature is $R = L/\cos(\alpha/2)$. Therefore,

$$P = \frac{\gamma}{L} \cos \frac{\alpha}{2} \qquad (13.34)$$

Assuming a constant boundary mobility, M_B,

$$v = \frac{dL}{dt} = -M_B P = -\frac{M_B \gamma^B}{L} \cos \frac{\alpha}{2}$$

or

$$L \, dL = -M_B \gamma^B \cos \frac{\alpha}{2} \, dt \qquad (13.35)$$

Integrating Eq. 13.35 yields

$$L^2(0) - L^2(t) = 2M_B\gamma^B \cos\frac{\alpha}{2}\, t \qquad (13.36)$$

(b) The data indicate that the boundary motion corresponding to Eq. 13.36 does not begin immediately. This could be the result of grain boundary pinning by solute atoms or small precipitates or by the existence of a grain-boundary groove in the initial position of the boundary from which the boundary must break away. It could also be the result of time taken for the boundary to adopt the curved shape—a process that would presumably begin from the specimen edge.

(c) Equation 13.35 indicates that the slope of the linear regions of the two curves in Fig. 13.16 is proportional to the boundary mobility, M_B. Assuming that the boundary energy, γ^B, is independent of temperature and that M_B follows an Arrhenius law, $M_B = M_B^\circ \exp[-E^B/(kT)]$, the activation energy for migration can be calculated. Letting $T_1 = 1073$ K and $T_2 = 1273$ K, the corresponding mobilities are proportional to the slopes of the curves in Fig. 13.16:

$$M_B(T_1) \propto \frac{4}{55-6} = 0.082 \quad \text{and} \quad M_B(T_2) \propto \frac{4}{19-2} = 0.23 \qquad (13.37)$$

From the Arrhenius law,

$$\ln M_B(T_1) - \ln M_B(T_2) = -\frac{E^B}{k}\left(\frac{1}{T_1} - \frac{1}{T_2}\right) \qquad (13.38)$$

and hence

$$
\begin{aligned}
E^B &= \frac{[\ln M_B(T_2) - \ln M_B(T_1)]\, k}{(1/T_1) - (1/T_2)} \\
&= \frac{1.054 \times 1.38 \times 10^{-19}\ \text{J K}^{-1}}{1.46 \times 10^{-4}\ \text{K}^{-1}} = 1.0 \times 10^{-15}\ \text{J}
\end{aligned}
\qquad (13.39)
$$

(d) The most likely explanation for jerky motion of the boundary is localized pinning by precipitates or small inclusions from which the boundary must repeatedly escape.

(e) Figure 13.18 indicates that the boundary initially moves steadily at a relatively slow rate, then undergoes a transition to steady motion at a higher rate. Such behavior is consistent with an impurity-drag breakaway effect, which could be due to certain impurities in the material from the different supplier. Note that according to Eq. 13.34, the driving pressure increases as the boundary migrates and L decreases. The initial slow migration regime takes place when the boundary is less highly curved and is moving under a relatively small capillary driving force. Under these conditions, the boundaries in the impure material could have an impurity atmosphere that would have to move along with the boundary as it migrates with a relatively low mobility (extrinsic migration). As the boundary moves toward the apex of the wedge-shaped bicrystal specimen, it becomes more highly curved and the capillary driving force rises (ultimately becoming infinite). When the driving force becomes sufficiently high, the moving boundary can break free of the impurity atoms and move with a higher intrinsic mobility.

13.5 Consider the pinned second-phase particle illustrated in Fig. 13.7 and described by Eqs. 13.9–13.12. Assume that $\gamma^{P1} = \gamma^{P2}$, so that $\alpha = 90°$.

(a) Show that the force exerted by the interface on the particle (or alternately, the force exerted by the particle on the interface) when the interface is at a position where it meets the particle at the distance along

y denoted by $y = \eta$ is given (as a function of η) by

$$F(\eta) = 2F_{\max} \frac{\eta}{R} \sqrt{1 - \left(\frac{\eta}{R}\right)^2} \tag{13.40}$$

which is plotted in Fig. 13.19.

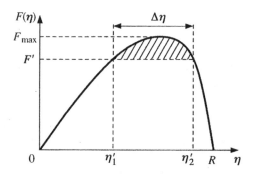

Figure 13.19: $F(\eta)$ vs. η. $F(\eta)$ is the pinning force exerted on an interface by a spherical particle embedded in the interface in the configuration illustrated in Fig. 13.7. η is the displacement of the interface in the y direction, corresponding to $\eta = R \sin \phi$ in Fig. 13.7.

(b) Consider the possible thermally activated unpinning of the particle when the interface is initially held at the position η_1' by a force F'. The activation energy E for the unpinning will be the work required to move the interface off the particle in the direction of y in the presence of the initial force F'. The force resisting this motion will correspond to $F(\eta)$. Show that E is given by

$$E = \frac{2}{3} F_{\max} R \left(1 - \frac{F'}{F_{\max}}\right)^{3/2} \tag{13.41}$$

The thermally activated unpinning is therefore favored by a small particle size and a large applied force, as might be expected intuitively.

Solution.

(a) When $\alpha = 90°$, Eqs. 13.10 and 13.12 yield

$$F = 2F_{\max} \cos \phi \, \cos(90 - \phi) = 2F_{\max} \cos \phi \sin \phi \tag{13.42}$$

From Fig. 13.7, $\cos \phi = \sqrt{R^2 - \eta^2}/R$, and $\cos(90 - \phi) = \sin \phi = \eta/R$. Substitution of these quantities into Eq. 13.42 then yields the desired relationship.

(b) The energy that must be supplied by thermal activation in the presence of the initial force F' is the shaded area in the figure, and therefore

$$E = \int_{F'}^{F_{\max}} \Delta\eta(F) \, dF \tag{13.43}$$

Using $\eta = R \sin \phi$, $\Delta\eta = R(\sin \phi_2 - \sin \phi_1)$. From Eq. 13.42,

$$F(\eta_1) = 2F_{\max} \cos \phi_1 \, \cos(90 - \phi_1) \qquad F(\eta_2) = 2F_{\max} \cos \phi_2 \, \cos(90 - \phi_2) \tag{13.44}$$

$F(\eta_1) = F(\eta_2)$ requires that $\phi_2 = 90 - \phi_1$. Therefore,

$$\Delta\eta = R(\sin\phi_2 - \sin\phi_1) = R(\cos\phi_1 - \sin\phi_1)$$
$$= R\sqrt{1 - 2\sin\phi_1\cos\phi_1} = R\sqrt{1 - F/F_{max}} \tag{13.45}$$

Putting Eq. 13.45 into Eq. 13.43 and integrating then produces the desired result.

13.6 Equation 13.22 holds for the quasi-steady rate at which B atoms diffuse to a spherical precipitate in a distribution of similar precipitates in a dilute supersaturated solution. Show that an equation of the same form holds for the quasi-steady rate of diffusion to a single spherical precipitate of constant radius R embedded in an infinite volume of a similar solution. [Note that a similar result is also obtained in Section 20.2.1 (Eq. 20.47) for the more realistic case where the particle is allowed to grow as the diffusion occurs.]

- Use spherical coordinates and solve the resulting boundary-value problem by making the change of variable $u = (c - c_o)r$, where c_o is the concentration when $r \to \infty$.

Solution. The basic differential equation is

$$\frac{\partial c}{\partial t} = \nabla^2 c = D_B\left(\frac{\partial^2 c}{\partial r^2} + \frac{2}{r}\frac{\partial c}{\partial r}\right) \tag{13.46}$$

and the boundary and initial conditions are

$$c(r = R, t > 0) = c_{eq}^{\alpha\beta} \qquad c(r \geq R, t = 0) = c_o \qquad c(r \to \infty, t > 0) = c_o \tag{13.47}$$

Making the change of variable $u = (c - c_o)r$, Eq. 13.46 becomes

$$\frac{\partial u}{\partial t} = D_B\frac{\partial^2 u}{\partial r^2} \tag{13.48}$$

A standard solution of Eq. 13.48 (see Chapter 4) is

$$u(r,t) = a_1\left[1 - \text{erf}\left(\frac{r - R}{2\sqrt{D_B t}}\right)\right] + a_2 \tag{13.49}$$

where a_1 and a_2 are constants. Therefore,

$$c - c_o = \frac{a_1}{r}\left[1 - \text{erf}\left(\frac{r - R}{2\sqrt{D_B t}}\right)\right] + \frac{a_2}{r} \tag{13.50}$$

and applying the boundary and initial conditions, $a_2 = 0$ and $a_1 = R(c_{eq}^{\alpha\beta} - c_o)$. The final solution is then

$$\frac{c - c_o}{c_{eq}^{\alpha\beta} - c_o} = \frac{R}{r}\left[1 - \text{erf}\left(\frac{r - R}{2\sqrt{D_B t}}\right)\right] \tag{13.51}$$

The total diffusion current entering the particle is then

$$I = 4\pi R^2 D_B\left(\frac{\partial c}{\partial r}\right)_{r=R} = 4\pi D_B R\left(c_o - c_{eq}^{\alpha\beta}\right)\left[1 + \frac{R}{2\sqrt{D_B t}}\right] \tag{13.52}$$

The second term inside the brackets is seen to be an initial transient that falls off as $t^{-1/2}$. It is associated with the establishment at early times of a steep concentration gradient in the diffusion field over a distance from the particle equal to about R.

Once the diffusion distance, $2(D_B t)^{1/2}$, substantially exceeds R, the transient becomes unimportant, and Eq. 13.52 then has the same form as Eq. 13.22.

13.7 Consider the analysis leading to Eqs. 13.23 and 13.24 for a spherical precipitate particle acting as an ideal or nonideal sink.

(a) Show that when $\widetilde{D}/(\kappa R) \ll 1$ and the particle is acting almost as a perfect sink, the concentration maintained at the particle/matrix interface, $c^{\alpha\beta}$, must inevitably be slightly larger than the equilibrium concentration, $c_{\text{eq}}^{\alpha\beta}$, and that, in fact,

$$c^{\alpha\beta} \approx c_{\text{eq}}^{\alpha\beta} + [\langle c \rangle - c_{\text{eq}}^{\alpha\beta}]\frac{\widetilde{D}}{\kappa R} \qquad (13.53)$$

(b) Next, show that when $\widetilde{D}/(\kappa R) \gg 1$ and the particle is acting as a very poor sink, the concentration at the interface will inevitably be slightly lower than the average concentration in the bulk, $\langle c \rangle$, and that, in fact,

$$c^{\alpha\beta} \approx \langle c \rangle - \frac{\langle c \rangle - c_{\text{eq}}^{\alpha\beta}}{\widetilde{D}/(\kappa R)} \qquad (13.54)$$

• Neglect higher-order terms in any expansions that are employed.

Solution. First, solve for $c^{\alpha\beta}$ by equating Eqs. 13.23 and 13.24 to obtain

$$c^{\alpha\beta} = \frac{\langle c \rangle[\widetilde{D}/(\kappa R)] + c_{\text{eq}}^{\alpha\beta}}{1 + \widetilde{D}/(\kappa R)} \qquad (13.55)$$

(a) When $\widetilde{D}/(\kappa R) \ll 1$, let $\widetilde{D}/(\kappa R) = \varepsilon$. Then

$$c^{\alpha\beta} = \frac{\langle c \rangle\varepsilon + c_{\text{eq}}^{\alpha\beta}}{1 + \varepsilon} \approx c_{\text{eq}}^{\alpha\beta} + [\langle c \rangle - c_{\text{eq}}^{\alpha\beta}]\varepsilon = c_{\text{eq}}^{\alpha\beta} + \frac{[\langle c \rangle - c_{\text{eq}}^{\alpha\beta}]\widetilde{D}}{\kappa R} \qquad (13.56)$$

(b) When $\widetilde{D}/(\kappa R) \gg 1$, let $\kappa R/\widetilde{D} = \varepsilon$. Then

$$c^{\alpha\beta} = \frac{\langle c \rangle + \varepsilon c_{\text{eq}}^{\alpha\beta}}{1 + \varepsilon} = \langle c \rangle + \varepsilon c_{\text{eq}}^{\alpha\beta} - \langle c \rangle\varepsilon = \langle c \rangle - \frac{[\langle c \rangle - c_{\text{eq}}^{\alpha\beta}]\kappa R}{\widetilde{D}} \qquad (13.57)$$

13.8 Using the model and results in Section 13.4.2, find a relationship showing how the particle radius grows with increasing time during the precipitation of the finite number of supersaturated interstitial atoms with diffusivity D_B available. Assume that the particles are of fixed composition throughout and act as perfect sinks and that $R \ll R_c$.

Solution. Starting with Eq. 13.22, the instantaneous rate at which the particle volume increases is

$$\frac{dV}{dt} = 4\pi R^2 \frac{dR}{dt} = 4\pi\widetilde{D}R[\langle c \rangle - c_{\text{eq}}^{\alpha\beta}]\Omega' \qquad (13.58)$$

where Ω' is the increase in particle volume per added interstitial. At any time, the volume of the precipitate is given by

$$V = \frac{4}{3}\pi R^3 = \frac{4}{3}\pi R_c^3[c_{\text{o}} - \langle c \rangle]\Omega' \qquad (13.59)$$

Solving Eq. 13.59 for $\langle c \rangle$ and putting the result into Eq. 13.58 yields the differential equation

$$\frac{R\,dR}{(c_{\mathrm{o}} - c_{\mathrm{eq}}^{\alpha\beta}) - R^3/R_{\mathrm{o}}^3\,\Omega'} = \widetilde{D}\Omega'\,dt \qquad (13.60)$$

The final size of the particle when the precipitation is complete is given by

$$R_{\infty}^3 = R_c^3(c_{\mathrm{o}} - c_{\mathrm{eq}}^{\alpha\beta})\Omega' \qquad (13.61)$$

and therefore Eq. 13.60 may be put into the more convenient form

$$\frac{R\,dR}{R_{\infty}^3 - R^3} = \frac{\widetilde{D}}{R_c^3}\,dt \qquad (13.62)$$

The solution of Eq. 13.62, subject to the initial condition $R = 0$ when $t = 0$, is

$$\ln\left(\frac{\sqrt{R^2 + R_{\infty}R + R_{\infty}^2}}{R_{\infty} - R}\right) - \sqrt{3}\tan^{-1}\left(\frac{2R + R_{\infty}}{\sqrt{3}\,R_{\infty}}\right) + \frac{\pi}{2\sqrt{3}} = \frac{3R_{\infty}\widetilde{D}}{R_c^3}\,t \quad (13.63)$$

A few calculations show that R increases with time at an ever-decreasing rate and approaches R_{∞} asymptotically as $t \to \infty$.

13.9 Consider a small-angle tilt boundary of the type shown in Fig. 13.1, which is acting as a sink for highly supersaturated vacancies at initial concentration given by $\bar{c}$. Assume that each edge dislocation acts as a perfect line sink and that the vacancies are diffusing to the boundary from a distance, L, larger than the spacing of the dislocations in the boundary, d. Show that the boundary will act as a perfect sink for all practical purposes whenever $L \gg d$.

- Neglect any effects due to the climb motion of the dislocations and assume quasi-steady-state diffusion. The vacancy isoconcentration contours around the dislocations in the boundary will then appear approximately as illustrated in Fig. 13.20.

At distances from the boundary along x greater than about half the dislocation spacing (i.e., $d/2$), the contours will be unaffected by the fine structure of the boundary and will essentially be planes running parallel to the boundary. Nearer to the dislocation cores, the contours will be concentric cylinders. A reasonable approximation is then to represent the diffusion field as shown in

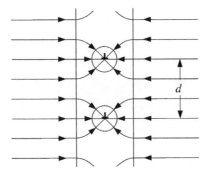

Figure 13.20: Approximate flux lines around edge dislocations in a small-angle tilt boundary acting as a sink for supersaturated vacancies.

Fig. 13.20. Here, at distances from each core less than $d/2$, the diffusion into each core is taken to be cylindrical. Beyond this distance, the diffusion into the boundary as a whole along x is taken to be planar (one-dimensional). At the "surface" of the core (i.e., the core radius of the dislocation, $r = R_o$), the equilibrium vacancy concentration is negligible compared with the high concentration, $\bar{c}$, at the large distance, L. Therefore, $c(r = R_o) = 0$. Assume a quasi-steady state in the boundary region. The concentration at $x = d/2$ will then take on an intermediate value, c', which can be determined. The total diffusion current into the boundary can then be determined.

Solution. Using Eq. 5.13, the quasi-steady-state current entering each dislocation by cylindrical diffusion is

$$I_C = \frac{2\pi D_V c'}{\ln\left[d/(2R_o)\right]} \tag{13.64}$$

while the current diffusing toward each dislocation from the distance L by planar diffusion is

$$I_P = \frac{2dD_V(\bar{c} - c')}{L - d/2} \tag{13.65}$$

Equating these currents then yields

$$c' = \frac{d\ln[d/(2R_o)]}{\pi(L - d/2) + d\ln[d/(2R_o)]}\,\bar{c} \tag{13.66}$$

Substituting this into the expression for I_C,

$$I_C = \frac{2\pi D_V d}{\pi(L - d/2) + d\ln[d/(2R_o)]}\,\bar{c} \tag{13.67}$$

The current that would diffuse into the boundary if it were a perfect sink would be

$$I_{\text{ideal}} = \frac{2dD_V\bar{c}}{L} \tag{13.68}$$

The sink efficiency (see Eq. 11.25) is then

$$\eta = \frac{I_C}{I_{\text{ideal}}} = \frac{1}{1 - d/(2L) + [d/(\pi L)]\ln[d/(2R_o)]} \tag{13.69}$$

Therefore, $\eta \approx 1$ when $L \gg d$. *Note:* a more exact treatment using the analogy between steady-state diffusion fields and electrostatic fields leads to a very similar result [22].

PART III

MORPHOLOGICAL EVOLUTION DUE TO CAPILLARY AND APPLIED MECHANICAL FORCES

A material can change its *morphology*—its external macroscopic shape and its internal microstructure—when suitable driving forces exist. As always, these driving forces result from differential decreases in total free energy.

Morphology depends to a major extent upon the shapes and positions of interfaces. Every real material has at least one interface—the exterior interface which separates it from its environment. If an interface separates a material from its vapor or a vacuum, it is typically called a *free surface*. Interfaces that exist in materials within a condensed phase (or between condensed phases) are termed *internal interfaces*.

Chapters 14 and 15 are concerned with the contributions made to morphological change by the motion of free surfaces and internal interfaces. Interfaces contribute an excess free energy to a material which depends on their areas and their crystallographic degrees of freedom, and they can also respond to mechanical forces. As a result, interfaces are commonly subjected to two independent forces which tend to drive their motion and produce morphological change. If the interface motion decreases the total excess free energy associated with the interface itself (i.e., through the product of its area and energy per unit area), the resulting force for motion is conventionally called a *capillary force*. For a curved isotropic interface, the capillary force is present due to the pressure difference across the interface caused by its local mean curvature. Such capillary forces are especially influential in fine-scale microstructures containing interfaces with large curvatures where pressure differences can become correspondingly large. In addition to a capillary force, a force for interface motion is produced whenever motion of the interface allows an applied force to perform work: such a force is an *applied force*.

The three chapters that comprise Part III are limited to cases where chemical driving forces are absent, and phase transformations therefore do not occur. Morphological change due to the evolution of free surfaces under the influence of capillary forces is treated in Chapter 14. Topics include surface smoothing and faceting, surface grooving at intersections with grain boundaries, and the incipient evolution of a cylinder into a row of spheres. Transport mechanisms considered include surface and volume diffusion and vapor transport. Morphological change due to the capillarity-driven motion of internal interfaces is treated in Chapter 15, which includes grain growth in polycrystals and the coarsening of a fine distribution of second-phase particles. In Chapter 16 we address cases of morphological change that involve free surfaces and internal interfaces and include capillary and applied mechanical forces. Attention is focused on two technologically important kinetic processes: the surface smoothing and removal of porosity which occurs during sintering of porous bodies and the change in body shape produced by diffusional creep.

CHAPTER 14

SURFACE EVOLUTION DUE TO CAPILLARY FORCES

The total interfacial contribution to the free energy of a system corresponds to the integral of the interfacial free energy per unit area, γ, over all interfaces. γ is a material property that is a function of the geometric degrees of freedom of an interface. Therefore, driving forces for interfacial motion in a particular material derive from two sources—reductions in interfacial area and reductions in interfacial energy per unit area due to changes in the geometrical/crystallographic attributes of the interfaces.

Analysis is simplified if γ is *isotropic*—i.e., independent of geometrical attributes such as interfacial inclination $\hat{n}$ and, for internal interfaces in crystalline materials, the crystallographic misorientation across the interface. All interfacial energy reduction then results from a reduction of interfacial area through interface motion. The rate of interfacial area reduction per volume transferred across the interface is the local geometric mean curvature. Thus, local driving forces derived from variations in mean curvature allow tractable models for the *capillarity-induced* morphological evolution of isotropic interfaces.

This chapter treats the morphological evolution of free surfaces under capillary driving forces, including phenomena associated with both isotropic and anisotropic surfaces. The kinetics of surface smoothing, the formation of surface grooves on free surfaces at intersections with grain boundaries, and the incipient evolution of a cylinder into a row of isolated spheres are considered for bodies with isotropic surfaces. Transport mechanisms include surface and volume diffusion and vapor transport. General methods of analyzing the evolution of anisotropic surfaces are discussed; particular attention is given to the phenomenon of faceting and crystal growth when the surface velocity is a function of the surface inclination.

In a two-phase composite material of isolated spherical particles embedded in a matrix, there is a driving force to transport material from particles enclosed by isotropic surfaces of larger constant mean curvature to particles of smaller constant mean curvature. This coarsening process and the motion of internal interfaces due to curvature are treated in Chapter 15.

14.1 ISOTROPIC SURFACES

An isotropic surface with undulations will generally evolve toward constant mean curvature as a result of the driving force arising from differences in curvature. As indicated by Eq. 3.76, material below a bump of large mean curvature has a larger diffusion potential than material below a bump of smaller mean curvature.[1] Considering only capillary driving forces, it can be shown that irregularities on a uniform flat surface flatten due to capillarity-induced evolution, as illustrated in Fig. 3.7. However, a cylinder with surface irregularities can either evolve back to a uniform cylinder or toward a row of isolated spheres.

The rate and characteristics of surface evolution depend on the particular transport mechanisms that accomplish the necessary surface motion. These can include surface diffusion, diffusion through the bulk, or vapor transport. Kinetic models of capillarity-induced interface evolution were developed primarily by W.W. Mullins [1–4]. The models involving surface diffusion, which relate interface velocity to fourth-order spatial derivatives of the interface, and vapor transport, which relate velocity to second-order spatial derivatives, derive from Mullins's pioneering theoretical work.

14.1.1 Flattening of Free Surfaces by Surface Diffusion

Consider a system composed of only one type of atom along with its equilibrium point defects, having an initial surface profile

$$y - y_\circ = h(x) \quad \text{or} \quad F(x,y) = y - h(x) = \text{constant} \tag{14.1}$$

where $h(x)$ represents deviations from a flat ($y = \text{constant}$) surface as in Fig. 14.1.[2]

The diffusion potential of an atom at the surface is proportional to local surface curvature as demonstrated in Section 3.4. The curvature can be determined from Eq. 14.1 and is a function of x. The local diffusion potential produces boundary conditions for diffusion through the bulk or transport via the vapor phase. For surface diffusion, gradients in the diffusion potential produce fluxes along the surface.

For surface diffusion, migration is constrained to a thin surface slab of thickness δ, as described in Section 9.1. Because the surface is an efficient point-defect source, point defects may be assumed to be at their local equilibrium populations.

---

[1] The following convention is employed: when a center of curvature is on the same side as the material in question, the curvature is positive. A ball has positive curvature; a spherical hole has negative curvature.

[2] Additional material regarding the mathematical descriptions of interfaces is given in Appendix C, particularly Eq. C.9.

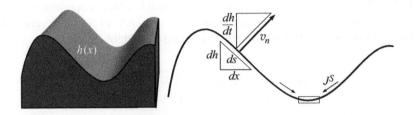

Figure 14.1:   Flattening of a ruled surface $h(x)$ by surface diffusion. The normal velocity is proportional to the accumulation of flux. The rate of vertical motion dh/dt is related to the normal velocity v_n by the local geometry of the surface.

Therefore, combining the force–flux equation (Eq. 2.21) and the diffusion potential at a curved interface (Eq. 3.76),

$$\vec{J}^S = c\langle v \rangle = -c\,M^S \nabla_{\text{surf}}\,\Phi^S = -c\,\Omega\gamma^S M^S \nabla_{\text{surf}}\,\kappa = -\frac{\gamma^S\,{}^\star D^S}{kT}\nabla_{\text{surf}}\kappa \qquad (14.2)$$

where $c = 1/\Omega$ is the concentration of atoms in a surface layer of thickness δ, and the surface mobility M^S is related to the surface diffusivity ${}^\star D^S$ by Eq. 3.43. ∇_{surf} in Eq. 14.2 is the surface gradient operator, which has derivatives within the interface (i.e., it is a two-dimensional gradient operator in the interface tangent plane). For a two-dimensional interface embedded in three dimensions $\{x(u,v), y(u,v), z(u,v)\}$, with interfacial coordinates (u,v) and curvature κ, the surface gradient of κ,

$$\nabla_{\text{surf}}\,\kappa(u,v) = \nabla\,|_{x(u,v),y(u,v),z(u,v)} = \left(\frac{\partial\kappa}{\partial u}, \frac{\partial\kappa}{\partial v}\right) \qquad (14.3)$$

is a vector perpendicular to the surface normal which points in the direction of fastest increase in κ. For the ruled surface described by Eq. 14.1,

$$J^S = -\frac{\gamma^S\,{}^\star D^S}{kT}\frac{\partial\kappa}{\partial s} \qquad (14.4)$$

where s is the arc length along the surface curve $y - h(x) = \text{constant}$, as illustrated in Fig. 14.1. The rate of particle accumulation at a particular surface site is the integral of the flux over the surface enclosing each surface site. Representing each surface site as a box with a square base of area $1/c^S$ (where c^S is the number of surface sites per area) and height δ, the particle accumulation rate is

$$\dot{\mathcal{N}}^S = -\int_{\text{box-walls}} \vec{J}^S \cdot d\vec{A} = -\int_{\text{box-volume}} \nabla \cdot \vec{J}^S \, dV \approx -\nabla \cdot \vec{J}^S \frac{\delta}{c^S} \qquad (14.5)$$

The atom addition rate per unit area is $\dot{\mathcal{N}}^S c^S = -\delta\nabla \cdot \vec{J}^S$, and therefore the normal velocity of the surface is

$$v_n = \Omega\dot{\mathcal{N}}^S c^S = -\Omega\,\delta\,\nabla \cdot \vec{J}^S = \frac{\Omega\gamma^S\delta\,{}^\star D^S}{kT}\nabla^2_{\text{surf}}\kappa = B^S\frac{d^2\kappa}{ds^2} \qquad (14.6)$$

where B^S collects the kinetic and material coefficients that multiply the surface Laplacian, $\nabla^2_{\text{surf}} \equiv \nabla_{\text{surf}} \cdot \nabla_{\text{surf}}$, the two-dimensional Laplacian operator with derivatives with respect to coordinates in the interface as in Eq. 14.3. The figure indicates a relation between the rate of change of the surface height, h, and v_n:

$$\frac{\partial h}{\partial t} = \frac{v_n}{\sqrt{1 + (\frac{\partial h}{\partial x})^2}} = \frac{B^S}{\sqrt{1 + (\frac{\partial h}{\partial x})^2}} \frac{\partial^2 \kappa}{\partial s^2} \tag{14.7}$$

Figure 14.1 indicates that the derivatives with respect to arclength, s, can be converted to derivatives with respect to x using $ds^2 = dh^2 + dx^2$. From the expression for κ in Eq. C.5,

$$\begin{aligned}
\frac{\partial h}{\partial t} &= \frac{B^S}{1 + (\frac{\partial h}{\partial x})^2} \frac{\partial}{\partial x} \left[\frac{1}{[1 + (\frac{\partial h}{\partial x})^2]^{1/2}} \frac{\partial \kappa}{\partial x} \right] \\
&= \frac{-B^S}{1 + (\frac{\partial h}{\partial x})^2} \frac{\partial}{\partial x} \left[\frac{1}{[1 + (\frac{\partial h}{\partial x})^2]^{1/2}} \frac{\partial}{\partial x} \left(\frac{\frac{\partial^2 h}{\partial x^2}}{[1 + (\frac{\partial h}{\partial x})^2]^{3/2}} \right) \right]
\end{aligned} \tag{14.8}$$

In the limit of small slopes, $|\partial h / \partial x| \ll 1$,

$$\frac{\partial h}{\partial t} = -B^S \frac{\partial^4 h}{\partial x^4} + 3B^S \left(\frac{\partial^2 h}{\partial x^2} \right)^3 \tag{14.9}$$

The second term on the right-hand side is negligible unless there are regions where the (small) slopes change rapidly. Any such regions disappear rapidly, due to the first term. Therefore, the linearized surface diffusion equation is

$$\frac{\partial h}{\partial t} = -B^S \frac{\partial^4 h}{\partial x^4} = \frac{-\Omega \gamma^S \delta^\star D^S}{kT} \frac{\partial^4 h}{\partial x^4} \tag{14.10}$$

The dependence of h on the length scales of the surface roughness can be analyzed with independent Fourier components having the form

$$h(x, t) = A(t) \sin \left(\frac{2\pi x}{\lambda} \right) \tag{14.11}$$

Substitution of Eq. 14.11 into Eq. 14.10 gives

$$\frac{dA(t)}{A(t)} = \frac{-\Omega \gamma^S \delta^\star D^S}{kT} \left(\frac{2\pi}{\lambda} \right)^4 dt \tag{14.12}$$

so that the time dependence of the Fourier amplitude, $A(t)$, is

$$A(t) = A(0) e^{-B^S (2\pi/\lambda)^4 t} \tag{14.13}$$

The amplitude decay exponent is proportional to $1/\lambda^4$; fine-scale roughness therefore disappears much more rapidly than longer-wavelength surface roughness. General roughness can be decomposed into a sum of Fourier terms, and since Eq. 14.10 is linear, each term decays independently. Physically, the short-wavelength modes disappear rapidly and the flattening rate of an isotropic surface by surface diffusion is limited by the longest wavelength. Equation 14.13 is an approximation for the limit of small slopes. However, numerical solutions of the nonlinear evolution equation, Eq. 14.8, for large-slope conditions indicate that the small-slope approximation can be applied in the general case without significant error [5].

The wavelength dependence in Eq. 14.13 can be used for experimental measurements of the surface and kinetic coefficients that constitute B^S. If an array of evenly spaced parallel grooves is introduced on a surface, the spacing dependence of the grooves' amplitude-decay factor can be measured [6]. An analysis for flattening of an isotropic surface by bulk diffusion as in Fig. 3.7 is presented in Exercise 14.1.

14.1.2 Surface Evolution by Vapor Transport

Solid/vapor interface motion can be produced by evaporation—the atoms that compose the solid phase are removed from the surface via the vapor phase; reverse motion can be produced by condensation where a vapor-phase flux is directed onto the solid phase. Figure 14.2 illustrates how simultaneous evaporation and condensation can result in surface smoothing.

Vapor transport differs from surface diffusional transport, where the flux is always in the surface plane. For both surface diffusion and vapor transport, the diffusion potential at the surface is proportional to the local value of $\gamma^S \kappa$ if the surface free energy is isotropic. For surface diffusion, the interface normal velocity is related to a derivative (i.e., the divergence of the flux). Also, the total volume is conserved during surface diffusion. For vapor transport, the interface normal velocity is directly proportional to the vapor flux, and the total number of atoms is not necessarily conserved.

Crystal growth from the vapor phase has been treated in Chapter 12. An expression for the net atom flux, J^V, gained at a macroscopically flat crystal surface during growth from the vapor has been obtained in Exercise 12.2 in the form of Eq. 12.27. To treat surfaces possessing nonuniform curvature, this relationship can be generalized in the form

$$J^V = K[P^{\mathrm{amb}} - P^{\mathrm{eq}}(\kappa)] \tag{14.14}$$

where P^{amb} is the ambient vapor pressure in the system, $P^{\mathrm{eq}}(\kappa)$ is the vapor pressure in equilibrium with a local region of the surface with curvature κ, and K is a rate constant. The function $P^{\mathrm{eq}}(\kappa)$ may be derived by equating the atom diffusion potential in the surface and in the equilibrium vapor. For the vapor, $dG = kT \ln P$, and therefore

$$\mu^V(\kappa) - \mu^V(\kappa = 0) = \Phi^V(\kappa) - \Phi^V(\kappa = 0) = kT \ln \left[\frac{P^{\mathrm{eq}}(\kappa)}{P^{\mathrm{eq}}(\kappa = 0)} \right]$$

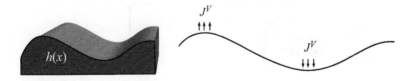

Figure 14.2: Flattening of surface $h(x)$ by vapor transport involving evaporation from regions of larger curvature and condensation at regions of lower curvature. The velocity along the surface normal is proportional to the normal flux at the solid/vapor interface.

while at the surface, according to Eq. 3.76, $\Phi^S(\kappa) - \Phi^S(\kappa = 0) = \Omega\gamma^S\kappa$. Equating these expressions results in

$$P^{\text{eq}}(\kappa) = P^{\text{eq}}(\kappa = 0)e^{\Omega\gamma^S\kappa/(kT)} \approx P^{\text{eq}}(\kappa = 0)\left(1 + \frac{\Omega\gamma^S\kappa}{kT}\right) \tag{14.15}$$

The first-order expansion employed will be valid under all usual conditions.

The normal growth velocity of a local region of the surface with curvature κ can then be found by using Eqs. 14.14 and 14.15:

$$v_n = \Omega J^V = \Omega K \left[P^{\text{amb}} - P^{\text{eq}}(\kappa = 0) - \frac{P^{\text{eq}}(\kappa = 0)\Omega\gamma^S}{kT}\kappa\right] \tag{14.16}$$

In the common situation where the surface contains undulations but is macroscopically flat, P^{amb} will be well approximated by $P^{\text{eq}}(\kappa = 0)$, so that

$$v_n = -\frac{K\Omega^2\gamma^S P^{\text{eq}}(\kappa = 0)}{kT}\kappa = -B^V\kappa \tag{14.17}$$

where the coefficient B^V collects the vapor-transport kinetic and material parameters (and has the same units as a diffusivity).

Surface Grooving at Surface/Grain-Boundary Intersections by Vapor Transport. When a grain boundary with isotropic surface energy γ^B intersects a free surface with isotropic surface energy γ^S, a *grain-boundary groove* forms in order to achieve a capillary-force balance, as illustrated in Fig. 14.3. The dihedral angle ψ is determined by Young's equation,

$$\cos\frac{\psi}{2} = \frac{\gamma^B}{2\gamma^S} \tag{14.18}$$

Grain-boundary grooves can develop during thermal annealing by mass transport arising from vapor transport, surface diffusion, or surface-to-surface transport by means of volume diffusion.

Equation 14.17 can be used to model grain-boundary grooving kinetics when vapor transport is the dominant mechanism. The normal velocity $\partial h/\partial t$ is related

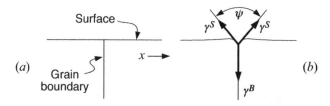

Figure 14.3: (a) Initial intersection of grain boundary with flat surface and (b) subsequent formation of surface groove with dihedral angle, ψ. Because a grain boundary migrates toward its center of curvature (discussed in Section 15.2), stationary grain boundaries intersect the surface at right angles.

to the vertical velocity as in Fig. 14.1 and Eq. 14.7:

$$\frac{\partial h}{\partial t} = -\frac{B^V}{\left[1 + (\frac{\partial h}{\partial x})^2\right]^{1/2}}\kappa$$

$$= B^V \frac{\frac{\partial^2 h}{\partial x^2}}{\left[1 + (\frac{\partial h}{\partial x})^2\right]^2}$$

(14.19)

In the limit of small slopes $|\frac{\partial h}{\partial x}| \ll 1$,

$$\frac{\partial h}{\partial t} = B^V \frac{\partial^2 h}{\partial x^2}$$

(14.20)

This surface evolution equation has the same form as the bulk mass diffusion equation; the concentration is replaced by the height of the surface, h, and the diffusivity is replaced by B^V.

For the grain-boundary grooving problem, the initial and boundary conditions derive from the initial shape of the surface and Young's equation at the groove notch:

$$h(x, t = 0) = 0 \qquad \frac{\partial h}{\partial x}(x = 0, t) = \tan(\frac{\pi - \psi}{2}) \qquad \frac{\partial h}{\partial x}(x = \infty, t) = 0 \qquad (14.21)$$

Only $x > 0$ is considered, since the solution will have mirror symmetry about the grain boundary plane $x = 0$.

These boundary conditions correspond to the constant surface-flux diffusion problem in Section 5.2.5, in which the surface concentration increased proportionally to $t^{1/2}$. Therefore, adapting the solution given by Eq. 5.69 to the grain-boundary grooving model,

$$h(0, t) = -2\tan\left(\frac{\pi - \psi}{2}\right)\sqrt{\frac{B^V t}{\pi}}$$

(14.22)

and the groove therefore deepens proportionally with $t^{1/2}$ as well.

Stability of a Cylinder. A cylinder of radius R_o with isotropic γ^S can reduce its total surface energy if it evolves into a row of spheres with radii greater than $3R_o/2$ (i.e., if each cylinder section with a length of at least $9R_o/2$ evolves into a sphere). Therefore, microstructures with cylindrical features can be morphologically unstable. However, kinetics requires that the total surface energy (or, surface area for the isotropic case) must decrease *continuously* if a cylinder evolves into a row of isolated particles. Continuous surface-energy reduction sets a fundamental or *critical* length scale for the wavelength, λ_{crit}, of an infinitesimal axial undulation on a uniform cylinder, which produces instability. Rayleigh derived an expression for λ_{crit} using *perturbation analysis* as follows [7].

Consider a radially symmetric perturbation (or an infinitesimal Fourier mode) on a uniform cylinder of initial radius R_o:

$$R(z, t) = R_{\text{cyl}}(t) + \epsilon(t)\cos\frac{2\pi z}{\lambda}$$

(14.23)

where $\epsilon(t)$ is the time-dependent perturbation amplitude (see Fig. 14.4). The cylin-

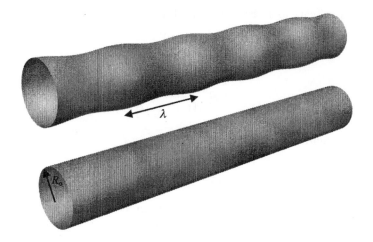

Figure 14.4:   Perturbation of cylindrical body by a sinusoid.   Long-wavelength perturbations reduce the surface area of the body and ultimately result in breakup of the body into a linear array of spheres.

der is isolated so that its volume per wavelength, $\pi R_{\mathrm{o}}^2 \lambda$, is constant. Therefore,

$$
\begin{aligned}
\pi R_{\mathrm{o}}^2 \lambda &= \int_0^\lambda \pi R^2(z,t)\, dz \\
&= \pi \int_0^\lambda \left[R_{\mathrm{cyl}}^2 + 2 R_{\mathrm{cyl}}(t)\, \epsilon(t) \cos \frac{2\pi z}{\lambda} + \epsilon^2(t) \cos^2 \frac{2\pi z}{\lambda} \right] dz \qquad (14.24) \\
&= \pi \lambda R_{\mathrm{cyl}}^2 + \frac{\pi \lambda \epsilon^2(t)}{2}
\end{aligned}
$$

The average cylinder radius, R_{cyl}, must shrink when $\epsilon(t)/R_{\mathrm{o}}$ is small according to

$$
R_{\mathrm{cyl}}(t) = R_{\mathrm{o}} - \frac{\epsilon^2(t)}{4 R_{\mathrm{o}}} + \cdots \qquad (14.25)
$$

The total surface energy per wavelength for the surface of revolution, Eq. 14.23, is

$$
\begin{aligned}
E^S &= \int_0^\lambda \gamma^S dA = \gamma^S \int_0^\lambda 2\pi R\, ds \\
&= 2\pi \gamma^S \int_0^\lambda R \sqrt{dz^2 + dR^2} = 2\pi \gamma^S \int_0^\lambda R \sqrt{1 + \left(\frac{dR}{dz} \right)^2}\, dz
\end{aligned} \qquad (14.26)
$$

Expanding in powers of $\epsilon(t)$,

$$
\begin{aligned}
E^S &= 2\pi \gamma^S \int_0^\lambda \left[R_{\mathrm{cyl}}(t) + \epsilon(t) \cos \frac{2\pi z}{\lambda} \right] \sqrt{1 + \left[\frac{2\pi \epsilon(t) \sin \left(\frac{2\pi z}{\lambda} \right)}{\lambda} \right]^2}\, dz \\
&= 2\pi \gamma^S R_{\mathrm{cyl}}(t) \int_0^\lambda dz + \frac{4\pi^3 \epsilon^2(t) \gamma^S R_{\mathrm{cyl}}(t)}{\lambda^2} \int_0^\lambda \sin^2 \frac{2\pi z}{\lambda}\, dz \qquad (14.27) \\
&= E_{\mathrm{o}}^S + \frac{\pi \gamma^S \epsilon^2(t)}{2\lambda R_{\mathrm{o}}} \left[(2\pi R_{\mathrm{o}})^2 - \lambda^2 \right] + \cdots
\end{aligned}
$$

Therefore, the energy decreases continuously with time if *the Rayleigh instability condition* is satisfied,

$$\lambda > \lambda_{\text{crit}} = 2\pi R_\circ \tag{14.28}$$

Any perturbation with a wavelength less than the circumference of the cylinder will not grow.

The particular characteristics of morphological evolution are determined by the dominant transport mechanism; their analyses derive from the diffusion potential, which depends on the local curvature. For a surface of revolution about the z-axis, the curvature is given by Eq. C.16; that is,

$$\kappa(z) = \frac{1}{R(z)\sqrt{1 + \left(\frac{\partial R}{\partial z}\right)^2}} - \frac{\frac{\partial^2 R}{\partial z^2}}{\left[1 + \left(\frac{\partial R}{\partial z}\right)^2\right]^{3/2}} \tag{14.29}$$

Substituting Eq. 14.23 into Eq. 14.29 and expanding for small $\epsilon/R_\circ$ yields

$$\kappa(z) \approx \frac{1}{R_\circ} + \frac{\epsilon}{R_\circ^2}\left[\left(\frac{2\pi R_\circ}{\lambda}\right)^2 - 1\right]\cos\frac{2\pi z}{\lambda} \tag{14.30}$$

14.1.3 Evolution of Perturbed Cylinder by Vapor Transport

Suppose that a perturbed cylinder with radius given by Eq. 14.23 evolves by vapor transport in an environment with an ambient vapor pressure in equilibrium with the unperturbed cylinder, $P^{\text{amb}} = P^{\text{eq}}(\kappa = 1/R_\circ)$. Then, using Eqs. 14.15, 14.16, and 14.17,

$$v_n = B^V\left(\frac{1}{R_\circ} - \kappa\right) \tag{14.31}$$

According to Eq. 14.25, $R_{\text{cyl}} \simeq R_\circ$, so Eq. 14.23 shows that v_n at $z = 0$ is approximately $d\varepsilon(t)/dt$. Therefore, using Eq. 14.30,

$$\frac{d\epsilon(t)}{dt} = \frac{B^V}{R_\circ^2}\left[1 - \left(\frac{2\pi R_\circ}{\lambda}\right)^2\right]\epsilon(t) = \frac{1}{\tau^V(\lambda)}\epsilon(t) \tag{14.32}$$

Small perturbations therefore evolve according to

$$\epsilon(t) = \epsilon(0)e^{t/\tau^V(\lambda)} \tag{14.33}$$

where the *amplification factor* $1/\tau^V = (B^V/R_\circ^2)[1 - (2\pi R_\circ/\lambda)^2]$. This first-order kinetic result is consistent with the previous Rayleigh result: only perturbations with wavelengths longer than λ_{crit} will grow.

14.1.4 Evolution of Perturbed Cylinder by Surface Diffusion

Suppose that the perturbed cylinder considered above evolves by surface diffusion. A first-order differential equation for the amplitude $\epsilon(t)$ follows from inserting Eq. 14.30 into the surface diffusion relation, Eq. 14.6, and again setting $v_n = d\epsilon(t)/dt$ at $z = 0$:

$$\frac{d\epsilon(t)}{dt} = \frac{4\pi^2 B^S}{R_\circ^2\lambda^2}\left[1 - \left(\frac{2\pi R_\circ}{\lambda}\right)^2\right]\epsilon(t) = \frac{1}{\tau^S(\lambda)}\epsilon(t) \tag{14.34}$$

In addition to the Rayleigh result, Eq. 14.34 predicts that a particular perturbation wavelength, $\lambda_{\max}$, grows the fastest and hence dominates the morphology of the evolving cylinder. This *kinetic wavelength* maximizes the right-hand side of Eq. 14.34, giving the result $\lambda_{\max} = \sqrt{2}\,\lambda_{\text{crit}}$.

14.1.5 Thermodynamic and Kinetic Morphological Wavelengths

Comparison of surface-diffusion and vapor-transport kinetics in Fig. 14.5 shows a difference in long-wavelength behavior. The amplification factor $1/\tau(\lambda)$ in the perturbation growth rate $\epsilon(t) = \epsilon(0)\exp[t/\tau(\lambda)]$ is monotonically increasing for vapor transport and approaches B^V/R_o^2 asymptotically for long wavelengths. For surface diffusion, $1/\tau(\lambda)$ goes to zero for long wavelengths and has a maximum at $\lambda = \sqrt{2}\,(2\pi R_o)$. For a cylinder with an initial small random roughness, evolution by surface diffusion results in a morphological scale associated with $\lambda_{\max}$. For vapor diffusion, no characteristic morphological scale is predicted.

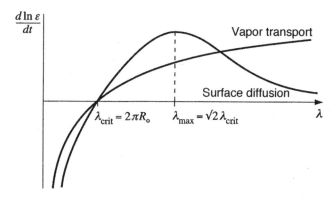

Figure 14.5:   Behavior of the perturbation-amplitude growth coefficients $1/\tau^V$ and $1/\tau^S$ for cylinder-perturbation growth by vapor transport and surface diffusion, respectively. For surface diffusion, a fastest-growing wavelength $\lambda_{\max}$ determines a morphological scale for the initial instability. Except for the Rayleigh critical wavelength, λ_{crit}, no characteristic length scale appears for vapor transport.

14.2 ANISOTROPIC SURFACES

14.2.1 Some Geometrical Aspects of Anisotropic Surfaces

An anisotropic surface's energy per unit area, $\gamma(\hat{n})$, depends on its inclination, $\hat{n}$. For isotropic surfaces, the surface energy is simply proportional to the area, but two additional degrees of freedom emerge for the anisotropic case. These correspond to the two parameters required to specify the surface inclination.[3] An anisotropic surface can often decrease its energy at constant area by tilting (i.e., changing its normal). The variation of the interfacial energy with inclination can be represented conveniently in the form of a polar plot (or γ-plot), as shown in two dimensions in Fig. 14.6. Here, the energy of each inclination is represented by a vector, $\gamma(\hat{n})\hat{n}$

[3]Geometrical constructions for describing anisotropic surfaces are reviewed in Section C.3.1.

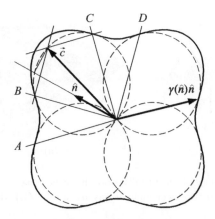

Figure 14.6: Construction for testing whether an interface of inclination $\hat{n}$ will prefer to be faceted into inclinations corresponding to points B and C. γ-plot same as in Fig. C.4a.

(i.e., a vector normal to that inclination and of magnitude equal to the interfacial energy at that inclination). If all of these vectors are referred to a single origin, the γ-plot is the surface passing through the tips of these vectors. Inclinations of particularly low energies will therefore appear as cusps or depressions in the plot.

Conceptually, treatment of the morphological evolution for an anisotropic surface is no different than for an isotropic surface—kinetics requires that $\int \gamma(\hat{n}) \, dA$ (compared to $\gamma \int dA$ for an isotropic surface) must decrease monotonically. However, because the evolving surface's geometry is linked to the local surface-energy density through $\hat{n}$, the analysis is considerably more complicated. Furthermore, when a surface is sufficiently anisotropic, inclinations $\hat{n}$ associated with large energies become unstable and cannot be in local equilibrium—the surface must develop corners or edges. The missing inclinations create points or curves on a surface where surface derivatives will be discontinuous. When the γ-plot has cusp singularities, planar facets may appear; such a surface can have portions that are smoothly curved or portions that are flat and these portions are separated by edges or corners where derivatives are discontinuous.

For surfaces with the two-dimensional γ-plot shown in Fig. 14.6, certain inclinations will be unstable and will be replaced by other inclinations (facets), even though this increases the total surface area. Whether a certain inclination is unstable and prone to facet into other inclinations can be determined by a simple geometrical construction using the γ-plot [8]. The surface will consist of two different types of facets, as in Fig. 14.7a. The energy of such a structure per unit area projected on the macroscopically flat surface, $\gamma_{\rm fac}$, is

$$\gamma_{\rm fac} = \gamma_1 f_1 + \gamma_2 f_2 \tag{14.35}$$

where γ_i is the surface energy of the ith-type facet and f_i is the fraction of the projected area contributed by facets of type i. If $\hat{n}$ is the unit normal to the flat surface and $\hat{n}_1$, $\hat{n}_2$, and $\hat{n}_3$ are unit vectors normal to type-1 facets, type-2 facets, and along the facet intersections, respectively, as in Fig. 14.7a,

$$\hat{n} = f_1 \hat{n}_1 + f_2 \hat{n}_2 \tag{14.36}$$

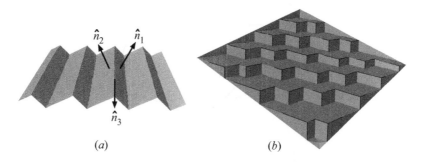

(a) (b)

Figure 14.7: Morphology of an initially smooth surface that has reduced its energy by faceting. **(a)** Morphology if two facet inclinations are stable. **(b)** Morphology if three facet inclinations are stable.

If a set of vectors, $\vec{n}_i^{\star}$, reciprocal to the vectors $\hat{n}_i$, is introduced so that

$$\vec{n}_1^{\star} = \frac{\hat{n}_2 \times \hat{n}_3}{\hat{n}_1 \cdot (\hat{n}_2 \times \hat{n}_3)} \qquad \vec{n}_2^{\star} = \frac{\hat{n}_3 \times \hat{n}_1}{\hat{n}_1 \cdot (\hat{n}_2 \times \hat{n}_3)} \qquad \vec{n}_3^{\star} = \frac{\hat{n}_1 \times \hat{n}_2}{\hat{n}_1 \cdot (\hat{n}_2 \times \hat{n}_3)} \qquad (14.37)$$

so that $\vec{n}_i^{\star} \cdot \hat{n}_j = \delta_{ij}$, Eq. 14.35 can be rewritten

$$\gamma_{\text{fac}} = \left(\gamma_1 \vec{n}_1^{\star} + \gamma_2 \vec{n}_2^{\star}\right) \cdot \hat{n} = \vec{c} \cdot \hat{n} \qquad (14.38)$$

where $\vec{c} = \gamma_1 \vec{n}_1^{\star} + \gamma_2 \vec{n}_2^{\star}$ has the properties

$$\vec{c} \cdot \hat{n}_1 = \gamma_1 \qquad \vec{c} \cdot \hat{n}_2 = \gamma_2 \qquad (14.39)$$

Whether faceting will occur can now be settled by a simple geometrical construction using the γ-plot shown in Fig. 14.6. If the surface to be tested has the inclination $\hat{n}$ and the inclinations corresponding to points B and C are chosen as the inclinations for the $i = 1$ and $i = 2$ facets, $\vec{c}$ must appear as shown in Fig. 14.6 in order to be consistent with Eq. 14.39. The energy of the surface of average inclination $\hat{n}$ that is faceted into inclinations corresponding to points B and C is then, according to Eq. 14.38, the projection of $\vec{c}$ on $\hat{n}$. This energy is smaller than the energy of the nonfaceted interface (indicated by the outer envelope of the γ-plot) and the surface will prefer to be faceted.

It may also be seen that the energies of all other surfaces with inclinations varying between those at B and C will fall on the dashed circle. All of these surfaces will therefore be faceted. On the other hand, a similar construction shows that all surfaces with inclinations between those at C and D will be stable against faceting into the inclinations at C and D. Points such as those at B and C where the dashed circle is tangent to the γ-plot therefore delineate the ranges of inclination between which the surface is either faceted or nonfaceted. The construction indicated in Fig. 14.6 is readily generalized to three dimensions: three facet planes could then be present, as in Fig. 14.7b, and $\vec{c}$ then terminates at the point of intersection of three planes rather than two lines.

Figure 14.8 shows a three-dimensional γ-plot comprised of eight equivalent spherical surface regions. The shape of this γ-plot is consistent with all surfaces represented by the plot being composed of various mixtures of the three types of facets,

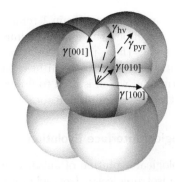

Figure 14.8: The γ-plot for a material with a Wulff shape corresponding to a cube when $\gamma[100] = \gamma[010] = \gamma[001]$. It consists of portions of eight identical spheres, shown here in cutaway view. These spheres share a common point at the origin, but each has a diametrically opposed point directed toward the eight $\langle 111 \rangle$ directions.

corresponding to the $\gamma[100]$, $\gamma[010]$, and $\gamma[001]$ vectors shown.[4] Any interface corresponding to a vector lying on a groove at the intersection of two spheres, such as γ_{hv}, will consist of *two types* of facets, corresponding to a pair of the vectors $\gamma[100]$, $\gamma[010]$, or $\gamma[001]$. Any interface corresponding to a vector going to a spherical region of the plot, such as γ_{pyr}, will consist of *three types* of facets, corresponding to $\gamma[100]$, $\gamma[010]$, and $\gamma[001]$.

Figure 14.9 shows a three-grain junction on the surface of polycrystalline Al_2O_3 after high-temperature annealing. Each grain surface has a different inclination

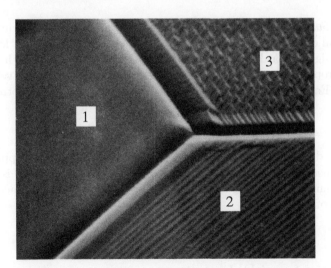

Figure 14.9: Surface morphology of three faceted grains in an annealed alumina polycrystal. From J.M. Dynys [9].

[4]In Fig. 14.6, which holds in two dimensions, the energies of all faceted surfaces with inclinations between B and C fall on the dashed tangent circle shown. In three dimensions, a comparable construction would show that faceting would occur on three facet planes, such as in Fig. 14.7*b*, and that the counterpart to the tangent circle would be a tangent sphere.

and exhibits a different facet morphology. Grain 1 remains flat, grain 2 shows two facet inclinations, and grain 3 exhibits three facet inclinations.

Other constructions employing the γ-plot are reviewed in Section C.3.1. These include the reciprocal γ-plot, which is also useful in treating the faceting problem above, and the Wulff construction, which is used to find the shape (Wulff shape) of a body of fixed volume that possesses minimum total surface energy.

14.2.2 Rate of Morphological Interface Evolution

The kinetics of the morphological evolution of anisotropic interfaces can be developed as an extension of the isotropic case. Isotropic interface evolution originates from a diffusion potential proportional to the local geometric curvature (mean curvature) multiplied by the surface energy per unit area. The local geometric curvature is the change of interface area, δA, with the addition of volume δV, $\kappa = \delta A / \delta V$ (see Section C.2.1). Therefore, the local energy increase due to the addition of an atom of volume Ω is $\Omega \gamma \kappa$. The anisotropic analog to the isotropic energy increase is the *weighted mean curvature* $\kappa_\gamma = \delta(\gamma A)/\delta V$, developed by J. Taylor [10]. In the anisotropic case, the diffusion potential is increased by, $\Omega \kappa_\gamma$, the local energy increase per adatom. It can be shown that

$$\kappa_\gamma = \nabla_{\text{surf}} \cdot \vec{\xi}(\hat{n}) \tag{14.40}$$

where $\vec{\xi}$ is the capillarity vector and ∇_{surf} is the surface divergence operator, similar to the surface gradient introduced in Eq. 14.2.[5] Two different types of derivatives are involved in this expression for κ_γ—the first produces $\vec{\xi}$ from a derivative in γ-space as seen in Eq. C.20; the second derivative used to obtain the divergence is taken along the evolving interface.

Evolution by Surface Diffusion and by Vapor Transport. Although calculation of the morphological evolution for particular cases can become tedious, the kinetic equations are straightforward extensions of the isotropic case [11]. For the movement of an anisotropic surface by surface diffusion, the normal interface velocity is an extension of Eq. 14.6 which holds for the isotropic case; for the anisotropic case,

$$v_n = \dot{\mathcal{N}}^S \Omega c^S = \frac{\Omega \delta^\star D^S}{kT} \nabla_{\text{surf}}^2 \kappa_\gamma \tag{14.41}$$

If the surface diffusivity is anisotropic, its surface derivatives must appear as well.

For movement by vapor transport of an anisotropic interface that is exposed to a vapor in equilibrium with a very large particle[6], the normal interface velocity is an extension of Eq. 14.17:

$$v_n = -\frac{K \Omega^2 P^{\text{eq}}(\kappa = 0)}{kT} \kappa_\gamma \tag{14.42}$$

The expression for weighted mean curvature for any surface in local equilibrium is simplified when the Wulff shape is completely faceted [10, 12]. In this case,

---

[5]The capillarity vector $\vec{\xi}$ and the weighted mean curvature κ_γ are discussed in more detail in Section C.3.2.

[6]Weighted mean curvature, which is uniform on a Wulff shape, goes to zero in the limit of large body volumes.

tractable expressions and simulations can be produced for morphological evolution by surface diffusion and vapor transport [13]. However, these models do not include edge and corner energies because they are inadmissible in the Wulff construction— nor do they include nucleation barriers for ledge and step creation, ledge–ledge interactions, and elastic effects associated with edges and corners.

Growth Rate for Inclination-Dependent Interface Velocity. For a crystalline particle growing from a supersaturated solution, the surface velocity often depends on atomic attachment kinetics. Attachment kinetics depends on local surface structure, which in turn depends on the surface inclination, $\hat{n}$, with respect to the crystal frame. In limiting cases, surface velocity is a function only of inclination; the interfacial speed in the direction of $\hat{n}$ is given by $v(\hat{n})$. The main aspects of a method for calculating the growth shapes for such cases when $v(\hat{n})$ is known is described briefly in this section.

Given an initial surface, $\vec{r}(t = 0)$, the surface morphology at some later time, t, can be computed from the growth law $v(\hat{n})$ with a simple construction [14, 15]. Let $\tau(\vec{r})$ be the time that the growing interface reaches a position $\vec{r}$; therefore, the level set $t_{\text{const}} = \tau(\vec{r})$ could be inverted to give the surface $\vec{r}(t_{\text{const}})$. The surface normal must be in the direction of the gradient of τ; $\hat{n} = \nabla\tau/|\nabla\tau|$, where $|\nabla\tau|$ must be proportional to $[v(\hat{n})]^{-1}$. Solving for the constant of proportionality, α, as a function of $\nabla\tau$,

$$\alpha \equiv v^{\text{ext}}(\nabla\tau) = |\nabla\tau|\, v(\hat{n}) = |\nabla\tau|\, v\left(\frac{\nabla\tau}{|\nabla\tau|}\right) \qquad (14.43)$$

$v^{\text{ext}}(\vec{p})$ is the homogeneous extension of the surface velocity $v(\hat{n})$ from $\hat{n}$ on the unit sphere to gradients of arbitrary magnitude $\vec{p} = \nabla\tau$ [16].

The extended normal velocity, $v^{\text{ext}}(\vec{p})$, can be used to construct *characteristics* that specify the surface completely at some time t [14]. The characteristics are rays that emanate from each position on the initial surface $\vec{r}(t = 0)$, given by

$$\begin{aligned}
\vec{r}(t) &= \vec{r}(t = 0) + t\,\nabla_{\vec{p}}\, v^{\text{ext}}(\vec{p}) \\
&= \vec{r}(t = 0) + t\left(\frac{\partial v^{\text{ext}}}{\partial p_1}, \frac{\partial v^{\text{ext}}}{\partial p_2}, \frac{\partial v^{\text{ext}}}{\partial p_3}\right)
\end{aligned} \qquad (14.44)$$

The surface normal $\hat{n}$ is constant along the characteristics, and therefore the surface velocity $v(\hat{n})$ is constant as well (see Exercise 14.5). The characteristics, defined as

$$\vec{\zeta} \equiv \nabla_{\vec{p}}\, v^{\text{ext}}(\vec{p}) = \nabla_{\vec{p}}\,[|\nabla\vec{p}|\, v(\hat{n})] = \nabla_{\vec{p}}\left[|\nabla\vec{p}|\, v\left(\frac{\vec{p}}{|\vec{p}|}\right)\right] \qquad (14.45)$$

do not depend on the magnitude $|\nabla\vec{p}|$. Therefore, the time-dependent morphology can be calculated directly from any initial surface $\vec{r}(t = 0)$ and a normal velocity $v(\hat{n})$ by the following procedure. First calculate $\zeta(\hat{n})$ for every point on the initial surface $\vec{r}(t = 0)$, then construct rays equal to $t\vec{\zeta}$ from each point. Using Eqs. 14.44 and 14.45, the surface positions at an arbitrary time t are

$$\vec{r}(t) = \vec{r}(t = 0) + t\vec{\zeta}(\hat{n}) \qquad (14.46)$$

The method is illustrated with a simple example in two dimensions. Suppose that the surface has the symmetry of a square and $v(\hat{n}) = v(n_1, n_2) = v(\cos\theta, \sin\theta)$

is given by

$$v(\hat{n}) = 1 + \beta(n_1^4 + \alpha n_1^2 n_2^2 + n_2^4)$$

$$v^{\text{ext}}(\vec{p}) = (p_1^2 + p_2^2)^{1/2} v \left[\frac{\vec{p}}{(p_1^2 + p_2^2)^{1/2}} \right] \tag{14.47}$$

$$\vec{\zeta}(\hat{n}) = \nabla_{\vec{p}} v^{\text{ext}}(\vec{p}) = \left(\begin{array}{c} n_1 \{1 + \beta[4n_1^2 - 3n_2^4 - 3n_2^4 + 2\alpha n_2^2(2 - 3n_1^2)]\} \\ n_2 \{1 + \beta[4n_2^2 - 3n_2^4 - 3n_1^4 + 2\alpha n_1^2(2 - 3n_2^2)]\} \end{array} \right)$$

where α and β are constants. The velocity $v(\hat{n})$ and its associated $\zeta(\hat{n})$ are illustrated in Fig. 14.10 for particular values of α and β.[7]

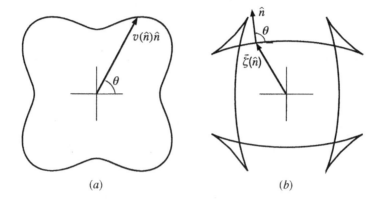

Figure 14.10:   Examples of $v(\hat{n})\hat{n}$ and $\vec{\zeta}(\hat{n})$ from Eq. 14.47 for $\beta = 1/2$ and $\alpha = 4$. **(a)** A polar plot of $v(\hat{n})\hat{n}$. The magnitude of the plot in each direction, $\hat{n} = (\cos\theta, \sin\theta)$, is the velocity in that direction. **(b)** $\vec{\zeta}(\hat{n})$ is plotted parametrically as a function of θ. The vector $\vec{\zeta}(\hat{n}) = \vec{\zeta}(\theta)$ is generally not in the direction of $\hat{n}(\theta)$. However, the surface of the $\vec{\zeta}(\theta)$-plot at any point is always normal to $\hat{n}(\theta)$, as shown in Eq. C.19, which although written for $\vec{\xi}(\hat{n})$ and $\gamma(\hat{n})$, also holds for $\vec{\zeta}(\hat{n})$ and $v(\hat{n})$.

Figure 14.11 shows the shape evolution due to $v(\hat{n})$ and its characteristics $\vec{\zeta}$ in Eq. 14.47 for an initially circular particle. After very long times, the only remaining orientations on the growth shape are those that lie on the interior portion of the $\vec{\zeta}$-surface; therefore, the portion of the $\vec{\zeta}$-surface with the spinodes (the swallowtail-shaped region) is removed.

For morphological evolution during dissolution of a crystal (or disappearance of voids in a crystalline matrix), the same characteristic construction applies, but the sense of the surface normal is switched compared to Fig. 14.11. An example of dissolution is illustrated in Fig. 14.12.

The asymptotic growth shapes (Fig. 14.11) are composed of inclinations associated with the slowest growth velocities, and the fastest inclinations grow out of existence by forming corners. On the contrary, for dissolution shapes (Fig. 14.12), the inclinations associated with the fastest dissolution remain and the slow-speed inclinations disappear into the corners. The asymptotic growth shape is the in-

---

[7]$\vec{\zeta}(\hat{n})$ is related to $v(\hat{n})$ in the same way that the capillarity vector, $\vec{\xi}$, is related to $\gamma(\hat{n})$ and is constructed in the same way. The Wulff construction applied to $v(\hat{n})$ produces the asymptotic growth shape. This and other relations between the Wulff construction and the common-tangent construction for phase equilibria are discussed by Cahn and Carter [16].

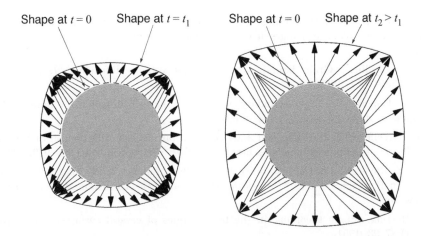

Figure 14.11:   Development of growth shape for an initially circular particle for the $v(\hat{n})$ illustrated in Fig. 14.10. Rays $t\vec{\zeta}(\hat{n})$ are drawn from each associated inclination on the initial surface. Fastest-growing inclinations accumulate at $45°$ and its equivalents and form corners.

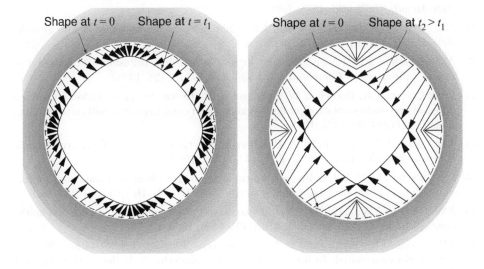

Figure 14.12:   Development of dissolution shape for initially circular particle for the $v(\hat{n})$ illustrated in Fig. 14.10. Rays $t\vec{\zeta}(-\hat{n})$ are drawn from each associated inclination on the initial surface. The slowest-growing inclinations accumulate at $90°$ and its equivalents and form corners.

terior of the $\vec{\zeta}$-surface and the asymptotic dissolution shape is composed of those inclinations between the cusps on the swallowtail-shaped region on the $\vec{\zeta}$-surface.

Bibliography

1. W.W. Mullins. Solid surface morphologies governed by capillarity. In N.A. Gjostein, editor, *Metal Surfaces: Structure, Energetics and Kinetics*, pages 17–66, Metals Park, OH, 1962. American Society for Metals.

2. W.W. Mullins. Theory of thermal grooving. *J. Appl. Phys.*, 28(3):333–339, 1957.

3. F.A. Nichols and W.W. Mullins. Surface- (interface-) and volume-diffusion contributions to morphological changes driven by capillarity. *Trans. AIME*, 233(10):1840–1847, 1965.

4. W.W. Mullins. Grain boundary grooving by volume diffusion. *Trans. AIME*, 218(4):354–361, 1960.

5. W.M. Robertson. Grain-boundary grooving by surface diffusion for finite surface slopes. *J. Appl. Phys.*, 42(1):463–467, 1971.

6. M.E. Keeffe, C.C. Umbach, and J.M. Blakely. Surface self-diffusion on Si from the evolution of periodic atomic step arrays. *J. Phys. Chem. Solids*, 55:965–973, 1994.

7. J.W.S. Rayleigh. On the instability of jets. *Proc. London Math. Soc.*, 1:4–13, 1878. Also in Rayleigh's *Collected Scientific Papers* and *Theory of Sound*, Vol. I, Dover, New York.

8. C. Herring. Some theorems on the free energies of crystal surfaces. *Phys. Rev.*, 82(1):87–93, 1951.

9. J.M. Dynys. *Sintering Mechanisms and Surface Diffusion for Aluminum Oxide*. PhD thesis, Department of Materials Science and Engineering, Massachusetts Institute of Technology, 1982.

10. J.E. Taylor. Overview No. 98. II—Mean curvature and weighted mean curvature. *Acta Metall.*, 40(7):1475–1485, 1992.

11. J.E. Taylor, C.A. Handwerker, and J.W. Cahn. Geometric models of crystal growth. *Acta Metall.*, 40(5):1443–1474, 1992.

12. A. Roosen and J.E. Taylor. Modeling crystal growth in a diffusion field using fully-faceted interfaces. *J. Computational Phys.*, 114(1):113–128, 1994.

13. W.C. Carter, A.R. Roosen, J.W. Cahn, and J.E. Taylor. Shape evolution by surface diffusion and surface attachment limited kinetics on completely facetted surfaces. *Acta Metall.*, 43(12):4309–4323, 1995.

14. J.E. Taylor, J.W. Cahn, and C.A. Handwerker. Overview No. 98. I—Geometric models of crystal growth. *Acta Metall.*, 40(7):1443–1474, 1992.

15. W.C. Carter and C.A. Handwerker. Morphology of grain growth in response to diffusion induced elastic stresses: Cubic systems. *Acta Metall.*, 41(5):1633–1642, 1993.

16. J.W. Cahn and W.C. Carter. Crystal shapes and phase equilibria: A common mathematical basis. *Metall. Trans.*, 27A(6):1431–1440, 1996.

17. J.W. Cahn, J.E. Taylor, and C.A. Handwerker. Evolving crystal forms: Frank's characteristics revisited. In R.G. Chambers, J.E. Enderby, A. Keller, A.R. Lang, and J.W. Steeds, editors, *Sir Charles Frank, OBE, FRS, An Eightieth Birthday Tribute*, pages 88–118, New York, 1991. Adam Hilger.

EXERCISES

14.1 Section 14.1.1 treated the smoothing of a sinusoidally roughened surface by means of surface diffusion to obtain Eq. 14.13. Show that the corresponding expression for smoothing by means of crystal bulk diffusion, as in Fig. 3.7, is

$$A(t) = A(0)e^{-{}^{*}D^{XL}\gamma^{S}\Omega\omega^{3}t/(kT\mathbf{f})} \tag{14.48}$$

where $\omega = 2\pi/\lambda$.

- Use the same small-slope approximations as in Section 14.1.1.

- Assume that self-diffusion occurs by a vacancy mechanism and take Eq. 13.3 as the volume diffusion equation.

- Assume that the diffusion field is in a quasi-steady state and that local equilibrium is maintained at the surface and in the volume at a long distance from the surface, where $\mu_V = 0$ and μ_A has the value characteristic of a flat surface.

- Note that one of the solutions to Laplace's equation is

$$\Phi_A = \mu_A - \mu_V = a_1 + a_2 \sin(\omega x)\, e^{\omega y} \qquad (14.49)$$

Solution. The height of the surface is given by $h = A\sin(\omega x)$ and the flux equation is given by Eq. 13.3. Therefore,

$$\frac{\partial h}{\partial t} = \vec{J}_A \cdot \hat{j}\,\Omega = -\frac{{}^{\star}D^{XL}}{kT\mathbf{f}} \left(\frac{\partial \Phi_A}{\partial y} \right)_{y=0} \qquad (14.50)$$

To evaluate Eq. 14.50 we must obtain an expression for Φ_A by solving the steady-state diffusion equation,

$$\nabla^2 \Phi_A = \frac{\partial^2 \Phi_A}{\partial x^2} + \frac{\partial^2 \Phi_A}{\partial y^2} = 0 \qquad (14.51)$$

in the volume, subject to appropriate boundary conditions. At the surface (i.e., at $y = 0$), $\mu_V = 0$, and from Eq. 3.76, $\Phi_A^S = \mu_A^\circ + \gamma^S \Omega \kappa$. In the deep interior, $\mu_V = 0$ and $\mu_A = \mu_A^\circ$, so that $\Phi_A = \mu_A^\circ$. Because

$$\kappa \cong -\frac{\partial^2 h}{\partial x^2} = A\omega^2 \sin \omega x \qquad (14.52)$$

the boundary conditions above and the diffusion equation are satisfied by a solution of the form of Eq. 14.49 with $a_1 = \mu_A^\circ$ and $a_2 = \gamma^S \Omega A\omega^2$. Therefore, using Eqs. 14.49 and 14.50,

$$\frac{\partial h}{\partial t} = -\frac{{}^{\star}D^{XL}\gamma^S \Omega}{kT\mathbf{f}} A\omega^3 \sin \omega x = -\frac{{}^{\star}D^{XL}\gamma^S \Omega}{kT\mathbf{f}} h\,\omega^3 \qquad (14.53)$$

Finally, because $(1/h)(\partial h/\partial t) = (1/A)(\partial A/\partial t)$, Eq. 14.53 may be integrated to produce Eq. 14.48.

14.2 Figure 14.13 illustrates a portion of an infinite thin plate of thickness h containing a circular hole of radius R. The plate is held at a high temperature where diffusional transport processes become active.

(a) At which specific location(s) will the shape of the plate first begin to change? Explain your reasoning in terms of driving forces for diffusion.

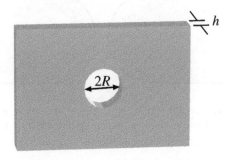

Figure 14.13:   Portion of infinite plate of thickness h containing a hole of radius R.

(b) What role do you expect the initial value of the ratio h/R to have in determining whether the hole in the plate will either shrink and disappear spontaneously or grow spontaneously? Explain your reasoning.

Solution.

(a) Equation 3.76 demonstrates that the diffusion potential of an atom at a surface depends on the local surface curvature. Consistent with the convention that a convex spherical surface has a curvature $+2/R$ (see Section 14.1), the curvature of the surface of the flat plate is zero and the initial curvature of the cylindrical surface inside the hole is $\kappa^I = 1/\infty - 1/R = -1/R$. The highest curvature is at the "rim" of the hole where the hole intersects the flat surface; the curvature here is $\kappa^I = 1/\epsilon - 1/R \to +\infty$. Therefore, there is a large diffusion-potential gradient for atoms at the rim of the hole. The first shape change would therefore be rounding of the sharp edges of the hole. The driving force for diffusion would be reduction of the total surface area, and this would commence by movement of atoms away from the rim of the hole toward both the flat plate surface and the cylindrical surface of the hole. The interior surface of the hole will continue to evolve at a slower rate, as described in part (b).

(b) Recall that curved interfaces can reduce their area by migration toward the center of curvature of the higher principal curvature. Consider two limiting cases, depicted in Fig. 14.14. Case 1, $R \gg h$: assuming complete rounding of the sharp hole edges, as in Fig. 14.14a, the curvature of the rounded hole will be $\kappa^I \approx 1/(h/2) - 1/R \approx 2/h$, and the surface tension force will cause the hole to increase in diameter. Case 2, $h \gg R$: as in Fig. 14.14b, the hole interior has curvature $\kappa^I \approx 1/(h/2) - 1/R \approx -1/R$, and the surface-tension force acts to reduce the diameter of the hole. One can make a simple calculation to investigate this problem further. Assume that the hole of diameter $2R$ lies somewhere in a fixed area A of the plate. Then the initial total surface area of the plate and hole (with sharp corners) will be

$$A^{\text{tot}} = 2A + \pi 2Rh - 2\pi R^2$$

Now the rate of change of A^{tot} with hole diameter R is

$$\frac{dA^{\text{tot}}}{dR} = 2\pi h - 4\pi R = 2\pi(h - 2R)$$

and the limiting condition for hole expansion or contraction is simply $h = 2R$.

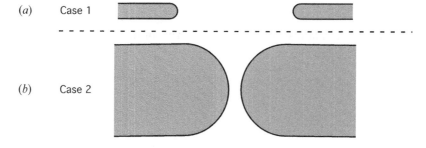

(a) Case 1

(b) Case 2

Figure 14.14:   Limiting cases of the evolving shape of a plate with a cylindrical hole, depicted in cross section. In Case 1, the hole expands; in Case 2, it will fill in.

14.3 Consider a pillbox-shaped grain embedded in an otherwise single-crystal sheet (not shown) of thickness h, as in Fig. 14.15. Such a grain will shrink and

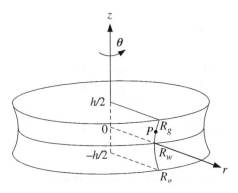

Figure 14.15:   Pillbox-shaped grain in a single-crystal sheet of thickness h.

eventually disappear. However, if grain boundary grooves develop on the two sheet surfaces and pin the boundary so that it is essentially stationary, the boundary can equilibrate locally and develop a minimum-energy form similar to that of a soap film held between two rigid circular wires.

(a) Show that such an equilibrated boundary would have the form

$$r(z) = R_w \cosh \frac{z}{R_w} \tag{14.54}$$

Here, the cylindrical coordinate system in Fig. 14.15 has been employed and R_w is the radius at the "waist" of the boundary.

(b) Calculate the force per unit length exerted on each groove by the pinned boundary when $R_g = h$. *Note:* $\beta = \cosh(\beta/2)$ has two solutions, $\beta = 1.1787$ and $\beta = 4.2536$.

(c) What happens to the grain when R_g decreases to $3/4h$? *Note:* $\alpha\beta = \cosh(\beta/2)$ has no solutions when $\alpha < 0.75$.

Solution.

(a) One way to solve this exercise is to show that the mean curvature of the boundary is zero when Eq. 14.54 is satisfied by inserting $r(z)$ into Eq. 14.29. There is then no pressure anywhere on the pinned boundary urging it to change its shape, and it possesses the shape of minimum energy. However, direct consideration of the two curvatures is instructive. Figure 14.16 shows a convenient choice for the two orthogonal planes which will be used to find the mean curvature by the method illustrated in Fig. C.2. Consider the curvature at a general point on the boundary such as P in Figs. 14.15 and 14.16. The first plane, Plane 1, selected is the constant-θ plane in Fig. 14.15, which lies in the plane of the paper in Fig. 14.16. The second plane, Plane 2 (which must be orthogonal to the first and intersect it along $\hat{n}$) is indicated by its trace, AB, in Fig. 14.16. Using Eq. C.5, the curvature of the boundary intersection with Plane 1 is

$$\kappa_1 = -\frac{\frac{d^2r}{dz^2}}{\left[1 + \left(\frac{dr}{dz}\right)^2\right]^{3/2}}$$
$$= \frac{-\cosh(z/R_w)}{R_w[1 + \sinh^2(z/R_w)]^{3/2}} = \frac{-1}{R_w \cosh^2(z/R_w)} \tag{14.55}$$

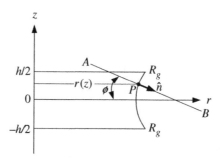

Figure 14.16:   Intersection of the pillbox-shaped grain in Fig. 14.15 with a constant-θ plane.

Plane 2 is tilted with respect to the $z = 0$ plane by the angle $\phi = \tan^{-1}(dr/dz)$. Therefore,

$$\cos\phi = \frac{1}{\sqrt{1 + \left(\frac{dr}{dz}\right)^2}} \tag{14.56}$$

The curvature of the line of intersection of the boundary with a plane parallel to the sheet surface is $\kappa_S = 1/r(z)$. The curvature of the line of intersection of the boundary with Plane 2 is then

$$\kappa_2 = \kappa_S \cos\phi = \frac{1}{r(z)} \frac{1}{\sqrt{1 + \left(\frac{dr}{dz}\right)^2}} \tag{14.57}$$

This follows from the fact that the curvature is the rate of change of the tangent vector as the curve is traversed (see Fig. C.1). Using Eq. 14.57,

$$\kappa_2 = \frac{1}{R_w \cosh(z/R_w)\sqrt{1 + \sinh^2(z/R_w)}} = \frac{1}{R_w \cosh^2(z/R_w)} \tag{14.58}$$

The mean curvature of the boundary at P is therefore $\kappa = \kappa_1 + \kappa_2 = 0$. Alternatively, Eq. C.18 can be used to demonstrate that the mean curvature is zero on the boundary.

(b) If f is the force per unit length, the work required to expand the boundary radially by dR_g is $dW = d\mathcal{G} = f\, 2\pi R_g dR_g$, where $d\mathcal{G}$ is the change in the energy of the boundary region lying between $z = 0$ and $z = h/2$. So

$$f = \frac{1}{2\pi R_g} \frac{d\mathcal{G}}{dR_g} \tag{14.59}$$

Integrating over this boundary region yields

$$\mathcal{G} = 2\pi\gamma^B \int_0^{h/2} r\sqrt{1 + (dr/dz)^2}\, dz = \frac{\pi\gamma^B}{2}\left(hR_w + R_w^2 \sinh\frac{h}{R_w}\right) \tag{14.60}$$

Using Eq. 14.54,

$$R_g = R_w \cosh\frac{h}{2R_w} \tag{14.61}$$

Because $\mathcal{G} = \mathcal{G}(R_w)$ and $R_g = R_g(R_w)$, Eq. 14.59 can be written

$$f = \frac{1}{2\pi R_g} \frac{d\mathcal{G}/dR_w}{dR_g/dR_w} \tag{14.62}$$

Using Eqs. 14.60 and 14.61 and defining $\beta \equiv h/R_w$,

$$f = \frac{\gamma^B h}{4 R_g} \frac{1 - \cosh\beta + (2/\beta)\sinh\beta}{\cosh(2/\beta) - (2/\beta)\sinh(2/\beta)} \tag{14.63}$$

When $R_g = h$, Eq. 14.61 becomes

$$\beta = \cosh(2/\beta) \tag{14.64}$$

and Eq. 14.63 becomes

$$f = \frac{\gamma^B}{4} \frac{2(1 - \beta^2) + (2/\beta)\sinh\beta}{\beta - (2/\beta)\sinh(2/\beta)} \tag{14.65}$$

The value of β in Eq. 14.65 must satisfy Eq. 14.64. The smaller solution of Eq. 14.64 for β gives the smaller boundary energy $\mathcal{G}$, and therefore putting $\beta = 1.1787$ into Eq. 14.65 gives the force per unit length acting on the groove as

$$f = 0.5296\gamma^B \tag{14.66}$$

The total force, F^{tot}, acting on the groove is

$$F^{\text{tot}} = 2\pi R_g F = 2\pi R_g (0.5296\gamma^B) = 3.3273 R_g \gamma^B \tag{14.67}$$

(c) When the groove shrinks below a critical size, no minimal solutions can be obtained for the grain boundary. Note that this happens well before the "waist" pinches down to zero. The boundary then becomes unstable. Assuming that the boundary mobility is large compared to the groove mobility, beyond the stability point, the boundary will pinch down and form two caps, which will subsequently "pop" through the top and bottom surfaces of the sheet. A circular "ghost" groove of the critical radius will be left on the surface.

14.4 Consider two faces of a faceted crystal advancing at different velocities during crystal growth as in Fig. 14.17. The growth rates of facets 1 and 2 are $\vec{v}_1$ and $\vec{v}_2$.

   (a) Find the condition on the velocities under which facet 2 will grow at the expense of facet 1.

   (b) Find the corresponding condition under which facet 1 will grow at the expense of facet 2.

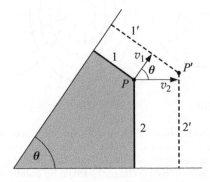

Figure 14.17: Two facets on a crystal growing at different velocities.

Solution. In general, during unit time the faceted interface advances as shown in Fig. 14.17. In the situation shown, facets 1 and 2 will both grow with time. However, a few simple constructions show that if $v_2 < v_1 \cos\theta$, facet 1 will shrink and facet 2 will grow. On the other hand, if $v_1 < v_2 \cos\theta$, facet 2 will shrink and facet 1 will grow.

14.5 Consider the growth from the vapor of a crystal possessing vicinal faces. The shape that the crystal assumes depends upon the number and rate at which ledges move across its surfaces. When one of the surfaces consists of a series of straight parallel ledges of height, h, running parallel to the z axis, the ledges move parallel to x, causing crystal growth along y. Let $k = k(x,t)$ be the local ledge density at the point x. Also, let $q = q(x,t)$ be the ledge flux equal to the number of ledges passing the point x per unit time. The local slope of the surface is then $(\partial y/\partial x)_t = -hk$ and the rate of crystal growth along y is $(\partial y/\partial t)_x = qh$. Assume that $q = q(k)$, which is often the case because the flux depends upon both the ledge density and the ledge velocity, which is, itself, dependent upon the ledge density. The rate of growth will then be a function of the inclination of the surface [i.e., $v = v(\hat{n})$], as in Section 14.2.2.

(a) Show that the moving ledges obey the equation of continuity,

$$\left(\frac{\partial k}{\partial t}\right)_x = -\left(\frac{\partial q}{\partial x}\right)_t \tag{14.68}$$

(b) Using $q = q(k)$ and the equation of continuity, show that

$$\frac{dk}{dt} = \left(\frac{\partial k}{\partial x}\right)_t \left(\frac{dx}{dt} - \frac{dq}{dk}\right) \tag{14.69}$$

(c) Now consider a point on the evolving surface where the slope is constant. Use the results above to show that this point moves along a trajectory which projects on the xy-plane as a straight line. This trajectory, called a *characteristic*, is shown in Fig. 14.18.

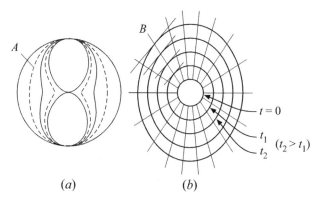

(a) $\qquad\qquad\qquad$ (b)

Figure 14.18: (a) Various polar plots of $v(\hat{n})$ for interfaces whose growth velocities, $v(\hat{n})$, are functions of their inclinations, n. (b) Shapes at increasing times of a body that was initially spherical and whose $v(\hat{n})$-plot is indicated by A in (a). Growth characteristics (outward rays) are shown which delineate the paths taken by points on the interface where the inclination remains constant. The tangent constructions along the characteristic indicated by B illustrate this constancy of inclination. From Cahn et al. [17].

Solution.

(a) In the usual way, the continuity relationship is

$$\left(\frac{\partial k}{\partial t}\right)_x = -\nabla \cdot q = -\left(\frac{\partial q}{\partial x}\right)_t \tag{14.70}$$

(b) Because $k = k(x, t)$,

$$\frac{dk}{dt} = \left(\frac{\partial k}{\partial x}\right)_t \frac{dx}{dt} + \left(\frac{\partial k}{\partial t}\right)_x \tag{14.71}$$

Because $q = q(k)$, $dq = (dq/dk)dk$, and using Eq. 14.70,

$$\left(\frac{\partial q}{\partial x}\right)_t = \frac{dq}{dk}\left(\frac{\partial k}{\partial x}\right)_t = -\left(\frac{\partial k}{\partial t}\right)_x \tag{14.72}$$

Therefore, combining Eqs. 14.71 and 14.72 yields

$$\frac{dk}{dt} = \left(\frac{\partial k}{\partial x}\right)_t \left(\frac{dx}{dt} - \frac{dq}{dk}\right) \tag{14.73}$$

(c) The evolution of the surface shape as a function of x and t is shown in Fig. 14.19. At a point where the slope is constant, k is constant and, since $q = q(k)$, q is also constant. Because $dq/dk = f(k)$, dq/dk must also be constant. Therefore, according to Eq. 14.73, $dx/dt = dq/dk = $ constant. The point of constant slope must therefore project as a straight line in the xt-plane. Now,

$$dy = \left(\frac{\partial y}{\partial x}\right)_t dx + \left(\frac{\partial y}{\partial t}\right)_x dt \tag{14.74}$$

and

$$\frac{dy}{dx} = \left(\frac{\partial y}{\partial x}\right)_t + \left(\frac{\partial y}{\partial t}\right)_x \frac{dt}{dx} = -hk + qh\frac{dt}{dx} \tag{14.75}$$

Because all the terms on the right side of Eq. 14.75 are constant, its projection in the xy-plane must therefore be a straight line.

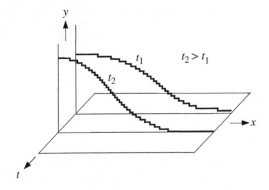

Figure 14.19: Stepped-surface evolution during crystal growth.

14.6 Prove that all of the results obtained in Exercise 14.5 for crystal growth (including the basic differential equation, its solution, and the expression for the sink efficiency) also hold for crystal evaporation.

Solution.   The basic differential equation is in a form that holds for both growth and evaporation. The diffusion boundary conditions are the same for the two cases, and therefore the solution is equally applicable. Finally, the same expression for the efficiency of the surface is obtained. (Note that the two changes of sign encountered in its derivation cancel.)

CHAPTER 15

COARSENING OF MICROSTRUCTURES DUE TO CAPILLARY FORCES

In Chapter 14 we focused on capillarity-driven processes that primarily alter the *shape* of a body. Two types of changes were considered: those driven by reduction of surface area, and those driven by altering the inclination of surfaces. In this chapter, changes in the *length scales* that characterize the microstructure are treated.

Coarsening is an increase in characteristic length scale during microstructural evolution. Total interfacial energy reduction provides the driving force for coarsening of a particle distribution. Coarsening plays an important role in microstructural evolution in two principal ways. When a particulate phase is embedded in a matrix of a second phase, flux from smaller to larger particles causes the average particle size to increase as the total heterophase interfacial energy decreases. The particles compete for solute and the larger particles have the advantage. This process degrades many material properties, depending on the presence of fine precipitates. In single-phase polycrystalline materials, larger grains tend to grow at the expense of the smaller grains as the the total grain-boundary free energy decreases. This process is also competitive and often produces unwanted coarse-grained structures.

15.1 COARSENING OF A DISTRIBUTION OF PARTICLES

15.1.1 Classical Mean-Field Theory of Coarsening

In 1961, the classical theory of particle coarsening was developed at about the same time, but independently, by Lifshitz and Slyozov [1] and Wagner [2]. Most of the

Kinetics of Materials. By Robert W. Balluffi, Samuel M. Allen, and W. Craig Carter.   **363**
Copyright © 2005 John Wiley & Sons, Inc.

theory's essential elements were worked out earlier by Greenwood [3]. This theory is often referred to as the *LSW theory* of particle coarsening and sometimes as the *GLSW theory*.

Consider a binary system at an elevated temperature composed of A and B atoms containing a distribution of spherical β-phase particles of pure B embedded in an A-rich matrix phase, α. The concentration of B atoms in the vicinity of each β-phase particle has an equilibrium value that increases with decreasing particle radius, as demonstrated in Fig. 15.1. Because of concentration differences, a flux of B atoms from smaller to larger particles develops in the matrix. This flux causes the smaller particles to shrink and the larger particles to grow.

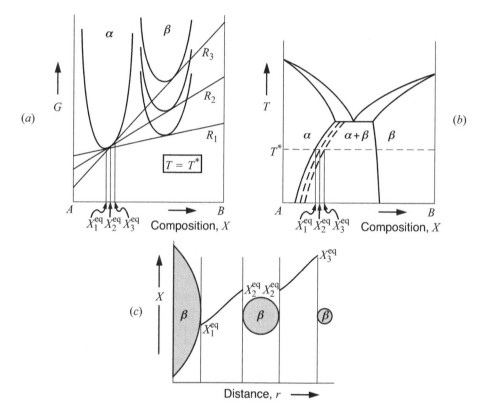

Figure 15.1:   Effect of β-phase particle size on the concentration, X^{eq}, of component B in the α phase in equilibrium with a β-phase particle in a binary system at the temperature T^*, assuming that β is pure B. **(a)** Schematic free-energy curves for α phase and three β-phase particles of different radii, $R_1 > R_2 > R_3$. The free energies (per mole) of the particles increase with decreasing radius due to the contributions of the interfacial energy, which increase as the ratio of interfacial area to volume increases. **(b)** Corresponding phase diagram. The concentration of B in the α phase in equilibrium with the β-phase particles, as determined by the common-tangent construction in (a), increases as R decreases, as shown in an exaggerated fashion for clarity. **(c)** Schematic concentration profiles in the α matrix between the three β-phase particles.

In the following, this model is used to analyze the kinetics for the two cases where the particle growth is either diffusion- or source-limited. Each of the two cases yields a different growth law for the particles in the distribution.

At any time t, a distribution of particle sizes will exist which can be quantified by defining a *particle-size distribution function*, $f(R,t)$ [units, (length)$^{-4}$], such that the number of particles per unit volume with radii between R and $R + dR$, $n(R, R + dR; t)$, is given by

$$n(R, R + dR; t) = f(R,t)\,dR \tag{15.1}$$

It is assumed that the total number of B atoms in solution remains constant during coarsening, and that a particle increases its volume by Ω as it absorbs a B atom from solution. Therefore, the total volume of the particles is constant,[1] and

$$\frac{d}{dt} \sum_{\text{part}} \frac{4\pi}{3} R^3 = 0 \tag{15.2}$$

Thus,

$$\sum_{\text{part}} R^2 \frac{dR}{dt} = 0 \tag{15.3}$$

where the sums are over all particles in the distribution.

Diffusion-Limited Coarsening.   During diffusion-limited coarsening, the heterophase interfaces surrounding the particles act as highly effective point-defect sources and sinks and maintain the concentrations of B in the α phase in their direct vicinities at the equilibrium values. The rate of coarsening is then controlled by the rate at which diffusion can take place between the particles. An approximate expression for this equilibrium concentration as a function of particle radius can be obtained by assuming that: α and β are fluid phases; β is pure component B and is an incompressible spherical particle of radius, R; and α is a dilute solution (see Section C.4.2). Using Eqs. A.4 and C.37, the concentration of B atoms in α at the α/β interface is

$$c^{\text{eq}}(R) = c^{\text{eq}}(\infty)e^{2\gamma\Omega/(kTR)} \simeq c^{\text{eq}}(\infty)\left[1 + \frac{2\gamma\Omega}{kTR}\right] \tag{15.4}$$

where $c^{\text{eq}}(\infty)$ is the solubility of B in α for a system with a planar α/β interface.

Because particles of different sizes are distributed throughout the bulk randomly, developing an exact model that couples diffusion to particle size evolution is daunting. However, a mean-field approximation is reasonable because diffusion near a spherical sink (see Section 13.4.2) has a short transient and a steady state characterized by steep concentration gradients near the surface. The particles act as independent sinks in contact with a mean-field as in Fig. 15.2.

In the mean-field approximation, each particle develops a spherically symmetric diffusion field with the same far-field boundary condition fixed by the mean concentration, $\langle c \rangle$. This mean concentration is lower than the smallest particles'

---

[1]This may not always be the case. For example, if the particles are formed by precipitation, coarsening may begin before complete precipitation has occurred [4].

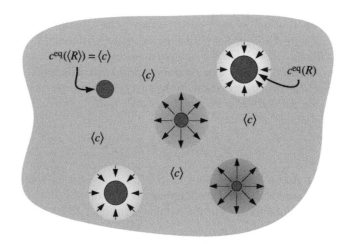

Figure 15.2: The mean-field approximation for diffusion-limited coarsening. Each particle is surrounded by a spherically symmetric diffusion field (fluxes are indicated by arrows). The concentration in the matrix at the interface of each particle is fixed by Eq. 15.4 and the concentration far-removed from particles is fixed at $\langle c \rangle$. The flux is zero near particles of average size, $\langle R \rangle$.

equilibrium concentration and higher than the largest particles' equilibrium concentration. Therefore, the large particles tend to grow and the small particles to shrink, as shown in Fig. 15.2. Using Eq. 13.22, the growth rate of a particle with radius R is

$$\frac{dR}{dt} = -\widetilde{D}\, \frac{c^{\text{eq}}(R) - \langle c \rangle}{R}\, \Omega \tag{15.5}$$

Combining Eqs. 15.3 and 15.5 gives

$$\sum_{\text{part}} R\left(c^{\text{eq}}(R) - \langle c \rangle\right) = 0 \tag{15.6}$$

Substituting the expression for $c^{\text{eq}}(R)$ into Eq. 15.6,

$$\sum_{\text{part}} R\left[\langle c \rangle - c^{\text{eq}}(\infty)\left(1 + \frac{2\gamma\Omega}{kTR}\right)\right] = 0 \tag{15.7}$$

Rearranging this equation gives

$$[\langle c \rangle - c^{\text{eq}}(\infty)] \sum_{\text{part}} R = \frac{2\gamma\Omega c^{\text{eq}}(\infty)}{kT} \sum_{\text{part}} 1 \tag{15.8}$$

On further rearrangement, Eq. 15.8 becomes

$$\langle c \rangle - c^{\text{eq}}(\infty) = \frac{2\gamma\Omega c^{\text{eq}}(\infty)}{kT\langle R \rangle} \tag{15.9}$$

where $\langle R \rangle$ is the average particle size (radius), given by

$$\langle R \rangle = \frac{\sum_{\text{part}} R}{N^{\text{tot}}} \tag{15.10}$$

where $N^{\text{tot}} = \sum_{\text{part}} 1$ is the total number of particles. By comparison with Eq. 15.4, Eq. 15.9 shows that $\langle c \rangle$ is the matrix concentration in equilibrium with particles of size $\langle R \rangle$:

$$\langle c \rangle = c^{\text{eq}}(\langle R \rangle) \tag{15.11}$$

Subtracting Eq. 15.4 from Eq. 15.9 yields

$$\langle c \rangle - c^{\text{eq}}(R) = \frac{2\gamma\Omega c^{\text{eq}}(\infty)}{kT}\left(\frac{1}{\langle R \rangle} - \frac{1}{R}\right) \tag{15.12}$$

Equation 15.5 can be combined with Eq. 15.12 to obtain

$$\frac{dR}{dt} = \frac{2\widetilde{D}\gamma\Omega^2 c^{\text{eq}}(\infty)}{kTR}\left(\frac{1}{\langle R \rangle} - \frac{1}{R}\right) \tag{15.13}$$

Therefore, when $R < \langle R \rangle$, dR/dt is negative, and when $R > \langle R \rangle$, dR/dt is positive. Equation 15.13 is an example of the results of a *mean-field theory*—the behavior of any particular particle depends only on its size compared to the mean particle size $\langle R \rangle$. Figure 15.3 presents a schematic plot of this equation for two different particle-size distributions, f_1 and f_2, such that $\langle R \rangle_2 = 1.5\langle R \rangle_1$.

The particle size for which the rate of particle-size growth will be a maximum must satisfy the condition

$$\frac{d}{dt}\left(\frac{dR}{dt}\right) = 0 = \frac{2\langle R \rangle - R_{\text{max}}}{R_{\text{max}}^3 \langle R \rangle} \tag{15.14}$$

Thus, the maximum rate of particle size increase occurs at $R_{\text{max}} = 2\langle R \rangle$.

Several qualitative observations may be made about the time dependence of the particle-size distribution. As predicted by Eq. 15.14, the growth of the largest particles will be slow and particles with smaller radii (but greater than $\langle R \rangle$) will grow more quickly, shifting the distribution of particles toward $2\langle R \rangle$. However, the average radius, $\langle R \rangle$, is an increasing function of time. Ultimately, the tail of the distribution at large R is expected to diminish with time and particles with $R \gg \langle R \rangle$ will be rarely observed. Figure 15.4 depicts this expected shift of the distribution.

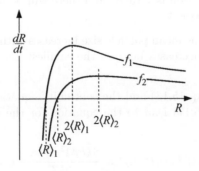

Figure 15.3:   Particle growth rate vs. particle size in Eq. 15.13. f_1 and f_2 represent the growth rate for two different particle-size distributions, such that $\langle R \rangle_2 = 1.5\langle R \rangle_1$.

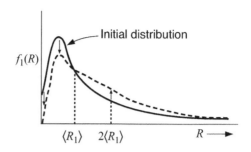

Figure 15.4: Initial distribution, $f_1(R)$, changing with time according to the growth law, Eq. 15.13. The smallest particles disappear with relatively large shrinkage rates, so the lower end of the distribution collapses to zero.

The time dependence of the particle-size distribution can be studied analytically by developing a differential equation based on the flux of particles that occurs in particle-size space as the distribution evolves. The flux of particle density passing the size R in this space is

$$J = f(R,t)\frac{\partial R}{\partial t} \tag{15.15}$$

The accumulation equation is then

$$\frac{\partial f(R,t)}{\partial t} = -\frac{\partial J}{\partial R} = -\frac{\partial}{\partial R}\left[f(R,t)\frac{\partial R}{\partial t}\right] \tag{15.16}$$

Equation 15.16 is the rate of change of f in the interval $(R, R+dR)$, and is related to the difference between the rates of particles entering from below and exiting the interval at $R + dR$. Combining Eqs. 15.13 and 15.16 gives

$$\frac{\partial f(R,t)}{\partial t} = -\frac{2\widetilde{D}\gamma\Omega^2 c^{\mathrm{eq}}(\infty)}{kT\langle R\rangle}\frac{\partial}{\partial R}\left[\frac{(R - \langle R\rangle)\,f(R,t)}{R^2}\right] \tag{15.17}$$

If an initial form of $f(R, t = 0)$ is assumed (e.g., a Gaussian distribution), it is possible to compute the form of $f(r,t)$ at later times using Eq. 15.17. Solutions of this type yield the following results [1, 2]:

- A steady-state (normalized) distribution function is approached asymptotically as $t \to \infty$. This steady-state distribution, illustrated in Fig. 15.5, is approached by all initial distributions. The most frequent particle size in the steady-state distribution is $1.13\langle R\rangle$ and there will be no particles larger than $1.5\langle R\rangle$, *the cut-off size*.

- During annealing, the mean particle size increases with time, and the number of particles, N^{tot}, decreases because the smallest particles disappear as the larger ones grow.

- The growth law (Eq. 15.13) and the continuity equation for the particle-size distribution (Eq. 15.17) lead to the equation for the evolution of the mean particle size:

$$\langle R(t)\rangle^3 - \langle R(0)\rangle^3 = \frac{8\widetilde{D}\gamma\Omega^2 c^{\mathrm{eq}}(\infty)}{9kT}\,t = K_D\,t \tag{15.18}$$

where K_D is the rate constant for diffusion-limited coarsening. See Exercise 15.2.

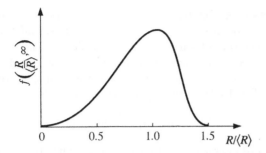

Figure 15.5:   Final steady-state normalized particle-size distribution for diffusion-limited coarsening.

Experimental Observations.   Generally, two quantities are measured in experimental studies of particle coarsening [5]. First, the mean particle size is studied as a function of time. For volume-diffusion-limited coarsening, the $t^{1/3}$-*law* corresponding to Eq. 15.18 is generally observed, in agreement with theoretical predictions. The second measured characteristic is the particle-size distribution, including its time dependence. The experimental time-dependent evolution of particle-size distributions does not always match predictions from classical coarsening theory [5, 6]; the distributions observed are generally broader than the classical theory predicts and particles are often larger than the predicted cut-off size, $1.5\langle R \rangle$.

A study of coarsening in semisolid Pb–Sn alloys verified the $t^{1/3}$-law kinetics predicted by the mean-field theory (see Fig. 15.6). However, some aspects of the classical theory are not observed in Fig. 15.6. Limitations of the classical mean-field theory are discussed in Section 15.1.2.

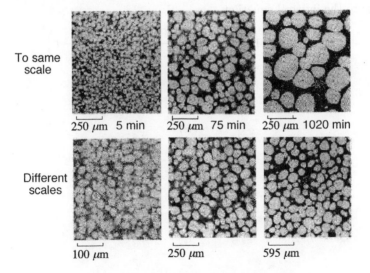

Figure 15.6:   Particle distributions observed in coarsening experiments on semisolid Pb–Sn alloys. The volume fraction of particles is 0.64. The upper row shows a steady increase in mean particle size with aging time. The lower row is scaled so that the apparent mean particle size is invariant—demonstrating that the particle distribution remains essentially constant during coarsening. From Hardy and Voorhees [7].

Source-Limited Coarsening. During source-limited coarsening, the interfaces surrounding the particles behave as poor sources and sinks, and the coarsening rate then depends upon the rate at which the diffusion fluxes between the particles can be created or destroyed (accommodated) at the particle interfaces. In a simple model, the same assumptions can be made about the source action at the particles as those that led to Eq. 13.24. The rate of particle growth can then be written

$$\frac{dR}{dt} = K\left[c(R) - c^{\mathrm{eq}}(R)\right]\Omega \tag{15.19}$$

where K is the same type of rate constant as in Eq. 13.24. If diffusion in the matrix between the particles is so rapid, or K is so small, that all gradients in composition are essentially eliminated on a short time scale compared to the growth time of the particle, $c(R) = \langle c \rangle$ and the rate-limiting process is the source action.

An expression for the average composition $\langle c \rangle$ can be obtained by first combining the growth law, Eq. 15.19, with Eq. 15.3 and assuming that the concentration field is uniform ($\langle c \rangle = c(R)$; see Fig. 15.7):

$$\sum_{\text{part}} R^2 (c^{\mathrm{eq}}(R) - \langle c \rangle) = 0 \tag{15.20}$$

Then, using Eq. 15.4,

$$\langle c \rangle = c^{\mathrm{eq}}(\infty)\left(1 + \frac{2\gamma\Omega}{kT}\frac{\langle R \rangle}{\langle R^2 \rangle}\right) \tag{15.21}$$

Equations 15.19 and 15.21 lead to an expression for the growth rate of any particle in a particular size distribution:

$$\frac{dR}{dt} = \frac{2Kc^{\mathrm{eq}}(\infty)\gamma\Omega^2}{kT}\left(\frac{\langle R \rangle}{\langle R^2 \rangle} - \frac{1}{R}\right) \tag{15.22}$$

This leads to the integral equation,

$$R^2(t) - R^2(0) = \frac{4Kc^{\mathrm{eq}}(\infty)\gamma\Omega^2}{kT}\left(\int_0^t \frac{\langle R \rangle R}{\langle R^2 \rangle}\,dt - t\right) \tag{15.23}$$

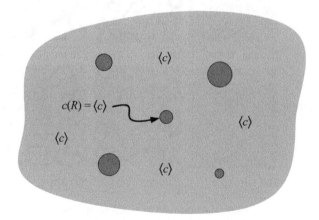

Figure 15.7: Solute concentration field for source-limited kinetics. The matrix concentration field is essentially uniform and $c = \langle c \rangle$ everywhere, including the regions adjacent to the particles where $c(R) = \langle c \rangle$.

Equation 15.23 can be averaged and Wagner derives the solution,

$$\langle R^2(t) \rangle - \langle R^2(0) \rangle = \frac{64Kc^{\text{eq}}(\infty)\gamma\Omega^2}{81kT}\, t = K_S t \tag{15.24}$$

where K_S is the rate constant for source-limited coarsening [2].

15.1.2 Beyond the Classical Mean-Field Theory of Coarsening

The mean-field theory has a number of shortcomings, including the approximations of a mean concentration around all particles and the establishment of spherically symmetric diffusion fields around every particle, similar to those that would exist around a single particle in a large medium. The larger the particles' total volume fraction and the more closely they are crowded, the less realistic these approximations are. No account is taken in the classical model of such volume-fraction effects. Ratke and Voorhees provide a review of this topic and discuss extensions to the classical coarsening theory [8].

Work by Voorhees and Glicksman concludes that the classical theory is correct in the limit of zero volume fraction of the coarsening phase and that both the kinetics and the size distributions are significantly dependent on the precipitate volume fraction, ϕ [9–12]. The temporal law for diffusion-limited coarsening, given by Eq. 15.18, remains valid for all volume fractions, but the rate constant K_D is a monotonically increasing function of ϕ, as in Fig. 15.8.

Volume-fraction effects on particle coarsening rates have been observed experimentally. For comparisons between theory and experiment, data from liquid+solid systems are far superior to those from solid+solid systems, as the latter are potentially strongly influenced by coherency stresses. Hardy and Voorhees studied Sn-rich and Pb-rich solid phases in Pb–Sn eutectic liquid over the range $\phi = 0.6$–0.9 and presented data in support of the volume-fraction effect, as shown in Fig. 15.9 [7].

Voorhees's experimental study of low-volume-fraction-solid liquid+solid Pb–Sn mixtures carried out under microgravity conditions during a space shuttle flight enabled a wider range of solid-phase volume fractions to be studied without significant influence of buoyancy (flotation and sedimentation) effects [13]. The rate of approach to the steady-state particle-size distribution in 0.1–0.2 volume-fraction

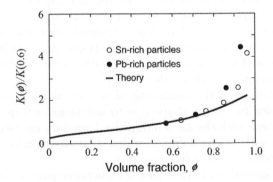

Figure 15.8: Rate constant $K_D(\phi)$ for particle coarsening vs. volume fraction of particles, expressed as a ratio of the rate constant $K_D(\phi)$ at volume fraction ϕ to that at $\phi = 0.6$. From Hardy and Voorhees [7].

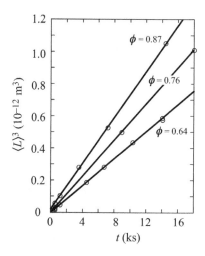

Figure 15.9:   Coarsening data from experiments on semisolid Pb–Sn alloys, showing the effect of volume fraction ϕ on the rate constant for coarsening, which is proportional to the slope of the curves. $\langle L \rangle$ is the mean linear intercept of the coarsening particles, measured on a random two-dimensional section of the microstructure. From Hardy and Voorhees [7].

solid specimens was very slow. The initial stages of coarsening involved significant solid-particle clustering that accelerated the coarsening kinetics compared to a system with a random spatial arrangement of particles.

Another complication is that the particles interact elastically, causing shape changes and other effects. Ardell and Nicholson were first to realize that elastic energy could influence particle distributions during coarsening [14]. Their study reported the formation of *modulated* (aligned) precipitate structures in Ni–Al alloys after prolonged aging, thereby demonstrating that alignment of coherent misfitting precipitates along the cube axes of a material could lower its elastic energy.

Another effect of elastic energy, described as *stress coarsening*, is seen in systems with coherent misfitting precipitates aged under an applied load, as studied originally by Tien and Copley [15]. In a nickel-based superalloy, single crystals coarsen when subjected to an applied uniaxial load, and precipitates form either platelets or rods, depending on whether the load is tensile or compressive. In the absence of load, the particles are cuboidal. This phenomenon, known as *rafting*, may allow the production of structural alloys with improved creep resistance. Even for a single coherent particle in a matrix, elastic anisotropy, elastic inhomogeneity, the applied stress state, and the stress-free transformation strain of the precipitates all influence the equilibrium precipitate shapes [16].

Numerical simulations of the coarsening of several particles are now possible, allowing the particles to change shape due to diffusional interparticle transport in a manner consistent with the local interphase boundary curvatures [17]. These studies display interparticle translational motions that are a significant phenomenon at high volume fractions of the coarsening phase.

Examples which show that elastic interactions between particles can stabilize distributions of fine particles against coarsening include the observation by Miyazaki et al. that large particles in aged Ni–Si and Ni–Al alloys split into pairs and even octets of smaller particles as aging proceeds [18]. In such cases, the reduction in elastic energy achieved by the splitting outweighs the attendant increase in interphase boundary area. Johnson has given a theoretical treatment of the coarsening in a two-precipitate system under applied load [19].

15.2 GRAIN GROWTH

Grain growth is a kinetic process by which the average grain size of a polycrystalline material increases during annealing at an elevated temperature. In this process (illustrated in Fig. 15.10), the larger grains tend to increase in size while the smaller grains shrink and eventually disappear, causing the total number of grains to decrease and the average grain size to increase. The driving force is the decrease in the total interfacial energy of the system that accompanies the growth.

The construction of realistic models for this process, which include detailed information about the distribution of the evolving grains' sizes and shapes, has proven to be elusive and has been reviewed [20, 21]. Grain growth in two-dimensional polycrystals (e.g., polycrystalline sheets with columnar grain structures) is taken up first because it is considerably simpler than growth in three dimensions. For two-dimensional growth, there are powerful topological rules, and considerable progress has been made in developing detailed models for the process. Following this, grain growth in three dimensions is considered. Here, relatively little success has been met in developing physical models, and progress has been achieved mainly by using computer simulation.

Grain growth in polycrystalline materials is similar in many respects to the growth of bubbles in froths or foams, to magnetic domain coarsening, and to the growth of cells in many types of cellular materials. Concepts and results developed for grain growth therefore also apply to these phenomena.

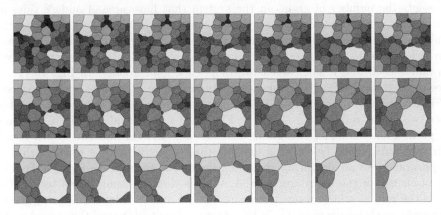

Figure 15.10: Simulation of isotropic and uniform grain growth in two dimensions. The $(N - 6)$ rule is obeyed for all grains except those intersecting the edge of the domain. Calculation using Surface Evolver [22]. Courtesy of Ellen J. Siem.

15.2.1 Grain Growth in Two Dimensions

Topology of Two-Dimensional Polycrystals. In two dimensions, three (and only three) grains meet at every grain boundary vertex, as in Fig. 15.10. Vertices where larger numbers of boundaries come together are unstable: for example, a conceivable four-grain vertex would immediately decompose into two three-grain vertices connected by a new boundary segment. It is expected that the three boundaries at each vertex will be locally equilibrated with respect to one another; when all boundary energies are isotropic and equal, this equilibration requires that the net capillary force exerted by the three boundaries on the vertex be zero. This is accomplished when the boundaries meet at equal 120° angles.

An important geometrical growth parameter is the average number of sides per grain, $\langle N \rangle$, in the ensemble, which can be determined with Euler's theorem, which states that

$$N^G - N^B + N^V = 1 \tag{15.25}$$

where N^G is the number of grains, N^B is the number of boundary segments, and N^V is the number of vertices in the system. If there are N^V vertices, the number of boundary segments must be

$$N^B = \frac{3}{2} N^V \tag{15.26}$$

Each boundary segment provides a side for two grains, therefore $\langle N \rangle = 2N^B/N^G$. Since N^G, N^B, and N^V in Eq. 15.25 are all much greater than unity, $N^G + N^V = N^B$ to a good approximation, and, using Eq. 15.26, $N^B = 3N^G$. Therefore,

$$\langle N \rangle = \frac{2N^B}{N^G} = 6 \tag{15.27}$$

Model for Two-Dimensional Grain Growth. During growth, the distribution of the sizes and shapes of grains evolves with time. The changes in the overall structure can then be described adequately by a distribution function $F(A, N, t)$, which represents the number of grains in the system that have area A and N sides at time t. This distribution evolves when grains grow or shrink without changing the number of their sides or discontinuously change the number of their sides by means of switching events, as illustrated in Fig. 15.11. In Fig. 15.11a, two vertices meet, two grains each lose a side, and two grains each gain a side. In Fig. 15.11b, a three-sided grain shrinks to disappearance and all three grains lose one side apiece. In Fig. 15.11c, a four-sided grain disappears, two grains each lose one side, and the other two grains retain their sides. In Fig. 15.11d, a five-sided grain disappears, one grain gains one side, two grains lose one side apiece, and two grains retain their sides. As shown below, it is unlikely that grains with $N \geq 6$ will shrink to disappearance, so all significant switching events are included in Fig. 15.11. The rate of change of the distribution, $F(A, N, t)$, is then the sum of two terms: the first is the rate at which grains possessing area A and N sides accumulate due to the growth or shrinkage of grains in the absence of switching, and the second is the

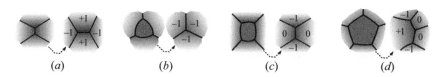

(a)   (b)   (c)   (d)

Figure 15.11:   Switching events during two-dimensional grain growth. From Thompson [21].

rate at which they accumulate due to the switching of grain sides. To determine the first term, an expression is needed for the rate of growth or shrinkage of a single grain in the absence of switching.

Growth or Shrinkage of a Single Grain in the Absence of Switching. A simple expression can be obtained if it is assumed that all boundaries are isotropic and of the same energy, γ. Consider first the single isolated interior grain shown in Fig. 15.12a. According to Eq. C.17, any portion of its boundary will be subjected to a normal pressure directed toward its concave side and given by $P = \gamma \kappa$ (see the arrows in Fig. 15.12a). Following Eq. 13.7, we assume that the resulting boundary velocity will vary linearly with the pressure according to

$$v = M_B P = M_B \gamma \kappa \tag{15.28}$$

where M_B is a mobility linking velocity with pressure. If the local velocities of the boundaries are integrated along the perimeter of a grain, an expression for the rate of change of area of the interior grain is obtained:

$$\frac{dA}{dt} = -\int_{GB} v \, ds = -\int_{GB} M_B \gamma \kappa \, ds \tag{15.29}$$

Assuming that M_B and γ and do not depend on orientation and curvature,

$$
\begin{aligned}
\frac{dA}{dt} &= -M_B \gamma \int_{GB} \kappa \, ds \\
&= -M_B \gamma \int_{GB} \frac{d\theta}{ds} ds \\
&= -2\pi M_B \gamma
\end{aligned}
\tag{15.30}
$$

In the relationships above, θ is the angle that the boundary normal makes with a fixed direction in the plane of the specimen. Because the curvature is the rate of change of the boundary normal as the line integral is carried out, $\kappa = d\theta/ds$. Also, θ varies between 0 and 2π in the integration, because the normal rotates by 2π as the boundary is traversed. Therefore, independent of the shape of the grain, Eq. 15.30 becomes

$$\frac{dA}{dt} = -\text{constant} \tag{15.31}$$

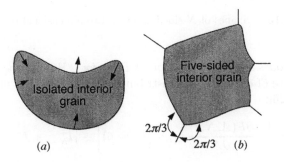

(a)   (b)

Figure 15.12: (a) Two-dimensional isolated grain. Arrows indicate local pressures acting on the boundary. (b) Five-sided grain in polycrystalline material with isotropic grain-boundary energies.

This result holds even if mass is leaving the grain along some segments of the boundary and entering the grain along other segments, as in Fig. 15.12*a*.

Now consider that the grain is one of many in a two-dimensional polycrystal, as in Fig. 15.12*b*. The rate of area change is again given by Eq. 15.29, but now the change in boundary orientation has discontinuities in $d\theta/ds$ at the vertices. If the grain has N sides, and there are therefore N vertices, the integral can be written as a sum of contributions from each boundary segment:

$$\frac{dA(N)}{dt} = -M_B\gamma\left(\int_{\text{seg }1} d\theta + \int_{\text{seg }2} d\theta + \cdots + \int_{\text{seg }N} d\theta\right)$$

$$= -M_B\gamma(2\pi - N\,\Delta\theta) \tag{15.32}$$

where $\Delta\theta$ is the change in the exterior angle at each vertex:

$$\Delta\theta = \theta_{\text{begin}}(\text{segment } i+1) - \theta_{\text{end}}(\text{segment } i)$$

Assuming that the boundaries are locally equilibrated at each vertex, the angles between boundaries at each vertex are $2\pi/3$, so $\Delta\theta = 4\pi/3 - \pi$. Therefore,

$$\frac{dA(N)}{dt} = M_B\gamma\frac{\pi}{3}(N - 6) \tag{15.33}$$

and grains with more than six sides will grow and those with fewer than six will shrink. This remarkable result of Mullins [23] generalizes an earlier result for the $(N - 6)$-*rule* by Smith and von Neumann [24]. The growth rate depends only on grain topology and not on grain size. Grains with more than six sides grow as shown in Fig. 15.13, because the fixed angles at each vertex force the boundary segments to be concave away from the grain center. The pressure due to this curvature therefore induces growth. Grains with fewer than six sides show the opposite behavior, whereas grains with six sides possess flat sides and are static.

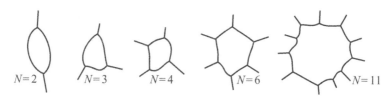

$N=2$ $N=3$ $N=4$ $N=6$ $N=11$

Figure 15.13:   Shapes of N-sided grains in two-dimensional polycrystals.

Completion of Model and Description of Its Behavior.   Having the result given by Eq. 15.33, the rate of change of the distribution, $F(A, N, t)$, can now be expressed in the form [25, 26]:

$$\frac{\partial F(A, N, t)}{\partial t} = \left[-\frac{\partial F(A, N, t)}{\partial A} M_B\gamma\frac{\pi}{3}(N - 6)\right] + \left[I_{N-1}^+ + I_{N+1}^- - I_N^+ - I_N^-\right] \tag{15.34}$$

The first term in brackets is the rate at which grains of area A and sides N accumulate in *area space* due to the growth or shrinkage of grains as described by Eq. 15.33. The second is the corresponding rate of accumulation of such grains due to switching. In this nomenclature, I^{+}_{N-1}, for example, is the rate at which grains of area A with $N-1$ sides gain a side by switching to become grains of area A with N sides.

The development of exact expressions for the switching terms in Eq. 15.34 is difficult and has not been achieved. Marder obtained approximate expressions by applying some simple rules that ignored spatial correlations [25]. For example, when a four-sided grain disappears, four grains are picked at random from the distribution and a side is removed from the two having the smallest areas. When a five-sided grain disappears, the grain that gains a side is the one that has the largest area. Using several additional approximations to obtain expressions for the switching terms and the rule that $\langle N \rangle = 6$, Marder solved Eq. 15.34 numerically to obtain the results shown in Fig. 15.14. The figure also contains experimental measurements of bubble growth in two-dimensional soap froths for purposes of comparison.

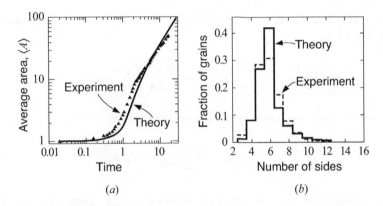

(a)   (b)

Figure 15.14:   (a) Average area of grains vs. time. (b) Fraction of grains possessing N sides. Figures compare Marder's theoretical results with experimental measurements of bubble growth in soap froths made by Glazier et al. From Marder [25] and Glazier et al. [27].

The agreement between theory and experiment is reasonably good, indicating that the model captures the essential physics of the phenomena. An important result, apparent in Fig. 15.14a, is that, after an initial transient, the model predicts a linear increase in the average grain area with time:

$$\langle A \rangle \propto t \tag{15.35}$$

As discussed below, this is expected if Eq. 15.33 applies and if statistical self-similarity holds during the growth process.[2] This work is reviewed by Thompson [21].

Computer Simulation of Two-Dimensional Grain Growth.   Because of the great difficulties in developing rigorous models for grain growth, computer simulation has been used to obtain more information about the details of two-dimensional grain growth [21]. A wide variety of techniques is employed, including, among others,

---

[2]Statistical self-similarity holds when the normalized grain-size distribution and the number-of-sides distribution remain constant during growth. A grain structure at a later time then looks statistically similar to itself at an earlier time, except for a uniform magnification, and the structures are therefore scaled with respect to each other.

Monte Carlo methods, direct tracking methods, and vertex models, where the evolution of the two-dimensional grain structure is described in terms of the motion of the vertices. After initial transients, all of these simulations exhibit statistical self-similarity during growth and an average grain area that increases linearly with time according to Eq. 15.35.

Further Aspects. The constant growth rate of the average grain area found above (Eq. 15.35) is to be expected when statistical self-similarity holds and individual grains grow or shrink according to Eq. 15.33 [28]. Consider a large area, A, containing G grains. Then $\langle A \rangle = A/G$ and $\langle A^2 \rangle = (1/G) \sum_{i=1}^{G} a_i^2$, where a_i is the area of the ith grain, and the sum is over all grains. The latter expression can also be written

$$\langle A \rangle \frac{\langle A^2 \rangle}{\langle A \rangle^2} = \frac{1}{A} \sum_{i=1}^{G} a_i^2 \qquad (15.36)$$

Differentiating Eq. 15.36, with $\langle A^2 \rangle / \langle A \rangle^2 = $ constant because of self-similarity, and using Eq. 15.33,

$$\frac{\langle A^2 \rangle}{\langle A \rangle^2} \frac{d\langle A \rangle}{dt} = \frac{1}{A} \sum_{i=1}^{G} 2 a_i \frac{da_i}{dt} = \frac{2\pi M_B \gamma}{3A} \sum_{i=1}^{G} a_i (N_i - 6) \qquad (15.37)$$

where N_i is the number of sides of the ith grain. However,

$$\sum_{n=1}^{G} a_i (N_i - 6) = \sum_{N=2}^{\infty} (A_N N - 6 A_N) \qquad (15.38)$$

where A_N is the total area occupied by N-sided grains. Note that the second sum begins with $N = 2$, because one-sided grains do not exist. Therefore, using Eq. 15.38 in Eq. 15.37 yields

$$\frac{d\langle A \rangle}{dt} = \frac{2\pi M_B \gamma}{3} \frac{\langle A \rangle^2}{\langle A^2 \rangle} \sum_{N=2}^{\infty} \frac{A_N}{A} (N - 6) = C = \text{constant} \qquad (15.39)$$

Here, A_N/A, the fraction of the total area occupied by N-sided grains, is constant because of self-similarity.

It is often convenient to express the growth law given by Eq. 15.39 in terms of a linear grain-size dimension. If $R_{\rm rms}$ is the effective root-mean-square radius of the roughly equiaxed grains,

$$\frac{d\langle A \rangle}{dt} = \frac{d}{dt} \left(\pi R_{\rm rms}^2 \right) = C \qquad (15.40)$$

and, therefore, upon integrating,

$$R_{\rm rms}^2(t) - R_{\rm rms}^2(0) = \frac{C}{\pi} t \qquad (15.41)$$

The growth laws expressed by Eqs. 15.35 and 15.41 are consistent with the behavior of a wide variety of experimental systems and models as well as computer simulation [21].

Finally, it should be noted that during grain growth in thin polycrystalline films, two-dimensional grain growth may be seriously impeded if surface grooves form at the intersections of the boundaries and the specimen surface(s) as described in Section 14.1.2 [21, 29]. The boundaries can then become trapped at these grooves leading to growth stagnation (see Exercise 14.3).

15.2.2   Grain Growth in Three Dimensions

Topological Aspects.   The geometry of grain growth in three dimensions is more difficult to visualize and to analyze than in two dimensions, and only limited success has been achieved. Figure 15.15 shows the relatively simple structure of an idealized polycrystal in which all the grains are identical. Each grain is a tetrakaidecahedron (possessing 14 faces) arranged on a b.c.c. lattice. The three-dimensional grain structure can be described in terms of grains, boundaries, edges, and vertices. Boundaries are the interfaces where two grains meet, edges are the lines along which three grains meet, and vertices are the points where four edges (and also four grains) meet.

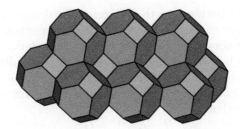

Figure 15.15:   Three-dimensional polycrystal with all grains identical. Each grain is a tetrakaidecahedron with 6 square and 8 hexagonal faces.

When the boundaries are equilibrated with respect to one another locally at all edges and vertices, the net capillary force exerted on each junction will be zero. When γ^B is isotropic and the same for all boundaries, this condition is satisfied when there is trihedral symmetry along each edge and tetrahedral symmetry at each vertex [i.e., at three-boundary junctions, two faces have a dihedral angle $\cos^{-1}(-1/2) = 120°$, and at four-edge junctions, each pair of edges makes an angle $\cos^{-1}(-1/3) = 109°28'$]. It is expected that these local equilibrium symmetry conditions, known as *Plateau's laws*, will be closely satisfied during grain growth.[3]

However, Plateau's laws are not obeyed by the grain boundary structure in the special polycrystal in Fig. 15.15. To achieve local equilibrium at all grain junctions, so that Plateau's laws are obeyed, the faces of each grain must be curved [31]. A boundary structure in which all junctions obey Plateau's laws is presented in Fig. 15.16, which shows a polycrystal consisting of six grains that meet at four vertices. Each grain fully occupies one face of the polycrystal. As in Fig. 15.15,

---

[3]These conditions are Young's equation, $\gamma_1/\sin\phi_1 = \gamma_2/\sin\phi_2 = \gamma_3/\sin\phi_3$, which requires all grain-boundary angles to be $2\pi/3$ for the uniform boundary-energy case and the requirement that quad- and higher-order junctions are unstable in two dimensions [30].

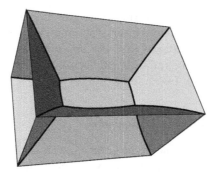

Figure 15.16:   An orthorhombic-shaped polycrystal composed of six grains that meet at four vertices. Each grain occupies one face of the polycrystal. Each vertex is shared by four grains. Three grains meet at each edge (triple line). Four edges meet at each vertex. All edges and vertices are in local equilibrium and the system obeys Plateau's laws. All boundary energies are isotropic and equal. Figure was computed with Surface Evolver [22].

three grains meet at each edge, four grains meet at each vertex, and four edges meet at each vertex.

The grain structure during growth will generally consist of a distribution of grains possessing a range of sizes (volumes) and various numbers of faces and edges. Plateau's laws will be obeyed at all junctions. To characterize the distribution of grain shapes and sizes in a large three-dimensional polycrystalline system, we take v, e, and f as the volume, number of edges, and number of faces, respectively, possessed by any particular grain. If E is the number of grain edges in the system, Euler's theorem states that

$$B - E + V = G \tag{15.42}$$

Each boundary in the system provides two faces for grains. Therefore, the average number of faces per grain, $\langle f \rangle_{\mathrm{syst}}$, is

$$\langle f \rangle_{\mathrm{syst}} = \frac{2B}{G} \tag{15.43}$$

where the symbol $\langle \bullet \rangle_{\mathrm{syst}}$ indicates an average over the system. Because three grains meet at each edge, each edge in the system provides three edges for grains. Therefore, the average number of edges per grain, $\langle e \rangle_{\mathrm{syst}}$, is

$$\langle e \rangle_{\mathrm{syst}} = \frac{3E}{G} \tag{15.44}$$

Also, because four edges emanate from each vertex and every edge connects two vertices,

$$E = \frac{4V}{2} = 2V \tag{15.45}$$

By combining Eqs. 15.42–15.45,

$$\langle e \rangle_{\mathrm{syst}} = 3 \langle f \rangle_{\mathrm{syst}} - 6 \tag{15.46}$$

Because each edge on a grain provides two edges for faces on that grain, the average number of edges per face on grains is $\langle e_f \rangle_{\mathrm{syst}} = 2 \langle e \rangle_{\mathrm{syst}} / \langle f \rangle_{\mathrm{syst}}$. Combining this result with Eq. 15.46 yields

$$\langle f \rangle_{\text{syst}} = \frac{12}{6 - \langle e_f \rangle_{\text{syst}}} \tag{15.47}$$

It is readily verified by inspection that Eqs. 15.46 and 15.47 are obeyed for the grain structure in Fig. 15.15.

Rate of Grain Growth.[4]   The $(N - 6)$-rule for two-dimensional grain growth in a system with isotropic and uniform boundary energies (Eq. 15.33) is an exact example of a principle which states that *equivalent* grains grow at identical rates. Equivalent grains are grains that belong to an *equivalence class* (i.e., grains having at least one topological feature in common). In the two-dimensional case, the equivalence class consists of grains with the same number of sides, N, and grains with the same number of sides grow at the same rate. The $(N-6)$-rule's derivation is simple because there are no out-of-plane contributions to the boundary curvature.[5]

No known simple equivalence class describes grain growth in three dimensions. Although no exact growth law for an equivalence class exists, approximate laws can be found that are obeyed on average for grains of the same equivalence class. Any particular grain will obey a growth law that depends on the curvature of its boundaries, subject to Plateau's laws, as well as a statistical growth law applicable for all members of its equivalence class. In two-dimensional growth, the particular growth law is identical to the statistical law when the topological feature defining the class is chosen to be the number of grain sides, N.

For three-dimensional growth, a statistical growth law can be justified if v and e are chosen as the characteristic topological features that define the equivalence class. The choice of e is favored because all edges terminate at vertices and have similar topologies. For symmetric grains of volume v and number of faces f, the average growth rate, $\langle dv/dt \rangle_{\{v,f\}}$, varies as $\sqrt{f}$ as f becomes large [32]. Here, the symbol $\{v, f\}$ indicates an average over all grains in the equivalence class possessing volume v and number of faces f. Equation 15.46, where $\langle e \rangle_{\text{syst}}$ is approximately proportional to $\langle f \rangle_{\text{syst}}$ for large f and e, indicates possibly similar behavior on average if e is chosen rather than f, and this suggests the expansion

$$\left\langle \frac{dv}{dt} \right\rangle_{\{e,v\}} = \left(\frac{dv}{dt} \right)_0 (aE^{1/2} + b + cE^{-1/2} + \cdots) \tag{15.48}$$

where $\langle dv/dt \rangle_{\{e,v\}}$ is the average volumetric growth rate of grains belonging to the equivalence class $\{e, v\}$ and the quantity $(dv/dt)_0$ is the shrinkage rate of a sphere of the same volume embedded in a matrix grain. The undetermined coefficients, $a, b, c, \ldots$, have no explicit dependence on e. To obtain the behavior for $e \gg 1$, the third- and higher-order terms are neglected, and in analogy with soap froths in two dimensions, b is set equal to 1. This represents an offset from the mean pressure of the froth. By enforcing conservation of total volume by summing over all classes,

$$\sum_{\{e,v\}} \frac{dv}{dt} = 0 \tag{15.49}$$

[4]We are grateful to David T. Wu for content and sharing his recent results in this section.
[5]The Gauss–Bonnet theorem, which relates integrals of Gaussian curvature ($1/(R_1 R_2)$ in three dimensions) over a surface to integrals of mean curvature ($1/R_1 + 1/R_2$ in three dimensions) over boundaries of the surface, is particularly simple in two dimensions. In two dimensions, the $(N - 6)$-rule *is equivalent* to the Gauss–Bonnet theorem.

A statistical growth law follows for grains with large edge number, e, given by

$$\left\langle \frac{dv}{dt} \right\rangle_{\{e,v\}} = \left[\frac{dv}{dt} \right]_0 \left(\sqrt{\frac{e}{e_{\text{crit}}}} - 1 \right) \tag{15.50}$$

where the critical edge number, e_{crit}, is given by

$$e_{\text{crit}} = \left(\frac{\sum_{\{e,v\}} v^{1/3} \sqrt{e}}{\sum_{\{e,v\}}} \right)^2 \tag{15.51}$$

Comparisons to direct simulations of grain growth by Wu et al. show that Eq. 15.50 is correct for small as well as large e [33, 34]. Unlike two-dimensional growth, the critical topological feature controlling the growth rate (in this case e rather than N) is not constant but depends upon the state of the grain-boundary structure.

Little progress has been made in analyzing the complicated problem of grain switching in three dimensions. In view of this, extensive recourse has been made to computer simulation. This work requires extensive computer power, and the results have not been as definitive as for two-dimensional grain growth. Evidence for statistical self-similarity has been obtained, and reasonable agreement with experiment has been found in certain cases. In experiments and in simulations, agreement with Eq. 15.41 has been found (see Thompson's review [21]).

Bibliography

1. I.M. Lifshitz and V.V. Slyozov. The kinetics of precipitation from supersaturated solid solution. *J. Phys. Chem. Solids*, 19(1–2):35–50, 1961.

2. C. Wagner. Theorie der Alterung von Niederschlagen durch Umlosen (Ostwaldreifung). *Z. Elektrochem.*, 65(7–8):581–591, 1961.

3. G.W. Greenwood. The growth of dispersed precipitates in solutions. *Acta Metall.*, 4(3):243–248, 1956.

4. J.S. Langer and A.J. Schwartz. Kinetics of nucleation in near-critical fluids. *Phys. Rev. A*, 21(3):948–958, 1980.

5. A.J. Ardell. *Experimental Confirmation of the Lifshitz–Wagner Theory of Particle Coarsening*, pages 111–116. Institute of Metals, London, 1969.

6. A.J. Ardell. Precipitate coarsening in solids: Modern theories, chronic disagreement with experiment. In G.W. Lorimer, editor, *Phase Transformations 87*, pages 485–490, London, 1988. Institute of Metals.

7. S.C. Hardy and P.W. Voorhees. Ostwald ripening in a system with a high volume fraction of coarsening phase. *Metall. Trans.*, 19A(11):2713–2721, 1988.

8. L. Ratke and P.W. Voorhees. *Growth and Coarsening*. Springer-Verlag, Berlin, 2002.

9. A.D. Brailsford and P. Wynblatt. The dependence of Ostwald ripening kinetics on particle volume fraction. *Acta Metall.*, 27(3):489–497, 1979.

10. P.W. Voorhees and M.E. Glicksman. Solution to the multi-particle diffusion problem with application to Ostwald ripening—I. Theory. *Acta Metall.*, 32(11):2001–2011, 1984.

11. P.W. Voorhees and M.E. Glicksman. Solution to the multi-particle diffusion problem with application to Ostwald ripening—II. Computer simulations. *Acta Metall.*, 32(11):2013–2030, 1984.

12. P.W. Voorhees. The theory of Ostwald ripening. *J. Stat. Phys.*, 38(1–2):231–254, 1985.

13. V.A. Snyder, J. Alkemper, and P.W. Voorhees. The development of spatial correlations during Ostwald ripening: A test of theory. *Acta Materialia*, 48(10):2689–2701, 2000.

14. A.J. Ardell and R.B. Nicholson. On the modulated structure of aged Ni–Al. *Acta Metall.*, 14(10):1295–1310, 1966.

15. J.K. Tien and S.M. Copley. The effect of uniaxial stress on the periodic morphology of coherent gamma prime precipitates in nickel-base superalloy crystals. *Metall. Trans.*, 2(1):215–219, 1971.

16. S.M. Allen and J.C. Chang. Elastic energy changes accompanying the gamma-prime rafting in nickel-base superalloys. *J. Mater. Res.*, 6(9):1843–1855, 1991.

17. P.W. Voorhees, G.B. McFadden, R.F. Boisvert, and D.I. Meiron. Numerical simulation of morphological development during Ostwald ripening. *Acta Metall.*, 36(1):207–222, 1988.

18. T. Miyazaki, H. Imamura, and T. Kozakai. The formation of γ precipitate doublets in Ni–Al alloys and their energetic stability. *Mat. Sci. Eng.*, 54(1):9–15, 1982.

19. W.C. Johnson. On the elastic stabilization of precipitates against coarsening under applied load. *Acta Metall.*, 32(3):465–475, 1984.

20. J. Stavans. The evolution of cellular structures. *Rep. Prog. Phys.*, 56:733–789, 1993.

21. C.V. Thompson. Grain growth and evolution of other cellular structures. *Solid State Phys.*, 55:269–314, 2001.

22. K. A. Brakke. The Surface Evolver. *Exp. Math.*, 1(2):141–165, 1992. Publicly available software.

23. W.W. Mullins. Two-dimensional motion of idealized grain boundaries. *J. Appl. Phys.*, 27(8):900–904, 1956.

24. C.S. Smith. Grain shapes and other metallurgical applications of topology. In *Metal Interfaces*, pages 65–108, Metals Park, OH, 1952. American Society for Metals. Also, see discussion by von Neumann following article, pages 108–110.

25. M. Marder. Soap-bubble growth. *Phys. Rev. A*, 36(1):438–440, 1987.

26. F.E. Fradkov. A theoretical investigation of two-dimensional grain growth in the "gas" approximation. *Phil. Mag. Lett.*, 58(6):271–275, 1988.

27. J.A. Glazier, S.P. Gross, and J. Stavans. Dynamics of two-dimensional soap froths. *Phys. Rev. A*, 36(1):306–312, 1987.

28. W.W. Mullins. On idealized 2 dimensional grain-growth. *Scripta Metall.*, 22(9):1441–1444, 1988.

29. W.W. Mullins. The effect of thermal grooving on grain boundary motion. *Acta Metall.*, 6(6):414–426, 1958.

30. J. A. F. Plateau. *Statique expérimentale et théorique des liquides soumis aux seules forces moléculaires.* Gauthier-Villars, Paris, 1873.

31. C.S. Smith. Grain shapes and other metallurgical applications of topology. In *A Search for Structure*, pages 17–25, Cambridge, MA, 1981. MIT Press.

32. S. Hilgenfeldt, A.M. Kraynik, S.A. Koehler, and H.A. Stone. An accurate von Neumann's law for three-dimensional foams. *Phys. Rev. Lett.*, 86:2685–2688, 2001.

33. D.T. Wu and C.E. Krill III. Unknown title (work in progress). *To be submitted*, 2005 (expected).

34. D.T. Wu and F. Wakai. Unknown title (work in progress). *To be submitted*, 2005 (expected).

EXERCISES

15.1 Consider how the particle-size distribution changes with time for source-limited kinetics. Suppose that at $t = t_0$, the particle-size distribution is given by $f(R, t = t_0) = AR(R_{\max} - R)$, where $R_{\max}$ is the maximum radius of any particle. Calculate the rate at which $f(R, t = t_0)$ is increasing at $R = R_{\max}$. At what particle size is the value of $f(r, t = t_0)$ constant?

Solution. First calculate expressions for $\langle R \rangle$ and $\langle R^2 \rangle$ for the growth rate:

$$\langle R \rangle = \frac{\int_0^{R_{\max}} R f(R, t) \, dR}{\int_0^{R_{\max}} f(R, t) \, dR} = \frac{R_{\max}}{2}$$

and

$$\langle R^2 \rangle = \frac{\int_0^{R_{\max}} R^2 f(R, t) \, dR}{\int_0^{R_{\max}} f(R, t) \, dR} = \frac{3R_{\max}^2}{10}$$

Putting these results into Eq. 15.22,

$$\frac{dR}{dt} = M \left(\frac{\langle R \rangle}{\langle R^2 \rangle} - \frac{1}{R} \right) = M \left(\frac{5}{3R_{\max}} - \frac{1}{R} \right)$$

where $M = $ constant. Using Eq. 15.16, we can now find how f changes in time:

$$\frac{\partial f}{\partial t} = -\frac{\partial}{\partial R} \left[f(R, t) \frac{\partial R}{\partial t} \right] = \frac{2AM}{3R_{\max}} (5R - 4R_{\max})$$

Therefore, at $R = R_{\max}$, $\partial f / \partial t = (2/3)AM$, and the distribution is constant at $R = 4R_{\max}/5$.

15.2 In Section 15.1.1 we pointed out that during diffusion-limited coarsening a final steady-state particle distribution corresponding to Fig. 15.5 is expected to develop after initial transients die away. Derive the particle growth law given by Eq. 15.18 using only Eq. 15.13 and the fact that no particles larger than $1.5\langle R \rangle$ appear in the steady-state particle distribution, as is evident in Fig. 15.5.

Solution. In the distribution, the size of the largest particles in the distribution, $R_{\text{cut-off}}$, always corresponds to $1.5\langle R \rangle$. The rate of growth of the largest particles is therefore given by Eq. 15.13:

$$\frac{dR_{\text{cut-off}}}{dt} = \frac{2\widetilde{D}\gamma\Omega c^{\text{eq}}(\infty)}{kT R_{\text{cut-off}}} \left(\frac{1}{\langle R \rangle} - \frac{1}{R_{\text{cut-off}}} \right) \tag{15.52}$$

Substituting $\langle R \rangle = (2/3)R_{\text{cut-off}}$ into Eq. 15.52 and integrating yields

$$R_{\text{cut-off}}^3(t) - R_{\text{cut-off}}^3(0) = \frac{3\widetilde{D}\gamma\Omega c^{\text{eq}}(\infty)}{kT} t \tag{15.53}$$

or, again using $\langle R \rangle = (2/3)R_{\text{cut-off}}$,

$$\langle R(t) \rangle^3 - \langle R(0) \rangle^3 = \frac{8\widetilde{D}\gamma\Omega c^{\text{eq}}(\infty)}{9kT} t \tag{15.54}$$

15.3 The $(N - 6)$-rule was derived for two-dimensional grain growth in a thin film. However, only the interior grains were treated. What is the rule for the growth of the edge and corner grains? Under the same assumptions that apply for the $(N - 6)$-rule, find how the growth of a side grain and a corner grain in a square specimen such as shown in Fig. 15.17 depends on the number of neighboring grains, N. It is reasonable to assume that the grain boundaries are maintained perpendicular to the edges of the sample at the locations of their intersections, as shown. Local interface-tension equilibration obtains and Young's equation is satisfied.

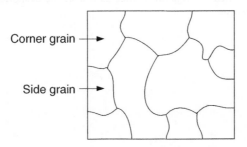

Corner grain

Side grain

Figure 15.17:   Two-dimensional grain growth on a square domain.

Solution. As shown in Fig. 15.17, for side grains and corner grains the number of triple junctions is one less than the number of neighboring grains, N. For the side grains, the inclination of the boundary normal changes by π from one end to the other:

$$\frac{dA}{dt} \propto \left[(N - 1)\frac{\pi}{3} - \pi\right] = \frac{\pi}{3}(N - 4) \tag{15.55}$$

Side grains with more than four neighboring grains therefore grow. For the corner grains, the change is $\pi/2$:

$$\frac{dA}{dt} \propto \left[(N - 1)\frac{\pi}{3} - \frac{\pi}{2}\right] = \frac{\pi}{3}\left(N - \frac{5}{2}\right) \tag{15.56}$$

Since there is no integer number of neighbors that can produce constant area for a corner grain, it is impossible to stabilize grain growth on a rectangular domain.

15.4 **(a)** A cylindrical grain of circular cross section embedded in a large single-crystalline sheet is shrinking under the influence of its grain-boundary energy. Find an expression for the grain radius as a function of time. Assume isotropic boundary energy, γ, and a constant grain-boundary mobility, M_B.

  (b) Derive a corresponding expression for a shrinking spherical grain embedded in a large single crystal in three dimensions.

Solution.

  (a) The grain area, A, is related to its radius, R, by $A = \pi R^2$. Therefore, using Eq. 15.30,

$$\frac{dA}{dt} = -2\pi M_B\gamma = 2\pi R\frac{dR}{dt} \tag{15.57}$$

Integration of Eq. 15.57 then yields

$$R^2(t) - R^2(0) = -2M_B\gamma\, t \tag{15.58}$$

(b) Here, the velocity of the spherical interface normal to itself is given by Eq. 15.28 and, therefore,

$$v = M_B \gamma \kappa = M_B \gamma \frac{2}{R} = -\frac{dR}{dt} \tag{15.59}$$

Integration of Eq. 15.59 then yields

$$R^2(t) - R^2(0) = -4M_B \gamma t \tag{15.60}$$

and the spherical grain shrinks twice as fast as the cylindrical grain because of its larger curvature.

CHAPTER 16

MORPHOLOGICAL EVOLUTION DUE TO CAPILLARY AND APPLIED FORCES: DIFFUSIONAL CREEP AND SINTERING

Capillary forces induce morphological evolution of an interface toward uniform diffusion potential—which is also a condition for constant mean curvature for isotropic free surfaces (Chapter 14). If a microstructure has many internal interfaces, such as one with fine precipitates or a fine grain size, capillary forces drive mass between or across interfaces and cause coarsening (Chapter 15). Capillary-driven processes can occur simultaneously in systems containing both free surfaces and internal interfaces, such as a porous polycrystal.

Applied forces can also induce mass flow between interfaces. When tensile forces are applied, atoms from an unloaded free surface will tend to diffuse toward internal interfaces that are normal to the loading direction; this redistribution of mass causes the system to expand in the tensile direction. Applied compressive forces can superpose with capillary forces to cause shrinkage. In this chapter, we introduce a framework to treat the combined effects of capillary and applied mechanical forces on mass redistribution between surfaces and internal interfaces.

Applications of this framework include diffusional creep in dense polycrystals and sintering of porous polycrystals. Diffusional creep and sintering derive from similar kinetic driving forces. Diffusional creep is associated with macroscopic shape change when mass is transported between interfaces due to capillary and mechanical driving forces. Sintering occurs in response to the same driving forces, but is identified with porous bodies. Sintering changes the shape and size of pores; if pores shrink, sintering also produces macroscopic shrinkage (densification).

Kinetics of Materials. By Robert W. Balluffi, Samuel M. Allen, and W. Craig Carter.   **387**
Copyright © 2005 John Wiley & Sons, Inc.

Microstructures are generally too complex for exact models. In a polycrystalline microstructure, grain-boundary tractions will be distributed with respect to an applied load. Microstructures of porous bodies include isolated pores as well as pores attached to grain boundaries and triple junctions. Nevertheless, there are several simple representative geometries that illustrate general coupled phenomena and serve as good models for subsets of more complex structures.

16.1   MORPHOLOGICAL EVOLUTION FOR SIMPLE GEOMETRIES

Both capillarity and stresses contribute to the diffusion potential (Sections 2.2.3 and 3.5.4). When diffusion potential differences exist between interfaces or between internal interfaces and surfaces, an atom flux (and its associated volume flux) will arise. These driving forces were introduced in Chapter 3 and illustrated in Fig. 3.7 (for the case of capillarity-induced surface evolution) and in Fig. 3.10 (for the case of shape changes due to capillary and applied forces).

For pores within an unstressed body, the diffusion potential at a pore surface will be lower than at nearby grain boundaries if the surface curvature is negative.[1] In this case, the material densifies as atoms flow from grain boundaries to the pore surfaces. Conversely, macroscopic expansion occurs if the pore surface has average positive curvature.

An applied stress, as in Fig. 16.1, can reverse the situation by modifying the diffusion potential on interfaces if their inclinations are not perpendicular to the loading direction. With applied stress and capillary forces, the flux equations for crystal diffusion and surface diffusion are given by Eqs. 13.3 and 14.2. For grain-boundary

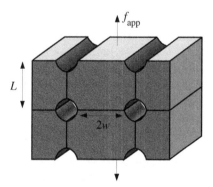

Figure 16.1:   A bundle of parallel wires bonded with grain-boundary segments. An applied force per unit length of wire f_{app} is applied to each wire in the bundle. The system shrinks if mass is transported from the boundaries of width $2w$ into the pores.

---

[1]The sign of the average pore-surface curvature will generally be negative if the dihedral angles are large and the number of neighboring grains is small. In two dimensions—if the pore-surface tension is equal to the grain-boundary surface tension—the average pore-surface curvature will be positive if there are more than six neighbors, and the pore can grow by absorbing vacancies from its abutting grain boundaries. This is equivalent to the $(N-6)$-rule (Eq. 15.33). If the grain boundaries have variable tensions, pore growth or shrinkage will depend on the particular abutting grain boundary energies. However, two-dimensional pores with more than $N_{\text{crit}} = 2\pi/(\pi - \langle\psi\rangle)$ abutting grains (where $\langle\psi\rangle$ is the average dihedral angle $\langle 2\cos^{-1}\gamma^B/(2\gamma^S)\rangle$) will grow on the average.

diffusion, the flux along a boundary under normal stress, σ_{nn}, is determined from Eqs. 2.21, 3.43, and 3.84,

$$\vec{J}^B = -L^B \nabla \Phi^B = -\frac{c^\star D^B}{kT} \nabla \Phi^B = \frac{c \Omega_A^\star D^B}{kT} \nabla \sigma_{nn} = \frac{{}^\star D^B}{kT} \nabla \sigma_{nn} \qquad (16.1)$$

As in surface diffusion (Eq. 14.6), flux accumulation during grain-boundary diffusion leads to atom deposition adjacent to the grain boundary. The resulting accumulation causes the adjacent crystals to move apart at the rate[2]

$$v_n = \frac{dL}{dt} = -\Omega_A \delta \nabla \cdot \vec{J}^B = -\frac{\Omega_A \delta^\star D^B}{kT} \nabla^2 \sigma_{nn} \qquad (16.2)$$

Three conditions are required for a complete solution to the problems illustrated in Figs. 3.10 and 16.1. If the grain boundary remains planar, dL/dt in Eq. 16.2 must be spatially uniform—the Laplacian of the normal surface stress under quasi-steady-state conditions must then be constant:

$$\nabla^2 \sigma_{nn} = \text{constant} = A \qquad (16.3)$$

Continuity of the diffusion potential at the intersection of the grain boundary and the adjoining surface requires that

$$\sigma_{nn} \big|_{\text{bndy int}} = -\gamma^S \kappa \big|_{\text{bndy int}} \qquad (16.4)$$

Finally, the total force across the boundary plane must be zero:

$$F_{\text{app}} = \iint_{\text{bndy}} \sigma_{nn} dA + \int_{\text{bndy int}} \gamma^S \cos\theta \, ds \qquad (16.5)$$

The physical basis for the three terms in Eq. 16.5 is illustrated by Fig. 16.2 for the geometry indicated by Fig. 3.10.[3]

16.1.1 Evolution of Bamboo Wire via Grain-Boundary Diffusion

For this case of an isotropic polycrystalline wire loaded parallel to its axis as illustrated in Fig. 3.10b, Eqs. 16.3 and 16.4 become[4]

$$\nabla^2 \sigma_{nn} = \frac{\partial^2 \sigma_{nn}}{\partial r^2} + \frac{1}{r}\frac{\partial \sigma_{nn}}{\partial r} = A = \text{constant}$$
$$\sigma_{nn}(r = R_b) = -\gamma^S \kappa \qquad (16.6)$$

Solving Eqs. 16.6 subject to the symmetry condition $(d\sigma_{nn}/dr|_{r=0} = 0)$,

$$\sigma_{nn}(r) = \frac{A}{4}(r^2 - R_b^2) - \gamma^S \kappa \qquad (16.7)$$

---

[2]This could be measured by observing the separation of inert markers buried in each crystal opposite one another across the boundary.

[3]The justification for the projected interface contribution is presented elsewhere [1–4]. The total force F_{app} is that measured by a wetting balance [5].

[4]By symmetry, there is no angular dependence of σ_{nn}.

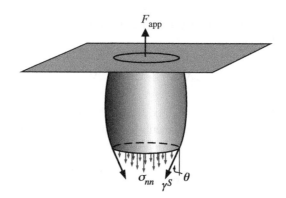

Figure 16.2: Force-balance diagram for a body with capillary forces and applied load F_{app}. The plane cuts the body normal to the applied force. There are two contributions from the body itself. One is the projection of the surface capillary force per unit length (γ^S) onto the normal direction and integrated over the bounding curve. The second is the normal stress σ_{nn} integrated over the cross-sectional area—in the case of fluids bounded by a surface of uniform curvature κ^S, $\sigma_{nn} = \gamma^S \kappa^S$ [4].

The constant A is determined from the force balance in Eq. 16.5,

$$F_{\mathrm{app}} = 2\pi \int_0^{R_b} \sigma_{nn}(r) r \, dr + 2\pi R_b \gamma^S \cos\theta \tag{16.8}$$

Using Young's force-balance equation (Eq. 14.18),

$$\cos\theta = \sin\frac{\psi}{2} = \sqrt{1 - \left(\frac{\gamma^B}{2\gamma^S}\right)^2} \tag{16.9}$$

at the grain boundary/surface intersection and the elongation rate (Eq. 16.2) becomes

$$\frac{dL}{dt} = -\frac{\Omega_A \delta^\star D^B A}{kT} = \frac{8\Omega_A \delta^\star D^B}{\pi R_b^4 kT}(F_{\mathrm{app}} + \Upsilon_{\mathrm{bamboo}})$$

$$\Upsilon_{\mathrm{bamboo}} = -\pi\gamma^S R_b \left(2\sqrt{1 - \left(\frac{\gamma^B}{2\gamma^S}\right)^2} - R_b\kappa\right) \tag{16.10}$$

When the grain boundaries are not spaced too closely, the quantity $\Upsilon_{\mathrm{bamboo}}$ is generally negative because $R_b\kappa \approx 1$ is less than $2\sqrt{1 - [\gamma^B/(2\gamma^S)]^2}$ and $\gamma^B/(2\gamma^S) \approx 1/6$ for metals. Υ, the *capillary shrinkage force*, arises from a balance between reductions of surface and grain-boundary area. If F_{app} is adjusted so that the elongation rate goes to zero, $F_{\mathrm{app}} = -\Upsilon_{\mathrm{bamboo}}$, and this provides an experimental method to determine γ^B/γ^S, and thus γ^S if ψ is measured. This is known as the Udin–Schaler–Wulff *zero-creep method* [6].

Scaling arguments can be used to estimate elongation behavior. Because κ and $1/R_b$ will scale with $\sqrt{\pi L/V_g}$ and the grain volume, V_g is constant, Eq. 16.10 implies that

$$\frac{dL}{dt} \propto L^2 \left(F_{\mathrm{app}} - \gamma^S \sqrt{\frac{\pi V_g}{L}}\right) \tag{16.11}$$

where $\Upsilon_{\text{bamboo}} \approx -\pi\gamma^S R_b$ is replaced by a term that depends on L alone. Elongation proceeds according to[5]

$$
\begin{aligned}
\left(\frac{1}{L(t=0)} - \frac{1}{L(t)}\right) &\propto F_{\text{app}}t & \text{if } F_{\text{app}} \gg \gamma^S \sqrt{\frac{\pi V_g}{L}} \\
\left(\sqrt{\frac{1}{L(t)}} - \sqrt{\frac{1}{L(t=0)}}\right) &\propto \gamma^S V_g^{1/2}t & \text{if } F_{\text{app}} \ll \gamma^S \sqrt{\frac{\pi V_g}{L}}
\end{aligned}
\tag{16.12}
$$

16.1.2 Evolution of a Bundle of Parallel Wires via Grain-Boundary Diffusion

For the boundary of width $2w$ in Fig. 16.1, Eq. 16.4 becomes

$$
\sigma_{nn}(x = \pm w) = -\gamma^S \kappa
\tag{16.13}
$$

where κ is evaluated at the pore surface/grain boundary intersection. Solving Eq. 16.3 subject to Eq. 16.13 and the symmetry condition $(d\sigma_{nn}/dx)|_{x=0} = 0$,

$$
\sigma_{nn}(x) = \frac{A}{2}(x^2 - w^2) - \gamma^S \kappa
\tag{16.14}
$$

where the grain-boundary center is located at $x = 0$. The constant A can be determined from Eq. 16.5,

$$
A = \frac{3}{2w^3}\left(2\gamma^S \sin\frac{\psi}{2} - 2\gamma^S \kappa w - f_{\text{app}}\right)
\tag{16.15}
$$

The shrinkage rate, Eq. 16.2, becomes

$$
\begin{aligned}
\frac{dL}{dt} &= -\frac{\Omega_A \delta^\star D^B A}{kT} = \frac{3\Omega_A \delta^\star D^B}{2w^3 kT}(f_{\text{app}} + \Upsilon_{\text{wires}}) \\
\Upsilon_{\text{wires}} &= 2\gamma^S(\kappa w - \sin\frac{\psi}{2})
\end{aligned}
\tag{16.16}
$$

If surface diffusion or vapor transport is rapid enough, the pores will maintain their quasi-static equilibrium shape, illustrated in Fig. 16.1 in the form of four cylindrical sections of radius R.[6] The dihedral angle at the four intersections with grain boundaries, ψ, will obey Young's equation. ψ is related to θ by $\sin(\psi/2) = \cos\theta$.

An exact expression can be calculated for the quasi-static capillary force, Υ_{wires}, as a function of the time-dependent length $L(t)$. Young's equation places a geometric constraint among $L(t)$, the cylinder's radius of curvature $R(t)$, and boundary width $w(t)$; conservation of material volume provides the second necessary equation. With $\Upsilon_{\text{wire}}(L)$ and $w(L)$, Eq. 16.16 can be integrated. This model could be extended to general two-dimensional loads by applying different forces onto the horizontal and vertical grain boundaries in Fig. 16.1. The three-dimensional case, with sections of spheres and a triaxial load, could also be derived exactly.

---

[5] An exact quasi-static [e.g., surfaces of uniform curvature (Eq. 14.29)] derivation exists for this model [4].
[6] The Rayleigh instability (Section 14.1.2) of the pore channel is neglected. Pores attached to grain boundaries have increased critical Rayleigh instability wavelengths [7].

16.1.3   Evolution of Bamboo Wire by Bulk Diffusion

Morphological evolution and elongation can also occur by mass flux (and its associated volume) from the grain boundary through the bulk to the surface as illustrated in Fig. 3.10a. For elongation of a crystalline material, vacancies could be created at the grain boundary and diffuse through the grain to the surface, where they would be removed. The quasi-steady-state rate of elongation can be determined by solving the boundary-value problem described in Section 3.5.3 involving the solution to Laplace's equation $\nabla^2 \Phi_A = 0$ within each grain of the idealized bamboo structure. For isotropic surfaces and grain boundaries, Φ_A is given by Eqs. 3.76 and 3.84. The expression for bulk mass flux is given by Eq. 13.3, and using the coordinate system shown in Fig. 3.10, symmetry requires that

$$\left(\frac{\partial \Phi_A}{\partial z} \right)_{z=L/2} = 0 \tag{16.17}$$

If the grain boundary remains planar, the flux into the boundary must be uniform,

$$\left(\frac{\partial \Phi_A}{\partial z} \right)_{z=0} = C = \text{constant} \tag{16.18}$$

Laplace's equation in cylindrical coordinates is

$$\frac{1}{r} \frac{\partial}{\partial r} \left(r \frac{\partial \Phi_A}{\partial r} \right) + \frac{\partial^2 \Phi_A}{\partial z^2} = 0 \tag{16.19}$$

Assuming that the solution to Eq. 16.19 is the product of functions of z and r and using the separation-of-variables method (Section 5.2.4),

$$\Phi_A = [c_1 \sinh(kz) + c_2 \cosh(kz)][c_3 J_\circ(kr) + c_4 Y_\circ(kr)] \tag{16.20}$$

where c_1, c_2, c_3, c_4, and k are constants to be determined, and $J_\circ$ and $Y_\circ$ are the zeroth-order Bessel functions of the first and second kinds. Because $\Phi_A(r = 0)$ must be bounded, $c_4 = 0$. Introducing a new variable $p(r, z)$ that will necessarily vanish on the free surface,

$$p(r, z) \equiv \Phi_A - \left(\mu^\circ + \Omega_A \gamma^S \kappa \right) \tag{16.21}$$

The general solution to Eq. 16.19 is the superposition of the homogeneous solutions,

$$p(r, z) = \sum_n J_\circ(k_n r) \left[b_n \sinh(k_n z) + c_n \cosh(k_n z) \right] \tag{16.22}$$

The bamboo segment can be approximated as a cylinder of average radius R_c, where

$$\pi R_c^2 L = \int_0^L \pi R^2(z) \, dz \tag{16.23}$$

The boundary condition (Eq. 3.76) is then approximated by

$$\Phi_A = \mu^\circ + \frac{\gamma^S \Omega_A}{R_c} \qquad \text{or, equivalently,} \qquad p(r = R_c, z) = 0 \tag{16.24}$$

The $k_n R_c$ quantities are the roots of the zeroth-order Bessel function of the first kind,

$$J_\circ(k_n R_c) = 0 \tag{16.25}$$

The symmetry condition Eq. 16.17 is satisfied if $b_n \cosh(k_n L/2) + c_n \sinh(k_n L/2) = 0$, and therefore,

$$p(r, z) = \sum_n b_n J_\circ(k_n r) \left[\sinh(k_n z) - \coth\left(\frac{k_n L}{2}\right) \cosh(k_n z) \right] \tag{16.26}$$

The planar grain-boundary condition given by Eq. 16.18 is satisfied if

$$\sum_n b_n k_n J_\circ(k_n r) = C \tag{16.27}$$

The coefficients, $b_n k_n$, of $J_\circ$ in this Bessel function series can be determined [8]:

$$b_n k_n = \frac{2C}{k_n R_c J_1(k_n R_c)} \tag{16.28}$$

The constant C can be determined by substituting Eqs. 16.26 and 16.28 into the force-balance condition (Eq. 16.5),

$$C = \frac{[F_{\text{app}} + \Upsilon_{\text{approx cyl}}] \Omega_A}{4\pi R_c^3 \sum_n \left[\coth(k_n L/2)/(k_n{}^3 R_c{}^3) \right]} \tag{16.29}$$

where

$$\Upsilon_{\text{approx cyl}} = -\pi\gamma^S R_c \left(2\sqrt{1 - \left(\frac{\gamma^B}{2\gamma^S}\right)^2} - 1 \right) = -\pi\gamma^S R_c \left(2\sin\frac{\psi}{2} - 1 \right) \tag{16.30}$$

The total atom current into the boundary is $I_A = -2\pi R_c^2 J_A$; therefore,

$$\begin{aligned}
\frac{dL}{dt} &= \frac{\Omega_A I_A}{\pi R_c^2} = -2J_A\Omega_A = \frac{2\,{}^\star D^{XL}}{\mathbf{f}kT} C \\
&= \frac{{}^\star D^{XL}\Omega_A B}{2\pi \mathbf{f}kT R_c^3} [F_{\text{app}} + \Upsilon_{\text{approx cyl}}]
\end{aligned} \tag{16.31}$$

$$B \equiv \left[\sum_n \frac{\coth(k_n L/2)}{k_n^3 R_c^3} \right]^{-1}$$

$B \approx 12$ for $L/R_c \approx 2$ [9].

The elongation-rate expressions for grain-boundary diffusion (Eq. 16.10) and bulk diffusion (Eq. 16.31) for a bamboo wire are similar except for a length scale. The approximate capillary shrinkage force $\Upsilon_{\text{approx cyl}}$ reduces to the exact force Υ_{bamboo} as the segment shapes become cylindrical, $R_b \approx R_c \approx 1/\kappa$. However, because the grain-boundary diffusion elongation rate is proportional to ${}^\star D^B/R_b^4$, while the bulk diffusion rate is proportional to ${}^\star D^{XL}/R_c^3$, grain-boundary transport will dominate at low temperatures and small wire radii.

16.1.4   Neck Growth between Two Spherical Particles via Surface Diffusion

Figure 16.3 illustrates neck growth between two particles by surface diffusion. Surface flux is driven toward the neck region by gradients in curvature. Neck growth (and particle bonding) occurs as a result of mass deposition in that region of smallest curvature. Because no mass is transported from the region between the particle centers, the two spheres maintain their spacing at $2R$ as the neck grows through rearrangement of surface atoms. This is surface evolution toward a uniform potential for which governing equations were derived in Section 14.1.1. However, the small-slope approximation that was used to obtain Eq. 14.10 does not apply for the sphere–sphere geometry. Approximate models, such as those used in the following treatment of Coblenz et al., can be used and verified experimentally [10].

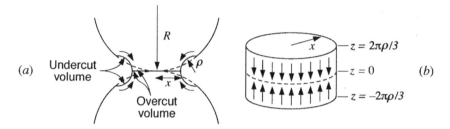

Figure 16.3:   **(a)** Model for formation of a neck between two spherical particles due to surface diffusion. **(b)** Approximation in which the surface diffusion zone within the saddle-shaped neck region of (a) is mapped onto a right circular cylinder of radius x. z is the distance parameter in the diffusion direction. Arrows parallel to the surface indicate surface-diffusion directions in both (a) and (b). From Coblenz et al. [10].

Because of the proximity effect of surface diffusion, the flux from the regions adjacent to the neck leaves an *undercut* region in the neck vicinity.[7] Diffusion along the uniformly curved spherical surfaces is small because curvature gradients are small and therefore the undercut neck region fills in slowly. This undercutting is illustrated in Fig. 16.3a. Because mass is conserved, the undercut volume is equal to the *overcut* volume. Conservation of volume provides an approximate relation between the radius of curvature, ρ, and the neck radius, x:

$$\rho = 0.26x \left(\frac{x}{R} \right)^{1/3} \tag{16.32}$$

This surface-diffusion problem can be mapped to a one-dimensional problem by approximating the neck region as a cylinder of radius x as shown in Fig. 16.3b. The fluxes along the surface in the actual specimen (indicated by the arrows in Fig. 16.3a) are mapped to a corresponding cylindrical surface (indicated by the arrows in Fig. 16.3b). The zone extends between $z = \pm 2\pi\rho/3$. The flux equation has the same form as Eq. 14.4, so that[8]

$$J^S \approx -\frac{{}^{\star}D^S \gamma^S}{kT} \frac{d\kappa}{dz} \tag{16.33}$$

---

[7]The proximity effect is reflected in the strong wavelength dependence of surface smoothing (i.e., $1/\lambda^4$ in Eq. 14.12).

[8]Equation 16.33 ignores the relatively small effect of the increase in energy due to the growing grain boundary.

The curvature has the value $2/R$ at $z = \pm 2\pi\rho/3$ and approximately $-1/\rho$ at $z = 0$ (neglecting terms of order ρ/R). The average curvature gradient $-3/(2\pi\rho^2)$ can be inserted into Eq. 16.33 for an approximation to the total accumulation at the neck (per neck circumference),

$$I^S \approx 2\,\delta J^S \approx \frac{3\,\delta^\star D^S \gamma^S}{\pi k T \rho^2} \tag{16.34}$$

The corresponding neck surface area is approximately ρ (per neck circumference), and therefore the neck growth rate is approximately

$$\frac{dx}{dt} \approx \frac{3\,\delta^\star D^S \gamma^S \Omega_A}{\pi k T \rho^3} \tag{16.35}$$

Putting Eq. 16.32 into Eq. 16.35 and integrating yields the neck growth law,

$$\left(\frac{x}{R}\right)^5 \approx \frac{271\delta^\star D^S \gamma^S \Omega_A}{R^4 k T} t \tag{16.36}$$

Equation 16.36 predicts that $x(t) \propto t^{1/5}$ and that the neck growth rate will therefore fall off rapidly with time. The time to produce a neck size that is a given particle-size fraction is a strong function of initial particle size—it increases as R^4. Equation 16.36 agrees with the results of a numerical treatment by Nichols and Mullins [11].[9]

16.2 DIFFUSIONAL CREEP

Mass diffusion between grain boundaries in a polycrystal can be driven by an applied shear stress. The result of the mass transfer is a high-temperature permanent (plastic) deformation called *diffusional creep*. If the mass flux between grain boundaries occurs via the crystalline matrix (as in Section 16.1.3), the process is called *Nabarro–Herring creep*. If the mass flux is along the grain boundaries themselves via triple and quadjunctions (as in Sections 16.1.1 and 16.1.2), the process is called *Coble creep*.

Grain boundaries serve as both sources and sinks in polycrystalline materials—those grain boundaries with larger normal tensile loads are sinks for atoms transported from grain boundaries under lower tensile loads and from those under compressive loads. The diffusional creep in polycrystalline microstructure is geometrically complex and difficult to analyze. Again, simple representative models are amenable to rigorous treatment and lead to an approximate treatment of creep in general.

16.2.1 Diffusional Creep of Two-Dimensional Polycrystals

A representative model is a two-dimensional polycrystal composed of equiaxed hexagonal grains. In a dense polycrystal, diffusion is complicated by the necessity

---

[9]Different growth-law exponents are obtained for other dominant transport mechanisms. Coblenz et al. present corresponding neck-growth laws for the vapor transport, grain-boundary diffusion, and crystal-diffusion mechanisms [10].

of simultaneous *grain-boundary sliding*—a thermally activated shearing process by which abutting grains slide past one another—to maintain compatibility between the grains. In the absence of sliding, gaps or pores will develop. Sliding is confined to the grain-boundary region and occurs by complex mechanisms that are not yet completely understood [12].

The need for such sliding can be demonstrated by analyzing the diffusional creep of the idealized polycrystal illustrated in Fig. 16.4 [12–15]. The specimen is subjected to the applied tensile stress, σ, which motivates diffusion currents between the boundaries at differing inclinations and causes the specimen to elongate along the applied stress axis. Figure 16.4 shows the currents associated with Nabarro–Herring creep. Currents along the boundaries can occur simultaneously, and if these dominate the dimensional changes, produce Coble creep. For the equiaxed microstructure in Fig. 16.4, there are only three different boundary inclinations with respect to a general loading direction; these are exhibited by the boundaries between grains A, B, and C indicated in Fig. 16.4. Mass transport between these boundaries will cause displacement of the centers of their adjoining grains. The normal displacements are indicated by L^A, L^B, and L^C in Fig. 16.4 and the shear displacements by S^A, S^B, and S^C. These combined grain-center displacements produce an equivalent net shape change of the polycrystal.

Compatibility relationships between the displacements must exist if the grain boundaries remain intact. Along the 1 axis, the displacement of grain C relative to grain B must be consistent with the difference between the displacement of grain C with respect to grain A and with that of grain B with respect to grain A. This requirement is met if

$$L^A + L^B - 2L^C = \sqrt{3}\,S^A - \sqrt{3}\,S^B \qquad (16.37)$$

Similarly, along the 2 axis,

$$\sqrt{3}\,L^A - \sqrt{3}\,L^B = -S^A - S^B + 2S^C \qquad (16.38)$$

Also, the volume must remain constant. Therefore,

$$\varepsilon_{11} + \varepsilon_{22} = 0 \qquad (16.39)$$

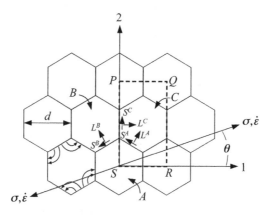

Figure 16.4:   Two-dimensional polycrystal consisting of identical hexagonal grains subjected to uniaxial applied stress, σ, giving rise to an axial strain rate $\dot{\varepsilon}$. From Beeré [14].

where ε_{11} and ε_{22} are the normal strains of the overall network connecting the centers of the grains in the $(1,2)$ coordinate system in Fig. 16.4.

These strains are related to the displacements through $\varepsilon_{11} = \partial u_1/\partial x_1$, $\varepsilon_{22} = \partial u_2/\partial x_2$, and $\varepsilon_{12} = (1/2)(\partial u_1/\partial x_2 + \partial u_2/\partial x_1)$, where the u_i are the displacements produced throughout the network of grain centers. For the representative unit cell $PQRS$ in Fig. 16.4,

$$
\begin{aligned}
\varepsilon_{11} &= \frac{\Delta C_1 - \Delta B_1}{d} \\
\varepsilon_{22} &= \frac{\Delta B_2 + \Delta C_2}{\sqrt{3}d} \\
\varepsilon_{12} &= \frac{\Delta B_1 + \Delta C_1}{2\sqrt{3}d} + \frac{\Delta C_2 - \Delta B_2}{2d}
\end{aligned}
\tag{16.40}
$$

where d is the width of a hexagonal grain, and ΔB_i and ΔC_i are the components of the displacements of the centers of the grains B and C relative to A and are given by

$$
\begin{aligned}
2\Delta B_1 &= -\sqrt{3}\,S^B - L^B \\
2\Delta B_2 &= -S^B + \sqrt{3}\,L^B \\
2\Delta C_1 &= -\sqrt{3}\,S^A + L^A \\
2\Delta C_2 &= S^A + \sqrt{3}\,L^A
\end{aligned}
\tag{16.41}
$$

Therefore,

$$
\begin{aligned}
\varepsilon_{11} &= \frac{\sqrt{3}\left(S^B - S^A\right)}{2d} + \frac{L^A + L^B}{2d} \\
\varepsilon_{22} &= \frac{S^A - S^B}{2\sqrt{3}d} + \frac{L^A + L^B}{2d} \\
\varepsilon_{12} &= \frac{L^A - L^B}{\sqrt{3}d}
\end{aligned}
\tag{16.42}
$$

Substituting Eqs. 16.42 into Eq. 16.39 yields

$$
\sqrt{3}\,L^A + \sqrt{3}\,L^B = S^A - S^B
\tag{16.43}
$$

Combining Eqs. 16.38, 16.37, and 16.43,

$$
\begin{aligned}
\sqrt{3}\,L^A &= S^C - S^B \\
\sqrt{3}\,L^B &= S^A - S^C \\
\sqrt{3}\,L^C &= S^B - S^A
\end{aligned}
\tag{16.44}
$$

and

$$
L^A + L^B + L^C = 0
\tag{16.45}
$$

which is equivalent to the constant-volume condition.

To show that boundary sliding must participate in the diffusional creep to maintain compatibility, suppose that all of the S^A, S^B, and S^C sliding displacements are zero. Equations 16.44 require that the L^A, L^B, and L^C must also vanish. Therefore, nonzero S^i's (sliding) are required to produce nonzero grain-center normal displacements.

This result can be demonstrated similarly by solving for the strain, ε, along the applied tensile stress axis shown in Fig. 16.4 in terms of only the L^i's or only the S^i's:

$$\varepsilon = \cos^2\theta\varepsilon_{11} + \sin^2\theta\varepsilon_{22} + 2\sin\theta\cos\theta\varepsilon_{12} \tag{16.46}$$

or, using Eq. 16.40–16.44,

$$\varepsilon = \frac{1}{d}\left[(L^A + L^B)(1 - 2\cos^2\theta) + \frac{L^A - L^B}{\sqrt{3}}2\sin\theta\cos\theta\right] \tag{16.47}$$

$$\varepsilon = \frac{1}{d}\left[\frac{S^A - S^B}{\sqrt{3}}(1 - 2\cos^2\theta) + \frac{2S^C - S^A - S^B}{3}2\sin\theta\cos\theta\right] \tag{16.48}$$

Equation 16.47 indicates that the creep strain may be regarded as diffusional transport accommodated by boundary sliding, and Eq. 16.48 indicates that it may equally well be regarded as boundary sliding accommodated by diffusional transport.[10] The creep rate, $\dot{\varepsilon}$, can be obtained by taking time derivatives of ε in Eqs. 16.47 and 16.48. The applied tensile stress, σ, shown in Fig. 16.4 will generate stresses throughout the polycrystal, and each boundary segment will, in general, be subjected to a shear stress (parallel to the boundary) and a normal stress (perpendicular to the boundary). The shear stresses will promote the grain-boundary sliding displacements, S^A, S^B, and S^C, while the normal stresses will promote the diffusion currents responsible for the L^A, L^B, and L^C displacements. A detailed analysis of the shear and normal stresses at the various boundary segments is available (see also Exercise 16.2) [12–14].

16.2.2 Diffusional Creep of Three-Dimensional Polycrystals

The analysis can be extended to a three-dimensional polycrystal with an equiaxed grain microstructure. As in two-dimensional creep, grain-boundary sliding must accompany the diffusional creep, and because these processes are interdependent, either sliding or diffusion may be rate limiting. In most observed cases, the rate is controlled by the diffusional transport [14, 15, 18, 19]. Exact solutions for corresponding tensile strain rates are unknown, but approximate expressions for the Coble and Nabarro–Herring creep rates under diffusion-controlled conditions where the boundaries act as perfect sources may be obtained from the solutions for the bamboo-structured wire in Section 16.1.1. The equiaxed polycrystal can be approximated as an array of bonded bamboo-structured wires with their lengths running parallel to the stress axis and with the lengths of their grains (designated by L in Fig. 3.10) equal to the wire diameter, $2R$. This produces a polycrystal with an approximate equiaxed grain size $d = L = 2R$. The Coble and Nabarro–Herring creep rates of this structure can be approximated by those given for the creep rates of the bamboo-structured wire by Eqs. 16.10 and 16.31 with $L = 2R = d$ and the sintering potential set to zero. In this approximation, the effects of internal normal stresses generated along the vertical boundaries (between the bonded wires) may be neglected because these stresses are zero on average. Using this approximation, for diffusion-controlled Coble creep,

$$\dot{\varepsilon} \cong A_1\frac{^\star D^B\delta\,\Omega_A}{kT\mathfrak{f}d^3}\,\sigma \tag{16.49}$$

---

[10]This duality has been recognized (e.g., Landau and Lifshitz [16] and Raj and Ashby [17]).

with $A_1 = 32$, and for diffusion-controlled Nabarro–Herring creep,

$$\dot{\varepsilon} \cong A_2 \, \frac{{}^\star D^{XL} \Omega_A}{kT\mathfrak{f}d^2} \, \sigma \tag{16.50}$$

with $A_2 = 12$.[11] Because the Coble creep rate is proportional to ${}^\star D^B/d^3$ and the Nabarro–Herring rate to ${}^\star D^{XL}/d^2$, Coble creep will be favored as the temperature and grain size are reduced.

Figure 16.5 shows a *deformation map* for polycrystalline Ag possessing a grain size of 32 μm strained at a rate of $10^{-8}\,\mathrm{s}^{-1}$ [20]. Each region delineated on the map indicates a region of applied stress and temperature where a particular kinetic mechanism dominates. Experimental data and approximate models are used to produce such deformation maps. The mechanisms include elastic deformation at low temperatures and low stresses, dislocation glide at relatively high stresses, dislocation creep at somewhat lower stresses and high temperatures, and Nabarro–Herring and Coble diffusional creep at high temperatures and low stresses. Coble creep supplants Nabarro–Herring creep as the temperature is reduced. An analysis of diffusional creep when the boundaries do not act as perfect sources and sinks has been given by Arzt et al. [19] and is explored in Exercise 16.1.

The creep rate when boundary sliding is rate-limiting has been treated and discussed by Beeré [13, 14]. If a viscous constitutive relation is used for grain-boundary sliding (i.e., the sliding rate is proportional to the shear stress across the boundary), the macroscopic creep rate is proportional to the applied stress, and the bulk polycrystalline specimen behaves as a viscous material. An analysis of the sliding-controlled creep rate of the idealized model in Fig. 16.4 is taken up in Exercise 16.2.

Variable boundary behavior complicates the results derived from the uniform equiaxed model presented above. Nonuniform boundary sliding rates may cause

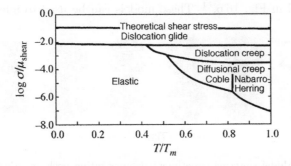

Figure 16.5: Deformation mechanism map for Ag polycrystal: σ = applied stress, μ = shear modulus, grain size = 32 μm, and strain rate = $10^{-8}\,\mathrm{s}^{-1}$. The diffusional creep field is divided into two subfields: the Coble creep field and the Nabarro–Herring creep field. From Ashby [20].

[11]The functional forms of these relationships agree with those obtained by other investigators using a variety of approximations. However, the values of the constants A_1 and A_2 vary in some cases by factors as large as three. See Ashby [20], Burton [18], Arzt et al. [19], and Pilling and Ridley [15].

individual grains to rotate. Also, grain-boundary migration and the formation of new grains by recrystallization will affect both microstructure and creep rate.[12]

Finally, mechanisms besides diffusional transport of mass between internal interfaces can contribute to diffusional creep. For instance, single crystals containing dislocations exhibit limited creep if the dislocations act as sources and sinks, depending on their orientation with respect to an applied stress (see Exercise 16.3).

16.3   SINTERING

Sintering is a kinetic process that converts a compacted particle mass (or powder) or fragile porous body into one with more structural integrity. Increased mechanical integrity stems from both *neck growth* (due to mass transport that increases the particle/particle "necks") and densification (due to mass transport that reduces porosity). The fundamental sintering driving force—capillarity—derives from reduction of total surface energy and is often augmented by applied pressure.

The kinetic transport mechanisms that permit sintering are solid-state processes, and therefore sintering is an important forming process that does not require melting. Materials with high melting temperatures, such as ceramics, can be molded into a complex shape from a powder and subsequently sintered into a solid body.[13]

16.3.1   Sintering Mechanisms

Neck growth can occur by any mass transport mechanism. However, processes that permit shrinkage by pore removal must transport mass from the interior of the particles to the pore surfaces—these mechanisms include grain-boundary diffusion, volume diffusion, and viscous flow. Other mechanisms simply rearrange volume at the pore surfaces and contribute to particle/particle neck growth without reduction in porosity and shrinkage—these mechanisms include surface diffusion and vapor transport. Particle compacts and porous bodies have complex geometries, but models for sintering and shrinkage can be developed for simpler geometries such as the one captured in Fig. 16.6.[14] These models can be used to infer behaviors of

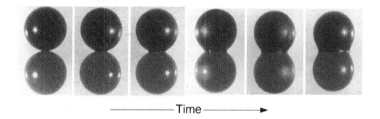

—————— Time ——————▶

Figure 16.6:   Model sintering experiment demonstrating neck growth during sintering by viscous flow of initially spherical 3 mm diameter glass beads at $1000°$C over 30 minutes. Courtesy of Hans-Eckart Exner.

[12]These phenomena, and their effects on the creep rate, are described in more detail by Sutton and Balluffi [12], Beeré [13, 14], and Pilling and Ridley [15].
[13]General reviews of sintering appear in introductory ceramics texts [21, 22], and a more complete exposition is given in German's book on sintering [23].
[14]Further details about such models can be found in *Reviews in Powder Metallurgy and Physical Ceramics* or in *Physical Metallurgy* [24, 25].

complex systems of which these simpler geometries are component parts.

Figure 16.7 summarizes the atom-transport paths that can contribute to neck growth and also, in some cases, densification. If the particles are crystalline, a grain boundary will generally form at the contact region (the neck). A dihedral angle ψ will form at the neck/surface junction, and for the isotropic case, conform to Young's equation, $\gamma^B = 2\gamma^S \cos(\psi/2)$. The seven different transport paths in Fig. 16.7 are listed in Table 16.1 with their kinetic mechanisms. Atoms generally flow to the neck region, where the surface has a large negative principal curvature and therefore a low diffusion potential compared to neighboring regions. Densification will accompany neck growth if the centers of the abutting spheres move toward one another. For example, with mechanism **BS·B**, atoms are removed from the boundary region causing such motion.

The dominant mechanism and transport path—or combinations thereof—depend upon material properties such as the diffusivity spectrum, surface tension, temperature, chemistry, and atmosphere. The dominant mechanism may also change as the microstructure evolves from one sintering stage to another. *Sintering maps* that indicate dominant kinetic mechanisms for different microstructural scales and environmental conditions are discussed in Section 16.3.5.

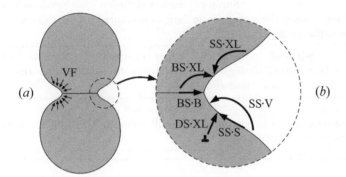

Figure 16.7:   (a) Sintering of two abutting single-crystal spherical particles of differing crystal orientations. A grain boundary has formed across the neck region. (b) Detail of neck perimeter. Seven possible sintering mechanisms for the growth of the neck are illustrated (see the text and Table 16.1).

16.3.2 Sintering Microstructures

Powder compressed into a desired shape at room temperature provides an initial microstructure for a typical sintering process. Such a microstructure may be composed of equiaxed particles or the particles may vary in size and shape. Particle packing may be regular and nearly "crystalline," highly irregular, or mixtures of both. Sintering microstructures are generally complex, but some aspects of their microstructural evolution can be understood by investigating primary process models such as those described in Section 16.1 and the simple neck-growth models presented in Section 16.1.4. However, some microstructural evolution processes are not easily captured by simple models. Additional modeling difficulties arise for irregular packings, variability in particle size and shape, and inhomogeneous chemistry.

Table 16.1:   Mass Transport Mechanisms for Sintering

Mechanism	Source	Sink	Transport Mechanism	Densifying or Nondensifying
SS·XL	Surface	Surface	Crystal diffusion	Nondensifying

Atoms diffuse through the crystal from larger-curvature surface regions to lower-curvature regions.

BS·XL	Boundary	Surface	Crystal diffusion	Densifying

Atoms diffuse through the crystal from the grain boundary to low-curvature surface regions.

BS·B	Boundary	Surface	Boundary diffusion	Densifying

Atoms diffuse along the boundary to the surface; subsequently, they are transported along the surface by one or more of the **SS·XL**, **SS·S**, or **SS·V** paths.

DS·XL	Dislocation	Surface	Crystal diffusion	Either

Atoms diffuse through the crystal from climbing dislocations. Equivalently, vacancies diffuse from the surface.

SS·S	Surface	Surface	Surface diffusion	Nondensifying

Atoms diffuse along the surface from larger-curvature surface regions to lower-curvature surface regions.

SS·V	Surface	Surface	Vapor transport	Nondensifying

Atoms are transported through the vapor phase from larger-curvature surface regions to lower-curvature surface regions.

VF	—	—	Viscous flow	Either

Atoms are transported by viscous flow by differences in the capillary pressure at nonuniformly curved surfaces.

Nevertheless, there are parallel stages in any powder sintering process that can be used to catalog behavior. Each powder sintering process begins with particle/particle neck formation and a porous phase between the weakly attached particles. As these necks grow, the particle/pore interface becomes more uniformly curved but remains interconnected throughout the compact. Before the porous phase is removed, it becomes disconnected and isolated at pockets where four grain boundaries intersect.

Initial, Intermediate, and Final Stages of Powder Sintering. Following Coble's pioneering work, the microstructural evolution of a densifying compact is separated into an *initial stage*, an *intermediate stage*, and a *final stage* of sintering [26]. Figure 16.8 illustrates some of the microstructural features of each stage.

The initial stage comprises neck growth along the grain boundary between abutting particles. The intermediate stage occurs during the period when the necks between the particles are no longer small compared to the particle radii and the porosity is mainly in the form of tubular pores along the three-grain junctions in the compact. The geometries of both the initial and intermediate stages therefore have intergranular porosity percolating through the compact.

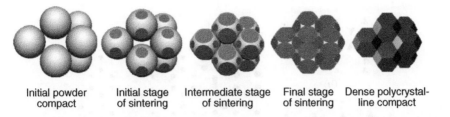

| Initial powder compact | Initial stage of sintering | Intermediate stage of sintering | Final stage of sintering | Dense polycrystal-line compact |

Figure 16.8: Stages of powder sintering. Initial stage involves neck growth. Intermediate state is marked by continuous porosity along three-grain junctions. Final stage involves removal of isolated pores at four-grain junctions. Figure calculated using Surface Evolver [27]; figure concept by Coble [26]. Courtesy of Ellen J. Siem.

The transition from the intermediate to the final stage occurs when the interconnected tubular porosity along the grain junctions (edges) breaks up because of the Rayleigh instability (see Section 14.1.2) and leaves isolated pores of equiaxed shape at the grain corners [7]. The final stage occurs when the porosity is isolated and located at multiple-grain junctions. Final pore elimination occurs by mass transfer from the grain boundaries to the pores attached to the grain boundaries, similar to the transport in the wire-bundle model treated in Section 16.1.2. If grain growth occurs during any stage, the pores may break away from the grain boundaries. In such cases, the pores will be isolated from the grain boundaries in the final stage, and further densification will be limited by the rate of crystal diffusion of atoms from dislocation sources by the mechanism **DS·XL** illustrated in Fig. 16.7. Failure to reach full density is often caused by such pore breakaway.

16.3.3 Model Sintering Experiments

Experiments have been designed to reveal details of the sintering mechanisms indicated by Fig. 16.7 and the sintering stages illustrated by Fig. 16.8. Such sintering experiments include sphere–sphere model experiments similar to that depicted in Fig. 16.6 [28], sintering of rows of spheres [29], sintering of spheres and wires to flat plates [30], and sintering of bundles of wires such as that depicted in Fig. 16.9 [31].

With their simple geometry, these model experiments reveal fundamental processes during the various stages of sintering. Initial-stage processes are illuminated by the sphere–sphere experiments, and transitions between the intermediate and final stages are captured in the wire-bundle experiments. Figure 16.9*d*, in particular, demonstrates the important role of grain-boundary attachment for pore removal—essentially all of the grain-boundary segments trapped between the pores have broken free and left the specimen. However, one boundary remains and continues to feed atoms to the pores to which it is connected.

16.3.4 Scaling Laws for Sintering

Because the surface energy per volume is larger for small particles and because the fundamental driving force for sintering is surface-energy reduction, compacts composed of smaller powders will typically sinter more rapidly. Smaller powders are more difficult to produce and handle; therefore, predictions of sintering rate dependence on size are used to make choices of initial particle size. Herring's

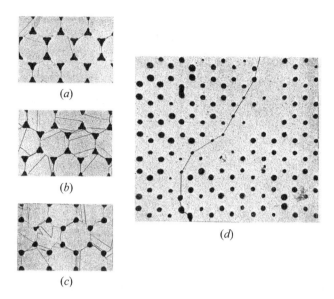

Figure 16.9:   Cross section of bundle of parallel 128 μm diameter Cu wires after sintering at 900°C: **(a)** 50 h, **(b)** 100 h, **(c)** 300 h; and at 1075°C: **(d)** 408 h. From Alexander and Balluffi [31].

scaling laws provide a straightforward method to predict sintering rate dependence on length scale [32].

Suppose that two sintering systems, S and B, are identical in all aspects except their size.[15] Each length dimension of system B is λ times as large as the corresponding dimension of system S. Under identical conditions and provided that the same sintering mechanism is operative, the ratio of sintering rates can be determined from the relative sizes of the specimens.

In general, a sintering rate is proportional to the mass-transport current, I, due to sintering driving forces and is inversely proportional to the transported material volume, ΔV, required to produce a given shape change (e.g., the volume associated with neck growth). The current I is the vector product of the atomic flux $\vec{J}$ and the area $\vec{A}$ through which the current flows during sintering. Therefore, the rates at which bodies S and B undergo geometrically similar changes will be in the ratio

$$\frac{\text{rate}_B}{\text{rate}_S} = \frac{I_B}{\Delta V_B} \frac{\Delta V_S}{I_S} \tag{16.51}$$

The current, I, is proportional to the diffusion potential gradient, $\nabla\Phi$, and to the cross-sectional area, $\vec{A}$, through which this flux flows. Therefore,

$$\frac{I_B}{I_S} = \frac{\nabla\Phi_B \cdot \vec{A_B}}{\nabla\Phi_S \cdot \vec{A_S}} \tag{16.52}$$

Suppose that the plane $\vec{A}$ is the bisector between the source of atoms and their sink (i.e., the sources and sinks listed in Table 16.1). The component of diffusion

---

[15] The systems may be similar powder compacts of the same powder material but differing particle sizes, or they may be model systems such as those illustrated in Figs. 16.6 and 16.9 but with all corresponding length dimensions scaled similarly.

potential gradient normal to the plane $\vec{A}$ (the component projected onto the normal $\hat{n}$ of $\vec{A}$) is proportional to the difference in diffusion potential between the source and the sink, $\Delta\Phi$, divided by the distance Δl between the source and sink. Therefore,

$$\frac{\nabla\Phi_B \cdot \hat{n}}{\nabla\Phi_S \cdot \hat{n}} = \frac{\Delta\Phi_B}{\Delta l_B}\frac{\Delta l_S}{\Delta\Phi_S} \tag{16.53}$$

Combining these relationships with Eq. 16.51,

$$\frac{\text{rate}_B}{\text{rate}_S} = \frac{\Delta\Phi_B}{\Delta\Phi_S}\frac{\Delta l_S}{\Delta l_B}\frac{A_B}{A_S}\frac{\Delta V_S}{\Delta V_B} \tag{16.54}$$

For free sintering, the diffusion potential is proportional to curvature; $\Delta\Phi_B/\Delta\Phi_S$ will scale as $1/\lambda$. The ratio $\Delta l_S/\Delta l_B$ also scales as $1/\lambda$.

For surface diffusion, one of the cross-sectional area's dimensions is δ, the thickness of the high-diffusivity surface layer, independent of system size. The remaining cross-sectional area length scales as λ, and therefore A_B/A_S must scale as λ. $\Delta V_S/\Delta V_B$ scales as $1/\lambda^3$. Therefore, substituting into Eq. 16.54,

$$\frac{\text{rate}_B}{\text{rate}_S} = \frac{1}{\lambda} \times \frac{1}{\lambda} \times \lambda \times \frac{1}{\lambda^3} = \lambda^{-4} \quad \text{(surface diffusion)} \tag{16.55}$$

If sintering occurs by diffusion through the bulk crystal (mechanism **BS·XL**), all the ratios will be the same as for surface diffusion except for the cross-sectional area A_B/A_S, which will scale as λ^2. Therefore,

$$\frac{\text{rate}_B}{\text{rate}_S} = \frac{1}{\lambda} \times \frac{1}{\lambda} \times \lambda^2 \times \frac{1}{\lambda^3} = \lambda^{-3} \quad \text{(crystal diffusion)} \tag{16.56}$$

If sintering occurs by grain-boundary diffusion, the ratio of rates will be the same as for the surface-diffusion case, λ^{-4}. A λ^{-1} scaling law can be derived for viscous flow and a λ^{-2} law applies for vapor transport [32].

To show that the rate of two-particle neck growth by surface diffusion in Section 16.1.4 is consistent with the λ^{-4} scaling law, Eq. 16.36 can be written in terms of its fundamental length scales and differentiated:

$$\frac{1}{x}\frac{dx}{dt} \propto \frac{R}{x^5} \tag{16.57}$$

Therefore,

$$\frac{\text{rate}_B}{\text{rate}_S} = \frac{(1/x_B)\frac{dx_B}{dt}}{(1/x_S)\frac{dx_S}{dt}} = \frac{R_B/x_B^5}{R_S/x_S^5} = \frac{\lambda R_S/(\lambda^5 x_S^5)}{R_S/x_S^5} = \lambda^{-4} \tag{16.58}$$

A similar result may be obtained (Exercise 16.6) using the result derived in Eq. 16.16 for the neck growth for a bundle of parallel wires by grain-boundary diffusion.

16.3.5 Sintering Mechanisms Maps

Any of the various mechanisms for sintering identified in Table 16.1 may contribute to the sintering rate. Which of the mechanisms contributes most to sintering depends on, among other things, particle size and temperature. Sometimes certain mechanisms can be ruled out immediately. For example, viscous flow (**VF**) cannot contribute for crystalline materials, and the nondensifying mechanisms (e.g., **SS·S** and **SS·V**) cannot contribute to pore removal in the final stages of sintering.

Processing decisions depend on the particular mechanism, or combination of mechanisms, that contribute to sintering. Sintering maps such as Fig. 16.10 provide information for such decisions. These plots can be created by employing approximate models for the sintering rates for specific systems by the various mechanisms. These models, combined with experimental data, can be used to plot regimes for which a particular mechanism makes the largest contribution to the sintering rate [33].

Sintering maps for different systems vary considerably—even for the same material, but having different initial particle sizes. For example, a map corresponding to Fig. 16.10 for silver particles of a smaller radius (i.e., 10 μm) shows a considerably reduced field for the **BS·XL** mechanism. On the other hand, a map for 10 μm UO$_2$ particles shows a vapor-transport (**SS·V**) regime [33]. Scaling laws are limited to regions of the sintering map where the dominant mechanism is unchanged (Section 16.3.4). Although sintering mechanism maps are no better than the models and data used to construct them, they provide useful insights.

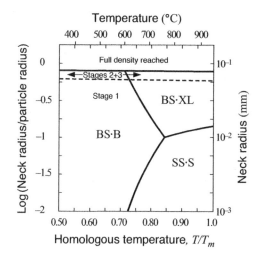

Figure 16.10: Sintering mechanism map for silver powder of radius 100 μm plotted with coordinates of reduced temperature and neck radius. The assumed conditions are that grain boundaries remain between abutting particles and that no trapped gases are present to impede isolated pore shrinkage. Each region represents the regime where the indicated mechanism is dominant (see Table 16.1). The dashed line indicates transitions between initial-stage and intermediate- and final-stage sintering. Although all possible mechanisms were considered, the three shown were dominant in their respective regimes. From Ashby [33].

Bibliography

1. R.B. Heady and J.W. Cahn. An analysis of capillary force in liquid-phase sintering of spherical particles. *Metall. Trans.*, 1(1):185–189, 1970.

2. J.W. Cahn and R.B. Heady. Analysis of capillary force in liquid-phase sintering of jagged particles. *J. Am. Ceram. Soc.*, 53(7):406–409, 1970.

3. W.C. Carter. The forces and behavior of fluids constrained by solids. *Acta Metall.*, 36(8):2283–2292, 1988.

4. R.M. Cannon and W.C. Carter. Interplay of sintering microstructures, driving forces, and mass transport mechanisms. *J. Am. Ceram. Soc.*, 72(8):1550–1555, 1989.

5. F. M. Orr, L. W. Scrivin, and T. Y. Chu. Mensici around plates and pins dipped in liquid—interpretation of Wilhemy plate and solderability measurements. *J. Colloid Interf. Sci.*, 60:402–405, 1977.

6. H. Udin, A. J. Shaler, and J. Wulff. The surface tension of copper. *Trans. AIME*, 186:186–190, 1949.

7. W.C. Carter and A.M. Glaeser. The morphological stability of continuous intergranular phases—thermodynamics considerations. *Am. Ceram. Soc. Bull.*, 63(8):993–993, 1984.

8. F.B. Hildebrand. *Advanced Calculus for Engineers*. Prentice-Hall, Englewood Cliffs, NJ, 2nd edition, 1976.

9. C. Herring. Diffusional viscosity of a polycrystalline solid. *J. Appl. Phys.*, 21:437–445, 1950.

10. W.S. Coblenz, J.M. Dynys, R.M. Cannon, and R.L. Coble. Initial stage solid-state sintering models: A critical analysis and assessment. In G. C. Kuczynski, editor, *Proceedings of the Fifth International Conference on Sintering and Related Phenomena*, pages 141–157, New York, 1980. Plenum Press.

11. F.A. Nichols and W.W. Mullins. Surface- (interface-) and volume-diffusion contributions to morphological changes driven by capillarity. *Trans. AIME*, 233(10):1840–1847, 1965.

12. A.P. Sutton and R.W. Balluffi. *Interfaces in Crystalline Materials*. Oxford University Press, Oxford, 1996.

13. W. Beeré. Stress redistribution during Nabarro-Herring and superplastic creep. *Metal Sci.*, 10(4):133–139, 1976.

14. W. Beeré. Stresses and deformation at grain boundaries. *Phil. Trans. Roy. Soc. London A*, 288(1350):177–196, 1978.

15. J. Pilling and N. Ridley. *Superplasticity in Crystalline Solids*. Institute of Metals, London, 1989.

16. L.D. Landau and E.M. Lifshitz. *Statistical Physics*. Pergamon Press, New York, 1963.

17. R. Raj and M.F. Ashby. On grain boundary sliding and diffusional creep. *Metall. Trans.*, 2:1113–1127, 1971.

18. B. Burton. *Diffusional Creep of Polycrystalline Materials*. Diffusion and Defect Monograph Series, No. 5. Trans Tech Publications, Bay Village, OH, 1977.

19. E. Arzt, M.F. Ashby, and R.A. Verrall. Interface controlled diffusional creep. *Acta Metall.*, 31(12):1977–1989, 1983.

20. M.F. Ashby. A first report on deformation mechanism maps. *Acta Metall.*, 20(7):887–897, 1972.

21. Y.-M. Chiang, D. Birnie, and W.D. Kingery. *Physical Ceramics*. John Wiley & Sons, New York, 1996.

22. W.D. Kingery, H.K. Bowen, and D.R. Uhlmann. *Introduction to Ceramics*. John Wiley & Sons, New York, 1976.

23. R.M. German. *Sintering Theory and Practice*. John Wiley & Sons, New York, 1996.

24. H.E. Exner. Principles of single phase sintering. *Reviews on Powder Metallurgy and Physical Ceramics*, 1:1–237, 1979.

25. R.W. Cahn. Recovery and recrystallization. In R.W. Cahn and P. Haasen, editors, *Physical Metallurgy*, pages 1595–1671. North-Holland, Amsterdam, 1983.

26. R.L. Coble. Sintering crystalline solids I. Intermediate and final state diffusion models. *J. Appl. Phys.*, 32(5):787, 1961.

27. K. A. Brakke. The Surface Evolver. *Exp. Math.*, 1(2):141–165, 1992. Publicly available software.

28. W. Zhang, P. Sachenko, J.H. Schneibel, and I. Gladwell. Coalescence of two particles with different sizes by surface diffusion. *Phil. Mag. A*, 82(16):2995–3011, 2002.

29. W.D. Kingery and M. Berg. Study of the initial stages of sintering solids by viscouus flow, evaporation-condensation, and self-diffusion. *J. Appl. Phys.*, 26(10):1205–1212, 1955.

30. G.C. Kuczynski. Theory of solid state sintering. In W. Leszynski, editor, *Powder Metallurgy*, pages 11–30, New York, 1961. Interscience Publishers.

31. B.H. Alexander and R.W. Balluffi. Mechanism of sintering of copper. *Acta Metall.*, 5(11):666–677, 1957.

32. C. Herring. Effect of change of scale on sintering phenomena. *J. Appl. Phys.*, 21:301–303, 1950.

33. M.F. Ashby. A first report on sintering diagrams. *Acta Metall.*, 22(3):275–289, 1974.

EXERCISES

16.1 An analysis of the rate of elongation of a wire possessing a bamboo-type grain structure is given in Section 16.1.3. An essential aspect of the analysis is the assumption that the stress-induced atomic transport producing the elongation is diffusion-limited. Now, construct the main framework of a model for the same system in which the atomic transport is source-limited, as indicated below, and explain how the model works.

- Assume that the grain boundaries are much poorer sources than the wire surface and that it is the poor source action of the boundaries that causes the process to be source-limited.

- Use the simple rate-constant type of formulation employed in Sections 13.4.2 and 15.1.1 to analyze source-limited precipitate growth and particle coarsening, respectively.

- Assume, as in Section 16.1.1, that the diffusion occurs via vacancies and that the rate-limiting process is the rate of creation and destruction of vacancies at the grain boundaries.

Solution. When the process is diffusion-limited and the rate-constant formalism is used, the net rate at which vacancies are destroyed at a boundary (i.e., the rate of destruction minus the rate of creation) is

$$I^B = 2 \int_0^R K^B \left[c_V^B(r) - c_V^{B,\text{eq}}(r) \right] 2\pi r \, dr \tag{16.59}$$

The corresponding rate at the free surface of the cell, which is of length L, is

$$I^S = \int_0^L K^S \left[c_V^S - c_V^{S,\text{eq}} \right] 2\pi R \, dz \tag{16.60}$$

I^B is negative, since $c_V^{B,\text{eq}} > c_V^B$ and vacancies are being created, whereas I^S is positive, since $c_V^S > c_V^{S,\text{eq}}$ and vacancies are being destroyed. The crystal diffusive current at each interface is equal to the net destruction rate, and therefore

$$I_{\text{diff}}^B = I^B \quad \text{and} \quad I_{\text{diff}}^S = I^S \tag{16.61}$$

In addition, in the quasi-steady state,

$$I^B = I_{\text{diff}}^B = -I^S = -I_{\text{diff}}^S \tag{16.62}$$

and the elongation rate, $\dot{\varepsilon}$, is given by

$$\frac{dL}{dt} = -\frac{I^B \Omega}{\pi R^2} \tag{16.63}$$

The diffusion within the grains is relatively slow, so that $I_{\text{diff}}^B = I^B$ and $I_{\text{diff}}^S = I^S$ are small compared to the vacancy creation and destruction rates in the equations above. Therefore, $c_V^B \cong c_V^{B,\text{eq}}$ and $c_V^S \cong c_V^{S,\text{eq}}$ and the rate is diffusion-limited.

When the rate is source-limited, the vacancy diffusion rate within the grains is relatively large. $K^B \ll K^S$ and the relatively slow source action at the boundary has a negligible effect on the vacancy gradients within the grains. The vacancy concentration is then maintained everywhere at an essentially constant level corresponding to $c_V^{S,\text{eq}}$ as a result of the relatively fast source action at the wire surface. I^B is then given by Eq. 16.59 with $c_V^B \cong c_V^{S,\text{eq}}$. Because the stress and the concentration $c_V^{B,\text{eq}}$ are uniform over the boundary, the elongation rate is

$$\frac{dL}{dt} = -2\Omega K^B \left[c_V^{S,\text{eq}} - c_V^{B,\text{eq}} \right] \tag{16.64}$$

The equilibrium vacancy concentration is given, in general, by Eq. 3.65, which for present purposes may be written in the form

$$c_V^{\text{eq}} = \frac{1}{\Omega} e^{-G_V^f/(kT)} = \frac{1}{\Omega} e^{-[G_V^f(\infty) + \Delta G_V^f]/(kT)} \tag{16.65}$$

where $G_V^f(\infty)$ is the work required to form a vacancy at a flat stress-free surface and ΔG_V^f is any additional work. The wire volume is increased when a vacancy is formed, and in the present model ΔG_V^f is negative when a vacancy is formed at the grain boundary because of the work done by the applied tensile stress during its formation. On the other hand, ΔG_V^f is positive at the wire surface because of the work that must be done to increase the surface area. Therefore, $c_V^{B,\text{eq}} > c_V^{S,\text{eq}}$ and dL/dt is positive.

16.2 Consider the diffusional creep of the idealized two-dimensional polycrystal illustrated in Fig. 16.4 and discussed in Section 16.2. Each boundary will be subjected to a normal stress, σ_n, and a shear stress, σ_s, as illustrated in Fig. 16.11. Suppose that all boundaries shear relatively slowly at a rate corresponding to

$$\frac{dS}{dt} = K\sigma_s \tag{16.66}$$

where K is a boundary shear rate constant, whereas diffusional transport between the different boundary segments is extremely rapid. The creep will then proceed at a rate controlled by the rate of the grain-boundary sliding and not by diffusional transport through the grains or along the grain boundaries. Using the results in Section 16.2, find an expression for the sliding-limited creep rate of the specimen illustrated in Figs. 16.4 and 16.11.

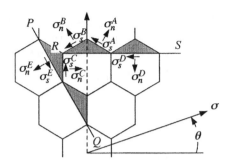

Figure 16.11:   Normal stresses and shear stresses present in two-dimensional polycrystal subjected to uniaxial applied stress, σ. The geometry is the same as in Fig. 16.4.

Determine the shear stresses acting on the three types of boundary segments present. When diffusion is extremely rapid, all differences in the diffusion potential will be eliminated, and all three normal stresses at the three different types of boundary segments will be uniform along each segment and equal to one another. Therefore,

$$\sigma_n^A = \sigma_n^B = \sigma_n^C = \sigma_n \tag{16.67}$$

Also, if each grain is not to rotate,

$$\sigma_s^A + \sigma_s^B + \sigma_s^C = 0 \tag{16.68}$$

Since the stresses in each grain are the same, the normal and shear stresses along planes PQ and RS can be found from the forces exerted on them by the applied stress, σ, with the results

$$\sigma_s^D = \sin\theta\cos\theta\,\sigma$$
$$\sigma_n^D = \sin^2\theta\,\sigma$$
$$\sigma_s^E = \left[-\frac{\sqrt{3}}{4}\cos^2\theta + \frac{1}{2}\sin\theta\cos\theta + \frac{\sqrt{3}}{4}\sin^2\theta\right]\sigma \tag{16.69}$$
$$\sigma_n^E = \left[\frac{3}{4}\cos^2\theta + \frac{\sqrt{3}}{2}\sin\theta\cos\theta + \frac{1}{4}\sin^2\theta\right]\sigma$$

Next, each triangular shaded region in Fig. 16.11 must be in mechanical equilibrium (i.e., the sum of the forces on it parallel and normal to PQ, or RS, must be zero). This leads to the conditions

$$\begin{aligned}
0 &= -2\sqrt{3}\,\sigma_n^D - \sigma_s^B + 2\sqrt{3}\,\sigma_n + \sigma_s^A \\
0 &= 2\,\sigma_s^D + \sigma_s^B + \sigma_s^A \\
0 &= -2\,\sigma_s^E + \sigma_s^A + \sigma_s^C \\
0 &= 2\sqrt{3}\,\sigma_n^E + \sigma_s^A - 2\sqrt{3}\,\sigma_n - \sigma_s^C
\end{aligned} \tag{16.70}$$

These linear equations are sufficient to allow the determination of the shear stress acting on each boundary segment.

Solution. Expressions for the shear stresses at the three types of boundary segments may be obtained by the simultaneous solution of the equations given above for the boundary stresses. The results are

$$\sigma_s^A = \left[-\sqrt{3}\cos^2\theta - \sin\theta\cos\theta + \frac{\sqrt{3}}{2} \right]\sigma$$

$$\sigma_s^B = \left[\sqrt{3}\cos^2\theta - \sin\theta\cos\theta - \frac{\sqrt{3}}{2} \right]\sigma \qquad (16.71)$$

$$\sigma_s^C = 2\sin\theta\cos\theta\,\sigma$$

The expression for the creep rate due to boundary sliding is obtained by differentiating Eq. 16.48,

$$\dot{\varepsilon} = \frac{1}{d}\left[\frac{1}{\sqrt{3}}\frac{d(S_s^A - S_s^B)}{dt}(1 - 2\cos^2\theta) + \frac{1}{3}\frac{d(2S_s^C - S_s^A - S_s^B)}{dt}2\sin\theta\cos\theta \right] \quad (16.72)$$

Substituting Eqs. 16.66 and 16.71 into Eq. 16.72 then produces the surprisingly simple result

$$\dot{\varepsilon} = \frac{K}{d}\sigma \qquad (16.73)$$

The creep rate is therefore proportional to the applied stress, and the polycrystal acts effectively as an ideally viscous material.

16.3 Diffusional creep can also occur by means of the stress-motivated transport of atoms between climbing dislocations in a material. This is illustrated in a highly idealized manner in Fig. 16.12, where a regular array of edge dislocations possessing four different Burgers vectors is present in a stressed material. The net Burgers vector content is zero. The stress exerts climb forces on the dislocations so that dislocations with Burgers vectors lying along $\pm x$ and $\pm y$ directions act alternately as sources and sinks. The arrows indicate the atomic fluxes associated with the climb. Each source dislocation is surrounded by four nearest-neighbor sink dislocations, and vice versa for the sink disloca-

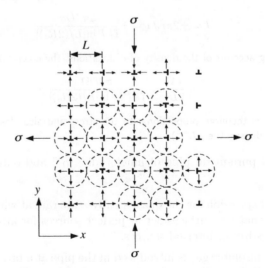

Figure 16.12:   Idealized array of edge dislocations subjected to applied stress, σ. Arrows show stress-induced diffusion current around each climbing dislocation.

tions. The climb of the dislocations in this arrangement adds atomic planes lying perpendicular to x and removes an equal number of planes lying perpendicular to y, causing the specimen to lengthen and shorten in the direction of the applied stresses. Find an expression for the instantaneous quasi-steady-state creep rate of this idealized structure, assuming that the dislocations act as perfect sources or sinks.

- Surround each dislocation by a cylindrical cell in which the diffusion to/from the dislocation is assumed to be cylindrical (see Fig. 16.12), and use a mean-field approximation similar to the one used in the analysis of particle coarsening.

Solution. The flux equation is given by Eq. 13.3, and the diffusion equation in the quasi-steady state is $\nabla^2 \Phi_A = 0$. The derivation of Eq. 13.4 shows that the climb force exerted on the sink dislocations will cause the value of Φ_A at their core radii, $R_\circ$, to have the value

$$\Phi_A^D(\text{sinks}) = \mu_A^\circ - \sigma \Omega \tag{16.74}$$

while at the source dislocations,

$$\Phi_A^D(\text{sources}) = \mu_A^\circ + \sigma \Omega \tag{16.75}$$

At the surface of the cell at $r = L/2$, we use the mean-field boundary value

$$\Phi_A\left(\frac{L}{2}\right) = \mu_A^\circ \tag{16.76}$$

The general solution of the diffusion equation in cylindrical coordinates is $\Phi_A = a_1 \ln r + a_2$, and using the boundary conditions above to determine the constants a_1 and a_2,

$$\Phi_A - \mu_A^\circ = \pm \frac{\sigma \Omega}{\ln[L/(2R_\circ)]} \ln\left(\frac{L}{2r}\right) \tag{16.77}$$

Using the flux equation, the diffusion current into a dislocation (per unit length) is

$$I = \pm 2\pi r J_A = \pm \frac{2\pi \,^\star\!D\sigma}{fkT \ln[L/(2R_\circ)]} \tag{16.78}$$

and after taking account of the density of dislocations, the creep rate along x is

$$\dot{\varepsilon} = \frac{I\Omega}{2L^2} = \frac{\pi \,^\star\!D\Omega}{fkTL^2 \ln[L/(2R_\circ)]} \sigma \tag{16.79}$$

The creep rate is therefore proportional to the stress and also closely proportional to the dislocation density (i.e., L^{-2}).

16.4 A thin-walled pure-metal pipe of inner radius R^{in} and outer radius R^{out} is heated.

(a) Find an expression for the quasi-steady-state rate at which it will shrink. Assume that the surfaces act as perfect sources for atoms and that the interior is free of internal sources.

(b) An inert insoluble gas is introduced in the pipe at a pressure P. Find the value of P that will stop the pipe from shrinking. The external pressure is small enough so that it may be ignored.

Solution.

(a) In the quasi-steady-state Laplace equation, $\nabla^2\Phi(r) = 0$ holds for the diffusion potential and Eq. 13.3 holds for the diffusion flux. The boundary conditions on Φ at the surfaces are

$$\Phi(r = R^{\text{out}}) = \Phi^{\text{out}} = \mu^\circ + \frac{\Omega\gamma^S}{R^{\text{out}}}$$
$$\Phi(r = R^{\text{in}}) = \Phi^{\text{in}} = \mu^\circ - \frac{\Omega\gamma^S}{R^{\text{in}}} \tag{16.80}$$

Using the solution of the Laplace equation for diffusion in cylindrical coordinates given by Eq. 5.10, fitting it to the boundary conditions given by Eq. 16.80, and employing Eq. 13.3 for the flux, the total diffusion current of atoms (per unit pipe length) passing radially from R^{in} to R^{out} is

$$I = \frac{2\pi^{\star}D(\Phi^{\text{in}} - \Phi^{\text{out}})}{\Omega kT \mathbf{f} \ln(R^{\text{out}}/R^{\text{in}})} \tag{16.81}$$

which may be compared with Eq. 5.13. Now,

$$\frac{dR^{\text{out}}}{dt} = \frac{\Omega I}{2\pi R^{\text{out}}} \tag{16.82}$$

and for the thin-walled pipe, $\ln(R^{\text{out}}/R^{\text{in}}) = \ln(1 + \Delta R/R^{\text{in}}) \approx \Delta R/\langle R\rangle$, where $\Delta R = R^{\text{out}} - R^{\text{in}}$ and $R^{\text{out}} \approx R^{\text{in}} \approx \langle R\rangle$. Using these results and Eq. 16.81,

$$\frac{d\langle R\rangle}{dt} = -\frac{2^{\star}D\Omega\gamma^S}{kT\mathbf{f}\langle R\rangle\,\Delta R} \tag{16.83}$$

(b) The internal pressure causes the diffusion potential at R^{in} to be $\Phi^{\text{in}} = \mu^\circ - \Omega\gamma^S/R^{\text{in}} + \Omega P$. Equation 16.81 then becomes

$$I = \frac{2\pi^{\star}D(\Phi^{\text{in}} - \Phi^{\text{out}})}{\Omega kT \mathbf{f} \ln(R^{\text{out}}/R^{\text{in}})} = \frac{2\pi^{\star}D(P - 2\gamma^S/\langle R\rangle)}{kT\mathbf{f} \ln(R^{\text{out}}/R^{\text{in}})} \tag{16.84}$$

and shrinkage will stop when $P = 2\gamma^S/\langle R\rangle$.

16.5 Suppose that a body made up of fine particles can sinter by either the crystal diffusion mechanism **BS $\cdot$ XL** or the grain-boundary diffusion mechanism **BS $\cdot$ B** as illustrated in Fig. 16.7. How will the relative sintering rates due to these two mechanisms vary as:

(a) The particle size is decreased?

(b) The temperature is decreased?

Solution.

(a) Let

$$\text{Ratio} = \frac{\text{sintering rate due to grain-boundary diffusion}}{\text{sintering rate due to crystal diffusion}}$$

The sintering rate due to boundary diffusion and crystal diffusion will be proportional to $^{\star}D^B$ and $^{\star}D^{XL}$, respectively. The scaling laws show that the sintering rate due to boundary diffusion will decrease by the factor λ^{-4} when the particle size is increased by the factor λ. The corresponding factor for sintering by crystal diffusion is λ^{-3}. Therefore,

$$\text{Ratio} \propto \frac{^{\star}D^B}{^{\star}D^{XL}}\frac{\lambda^3}{\lambda^4} = \frac{^{\star}D^B}{^{\star}D^{XL}\lambda} \tag{16.85}$$

Sintering by boundary diffusion will become more important as the particle size decreases.

(b) Because $^{\star}D^B$ increases relative to $^{\star}D^{XL}$ as the temperature decreases, sintering by boundary diffusion will become more important as the temperature decreases.

16.6 Show that a λ^{-4} scaling law holds for the sintering of a bundle of parallel wires by means of grain-boundary diffusion, which was analyzed in Section 16.1.2.

Solution. The rate of sintering is given by Eq. 16.16. Using the formalism of Section 16.3.4, where all dimensions in the B system are λ times larger than in the S system,

$$\begin{aligned}\frac{\text{rate}_B}{\text{rate}_S} &= \frac{(1/L_B)\frac{dL_B}{dt}}{(1/L_S)\frac{dL_S}{dt}} \propto \frac{[1/(L_B w_B^3)]\left[f_{\text{app}} + 2\gamma^S(\kappa_B w_B - \sin\frac{\psi}{2})\right]}{[1/(L_S w_S^3)]\left[f_{\text{app}} + 2\gamma^S(\kappa_S w_S - \sin\frac{\psi}{2})\right]}\\[2mm] &= \frac{[1/(\lambda L_S \lambda^3 w_S^3)]\left\{f_{\text{app}} + 2\gamma^S[(\kappa_S/\lambda)\lambda w_S - \sin\frac{\psi}{2}]\right\}}{[1/(L_S w_S^3)]\left[f_{\text{app}} + 2\gamma^S(\kappa_S w_S - \sin\frac{\psi}{2})\right]} = \lambda^{-4}\end{aligned} \tag{16.86}$$

16.7 Consider a grain boundary containing a uniform distribution of small pores (as shown in Fig. 16.13) that is subjected to a normal tensile stress σ_{nn}^∞ at a large distance from the boundary.

The pores will either grow or shrink by transferring atoms via grain-boundary diffusion to or from the grain boundary acting as a sink or source, respectively, depending upon the magnitude of the applied stress. Find an expression for the rate of growth of the pore volume in a form proportional to the quantity $(F_{\text{app}} + \Upsilon)$, where F_{app} is the force applied to each pore cell (shown dashed in Fig. 16.13) and Υ is the corresponding capillary force given by

$$\Upsilon = -\frac{2\pi\gamma^S R_c^2}{R} \tag{16.87}$$

- Construct a cylindrical cell of radius R_c centered on a single pore as illustrated in Fig. 16.13 and solve the diffusion problem within it using cylindrical coordinates and the same basic method employed to obtain

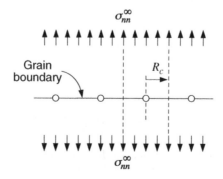

Figure 16.13: Distribution of pores in grain boundary subjected to tensile stress σ_{zz}^∞.

the sintering rate of a bundle of parallel wires in Section 16.1.2. Assume that the pore maintains the equilibrium shape illustrated in Fig. 16.14. The upper and lower surfaces are spherical with curvature $-2/R$.

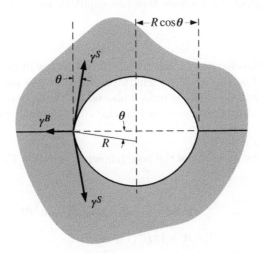

Figure 16.14:   Cross section of equilibrium shape of pore on grain boundary as in Fig. 16.13, assuming that $\gamma^B/\gamma^S = 1/3$, so that $\theta = 9.6°$. R is the radius of curvature of the top and bottom surfaces. The cross section in the grain-boundary plane is a circle of radius $R \cos \theta$.

Solution. Equation 16.3 applies and therefore

$$\nabla^2 \sigma_{nn} = \text{constant} = A = \frac{1}{r} \frac{d}{dr} \left(r \frac{d\sigma_{nn}}{dr} \right) \qquad (16.88)$$

Integrating twice and applying the boundary conditions corresponding to

$$\left[\frac{d\sigma_{nn}}{dr} \right]_{r=R_c} \qquad (16.89)$$

and Eq. 16.4

$$\sigma_{nn} = \frac{A}{4} \left[(r^2 - R^2 \cos^2 \theta) + 2R_c^2 \ln \left(\frac{R \cos \theta}{r} \right) \right] - \gamma^S \kappa \qquad (16.90)$$

A is now obtained by applying the integral force condition given by Eq. 16.5 and using the curvature relation $\kappa = -2/R$. Therefore,

$$A = -\frac{8(F_{\text{app}} + \Upsilon)}{\pi \left\{ R_c^4 \left[4 \ln \left(\frac{R_c}{R \cos \theta} \right) - 3 \right] + R^2 \cos^2 \theta (4R_c^2 - R^2 \cos^2 \theta) \right\}} \qquad (16.91)$$

with Υ given by

$$\Upsilon = -\frac{2\pi \gamma^S R_c^2}{R} \qquad (16.92)$$

and $F_{\text{app}} = \pi R_c^2 \sigma_{nn}^\infty$. The rate at which volume is transferred to the pore by grain-boundary diffusion is then

$$\frac{dV}{dt} = -\frac{\Omega \delta D^B \pi (R_c^2 - R^2 \cos^2 \theta)}{kT} A \qquad (16.93)$$

and dV/dt is seen to be proportional to $(F_{\text{app}} + \Upsilon)$.

16.8 Since pore shrinkage is driven by a decrease in interfacial energy, it may be expected on general principles that the capillary force, Υ, must correspond to

$$\Upsilon = -\frac{dG_I}{dL} \tag{16.94}$$

where dG_I is the change in the total interfacial energy of the system and dL is the corresponding change in its length produced by the pore shrinkage. Now demonstrate that Eq. 16.94 is indeed obeyed for the shrinkage of the pores on the grain boundary considered in Exercise 16.7, where Υ was found to equal $-2\pi\gamma^S R_c^2/R$.

- The volume and area of the pore (shown in detail in Fig. 16.14) are given by

$$V = \frac{2\pi R^3}{3}(2 - 3\cos\theta + \sin^3\theta)$$
$$A = 4\pi R^2(1 - \sin\theta) \tag{16.95}$$

Solution. Focus on one cell as illustrated in Fig. 16.13 and take the cell height to be L. Using Eq. 16.95, the total interfacial energy of the cell is given by

$$G_I = 4\pi R^2(1 - \sin\theta)\gamma^S + \pi(R_c^2 - R^2\cos^2\theta)\gamma^B \tag{16.96}$$

Then, since $\sin\theta = \gamma^B/(2\gamma^S)$,

$$-\frac{dG_I}{dL} = 2\pi\gamma^S R[2\sin\theta\cos^2\theta - 4(1 - \sin\theta)]\frac{dR}{dL} \tag{16.97}$$

However, the volume in the cell must remain constant, so

$$dV = d\left[\pi R_c^2 L - \frac{2\pi R^3}{3}(2 - 3\cos\theta + \sin^3\theta)\right] = 0 \tag{16.98}$$

and

$$\frac{dR}{dL} = \frac{R_c^2}{2R^2(2 - 3\cos\theta + \sin^3\theta)} \tag{16.99}$$

Substituting Eq. 16.99 into Eq. 16.97 and employing Eq. 16.94,

$$\Upsilon = -\frac{dG_I}{dL} = \frac{2\pi\gamma^S R_c^2[\sin\theta\cos^2\theta - 2(1 - \sin\theta)]}{R(2 - 3\cos\theta + \sin^3\theta)} = -\frac{2\pi\gamma^S R_c^2}{R} \tag{16.100}$$

in agreement with Eq. 16.92.

PART IV

PHASE TRANSFORMATIONS

Phase transformations are of central importance in materials science and engineering. An understanding of the thermodynamics of phase equilibria is the foundation for understanding their kinetics. Necessary conditions for equilibrium include: uniformity and equality of the diffusion potential for each chemical species that can be exchanged between the phases; equality of temperature; and equality of pressure if the two phases can freely exchange volume.[16] Deviations from these equilibrium conditions set the stage for kinetic processes. Parts II and III chiefly treated kinetic processes that derived from nonuniformity of a potential, such as chemical potential or temperature, within a single phase. Phase transformations occur when a region of the material can reduce the total free energy by changing its symmetry, equilibrium composition, equilibrium density, or any other quantity that defines a phase.

The transforming material portion may be adjacent to its prospective phase, which is the case for growth of a new phase; or the portion may be isolated, which is the case for nucleation of a new phase. In any case, the spatial variation of

[16]Strict equality of pressure is required absent capillarity effects: if a deforming heterophase interface stores energy during volume transfer, the two phases will have an equilibrium pressure difference.

the phase-defining quantities—order parameters—permit a convenient means to identify heterophase interfaces.

The definition of what constitutes a phase is troublesome. Gibbs required 40 pages of preamble before introducing *phase* with "such bodies as differ in composition or state, different phases of the matter considered, regarding all bodies which differ only in quantity and form as different examples of the same phase" [1]. This clearly eliminates two bodies that are identical except for their morphology and size—and perhaps one may credit Gibbs with the foresight to exclude crystallographic misorientation and symmetry operations from distinguishing phases. However, special cases, such as the distinction of a nearly cubic tetragonal variant from a strained cubic phase, must be handled carefully.

Contiguous phases must be separated by an interface, and therefore considerations of interface and morphological evolution play a role in phase transformation kinetics. However, every interface need not separate two phases—grain and antiphase boundaries separate crystallographic or symmetric variants of a single phase. Nevertheless, it is instructive to treat such interface motion—where a single phase alters its orientation—analogously with phase transformations. Such a treatment naturally introduces two different kinds of order parameters. Regions of material defined by one kind of order parameter—such as spin density, symmetry, and orientation—may alter without a corresponding flux. Such flux-less order parameters are called *nonconserved variables*. *Conserved variables*, such as composition, require flux for a material to change locally.

The thermodynamics of phase equilibria is reviewed in Chapter 17 and the fundamental thermodynamic differences between conserved and nonconserved order parameters are reinforced with a geometrical construction. These order parameters are used in the kinetic analyses of *continuous* and *discontinuous* phase transformations.

Continuous transformations are treated in detail in Chapter 18. Spinodal decomposition and certain types of order–disorder transformations follow from similar principles but differ only in the kinetics of conserved and nonconserved variables.

The remainder of the book treats discontinuous transformations. Nucleation, which is necessary for the production of a new phase, is treated in Chapter 19. The growth of new phases under diffusion- and interface-limited conditions is treated in Chapter 20. Concurrent nucleation and growth is treated in Chapter 21. Specific examples of discontinuous transformations are discussed in detail; these include solidification (Chapter 22), precipitation from solid solution (Chapter 23), and martensite formation (Chapter 24).

Bibliography

1. J.W. Gibbs. On the equilibrium of heterogeneous substances (1876). In *Collected Works*, volume 1. Longmans, Green, and Co., New York, 1928.

CHAPTER 17

GENERAL FEATURES OF PHASE TRANSFORMATIONS

The conditions under which a portion of a material system will undergo a phase transformation are determined by the system's current and equilibrium thermodynamic states. The equilibrium state is distinguished by the minimum of an energy function that is particular to physical constraints imposed on the system. For instance, under conditions of constant temperature, T, and applied pressure, P, if the total energy quantity

$$\mathcal{G} \equiv \mathcal{U} - T\mathcal{S} + P\mathcal{V} \equiv \mathcal{F} + P\mathcal{V} \equiv \mathcal{H} - T\mathcal{S} \tag{17.1}$$

can be decreased via any internal change in the system, the system is thermodynamically unstable with respect to that change, however large. If $\mathcal{G}$ cannot be decreased by any possible small variation, the system is in local equilibrium; if $\mathcal{G}$ achieves its global minimum, the system will remain in complete equilibrium as long as it is constrained to the same constant P and T. Under different constraints, other minimizing energy functionals can be derived through Legendre transformations.

The necessary conditions for a spontaneous phase transformation relate directly to the system's energy differences upon transformation. In addition to the transformed volume's chemical-energy change, its interfacial energy and the elastic energy to accommodate interfacial misfit contribute to the total free-energy change. Calculations of energy differences are simplified when the host material can be approximated as a reservoir with constant properties, so that the transformed volume and its interface need to be considered.

If in addition to a thermodynamic driving force, a system has kinetic mechanisms available to produce a phase transformation (e.g., diffusion or atomic structural relaxation), the rate and characteristics of phase transformations can be modeled through combinations of their cause (thermodynamic driving forces) and their kinetic mechanisms. Analysis begins with identification of parameters (i.e., *order parameters*) that characterize the internal variations in state that accompany the transformation. For example, site fraction and magnetization can serve as order parameters for a ferromagnetic crystalline phase.

Analysis proceeds by considering the temporal evolution of small variations in order parameter fields. However, a variation may be "small" in different ways. J. Willard Gibbs distinguished between a variation that "initially is small in degree, but may be great in its extent in space" and one that is "initially small in extent but great in degree" [1]. In the context of phase transformations, *degree* applies to the magnitude of an order parameter that characterizes a phase—specifically, whether it may vary continuously or not. *Extent* refers to the spatial region over which such variation occurs—specifically, whether the change is confined to a (typically small) finite material portion or throughout the entire material system.

Gibbs's classification serves as the fundamental basis for division of phase transformation processes into two broad categories: *continuous phase transformations* and *discontinuous phase transformations*. During a continuous transformation, order parameter fields evolve smoothly in time and evolution is not confined to a small region. Discontinuous transformations initiate with an abrupt variation in an order parameter field and are localized events (i.e., they involve nucleation). Subsequent to initiation, phase transformations can continue by *growth*, which occurs as host material adjacent to the interface transforms into the new phase. Growth is treated in Chapter 21. After driving forces for growth have been exhausted, the system can continue to evolve by coarsening through reducing the energetic contribution from interfaces at constant phase fraction (Chapter 15).

Model energy functionals will be obtained through consideration of the energetic contribution of order parameter fields, and this is preceded by a survey of order parameters.

17.1 ORDER PARAMETERS

17.1.1 One-Component or Fixed Stoichiometry Systems

Figure 17.1 shows the molar free energy, F, as a function of temperature for a pure, or stoichiometric, material at fixed volume. The material has a first-order phase transformation at the temperature where the molar free energies cross. The equilibrium free energy is a function of temperature only. The corresponding order parameter, ξ, which is also a function of T as illustrated in Fig. 17.1b, is a subsidiary parameter introduced by the series expansion

$$F(T,\xi) = a_0(T) + a_1(T)\xi + a_2(T)\xi^2 + \cdots \tag{17.2}$$

commonly known as a *Landau expansion* [2, 3]. The physical quantity corresponding to ξ might be a molar heat capacity, enthalpy density, or any derivative of the molar free energy and its metastable extensions. F is taken as a function of both T and ξ. At any temperature, the equilibrium value of ξ, $\xi^{\text{eq}}(T)$, is determined from the

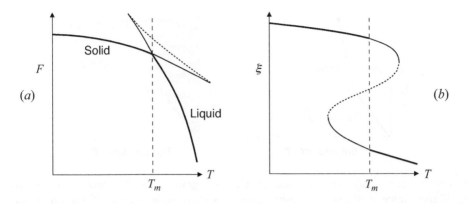

Figure 17.1:   **(a)** Molar free energy as a function of temperature for a melting transition. **(b)** Temperature dependence of an order parameter ξ for the transition in (a).

condition

$$\left(\frac{\partial F}{\partial \xi}\right)_T = 0 \tag{17.3}$$

and using Eq. 17.2, the equilibrium free energy, F^{eq}, can be expressed as a function of T alone, in the form $F^{\text{eq}}[T, \xi^{\text{eq}}(T)]$, and $\xi(\vec{x}) - \xi^{\text{eq}}$ becomes a measure of the local departure from equilibrium. The function $F(T, \xi)$ defined in Eq. 17.2 can be used to model the free energies of systems that are in the process of moving toward equilibrium—i.e., undergoing a phase transformation represented by particular values of ξ.

Introduced in this manner, the order parameter, ξ, is a "hidden" thermodynamic variable: its equilibrium values, $\xi^{\text{eq}}(T)$, are not independent but are fixed by Eq. 17.3. Therefore, an order parameter is characteristic of the transformation process because it cannot be fixed by an experimental condition.

Whether the phase transition is first- or second-order depends on the relative magnitudes of the coefficients in the Landau expansion, Eq. 17.2. For a first-order transition, the free energy has a discontinuity in its first derivative, as at the temperature T_m in Fig. 17.1a, and higher-order derivative quantities, such as heat capacity, are unbounded. In second-order transitions, the discontinuity occurs in the second-order derivatives of the free energy, while first derivatives such as entropy and volume are continuous at the transition.

The order of a transition can be illustrated for a fixed-stoichiometry system with the familiar P–T diagram for solid, liquid, and vapor phases in Fig. 17.2. The curves in Fig. 17.2 are sets of P and T at which the molar volume, V, has two distinct equilibrium values—the discontinuous change in molar volume as the system's equilibrium environment crosses a curve indicates that the phase transition is first order. Critical points where the change in the order parameter goes to zero (e.g., at the end of the vapor–liquid coexistence curve) are second-order transitions.

Connections to other types of phase diagrams can be obtained if order parameters are exchanged for intensive variables. Figure 17.2 is replotted with the order parameter V as the ordinate in Fig. 17.3b. The diagram predicts the phases that would exist for a molar volume fixed by a rigid container at different temperatures. The tie-lines connect equilibrium molar volumes at the same temperature

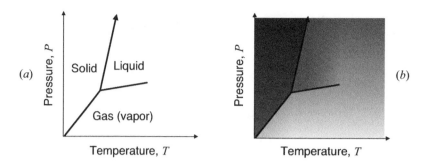

Figure 17.2:   (a) Single-component phase diagram.   (b) Shading represents the equilibrium value of a molar extensive quantity such as molar volume V (i.e., light gray represents a large value and dark gray a small value) that apply to each phase at that particular P and T. For phase transitions, the grayscale (or V) could be used as an order parameter indicating phase.

and pressure. An analog to a ternary diagram could be obtained by substituting molar entropy for the T-axis in Fig. 17.3.

Order parameters may also refer to underlying atomic structure or symmetry. For example, a piezoelectric material cannot have a symmetry that includes an inversion center. To model piezoelectric phase transitions, an order parameter, η, could be associated with the displacement of an atom in a fixed direction away from a crystalline inversion center. Below the transition temperature T_c, the molar Gibbs free energy of a crystal can be modeled as a Landau expansion in even powers of η (because negative and positive displacements, η, must have the same contribution to molar energy) with coefficients that are functions of fixed temperature and pressure,

$$G(T, P, \eta) = G_o(T, P) + a(P)(T - T_c)\eta^2 + B(P)\eta^4 \qquad (17.4)$$

The equilibrium state is entirely determined by the minima of G as a function of pressure and temperature. The equilibrium order parameter η^{eq} is determined by the minima of G and the equilibrium molar free energy can be calculated explicitly,

$$\eta^{\text{eq}} = \begin{cases} \pm\sqrt{a(P)(T_c - T)/[2B(P)]} & \text{if } T < T_c \\ 0 & \text{if } T \geq T_c \end{cases} \qquad (17.5)$$

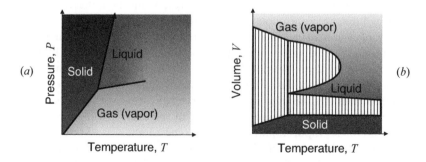

Figure 17.3:   (a) Single-component P–T phase diagram. (b) Phase diagram obtained from (a) by plotting the molar volume, V, as an order parameter in place of pressure. The grayscale could indicate variations in an order parameter such as S.

$$G^{\text{eq}}(T, P) - G_o(T, P) = \begin{cases} -[a(P)(T - T_c)]^2/[4B(P)] & \text{if } T < T_c \\ 0 & \text{if } T \geq T_c \end{cases} \quad (17.6)$$

where Eq. 17.6 is the free-energy change when a mole undergoes a phase transformation from the nonpiezoelectric phase ($\eta = 0$) to a piezoelectric phase ($\eta^2 \geq 0$). Atomic displacements of opposite sign, $\pm \eta$, correspond to different polarizations of the same piezoelectric phase.

Naturally, the fixed composition phase transformations treated in this section can be accompanied by local fluctuations in the composition field. Because of the similarity of Fig. 17.3 to a binary eutectic phase diagram, it is apparent that composition plays a similar role to other order parameters, such as molar volume. Before treating the composition order parameter explicitly for a binary alloy, a preliminary distinction between *types* of order parameters can be obtained. Order parameters such as composition and molar volume are derived from extensive variables; any kinetic equations that apply for them must account for any conservation principles that apply to the extensive variable. Order parameters such as the atomic displacement η in a piezoelectric transition, or spin in a magnetic transition, are not subject to any conservation principles. Fundamental differences between *conserved* and *nonconserved* order parameters are treated in Sections 17.2 and 18.3.

17.1.2 Two-Component Systems

For a binary A–B alloy, another independent parameter, X_B (or $X_A = 1 - X_B$) must be added to the fixed-stoichiometry order parameters in the preceding section. The phenomenological form of the Landau expansion, Eq. 17.2, can be extended to include X_B and has been used to catalog the conditions for many transitions in two-component systems [3].

The methods of constructing homogeneous molar free energies for phase diagrams can also be used to construct first-order approximations of free-energy densities when a composition order parameter field is heterogeneous. Multicomponent phase diagrams can be accurately predicted from empirical, computed, or theoretical models of the composition dependence of molar enthalpies and entropies of formation and mixing. Macroscopic models for molar free energies of mixing can be obtained from combinations of atomistic bond energies, crystal structure, and configurational and vibrational entropy. A simple example of ordering or clustering of an A–B alloy on a b.c.c. lattice illustrates how composition and structural order parameters arise naturally in the construction of the homogeneous molar free energy.

Decomposition and Order–Disorder Transformations on a B.C.C. Lattice.[1] Suppose that two species, A and B, occupy a b.c.c. lattice. If unlike bonds have lower molar enthalpies than like bonds [i.e., $H_{AB} < \frac{1}{2}(H_{AA} + H_{BB})$], then at low temperatures, *ordered* structures result in which the nearest-neighbors of A atoms are

---

[1]In this section, the terms *ordered structures*, *order parameters*, and *ordering transformations* appear and may present some confusion. Unfortunately, these historic terms are in common use. An ordered structure typically indicates a regular site occupation pattern at the microscopic scale. Ordering transformations are those associated with such regular microscopic patterns. Order parameters are coarse-grained measures that collectively indicate phase plus additional information indicating geometric configurations of the same phase—for example, at antiphase boundaries.

predominately B atoms.[2] At high temperatures, disordered structures should appear because of the significant entropic contributions of the numerous disordered configurations but at the expense of slightly increased molar enthalpies. In this case, the high-temperature phase is b.c.c. ($A2$, in *Structurbericht* notation) and, on cooling, there is a phase transition to the $B2$ structure (primitive-cubic ionic CsCl is the prototype of the $B2$ structure shown in Fig. 17.4). At this phase transition, there is a decrease of symmetry on the transformation to the ordered phase—the $a/2\langle 111 \rangle$ translational symmetry of the $A2$ phase is lost. An order parameter that indicates this symmetry loss would be a candidate to characterize an order–disorder transition on the b.c.c. lattice. An example of a structure that undergoes such an ordering transformation is β-brass (a Cu–Zn alloy).

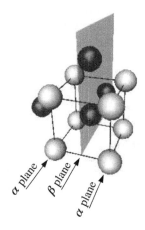

Figure 17.4: $B2$ ordering on a b.c.c. lattice where grayscale is associated with the composition at each site. $B2$ is a particular type of microscopic ordering (of which CsCl is the prototype); the compositions are the same on all α-planes but differ from the composition on all the β-planes.

Considering Fig. 17.4, the development of the $B2$ structure creates two sublattices from the original $A2$ structure. One of the $B2$ sublattices consists of the b.c.c. unit-cell centers (indicated by β in Fig. 17.4) are displaced from the b.c.c. corners (α in Fig. 17.4) by $a/2\langle 111 \rangle$. An ordering transformation produces sublattices, α and β, with differing site fractions, X_B^α and X_B^β. Their difference becomes a structural order parameter:

$$\eta = \frac{1}{2}\left(X_B^\alpha - X_B^\beta\right) \tag{17.7}$$

η evolves from zero toward equilibrium finite values $\pm\eta^{\text{eq}}$. Symmetric stable values exist because the b.c.c. corner sites are equivalent to the b.c.c. center sites—any result must be invariant to exchange of α with β sublattices. Compositions on the two sublattices must be coupled to the local average composition,

$$X_B = \frac{1}{2}\left(X_B^\alpha + X_B^\beta\right) \tag{17.8}$$

---

[2]The molar enthalpy and the molar internal energy of bonding, H_{AB} and U_{AB}, are related to the bond energies E_{AB} at constant pressure and at constant volume, respectively.

which is unchanged from the system's average composition upon ordering. The order parameters η and X_B are derivable from microscopic values of X_B^α and X_B^β.

However, if unlike bonds have higher molar enthalpies, low-temperature equilibrium configurations will consist of two-phase A- and B-rich $A2$ regions. At higher temperatures, entropic contributions will favor complete mixing. Therefore, systems that favor like bonds will tend to have a *decomposition* transformation upon cooling. As a result of a decomposition transformation, composition (i.e., X_B in a binary alloy, Eq. 17.8) plays the role of an order parameter and evolves into stable values indicated by tie-lines on a phase diagram. However, η in Eq. 17.7 remains zero during a decomposition reaction.

A model for the homogeneous free energy of mixing and $B2$ ordering can be obtained with *site occupation probabilities*. In a *homogeneous* system, the probability, p_B^α, of finding a B-species occupying an α-site is X_B^α; similarly, from Eqs. 17.7 and 17.8,

$$
\begin{aligned}
p_B^\alpha &= X_B^\alpha = X_B + \eta \\
p_B^\beta &= X_B^\beta = X_B - \eta \\
p_A^\alpha &= 1 - X_B^\alpha = 1 - X_B - \eta \\
p_A^\beta &= 1 - X_B^\beta = 1 - X_B + \eta
\end{aligned}
\tag{17.9}
$$

These probabilities must all lie between zero and one; this sets bounds on physical values for the structural order parameter

$$
\begin{aligned}
-X_B < \eta < X_B \quad &\text{if} \quad 0 < X_B < \frac{1}{2} \\
X_B - 1 < \eta < 1 - X_B \quad &\text{if} \quad \frac{1}{2} < X_B < 1
\end{aligned}
\tag{17.10}
$$

The simplest approximation to the total entropy of mixing is the mean-field Bragg–Williams–Gorsky configurational entropy, proportional to $\sum p_i^\omega \ln p_i^\omega$ [4]:[3]

$$
\Delta S = -\mathcal{N}k \left(p_A^\alpha \ln p_A^\alpha + p_A^\beta \ln p_A^\beta + p_B^\alpha \ln p_B^\alpha + p_B^\beta \ln p_B^\beta \right)
\tag{17.11}
$$

If it can be assumed that mixing, decomposition, and ordering transformations occur at constant volume, an estimate of the internal energy, U, can be used to obtain the appropriate minimizing molar free energy $F = U - TS$. A model for the total internal energy can be obtained from the nearest-neighbor bond energies of the three possible bond types E_{AA}, E_{BB}, and E_{AB}, and their populations calculated from site probabilities. Because each atom contributes four bonds,

$$
\begin{aligned}
\mathcal{N}_{AA} &= 4\mathcal{N} p_{AA} = 4\mathcal{N} p_A^\alpha p_A^\beta = 4\mathcal{N} \left[(1 - X_B)^2 - \eta^2 \right] \\
\mathcal{N}_{BB} &= 4\mathcal{N} p_{BB} = 4\mathcal{N} p_B^\alpha p_B^\beta = 4\mathcal{N} (X_B^2 - \eta^2) \\
\mathcal{N}_{AB} &= 4\mathcal{N} p_{AB} = 4\mathcal{N} (p_A^\alpha p_B^\beta + p_B^\alpha p_A^\beta) = 8\mathcal{N} \left(X_B - X_B^2 + \eta^2 \right) \\
\mathcal{U} &= \mathcal{N}_{AA} E_{AA} + \mathcal{N}_{BB} E_{BB} + \mathcal{N}_{AB} E_{AB} \\
&= 4\mathcal{N} \{ [E_{AA} + E_{BB} - 2E_{AB}][X_B(X_B - 1) - \eta^2] \\
&\quad + E_{AA}(1 - X_B) + E_{BB} X_B \}
\end{aligned}
\tag{17.12}
$$

---

[3]The reference state is the composition-weighted linear combination of pure A and B components. This approximation neglects vibrational entropy. Higher-order mean-field approximations to configurational entropy, known as the *cluster-variation method*, are known [5, 6].

The last two terms in $\mathcal{U}$ can be used as a reference energy—the total internal energy of a system of composition X_B that is divided into pure A and pure B portions. The change in internal energy due to mixing and ordering relative to this reference state is

$$\begin{aligned} \Delta\mathcal{U} &= \mathcal{U} - E_{AA}(1 - X_B) - E_{BB}X_B \\ &= 4\mathcal{N}\left\{[E_{AA} + E_{BB} - 2E_{AB}][X_B(X_B - 1) - \eta^2]\right\} \end{aligned} \tag{17.13}$$

The nearest-neighbor interaction coefficient, $W \equiv E_{AA} + E_{BB} - 2E_{AB}$, and the temperature together characterize the molar free energy of mixing:

$$\begin{aligned} \Delta F =&\, 4\mathcal{N}W\left[X_B(X_B - 1) - \eta^2\right] \\ &+ kT\left[\begin{array}{l}(X_B + \eta)\ln(X_B + \eta) + (X_B - \eta)\ln(X_B - \eta) \\ +(1 - X_B + \eta)\ln(1 - X_B + \eta) + (1 - X_B - \eta)\ln(1 - X_B - \eta)\end{array}\right] \\ =&\, 4\mathcal{N}W\left[X_B X_A - \eta^2\right] \\ &+ kT\left[\begin{array}{l}(X_B + \eta)\ln(X_B + \eta) + (X_B - \eta)\ln(X_B - \eta) \\ +(X_A + \eta)\ln(X_A + \eta) + (X_A - \eta)\ln(X_A - \eta)\end{array}\right] \end{aligned} \tag{17.14}$$

The equilibrium order parameters X_B^{eq} and η^{eq} minimize ΔF subject to any system constraints. Supposing that the system's composition is fixed, the method of Lagrange multipliers leads to a common-tangent construction for ΔF with respect to X_B—or equivalently, equality of chemical potentials of both A and B. Two compositions, $X_B^{\text{eq}-}$ and $X_B^{\text{eq}+}$, will coexist at equilibrium for average compositions X_B in the composition range $X_B^{\text{eq}-} < X_B < X_B^{\text{eq}+}$ if they satisfy

$$\left.\frac{\partial\Delta F}{\partial X_B}\right|_{X_B^{\text{eq}-}} = \left.\frac{\partial\Delta F}{\partial X_B}\right|_{X_B^{\text{eq}+}} = \frac{\Delta F(X_B^{\text{eq}-}) - \Delta F(X_B^{\text{eq}+})}{X_B^{\text{eq}-} - X_B^{\text{eq}+}} \tag{17.15}$$

The condition expressed by Eq. 17.15 pertains only if $X_B^{\text{eq}-} \neq X_B^{\text{eq}+}$; it is automatically satisfied for fixed uniform compositions with variable order parameter (this situation appears below for $B2$ ordering). Because η is not constrained,

$$\left.\frac{\partial\Delta F}{\partial\eta}\right|_{\eta_B^{\text{eq}}} = 0 \tag{17.16}$$

The symmetry of the particular model, Eq. 17.14, simplifies the equilibrium condition, Eq. 17.15:

$$\left.\frac{\partial\Delta F}{\partial X_B}\right|_{\frac{1}{2}-\Delta X_B^{\text{eq}}} = \left.\frac{\partial\Delta F}{\partial X_B}\right|_{\frac{1}{2}+\Delta X_B^{\text{eq}}} = 0 \tag{17.17}$$

where $X_B^{\text{eq}\pm} = \frac{1}{2} \pm \Delta X_B^{\text{eq}}$. For this nearest-neighbor b.c.c. model, equilibrium order parameters can be reduced to solutions of

$$\frac{(\frac{1}{2} - \Delta X_B^{\text{eq}})^2 - (\eta^{\text{eq}})^2}{(\frac{1}{2} + \Delta X_B^{\text{eq}})^2 - (\eta^{\text{eq}})^2} = e^{-8\Delta X_B^{\text{eq}}W/(kT)} \tag{17.18}$$

and

$$\frac{(\frac{1}{2} + \eta^{\text{eq}})^2 - (\Delta X_B^{\text{eq}})^2}{(\frac{1}{2} - \eta^{\text{eq}})^2 - (\Delta X_B^{\text{eq}})^2} = e^{-8\eta^{\text{eq}}W/(kT)} \tag{17.19}$$

For the disordered ($A2$) phases, $\eta^{\mathrm{eq}} = 0$, Eq. 17.19 is satisfied automatically, and equilibrium tie-lines are present if $W < 0$ and $T < T^{\mathrm{crit}} \equiv |W/k|$, as illustrated in Fig. 17.5. In the nearest-neighbor model, ordered $B2$ solutions can appear at any uniform composition X_B. Nonzero equilibrium structural order parameters appear only if $W > 0$ at temperatures T satisfying

$$T = 8\eta T^{\mathrm{crit}} \left\{ \ln \left[\frac{(X_B + \eta)(X_A + \eta)}{(X_A - \eta)(X_B - \eta)} \right] \right\}^{-1} \tag{17.20}$$

as illustrated in Fig. 17.5.

The development of the miscibility gap for $W < 0$ and the antiphases ($\pm\eta^{\mathrm{eq}}$) for $W > 0$ have entirely different kinetic implications. For decomposition, mass flux is necessary for the evolution of two phases with differing compositions. Furthermore, interfaces between these two phases necessarily develop. The evolution of ordered phases from disordered phases (i.e., the onset of nonzero structural order parameters) can occur with no mass flux; macroscopic diffusion is not necessary. Because the η_+^{eq}-phase is thermodynamically equivalent to the η_-^{eq}-phase, the development of η_+^{eq}-phase in one material location is simultaneous with the evolution of η_-^{eq}-phase at another location. The impingement of these two phases creates an *antiphase domain boundary*. These interfaces are regions of local heterogeneity and increase the free energy above the homogeneous value given by Eq. 17.14. The kinetic implications of macroscopic diffusion and of the development of interfaces are treated in Chapter 18.

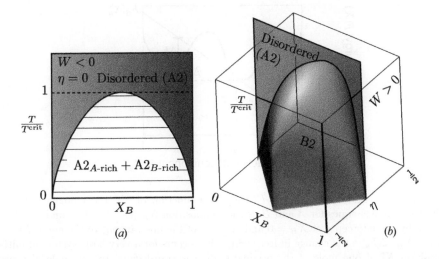

Figure 17.5:   Phase diagrams for the nearest-neighbor bond-energy model for the b.c.c. lattice. **(a)** The temperature dependence of the decomposition reaction for $W = E_{AA} + E_{BB} - 2E_{AB} < 0$. A miscibility gap appears at temperatures below $T^{\mathrm{crit}} = W/k$ but $\eta^{\mathrm{eq}} = 0$ at all compositions and temperatures. **(b)** $B2$ ordering transformations occur if $W > 0$ and $T/T^{\mathrm{crit}} < 4X_A X_B$ and are indicated by the onset of nonzero values of η^{eq}. The η^{eq} are symmetric about $\eta = 0$. The two values $\pm\eta^{\mathrm{eq}}$ indicate equivalent ordered states where the α- and β-planes in Fig. 17.4 are exchanged.

The composition order parameter X_B is an example of a *conserved order parameter*; it cannot increase or decrease locally without an opposite change elsewhere. The structural order parameter η is an example of a *nonconserved order parameter*—no spatial correlations in their variations are necessary. Energetic changes associated with the two different types of order parameters have already been distinguished by the difference between the common-tangent condition (Eq. 17.18) and the global-minimum condition expressed by Eq. 17.19. The changes in molar free energies for the two different types of order parameters are treated in the next section.

17.2   MOLAR FREE-ENERGY CHANGES FOR CONSERVED AND NONCONSERVED ORDER PARAMETERS

The composition or the number fraction of component B, X_B, is an example of an order parameter that is conserved in a closed system. Figure 17.6 shows a molar free energy versus composition curve for a binary solution. The molar free energy for a solution at any composition X_B can be written in terms of its *partial* molar quantities, $F_A(X_B)$ and $F_B(X_B)$:[4]

$$F(X_B) = \frac{\mathcal{N}_A}{\mathcal{N}_A + \mathcal{N}_B} F_A(X_B) + \frac{\mathcal{N}_B}{\mathcal{N}_A + \mathcal{N}_B} F_B(X_B)$$
$$= (1 - X_B)F_A(X_B) + X_B F_B(X_B) \tag{17.21}$$

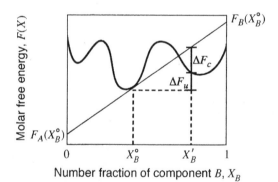

Figure 17.6:   Molar free energy, $F(X_B)$, as a function of X_B for a binary solution. Partial molar free energies for A and B at the composition X_B° [i.e., $F_A(X_B^\circ)$ and $F_B(X_B^\circ)$] are given by the intercepts at $X_B = 0$ and $X_B = 1$ of the line tangent to the curve $F(X_B)$ at $X_B = X_B^\circ$. ΔF_u is the change in free energy that occurs for a very large system of initial composition X_B° if one mole of the original system is transformed to one mole of a new solution of composition X_B' and the system is open and unconstrained so that matter can be transferred in or out. ΔF_c is the corresponding change that occurs if the system is closed so that its total composition is fixed.

---

[4]The Helmholtz free energy, used here as an example of a molar free energy, is the appropriate minimizing functional for a system at fixed volume in equilibrium with a reservoir at fixed temperature. For different types of system constraints, F would be replaced with another appropriate molar free energy.

Consider a general system at composition X_B°. If any of its subsystems (i.e., material portions) could transform to a new composition X_B' *without affecting the rest of the system*, the change in the system's total free energy $\mathcal{F}$ would be simply

$$\Delta F = F(X_B') - F(X_B^\circ) \tag{17.22}$$

for each transformed mole as indicated by ΔF_u in Fig. 17.6. Such a change would be possible in a system that can exchange B with its environment—an open system for which no conservation principle applies. In a closed system, the total number of B, $\mathcal{N}_B$, is effectively conserved and such a change is not possible. Therefore, for any nonconserved parameter ξ, the change in molar free energy is

$$\Delta F = F(\xi') - F(\xi_\circ) \tag{17.23}$$

In a closed system, where the numbers of A and B atoms are conserved, a change in any subsystem must affect the rest of the system—atoms must be exchanged internally to accomplish the transformation. For each mole transformed, the change in F for the X_B' moles of the B component is $[F_B(X_B') - F_B(X_B^\circ)]X_B'$, with a similar term for the $(1 - X_B')$ moles of the A component,

$$\Delta F = [F_B(X_B') - F_B(X_B^\circ)]X_B' + [F_A(X_B') - F_A(X_B^\circ)](1 - X_B') \tag{17.24}$$

which can be rewritten

$$\Delta F = X_B' F_B(X_B') + (1 - X_B')F_A(X_B') - X_B^\circ F_B(X_B^\circ) \\ - (1 - X_B^\circ)F_A(X_B^\circ) + (X_B^\circ - X_B')[F_B(X_B^\circ) - F_A(X_B^\circ)] \tag{17.25}$$

or

$$\Delta F = F(X_B') - F(X_B^\circ) - (X_B' - X_B^\circ) \left.\frac{\partial F}{\partial X_B}\right|_{X_B = X_B^\circ} \tag{17.26}$$

which is numerically equal to the distance indicated in Fig. 17.6 by ΔF_c. Note that ΔF_c is negative if at $X_B = X_B'$, the curve for $F(X_B)$ lies below the tangent; ΔF_c is positive otherwise. Equation 17.26 holds for any composition X_B' when the composition X_B° is fixed.

Consider a small *fluctuation* of composition (or any other order parameter that is conserved for a system), $\delta X_B = X_B' - X_B^\circ$. Expanding $F(X_B)$ in δX_B yields

$$F(X_B') = F(X_B^\circ) + \delta X_B \left.\frac{\partial F}{\partial X_B}\right|_{X_B = X_B^\circ} + \frac{1}{2}(\delta X_B)^2 \left.\frac{\partial^2 F}{\partial X_B^2}\right|_{X_B = X_B^\circ} + \cdots \tag{17.27}$$

Substitution of Eq. 17.27 into Eq. 17.26 gives

$$\Delta F = \frac{1}{2}(\delta X_B)^2 \left.\frac{\partial^2 F}{\partial X_B^2}\right|_{X_B = X_B^\circ} \tag{17.28}$$

for the change in the molar free energy. Equation 17.28 indicates that when X_B is conserved, the variation in molar free energy is proportional to $(\delta X_B)^2$. Therefore, a barrier to the growth of small variations exists whenever the second derivative in Eq. 17.28 is positive.

On the other hand, if ξ is nonconserved, using Eq. 17.23 and expanding $F(\xi^\circ + \delta\xi)$ shows that the lowest-order term for the change in the molar free energy is

$$\Delta F = \delta\xi \left. \frac{\partial F}{\partial \xi} \right|_{\xi = \xi^\circ} \tag{17.29}$$

A decrease in the free energy can always be achieved by picking a variation $\delta\xi$ with a sign that makes the product in Eq. 17.29 negative. Therefore, if ξ is nonconserved, there is no barrier to evolution to a local minimum in the appropriate free-energy functional, and the system will evolve if there are kinetic mechanisms that enable such variations.

17.3 CLASSIFICATION OF PHASE TRANSFORMATIONS: CONTINUOUS VERSUS DISCONTINUOUS TRANSFORMATIONS

Figure 17.7a shows a typical molar free energy versus composition curve for a binary system exhibiting a miscibility gap. The corresponding equilibrium diagram is shown in Fig. 17.7b. If a homogeneous alloy in Region I of Fig. 17.7b is rapidly cooled into the miscibility gap and held there at T_0, it decomposes into solute-rich and solute-lean phases at the ends of the tie-line shown. If the alloy composition is only slightly higher than $X_B^{\alpha'}$ so that it is in the range where $d^2F/dX_B^2 > 0$, Eq. 17.28 shows that a free-energy barrier exists for any local region in the system to enrich itself progressively in solute-atom concentration. In addition, Fig. 17.6 shows that if such a region could undergo a fluctuation corresponding to a large increase in local solute concentration, the free energy would decrease. In contrast, if the alloy composition is higher and in the range where $d^2F/dX_B^2 < 0$, a region of the specimen can progressively enrich itself and eventually form the equilibrium solute-rich phase as the free energy decreases continuously. The dividing line between these two types of behavior is set by the condition $d^2F/dX_B^2 = 0$ and is known as the *chemical spinodal*. This line is mapped in Fig. 17.7b.

Phase transformations proceed only when there is an ultimate decrease in the free energy, and the considerations above show that two quite different types of

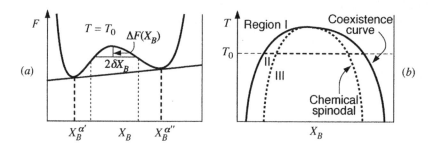

Figure 17.7:   (a) Molar free energy, $F(X_B)$ vs. composition curve for a binary system exhibiting a miscibility gap. The corresponding equilibrium diagram is shown in (b). The compositions of the α' and α'' phases in equilibrium across the miscibility gap in (b) are obtained from common-tangent constructions at each temperature such as the one shown in (a). The points on the chemical spinodal in (b) correspond to points on the curve in (a) where $\partial^2 F/\partial X_B^2 = 0$.

phase transformation can take place depending upon whether the composition is inside or outside the spinodal line in Fig. 17.7b. If the composition is outside the chemical spinodal (but still within the miscibility gap) there is a bulk free-energy barrier to the formation of the solute-rich phase in any local region, and the solute-rich phase can form only if such a region somehow becomes sufficiently enriched in solute atoms so that the free-energy barrier is overcome. Once the barrier is surmounted, the region can continue to develop into the equilibrium phase with a continuous decrease in the free energy. The only way that this can occur at a sensible rate is by means of a fluctuation that produces a relatively large enrichment of composition in a small region, i.e., nucleation of the solute-rich phase must take place in a region of atomic dimensions (see Chapter 19). On the other hand, if the composition is within the chemical spinodal, there is no bulk free-energy barrier to the progressive formation of solute-rich regions, and the transformation takes place over large regions of the system with a continuous decrease in the free energy.

This leads to the general classification of continuous and discontinuous transformations.

Continuous transformations: In this type of transformation, the system is initially unstable, and an infinitesimal variation will initiate the transformation and the decrease of bulk free energy. The transformation can begin over large regions, and all volumes of a system can transform simultaneously. The beginning of the transformation involves a change that is small in degree but large in extent. Primary examples include spinodal decomposition and some order–disorder transformations in alloys that exhibit long-range order.

Discontinuous transformations: In this type of transformation, there is a free-energy barrier to infinitesimal variations and the system is initially metastable. However, a sufficiently large variation can cause the free energy to decrease. The transformation therefore can be initiated at a finite rate only by a variation that is large in degree but small in extent (i.e., nucleation is required). Examples include the formation of B-rich precipitates from a supersaturated A–B solution.

Bibliography

1. J.W. Gibbs. On the equilibrium of heterogeneous substances (1876). In *Collected Works*, volume 1. Longmans, Green, and Co., New York, 1928.

2. L.D. Landau and E.M. Lifshitz. *Statistical Physics*. Pergamon Press, New York, 1963.

3. S.M. Allen and J.W. Cahn. Phase diagram features associated with multicritical points in alloy systems. *Bull. Alloy Phase Diagrams*, 3(3):287–295, 1982.

4. G. Inden. Ordering and segregation reactions in BCC binary alloys. *Acta Metall.*, 22(8):945–951, 1974.

5. R. Kikuchi. A theory of cooperative phenomena. *Phys. Rev. B*, 81(6):988–1003, 1951.

6. D. de Fontaine. Studies of the thermodynamics of ordering by the cluster variation method. In H.I. Aaronson, D.E. Laughlin, R.F. Sekerka, and C.M. Wayman, editors, *Proceedings of an International Conference on Solid→Solid Phase Transformations*, pages 25–47, Warrendale, PA, 1981. The Metallurgical Society of AIME.

CHAPTER 18

SPINODAL DECOMPOSITION AND ORDER–DISORDER TRANSFORMATIONS

Spinodal decomposition and certain order–disorder transformations are the two categories of continuous phase transformations. Both arise from an order parameter instability: in the case of spinodal decomposition, it is a conserved order parameter; for continuous ordering, it is a nonconserved order parameter.

As spinodal and ordering transformations progress, order parameter gradients arise and the extra energy associated with these gradients must be reflected in a modified diffusion potential for the transformations. The evolution of spinodal decomposition and continuous ordering transformations is described by the Cahn–Hilliard equation and the Allen–Cahn equation, respectively. Both incorporate gradient-energy terms via the modified diffusion potential.

Fine-scale, spatially periodic microstructures are characteristic of spinodal decomposition. In elastically anisotropic crystalline solutions, spinodal microstructures are aligned along elastically soft directions to minimize elastic energy. Microstructures resulting from continuous ordering contain interfaces called *antiphase boundaries* which coarsen slowly in comparison to the rate of the ordering transformation.

18.1 THE INTERDIFFUSIVITY AT UNSTABLE COMPOSITIONS

Because composition is a locally-conserved order parameter, it cannot change in one location without affecting its neighborhood—fluxes are required to change a composition field. For example, in a binary alloy, the concentration field c_B is re-

Kinetics of Materials. By Robert W. Balluffi, Samuel M. Allen, and W. Craig Carter.

lated to the composition field $(c_B = X_B/\langle\Omega\rangle)$ and the flux of B is linearly related to its chemical potential gradient by Eq. 3.7.[1] The fluxes that produce a concentration change at some point necessarily affect concentrations in the neighborhood of that point. This coupling between the space and time behavior of concentration is encoded in Fick's first and second laws for a binary alloy by

$$\vec{J}_B = -\widetilde{D}\nabla c_B \tag{18.1}$$

where the flux, $\vec{J}_B$, is relative to the V-frame described in Section 3.1.3, and

$$\frac{\partial c_B}{\partial t} = \widetilde{D}\nabla^2 c_B \tag{18.2}$$

which is linearized by expanding the interdiffusivity $\widetilde{D}$ about an average composition. The exact form is given by Eq. 3.25:

$$
\begin{aligned}
\widetilde{D} &\approx \langle\Omega\rangle(c_A D_B + c_B D_A) = X_A D_B + X_B D_A \\
&= (X_A \, {}^\star D_B + X_B \, {}^\star D_A)\left(1 + \frac{\partial \ln\gamma_A}{\partial \ln X_A}\right)
\end{aligned}
\tag{18.3}
$$

If $\widetilde{D} > 0$, Eq. 18.2 implies that a system with a uniform composition $X_B(\vec{x}) = \langle\Omega\rangle c_B^{\text{const}}$ would not evolve. However, if $\widetilde{D} < 0$, small composition fluctuations would grow without bound: a system described by Eq. 18.2 would develop into many unphysical and finely spaced large fluctuations in composition.

The stability conditions for a conserved order parameter are developed in Section 17.2, and Eq. 17.28 relates this condition to the sign of $\partial^2 F/\partial X_B^2$. If a system is at a nearly uniform composition X_B where the second derivative is negative, the system is unstable and should evolve. As shown below, the sign of $\widetilde{D}$, determined by the sign of $\partial^2 F/\partial X_B^2$, predicts the decomposition of a uniform unstable composition. However, Eqs. 18.1 and 18.2 do not account for interfacial energies between the developing phases. The correction to Eqs. 18.1 and 18.2 due to interfacial energies is accomplished through the addition of a gradient-energy term and is addressed in Section 18.2.

To show how the sign of $\widetilde{D}$ is related to the sign of $\partial^2 F/\partial X_B^2$, consider chemical diffusion in a substitutional A–B crystalline material, as described in Section 3.1.3. The fluxes $\vec{J}_A^C$ and $\vec{J}_B^C$ in the C-frame are given by Eq. 3.7 and can be written in the forms

$$\vec{J}_A^C = -L_A\nabla\mu_A \qquad \vec{J}_B^C = -L_B\nabla\mu_B \tag{18.4}$$

where $L_A = L_{AA} - (c_A/c_B)L_{AB}$ and $L_B = L_{BB} - (c_B/c_A)L_{BA}$. Using Eq. 3.15, the corresponding fluxes in the V-frame are related to the fluxes in the C-frame by

$$\vec{J}_A^V = J_A^C + c_A v_C^V \qquad \vec{J}_B^V = J_B^C + c_B v_C^V \tag{18.5}$$

For simplicity, we assume in this chapter that the atomic volumes of A and B are equal and constant (i.e., $\Omega_A \approx \Omega_B \approx \Omega = $ constant).[2] Then, using Eqs. 3.21, 18.4, and 18.5, solving for $\vec{J}_B^V$, and simplifying the result using Eq. 3.9 yields

$$J_B^V = \Omega(c_A L_B\nabla\mu_B - c_B L_A\nabla\mu_A) \tag{18.6}$$

---

[1]See Section A.2 for this and similar conversions between composition and concentration.
[2]These approximations were eliminated by Hilliard [1], and a corresponding derivation that utilizes methods developed in Section 3.1.3 appears in Exercise 18.1.

Equation 18.6 can be simplified using the identity

$$
\begin{aligned}
c_A L_B \nabla \mu_B - c_B L_A \nabla \mu_A = \Omega[&(c_A^2 L_B + c_B^2 L_A)(\nabla \mu_B - \nabla \mu_A) \\
&+ (c_A L_B - c_B L_A)(c_A \nabla \mu_A + c_B \nabla \mu_B)]
\end{aligned}
\tag{18.7}
$$

In addition, because of the Gibbs–Duhem relation, $c_A \, d\mu_A + c_B \, d\mu_B = 0$, the chemical potential gradients are interdependent:

$$
c_A \, \nabla \mu_A + c_B \, \nabla \mu_B = 0
\tag{18.8}
$$

With the aid of Eqs. 18.7 and 18.8, the flux of B in the volume-fixed frame is then

$$
\begin{aligned}
\vec{J}_B^V &= -\Omega^2 \left(c_A^2 L_B + c_B^2 L_A \right) \nabla (\mu_B - \mu_A) \\
&\equiv -M \nabla (\mu_B - \mu_A) \\
&= -\frac{M}{N_o} \nabla \frac{\partial F}{\partial X_B} = -\frac{M}{N_o \Omega} \nabla \frac{\partial F}{\partial c_B} \\
&= -\frac{M}{N_o \Omega} \frac{\partial^2 F}{\partial c_B^2} \nabla c_B = -\frac{M \Omega}{N_o} \frac{\partial^2 F}{\partial X_B^2} \nabla c_B \\
&\equiv -\widetilde{D} \nabla c_B
\end{aligned}
\tag{18.9}
$$

where $M \equiv \Omega^2 \left(c_A^2 L_B + c_B^2 L_A \right)$ must be positive to satisfy the basic postulate of irreversible thermodynamics, $\dot{\sigma} \geq 0$ (Eq. 2.16). Comparison of Eqs. 18.2 and 18.9 shows that $\widetilde{D}$ is negative when the system is unstable to small concentration fluctuations, $\partial^2 F / \partial X_B^2 < 0$. This condition occurs during *spinodal decomposition*, and the negative interdiffusivity gives rise to *uphill diffusion*.[3]

However, a negative interdiffusivity makes the diffusion equation ill-posed. With a positive diffusion coefficient, the amplitude of a composition wave decays at a rate proportional to the square of the inverse wavelength (see Section 5.2.4). Conversely, if the diffusion coefficient is negative, the smallest wavelengths will *grow* the most rapidly. Consider two concentration profiles that are nearly the same except for small differences in their highest-order Fourier coefficients; after a short time, the profiles will be very different if the diffusion coefficient is negative. Therefore, when $\widetilde{D} < 0$, the behavior of the concentration profile is not robust with respect to small variations in the initial data and the problem is ill-posed. However, as shown in the following section, interfacial energy terms can be included by associating additional energy with the concentration gradients, and their inclusion *regularizes* the diffusion equation when $\widetilde{D} < 0$ (i.e., the resulting equation is not ill-posed).

18.2 FREE ENERGY OF INHOMOGENEOUS SYSTEMS: DIFFUSE INTERFACES AND THE POTENTIAL FOR TRANSFORMATION

The conditions for continuous phase transformations (described in Chapter 17) derive from considerations of the molar free energy of a *homogeneous*, spatially uniform system with no interfaces. For phase transformations involving nonconserved

---

[3] *Spinodals* occur at points where second derivatives vanish and for molar free energy and compositions give rise to the sharp "cusps" in Fig. 17.1 [2]. *Spinodal* derives from the Latin *spina*, for thorn (the plural, *spinae*, meant difficulties or perplexities).

order parameters, such as for $A2 \rightarrow B2^{\pm}$ $((\eta^{\text{eq}} = 0) \rightarrow (\eta = \pm\eta^{\text{eq}}))$, there is no bias to form one ordered $B2$ variant over another (the two equivalent variants are indicated by $B2^{\pm}$; see Fig. 17.4). The two equivalent variants emerge at random locations, and interfaces develop as one impinges upon the other. For conserved order parameters, such as composition, interfaces between phases on phase-diagram tie-lines necessarily appear.

In the absence of interfaces, a linear kinetic theory could be developed where the transformation driving force derives from decreases in homogeneous molar free energy as derived in Eqs. 17.28 and 17.29 for the conserved and nonconserved cases. However, at the onset of a continuous phase transition, the system is virtually *all* interface between new phases or variants. For example, when equivalent variants emerge in adjacent regions during ordering, gradients in the order parameter are generated; these constitute emerging diffuse antiphase boundaries. Neglecting the contribution of these interfaces leads to ill-posed linearized kinetics, as indicated by the negative interdiffusivity in Eq. 18.9.

The theory for the free energy of inhomogeneous systems incorporates contributions from interfacial free energy through the *diffuse interface method* [3]. Interfaces are defined by the locations where order parameters change and can be located by the regions with significant order-parameter gradients. Interfacial energy appears in the diffuse-interface methods because order-parameter gradients contribute extra energy.

18.2.1   Free Energy of an Inhomogeneous System

Let $\xi(\vec{r})$ represent either a conserved or a nonconserved order parameter, such as $c_B(\vec{r})$ or $\eta(\vec{r})$. Also, let the field $f(\vec{r}) = f(\xi(\vec{r}), \nabla\xi(\vec{r}))$ be the free-energy density (energy/volume) at position $\vec{r}$. The homogeneous free-energy density, $f^{\text{hom}}(\xi) = f(\xi, \nabla\xi = 0)$, is the free-energy density in the absence of gradients and is related to molar free energies, $F(\xi) = N_{\text{o}}\langle\Omega\rangle f^{\text{hom}}(\xi)$, used to construct phase diagrams such as Figs. 17.5 and 17.7. Expanding the free-energy density about its homogeneous value in powers of gradients,[4]

$$f(\xi, \nabla\xi) = f(\xi, 0) + \vec{L} \cdot \nabla\xi + \nabla\xi \cdot \boldsymbol{K}\, \nabla\xi + \dots \tag{18.10}$$

where

$$\vec{L} = [L_{x_1}, L_{x_2}, L_{x_3}] = \frac{\partial f}{\partial(\partial\xi/\partial x_i)} \tag{18.11}$$

is a vector evaluated at zero gradient, and $\boldsymbol{K}$ is a tensor property known as the gradient-energy coefficient with components

$$K_{ij} = \frac{1}{2}\frac{\partial^2 f}{\partial(\partial\xi/\partial x_i)\partial(\partial\xi/\partial x_j)} \tag{18.12}$$

The free-energy density should not depend on the choice of coordinate system [i.e., $f(\xi, \nabla\xi)$ should not depend on the gradient's direction] and therefore $\vec{L} = 0$ and $\boldsymbol{K}$ will be a symmetric tensor.[5] Furthermore, if the homogeneous material is isotropic

---

[4]There are expansions that contain higher-order spatial derivatives, but the resulting free energy is the same as that derived here [1, 4].

[5]If the homogeneous material has an inversion center (center of symmetry), $\vec{L}$ is automatically zero.

or cubic, $\boldsymbol{K}$ will be a diagonal tensor with equal components K. The free-energy density will be, to second order,

$$f(\xi, \nabla\xi) = f^{\mathrm{hom}}(\xi) + K\nabla\xi \cdot \nabla\xi = f^{\mathrm{hom}}(\xi) + K|\nabla\xi|^2 \qquad (18.13)$$

The free-energy density is thus approximated as the first two terms in a series expansion in order-parameter gradients: the first term is related to homogeneous molar free energy and the second is proportional to the gradient squared.

In the expansion that leads to Eq. 18.13, it is assumed that the free energy varies smoothly from its homogeneous value as the magnitude of the order-parameter gradient increases from zero. This assumption is usually correct, but there may be cases that include a lower-order term proportional to $|\nabla\xi|$ if the free-energy density has a *cusp* at zero gradient. Such cusps appear in the interfacial free energy at a faceting orientation; they are also present at small tilt-misorientation grain-boundary energies [5]. Models with crystallographic orientation as an order parameter incorporate gradient magnitudes, $|\nabla\xi|$, into the inhomogeneous free-energy density [6].

18.2.2   Structure and Energy of Diffuse Interfaces

There are two energetic contributions to interfaces in systems that undergo decomposition and ordering transformations such as illustrated in Figs. 17.5 and 17.7. One is due to the gradient-energy term in Eq. 18.13; this contribution tends to spread the interface region and thereby reduce the gradient as the order parameter changes between its stable values in adjacent phases. A second contribution derives from the increased homogeneous free-energy density associated with the "hump" in Fig. 18.1, and this term tends to narrow the interface region. Thus, systems modeled with Eq. 18.13 contain *diffuse interfaces* where the order parameter varies smoothly as in Fig. 18.2. Equilibrium order-parameter profiles and energies can be determined by minimizing $\mathcal{F}$, the volume integral of Eq. 18.13 [1, 4].

Figure 18.2a shows a planar interface between two equilibrium phases possessing different conserved order parameters corresponding to local free-energy density minima in their order parameters as in Fig. 17.7a. Figure 18.2b shows a corresponding profile of the distribution of order between two identical ordered domains possessing different nonconserved order parameters corresponding to local free-energy density

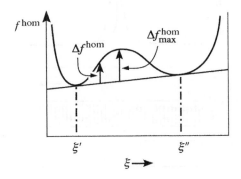

Figure 18.1:   Properties of diffuse interfaces expressed in terms of the function $\Delta f^{\mathrm{hom}}(\xi)$, which has the maximum value $\Delta f^{\mathrm{hom}}_{\mathrm{max}}$.

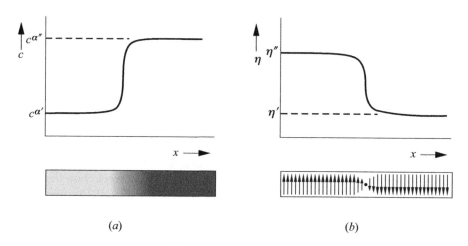

(a)   (b)

Figure 18.2:   **(a)** Composition and **(b)** order variations across diffuse, planar interfaces. The profiles $c(x)$ and $\eta(x)$ are continuous. In (a), the grayscale image represents the spatial variation of a conserved variable, and the quantities $c^{\alpha'}$ and $c^{\alpha''}$ are the equilibrium values in the bulk phases at large distances from the interface (see Fig. 17.7). In (b), the drawing below the profile illustrates the spatial variation of a nonconserved variable such as local magnetization in the region around a domain wall.

minima. Both kinds of interfaces can coexist, so that the variations of c_B and η are coupled as in Fig. 18.3. In all cases, the distribution of the order parameter (or order parameters) minimizes the total free energy of the system, $\mathcal{F}$. The coupled-parameter case can be treated as an extension to the theory so that the free energy is a function of both c_B and η.

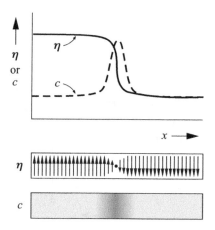

Figure 18.3:   Coupled system of order and concentration parameters representing an antiphase boundary with segregation.

Minimizing $\mathcal{F} = \int f(\xi, \nabla\xi) dV$ produces equilibrium interface profiles $\xi(\vec{r})$. An equilibrated planar interface is characterized by its excess energy per unit area, γ,

$$\gamma = 2 \int_{\xi'}^{\xi''} \sqrt{K \, \Delta f^{\text{hom}}(\xi)} \, d\xi \approx \sqrt{K \, \Delta f_{\text{max}}^{\text{hom}}} \, \Delta\xi \qquad (18.14)$$

and a characteristic width δ,

$$\delta \approx \sqrt{K/\Delta f_{\text{max}}^{\text{hom}}} \qquad (18.15)$$

where Δf^{hom} is the increase in free-energy density relative to a homogeneous system at its equilibrium values of ξ (i.e., relative to the common-tangent line) and $\Delta f_{\text{max}}^{\text{hom}}$ is the maximum value of Δf^{hom} (indicated in Fig. 18.1). γ and δ can be measured and their values uniquely determine the model parameters, $\Delta f_{\text{max}}^{\text{hom}}$ and K.

18.2.3 Diffusion Potential for Transformation

The local diffusion potential for a transformation, $\Phi(\vec{r})$, at a time $t = t_\circ$, can be determined from the rate of change of total free energy, $\mathcal{F}$, with respect to its current order-parameter field, $\xi(\vec{r}, t_\circ)$. At time $t = t_\circ$, the total free energy is

$$\mathcal{F}(t_\circ) = \int_{\mathcal{V}} \left[f^{\text{hom}}(\xi(\vec{r}, t_\circ)) + K\nabla\xi \cdot \nabla\xi \right] dV \qquad (18.16)$$

which defines $\mathcal{F}$ as a functional of $\xi(\vec{r}, t_\circ)$.[6] If the order parameter is changing with local "velocity" $\dot{\xi}$ [i.e., such that $\xi(\vec{r}, t) = \xi(\vec{r}, t_\circ) + \dot{\xi}(\vec{r}, t_\circ)t$], the rate of change of $\mathcal{F}$ can be summed from all the contributions to $f(\xi, \nabla\xi)$ due to changes in the order-parameter field and its gradient,

$$\left. \frac{d\mathcal{F}}{dt} \right|_{t_\circ} = \int_{\mathcal{V}} \left[\frac{\partial f^{\text{hom}}(\xi(\vec{r}, t_\circ))}{\partial\xi} \dot{\xi} + 2K\nabla\xi \cdot \nabla\dot{\xi} \right] dV \qquad (18.17)$$

Using the relation

$$\nabla \cdot \left(\dot{\xi}\nabla\xi \right) = \dot{\xi}\nabla^2\xi + \nabla\xi \cdot \nabla\dot{\xi} \qquad (18.18)$$

Eq. 18.17 can be written

$$\left. \frac{d\mathcal{F}}{dt} \right|_{t_\circ} = \int_{\mathcal{V}} \left(\frac{\partial f^{\text{hom}}(\xi)}{\partial\xi} - 2K\nabla^2\xi \right) \dot{\xi} \, dV + 2K \int_{\mathcal{V}} \nabla \cdot \left(\dot{\xi}\nabla\xi \right) dV \qquad (18.19)$$

Applying the divergence theorem to the second integral in Eq. 18.19,

$$\left. \frac{d\mathcal{F}}{dt} \right|_{t_\circ} = \int_{\mathcal{V}} \left(\frac{\partial f^{\text{hom}}(\xi)}{\partial\xi} - 2K\nabla^2\xi \right) \dot{\xi} \, dV + 2K \int_{\partial\mathcal{V}} \dot{\xi}\nabla\xi \cdot d\vec{A} \qquad (18.20)$$

where $\partial\mathcal{V}$ is the oriented surface bounding the volume $\mathcal{V}$. The boundary integral on the right-hand side of Eq. 18.20 is negligible. It vanishes identically if $\dot{\xi}(\partial\mathcal{V}) = 0$,

---

[6]Some readers will recognize this development as the calculus of variations [7]. A functional is a *function of a function*; in this case, $\mathcal{F}$ takes the function $\xi(\vec{r}, t_\circ)$ and maps it to a scalar value that is numerically equal to the total free energy of the system.

which is the case if $\xi(\partial V)$ has fixed boundary values (Dirichlet boundary conditions), or if the projections of the gradients onto the boundary vanish (Neumann boundary conditions). If neither Dirichlet or Neumann conditions apply, the boundary integral will usually be insignificant compared to the volume integral for large systems (e.g., if the volume-to-surface ratio is greater than any intrinsic length scale).

Therefore, if the order parameter changes by a small amount $\delta\xi = \dot{\xi}\,\delta t$, the change in total free energy is the sum of local changes:

$$\delta\mathcal{F} = \int_{\mathcal{V}} \left(\frac{\partial f^{\text{hom}}}{\partial \xi} - 2K\nabla^2\xi \right) \delta\xi(\vec{r})\,dV \tag{18.21}$$

The quantity

$$\Phi(\vec{r}) = \frac{\partial f^{\text{hom}}}{\partial \xi} - 2K\nabla^2\xi \tag{18.22}$$

is the localized density of free-energy change due to a variation in the order-parameter field, $\delta\xi$, and is therefore the potential to change ξ. Equation 18.22 is the starting point for the development of kinetic equations for conserved and nonconserved order-parameter fields.

18.3 EVOLUTION EQUATIONS FOR CONSERVED AND NON-CONSERVED ORDER PARAMETERS

18.3.1 Cahn–Hilliard Equation

The Cahn–Hilliard equation applies to conserved order-parameter kinetics. For the binary A–B alloy treated in Section 18.1, the quantity Φ in Eq. 18.22 is the change in homogeneous and gradient energy due to a change of the local concentration c_B and is related to flux by

$$
\begin{aligned}
\vec{J}_B &= -L\nabla(\mu_B - \mu_A) \\
&= -L\nabla\frac{\delta\mathcal{F}}{\delta c_B} \\
&= -L\nabla\left\{ \frac{\partial f^{\text{hom}}[c_B(\vec{r})]}{\partial c_B} - 2K_c\nabla^2 c_B \right\} \\
&= -\frac{\widetilde{D}}{\langle\Omega\rangle\frac{\partial^2 F^{\text{hom}}}{\partial X_B^2}}\nabla\left\{ \frac{\partial f^{\text{hom}}[c_B(\vec{r})]}{\partial c_B} - 2K_c\nabla^2 c_B \right\} \\
&= -\frac{\widetilde{D}}{\frac{\partial^2 f^{\text{hom}}}{\partial c_B^2}}\nabla\left\{ \frac{\partial f^{\text{hom}}[c_B(\vec{r})]}{\partial c_B} - 2K_c\nabla^2 c_B \right\}
\end{aligned}
\tag{18.23}
$$

where the subscript is affixed to the gradient energy coefficient as a reminder that the homogeneous system is expanded in composition and its gradient.

Therefore, the accumulation gives a kinetic equation for the concentration $c_B(\vec{r}, t)$ in an A–B alloy:

$$\frac{\partial c_B}{\partial t} = \nabla \cdot \left\{ \frac{\widetilde{D}}{\frac{\partial^2 f^{\text{hom}}}{\partial c_B^2}}\nabla\left[\frac{\partial f^{\text{hom}}}{\partial c_B} - 2K_c\nabla^2 c_B \right] \right\} \tag{18.24}$$

which is the Cahn–Hilliard equation [3]. The Cahn–Hilliard equation is often linearized for concentration around the average value of the inherently positive kinetic coefficient $M_\circ = \langle M \rangle = \langle \widetilde{D}/[\Omega(\partial^2 F^{\text{hom}})/(\partial X_B^2)] \rangle$, defined in Eq. 18.9:

$$\frac{\partial c_B}{\partial t} = M_\circ \left[\frac{\partial^2 f^{\text{hom}}}{\partial c_B^2} \nabla^2 c_B - 2K_c \nabla^4 c_B \right] \qquad (18.25)$$

The first term on the right-hand side in Eq. 18.25 is diffusive. The second term accounts for interfacial-energy penalties from concentration gradients.

18.3.2 Allen–Cahn Equation

The Allen–Cahn equation applies to the kinetics of a diffuse-interface model for a nonconserved order parameter—for example, the order–disorder parameter $\eta(\vec{r}, t)$ that characterizes the $A2 \rightarrow B2^{\pm}$ phase transformation treated in Section 17.1.2. The increase in local free-energy density, $\Phi(\vec{r})$ from Eq. 18.22, does not require any macroscopic flux.[7] In a linear model, the local rate of change is proportional to its energy-density decrease,

$$\begin{aligned} \frac{\partial \eta}{\partial t} &= -M_\eta \Phi(\vec{r}) \\ &= -M_\eta \left\{ \frac{\partial f^{\text{hom}}[\eta(\vec{r})]}{\partial \eta} - 2K_\eta \nabla^2 \eta \right\} \end{aligned} \qquad (18.26)$$

where M_η is a positive kinetic coefficient related to the microscopic rearrangement kinetics. According to the Allen–Cahn equation, Eq. 18.26, η will be attracted to the local minima of f^{hom}. Depending on initial variations in η, a system may seek out multiple minima at a rate controlled by M_η. The second term on the right-hand side in Eq. 18.26 will govern the profile of η at the antiphase boundary and will cause interfaces to move toward their centers of curvature [8].

18.3.3 Numerical Simulation and the Phase-Field Method

Numerical models of conserved order-parameter evolution and of nonconserved order-parameter evolution produce simulations that capture many aspects of observed microstructural evolution. These equations, as derived from variational principles, constitute the phase-field method [9]. The phase-field method depends on models for the homogeneous free-energy density for one or more order parameters, kinetic assumptions for each order-parameter field (i.e., conserved order parameters leading to a Cahn–Hilliard kinetic equation), model parameters for the gradient-energy coefficients, subsidiary equations for any other fields such as heat flow, and trustworthy numerical implementation.

The phase-field simulations reproduce a wide range of microstructural phenomena such as dendrite formation in supercooled fixed-stoichiometry systems [10], dendrite formation and segregation patterns in constitutionally supercooled alloy systems [11], elastic interactions between precipitates [12], and polycrystalline solidification, impingement, and grain growth [6].

---

[7]This ordering transition occurs at constant composition and is accomplished by microscopic re-arrangement of atoms into two sublattices.

The simple two-dimensional phase-field simulations in Figs. 18.4 and 18.5 were obtained by numerically solving the Cahn–Hilliard (Eq. 18.25) and the Allen–Cahn equations (Eq. 18.26). Each simulation's initial conditions consisted of unstable order-parameter values from the "top of the hump" in Fig. 18.1 with a small spatial

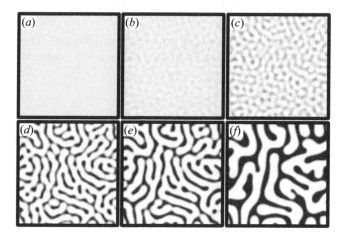

Figure 18.4:   Example of numerical solution for the Cahn–Hilliard equation, Eq. 18.25, demonstrating the kinetics of spinodal decomposition. The system is initially near an unstable concentration, **(a)**, and initially decomposes into two distinct phases with compositions c^α (black) and c^β (white) with a characteristic length scale, **(c)** and **(d)**. Subsequent evolution coarsens the length scale while maintaining fixed phase fractions. The effective time interval between images increases from (a)–(f).

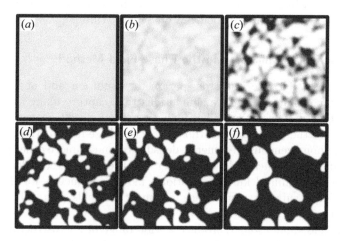

Figure 18.5:   Example of numerical solution for the Allen–Cahn equation, Eq. 18.26, for an order–disorder transition such as $A2 \rightarrow B2^{\pm}$. Initial data are near the disordered state, $\eta = 0$ (gray) in **(a)**. The system evolves into two types of domains (shown in black and white) with antiphase boundaries (APBs) separating them. The phase fractions are not fixed. The local rate of antiphase boundary migration is proportional to interface curvature [8, 13]. The effective time interval between images increases from (a)–(f).

variation. In each simulation, the magnitude of the order parameter is indicated by grayscale. Initial medium gray values correspond to the unstable initial conditions.

The characteristics of the initial evolution during spinodal decomposition or order–disorder transformations can be predicted by the perturbation analyses presented in the following section.

18.4 INITIAL STAGES OF DECOMPOSITION AND ORDER–DISORDER TRANSFORMATIONS

18.4.1 Cahn–Hilliard: Critical and Kinetic Wavelengths

A homogeneous free-energy density function $f^{\text{hom}}(c_B)$ that has a phase diagram similar to Fig. 17.7b has the form

$$f^{\text{hom}}(c_B) = \frac{16 f^{\text{hom}}_{\text{max}}}{(c^\beta - c^\alpha)^4} \left[(c_B - c^\alpha)(c_B - c^\beta) \right]^2 \tag{18.27}$$

with stable (common-tangent) concentrations located at its minima c^α and c^β and a maximum of height $f^{\text{hom}}_{\text{max}}$ at $c_B = c_0 \equiv (c^\alpha + c^\beta)/2$. Suppose that an initially uniform solution at $c_B = c_0$ is perturbed with a small one-dimensional concentration wave, $c_B(z,t) = c_0 + \epsilon(t)\sin\beta z$, where $\beta = 2\pi/\lambda$. Substituting $c_B(\vec{r},t)$ into Eq. 18.25 and keeping the lowest-order terms in $\epsilon(t)$ yields

$$\frac{d\epsilon(t)}{dt} = \frac{M_\circ \beta^2}{(c^\beta - c^\alpha)^2} \left[16 f^{\text{hom}}_{\text{max}} - 2K_c \beta^2 (c^\beta - c^\alpha)^2 \right] \epsilon(t) \tag{18.28}$$

so that

$$\epsilon(t) = \epsilon(0) e^{\frac{M_\circ \beta^2}{(c^\beta - c^\alpha)^2} \left[16 f^{\text{hom}}_{\text{max}} - 2K_c \beta^2 (c^\beta - c^\alpha)^2 \right] t}$$

$$\equiv \epsilon(0) e^{R(\beta)t} \tag{18.29}$$

where the sign of the *amplification factor* $R(\beta)$ indicates whether a fluctuation will grow or not [i.e., only composition fluctuations with wavelengths that satisfy $R(\beta) > 0$], or

$$\lambda > \lambda_{\text{crit}} \equiv \frac{\pi}{2}(c^\beta - c^\alpha)\sqrt{\frac{2K_c}{f^{\text{hom}}_{\text{max}}}} \tag{18.30}$$

will have $d\epsilon/dt > 0$ and will grow. Taking the derivative of the amplification factor in Eq. 18.28 with respect to β and setting it equal to zero, the fastest-growing wavelength is

$$\lambda_{\text{max}} = \sqrt{2}\lambda_{\text{crit}} = \frac{\sqrt{2}\pi}{2}(c^\beta - c^\alpha)\sqrt{\frac{2K_c}{f^{\text{hom}}_{\text{max}}}} \tag{18.31}$$

The characteristic length scale in the early stage of spinodal decomposition will correspond approximately to this wavelength.[8]

---

[8]Readers may recognize an analogy to the critical and fastest-growing wavelengths derived for surface diffusion and illustrated in Fig. 14.5. Both the surface diffusion equation and the Cahn–Hilliard equation are fourth-order partial-differential equations. The Allen–Cahn equation has analogies to the vapor transport equation. These analogies can be formalized with variational methods [14].

18.4.2   Allen–Cahn: Critical Wavelength

A homogeneous free-energy density function $f^{\text{hom}}(\eta)$ that has an order–disorder transition similar to Fig. 18.6b has the form

$$f^{\text{hom}}(\eta) = f^{\text{hom}}_{\text{max}}\left[(1+\eta)(1-\eta)\right]^2 \tag{18.32}$$

with local minima at $\eta = \pm1$ and a local maximum at $\eta = 0$.

Suppose that the system is initially uniform with an unstable disordered structure (i.e., $\eta = 0$). For instance, the system may have been quenched from a high-temperature, disordered state. $\eta = \pm1$ represents the two equivalent equilibrium ordered variants. If the system is perturbed a small amount by a one-dimensional perturbation in the z-direction, $\eta(\vec{r}) = \delta(t)\sin(\beta z)$. Substituting this ordering perturbation into Eq. 18.26 and keeping the lowest-order terms in the amplification factor, $\delta(t)$,

$$\frac{d\delta(t)}{dt} = M_\eta(4f^{\text{hom}}_{\text{max}} - 2K_\eta\beta^2)\delta(t) \tag{18.33}$$

so that

$$\delta(t) = \delta(0)e^{M_\eta(4f^{\text{hom}}_{\text{max}} - 2K_\eta\beta^2)t} = \delta(0)e^{R(\beta)t} \tag{18.34}$$

The perturbations therefore grow if

$$\lambda > \lambda_{\text{crit}} = \pi\sqrt{\frac{2K_\eta}{f^{\text{hom}}_{\text{max}}}} \tag{18.35}$$

which is about four times larger than the interface width given by Eq. 18.15.

Note that the amplification factor is a weakly increasing function of wavelength (asymptotically approaching $4M_\eta f^{\text{hom}}_{\text{max}}$ at long wavelengths). This predicts that the longest wavelengths should dominate the morphology. However, the probability of finding a long-wavelength perturbation is a decreasing function of wavelength, and this also has an effect on the kinetics and morphology.

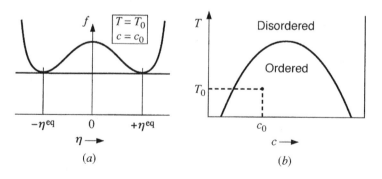

Figure 18.6:   (a) Free energy vs. nonconserved order parameter, η, at point (T_0, c_0) where the ordered phase is stable. **(b)** Corresponding phase diagram. The curve is the locus of order–disorder transition temperatures above which η^{eq} becomes zero. The equilibrium values of the order parameters, $\pm\eta^{\text{eq}}$, are the values that would be achieved at equilibrium in two equivalent variants lying on different sublattices and separated by an antiphase boundary as in Fig. 18.7c.

It is instructive to contrast the nature of the evolving early-stage morphologies predicted by Eq. 18.25 (for spinodal decomposition) and Eq. 18.26 (for ordering) and illustrated by the simulations in Figs. 18.4 and 18.5. In spinodal decomposition, the solution to the diffusion equation gives rise to a composition wave of wavelength λ_{max} given by Eq. 18.31. The decomposed microstructure is a mixture of two phases with different compositions separated by diffuse interphase boundaries (see Fig. 18.7b).

In continuous ordering, the solution to the diffusion equation gives rise to a wave of constant composition in which the order parameter varies. The theory does not predict that the order wave will have a "fastest-growing" wavelength—rather, it indicates that the longer the wavelength, the faster the wave should develop. The evolving structure will consist of coexisting antiphase domains, one with positive η and one with negative η, separated by diffuse antiphase boundaries (see Fig. 18.7c).

The crystal symmetry changes that accompany order–disorder transitions, discussed in Section 17.1.2, give rise to diffraction phenomena that allow the transitions to be studied quantitatively. In particular, the loss of symmetry is accompanied by the appearance of additional Bragg peaks, called *superlattice reflections*, and their intensities can be used to measure the evolution of order parameters.

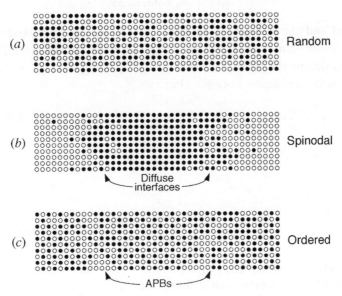

Figure 18.7: Interfaces resulting from two types of continuous transformation. (a) Initial structure consisting of randomly mixed alloy. (b) After spinodal decomposition. Regions of B-rich and B-lean phases separated by diffuse interfaces formed as a result of long-range diffusion. (c) After an ordering transformation. Equivalent ordering variants (domains) separated by two antiphase boundaries (APBs). The APBs result from A and B atomic rearrangement onto different sublattices in each domain.

18.5 COHERENCY-STRAIN EFFECTS

The driving force for transformation, Φ in Eq. 18.22, was derived from the total Helmholtz free energy, and it was assumed that molar volume is independent of concentration or structural order parameter. However, if an order-parameter fluctuation produces internal volume fluctuations, the differential expansions or contractions will produce internal strains, and additional strain-energy terms must be considered in the energetics leading to Eq. 18.21.

In crystalline solutions, the developing interfaces are initially *coherent*—strains are continuous across interfaces. Unless defects such as anticoherency dislocations intervene, the interfaces will remain coherent until a critical stress is attained and the dislocations are nucleated. For small-strain fluctuations, the system can be assumed to remain coherent and the resulting *elastic coherency energy* can be derived.[9]

For example, consider a binary alloy in which the stress-free molar volume is a function of concentration, $V(c_B)$. The linear expansion due to the composition change can be inferred from diffraction experiments under stress-free conditions (*Vegard's effect*) and is characterized by Vegard's parameter, α_c [e.g., in cubic or isotropic crystals $\epsilon_{xx}^{\sigma=0} = \epsilon_{yy}^{\sigma=0} = \epsilon_{zz}^{\sigma=0} = \alpha_c(c - c_0)$]. The assumption of coherency implies that the total strain in the interfacial planes is zero. If a planar composition fluctuation perturbation of the form

$$c_B(z) = c_0 + \delta c \cos \beta z \tag{18.36}$$

is postulated and the material is elastically isotropic, the total strain, $\epsilon_{ij}^{\text{tot}}$, will correspond to the sum of the strain due to the composition change, $\epsilon_{ij}^{\sigma=0}$, and the elastic strain due to the coherency stresses, $\epsilon_{ij}^{\text{elas}}$ (i.e., $\epsilon_{ij}^{\text{tot}} = \epsilon_{ij}^{\sigma=0} + \epsilon_{ij}^{\text{elas}}$). Then, using the equations of linear isotropic elasticity,[10]

$$\epsilon_{zz}^{\text{tot}} = \alpha_c \delta c \frac{1+\nu}{1-\nu} \cos \beta z \quad \text{and} \quad \epsilon_{xx}^{\text{tot}} = \epsilon_{yy}^{\text{tot}} = \epsilon_{xy}^{\text{tot}} = \epsilon_{yz}^{\text{tot}} = \epsilon_{zx}^{\text{tot}} = 0 \tag{18.37}$$

The elastic strains, $\epsilon_{ij}^{\text{elas}}$, required to satisfy Eq. 18.37 are

$$\epsilon_{zz}^{\text{elas}} = \alpha_c \, \delta c \frac{2\nu}{1-\nu} \cos \beta z$$
$$\epsilon_{xx}^{\text{elas}} = \epsilon_{yy}^{\text{elas}} = -\alpha_c \, \delta c \cos \beta z \tag{18.38}$$
$$\epsilon_{xy}^{\text{elas}} = \epsilon_{yz}^{\text{elas}} = \epsilon_{zx}^{\text{elas}} = 0$$

where ν is Poisson's ratio. The corresponding elastic stresses are given by $\boldsymbol{\sigma} = C\boldsymbol{\epsilon}^{\text{elas}}$, where C is the fourth-rank stiffness tensor:

$$\sigma_{zz} = 0$$
$$\sigma_{xx} = \sigma_{yy} = -\alpha_c \, \delta c \frac{E}{1-\nu} \cos \beta z \tag{18.39}$$
$$\sigma_{xy} = \sigma_{yz} = \sigma_{zx} = 0$$

---

[9]J.W. Cahn's early contributions to elastic coherency theory were motivated by his work on spinodal decomposition. His subsequent work with F. Larché created a rigorous thermodynamic foundation for coherency theory and stressed solids in general. A single volume, *The Selected Works of John W. Cahn* [15], contains papers that provide background and advanced reading for many topics in this textbook. This derivation follows from one in a publication included in that collection [16].

[10]Methods to calculate coherency stresses in anisotropic materials, and an example calculation for cubic materials, have been published [17].

where E is Young's elastic modulus. Therefore, the elastic coherency strain-energy contribution to the total energy is

$$\mathcal{F}^{\text{elas}} = \frac{1}{2} \int_{\mathcal{V}} \sum_{i,j} \sigma_{ij} \epsilon_{ij}^{\text{elas}} \, dV$$

$$= \frac{\alpha_c^2 E}{1 - \nu} \int_{\mathcal{V}} (\delta c)^2 \cos^2 \beta z \, dV \qquad (18.40)$$

$$= \frac{\alpha_c^2 E}{1 - \nu} \int_{\mathcal{V}} (c_B - c_0)^2 \, dV$$

The final equality in Eq. 18.40 demonstrates that the contribution to coherency energy is independent of wavelength and direction.

The coherency energy modifies Eqs. 18.21 and 18.22 as follows:

$$\delta \mathcal{F} = \int_{\mathcal{V}} \left[\frac{\partial f^{\text{hom}}}{\partial c_B} + \frac{2\alpha_c^2 E}{1 - \nu}(c_B - c_0) - 2K_c \nabla^2 c_B \right] \delta c_B \, dV$$

$$\Phi(\vec{r}) = \left[\frac{\partial f^{\text{hom}}}{\partial c_B} + \frac{2\alpha_c^2 E}{1 - \nu}(c_B - c_0) - 2K_c \nabla^2 \xi \right] \qquad (18.41)$$

and the Cahn–Hilliard equation linearized in c_B, corresponding to Eq. 18.25, including coherency effects for elastic materials, is

$$\frac{\partial c_B}{\partial t} = M_\circ \left[\left(\frac{\partial^2 f^{\text{hom}}}{\partial c_B^2} + \frac{2\alpha_c^2 E}{1 - \nu} \right) \nabla^2 c_B - 2K_c \nabla^4 c_B \right]$$

$$\equiv M_\circ \left[\left(\frac{\partial^2 f^{\text{hom}}}{\partial c_B^2} + 2\alpha_c^2 Y \right) \nabla^2 c_B - 2K_c \nabla^4 c_B \right] \qquad (18.42)$$

where the first equality gives the isotropic elastic contribution explicitly and the second defines a general coherency modulus, Y, for anisotropic materials [1, 17].

The coherency strain energy introduces an additional barrier to spinodal decomposition, which causes a shift on the temperature–composition phase diagram of the *chemical spinodal*, defined by $\partial^2 f^{\text{hom}}/\partial c_B^2 = 0$, to the *coherent spinodal*, defined by

$$\frac{\partial^2 f^{\text{hom}}}{\partial c_B^2} + 2\alpha_c^2 Y = 0 \qquad (18.43)$$

as in Fig. 18.8.

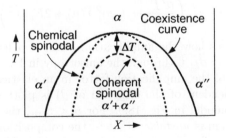

Figure 18.8: Relation between chemical and coherent spinodals.

An additional effect of undercooling on the kinetics and microstructure of spinodal decomposition arises from the temperature dependence of $\partial^2 f^{\text{hom}}/\partial c_B^2$, which is approximately linear near the chemical spinodal temperature [1]. Because $2\alpha_c^2 Y$ is always positive, the coherent spinodal lies below the chemical spinodal in the T–X diagram. The depression ΔT of the coherent spinodal below the consolute point can be calculated (Eq. 18.43) for various systems. In Al–Zn, ΔT is approximately 20 K; in Au–Ni, it is approximately 400 K.

In crystalline solids, *only* coherent spinodal decomposition is observed. The process of forming incoherent interfaces involves the generation of anticoherency dislocation structures and is incompatible with the continuous evolution of the phase-separated microstructure characteristic of spinodal decomposition. Systems with elastic misfit may first transform by coherent spinodal decomposition and then, during the later stages of the process, lose coherency through the nucleation and capture of anticoherency interfacial dislocations [18].

18.5.1 Generalizations of the Cahn–Hilliard and Allen–Cahn Equations

Microstructural length scales that initially arise from uniform, but unstable, order parameters are readily understood by the perturbation analyses that lead to the amplification factor $R(\beta)$ in Eqs. 18.28 and 18.34. When a system is anisotropic such as in a elastically coherent material, the perturbation's behavior may depend on its direction with respect to the material's symmetry axes.

Order-parameter fluctuations can be generalized by introducing the wave-vector $\vec{\beta}$ in a Fourier representation,

$$\xi(\vec{x}) - \xi_\circ = \int A(\vec{\beta}) e^{i\vec{\beta} \cdot \vec{x}} \, d\vec{\beta} \qquad (18.44)$$

where

$$A(\vec{\beta}) = \frac{1}{2\pi} \int [\xi(\vec{x}) - \xi_\circ] \, e^{-i\vec{\beta} \cdot \vec{x}} \, d\vec{x} \qquad (18.45)$$

are the amplitudes associated with each Fourier mode $\vec{\beta}$. Each Fourier mode is independent in a linear case. For example, when Eq. 18.44 is inserted into the linearized Cahn–Hilliard equation, Eq. 18.42,

$$\frac{\partial A(\vec{\beta})}{\partial t} = -M_\circ \left[\frac{\partial^2 f^{\mathrm{hom}}}{\partial c_B^2} + 2\alpha_c^2 Y(\hat{n}) + 2K_c |\vec{\beta}|^2 \right] |\vec{\beta}|^2 \equiv R(\vec{\beta}) A(\vec{\beta}) \qquad (18.46)$$

where

$$R(\vec{\beta}) = -M_\circ \left(\frac{\partial^2 f^{\mathrm{hom}}}{\partial c_B^2} + 2\alpha_c^2 Y(\hat{n}) + 2K_c |\vec{\beta}|^2 \right) |\vec{\beta}|^2 \qquad (18.47)$$

Equation 18.47 indicates how the amplification factor depends on the mode $\vec{\beta}$ as well as the direction $\hat{n} = \vec{\beta}/|\vec{\beta}|$ through the anisotropy in Y.

The β dependence of the amplification factor in an elastically isotropic crystal (for which R is independent of the direction of $\vec{\beta}$) is plotted for a temperature inside the coherent spinodal in Fig. 18.9. For $\beta < \beta_{\mathrm{crit}}$, the amplification factor $R(\beta) > 0$ and the system is *unstable*—that is, the composition waves in Eq. 18.44 will grow exponentially. The wavenumber β_{max}, at which $\partial R(\beta)/\partial \beta = 0$, receives maximum amplification and will dominate the decomposed microstructure. Outside the coherent spinodal, where $\partial^2 f^{\mathrm{hom}}/\partial c_B^2 + 2\alpha_c^2 Y(\hat{n}) > 0$, all wavenumbers will have $R(\beta) < 0$ and the system will be *stable* with respect to the growth of composition waves.

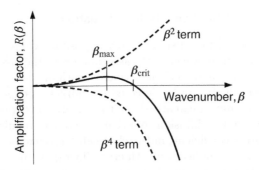

Figure 18.9:   Amplification factor vs. wavenumber plot for an elastically isotropic crystal at a temperature inside the coherent spinodal where $\partial^2 f^{\mathrm{hom}} / \partial c_B^2 + 2\alpha_c^2 Y < 0$.

The solution of Eq. 18.46 is

$$A(\vec{\beta}, t) = A(\vec{\beta}, 0) e^{R(\vec{\beta})t} \qquad (18.48)$$

which is a generic form for linear perturbation analysis. At least two sources of linearization lead to Eq. 18.48. As in the steps leading from Eq. 18.24 to Eq. 18.25, averaging is performed so that the kinetic equations are linear, and the perturbation modes are independent and linear in a small parameter.

The linear perturbation analyses reliably predict *initial* behavior and characteristic length scales. Equation 18.48 does not predict behavior at longer times, as nonlinearities and Fourier mode-coupling intervene. Numerical methods permit simulation of specific features and trends, such as the *coarsening* of the microstructural length scales in Figs. 18.4 and 18.5, which can be characterized, visualized, and understood. Furthermore, direct insight into the evolution path is obtained through physical considerations of energy functionals, Eqs. 18.21 and 18.41.

Because the kinetic equations are derived from variational principles for the total free energy, the total free energy always decreases.[11] The equilibrium state naturally has the lowest total free energy. For the total free energy given by Eq. 18.21, equilibrium corresponds to the phase composition and fractions predicted by the homogeneous free energy that minimizes total interfacial energy. For a conserved order parameter, energy-minimizing interfacial configurations have uniform mean curvature such as a planar or spherical interface.[12] For a nonconserved order parameter, the energy-minimizing configuration is *no* interface (i.e., a single variant). However, there are many locally minimizing microstructures in either case at which kinetic processes halt. Nevertheless, the coarsening observed in Figs. 18.4 and 18.5 can be rationalized by considerations of the differences between the early microstructures predicted by perturbation analyses, Eq. 18.48, and the microstructures that minimize the total free energy functional.

---

[11]Functionals that are monotonic, such as the appropriate total free energy, are called *Lyapanov functions*, and their existence simplifies global analysis. In the isothermal and constant-volume cases treated in this chapter, the total Helmholtz free energy is the Lyapanov function. However, other Lyapanov functions apply as the system constraints are generalized [9].

[12]Because K does not depend on the direction of the order-parameter gradient in Eq. 18.21, the interfacial energy is isotropic, and energy-minimizing partitions of space are constant-curvature surfaces. If the interfacial energy is anisotropic, energy-minimizing interfacial configurations have constant weighted mean curvature (see Appendix C).

18.5.2 Diffraction and the Cahn–Hilliard Equation

Microstructural characteristics of spinodal decomposition are *periodicity* and *alignment*. Periodicity arises from wavelengths associated with the fastest-growing initial mode. At later times, the characteristic periodic length increases due to microstructural coarsening. Periodicity can be detected by diffraction experiments.

Crystallographic alignment can arise from the orientation dependence of the elastic strain energy term in the diffusion equation. Alignment requires that a material has a nonzero Vegard's coefficient and is elastically anisotropic [i.e., the factor Y (Section 18.5) must vary significantly with crystallographic direction]. Under these conditions, composition waves directed along crystallographic directions that are elastically soft (i.e., along which Y is a minimum) will grow fastest, leading to alignment of the product microstructure with these directions. For cubic crystals, this alignment is along $\langle 100 \rangle$ and, less frequently, $\langle 111 \rangle$ directions.

Periodic microstructures can be corroborated by observations of wavevectors β in transmission electron microscope (TEM) images, particularly if the sample is oriented with the modulation waves directed perpendicular to the electron-beam direction (e.g., with the beam along [001] for a crystal with $\langle 100 \rangle$ modulations).

If there is alignment, contrast in TEM images is strong, because of the periodic strain field in the crystal. Selected-area diffraction shows evidence of such alignment by the location of "satellite" intensities around the Bragg peaks arising from the modulation of atomic scattering factors, lattice constant, or both [19]. In Fig. 18.10, the electron diffraction effects, expected from an f.c.c. crystal with $\langle 100 \rangle$ composition waves, are depicted with a [001] beam direction.

Examples of observations of spinodal microstructures include:

- Kubo and Wayman made TEM observations of an aligned $\langle 100 \rangle$ spinodal decomposition product in thin foils of long-range ordered β-brass [20]. (Interestingly, bulk material did not decompose, while thin foils with [001] foil normals did. The difference was attributed to a relaxation of elastic constraint in the thin foil.)

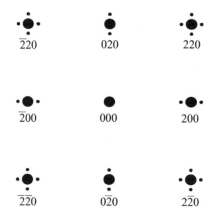

Figure 18.10: The (001) section of a reciprocal lattice of spinodally decomposed f.c.c. alloy, as observed by TEM. Note the systematic absences of satellites for which $\vec{g} \cdot \vec{R} = 0$ ($\vec{g}$ is the diffraction vector and $\vec{R}$ is the local atomic displacement vector).

- Miyazaki made TEM observations of an aligned $\langle 100 \rangle$ decomposition product in Fe–Mo alloys, with diffraction patterns similar to those in Fig. 18.10 [21].

- Allen reported TEM observations of a nonaligned decomposition products in long-range ordered Fe–Al alloy [22]. Such morphologies are called *isotropic spinodal microstructures*. Similar structures are observed in Al–Zn and Fe–Cr alloys. Such structures can be produced in systems that are elastically isotropic or in which the lattice constant does not change appreciably with composition.

- Brenner et al. reported an atom-probe field-ion microscope study of decomposition in an Fe–Cr–Co alloy (see Fig. 18.11) [23]. The *atom probe* allows direct compositional analysis of the peaks and valleys of the composition waves. It is probably the best tool for verifying a spinodal mechanism in metals, because the growth in amplitude of the composition waves can be studied as a function of aging time, with near-atomic resolution. In spinodal alloys, there is a continuous increase in the amplitude of the composition waves with aging time. On the other hand, for a transformation by nucleation and growth, the particles formed earliest generally exhibit a compositional discontinuity with the matrix.

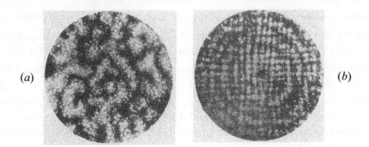

(a)   *(b)*

Figure 18.11: Spinodal decomposition observed by atom-probe field-ion microscopy. **(a)** Isotropic morphology observed in Fe–Mo alloys. **(b)** Aligned morphology observed in Fe–Cr–Co alloys. From Brenner et al. [23].

Bibliography

1. J.E. Hilliard. *Spinodal Decomposition*, pages 497–560. American Society for Metals, Metals Park, OH, 1970.

2. J.W. Cahn and W.C. Carter. Crystal shapes and phase equilibria: A common mathematical basis. *Metall. Trans.*, 27A(6):1431–1440, 1996.

3. J.W. Cahn and J.E. Hilliard. Free energy of a nonuniform system—I. Interfacial free energy. *J. Chem. Phys.*, 28(2):258–267, 1958.

4. J.W. Cahn. On spinodal decomposition. *Acta Metall.*, 9(9):795–801, 1961.

5. W.T. Read and W. Shockley. Dislocations models of grain boundaries. In *Imperfections in Nearly Perfect Crystals*. John Wiley & Sons, New York, 1952.

6. J.A. Warren, R. Kobayashi, A.E. Lobovsky, and W.C. Carter. Extending phase field models of solidification to polycrystalline materials. *Acta Mater.*, 51(20):6035–6058, 2003.

7. I.M. Gelfand and S.V. Fomin. *Calculus of Variations.* Prentice-Hall, Englewood Cliffs, NJ, 1963.

8. S.M. Allen and J.W. Cahn. Microscopic theory for antiphase boundary motion and its application to antiphase domain coarsening. *Acta Metall.*, 27(6):1085–1095, 1979.

9. O. Penrose and P.C. Fife. On the relation between the standard phase-field model and a thermodynamically consistent phase-field model. *Physica D*, 69(1–2):107–113, 1993.

10. R. Kobayashi. Modeling and numerical simulations of dendritic crystal-growth. *Physica D*, 63(3–4):410–423, 1993.

11. J.A. Warren and W.J. Boettinger. Prediction of dendritic growth and microsegregation patterns in a binary alloy using the phase-field method. *Acta Metall.*, 43(2):689–703, 1995.

12. S.Y. Hu and L.Q. Chen. A phase-field model for evolving microstructures with strong elastic inhomogeneity. *Acta Mater.*, 49(11):1879–1890, 2001.

13. P.C. Fife and A.A. Lacey. Motion by curvature in generalized Cahn-Allen models. *J. Stat. Phys.*, 77(1–2):173–184, 1994.

14. W.C. Carter, J.E. Taylor, and J.W. Cahn. Variational methods for microstructural-evolution theories. *JOM*, 49:30–36, 1997.

15. W.C. Carter and W.C. Johnson, editors. *The Selected Works of John W. Cahn.* The Minerals, Metals and Materials Society, Warrendale, PA, 1998.

16. J.W. Cahn. On spinodal decomposition. *Acta Metall.*, 9(9):795–801, 1961.

17. W.C. Carter and C.A. Handwerker. Morphology of grain growth in response to diffusion induced elastic stresses: Cubic systems. *Acta Metall.*, 41(5):1633–1642, 1993.

18. R.J. Livak and G. Thomas. Loss of coherency in spinodally decomposed Cu–Ni–Fe alloys. *Acta Metall.*, 22(5):589–599, 1974.

19. D. de Fontaine. A theoretical and analogue study of diffraction from one-dimensional modulated structures. In J.B. Cohen and J.E. Hillard, editors, *Local Atomic Arrangements Studied by X-Ray Diffraction*, pages 51–88. The Metallurgical Society of AIME, Warrendale, PA, 1966.

20. H. Kubo, I. Cornelis, and C.M. Wayman. Morphology and characteristics of spinodally decomposed β-brass. *Acta Metall.*, 28(3):405–416, 1980.

21. T. Miyazaki, S. Takagishi, H. Mori, and T. Kozakai. The phase decomposition of iron–molybdenum binary alloys by spinodal mechanism. *Acta Metall.*, 28(8):1143–1153, 1980.

22. S.M. Allen. Phase separation of Fe–Al alloys with Fe_3Al order. *Phil. Mag.*, 36(1):181–192, 1977.

23. S.S. Brenner, P.P. Camus, M.K. Miller, and W.A. Soffa. Phase separation and coarsening in Fe–Cr–Co alloys. *Acta Metall.*, 32(8):1217–1227, 1984.

24. K.B. Rundman and J.E. Hilliard. Early stages of spinodal decomposition in an aluminum–zinc alloy. *Acta Metall.*, 15(6):1025–1033, 1967.

25. H. Baker, editor. *ASM Handbook: Alloy Phase Diagrams*, volume 3, page 2•56. ASM International, Materials Park, OH, 1992.

EXERCISES

18.1 Equation 18.9 was derived assuming equal and constant atomic volumes in the A–B solid solution. Derive a corresponding relation for the interdiffusion flux in the V-frame, $\vec{J}_B^V$, assuming that that Ω_A and Ω_B remain independent of composition, but for which $\Omega_A \neq \Omega_B$. Find a relation between the interdiffusivity, $\tilde{D}$, alloy composition, atomic volumes, and L_A and L_B for this more general case.

Solution. Using Eqs. 3.9, the fluxes of components A and B in a local crystal frame (local C-frame) can be written

$$\vec{J}_A^C = -c_A \left[\frac{L_{AA}}{c_A} - \frac{L_{AB}}{c_B} \right] \nabla \mu_A = -c_A L_A \nabla \mu_A$$
$$\vec{J}_B^C = -c_B \left[\frac{L_{BB}}{c_B} - \frac{L_{BA}}{c_A} \right] \nabla \mu_B = -c_B L_B \nabla \mu_B \qquad (18.49)$$

where L_A and L_B are intrinsic mobilities. The flux of B in the V-frame is then

$$\vec{J}_B^V = \vec{J}_B^C + c_B \vec{v}_C^V \qquad (18.50)$$

where $\vec{v}_C^V$ is the velocity of the local C-frame in the V-frame as measured by the motion of an embedded inert marker at the origin of the C-frame. Using Eqs. 3.15, 3.23, and A.10, $\vec{v}_C^V = -[\Omega_A \vec{J}_A^C + \Omega_B \vec{J}_B^C]$ and therefore

$$\vec{J}_B^V = c_A \Omega_A \vec{J}_B^C - c_B \Omega_A \vec{J}_A^C \qquad (18.51)$$

Substituting Eqs. 18.49 into Eq. 18.51 yields

$$\vec{J}_B^V = -\Omega_A c_A c_B \left[L_B \nabla \mu_B - L_A \nabla \mu_A \right] \qquad (18.52)$$

Equation 18.52 may be put into another form by using the identity

$$L_B \nabla \mu_B - L_A \nabla \mu_A = (L_B c_A + L_A c_B) (\Omega_A \nabla \mu_B - \Omega_B \nabla \mu_A) \\ + (L_B \Omega_B - L_A \Omega_A) (c_A \nabla \mu_A + c_B \nabla \mu_B) \qquad (18.53)$$

Substituting Eq. 18.53 into Eq. 18.52 and using Eq. 18.8 gives the further expression

$$\vec{J}_B^V = -\Omega_A c_A c_B \left(L_B c_A + L_A c_B \right) \left(\Omega_A \nabla \mu_B - \Omega_B \nabla \mu_A \right) \qquad (18.54)$$

The second term in parentheses in Eq. 18.54 may be developed further by considering the free-energy density given by

$$f = c_A \mu_A + c_B \mu_B \qquad (18.55)$$

Differentiating Eq. 18.55 and using $(c_A \nabla \mu_A + c_B \nabla \mu_B) = 0$,

$$\frac{\partial f}{\partial c_B} = \mu_B - \frac{\Omega_B}{\Omega_A} \mu_A \qquad (18.56)$$

Applying the gradient operator to Eq. 18.56 then yields

$$\Omega_A \nabla \mu_B - \Omega_B \nabla \mu_A = \Omega_A \nabla \left(\frac{\partial f}{\partial c_B} \right) \qquad (18.57)$$

Substitution of Eq. 18.57 into Eq. 18.54 produces the relation

$$\vec{J}_B^V = -\Omega_A^{\,2} c_A c_B \left(L_B c_A + L_A c_B \right) \nabla \left(\frac{\partial f}{\partial c_B} \right) = -L \nabla \left(\frac{\partial f}{\partial c_B} \right) \qquad (18.58)$$

where the coefficient, L, is

$$L = \Omega_A{}^2 c_A c_B \left(L_B c_A + L_A c_B\right) \tag{18.59}$$

Also, because $\nabla(\partial f/\partial c_B) = (\partial^2 f/\partial c_B{}^2)\nabla c_B$,

$$\vec{J}_B^{\,\mathcal{V}} = -\left(L\,\frac{\partial^2 f}{\partial c_B^2}\right)\nabla c_B \tag{18.60}$$

and a comparison of Eq. 18.60 with Eq. 3.27 shows that

$$\widetilde{D} = L\,\frac{\partial^2 f}{\partial c_B^2} \tag{18.61}$$

18.2 The Al–Zn system was the first studied extensively in an attempt to verify the theory for spinodal decomposition [24]. The equilibrium diagram for this system, shown in Fig. 18.12, shows a monotectoid in the Al-rich portion of the diagram. The top of the miscibility gap at 40 at. % Zn is the critical consolute point of the incoherent phase diagram.

In concentrated Al–Zn alloys, the kinetics of precipitation of the equilibrium β phase from α are too rapid to allow the study of spinodal decomposition. An Al–22 at. % Zn alloy, however, has decomposition temperatures low enough to permit spinodal decomposition to be studied. For Al–22 at. % Zn, the chemical spinodal temperature is 536 K and the coherent spinodal temperature is 510 K. The early stages of decomposition are described by the diffusion equation

$$\frac{\partial c}{\partial t} = \widetilde{D}\left[\left(1 + \frac{2\alpha_c^2 Y}{f''}\right)\frac{\partial^2 c}{\partial x^2} - \frac{2K}{f''}\frac{\partial^4 c}{\partial x^4}\right] \tag{18.62}$$

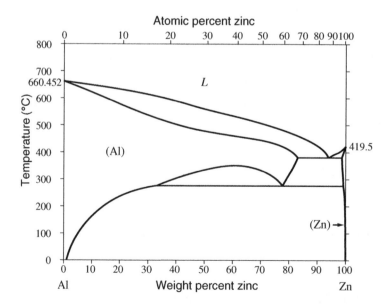

Figure 18.12:   Equilibrium diagram for Al–Zn alloys. From *ASM Handbook: Alloy Phase Diagrams*, Vol. 3 [25].

(a) What will be the characteristic periodicity in the microstructure in the early stages of decomposition at 338 K for an Al–22 at. % Zn alloy?

(b) Suppose that the specimen described in part (a) is suddenly heated to 473 K. Explain how the microstructure established at 338 K will change upon heating:

    i. At very short times

    ii. At intermediate times

    iii. At very long times

Data. Assume for Al–Zn alloys that $\alpha_c^2 Y$ is isotropic, the enthalpy of mixing of Al–Zn solutions is independent of temperature, and the entropy of mixing, s, is ideal; that is,

$s = -nk[c \ln c + (1 - c) \ln(1 - c)]$

$n = 6 \times 10^{22}$ cm^{-3} (number of atoms per unit volume)

$\tilde{D} = [c(1 - c)]f'' D^\circ e^{-Q/(N_o kT)}$

$Q = 104$ kJ mol^{-1}

$K = 1.6 \times 10^{-12}$ J cm^{-1}

$\tilde{D} = -2.6 \times 10^{-18}$ cm^2 s^{-1} (at 338 K and 22 at. % Zn)

$f'' = -1.17$ kJ cm^{-3} (at 338 K and 22 at. % Zn)

Solution. We will need an expression for $f''(T)$. Because $f = e - Ts$, $\partial f/\partial T = -s$ and $\partial(f'')/\partial T = -s''$. Also, for ideal entropy of mixing, s'' is independent of T, so f'' should vary linearly with T. From the fact that $f'' = 0$ at the chemical spinodal temperature and the value of f'' provided at 338 K, we obtain

$$f''(T) = \frac{1.17 \times 10^9}{198}(T - 536) \, \text{J m}^{-3} \tag{18.63}$$

We can evaluate the elastic energy term $2\alpha_c^2 Y$ because we know $f'' + 2\alpha_c^2 Y = 0$ at 510 K. Thus,

$$2\alpha_c^2 Y = -f''(510) = 1.536 \times 10^8 \, \text{J m}^{-3} \tag{18.64}$$

(a) The periodicity of the microstructure at 338 K (early times) is determined by $\beta_m(338 \text{ K})$. From $\partial R(\beta)/\partial \beta = 0$ at β_m,

$$\beta_m(T) = \sqrt{-\frac{f''(T) + 2\alpha_c^2 Y}{4K}} \tag{18.65}$$

$$\beta_m(338) = \sqrt{-\frac{-1.17 \times 10^9 + 1.536 \times 10^8 \, \text{J m}^{-3}}{4 \times 1.6 \times 10^{-10} \, \text{J m}^{-1}}} = 1.26 \times 10^9 \, \text{m}^{-1} \tag{18.66}$$

This corresponds to a modulation wavelength at 338 K of

$$\lambda_m = \frac{2\pi}{\beta_m} = 4.986 \text{ nm} \tag{18.67}$$

(b) Take the microstructure produced at 338 K with $\beta_m = 1.26 \times 10^9$ and heat to 473 K (still within the coherent spinodal). Let's compute β_m and β_c at 473 K:

$$\beta_m(473) = \sqrt{-\frac{-3.723 \times 10^8 + 1.536 \times 10^8 \, \text{J m}^{-3}}{4 \times 1.6 \times 10^{-10} \, \text{J m}^{-1}}} = 5.845 \times 10^8 \, \text{m}^{-1} \tag{18.68}$$

Recall that β_c is the value of β where $R(\beta) = 0$, so

$$\beta_c(T) = \sqrt{-\frac{f''(T) + 2\alpha_c^2 Y}{2K}} = \sqrt{2}\,\beta_m \qquad (18.69)$$

and thus,

$$\beta_c(473) = 8.266 \times 10^8 \text{ m}^{-1} \qquad (18.70)$$

Note that $\beta_m(338\text{ K}) > \beta_c(473\text{ K})$, so this is an example of *reversion*: the fine-scale decomposition structure produced at 338 K, on heating to 473 K, finds itself with a *negative* $R(\beta)$ at 473 K. Therefore, the fine-scale microstructure with approximately 5 nm periodicity *dissolves* at early times at 473 K. At intermediate times at 473 K, a microstructure dominated by waves with $\beta_m(473\text{ K})$ is expected. These have a wavelength of

$$\lambda_m(473\text{ K}) = 10.75 \text{ nm} \qquad (18.71)$$

which is more than double that of the original structure.

(c) At long times of aging at 473 K, the structure gradually coarsens—that is, the wavelength increases from its intermediate-time value of approximately 11 nm.

To quantitatively assess the relative rates of the reversion at 473 K with the decomposition that follows, we can compute the ratio of the two $R(\beta)$s:

$$\frac{R(\beta_1)}{R(\beta_2)} = \frac{\beta_1^2[(f'' + 2\alpha_c^2 Y) + 2K\beta_1^2]}{\beta_2^2[(f'' + 2\alpha_c^2 Y) + 2K\beta_2^2]} \qquad (18.72)$$

Taking $\beta_1 = \beta_m(338\text{ K})$ and $\beta_2 = \beta_m(473\text{ K})$ and using $f''(473\text{ K})$ at the reaction temperature of interest,

$$\frac{R(\beta_1)}{R(\beta_2)} = -12.313 \qquad (18.73)$$

So we conclude that the fine-scale structure "disappears" about 12 times faster than the coarser structure forms at 473 K.

18.3 Both Fe–Al and Fe–Mo alloys can undergo spinodal decomposition, yet the resultant microstructures have important differences. Figure 18.13 shows transmission electron micrographs of these two alloys taken with the electron beam parallel to $\langle 001 \rangle$, exhibiting typical spinodal microstructures.

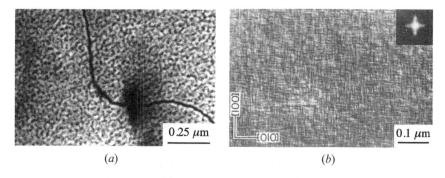

(a)   *(b)*

Figure 18.13:   TEM of **(a)** Fe–Al and **(b)** Fe–Mo alloy specimens after spinodal decomposition. From Allen [22] and Miyazaki et al. [21].

(a) What characteristic feature common to both microstructures is suggestive of spinodal decomposition? What is the theoretical reason for this characteristic feature?

(b) What is the most significant morphological difference between the spinodal microstructures in the two alloys? By using appropriate expressions from the theory of spinodal decomposition, identify *two* different physical properties of the alloy systems, whose behavior would provide an explanation for this difference. Fully explain your reasoning.

Solution.

(a) The characteristic common to both microstructures is *periodicity*. This arises from selective amplification of composition waves inside the coherent spinodal. The linear theory of spinodal decomposition predicts exponential growth of waves with nm-scale wavelengths, and the wavenumber corresponding to the maximum rate of decomposition, β_m, is

$$\beta_m(T) = \sqrt{-\frac{f''(T) + 2\alpha_c^2 Y}{4K}} \tag{18.74}$$

(b) The microstructure of the Fe–Al alloy, Fig. 18.13a, shows little evidence of crystallographic alignment, and at least two factors could be responsible for this lack of a preferred direction for the fastest-growing waves. First, it could be that the lattice constant of Fe–Al alloys is not very composition dependent, making $\alpha_c = (1/a)\,da/dc \cong 0$. Second, it could be for this alloy that the elastic modulus Y is independent of orientation; that is, the alloy is elastically isotropic. Either alternative would make the factor $2\alpha_c^2 Y$ in Eq. 18.74 very small relative to f''.

The microstructure of the decomposed Fe–Mo alloy, Fig. 18.13b, shows strong alignment of the developing two-phase microstructure along $\langle 100 \rangle$ directions. Such alignment is common in cubic crystals, and it arises from the anisotropy of the effective modulus, Y, in the diffusion equation. From Eq. 18.74 it is apparent that the crystallographic directions in which Y is a *minimum* will correspond to the wavevector of the fastest-growing waves.

18.4 If the progress of spinodal decomposition is measured isothermally at a series of temperatures, a plot of the time required to reach a given amount of decomposition at the various temperatures can be constructed [as described in Section 21.2, such a plot is called a *time–temperature–transformation* (TTT) *diagram*]. For spinodal decomposition, a TTT diagram has a "C" shape, similar to the shape of the corresponding TTT diagram for a nucleation and growth transformation (see Section 21.2).

Derive an expression for the temperature at which the rate of spinodal decomposition is a maximum (i.e., find the temperature of the nose of the C-curve).

Solution. The strategy is simple. We have an expression (Eq. 18.46) for the amplification factor, which is a function of wavevector (β) and temperature (T), which for a one-dimensional wavevector is

$$R(\beta, T) = -M_\circ \beta^2 \left(\frac{\partial^2 f}{\partial c^2} + 2\alpha_c^2 Y + 2K_c \beta^2 \right) \tag{18.75}$$

First, we take the derivative of $R(\beta, T)$ with respect to β to find the maximum wavevector β_m:

$$\beta_m^2 = -\frac{\frac{\partial^2 f}{\partial c^2} + 2\alpha_c^2 Y}{4K_c} \tag{18.76}$$

from which it is seen that β_m is function of temperature (through f''). Then, by plugging Eq. 18.76 into Eq. 18.75, an explicit expression for the temperature dependence $R(T)$ is obtained. Taking the derivative with respect to T of the resulting expression, an expression is obtained for the temperature at which the rate of spinodal decomposition is a maximum.

The temperature dependence of $R(\beta, T)$ arises from two sources: $\frac{\partial^2 f}{\partial c^2}$ and the mobility M_o. Using a regular solution model, $\frac{\partial^2 f}{\partial c^2}$ can be expressed

$$\frac{\partial^2 f}{\partial c^2} = -\frac{N_o k}{\Omega c(1 - c)}(T_s - T) \tag{18.77}$$

where T_s is the consolute temperature and c is normalized composition. The mobility has an Arrhenius temperature dependence:

$$M_o = A\,e^{-Q/(kT)} \tag{18.78}$$

After some algebraic manipulation, we get the following expression, which can be solved for the temperature of the nose, T_n:

$$\frac{Q}{c(1-c)}(T_n - T_s) + \frac{2Q\Omega}{N_o k}\alpha_c^2 Y + \frac{2(N_o k)^2}{\Omega^2 c^2(1-c)^2}T_n^2(T_n - T_s) + \frac{4N_o k}{\Omega c(1-c)}T_n^2 \alpha_c^2 Y = 0 \tag{18.79}$$

Note that the nose of the C-curve is composition-dependent.

18.5 Suppose that two equivalent variants of an ordered structure are present in a binary A–B system in the form of two domains (1 and 2) separated by an antiphase boundary as on the left and center of Fig. 18.7c. Only two sublattices are present in the structure.

Show that the long-range order parameter for domain 1 is the negative of the long-range order parameter of domain 2.

Solution. Using Eqs. 17.7 and 17.8,

$$\begin{aligned}
2\eta_1 &= [X_B^\alpha]_1 - \left[X_B^\beta\right]_1 \quad \text{(domain 1)} \\
2\eta_2 &= [X_B^\alpha]_2 - \left[X_B^\beta\right]_2 \quad \text{(domain 2)}
\end{aligned} \tag{18.80}$$

But because the variants are equivalent,

$$\begin{aligned}
[X_B^\alpha]_1 &= \left[X_B^\beta\right]_2 \\
\left[X_B^\beta\right]_1 &= [X_B^\alpha]_2
\end{aligned} \tag{18.81}$$

Combining Eqs. 18.80 and 18.81,

$$2\eta_2 = \left[X_B^\beta\right]_1 - [X_B^\alpha]_1$$

$$\eta_2 = -\eta_1$$

CHAPTER 19

NUCLEATION

The formation of a new phase by a discontinuous phase transformation (such as the formation of a solid from a liquid or the precipitation of a solute-rich solid phase from a supersaturated solid solution) requires the nucleation of the new phase in highly localized regions of the system. In this chapter we present the general theory of this nucleation, including classical and nonclassical models. Rates of nucleation are analyzed under quasi-steady-state and non-steady-state conditions. The influence of the nature of the nucleus/matrix interface, as well as effects due to the nucleus shape and presence of elastic strain energy, are included. Both homogeneous and heterogeneous nucleation are treated. Homogeneous nucleation takes place in uniform regions of a system in the absence of special sites such as at crystal defects or impurity particles which may aid the nucleation process. On the other hand, heterogeneous nucleation takes place at such special sites. These modes of nucleation generally compete with one another, and the predominant mode is the one that proceeds more rapidly.

Discontinuous transformations will generally occur in the series of stages illustrated in Fig. 19.1.

- Stage I is the incubation period in which the matrix phase is metastable and no stable particles of the new phase have formed. Nevertheless, small particles (termed *clusters* or *embryos*) which are precursors to the final stable phase continuously form and decompose in the matrix. The distribution of these clusters evolves with increasing time to produce larger clusters which are more

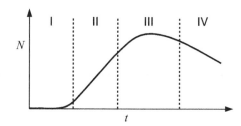

Figure 19.1: Number of particles formed during a discontinuous transformation, N, as a function of time at constant temperature.

stable and therefore less likely to revert back to the matrix. Eventually, some of the largest of these clusters evolve into particles (i.e., stable nuclei) of the new phase, remain in the system permanently, and continue to grow. At this point nucleation is well under way.

- Stage II is the *quasi-steady-state nucleation* regime. During this period, the distribution of clusters has built up into a quasi-steady state and stable nuclei are being produced at a constant rate.

- Stage III shows a decreased rate of nucleation to the point where the number of stable particles in the system becomes almost constant. This is commonly due to a decrease in the supersaturation (or free energy) which is driving the nucleation in untransformed regions of the system.

- Stage IV is the late period where the nucleation of new particles becomes negligible. However, many of the previously nucleated larger particles grow at the expense of the smaller particles in the system (i.e., coarsening occurs as described in Section 15.1). This causes the total number of particles to decrease.

Nucleation theory deals for the most part with Stages I and II. The treatment which follows starts with the relatively simple classical theory of the homogeneous nucleation of a new phase in a one-component condensed system without strain energy. This sets the stage for a description of the complications that occur when two components are present and for cases in which significant elastic strain energy is associated with the formation of a nucleus. Heterogeneous nucleation in crystalline systems is taken up with emphasis on grain boundaries and dislocations as heterogeneous nucleation sites.

19.1 HOMOGENEOUS NUCLEATION

19.1.1 Classical Theory of Nucleation in a One-Component System without Strain Energy

Consider a one-component system that consists initially of a total of $\mathcal{N}$ atoms of a parent α phase that is metastable with respect to the formation of an equilibrium

β phase.[1] According to the classical model, in order to nucleate the β phase, it is necessary for some of the α phase to be converted into small clusters of the β phase and, in turn, for at least some of these clusters to survive possible conversion back to the α phase and grow into much larger stable clusters corresponding to the bulk β phase. The small β clusters will have a large surface-to-volume ratio, and their interfacial energies will therefore be relatively large. This relatively large interfacial energy will make the formation of small clusters difficult and act as a barrier to the nucleation.

The Critical Nucleus. Imagine that a β cluster is produced by removing a cluster of atoms from the α-phase matrix leaving a cavity, transforming this cluster into the β phase, and then inserting the cluster into the cavity. In the classical model, the total energy of the embedded β cluster is assumed to be separable into bulk free energy and interfacial free energy terms. If the energy of the cluster/matrix interface is isotropic and no elastic strain energy is present (see Section 19.1.3), the energy change due to the formation of the cluster is then written

$$\Delta \mathcal{G}_{\mathcal{N}} = \Delta \mathcal{G}_{\mathcal{N}}^{\text{bulk}} + \Delta \mathcal{G}_{\mathcal{N}}^{\text{interfacial}} = \mathcal{N}(\mu^{\beta} - \mu^{\alpha}) + \eta \mathcal{N}^{2/3} \gamma \qquad (19.1)$$

where $\mathcal{N}$ is the number of atoms in the cluster, μ^{β} and μ^{α} are the chemical potentials of the bulk β and α phases, η is a shape factor, and γ is the interfacial energy per unit interfacial area. Since the interfacial energy is isotropic, the cluster will adopt a spherical shape to minimize its energy, and therefore $\eta = (36\pi)^{1/3} \Omega^{2/3}$. In the classical model, the bulk free-energy change due to the formation of the cluster is assumed to be the same as if it were a bulk phase. Furthermore, γ is assumed to be the same as that for a large flat α/β interface. In these approximations, the small cluster is therefore assumed to have the same properties as the bulk β phase.

When the interfacial energy depends on the inclination of the interface and γ is therefore anisotropic, Eq. 19.1 does not apply. In this case, the cluster will minimize its total interfacial energy by adopting its Wulff shape (Section C.3.1), which may be fully faceted or made up of faceted and smoothly curved regions. Several characteristic shapes are shown in Fig. 19.2. The interfacial-energy term in Eq. 19.1 can then be expressed as the sum of the interfacial energies of the various faceted or smoothly curved patches that make up the entire closed interface and

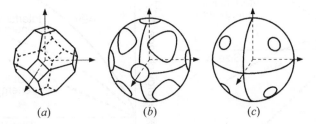

(a)   (b)   (c)

Figure 19.2:   Calculated nucleus shapes for homogeneous nucleation of an f.c.c. phase in an f.c.c. matrix. Interface is everywhere coherent with respect to a reference structure which can be taken as either the nucleus or matrix crystal. From LeGoues et al. [1].

---

[1]In systems with fixed stoichiometry such as CO_2 or polyvinyl chloride, N would be the number of molecules. In this chapter we refer to atoms, but molecular systems can be treated analogously.

written in the form

$$\Delta\mathcal{G}_{\mathcal{N}}^{\text{interfacial}} = \sum_i \gamma_i A_i = \mathcal{N}^{2/3} \sum_i \gamma_i \eta_i \qquad (19.2)$$

where the summation is carried out over all of the discrete areas, A_i, making up the enclosed interface, γ_i is the energy per unit area of the ith area, and η_i is the shape factor for the ith area. The interfacial energy is then again proportional to $\mathcal{N}^{2/3}$ but with γ replaced by $\sum_i \gamma_i \eta_i$.

The bulk term in Eq. 19.1 is negative, and a schematic plot of $\Delta\mathcal{G}_{\mathcal{N}}$ vs. $\mathcal{N}$ is shown in Fig. 19.3a. It is seen that because of the positive interfacial energy term, $\Delta\mathcal{G}_{\mathcal{N}}$ possesses a maximum before it becomes negative. The increase in free energy at the maximum will act as a barrier to the formation of large stable β clusters. For successful nucleation, small clusters must somehow form in the face of this positive free energy of formation and then grow to a larger size beyond the maximum where they continue to grow as the free energy decreases. For this reason, $\mathcal{N}_c$ and $\Delta\mathcal{G}_c$ are generally called the *critical cluster* (or nucleus) *size* and the *critical free energy of nucleation*, respectively. The values of $\mathcal{N}$ and $\Delta\mathcal{G}_{\mathcal{N}}$ at the maximum, designated by $\mathcal{N}_c$ and $\Delta\mathcal{G}_c$, respectively, are found from the condition $d\Delta\mathcal{G}_{\mathcal{N}}/d\mathcal{N} = 0$ and, for the isotropic γ case, are given by

$$\mathcal{N}_c = -\frac{8}{27}\left(\frac{\eta\gamma}{\mu_\beta - \mu_\alpha}\right)^3 \qquad (19.3)$$

$$\Delta\mathcal{G}_c = \frac{4}{27}\frac{(\eta\gamma)^3}{(\mu_\beta - \mu_\alpha)^2} = \frac{1}{3}\eta\gamma\mathcal{N}_c^{2/3} \qquad (19.4)$$

A kinetic model for the rate at which stable particles form in the presence of this barrier is developed below.

The classical model is an obvious approximation since the interface may be significantly diffuse and occupy a substantial fraction of the cluster volume. A clean separation between bulk and interfacial energies therefore becomes problematic. The classical model for the critical nucleus may be expected to be a reasonable

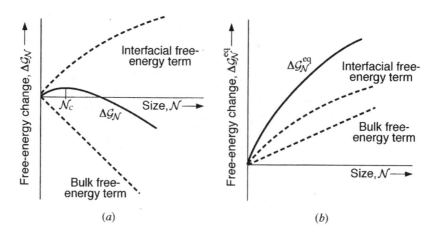

Figure 19.3:   Solid curves show cluster free energy $\Delta\mathcal{G}_{\mathcal{N}}$ vs. cluster size $\mathcal{N}$. **(a)** Nonequilibrium system during nucleation. **(b)** System at equilibrium.

approximation whenever the region strongly affected by the interface is fairly thin compared to the nucleus size. This tends to be the case when the supersaturation is small so that $(\mu^\beta - \mu^\alpha)$ is small and $\mathcal{N}_c$ is large according to Eq. 19.3. When this is not the case, nonclassical models (Section 19.1.6), which abandon the approximation of separate bulk and interfacial free-energy terms, must be employed. It will be shown that the rate-limiting kinetic events that produce critical nuclei take place over a small range of cluster sizes centered on the critical nucleus size, which may contain on the order of 100 atoms. Exact descriptions of the energies of the much smaller clusters in the distribution where the classical model is least realistic are therefore not essential. Furthermore, the classical model has the virtue of simplicity and allows the construction of a simple analytical model for nucleation which reproduces many of its main features in appropriate systems. The classical model has been shown to account reasonably well for several experimentally measured nucleation rates in selected systems (see Section 19.1.7).

Quasi-Steady-State Nucleation Rate.   Consider first the α phase containing a total of N atoms under stable equilibrium conditions. In general, this phase will contain an equilibrium distribution of small β clusters. If it is imagined that the clusters are produced by the procedure above, the plot of free-energy change vs. cluster size will appear as shown in Fig. 19.3b. In this case, where equilibrium prevails, Eq. 19.1 will again apply, but the bulk free-energy term will now be positive since the α phase is stable and $\mu^\beta > \mu^\alpha$. Values of $\Delta \mathcal{G}_{\mathcal{N}}^{\text{eq}}$ will therefore increase monotonically with $\mathcal{N}$ as seen in Fig. 19.3b. The contribution of the clusters to the free energy of the system can then be written

$$\Delta \mathcal{G}^{\text{eq}} = \sum_{\mathcal{N}} N_{\mathcal{N}}^{\text{eq}} \Delta \mathcal{G}_{\mathcal{N}}^{\text{eq}}$$
$$+ kT \left[\left(N - \sum_{\mathcal{N}} \mathcal{N} N_{\mathcal{N}}^{\text{eq}} \right) \ln \frac{N - \sum_{\mathcal{N}} \mathcal{N} N_{\mathcal{N}}^{\text{eq}}}{N - \sum_{\mathcal{N}} \mathcal{N} N_{\mathcal{N}}^{\text{eq}} + \sum_{\mathcal{N}} N_{\mathcal{N}}^{\text{eq}}} + \sum_{\mathcal{N}} N_{\mathcal{N}}^{\text{eq}} \ln \frac{\sum_{\mathcal{N}} N_{\mathcal{N}}^{\text{eq}}}{N - \sum_{\mathcal{N}} \mathcal{N} N_{\mathcal{N}}^{\text{eq}} + \sum_{\mathcal{N}} N_{\mathcal{N}}^{\text{eq}}} \right]$$

$$(19.5)$$

The first term in Eq. 19.5 consists of the free energies to form the clusters (where the free energy to form a particle is given by Eq. 19.1), and the second term is the free energy derived from the entropy of mixing of the $(N - \sum_{\mathcal{N}} \mathcal{N} N_{\mathcal{N}}^{\text{eq}})$ α atoms and all of the $\sum_{\mathcal{N}} N_{\mathcal{N}}^{\text{eq}}$ β clusters. The equilibrium fraction of each cluster can now be found by minimizing $\Delta \mathcal{G}$ with respect to $N_{\mathcal{N}}^{\text{eq}}$, with the result

$$\frac{N_{\mathcal{N}}^{\text{eq}}}{N - \sum_{\mathcal{N}} \mathcal{N} N_{\mathcal{N}}^{\text{eq}} + \sum_{\mathcal{N}} N_{\mathcal{N}}^{\text{eq}}} \left(\frac{N - \sum_{\mathcal{N}} \mathcal{N} N_{\mathcal{N}}^{\text{eq}} + \sum_{\mathcal{N}} N_{\mathcal{N}}^{\text{eq}}}{N - \sum_{\mathcal{N}} \mathcal{N} N_{\mathcal{N}}^{\text{eq}}} \right)^{\mathcal{N}} = e^{-\Delta \mathcal{G}_{\mathcal{N}}^{\text{eq}}/kT} \approx \frac{N_{\mathcal{N}}^{\text{eq}}}{N}$$

$$(19.6)$$

The simplification of Eq. 19.6 stems from the fact that $N >> \sum_{\mathcal{N}} \mathcal{N} N_{\mathcal{N}}^{\text{eq}}$. This result demonstrates that small β clusters will be present in the α phase under equilibrium conditions and that their concentrations will decrease rapidly with increasing size, since $\Delta \mathcal{G}_{\mathcal{N}}^{\text{eq}}$ increases monotonically with increasing $\mathcal{N}$ as indicated in Fig. 19.3b.

Consider now the nonequilibrium system when $\mu_\beta < \mu_\alpha$ and the nucleation of the stable β phase therefore becomes possible. Under this condition, $\Delta \mathcal{G}_{\mathcal{N}}$ goes through a maximum with increasing $\mathcal{N}$ as in Fig. 19.3a, and the formation of sufficiently large clusters causes the free energy of the system to decrease. Thermodynamics allows such clusters to grow without limit. A distribution of clusters can then develop in which relatively large β clusters in the distribution are gradually built up by the net gain of atoms and eventually become large enough to leave the nucleation process in the form of stable β particles. This process corresponds to the formation of stable nuclei of the β phase.

The rate of formation of stable nuclei can now be obtained by considering the rate at which clusters grow by adding (and sometimes losing) atoms as they advance through cluster space. If $N_{\mathcal{N}}$ is the number of clusters of size $\mathcal{N}$ in the system, the flux of clusters which are growing from size $\mathcal{N}$ to size $\mathcal{N}+1$ may be written as

$$J_{\mathcal{N}}(t) = \beta_{\mathcal{N}}\, N_{\mathcal{N}}(t) - \alpha_{\mathcal{N}+1}\, N_{\mathcal{N}+1}(t) \tag{19.7}$$

where $\beta_{\mathcal{N}}$ is the rate at which single atoms from the α phase join a β cluster of size $\mathcal{N}$, and $\alpha_{\mathcal{N}+1}$ is the rate at which single atoms are lost to the α phase from a β cluster of size $\mathcal{N}+1$. Other growth processes, such as the impingement of clusters, will be rare and are therefore ignored.

The coefficients in Eq. 19.7 may be taken as constants independent of the form of the distribution of cluster sizes. The relationship between the two coefficients $\beta_{\mathcal{N}}$ and $\alpha_{\mathcal{N}+1}$ may then be obtained by imposing an artificial constraint on the system: no clusters are allowed to grow beyond a limiting size, $\mathcal{N}_{\mathrm{lim}}$, which is considerably larger than the critical size $\mathcal{N}_c$. At the same time, all clusters of size below this limit are allowed to equilibrate with respect to one another so that detailed balance is achieved between them and all fluxes in cluster space go to zero. A new distribution of cluster sizes, $N_{\mathcal{N}}^{\mathrm{ceq}}$, will be produced in this constrained system. It is assumed that the same free-energy minimization procedure used previously to find the size distribution under true equilibrium conditions, and which led to Eq. 19.6, may be used for the constrained system, resulting in

$$\frac{N_{\mathcal{N}}^{\mathrm{ceq}}}{N} = e^{-\Delta \mathcal{G}_{\mathcal{N}}/(kT)} \tag{19.8}$$

where $\Delta \mathcal{G}_{\mathcal{N}}$ corresponds to the curve shown in Fig. 19.3a.[2] The form of this distribution is shown in Fig. 19.4b and is seen to possess a minimum at $\mathcal{N}_c$ due to the corresponding maximum in Fig. 19.3a. For the constrained system, $J_{\mathcal{N}}(t)$ in Eq. 19.7 is equal to zero and therefore, using Eq. 19.8 and expanding to first order,

$$\begin{aligned}
\alpha_{\mathcal{N}+1} &= \frac{N_{\mathcal{N}}^{\mathrm{ceq}}}{N_{\mathcal{N}+1}^{\mathrm{ceq}}}\beta_{\mathcal{N}} = \beta_{\mathcal{N}} e^{(\Delta \mathcal{G}_{\mathcal{N}+1} - \Delta \mathcal{G}_{\mathcal{N}})/(kT)} \\
&= \beta_{\mathcal{N}} e^{\frac{\partial \Delta \mathcal{G}_{\mathcal{N}}}{\partial \mathcal{N}}/(kT)} = \beta_{\mathcal{N}}\left(1 + \frac{1}{kT}\frac{\partial \Delta \mathcal{G}_{\mathcal{N}}}{\partial \mathcal{N}} \right)
\end{aligned} \tag{19.9}$$

Combining Eqs. 19.7 and 19.9 and approximating $N_{\mathcal{N}+1}(t) = N_{\mathcal{N}}(t) + \partial N_{\mathcal{N}}(t)/\partial \mathcal{N}$,

$$J_{\mathcal{N}}(t) = -\beta_{\mathcal{N}}\left[\frac{\partial N_{\mathcal{N}}(t)}{\partial \mathcal{N}} + \frac{1}{kT}\, N_{\mathcal{N}}(t)\frac{\partial \Delta \mathcal{G}_{\mathcal{N}}}{\partial \mathcal{N}} \right] \tag{19.10}$$

Comparing its form with Eq. 3.83 shows that Eq. 19.10 is equivalent to the one-dimensional diffusion of particles in cluster-size space under the influence of both a concentration gradient and a force field derived from a potential gradient.

---

[2]This "constrained equilibrium hypothesis" assumption, although seemingly plausible, is unproven [2].

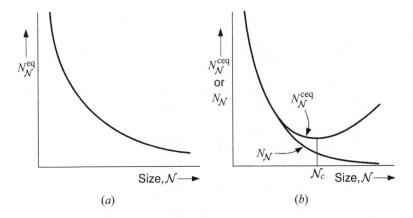

Figure 19.4:   Cluster size distribution. **(a)** System at equilibrium ($N_{\mathcal{N}}^{\rm eq}$). **(b)** System in constrained equilibrium ($N_{\mathcal{N}}^{\rm ceq}$) and for nonequilibrium system during nucleation ($N_{\mathcal{N}}$).

The nucleation rate under quasi-steady-state conditions can now be found by substituting Eq. 19.9 into Eq. 19.7 and rearranging and expanding to first order to obtain

$$J_{\mathcal{N}}(t) = \beta_{\mathcal{N}} N_{\mathcal{N}}^{\rm ceq} \left[\frac{N_{\mathcal{N}}(t)}{N_{\mathcal{N}}^{\rm ceq}} - \frac{N_{\mathcal{N}+1}(t)}{N_{\mathcal{N}+1}^{\rm ceq}} \right] = -\beta_{\mathcal{N}} N_{\mathcal{N}}^{\rm ceq} \frac{d}{d\mathcal{N}} \frac{N_{\mathcal{N}}(t)}{N_{\mathcal{N}}^{\rm ceq}} \qquad (19.11)$$

Equation 19.11 can then be integrated under quasi-steady-state conditions with $J_{\mathcal{N}} = J = $ constant in the form

$$-\frac{1}{J} \int_1^0 d\left(\frac{N_{\mathcal{N}}}{N_{\mathcal{N}}^{\rm ceq}} \right) = \frac{1}{N} \int_1^\infty \frac{1}{\beta_{\mathcal{N}}} e^{-\Delta \mathcal{G}_{\mathcal{N}}/(kT)} \, d\mathcal{N} \qquad (19.12)$$

The limits of integration are obtained by considering the differences between the actual quasi-steady-state cluster distribution, $N_{\mathcal{N}}$, and the constrained equilibrium distribution (i.e., $N_{\mathcal{N}}^{\rm ceq}$), as in Fig. 19.4b. In general, $N_{\mathcal{N}} < N_{\mathcal{N}}^{\rm ceq}$ because of the effect of the constraint. However, at very small sizes, well below $\mathcal{N}_c$, detailed balance (see Section 2.2.4) will be closely maintained in the quasi-steady-state distribution and the two distributions will therefore be nearly identical. Therefore, $N_{\mathcal{N}}/N_{\mathcal{N}}^{\rm ceq} \to 1$ as $\mathcal{N} \to 1$. On the other hand, at large sizes beyond $\mathcal{N}_c$, $N_{\mathcal{N}}^{\rm ceq}$ becomes relatively large as $\Delta \mathcal{G}_{\mathcal{N}}$ decreases, while $N_{\mathcal{N}}$ falls essentially to zero. Therefore, $N_{\mathcal{N}}/N_{\mathcal{N}}^{\rm ceq} \to 0$ as $\mathcal{N} \to \infty$. The integration on the right-hand side is simplified by the fact that values of the integrand are significant only near $\mathcal{N} = \mathcal{N}_c$ because of the maximum in $\Delta \mathcal{G}_{\mathcal{N}}$ at $\mathcal{N}_c$. Therefore, $\beta_{\mathcal{N}}$ will vary slowly in this range and can be replaced by $\beta_{\mathcal{N}} = \beta_c = $ constant. Also, $\Delta \mathcal{G}_{\mathcal{N}}$ can be replaced by an expansion around $\mathcal{N}_c$ of the form

$$\Delta \mathcal{G}_{\mathcal{N}} = \Delta \mathcal{G}_c + \frac{(\mathcal{N} - \mathcal{N}_c)^2}{2} \left(\frac{\partial^2 \Delta \mathcal{G}_{\mathcal{N}}}{\partial \mathcal{N}^2} \right)_{\mathcal{N}=\mathcal{N}_c} \qquad (19.13)$$

and from Eqs. 19.1 and 19.4,

$$\left(\frac{\partial^2 \Delta \mathcal{G}_{\mathcal{N}}}{\partial \mathcal{N}^2} \right)_{\mathcal{N}=\mathcal{N}_c} = -\frac{2}{3} \frac{\Delta \mathcal{G}_c}{\mathcal{N}_c^2} \qquad (19.14)$$

Therefore, putting these relationships into Eq. 19.12 yields

$$-\frac{1}{J}\int_1^0 d\left(\frac{N_\mathcal{N}}{N_\mathcal{N}^{\mathrm{ceq}}}\right) = \frac{1}{\beta_c N e^{-\Delta\mathcal{G}_c/(kT)}} \int_1^\infty e^{-\Delta\mathcal{G}_c\,(\mathcal{N}-\mathcal{N}_c)^2/(3kT\mathcal{N}_c^2)}\, d\mathcal{N} \qquad (19.15)$$

The lower limit of integration on the right-hand side can be replaced by $-\infty$ without significant error, and carrying out the integration,

$$J = \left(\frac{\Delta\mathcal{G}_c}{3\pi\mathcal{N}_c^2 kT}\right)^{1/2} \beta_c\, N e^{-\Delta\mathcal{G}_c/(kT)} \qquad (19.16)$$

or

$$J = Z\beta_c N e^{-\Delta\mathcal{G}_c/(kT)} \qquad (19.17)$$

where

$$Z \equiv \sqrt{\frac{\Delta\mathcal{G}_c}{3\pi\mathcal{N}_c^2 kT}} \qquad (19.18)$$

Equation 19.17 may be interpreted in a simple way. If the equilibrium concentration of critical clusters of size $\mathcal{N}_c$ were present, and if every critical cluster that grew beyond size $\mathcal{N}_c$ continued to grow without decaying back to a smaller size, the nucleation rate would be equal to $J = \beta_c N \exp[-\Delta\mathcal{G}_c/(kT)]$. However, the actual concentration of clusters of size $\mathcal{N}_c$ is smaller than the equilibrium concentration, and many supercritical clusters decay back to smaller sizes. The actual nucleation rate is therefore smaller and is given by Eq. 19.17, where the first term (Z) corrects for these effects. This dimensionless term is often called the *Zeldovich factor* and has a magnitude typically near 10^{-1}.

Non-Steady-State Nucleation: The Incubation Time. Although in principle, non-steady-state nucleation in single-component systems can be analyzed by solving the time-dependent nucleation equation (Eq. 19.10) under appropriate initial and boundary conditions, no exact solutions employing this approach have been obtained. Instead, various approximate solution have been derived, several of which have been reviewed by Christian [3]. Of particular interest is the incubation time described in Fig. 19.1. During this period, clusters will grow from some initial distribution, usually essentially free of nuclei, to a final steady-state distribution as illustrated in Fig. 19.5.

Approximate solutions of the time-dependent nucleation equation discussed by Christian indicate that the time-dependent nucleation rate in Region I for a single-component system may be approximated by

$$J(t) \approx J e^{-t/\tau} \qquad (19.19)$$

where J is the final quasi-steady-state rate and τ is the incubation time [3]. Assuming that this is the case, a reasonably good estimate for the magnitude of τ may be obtained using a physical argument introduced by Russell [4, 5]. Here it is argued that the curve of $\Delta\mathcal{G}_\mathcal{N}$ vs. $\mathcal{N}$ is essentially flat in the vicinity of $\mathcal{N} = \mathcal{N}_c$, as illustrated in Fig. 19.6, and that there is a range of cluster size, δ, over which the change in $\Delta\mathcal{G}_\mathcal{N}$ is less than kT. Over this range $\Delta\mathcal{G}_\mathcal{N}$ in Eq. 19.10 may be taken as constant, and this equation then becomes

$$J_\mathcal{N}(t) = -\beta_\mathcal{N}\frac{\partial N_\mathcal{N}(t)}{\partial \mathcal{N}} \qquad (19.20)$$

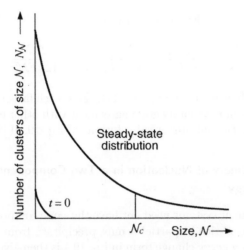

Figure 19.5:   Cluster-size distribution during transient nucleation.

which is of the form of the simple mass diffusion equation when only a concentration gradient is present. In this range, clusters will therefore grow ("move") in cluster space by a random-walk process just as during the mass diffusion of particles. Well away from $\mathcal{N}_c$, drift arising from the force field of the potential (i.e., $\Delta\mathcal{G}_\mathcal{N}$) dominates. The transition from predominant random walking to predominant drifting occurs when the potential deviates from flatness by approximately kT on either side of $\mathcal{N}_c$ (see Fig. 19.6). Because of drift, clusters of size $\mathcal{N} < (\mathcal{N}_c - \delta/2)$ have a high probability of shrinking, whereas clusters of size $\mathcal{N} > (\mathcal{N}_c + \delta/2)$ have a high probability of growing to stable nucleus size. The time required to form significant numbers of nuclei (i.e., the incubation time) will therefore be approximately the time required for clusters to random walk the distance δ in cluster space, provided that the time required to reach size $\mathcal{N} < (\mathcal{N}_c - \delta/2)$ is shorter than the random-walk time. Other calculations indicate that this is indeed the case [3, 6]. By analogy with the random walk for simple mass diffusion where, according to the one-dimensional form of Eq. 7.35, $\langle R^2 \rangle = 2Dt$,

$$\tau \approx \frac{\delta^2}{2\beta_c} \tag{19.21}$$

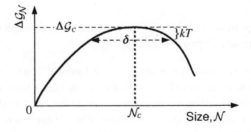

Figure 19.6:   Variation of free energy with size of fluctuation in the nucleation regime.

Furthermore, it is shown in Exercise 19.4 that

$$\frac{1}{\delta^2} = -\frac{1}{8kT} \left(\frac{\partial^2 \Delta \mathcal{G}_\mathcal{N}}{\partial \mathcal{N}^2} \right)_{\mathcal{N}=\mathcal{N}_c} = \frac{\Delta \mathcal{G}_c}{12kT\mathcal{N}_c^2} \tag{19.22}$$

and is therefore closely equal to the square of the Zeldovich factor given by Eq. 19.18. The results above are in reasonably good agreement with other estimates of τ based on approximate analytic and numerical solutions of Eq. 19.10 [3, 6].

19.1.2   Classical Theory of Nucleation in a Two-Component System without Strain Energy

Nuclei in two-component systems need not have the same composition as the parent phase. For example, B-rich β particles may precipitate from an A-rich α-phase matrix. The bulk free-energy change term in Eq. 19.1 is then given by $(\mathcal{N}/N_\circ)\,\Delta G_c$, (where the quantity ΔG_c is shown in Fig. 17.6) rather than $\mathcal{N}(\mu^\beta - \mu^\alpha)$. The rate of nucleation of the β phase can be determined by using a two-flux analysis where B atoms are added to a cluster by a two-step process consisting of a jump of a B atom onto the cluster from a nearest-neighbor matrix site followed by a replacement jump in the matrix in which a second B atom farther out in the matrix jumps into the site just evacuated by the first B atom [6]. The analysis for the steady-state nucleation rate is similar to that described previously, and the resulting expression for the rate is similar to Eq. 19.17. However, the β_c frequency is replaced by an effective frequency that reduces to the smaller of either the frequency of the matrix→cluster jumping or the matrix→matrix replacement jumping. (Note that the controlling rate is always the slower rate in a two-step process.) The concentration of B atoms in the vicinity of the nucleus is expected to be close to its average concentration in the matrix. Further details are given by Russell [6].

19.1.3   Effect of Elastic Strain Energy

When clusters form in solids, an elastic-misfit strain energy is generally present because of volume and/or shape incompatibilities between the cluster and the matrix. This energy must be added to the bulk chemical free energy in the expression for $\Delta \mathcal{G}_\mathcal{N}$. Since the strain-energy term is always positive, it acts, along with the interfacial energy term, as a barrier to the nucleation. The magnitude of the elastic-energy term generally depends upon factors such as the cluster shape, the mismatch between the cluster and the matrix (see below), and whether the interface between the matrix and cluster is coherent, semicoherent, or incoherent, as described in Section B.6.

   The elastic energy of a β cluster in an α matrix can be calculated by carrying out the following four-stage process [7]:

(i) Assume the cluster and the matrix to be linearly elastic continua. Cut the cluster (modeled as an elastic inclusion) out of the α matrix, leaving a cavity behind, and relax all stresses in both the inclusion and matrix. The inclusion will then have a generally different shape than the cavity. The homogeneous strain required to transform the cavity shape to the inclusion shape is called the *transformation strain*, ε_{ij}^T.

(ii) Apply surface tractions to the inclusion so that it fits back into the cavity. The tractions necessary to accomplish this, $-\sigma_{ij}^{T} n_j$, will be those required to produce the strains $-\varepsilon_{ij}^{T}$.

(iii) Insert the inclusion back into the cavity and join the inclusion and matrix along the inclusion/matrix interface in a manner that reproduces the type of interface (i.e., coherent, semicoherent, or incoherent) that existed initially between the β cluster and the matrix.

(iv) Remove the applied tractions by applying equal and opposite tractions (i.e., $\sigma_{ij}^{T} n_j$). This step restores the system to its original state. The tractions $\sigma_{ij}^{T} n_j$ that act on the system at the α/β interface will give rise to "constrained" displacements w_i^c, and thus strains ε_{ij}^c, in both the inclusion and the matrix which can be computed using the strain-displacement relationships of elasticity theory. Corresponding stresses σ_{ij}^c can then be computed from Hooke's law. The final strains and stresses are then ε_{ij}^c and σ_{ij}^c in the matrix and $(\varepsilon_{ij}^c - \varepsilon_{ij}^{T})$ and $(\sigma_{ij}^c - \sigma_{ij}^{T})$ in the particle. Finally, the elastic energy can be calculated from a knowledge of these stresses and strains, since for any elastic body the elastic energy is given by $1/2 \int_V \sigma_{ij} \varepsilon_{ij}\, dV$.

In problems of this type, the quantities that are given are the inclusion shape, the stress-free transformation strains ε_{ij}^{T}, the elastic properties of the two phases, and the degree of coherence between the inclusion and the matrix. When the elastic properties of the inclusion and matrix are the same, the system is said to be *elastically homogeneous*. Otherwise, it is *elastically inhomogeneous*. The main difficulty is the calculation of the constrained strains, ε_{ij}^c. Having these, the calculation of the elastic strain energy in the inclusion and matrix is straightforward. The original reference to such calculations is Eshelby [7]. An overview is given by Christian [3].

Some of the main results are given below for simple shapes such as spheres, discs, and needles which can be derived from a general ellipsoid of revolution by varying the relative lengths of its semiaxes. Only the limiting cases when the α/β interfaces are completely coherent or completely incoherent are included. Inclusions with semicoherent interfaces and interfaces where various patches possess different degrees of coherence will exhibit intermediate behavior which is much more complicated. Also, results for faceted interfaces are not included. In most cases, the energy of a faceted cluster can reasonably be approximated by using the result for a smoothly shaped cluster whose shape best approximates that of the faceted cluster.

Incoherent Clusters. As described in Section B.1, for incoherent interfaces all of the lattice registry characteristic of the reference structure (usually taken as the crystal structure of the matrix in the case of phase transformations) is absent and the interface's core structure consists of all "bad material." It is generally assumed that any shear stresses applied across such an interface can then be quickly relaxed by interface sliding (see Section 16.2) and that such an interface can therefore sustain only normal stresses. Material inside an enclosed, truly incoherent inclusion therefore behaves like a fluid under hydrostatic pressure. Nabarro used isotropic elasticity to find the elastic strain energy of an incoherent inclusion as a function of its shape [8]. The transformation strain was taken to be purely dilational, the particle was assumed incompressible, and the shape was generalized to that of an ellipsoid of revolution with semiaxes a, a, c so that its shape was given by

$$\frac{x^2}{a^2} + \frac{y^2}{a^2} + \frac{z^2}{c^2} = 1 \qquad (19.23)$$

The shape could therefore be varied between that of a thin disc ($c \ll a$) and that of a needle ($c \gg a$). The strain energy (per unit volume of inclusion) is expressed in the form

$$\Delta g_\varepsilon = 6\mu\epsilon^2\, E\!\left(\frac{c}{a}\right) \qquad (19.24)$$

where ϵ is the dilational transformation strain and $E(c/a)$ is a dimensionless shape-dependent function that has the form sketched in Fig. 19.7. From this plot, and the dependence of Δg_ε on $E(c/a)$ given in Eq. 19.24, it is apparent that the elastic strain energy of an incoherent particle can be made arbitrarily small if the particle has the form of a thin disc. Of course, such a shape would have very large interfacial area and corresponding interfacial free energy. The preferred shape for the nucleation is therefore that which minimizes the sum of the strain and interfacial energies.

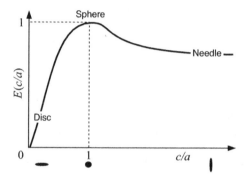

Figure 19.7: Elastic strain energy function $E(c/a)$ for an incoherent ellipsoid inclusion of aspect ratio c/a.

Coherent Clusters. As described in Section B.6, for coherent interfaces all of the coherence (lattice registry) of the reference lattice is retained. For $\alpha \to \beta$ phase transformations, the reference lattice is generally taken as the α-phase lattice, and the interface will contain an array of coherency dislocations as in Fig. B.8, which accounts for the surrounding stress field. A further example showing a spherical β cluster enclosed by a coherent interface is illustrated in Fig. 19.8a. As long as the α/β interface remains coherent during the growth of a β cluster, any shear stresses across it will be unrelaxed, since no interface sliding is possible in complete contrast to the case of the incoherent interface discussed above.

Eshelby treated systems that are both elastically homogeneous and elastically isotropic [7]. Some results for the ellipsoidal inclusion described by Eq. 19.23 are given below.

Case 1. *Pure dilational transformation strain with $\varepsilon_{xx}^T = \varepsilon_{yy}^T = \varepsilon_{zz}^T$.*

In an elastically homogeneous system, the elastic strain energy per unit volume of the inclusion Δg_ε is independent of inclusion shape and is given by

$$\Delta g_\varepsilon = \frac{2\mu\,(1+\nu)}{1-\nu}\left(\varepsilon_{xx}^T\right)^2 \qquad (19.25)$$

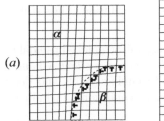 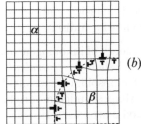

Figure 19.8: Interfacial structure for **(a)** coherent and **(b)** semicoherent interfaces between matrix phase α and particle phase β. The reference structure is the crystal lattice. Only coherency dislocations are present in (a); in (b), anticoherency dislocations relieve the elastic strain around the particle.

where ν is Poisson's ratio and μ is the shear modulus.[3] Another feature of this case is that purely dilational strain centers do not interact elastically, so that the strain fields of preexisting inclusions do not affect the strain energy of new ones that form. This is sometimes referred to as the *Bitter–Crum theorem* [9]. Finally, there is the degree of *accommodation*—this refers to the fraction of the total elastic strain energy residing in the matrix. For this example, it can be shown that two-thirds of Δg_ε always resides in the matrix.[4]

The case of a pure dilational transformation strain in an *inhomogeneous* elastically *isotropic* system has been treated by Barnett et al. [10]. For this case, the elastic strain energy does depend on the shape of the inclusion. Results are shown in Fig. 19.9, which shows the ratio of Δg_ε(inhomo) for the inhomogeneous problem to Δg_ε(homo) for the homogeneous case, vs. c/a. It is seen that when the inclusion is stiffer than the matrix, Δg_ε(inhomo) is a minimum

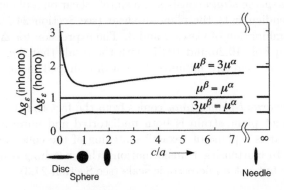

Figure 19.9: Effect of elastic inhomogeneity on elastic strain energy of a coherent ellipsoidal inclusion of aspect ratio c/a. Stress-free transformationstrains are $\varepsilon_{xx}^T = \varepsilon_{yy}^T = \varepsilon_{zz}^T$. From Barnett et al. [10].

---

[3]It is noted that Eqs. 19.24 and 19.25 do not agree exactly for the case of a sphere. Equation 19.25 correctly contains the factor $(1 + \nu)/[3(1 - \nu)] \approx 2/3$, introduced by Eshelby as an image term to make the surface of the matrix traction-free [7].

[4]Further discussion of accommodation can be found in Christian's text, p. 465 [3].

for a spherical inclusion and, when the inclusion is less stiff than the matrix, it is a minimum for a disc.

The elastic energy of inhomogeneous, anisotropic, ellipsoidal inclusions can be studied using Eshelby's *equivalent-inclusion method*. Chang and Allen studied coherent ellipsoidal inclusions in cubic crystals and determined energy-minimizing shapes under a variety of conditions, including the presence of applied uniaxial stresses [11].

Case 2. *Unequal dilational strains:* $\varepsilon_{xx}^T \equiv \varepsilon_x$, $\varepsilon_{yy}^T \equiv \varepsilon_y$, *and* $\varepsilon_{zz}^T \equiv \varepsilon_z$.

Here

$$\Delta g_\varepsilon = \frac{\mu}{1-\nu} \left\{ \begin{array}{c} \varepsilon_x^2 + \varepsilon_y^2 + 2\nu\varepsilon_x\varepsilon_y \\ -[\pi c/(32a)][13\left(\varepsilon_x^2 + \varepsilon_y^2\right) + 2\left(16\nu - 1\right)\varepsilon_x\varepsilon_y \\ -8\left(1 + 2\nu\right)\left(\varepsilon_x + \varepsilon_y\right)\varepsilon_z - 8\varepsilon_z^2] \end{array} \right\} \quad (19.26)$$

In this case the second and third terms become vanishingly small for a disc as it gets very thin, but the first term, which is independent of shape, remains. In addition, it may be seen that Eq. 19.25 is a special case of Eq. 19.26.

Case 3. *Pure shear transformation strain:* $\varepsilon_{13}^T = \varepsilon_{31}^T \equiv S/2$; *all other* $\varepsilon_{ij}^T = 0$.

Here

$$\Delta g_\varepsilon = \frac{\pi\mu}{8}\frac{2-\nu}{1-\nu}S^2\frac{c}{a} \quad (19.27)$$

Thus, for this case, Δg_ε becomes vanishingly small for a disc as it gets very thin.

Case 4. *Invariant-plane strain with* $\varepsilon_{13}^T = \varepsilon_{31}^T \equiv S/2$, $\varepsilon_{zz}^T \equiv \varepsilon_z$, *and all other* $\varepsilon_{ij}^T = 0$.

An *invariant-plane strain* consists of a simple shear on a plane, plus a normal strain perpendicular to the plane of shear (see Section 24.1 and Fig. 24.1). This is a combination of Cases 2 and 3. The expression for Δg_ε then follows directly from Eqs. 19.26 and 19.27, with the result that Δg_ε is proportional to c/a. Δg_ε is therefore minimized for a disc-shaped inclusion lying in the plane of shear.

The term *invariant-plane strain* comes from the fact that the plane of shear in an invariant plane strain is both undistorted and unrotated. Hence the plane of shear is a plane of "exact" matching of the coherent inclusion and the matrix. In martensitic transformations, this matching is met closely on a macroscopic but not a microscopic scale (see Section 24.3).

Additional factors that should often be considered in the treatment of strain energies (although commonly ignored) are: elastic anisotropy, which can be considerable, even for cubic crystals; elastic inhomogeneity, which can be treated by the Eshelby equivalent-inclusion method [12]; nonellipsoidal inclusion shapes; and elastic interactions between inclusions that can be significant, producing, for example, alignment of adjacent precipitates along elastically soft directions in anisotropic crystals [13].

19.1.4   Nucleus Shape of Minimum Energy

Both the interfacial energy and any strain energy associated with the formation of the critical nucleus act as barriers to homogeneous nucleation. Both energies are generally functions of the nucleus shape, and to find the nucleus of minimum energy, it is necessary to find the shape that minimizes the sum of these energies. As mentioned above, in the simple case where there is no strain energy, such as during solidification, the shape is given by the Wulff shape (described in Section C.3.1). However, in solid/solid transformations such as precipitation, where strain energy is generally present, the problem becomes considerably more complex.

The many variables that play a role include the anisotropic interfacial energy, which will be affected by the degree of coherency, and the elastic strain energy variables, which include the transformation strain, the degree of coherency, and the elastic properties (including elastic anisotropy). No analytical treatments of this complex minimization problem therefore exist. However, it is generally anticipated that the interfacial energy will be the dominant factor in most cases. Because the strain energy is proportional to the nucleus volume while the interfacial energy is proportional to the nucleus area, the interfacial energy should tend to dominate at the large surface-to-volume ratio characteristic of the small critical nucleus.

Both interfacial energy and strain energy have been incorporated in an analysis that gives some quantitative insight into the role that strain energy may play in determining the critical nucleus shape [14]. The nucleus is again taken to be ellipsoidal, so that the strain energy can be expressed as a function of c/a, as, for example, in Fig. 19.9. For simplicity, the interfacial energy is assumed to be isotropic. The free energy to form an ellipsoidal cluster may then be written

$$\Delta \mathcal{G} = \frac{4}{3}\pi a^3 \xi \left[\Delta g_B + \Delta g_\varepsilon(\xi) \right] + \pi a^2 \gamma \left[2 + \Lambda(\xi) \right] \tag{19.28}$$

where Δg_B is the bulk free-energy change per unit volume in the transformation, Δg_ε is a function of ξ where $\xi = c/a$, and $\Lambda(\xi)$ is a shape factor given by

$$\Lambda(\xi) = \begin{cases} (2\xi^2/\sqrt{1-\xi^2}) \tanh^{-1} \sqrt{1-\xi^2} & (\xi < 1) \\ 2 & (\xi = 1) \\ (2\xi/\sqrt{1+\xi^{-2}}) \sin^{-1} \sqrt{1-\xi^{-2}} & (\xi > 1) \end{cases} \tag{19.29}$$

The energy of the critical nucleus is now found by minimizing $\Delta \mathcal{G}$ with respect to a and ξ. The first minimization produces the results

$$a_c = -\frac{\gamma[2 + \Lambda(\xi)]}{2\xi[\Delta g_B + \Delta g_\varepsilon(\xi)]} \tag{19.30}$$

and

$$\Delta \mathcal{G}(\xi) = \frac{\pi\gamma^3[2 + \Lambda(\xi)]^3}{12\xi^2[\Delta g_B + \Delta g_\varepsilon(\xi)]^2} \tag{19.31}$$

Equation 19.31 may be divided by the expression $\Delta \mathcal{G}(1) = 16\pi\gamma^3/[3(\Delta g_B)^2]$, which is the form Eq. 19.31 would assume if the cluster were a sphere ($\xi = 1$) and the strain energy were zero. Therefore,

$$\Delta \mathcal{G}(\xi) = \Delta \mathcal{G}(1) \, \frac{[2 + \Lambda(\xi)]^3}{\left\{ 8\xi \left[1 + (\Delta g_\varepsilon(\xi)/\Delta g_B) \right] \right\}^2} \tag{19.32}$$

To find the effect of the strain energy on nucleus shape, the ratio $\Delta\mathcal{G}(\xi)/\Delta\mathcal{G}(1)$ from Eq. 19.32 is now plotted vs. ξ for various fixed values of the energy ratio $\Delta g_\varepsilon(1)/\Delta g_B$, where $\Delta g_\varepsilon(1)$ is the strain energy for the spherical nucleus ($\xi = 1$). Some results are shown in Fig. 19.10 for a coherent case corresponding to the lowest curve in Fig. 19.9, where the elastic energy decreased as the nucleus became disc-like. The minima in the curves correspond to the critical nuclei of minimum energy, and the critical nuclei remain spherical until the elastic energy is larger than about 85% of the absolute bulk free-energy change. ξ then decreases and the nucleus becomes progressively more disc-like. Similar results were found for other cases [14]. In general, the nucleus shape will not be strongly affected by the strain energy until $|\Delta g_\varepsilon|$ becomes comparable to $|\Delta g_B|$. But in most cases, $\Delta\mathcal{G}(\xi)$ will be so large that no significant homogeneous nucleation is possible. Therefore, strain energy will not affect the nucleus shape significantly in most actual cases. However, there will be exceptional cases where the interfacial energy is particularly small, as in the case of coherent clusters with close lattice matching, where $\Delta\mathcal{G}(1)$, and therefore $\Delta\mathcal{G}(\xi)$, are small enough so that significant nucleation can occur in the presence of strain energies large enough to affect the nucleus shape.

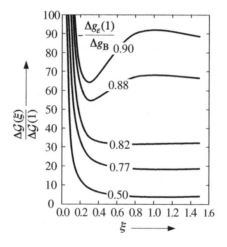

Figure 19.10:   Free energy to form ellipsoidal nucleus, $\Delta\mathcal{G}(\xi)$, as a function of the aspect ratio $\xi = c/a$ for various fixed values of the ratio $-\Delta g_\varepsilon(1)/\Delta g_B$. $\Delta\mathcal{G}(\xi)$ is normalized by $\Delta\mathcal{G}(1)$, the value $\Delta\mathcal{G}(\xi)$ would assume for a spherical nucleus ($\xi = 1$) in the absence of any strain energy. $\Delta g_\varepsilon(1)$ is the strain energy for a spherical nucleus. The elastic energy as a function of ξ corresponds to the lowest curve in Fig. 19.9. After Lee et al. [14].

19.1.5   More Complete Expressions for the Classical Nucleation Rate

With the background above, more complete expressions for the classical nucleation rate can be explored.

Single-Component System with Isotropic Interfaces and No Strain Energy.   This relatively simple case could, for example, correspond to the nucleation of a pure solid in a liquid during solidification. For steady-state nucleation, Eq. 19.16 applies with $\Delta\mathcal{G}_c$ given by Eq. 19.4 and it is necessary only to develop an expression for β_c. In a condensed system, atoms generally must execute a thermally activated jump over a local energy barrier in order to join the critical nucleus from the matrix. Therefore, $\beta_c = z_c X_B^\alpha \, \nu_c \exp[-G_c^m/(kT)]$ so that

$$J = \left(\frac{\Delta \mathcal{G}_c}{3\pi \mathcal{N}_c^2 kT} \right)^{1/2} z_c X_B^\alpha \, \nu_c \, N e^{-(\Delta \mathcal{G}_c + G_c^m)/(kT)} \tag{19.33}$$

Here, $z_c X_B^\alpha$ is the number of sites in the matrix from which atoms can jump onto the critical nucleus, ν_c is the effective vibrational frequency for such a jump, and G_c^m is the free energy of activation for the jump.

Two-Component System with Isotropic Interfaces and Strain Energy Present. An example of this case is the solid-state precipitation of a B-rich β phase in an A-rich α-phase matrix. For steady-state nucleation, Eq. 19.16 again applies. However, for a generalized ellipsoidal nucleus, the expression for $\Delta \mathcal{G}$ will have the form of Eq. 19.28. Also, β must be replaced by an effective frequency, as discussed in Section 19.1.2.

For nuclei that are coherent with the surrounding crystal, the lattice is continuous across the α/β interface. The jumps controlling the β_c frequency factor will then be essentially matrix-crystal jumps and β_c will be equal to the product of the number of solute atoms surrounding the nucleus in the matrix, $z_c X_B^\alpha$, and the solute atom jump rate, Γ, in the α crystal. The jump frequency can reasonably be approximated by $\Gamma \approx {}^\star D_1/a^2$ (see Eq. 7.52, where ${}^\star D_1$ is the solute tracer diffusivity and a is the jump distance). Therefore,

$$\beta_c \simeq \frac{z_c X_B^\alpha \, {}^\star D_1}{a^2} \tag{19.34}$$

For an incoherent nucleus, the jump rate across the cluster/matrix interface will be much faster than the lattice jump rate. Therefore, the β_c frequency factor is controlled by the lattice-replacement jumping and Eq. 19.34 holds.

In many cases, $\Delta \mathcal{G}_\mathcal{N}$ may be affected by the presence of supersaturated lattice vacancies resulting from the rapid cooling necessary to induce the precipitation. Incoherent interfaces are generally efficient sources for vacancies (in contrast to the coherent interfaces considered above), and in cases where ε_{xx}^T is positive, excess vacancies will annihilate themselves at the cluster/matrix interfaces and therefore eliminate the elastic strain energy that would otherwise have developed [6]. Furthermore, the excess vacancies may continue to annihilate beyond this point until the rate of buildup of elastic strain energy due to their annihilation is just equal to the rate at which energy is given up by the vacancy annihilation. In such a case, the excess vacancies provide a driving force aiding the nucleation and $\Delta \mathcal{G}_\mathcal{N}$ takes the form

$$\Delta \mathcal{G}_\mathcal{N} = \Omega \mathcal{N} (\Delta g_B + \Delta g_V) + \eta \gamma \mathcal{N}^{2/3} \tag{19.35}$$

where Ω is the atomic volume and Δg_V is the free-energy change due to the vacancy annihilation. For an elastically homogeneous spherical cluster where the transformation strain in the absence of any vacancy relaxation would be a uniform dilation, ε_{xx}^T, it may be shown (Exercise 19.5) that

$$\Delta g_V = -\frac{3\varepsilon_{xx}^T}{\Omega} kT \ln \left(\frac{X_V}{X_V^{\mathrm{eq}}} \right) + \frac{9(1-\nu)}{4\Omega^2 E} \left[kT \ln \left(\frac{X_V}{X_V^{\mathrm{eq}}} \right) \right]^2 \tag{19.36}$$

where E is Young's modulus. On the other hand, when ε_{xx}^T is negative, Δg_V will be positive and excess vacancies will hinder the nucleation.

19.1.6 Nonclassical Models for the Critical Nucleus

When the cluster interface is sufficiently diffuse that it occupies much of the cluster volume, the classical nucleation model breaks down. This will be the case, for example, in a precipitation system when the composition is near a spinodal and the interface becomes diffuse, as described in Section 18.2.2. It is then no longer possible to separate the nucleus energy into volume and interfacial terms, and the nucleus must be modeled as a single inhomogeneous body. The problem becomes one of determining the energy of a small critical cluster (nucleus) that is inhomogeneous in both composition and structure. In the special case when the precipitate and matrix have the same well-matched structures, the nucleus will be coherent with respect to a reference structure that can be taken to be either the matrix or precipitate lattice and there will be only compositional inhomogeneity with which to contend. The Cahn–Hilliard gradient-energy continuum approach to the energy of inhomogeneous systems described in Section 18.2 can then be used [15, 16]. When there is a difference in structure, a discrete atomistic calculation will be required.

An extensive formulation of classical and nonclassical models for homogeneous nucleation, as well as experimental tests of their validity, have been carried out for the Co–Cu precipitation system in which coherent Co-rich nuclei form [15].

19.1.7 Discussion

According to the classical model, the rate of nucleation during precipitation is sensitive to the magnitude of the interfacial energy because the critical nucleus energy, $\Delta\mathcal{G}_c$, varies as γ^3 (Eq. 19.4) and the nucleation rate varies as $\exp[-\Delta\mathcal{G}_c/(kT)]$ (Eq. 19.17). The interfacial energy of incoherent solid/solid interfaces is typically about $500\ \mathrm{mJ\,m^{-2}}$, whereas that of an interface that is coherent is lower by a factor of 3 or more. Homogeneous nucleation is therefore expected only in cases where the nucleus interface is coherent and the interfacial energy is relatively low. Otherwise, heterogeneous nucleation will predominate. This is consistent with experimental results obtained by Aaronson and Lee [17].

The nucleation rate is also sensitive to the magnitude of the driving energy since, according to Eq. 19.4, $\Delta\mathcal{G}_c$ is proportional to the inverse square of this quantity. When the temperature is changed and the system becomes metastable, the driving force increases with continued temperature change until the rate of nucleation increases explosively, as indicated in Fig. 19.11.

It is often useful to estimate values of $\Delta\mathcal{G}_c$ that may be required to produce an observable nucleation rate. For example, for the nucleation of a solid in a liquid, Eq. 19.33 applies and reasonable values for the various factors in the equation are: $(\Delta\mathcal{G}_c/3\pi\mathcal{N}_c^2 kT)^{1/2} \approx 10^{-1}$; $z_c X_B^\alpha \approx 10^2$; $\nu_c \approx 10^{13}\ \mathrm{s^{-1}}$; $N \approx 10^{23}\ \mathrm{cm^{-3}}$; $\exp[-G_c^m/(kT)] \approx 10^{-3}$; and $J \approx 1\ \mathrm{cm^{-3}\,s^{-1}}$. Therefore, $\Delta\mathcal{G}_c \approx 76kT$ and $\Delta\mathcal{G}_c$ must be no larger than approximately $76kT$ for observable rates of nucleation to occur.

The explosive onset of nucleation has made the experimental measurement of nucleation rates difficult, as measurable rates can be obtained only under a very limited range of experimental conditions. An additional difficulty has been counting the actual number of particles formed, since substantial concurrent particle coarsening often occurs (see Fig. 19.1). A common procedure has therefore been to find the driving force (which is relatively easy to quantify) that is necessary to produce

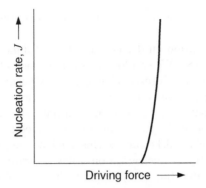

Figure 19.11:   Dependence of the nucleation rate J on the driving force for nucleation.

measurable amounts of nucleation and then to look for consistency between the value of $\Delta \mathcal{G}_c$ obtained from the data and that predicted from theory. Since the nucleation rate is so sensitive to the value of $\Delta \mathcal{G}_c$, many of the other factors in the overall expression for the nucleation rate need not be known with high precision. The various approximations used above to obtain expressions for these factors therefore do not lead to serious errors.

Despite these difficulties, Aaronson and LeGoues have measured the rate of the homogeneous nucleation of coherent Co-rich particles in the Co–Cu system by electron microscopy and compared their results with predictions of both the classical model and two nonclassical models [15]. Even though the thickness of the critical nucleus interface was roughly half the nucleus radius, as discussed in Section 23.4.1, relatively good agreement was obtained between the predictions of all three models. Furthermore, the predicted absolute nucleation rate was within a few orders of magnitude of the measured rate. This degree of agreement must be considered as relatively good in view of the many uncertainties involved.

19.2 HETEROGENEOUS NUCLEATION

Heterogeneous nucleation occurs in competition with homogeneous nucleation. Heterogeneous nucleation in solids is favored by the presence of special sites in the material that are capable of significantly lowering $\Delta \mathcal{G}_c$. Homogeneous nucleation is favored by the fact that the number of sites for homogeneous nucleation is generally equal to the number of atomic sites in the specimen and is therefore *far* greater than the number of heterogeneous sites. The mechanism with the faster kinetics dominates. We shall consider two types of heterogeneous nucleation processes: nucleation at grain boundaries in polycrystalline solids and nucleation on dislocations.

19.2.1 Nucleation on Grain Boundaries, Grain Edges, and Grain Corners

Grain boundaries are two-dimensional (planar) defects separating three-dimensional grains. Grain edges are one-dimensional (linear) defects found at the intersection of three grain boundaries. Grain corners are zero-dimensional (point) defects where four grains touch and where four grain edges meet (see Fig. 15.16). The number of each type of site per unit volume in a polycrystal decreases as its dimensionality decreases.

Our treatment of nucleation on defects in polycrystalline materials follows that first developed by Cahn [18]. We employ the simple classical model for the critical nucleus and assume isotropic interfacial energies. Consider the nucleation of a β-phase particle on a grain boundary between two grains of an α-phase matrix. Since $\gamma^{\alpha\alpha}$ and $\gamma^{\alpha\beta}$ are isotropic, the nucleus will have the shape of two truncated spheres joined in the plane of the grain boundary (referred to as a *lenticular* shape), as in Fig. 19.12. (Exercise 19.11 proves this nucleus shape, and Exercise 19.12 treats the related geometry of nucleation on a flat substrate.) A circular patch of grain boundary is eliminated but is replaced by the two spherical cap-shaped α/β interfaces. If the energy of this nucleus is lower than that of a spherical nucleus homogeneously nucleated within an α-phase grain, the boundary will act as an effective heterogeneous nucleation site.

The dihedral angle ψ is given by Young's equation:

$$\gamma^{\alpha\alpha} = 2\gamma^{\alpha\beta} \cos\psi \tag{19.37}$$

subject to

$$0 \le (\gamma^{\alpha\alpha}/2\gamma^{\alpha\beta}) \le 1 \tag{19.38}$$

Note the limiting physical situations implied by Eq. 19.38. When $\gamma^{\alpha\alpha}$ goes to zero, the grain boundary loses its ability to catalyze the reaction, and homogeneous nucleation will be favored ($\psi = \pi/2$). When $\gamma^{\alpha\alpha}$ rises to $2\gamma^{\alpha\beta}$, the grain boundary will be a perfect catalyzer of the reaction, because the grain boundary can be replaced by a continuous film of the β phase with no increase in energy ($\psi = 0$). In this instance, the nucleation barrier vanishes, a situation called *barrierless nucleation*. The β phase is said to *completely wet* the grain boundary when $\gamma^{\alpha\alpha} \ge 2\gamma^{\alpha\beta}$.

The nucleation barrier for the lenticular particle shown in Fig. 19.12 can be derived using the geometric relations for its volume V and interfacial area A:

$$V = \frac{2\pi R^3}{3}\left(2 - 3\cos\psi + \cos^3\psi\right) \tag{19.39}$$

and

$$A = 4\pi R^2 \left(1 - \cos\psi\right) \tag{19.40}$$

The semithickness c and radius r of the particle are given by

$$r = R\sin\psi \tag{19.41}$$

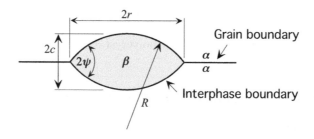

Figure 19.12:   Geometrical parameters defining size and shape of a lenticular β particle situated on a grain boundary in phase α.

and

$$c = R\left(1 - \cos\psi\right) \tag{19.42}$$

The free-energy change $\Delta\mathcal{G}^B$ for nucleation on a boundary site can then be expressed as

$$\Delta\mathcal{G}^B = \left(\frac{2\pi R^3}{3}\Delta g_B + 2\pi R^2 \gamma^{\alpha\beta}\right)\left(2 - 3\cos\psi + \cos^3\psi\right) \tag{19.43}$$

Note that in deriving Eq. 19.43, the quantity $\gamma^{\alpha\alpha}$ has been eliminated, using Eq. 19.37. It should be apparent from Eq. 19.43 that the value of the critical radius R_c for heterogeneous nucleation on a grain boundary is equal to that for homogeneous nucleation under the same conditions. The term in square brackets in Eq. 19.43 is equal to one-half the free-energy change for homogeneous nucleation. So the ratio of the critical free-energy change $\Delta\mathcal{G}_c^B$ for boundary nucleation to that for homogeneous nucleation $\Delta\mathcal{G}_c^H$ is

$$\frac{\Delta\mathcal{G}_c^B}{\Delta\mathcal{G}_c^H} = \frac{1}{2}\left(2 - 3\cos\psi + \cos^3\psi\right) \tag{19.44}$$

This ratio is the same as that of the volume of the grain boundary particle to that of a sphere having the same radius of curvature. The dihedral angle ψ is the sole parameter in determining this ratio.

Relations similar to Eq. 19.44 can be derived for the nucleation barrier for grain edges and corners, $\Delta\mathcal{G}_c^E$ and $\Delta\mathcal{G}_c^C$, respectively [18]. The extent to which the heterogeneous sites are favored relative to homogeneous nucleation and to each other can be seen by plotting the ratios $\Delta\mathcal{G}_c^B/\Delta\mathcal{G}_c^H$, $\Delta\mathcal{G}_c^E/\Delta\mathcal{G}_c^H$, and $\Delta\mathcal{G}_c^C/\Delta\mathcal{G}_c^H$ vs. $\cos\psi$, as shown in Fig. 19.13.

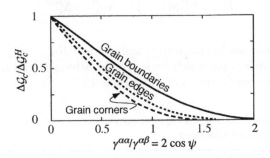

Figure 19.13:   Ratio of critical free-energy change for heterogeneous nucleation on grain boundaries, edges, and corners, $\Delta\mathcal{G}_c$, to that for homogeneous nucleation, $\Delta\mathcal{G}_c^H$, as a function of dihedral angle ψ. From Cahn [18].

Figure 19.13 demonstrates that for a given value of ψ, $\Delta\mathcal{G}_c$ decreases as the dimensionality of the heterogeneous site decreases. However, the number of sites available for nucleation also decreases as the dimensionality decreases. Thus, the kinetic equations for nucleation theory must be used to predict which mechanism will dominate. To accomplish this, some assumptions about the polycrystalline microstructure must be made. Let:

L = average grain diameter

δ = grain boundary thickness

n = number of atoms per unit volume

The densities of the heterogeneous sites can then be approximated by

$n^B = n(\delta/L)$ = number of boundary sites per unit volume

$n^E = n(\delta/L)^2$ = number of edge sites per unit volume

$n^C = n(\delta/L)^3$ = number of corner sites per unit volume

We now compare the rate of boundary nucleation to the rate of homogeneous nucleation, using Eq. 19.17:

$$J^B = Z\beta_c\, e^{-\Delta\mathcal{G}_c^B/(kT)} n\left(\frac{\delta}{L}\right) \tag{19.45}$$

and

$$J^H = Z\beta_c\, e^{-\Delta\mathcal{G}_c^H/(kT)}\, n \tag{19.46}$$

The ratio of these rates is

$$\frac{J^B}{J^H} = \frac{\delta}{L}\, e^{-(\Delta\mathcal{G}_c^B - \Delta\mathcal{G}_c^H)/(kT)} \tag{19.47}$$

Thus,

$$\ln\left(\frac{J^B}{J^H}\right) = \ln\left(\frac{\delta}{L}\right) + \frac{\Delta\mathcal{G}_c^H - \Delta\mathcal{G}_c^B}{kT} \tag{19.48}$$

Defining $R_B \equiv kT\ln(L/\delta)$, the rates J^B and J^H are equal when

$$R_B \equiv kT\ln\left(\frac{L}{\delta}\right) = \Delta\mathcal{G}_c^H - \Delta\mathcal{G}_c^B \tag{19.49}$$

and the homogeneous nucleation rate is higher when $R_B > \Delta\mathcal{G}_c^H - \Delta\mathcal{G}_c^B$. Similar analyses yield conditions for which each type of heterogeneous nucleation will be dominant, with the results summarized in Table 19.1.

Table 19.1:   Conditions for Heterogeneous Nucleation at Grain Boundaries, Edges, and Corners

Dominant Mode	Conditions
Homogeneous nucleation	$R_B > \Delta\mathcal{G}_c^H - \Delta\mathcal{G}_c^B$
Boundary nucleation	$\Delta\mathcal{G}_c^H - \Delta\mathcal{G}_c^B > R_B > \Delta\mathcal{G}_c^B - \Delta\mathcal{G}_c^E$
Edge nucleation	$\Delta\mathcal{G}_c^B - \Delta\mathcal{G}_c^E > R_B > \Delta\mathcal{G}_c^E - \Delta\mathcal{G}_c^C$
Corner nucleation	$\Delta\mathcal{G}_c^E - \Delta\mathcal{G}_c^C > R_B$

The results can be presented graphically, as in Fig. 19.14. The plot shows the kinetically dominant type of nucleation as a function of grain size (via R_B), $\Delta\mathcal{G}_c^H$, and $\gamma^{\alpha\alpha}/\gamma^{\alpha\beta}$. By setting the nucleation rate, J, at a fixed value, a curve such as $abcde$ can be plotted to indicate, for given value of L/δ, the dominant modes of nucleation at the designated nucleation rate at various values of $\gamma^{\alpha\alpha}/\gamma^{\alpha\beta}$.

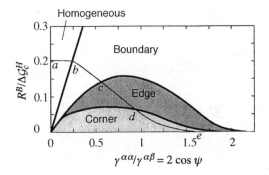

Figure 19.14:   Regimes in which grain corner, edge, boundary, and homogeneous nucleation are predicted to be dominant.   From Cahn [18].

19.2.2   Nucleation on Dislocations

Dislocations in crystals have an excess line energy per unit length that is associated with the elastic strain field of the dislocation and the bad material in its core. In many cases, the formation of a particle of the new phase at the dislocation can reduce this energy, enabling it to act as a favorable site for heterogeneous nucleation. The original treatment of heterogeneous incoherent nucleation on dislocations was by Cahn [19]. The general topic, including coherent nucleation on dislocations, has been reviewed by Larché [20].

Incoherent Nucleation.   Consider first incoherent nucleation on dislocations [19]. For linearly elastic isotropic materials, the energy per unit length E_l inside a cylinder of radius r having a dislocation at its center is given by

$$E_l = \frac{\mu b^2}{4\pi} \ln\left(\frac{r}{R_\circ}\right) \qquad \text{(screw dislocation)} \qquad (19.50)$$

and

$$E_l = \frac{\mu b^2}{4\pi (1 - \nu)} \ln\left(\frac{r}{R_\circ}\right) \qquad \text{(edge dislocation)} \qquad (19.51)$$

where b is the Burgers vector and $R_\circ$ is the usual effective core radius.

Poisson's ratio ν is approximately 0.3 for many solids, so to a fair approximation, the energy difference between edge and screw dislocations can be ignored. Following Cahn,

$$E_l = \frac{Bb}{2} \ln\left(\frac{r}{R_\circ}\right) \qquad (19.52)$$

where $B \approx \mu b/(2\pi)$.

Allowing the entire region inside a radius r to transform to incoherent β will allow essentially all of the dislocation energy originally inside the transformed region to be "released." Thus, the dislocation catalyzes incoherent nucleation by eliminating some of the dislocation's total energy. It is important to note that the dislocation will still effectively exist in the material along with its strain energy outside the transformed region, even though the incoherent β has replaced the core region. For example, a Burgers circuit around the dislocation in the matrix material surrounding the incoherent β-phase cylinder will still have a closure failure equal to b. On forming the incoherent cylinder of radius r, the total free energy change per unit length is

$$\Delta \mathcal{G}'(r) = \pi r^2 \Delta g_B + 2\pi r \gamma - \frac{Bb}{2} \ln r + \cdots \quad \text{(terms independent of } r\text{)} \quad (19.53)$$

Extreme values of $\Delta \mathcal{G}'(r)$ are given by the condition

$$\frac{\partial \Delta \mathcal{G}'(r)}{\partial r} = 2\pi (r \Delta g_B + \gamma) - \frac{Bb}{2r} = 0 \quad (19.54)$$

Plotting $\Delta \mathcal{G}'(r)$ vs. r in Fig. 19.15, two types of behavior are evident, depending on the value of the parameter, α, where

$$\alpha = -\frac{Bb \Delta g_B}{\pi \gamma^2} \quad (19.55)$$

For $\alpha > 1$, nucleation is barrierless—i.e., the transformation is controlled solely by growth kinetics. However, for $\alpha < 1$, a barrier exists. The local minimum of $\Delta \mathcal{G}'(r)$ at point A in the plot corresponds to a metastable cylinder of β of radius $r_\circ$ forming along the dislocation line. (In a sense, this is analogous to the Cottrell atmosphere described in Section 3.5.2.) In Eq. 19.54, the metastable cylinder's radius is

$$r_\circ = -\frac{\gamma}{2\Delta g_B} \left(1 - \sqrt{1 - \alpha}\right) \quad (19.56)$$

The nucleation barrier for $\alpha < 1$ is then related to the difference in $\Delta \mathcal{G}'(r)$ between the states A and B in Fig. 19.15, where the radius r_c corresponding to the unstable state at B is given from Eq. 19.54 as

$$r_c = -\frac{\gamma}{2\Delta g_B} \left(1 + \sqrt{1 - \alpha}\right) \quad (19.57)$$

However, the dislocation is practically infinitely long compared to the size of any realistic critical nucleus. If the nucleus were of uniform radius along a long length of the dislocation, $\Delta \mathcal{G}_c$ would be very large. A critical nucleus will form from a *local* fluctuation in the form of a "bulge" of the cylinder associated with the metastable state A, as illustrated in Fig. 19.16. The problem is thus to find the particular bulged-out shape that corresponds to a *minimum* activation barrier for nucleation.

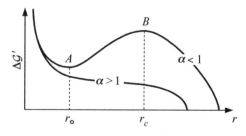

Figure 19.15:   Possible free energy vs. size behavior for the formation of an incoherent cylindrical precipitate along the core of a dislocation. From Cahn [19].

Figure 19.16:   Possible shape for incoherent critical nucleus forming along the core of a dislocation.

Let the function $r(z)$ specify the shape of the nucleus. The energy to go from the metastable state A to the unstable state B (see Fig. 19.15) can be expressed

$$\Delta \mathcal{G} = \int \left[\Delta \mathcal{G}'(r) - \Delta \mathcal{G}'(r_\circ) \right] dz \tag{19.58}$$

From earlier equations,

$$\Delta \mathcal{G} = \int_{-\infty}^{\infty} \left\{ \pi \Delta g_B \left(r^2 - r_\circ^2 \right) - \frac{Bb}{2} \ln \left(\frac{r}{r_\circ} \right) + 2\pi\gamma \left[r\sqrt{1 + \left(\frac{dr}{dz} \right)^2} - r_\circ \right] \right\} dz \tag{19.59}$$

The unknown shape $r(z)$ is determined by minimizing ΔG using variational calculus techniques. The solution to the Euler equation for this problem is somewhat complicated, requiring some substitutions and lengthy algebra [19]. From the resulting equations, one can plot the ratio of the activation barrier for nucleation on dislocations $\Delta \mathcal{G}_c^D$ to that for homogeneous nucleation $\Delta \mathcal{G}_c^H$ vs. α, in a manner analogous to the plot given in Fig. 19.13, which compared nucleation on various sites in polycrystals. The resulting plot in Fig. 19.17 shows a dramatic decrease in the relative value of $\Delta \mathcal{G}_c^D$ as $\alpha \to 1$.

Cahn also considered briefly the nucleation kinetics and showed that for reasonable values of the parameters in the theory, nucleation on dislocations in solids can be copious [19]. Typically, this occurs when α is in the range 0.4–0.7.

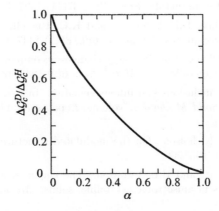

Figure 19.17:   Lowering of the activation barrier for heterogeneous incoherent nucleation at dislocations with increasing values of the parameter α (see Eq. 19.55). From Cahn [19].

Coherent Nucleation. The elastic interaction between the strain field of the nucleus and the stress field in the matrix due to the dislocation provides the main catalyzing force for heterogeneous nucleation of coherent precipitates on dislocations. This elastic interaction is absent for incoherent precipitates.

For coherent particles with dilational strains, there is a strong interaction with the elastic stress field of edge dislocations [20]. If a particle has a positive dilational transformation strain $(\varepsilon_{xx}^T + \varepsilon_{yy}^T + \varepsilon_{zz}^T > 0)$, it can relieve some of the dislocation's strain energy by forming in the region near the core that is under tensile strain. Conversely, when this strain is negative, the particle will form on the compressive side. Interactions with screw dislocations are generally considerably weaker, but can be important for transformation strains with a large shear component. Determinations of the various strain energies use Eshelby's method of calculating these quantities [20].

Bibliography

1. F.K. LeGoues, H.I. Aaronson, Y.W. Lee, and G.J. Fix. Influence of crystallography upon critical nucleus shapes and kinetics of homogeneous f.c.c.-f.c.c. nucleation. I. The classical theory regime. In *International Conference on Solid→Solid Phase Transformations*, pages 427–431, Warrendale, PA, 1982. The Minerals, Metals and Materials Society.

2. D.T. Wu. Nucleation theory. *Solid State Phys.*, 50:37–187, 1997.

3. J.W. Christian. *The Theory of Transformations in Metals and Alloys*. Pergamon Press, Oxford, 1975.

4. K.C. Russell. Linked flux analysis of nucleation in condensed phases. *Acta Metall.*, 16(5):761–769, 1968.

5. K.C. Russell. Grain boundary nucleation kinetics. *Acta Metall.*, 17(8):1123–1131, 1969.

6. K.C. Russell. Nucleation in solids: The induction and steady-state effects. *Adv. Colloid Interface Sci.*, 13(3–4):205–318, 1980.

7. J.D. Eshelby. On the determination of the elastic field of an ellipsoidal inclusion, and related problems. *Proc. Roy. Soc. A*, 241(1226):376–396, 1957.

8. F.R.N. Nabarro. The influence of elastic strain on the shape of particles segregating in an alloy. *Proc. Phys. Soc.*, 52(1):90–104, 1940.

9. F. Bitter. On impurities in metals. *Phys. Rev.*, 37(11):1527–1547, 1931.

10. D.M. Barnett, J.K. Lee, H.I. Aaronson, and K.C. Russell. The strain energy of coherent ellipsoidal precipitates. *Scripta Metall.*, 8(12):1447–1450, 1974.

11. S.M. Allen and J.C. Chang. Elastic energy changes accompanying the gamma-prime rafting in nickel-base superalloys. *J. Mater. Res.*, 6(9):1843–1855, 1991.

12. J.D. Eshelby. Elastic inclusions and inhomogeneities. In I.N. Sneddon and R. Hill, editors, *Progress in Solid Mechanics*, volume 2, pages 89–140, Amsterdam, 1961. North-Holland.

13. A.J. Ardell and R.B. Nicholson. On the modulated structure of aged Ni–Al. *Acta Metall.*, 14(10):1295–1310, 1966.

14. J.K. Lee, D.M. Barnett, and H.I. Aaronson. The elastic strain energy of coherent ellipsoidal precipitates in anisotropic crystalline solids. *Metall. Trans. A*, 8(6):963–970, 1977.

15. H.I. Aaronson and F.K. LeGoues. An assessment of studies on homogeneous diffusional nucleation kinetics in binary metallic alloys. *Metall. Trans. A*, 23(7):1915–1945, 1992.

16. J.W. Cahn and J.E. Hilliard. Free energy of a non-uniform system—III. Nucleation in a two-component incompressible fluid. *J. Chem. Phys.*, 31(3):688–699, 1959.

17. H.I. Aaronson and J.K. Lee. *The Kinetic Equations of Solid→Solid Nucleation Theory and Comparisons with Experimental Observations*, pages 165–229. The Minerals, Metals and Materials Society, Warrendale, PA, 2nd edition, 1999.

18. J.W. Cahn. The kinetics of grain boundary nucleated reactions. *Acta Metall.*, 4(5):449–459, 1956.

19. J.W. Cahn. Nucleation on dislocations. *Acta Metall.*, 5(3):169–172, 1957.

20. F.C. Larché. Nucleation and precipitation on dislocations. In F.R.N. Nabarro, editor, *Dislocations in Solids*, volume 4, pages 137–152, Amsterdam, 1979. North-Holland.

EXERCISES

19.1 An equilibrium temperature–composition diagram for an A–B alloy is shown in Fig. 19.18a. A nucleation study is carried out at 800 K using an alloy of 30 at. % B. The alloy is initially homogenized at 1200 K, then quenched to 800 K where the steady-state homogeneous nucleation rate is determined to be 10^6 m^{-3} s^{-1}. Since this rate is so small as to be barely detectable, it is desired to change the alloy composition (i.e., increase the supersaturation) so that with the same heat treatment the nucleation rate is increased to 10^{21} m^{-3} s^{-1}. Estimate the new alloy composition required to achieve this at 800 K. Use the free energy vs. composition curves in Fig. 19.18b, and assume that the interphase boundary energy per unit area, γ, is 75 mJ m^{-2}. List important assumptions in your analysis.

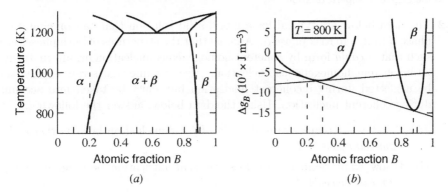

Figure 19.18: (a) Equilibrium diagram for A–B alloy. (b) Plot of free-energy density, Δg_B, vs. atomic fraction of component B at $T = 800$ K.

Solution. Important assumptions include that the interfacial free energy is isotropic, that elastic strain energy is unimportant, and that the nucleation rates mentioned are for steady-state nucleation. The critical barrier to nucleation, $\Delta \mathcal{G}_c$, can be calculated for the 0.3 atomic fraction B alloy using the *tangent-to-curve* construction on the curves in Fig. 19.18b to provide the value $\Delta g_B = -9 \times 10^7$ J m^{-3} for the chemical driving force for this supersaturation at 800 K. $\Delta \mathcal{G}_c$ is given for a spherical critical nucleus by

$$\Delta \mathcal{G}_c = \frac{16\pi\gamma^3}{3(\Delta g_B)^2} = \frac{16\pi(75 \times 10^{-3})^3}{3(-9 \times 10^7)^2} = 8.73 \times 10^{-19}\,\text{J} \qquad (19.60)$$

Note that at this temperature, $kT = 1.38 \times 10^{-23} \times 800 = 1.10 \times 10^{-20}$, so that at 800 K and $X_B = 0.3$, $\Delta\mathcal{G}_c \approx 79kT$. Based on the criterion that for significant nucleation $\Delta\mathcal{G}_c \leq 76kT$ (Section 19.1.7), it is reasonable that the nucleation rate is "barely detectable" in the alloy with $X_B = 0.3$.

The steady-state nucleation rate will be proportional to $\exp[-\Delta\mathcal{G}_c/(kT)]$ so we know that at 800 K and $X_B = 0.3$,

$$10^6 = C' \exp(-79) \tag{19.61}$$

where the constant C' is equal to $N\beta Z$ in the classical theory for steady-state nucleation. We need to find the critical nucleation barrier necessary to achieve the nucleation rate of 10^{21}, and this will be

$$\frac{10^6}{10^{21}} = \frac{\exp(-79)}{\exp[-\Delta\mathcal{G}_c/(kT)]} \tag{19.62}$$

or

$$\ln 10^{-15} = -79 + \frac{\Delta\mathcal{G}_c}{kT} \quad \text{or} \quad -34.54 + 79 = \frac{\Delta\mathcal{G}_c}{kT} \tag{19.63}$$

and thus for the higher nucleation rate we must have $\Delta\mathcal{G}_c \approx 44.5kT = 4.91 \times 10^{-19}\,\text{J}$. Next, solve for the chemical driving force required to get $\Delta\mathcal{G}_c$ down to this value, as follows:

$$\Delta g_B = \sqrt{\frac{16\pi\gamma^3}{3\Delta\mathcal{G}_c}} = \sqrt{\frac{16\pi \times (75 \times 10^{-3})^3}{3 \times 4.91 \times 10^{-19}}} = -12 \times 10^7\,\text{J}\,\text{m}^{-3} \tag{19.64}$$

Finally, use the free-energy density vs. composition curves and work the tangent-to-curve construction in reverse. Using the result that $\Delta g_B = -12 \times 10^7\,\text{J}\,\text{m}^{-3}$, the corresponding tangent to the α-phase curve will be at about 33 at. % B.

This calculation serves as a good example of the high sensitivity of nucleation rate to the degree of supersaturation.

19.2 The data below are typical for a metal solid solution that can precipitate a phase β from a matrix phase α. Assume that the structures of both phases are such that β *could* form by coherent homogeneous nucleation or, alternatively, by incoherent homogeneous nucleation. Also, assume that strain energy can be neglected during incoherent nucleation but must be taken into account during coherent nucleation. Using the data below, answer the following:

(a) Below what temperature does *incoherent* nucleation become *thermodynamically possible*?

(b) Below what temperature does *coherent* nucleation become *thermodynamically possible*?

(c) Which type of nucleation, coherent or incoherent, do you expect to occur at 510 K? Justify your answer.

Data
$\gamma^c = 160\,\text{mJ}\,\text{m}^{-2}$ (coherent interface)
$\gamma^i = 800\,\text{mJ}\,\text{m}^{-2}$ (incoherent interface)
$\Delta g_\varepsilon = 2.6 \times 10^9\,\text{J}\,\text{m}^{-3}$ (coherent particle)
$\Delta g_B = 8 \times 10^6\,(T - 900\text{K})\,\text{J}\,\text{m}^{-3}\,\text{K}^{-1}$ (driving force for precipitation)

Solution.

(a) Nucleation becomes thermodynamically possible if the thermodynamic driving force for the transformation is negative. For sufficiently large volumes nucleated incoherently in the absence of strain energy, and where the interfacial energy has become unimportant, the total energy change will be negative if $\Delta g_B < 0$. Therefore, we need $\Delta g_B = 8 \times 10^6 \, (T - 900 \, \text{K}) \, \text{J} \, \text{m}^{-3} \, \text{K}^{-1} < 0$, or $T < 900 \, \text{K}$.

(b) For coherent nucleation to be thermodynamically possible, $\Delta g_B + \Delta g_\varepsilon < 0$. Therefore, we need $8 \times 10^6 \, (T - 900 \, \text{K}) + 2.6 \times 10^9 < 0$, or $T < 575 \, \text{K}$.

(c) Assuming that the number of available sites for nucleation is the same for both coherent and incoherent mechanisms, the nucleation mechanism one expects to observe will be determined by the critical free-energy barrier, $\Delta \mathcal{G}_c$. Because the nucleation rates are proportional to $\exp[-\Delta \mathcal{G}_c/(kT)]$, the mechanism with the lowest value of $\Delta \mathcal{G}_c$ will dominate and be observable if $\Delta \mathcal{G}_c \leq 76kT$, approximately.

Assuming spherical nuclei, $\Delta \mathcal{G}_c = 16\pi\gamma^3/[3(\Delta g_B + \Delta g_\varepsilon)^2]$, where

$$
\begin{aligned}
\gamma = \gamma^i, \quad \Delta g_\varepsilon = 0 \quad &\text{(incoherent nucleation)} \\
\gamma = \gamma^c, \quad \Delta g_\varepsilon \neq 0 \quad &\text{(coherent nucleation)}
\end{aligned}
\tag{19.65}
$$

Using the given data, at $T = 510 \, \text{K}$, $\Delta g_B = -3.12 \times 10^9 \, \text{J} \, \text{m}^{-3}$. With this,

$$
\begin{cases}
\Delta \mathcal{G}_c = 8.81 \times 10^{-19} \, \text{J} \quad \text{(incoherent nucleation)} \\
\\
\Delta \mathcal{G}_c = 2.54 \times 10^{-19} \, \text{J} \quad \text{(coherent nucleation)}
\end{cases}
\tag{19.66}
$$

and, similarly,

$$
\begin{cases}
\Delta \mathcal{G}_c/(76kT) = 1.65 > 1 \quad \text{(incoherent nucleation)} \\
\\
\Delta \mathcal{G}_c/(76kT) = 0.475 < 1 \quad \text{(coherent nucleation)}
\end{cases}
\tag{19.67}
$$

Consequently, *coherent* nucleation is expected.

19.3 Martensitic transformations involve a shape deformation that is an invariant-plane strain (simple shear plus a strain normal to the plane of shear). The elastic coherency-strain energy associated with the shape change is often minimized if the martensite forms as thin plates lying in the plane of shear. Such a morphology can be approximated by an oblate spheroid with semiaxes (r, r, c), with $r \gg c$. The volume V and surface area S for an oblate spheroid are given by the relations

$$
V = \frac{4\pi}{3}r^2c \quad \text{and} \quad S = 2\pi r^2
\tag{19.68}
$$

The coherency strain energy per unit volume transformed is

$$
\Delta g_\varepsilon = \frac{Ac}{r}
\tag{19.69}
$$

(a) Find expressions for the size and shape parameters for a coherent critical nucleus of martensite. Use the data below to calculate values for these parameters.

(b) Find the expression for the activation barrier for the formation of a coherent critical nucleus of martensite. Use the data below to calculate the value of this quantity.

(c) Comment on the likelihood of coherent nucleation of martensite under these conditions.

(d) Make a sketch of the free-energy surface $\Delta\mathcal{G}(r, c)$ and indicate the location of the critical nucleus configuration (r_c, c_c) on the surface.

Data

$\Delta g_B = -170$ MJ m^{-3} (chemical driving force at observed transformation temperature)

$\gamma = 150$ mJ m^{-2} (interphase boundary energy per unit area)

$A = 2.4 \times 10^3$ MJ m^{-3} (strain energy proportionality factor)

Solution.

(a) Write the free energy to form a nucleus in the usual way as the sum of a bulk free-energy term, a strain-energy term, and an interfacial-energy term so that

$$\Delta\mathcal{G} = \frac{4}{3}\pi r^2 c \Delta g_B + \frac{4}{3}\pi r c^2 A + 2\pi r^2 \gamma \tag{19.70}$$

Now $\Delta\mathcal{G} = \Delta\mathcal{G}(c, r)$ and the critical values of c and r are then found by applying the simultaneous conditions

$$\frac{\partial \Delta\mathcal{G}}{\partial r} = \frac{\partial \Delta\mathcal{G}}{\partial c} = 0 \tag{19.71}$$

Substituting Eq. 19.70 into Eqs. 19.71 and solving for r_c and c_c yields

$$r_c = \frac{4K\gamma}{(\Delta g_B)^2} \qquad c_c = -\frac{2\gamma}{\Delta g_B} \tag{19.72}$$

Using the data provided, these quantities evaluate to

$$r_c = 50\,\text{nm} \qquad c_c = 1.76\,\text{nm} \qquad \frac{c}{r} = 0.035 \tag{19.73}$$

(b) Substituting Eqs. 19.72 into Eq. 19.70 then yields

$$\Delta\mathcal{G}_c = \frac{32\pi}{3}\frac{A^2\gamma^3}{(\Delta g_B)^4} \tag{19.74}$$

Using the data provided, this quantity is equal to

$$\Delta\mathcal{G}_c = 7.8 \times 10^{-16}\,\text{J} \tag{19.75}$$

(c) Nucleation would proceed at observable rates if $\Delta\mathcal{G}_c \leq 76kT$. Assuming a nucleation temperature of 350 K,

$$\frac{\Delta\mathcal{G}_c}{kT} = \frac{7.8 \times 10^{-16}\,\text{J}}{1.38 \times 10^{-23}\,\text{J K}^{-1} \times 350\,\text{K}} = 1.6 \times 10^5 \tag{19.76}$$

which is huge compared to 76! So homogeneous nucleation would be very unlikely. Note that the size parameter r_c is particularly large and thus the critical nucleus volume is large, consistent with the large value of $\Delta\mathcal{G}_c$.

(d) The saddle point on the free-energy surface, (r_c, c_c), is indicated in Fig. 19.19.

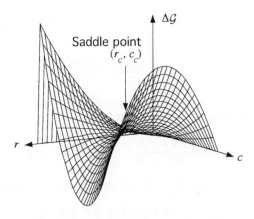

Figure 19.19:   Saddle point on free-energy surface.

19.4 Derive Eq. 19.22; i.e.,

$$\frac{1}{\delta^2} = -\frac{1}{8kT}\left(\frac{\partial^2 \Delta \mathcal{G}_\mathcal{N}}{\partial \mathcal{N}^2}\right)_{\mathcal{N}=\mathcal{N}_c}$$

Solution. Approximate the curve of $\Delta \mathcal{G}_\mathcal{N}$ vs. $\mathcal{N}$ in Fig. 19.6 by a circle of radius R. Then

$$kT = R - \frac{1}{2}\sqrt{4R^2 - \delta^2} \qquad (19.77)$$

Expanding Eq. 19.77 and neglecting the higher-order terms,

$$\frac{1}{\delta^2} = \frac{1}{8kT}\frac{1}{R} \qquad (19.78)$$

The standard expression for the curvature, $1/R$, is

$$\frac{1}{R} = -\frac{d^2 \Delta \mathcal{G}_\mathcal{N}/d\mathcal{N}^2}{[1 + (d\Delta \mathcal{G}_\mathcal{N}/d\mathcal{N})^2]^{3/2}} \cong -\frac{d^2 \Delta \mathcal{G}_\mathcal{N}}{d\mathcal{N}^2} \qquad (19.79)$$

Combining Eqs. 19.78 and 19.79, the desired result is then obtained.

19.5 Derive Eq. 19.36 for the free-energy change due to the annihilation of excess vacancies at nucleating incoherent clusters during precipitation.

Hint: The chemical potential of excess vacancies is given by Eq. 3.66.

Solution. First, calculate the energy change contributed by the excess vacancies which are eliminated to relieve the strain due to the dilatation ε_{11}^T. If V is the cluster volume, $\Delta V = 3\varepsilon_{11}^T V$. The number of vacancies required is then $N = 3\varepsilon_{11}^T V/\Omega$ and the free-energy change due to the removal of these vacancies is therefore

$$\Delta \mathcal{G}' = -\frac{3\varepsilon_{11}^T V}{\Omega}kT\ln\frac{X_V}{X_V^{\mathrm{eq}}} \qquad (19.80)$$

Next, calculate the free-energy change due to the destruction of the additional vacancies which are removed to the point where the rate of buildup of elastic strain due to their annihilation is just equal to the rate at which energy is given up by the vacancy annihilation. If N vacancies are destroyed in this fashion, the volume of matrix removed is $N\Omega$ and the dilational strain that is induced is then $N\Omega/3V$. Using Eq. 19.25, the strain energy that is created is

$$\Delta\mathcal{G}''_\varepsilon = \frac{2\mu(1+\nu)N^2\Omega^2}{9(1-\nu)V} \tag{19.81}$$

The energy released by the annihilated vacancies is $\Delta\mathcal{G}''_V = NkT\ln(X_V/X_V^{\rm eq})$, and the total energy change is then

$$\Delta\mathcal{G}'' = \frac{2\mu(1+\nu)N^2\Omega^2}{9(1-\nu)V} - NkT\ln\left(\frac{X_V}{X_V^{\rm eq}}\right) \tag{19.82}$$

$\Delta\mathcal{G}''$ is minimized when $\partial\Delta\mathcal{G}''/\partial N = 0$, and carrying out this operation, the minimum value is

$$\Delta\mathcal{G}''_{\rm min} = -\frac{9(1-\nu)V}{4E\Omega^2}\left[kT\ln\left(\frac{X_V}{X_V^{\rm eq}}\right)\right]^2 \tag{19.83}$$

Adding Eqs. 19.80 and 19.83, the total energy change (per unit cluster volume) is then finally

$$\Delta g = -\frac{3\varepsilon_{xx}^T}{\Omega}kT\ln\left(\frac{X_V}{X_V^{\rm eq}}\right) + \frac{9(1-\nu)}{4\Omega^2 E}\left[kT\ln\left(\frac{X_V}{X_V^{\rm eq}}\right)\right]^2 \tag{19.84}$$

19.6 Figure 19.20 shows a cross section through the center of a critical nucleus that has cylindrical symmetry around the vertical axis EF. AB and CD are the traces of flat facets that possess the interfacial energy (per unit area) γ^f, and AC and BD are the traces of the spherical portion of the interface that possesses the corresponding energy γ.

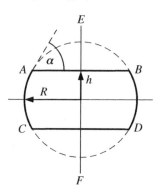

Figure 19.20:   Critical nucleus shape.

(a) Construct a Wulff plot that is consistent with the critical nucleus shape (Wulff shape) in Fig. 19.20.

(b) Show that the free energy to form this critical nucleus can be written

$$\frac{\Delta\mathcal{G}_c}{\Delta\mathcal{G}_c({\rm sphere})} = \frac{1}{2}\left(3\cos\alpha - \cos^3\alpha\right) \tag{19.85}$$

where $\Delta\mathcal{G}_c({\rm sphere})$ is the free energy to form a critical nucleus for the same transformation but which is spherical and possesses the interfacial energy γ. Assume the classical model for the nucleus.

Solution.

(a) The Wulff plot in Fig. 19.21 possesses two deep cusps necessary to produce the two low-energy facets.

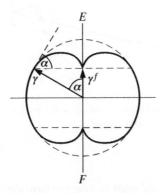

Figure 19.21:   Deeply cusped Wulff plot.

(b) Standard relationships for volumes and areas show that the volume of the nucleus is given by $V = 2\pi R^3 \left[\cos\alpha - (\cos^3\alpha)/3\right]$ while the area of the two facets is $2\pi R^2 \sin^2\alpha$ and the area of the spherical portion of the interface is $4\pi R^2 \cos\alpha$. The free energy to form the nucleus is therefore

$$\Delta\mathcal{G} = 2\pi R^3 \left(\cos\alpha - \frac{\cos^3\alpha}{3}\right)\Delta g_B + 2\pi R^2 \sin^2\alpha\,\gamma^f + 4\pi R^2\cos\alpha\,\gamma \quad (19.86)$$

But, according to Fig. 19.21, $\gamma^f/\gamma = \cos\alpha$ and therefore

$$\Delta\mathcal{G} = \left(2\pi R^3 \Delta g_B + 6\pi R^2\gamma\right)\left(\cos\alpha - \frac{\cos^3\alpha}{3}\right) \quad (19.87)$$

Minimizing $\Delta\mathcal{G}$ with respect to R in order to obtain $\Delta\mathcal{G}_c$ and using the result that $\Delta\mathcal{G}_c(\text{sphere}) = 16\pi\gamma^3/\left[3(\Delta g_B)^2\right]$, we obtain the result

$$\frac{\Delta\mathcal{G}_c}{\Delta\mathcal{G}_c(\text{sphere})} = \frac{1}{2}\left(3\cos\alpha - \cos^3\alpha\right) \quad (19.88)$$

19.7 Consider the possible heterogeneous nucleation of a solid phase from a liquid in a conically shaped pit in the wall of a mold as illustrated in Fig. 19.22. Let the energies of the liquid/mold, solid/mold, and liquid/solid interfaces be γ^{LM}, γ^{SM}, and γ^{LS}, respectively, and assume that $\gamma^{SM} = \gamma^{LM}$.

(a) Using the classical nucleation model, find an expression for the critical free energy for heterogeneous nucleation to occur within the pit, and compare it to the critical free energy for homogeneous nucleation in the bulk liquid. Assume that the pit is deep enough to allow a critical heterogeneous nucleus to form within it.

(b) For the solid, which has nucleated successfully within the pit, to grow without limit, it must be able to grow out of the pit and expand into the bulk liquid. Determine how deep the pit must be so that this can occur.

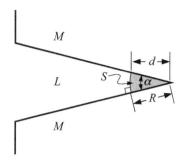

Figure 19.22:   Cross section of heterogeneous solid nucleus formed in a conical pit in the wall of a mold containing a metastable liquid.

Solution.

(a) The nucleus may be expected to have the form shown in Fig. 19.22. It consists of a cone with a spherical cap of constant curvature that meets the mold surface at $90°$ and therefore satisfies Young's equation for local interfacial equilibrium. The volume of the nucleus is then the volume of the cone of height d plus the volume of the spherical cap of radius R and is given by

$$
\begin{aligned}
V &= \frac{\pi}{3}R^3 \sin^2(\alpha/2)\cos(\alpha/2) + \frac{\pi R^3}{3}[2 - 3\cos(\alpha/2) + \cos^3(\alpha/2)] \\
&= \frac{2\pi R^3[1 - \cos(\alpha/2)]}{3}
\end{aligned}
\tag{19.89}
$$

The area of the cap, A, is given by $2\pi Rh$, where h is its height. Therefore,

$$
A = 2\pi Rh = 2\pi R^2[1 - \cos(\alpha/2)]
\tag{19.90}
$$

The free energy to form a nucleus as in Fig. 19.22 is then

$$
\Delta \mathcal{G} = \frac{2\pi R^3}{3}[1 - \cos(\alpha/2)]\Delta g_B + 2\pi R^2[1 - \cos(\alpha/2)]\gamma^{LS}
\tag{19.91}
$$

The critical nucleus radius, R_c, is found by setting $\frac{\partial \Delta \mathcal{G}}{\partial R} = 0$, with the result

$$
R_c = -\frac{2\gamma^{LS}}{\Delta g_B}
\tag{19.92}
$$

The critical radius given by Eq. 19.92 is equal to the critical radius for homogeneous nucleation in the bulk liquid. This is the expected result because $\gamma^{LM} = \gamma^{SM}$ (so that the liquid/solid interface makes an angle of $90°$ with the mold) and the inward pressure on the interface due to curvature, $\Delta P = 2\gamma^{LS}/R$ (Eq. 12.4), is then exactly balanced by the change in bulk free energy across the interface, $\Pi^{\text{phase trans}} = -\Delta g_B$ (Eq. 12.1). Substitution of Eq. 19.92 into Eq. 19.91 yields the critical free energy for nucleation:

$$
\Delta \mathcal{G}_c = \frac{8\pi(\gamma^{LS})^3[1 - \cos(\alpha/2)]}{3(\Delta g_B)^2}
\tag{19.93}
$$

$\Delta \mathcal{G}_c^H = (16\pi/3)(\gamma^{LS})^3/(\Delta g_B)^2$ is the critical free energy of homogeneous nucleation, so

$$
\frac{\Delta \mathcal{G}_c}{\Delta \mathcal{G}_c^H} = \frac{1 - \cos(\alpha/2)}{2}
\tag{19.94}
$$

and thus the heterogeneous nucleation will be much easier than homogeneous nucleation.

(b) Once the critical nucleus has formed within the pit, it will grow outward as shown in Fig. 19.23. When it reaches the mold surface, at a time t_1, it will continue to grow by bulging outward as long as its radius is always larger than the critical radius for homogeneous nucleation in the liquid, corresponding to R_c in Eq. 19.92. The minimum radius that will occur during this bulging out will correspond to that shown at time t_2 in Fig. 19.23, where the growing particle possesses a spherical cap with a radius that is just equal to half the width of the pit at the mold surface. This radius must be larger than R_c and the critical condition on the pit depth, D, for unlimited growth into the liquid is therefore

$$D > \frac{R_c}{\tan(\alpha/2)} \tag{19.95}$$

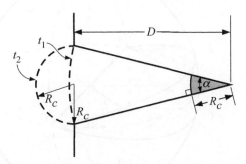

Figure 19.23: Critical pit depth, D, which will just allow the unlimited growth of a supercritical heterogeneous nucleus that has formed in the pit with a critical radius R_c.

19.8 In Exercise 19.7 we considered heterogeneous nucleation of a solid from a liquid in a conical pit in the wall of the mold holding the liquid. Consider now heterogeneous nucleation in the same solidification system but with the nucleation occurring in a crack in the wall of the mold as illustrated in Fig. 19.24 instead of in a conical pit. Again, let the energies of the liquid/mold, solid/mold, and liquid/solid interfaces be γ^{LM}, γ^{SM}, and γ^{LS}, respectively, and assume that $\gamma^{SM} = \gamma^{LM}$.

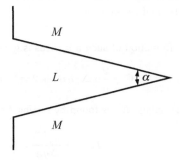

Figure 19.24: Cross section of a long crack in the wall of a mold containing a metastable liquid. The crack extends normal to the page.

(a) Describe the expected three-dimensional shape of the nucleus.

(b) Using the classical nucleation model, find an expression for the critical free energy for the heterogeneous nucleation to occur within the crack and compare it to the critical free energy for homogeneous nucleation in the bulk liquid. Assume that the crack is deep enough to allow a critical heterogeneous nucleus to form within it.

Solution.

(a) The expected shape of the nucleus is shown in Fig. 19.25. The interface $ABCD$ is spherical and of radius R and Young's equation for interface equilibrium is satisfied along the junction lines ABD and ACD.

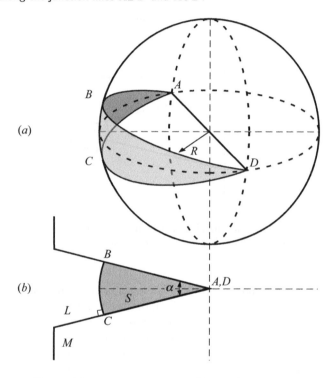

Figure 19.25: Shape of heterogeneous nucleus ($ABCD$) for solidification formed at the root of the crack in the mold wall. **(a)** Oblique view. **(b)** Cross section through the nucleus midplane. The crack extends normal to the page.

(b) The free energy of formation of such a nucleus is given by

$$\Delta\mathcal{G} = \frac{2}{3}\alpha\,\Delta g_B R^3 + 2\alpha\gamma^{LS}R^2 \qquad (19.96)$$

The critical nucleus radius, R_c, is found by setting $\frac{\partial\Delta\mathcal{G}}{\partial R} = 0$, with the result

$$R_c = -\frac{2\gamma^{LS}}{\Delta g_B} \qquad (19.97)$$

The critical radius given by Eq. 19.97 is equal to the critical radius for homogeneous nucleation in the bulk liquid. This result is similar to that obtained in Exercise 19.7

and might be expected since $\gamma^{LM} = \gamma^{SM}$, the liquid/solid interface makes an angle of $90°$ with the mold, and the inward pressure on the interface due to curvature, $P = 2\gamma^{LS}/R$ (Eq. 12.4), is then exactly balanced by the outward pressure, $P = -\Delta g_B$ (Eq. 12.1), due to the change in bulk free energy across the interface. Substitution of Eq. 19.97 into Eq. 19.96 yields the critical free energy for nucleation,

$$\Delta \mathcal{G}_c = \frac{8\alpha(\gamma^{LS})^3}{3(\Delta g_B)^2} \tag{19.98}$$

Because the critical free energy of homogeneous nucleation is

$$\Delta \mathcal{G}_c^H = \frac{16\pi(\gamma^{LS})^3}{3(\Delta g_B)^2} \tag{19.99}$$

it follows that

$$\frac{\Delta \mathcal{G}_c}{\Delta \mathcal{G}_c^H} = \frac{\alpha}{2\pi} \tag{19.100}$$

and as in Exercise 19.7, heterogeneous nucleation will be much easier than homogeneous nucleation. Note that for the solid that has nucleated successfully within the crack to grow without limit, it must be able to grow out of the crack and expand into the liquid. This will require that the crack be deep enough so that the nucleus will be supercritical when it emerges from the crack. (An analysis of this problem for a heterogeneous nucleus emerging from a conical pit in the mold wall is given in Exercise 19.7*b*.)

19.9 Experimental observations of precipitation from supersaturated, polycrystalline solid solutions seem to show that at small undercoolings $\Delta T = T^{\text{eq}} - T$ within a two-phase field, grain-boundary precipitation is observed, and at larger undercoolings, homogeneous precipitation occurs within the grain interiors. Interpret this observation by using nucleation theory.

Solution. Figure 19.14 delineates regions where homogeneous and heterogeneous nucleation will be dominant. The quantity $\Delta \mathcal{G}_c^H$ has the most pronounced temperature dependence of the important variables in the figure. The ratio $\gamma^{\alpha\alpha}/\gamma^{\alpha\beta}$ will normally be rather temperature-independent. Consequently, in Fig. 19.14, varying undercooling will correspond to changes parallel to the $R^B/\Delta \mathcal{G}_c^H$ axis. Because $\Delta \mathcal{G}_c^H \to \infty$ as the undercooling $\Delta T \to 0$, there will always be a range of temperatures close to T^{eq} where the rate of heterogeneous nucleation exceeds that of homogeneous nucleation. Conversely, high undercooling with correspondingly small values of $\Delta \mathcal{G}_c^H$ favors homogeneous nucleation.

19.10 Suppose that a material α with cubic symmetry is nucleating on a smooth, amorphous substrate. To demonstrate the crystallographic effects of nucleation, we will consider nucleation in two dimensions (i.e., the system lies in the plane of this page). Assume that the only surfaces which appear on the Wulff shape are the $\{10\}$-type faces, which have surface energy (energy per unit length) $\gamma^{10/v}$ at the interface with a vacuum. Furthermore, assume that there are only two crystallographic orientations of low-energy interfaces of α with the substrate: (1) the interface that is normal to $\{11\}$ which has interfacial energy $\gamma^{11/\text{sub}}$; (2) the interface that is normal to $\{10\}$ which has interfacial energy $\gamma^{10/\text{sub}}$. Let the interfacial free energy of the substrate/vacuum interface be $\gamma^{v/\text{sub}}$. Thus, nucleation could have two distinct morphologies, triangular and rectangular, as illustrated in Fig. 19.26.

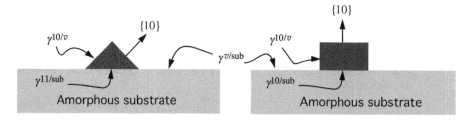

Figure 19.26:   Possible shapes for nucleation of a two-dimensional crystal on a one-dimensional substrate surface.

(a) Calculate the critical nucleus dimensions for each morphology in terms of the interfacial free energies and the chemical driving force Δg_a (energy per unit area). Assume that the nucleation shape for each case is the one that minimizes total surface energy for its volume.

(b) Find a relation between the various surface and interfacial free energies that would make the areas of the critical nuclei equal for the two cases.

Solution.

(a) For the triangular morphology, the free-energy change for forming a particle of edge length, l, will be

$$\Delta\mathcal{G}_{\text{tri}} = \frac{l^2}{2}\Delta g_a + 2\gamma^{10/v}l + \left(\gamma^{11/\text{sub}} - \gamma^{v/\text{sub}}\right)\sqrt{2}\,l \tag{19.101}$$

The critical size, l_c, is

$$\frac{\partial\Delta\mathcal{G}_{\text{tri}}}{\partial l} = 0 = 2\Delta g_a\frac{l_c}{2} + 2\gamma^{10/v} + \left(\gamma^{11/\text{sub}} - \gamma^{v/\text{sub}}\right)\sqrt{2} \tag{19.102}$$

or

$$l_c = -\frac{2\gamma^{10/v} + \sqrt{2}\left(\gamma^{11/\text{sub}} - \gamma^{v/\text{sub}}\right)}{\Delta g_a} \tag{19.103}$$

For the rectangular morphology, the free-energy change for forming a particle of height, b, and width, c, will be

$$\Delta\mathcal{G}_{\text{rect}} = bc\,\Delta g_a + (2b + c)\,\gamma^{10/v} + c\left(\gamma^{10/\text{sub}} - \gamma^{v/\text{sub}}\right) \tag{19.104}$$

The critical dimensions, b_c and c_c, are obtained from the minimization equations

$$\frac{\partial\Delta\mathcal{G}_{\text{rect}}}{\partial b} = 0 = c_c\Delta g_a + 2\gamma^{10/v} \tag{19.105}$$

and

$$\frac{\partial\Delta\mathcal{G}_{\text{rect}}}{\partial c} = 0 = b_c\Delta g_a + \left(\gamma^{10/v} + \gamma^{10/\text{sub}} - \gamma^{v/\text{sub}}\right) \tag{19.106}$$

and are given by

$$b_c = -\frac{\gamma^{10/v} + \gamma^{10/\text{sub}} - \gamma^{v/\text{sub}}}{\Delta g_a} \qquad\qquad c_c = \frac{-2\gamma^{10/v}}{\Delta g_a} \tag{19.107}$$

(b) The area of the triangular critical nucleus is

$$A_{\text{tri}} = \frac{1}{2}\, l_c^2 = \frac{1}{2}\left[-\frac{2\gamma^{10/v} + \sqrt{2}\left(\gamma^{11/\text{sub}} - \gamma^{v/\text{sub}}\right)}{\Delta g_a}\right]^2 \tag{19.108}$$

The area of the rectangular critical nucleus is

$$A_{\text{rect}} = b_c\, c_c = \frac{\gamma^{10/v} + \gamma^{10/\text{sub}} - \gamma^{v/\text{sub}}}{\Delta g_a}\,\frac{2\gamma^{10/v}}{\Delta g_a} \tag{19.109}$$

The two areas are equal when

$$(\gamma^{11/\text{sub}} - \gamma^{v/\text{sub}})^2 = 2\gamma^{10/v}\left[\gamma^{10/\text{sub}} - \sqrt{2}\gamma^{11/\text{sub}} + (\sqrt{2} - 1)\gamma^{v/\text{sub}}\right] \tag{19.110}$$

19.11 We wish to prove by means of the Wulff construction (Section C.3.1) that the equilibrium shape of the grain boundary nucleus in Fig. 19.12 is indeed composed of two spherical-cap-shaped interfaces.

The nucleus has cylindrical symmetry around an axis normal to the boundary and mirror symmetry across the grain-boundary plane. Figure 19.27 shows a cross section of the nucleus centered in a patch of boundary of constant circular area, $A_\circ$. The area of the nucleus projected on the boundary is indicated by A. The total interfacial energy of this configuration is then

$$\begin{aligned}
\mathcal{G}^{\text{int}} &= \int_{\alpha/\beta} \gamma^{\alpha/\beta}\, dA + \int_{\alpha/\alpha} \gamma^{\alpha/\alpha}\, dA \\
&= \int_{\alpha/\beta} \gamma^{\alpha/\beta}\, dA + \int_{A_\circ} \gamma^{\alpha/\alpha}\, dA - \int_{A} \gamma^{\alpha/\alpha}\, dA
\end{aligned} \tag{19.111}$$

where the integrals extend over the interface types indicated. The condition for minimum total interfacial energy is then

$$\delta\mathcal{G}^{\text{int}} = \delta\left(\int_{\alpha/\beta} \gamma^{\alpha/\beta}\, dA - \int_{A} \gamma^{\alpha/\alpha}\, dA\right) = 0 \tag{19.112}$$

since $A_\circ = $ constant. Equation 19.112 can also be written

$$\delta\mathcal{G}^{\text{int}} = \delta\left[2\left(\frac{1}{2}\int_{\alpha/\beta} \gamma^{\alpha/\beta}\, dA + \int_{A} \frac{-\gamma^{\alpha/\alpha}}{2}\, dA\right)\right] = 0 \tag{19.113}$$

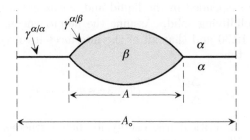

Figure 19.27: Cross section of grain boundary nucleus.

The first term in the curved inner bracket is the energy of either the upper or lower half of the α/β interface in Fig. 19.27, while the second represents half the energy of the area A. Their sum is therefore the total interfacial energy of the "half-nucleus" shape containing the fictitious boundary shown in Fig. 19.28. Using these results:

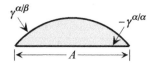

Figure 19.28:   Cross section of half-nucleus shape.

(a) Construct the Wulff plot for the half-nucleus and find the Wulff shape.

(b) Show that the upper and lower surfaces of the nucleus in Fig. 19.27 are indeed hemispherical as assumed.

(c) Show that Young's equilibrium relation at the triple junction at the nucleus and the grain boundary intersection is obeyed.

Solution. Cross sections of the Wulff plot and Wulff shape consistent with the symmetry of the problem are shown in Fig. 19.29. Since the α/β interface is isotropic, the top surface is spherical. Also, the construction is consistent with Young's equation, since from the figure, $\gamma^{\alpha\alpha} = 2\gamma^{\alpha\beta}\cos\theta$.

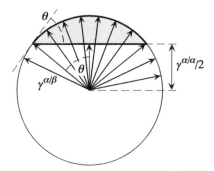

Figure 19.29:   Cross section of Wulff plot and related nucleus form.

19.12 During the solidification of a pure liquid, small, solid foreign particles (*inclusions*) may be suspended in the liquid and act as heterogeneous nucleation sites for the solidifying solid. Assume the heterogeneous nucleus geometry shown in Fig. 19.30 and that all of the interfaces involved possess isotropic energies. Show that

$$\frac{\Delta\mathcal{G}_c^P}{\Delta\mathcal{G}_c^H} = \frac{1}{4}\left(2 - 3\cos\theta + \cos^3\theta\right) \qquad (19.114)$$

where $\Delta\mathcal{G}_c^P$ is the critical free energy for heterogeneous nucleation at the particle and $\Delta\mathcal{G}_c^H$ is the critical free energy for homogeneous nucleation in the bulk liquid.

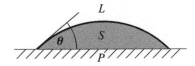

Figure 19.30: Heterogeneous nucleation of solid (S) on a solid particle (P) during the solidification of a bulk liquid (L).

Solution. Let γ^{LP}, γ^{LS}, and γ^{SP} be the energies (per unit area) of the liquid/particle, liquid/solid, and solid/particle interfaces, respectively. From Section 19.2.1 the volume of the solid nucleus is $V^S = \left(\pi R^3/3\right)\left(2 - 3\cos\theta + \cos^3\theta\right)$, the spherical liquid/solid cap area is $A^{LS} = 2\pi R^2(1 - \cos\theta)$, and the solid/particle area is $A^{SP} = \pi R^2 \sin^2\theta$. The free energy of nucleus formation on the particles is then

$$\Delta\mathcal{G}^P = V^S \Delta g_B + A^{LS}\gamma^{LS} + A^{SP}(\gamma^{SP} - \gamma^{LP}) \tag{19.115}$$

where Δg_B is the usual free energy of the bulk transformation. Also, equilibration at the interface junction requires that

$$\gamma^{LP} = \gamma^{LS}\cos\theta + \gamma^{SP} \tag{19.116}$$

Putting these relationships into Eq. 19.115 and minimizing $\Delta\mathcal{G}^P$ with respect to R,

$$\Delta\mathcal{G}_c^P = \frac{4\pi\left(\gamma^{LS}\right)^3}{3(\Delta g_B)^2}\left(2 - 3\cos\theta + \cos^3\theta\right) \tag{19.117}$$

On the other hand, for the homogeneous nucleation of a spherical solid nucleus in the bulk liquid, $\Delta\mathcal{G}_c^H = 16\pi\left(\gamma^{LS}\right)^3 / \left[3(\Delta g_B)^2\right]$. Combining this with Eq. 19.117 therefore produces Eq. 19.114.

19.13 A pure liquid being solidified contains a dispersion of very fine foreign particles which can act as heterogeneous nucleation sites as described in Exercise 19.12. The rate of nucleation of the solid falls off exponentially with time. Develop a simple model that might explain this phenomenon.

Solution. A simple model can be constructed based on the idea that the nucleation is heterogeneous and that the heterogeneous nucleation sites (i.e., the particles), are being neutralized as nucleation and growth proceed. Once a nucleation event has occurred at a given particle, the solid will grow and envelop the particle, and it is unlikely that an additional nucleation event will occur there. After the incubation period has passed, Eq. 19.17 may be written for heterogeneous nucleation in the form

$$J = Z\beta_c n_P N_p\, e^{-\Delta\mathcal{G}_c/(kT)} \tag{19.118}$$

where n_P is the number of nucleating particles per unit volume and N_P is the number of nucleating sites per particle. The number of active nucleating particles will then decrease according to

$$\frac{dn_P}{dt} = -J = -An_P \tag{19.119}$$

where $A = \text{constant}$. Equation 19.119 integrates to

$$n_P = n_P(0)\, e^{-At} \tag{19.120}$$

and therefore

$$J = An_P = A\, n_P(0)\, e^{-At} \tag{19.121}$$

CHAPTER 20

GROWTH OF PHASES IN CONCENTRATION AND THERMAL FIELDS

This chapter focuses on the growth of phases in transformations when long-range diffusion of mass and/or conduction of heat at the interface bounding the growing phase is necessary to sustain the growth. This occurs whenever there is a change of composition at the interface or a latent heat of transformation that must be supplied or removed. This type of growth is nonconservative with respect to mass and energy and is distinguished from purely conservative growth that occurs when no such long-range transport is required.[1]

In the treatment of this growth it is necessary to determine the concentration and/or thermal fields that are present in the bulk regions adjoining the moving interface under conditions where certain boundary conditions must be satisfied at the interface. Also, the rate of interface movement may be controlled by the rate at which the mass is transported to the interface by diffusion (diffusion-limited) or by the rate at which it can be incorporated at the interface (interface source-limited) as described in Section 13.4. A further complication may arise from the possibility that the interface, which is moving in adjoining concentration and/or thermal fields, will change its form and evolve into a cellular or dendritic structure (i.e., it will become morphologically unstable).

---

[1]Other types of growth are treated elsewhere in this book.

Kinetics of Materials. By Robert W. Balluffi, Samuel M. Allen, and W. Craig Carter.   **501**
Copyright © 2005 John Wiley & Sons, Inc.

20.1 GROWTH OF PLANAR LAYERS

We begin by analyzing the growth of planar layers when the growth rate is controlled by heat conduction, mass diffusion, or both simultaneously. Growth under interface source-limited conditions is considered as well. We assume throughout that the interface is morphologically stable and therefore remains planar. Morphological instability is analyzed in Section 20.3.

20.1.1 Heat Conduction-Limited Growth

Consider the melting of a pure material as depicted in Fig. 20.1. The melting is advancing from the left, where the temperature is maintained at T^{L0}, above the melting point. It is assumed that atoms exchange rapidly between the liquid and solid at the interface so that local equilibrium is achieved and the interface temperature is therefore maintained at the equilibrium melting point. Under this condition, the rate of melting will be controlled by the rate of heat conduction and not by processes at the interface. The temperature in the solid at a long distance on the right is maintained at $T^{S\infty}$. The interface will move as heat flows down the temperature gradient in the liquid to the interface, where it supplies the necessary latent heat of melting per unit mass, H_m, for the transformation. The rate of melting will then depend upon how rapidly this heat can be supplied. The differential equation for heat conduction (Eq. 4.11) applies in each phase. However, the moving interface modifies the conduction problem by making the position of the interface, where the boundary value of T is held at T_m, a function of the net flux to the interface and thus on the solution itself. Such a problem is called a *moving-boundary problem*, and the dependence of the boundary condition on the solution makes the problem nonlinear. Assuming a constant thermal diffusivity, the equations governing the temperature fields in the liquid and solid are

$$\frac{\partial T^L}{\partial t} = \kappa^L \frac{\partial^2 T^L}{\partial x^2} \qquad \frac{\partial T^S}{\partial t} = \kappa^S \frac{\partial^2 T^S}{\partial x^2} \tag{20.1}$$

with the conditions

$$\begin{aligned} T^L(x=0,t) &= T^{L0} & T^L[\chi(t),t] &= T_m \\ T^S(x=\infty,t) &= T^{S\infty} & T^S[\chi(t),t] &= T_m \end{aligned} \tag{20.2}$$

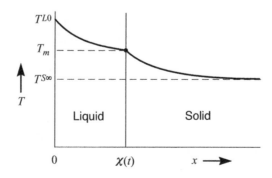

Figure 20.1:   Temperature distribution for the melting of a pure material. Melting is advancing from the left.

At this point, the interface position, $\chi(t)$, is an unknown function of time. However, it can be determined by imposing the requirement that the net rate at which heat flows into the boundary must be equal to the rate at which heat is delivered to supply the latent heat needed for the melting. Any small difference between the densities ρ^S and ρ^L may be neglected and the resulting uniform density is represented by ρ. Then, if the boundary advances a distance δx, an amount of heat $\rho H_m \, \delta x$ must be supplied per unit area. Also, if the time required for this advance is δt, the heat that has entered the boundary from the liquid is $J^L \delta t$ (at $x = \chi$) and the heat that has left the interface through the solid is $J^S \delta t$ (at $x = \chi$). Therefore,

$$\left(J^L - J^S\right)_{x=\chi} \delta t = -K^L \left(\frac{\partial T^L}{\partial x}\right)_{x=\chi} \delta t + K^S \left(\frac{\partial T^S}{\partial x}\right)_{x=\chi} \delta t = \rho H_m \, \delta x \quad (20.3)$$

or equivalently,

$$\frac{d\chi}{dt}\rho H_M = -K^L \left(\frac{\partial T^L}{\partial x}\right)_{x=\chi} + K^S \left(\frac{\partial T^S}{\partial x}\right)_{x=\chi} \quad (20.4)$$

This type of relationship, accounting for the flux into and out of the interface, is generally known as a *Stefan condition* [1]. The Stefan condition introduces a new variable, the interface position, and one new equation.

A solution of the conduction equation in the liquid is

$$T^L = a_1 \, \mathrm{erf}\left(\frac{x}{\sqrt{4\kappa^L t}}\right) + a_2 \quad (20.5)$$

where a_1 and a_2 are constants. Fitting this solution to the boundary conditions given by Eq. 20.2,

$$T_m - T^{L0} = a_1 \, \mathrm{erf}\left(\frac{\chi}{\sqrt{4\kappa^L t}}\right) \quad (20.6)$$

which can only be satisfied at all times if

$$\chi(t) = A\sqrt{t} \quad (20.7)$$

where A is a constant (to be determined). This yields the important result that the liquid/solid interface will advance parabolically (i.e., as $\sqrt{t}$). Solving for a_1, the solution of Eq. 20.5 becomes

$$\frac{T^{L0} - T^L}{T^{L0} - T_m} = \frac{\mathrm{erf}\left(x/\sqrt{4\kappa^L t}\right)}{\mathrm{erf}\left(A/\sqrt{4\kappa^L}\right)} \quad (20.8)$$

Using similar procedures, the solution of the conduction equation in the solid is

$$\frac{T^S - T^{S\infty}}{T_m - T^{S\infty}} = \frac{\mathrm{erfc}\left(x/\sqrt{4\kappa^S t}\right)}{\mathrm{erfc}\left(A/\sqrt{4\kappa^S}\right)} \quad (20.9)$$

Note that for these solutions no liquid exists at $t = 0$ and $T^S = T^{S\infty}$ everywhere. Finally, an equation for determining the constant A can now be obtained by substituting Eqs. 20.7, 20.8, and 20.9 and $K = c_P \rho \kappa$ into the Stefan condition (Eq. 20.4), with the result

$$
A = \frac{1}{\rho H_m \sqrt{\pi}} \left\{ \frac{c_P^L \sqrt{4\kappa^L} \left(T^{L0} - T_m\right) e^{-A^2/(4\kappa^L)}}{\mathrm{erf}\left(A/\sqrt{4\kappa^L}\right)} \right.
$$
$$
\left. - \frac{c_P^S \sqrt{4\kappa^S} \left(T_m - T^{S\infty}\right) e^{-A^2/(4\kappa^S)}}{\mathrm{erfc}\left(A/\sqrt{4\kappa^S}\right)} \right\} \tag{20.10}
$$

Transcendental equations, such as for A in Eq. 20.10, appear frequently in moving interface problems and can be solved using numerical methods.

20.1.2   Diffusion-Limited Growth

Many situations arise in solid materials where adjoining layers of different phases grow (or shrink) under diffusion-limited conditions. In these cases, the atom transfer across the interfaces between the layers must be sufficiently rapid so that the concentrations in the adjoining phases at each interface are maintained in local equilibrium. Also, the thermal conduction rates in the system must be rapid enough (compared with the mass-diffusion rates) so that no significant thermal gradients are present due to the latent heat emitted or absorbed at the interfaces. The analysis of the growth kinetics is therefore mathematically similar in many respects to the analysis of the melting kinetics in the preceding section. However, in the melting analysis it was assumed that $\rho^S = \rho^L$ and that any effects due to a difference in the densities of the two phases were neglected. Under this assumption, the total volume of the system remained constant and the two-phase conduction problem could be solved within a single coordinate system in a relatively simple manner. However, in the case of solid phases, larger differences in density may exist. In addition, important cases exist where a component may be delivered to a free surface via the vapor phase and produce a new growing phase at the surface as, for example, during oxidation. In such cases, the total volume of the system is not constant, and substantial expansions or contractions of the phases present will occur which cannot be ignored. These changes in volume will cause the phases to be displaced with respect to each other, causing additional flux terms to appear in the analysis.

To deal with this more complex problem, we follow Sekerka et al. [2] and Sekerka and Wang [3] and first establish a general analysis that allows for these changes of volume. The previous melting problem was solved by first obtaining independent solutions to the diffusion equation in each phase and then coupling them via the Stefan flux condition at the interface. A similar approach can be employed for the present problem. To accomplish this, it is necessary to identify suitable frames for analyzing the diffusion in each phase and then to find the relations between them necessary to construct the Stefan condition.

Framework for Describing the Diffusion.   We again make the acceptable approximation that the atomic volume of each component within a given phase is independent of concentration. No volume changes will therefore occur within each phase as a result of diffusion within the phase. As shown in Sections 3.1.3 and 3.1.4, chemical diffusion within each phase can then be described by employing a V-frame for that phase and then employing the interdiffusivity, $\widetilde{D}$. However, the total volume of the system will generally change when components diffuse into or out of adjoining phases across interfaces since the atomic volumes of the components will differ in the adjoining phases. Also, the various phases will grow or shrink and be bodily displaced with respect to each other. To cope with this situation, the chemical diffusion within each phase is analyzed within its own V-frame.[2] The relative displacements of the different phases (frames) are then determined by applying the Stefan condition at the various interfaces.

Stefan Condition at an α/β Interface. Consider the interface between the moving α and β phases shown in Fig. 20.2b. The α and β phases (along with their V-frames) will be bodily displaced with respect to each other, and the Stefan condition can be written as

$$\left[J_i^{V^\beta}\right]_{\alpha/\beta} - v_{\alpha/\beta}^{V^\beta} c_i^{\beta\alpha} = \left[J_i^{V^\alpha}\right]_{\alpha/\beta} - \left(v_{\alpha/\beta}^{V^\beta} - v_{V^\alpha}^{V^\beta}\right) c_i^{\alpha\beta} \quad (i = A, B) \qquad (20.11)$$

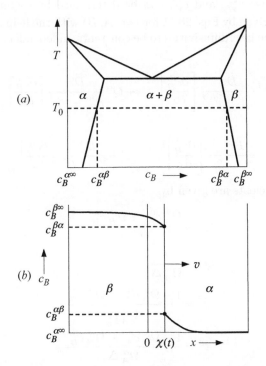

Figure 20.2: (a) Phase diagram for $A–B$ binary system. (b) Composition profile produced by bonding two thick slabs of pure A and pure B face to face and then annealing at T_0.

---

[2]Note that the use of a V-frame for diffusion within a phase merely requires that an equation such as Eq. 3.21 is satisfied. The fact that the volume of the phase may be changing due to the gain or loss of atoms at its interfaces is irrelevant.

where the relation $v_{V\alpha}^{V\beta} = v_{\alpha/\beta}^{V\beta} - v_{\alpha/\beta}^{V\alpha}$ has been used and where

$\left[J_i^{V\alpha} \right]_{\alpha/\beta}$ = flux of i into the α phase at the interface measured in

the α-phase V-frame

$\left[J_i^{V\beta} \right]_{\alpha/\beta}$ = flux of i into the β phase at the interface measured in

the β-phase V-frame

$v_{\alpha/\beta}^{V\alpha}$ = velocity of the interface measured in the α-phase V-frame

$v_{\alpha/\beta}^{V\beta}$ = velocity of the interface measured in the β-phase V-frame

$v_{V\alpha}^{V\beta}$ = velocity of the α-phase V-frame measured in the β-phase V-frame

$c_i^{\beta\alpha}$ = concentration of i at the α/β interface on the β side (β therefore

precedes α in the superscript),

$c_i^{\alpha\beta}$ = concentration of i at the α/β interface on the α side (α therefore

precedes β in the superscript).

The two velocities $v_{\alpha/\beta}^{V\beta}$ and $v_{V\alpha}^{V\beta}$ can be determined by solving simultaneously the two equations given by Eqs. 20.11 for $i = (A, B)$ with the help of Eqs. 3.24, 3.27, and A.8. Taking the interdiffusivities to be constants, independent of concentration, the results are [2, 3]

$$v_{\alpha/\beta}^{V\beta} = -Q^\beta \frac{\widetilde{D}^\beta}{c_B^{\beta\alpha} - c_B^{\alpha\beta}} \left[\frac{\partial c_B^\beta}{\partial x} \right]_{\alpha/\beta} + Q^\alpha \frac{\widetilde{D}^\alpha}{c_B^{\beta\alpha} - c_B^{\alpha\beta}} \left[\frac{\partial c_B^\alpha}{\partial x} \right]_{\alpha/\beta} \tag{20.12}$$

and

$$v_{V\alpha}^{V\beta} = \varepsilon^\beta \frac{\widetilde{D}^\beta}{c_B^{\beta\alpha} - c_B^{\alpha\beta}} \left[\frac{\partial c_B^\beta}{\partial x} \right]_{\alpha/\beta} + \varepsilon^\alpha \frac{\widetilde{D}^\alpha}{c_B^{\beta\alpha} - c_B^{\alpha\beta}} \left[\frac{\partial c_B^\alpha}{\partial x} \right]_{\alpha/\beta} \tag{20.13}$$

The Q and ε coefficients are given by

$$Q^\beta = \frac{\Omega_A^\beta c_A^{\alpha\beta} + \Omega_B^\beta c_B^{\alpha\beta}}{\Omega_A^\beta \Delta}$$

$$Q^\alpha = \frac{1}{\Omega_A^\alpha \Delta}$$

$$\varepsilon^\beta = \frac{1 - \Omega_A^\beta c_A^{\alpha\beta} - \Omega_B^\beta c_B^{\alpha\beta}}{\Omega_A^\beta \Delta} \tag{20.14}$$

$$\varepsilon^\alpha = \frac{1 - \Omega_A^\alpha c_A^{\beta\alpha} - \Omega_B^\alpha c_B^{\beta\alpha}}{\Omega_A^\alpha \Delta}$$

and the parameter Δ is given by

$$\Delta = \frac{c_B^{\beta\alpha} c_A^{\alpha\beta} - c_B^{\alpha\beta} c_A^{\beta\alpha}}{c_B^{\beta\alpha} - c_B^{\alpha\beta}} \tag{20.15}$$

In the special case where $\Omega_A^\beta = \Omega_A^\alpha$ and $\Omega_B^\beta = \Omega_B^\alpha$, $\varepsilon^\alpha = \varepsilon^\beta = 0$ and $Q^\alpha = Q^\beta = 1$. There is then no overall volume change or bodily movement between phases.

Stefan Condition at a Free Surface. Commonly, a component (i.e., B) with a high vapor pressure is diffused into the free surface of a β phase. Component A has a much lower vapor pressure and does not evaporate from the surface. The Stefan condition at the free surface is then

$$\left[J_A^{V^\beta} \right]_{\text{surf}} - v_{\text{surf}}^{V^\beta} c_A^{\beta\text{surf}} = 0 \tag{20.16}$$

Substituting Eqs. 3.24, A.8, and A.10 into Eq. 20.16 yields the further relation [2]

$$v_{\text{surf}}^{V^\beta} = \frac{\Omega_B^\beta}{1 - \Omega_B^\beta c_B^{\beta\text{surf}}} \widetilde{D}^\beta \left[\frac{\partial c_B^\beta}{\partial x} \right]_{\text{surf}} \tag{20.17}$$

Equation 20.17 serves as a useful boundary condition in cases where the concentration at the surface, $c_B^{\beta\text{surf}}$, is maintained constant.

Solution for Layer Growth in an Infinite α/β System. The formulation above can be used for solving a wide range of layer-growth problems in systems with constant interdiffusivities [2]. In the system in Fig. 20.2b, a thick slab of pure B was bonded to a corresponding slab of pure A in a binary A/B system having the phase diagram shown in Fig. 20.2a. The system was then annealed at T_0, and local equilibrium was quickly established across the α/β interface so that the concentrations in the α and β phases at the interface were maintained at $c_B^{\alpha\beta}$ and $c_B^{\beta\alpha}$, respectively. The original bonded interface, located at $x = 0$ (measured in the V^β-frame) has moved to the position $x = \chi(t)$. An expression $\chi(t)$ can now be found by using essentially the same procedure used to solve the previous melting problem.

First, solutions of the diffusion equation in the α and β phases (in the V^α- and V^β-frame, respectively) are found that match the boundary and initial conditions and then the Stefan condition is invoked. The solutions are of the error-function type and are given by

$$\frac{c_B^\alpha - c_B^{\alpha\infty}}{c_B^{\alpha\beta} - c_B^{\alpha\infty}} = \frac{1 - \text{erf}\left(x/\sqrt{4\widetilde{D}^\alpha t} \right)}{1 - \text{erf}\left(A/\sqrt{4\widetilde{D}^\alpha} \right)} \tag{20.18}$$

$$\frac{c_B^\beta - c_B^{\beta\infty}}{c_B^{\beta\alpha} - c_B^{\beta\infty}} = \frac{1 + \text{erf}\left(x/\sqrt{4\widetilde{D}^\beta t} \right)}{1 + \text{erf}\left(A/\sqrt{4\widetilde{D}^\beta} \right)} \tag{20.19}$$

with

$$\chi(t) = A\sqrt{t} \tag{20.20}$$

where $A = $ constant. A is now found by substituting Eqs. 20.18, 20.19, and 20.20 into Eq. 20.12 for the velocity of the interface (in the V^β-frame) with the result

$$\begin{aligned} A = & \frac{Q^\alpha \sqrt{4\widetilde{D}^\alpha}}{\sqrt{\pi}} \left(\frac{c_B^{\alpha\infty} - c_B^{\alpha\beta}}{c_B^{\beta\alpha} - c_B^{\alpha\beta}} \right) \frac{e^{-A^2/(4\widetilde{D}^\alpha)}}{1 - \text{erf}\left(A/\sqrt{4\widetilde{D}^\alpha} \right)} \\ & + \frac{Q^\beta \sqrt{4\widetilde{D}^\beta}}{\sqrt{\pi}} \left(\frac{c_B^{\beta\infty} - c_B^{\beta\alpha}}{c_B^{\beta\alpha} - c_B^{\alpha\beta}} \right) \frac{e^{-A^2/(4\widetilde{D}^\beta)}}{1 + \text{erf}\left(A/\sqrt{4\widetilde{D}^\beta} \right)} \end{aligned} \tag{20.21}$$

Therefore, the interface again moves parabolically, and A is given by a transcendental equation.

Solution for Layer Growth at a Free Surface. An example of layer growth at a surface is illustrated in Fig. 20.3, where a B-rich layer of β phase has formed on the surface of an initially pure A specimen at the temperature T_0. The phase diagram for the system is shown in Fig. 20.2a. Component B has a much higher vapor pressure than A and is supplied from the vapor phase (consisting of pure B) to the specimen surface located at $x = \chi_2(t)$ at a sufficient rate to maintain the concentration there at a constant value $c_B^{\beta\text{surf}}$ corresponding to pure B. At $t = 0$ the entire specimen was of composition $c_B^{\alpha\infty}$ (i.e., pure A) with its surface at $x = 0$. Due to the inward diffusion of B, a layer of β phase has formed with its surface moving to $x = \chi_2(t)$ and the α/β interface moving to $x = \chi_1(t)$. This problem can be solved in the same general way as the layer problem discussed above, but it is somewhat more complicated because there are now two interfaces instead of one. First, an error-function type solution to the diffusion equation is found for each phase within its own V-frame. To satisfy the boundary conditions, both interfaces must move parabolically with time with displacements given by $\chi_1 = A_1\sqrt{t}$ and $\chi_2 = A_2\sqrt{t}$, where A_1 and A_2 are constants. To find A_1 and A_2, Stefan conditions are invoked at the surface and at the α/β interface using Eqs. 20.17 and 20.12, respectively. The detailed solution is worked out in Exercise 20.4.

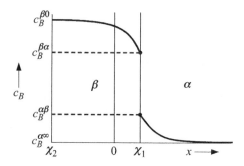

Figure 20.3: Layer of β phase formed on the surface of a semi-infinite slab of α phase.

20.1.3 Growth Limited by Heat Conduction and Mass Diffusion Simultaneously

The growth problem in alloys becomes considerably more complicated when the heat conduction is sufficiently slow so that the system is no longer closely isothermal throughout, and both temperature and composition gradients play important roles. We shall examine this problem only briefly. Consider, for example, the one-dimensional solidification of a liquid alloy of initial composition c_0 in a binary eutectic system with the phase diagram shown in Fig. 20.4a. Heat is removed in the $-x$ direction, the liquid/solid interface advances along $+x$, and the temperature gradients in both solid and liquid are positive. During cooling, the first solid to form at the temperature T_0 will be of solute concentration c_0^{SL}, which is lower than the solute concentration in the liquid. Solute must therefore be rejected into the liquid at the solid/liquid interface.

During further solidification, the solute concentration in the liquid in front of the advancing interface will then quickly build up to form a concentration "spike"

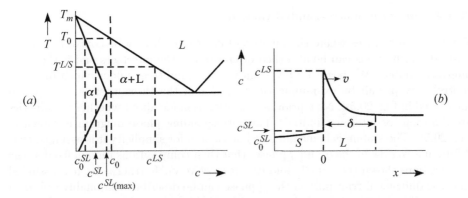

Figure 20.4:   One-dimensional solidification of liquid binary alloy. **(a)** Phase diagram of system. The solidus and liquidus are approximated by straight lines. **(b)** Solute concentration vs. distance in the vicinity of the liquid/solid interface (at $x = 0$) after the establishment of a quasi-steady-state concentration spike in the liquid in front of the advancing interface.

of the form shown in Fig. 20.4b. In the figure, further cooling and solidification has occurred and the temperature at the interface has dropped to $T^{L/S}(t)$. The compositions of the liquid and solid at the interface are maintained in local equilibrium at the interface and are now at $c^{LS}(T^{L/S})$ and $c^{SL}(T^{L/S})$, respectively.

In this example, the equilibrium concentrations maintained at the interface are functions of the interface temperature, which in turn is a function of time. In addition, the velocity of the interface, v (i.e., rate of solidification), depends simultaneously upon the mass diffusion rates and the rates of heat conduction in the two phases, as may be seen by examining the two Stefan conditions that apply at the interface. For the mass flow the condition is

$$\left[c^{LS}(T^{L/S}) - c^{SL}(T^{L/S})\right] v = -D^L \left[\frac{\partial c^L}{\partial x}\right]_{x=\chi(t)} + D^S \left[\frac{\partial c^S}{\partial x}\right]_{x=\chi(t)} \tag{20.22}$$

and for the heat flow

$$\rho H_m v = K^S \left[\frac{\partial T^S}{\partial x}\right]_{x=\chi(t)} - K^L \left[\frac{\partial T^L}{\partial x}\right]_{x=\chi(t)} \tag{20.23}$$

In addition, the following boundary conditions apply at the solid/liquid interface:

$$T^S[\chi(t)] = T^L[\chi(t)] = T^{L/S} \quad c^S[\chi(t)] = c^{SL} \quad c^L[\chi(t)] = c^{LS} \tag{20.24}$$

A complete general solution therefore requires solving for temperature and concentration fields in both phases which satisfy all boundary conditions as well as the two coupled Stefan conditions. Solving this problem is a challenging task [4, 5]; however, an analysis of the concentration spike under certain simplifying conditions when v is known is given in Section 22.1.1.

20.1.4 Interface Source-Limited Growth

We now turn to cases where the kinetics of layer growth may be controlled by the rates at which atoms can be absorbed or emitted at the interfaces and is therefore interface-limited. We begin by investigating the conditions under which interface control may prevail by examining the behavior of the three-layer $\alpha/\beta/\gamma$ system illustrated in Fig. 20.5a. No a priori assumption is made about whether the kinetics are diffusion- or interface-limited. The corresponding phase diagram is shown in Fig. 20.5b. The system is being annealed at T_0, and for simplicity it is assumed that diffusion is so slow in the α and γ phases that their compositions remain fixed at the compositions shown (i.e., at $c_{\mathrm{eq}}^{\alpha\beta}$ and $c_{\mathrm{eq}}^{\gamma\beta}$, respectively). Furthermore, it is assumed that the diffusional transport in the β phase can be described reasonably well by a constant diffusivity and simple linear profile, as shown, and that the atomic volume of each component is the same in each phase so that no changes in overall volume occur. Employing the same interfacial-reaction rate-constant model used to write Eq. 13.24, the following five equations will apply under general conditions:

$$J^\beta = K_1 \left(c_{\mathrm{eq}}^{\beta\alpha} - c^{\beta\alpha} \right) \tag{20.25}$$

$$J^\beta = \frac{d\chi_1}{dt} \left(c^{\beta\alpha} - c_{\mathrm{eq}}^{\alpha\beta} \right) \tag{20.26}$$

$$J^\beta = K_2 \left(c^{\beta\gamma} - c_{\mathrm{eq}}^{\beta\gamma} \right) \tag{20.27}$$

$$J^\beta = \frac{d\chi_2}{dt} \left(c^{\beta\gamma} - c_{\mathrm{eq}}^{\gamma\beta} \right) \tag{20.28}$$

$$J^\beta = -\frac{\widetilde{D}^\beta \left(c^{\beta\gamma} - c^{\beta\alpha} \right)}{\chi_2 - \chi_1} \tag{20.29}$$

Equations 20.25 and 20.27 show that $c^{\beta\alpha} \geq c_{\mathrm{eq}}^{\beta\alpha}$ and $c^{\beta\gamma} \leq c_{\mathrm{eq}}^{\beta\gamma}$, since J^β is negative and the rate constants are positive. By combining Eqs. 20.25 and 20.29,

$$c_{\mathrm{eq}}^{\beta\alpha} - c^{\beta\alpha} = \phi_1 \left(c^{\beta\alpha} - c^{\beta\gamma} \right) \tag{20.30}$$

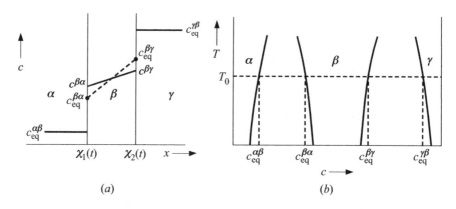

(a) (b)

Figure 20.5: (a) Three-layer $\alpha/\beta/\gamma$ system during annealing at T_0. (b) The corresponding phase diagram. χ_1 and χ_2 indicate the positions of the α/β and β/γ interfaces, respectively.

where

$$\phi_1 = \frac{\widetilde{D}^\beta}{K_1 \left(\chi_2 - \chi_1\right)} \tag{20.31}$$

Inspection of Eq. 20.30 demonstrates that $c^{\beta\alpha} \cong c_{\rm eq}^{\beta\alpha}$ when the parameter ϕ_1 is small and that $c^{\beta\alpha} \cong c^{\beta\gamma}$ when it is large, as expected. When ϕ_1 is small, diffusional transport away from the interface is discouraged by a relatively small $\widetilde{D}^\beta$ or large $(\chi_2 - \chi_1)$, while rapid atomic exchange at the interface is encouraged by a relatively large K_1. This favors the establishment of local equilibrium at the interface. On the other hand, when ϕ_1 is large, the situation is reversed, favoring diffusional transport and the elimination of any chemical gradient in the β layer. By combining Eqs. 20.27 and 20.29, and employing the parameter

$$\phi_2 = \frac{\widetilde{D}^\beta}{K_2 \left(\chi_2 - \chi_1\right)} \tag{20.32}$$

a similar exercise shows that $c^{\beta\gamma} \cong c_{\rm eq}^{\beta\gamma}$ when ϕ_2 is small, and $c^{\beta\alpha} \cong c^{\beta\gamma}$ when ϕ_2 is large.

Four basic situations can occur: both ϕ_i's small, both large, and two cases of one large and the other small. When ϕ_1 and ϕ_2 are both small, $c^{\beta\alpha} \cong c_{\rm eq}^{\beta\alpha}$ and $c^{\beta\gamma} \cong c_{\rm eq}^{\beta\gamma}$, and the β-phase layer should grow under diffusion-limited conditions. Putting these conditions into Eqs. 20.26 and 20.28 and summing them,

$$\chi \, d\chi = \widetilde{D}^\beta \left(c_{\rm eq}^{\beta\gamma} - c_{\rm eq}^{\beta\alpha}\right) \left(\frac{1}{c_{\rm eq}^{\beta\alpha} - c_{\rm eq}^{\alpha\beta}} - \frac{1}{c_{\rm eq}^{\beta\gamma} - c_{\rm eq}^{\gamma\beta}}\right) dt \tag{20.33}$$

where $\chi \equiv \chi_2 - \chi_1$. This yields parabolic growth upon integration, as expected.

When ϕ_1 and ϕ_2 are both large, $c^{\beta\gamma} = c^{\beta\alpha} = \bar{c}$ and the growth should be interface-limited. Putting these conditions into Eqs. 20.25 and 20.26 and equating them, doing the same with Eqs. 20.27 and 20.28, and then subtracting the results,

$$\frac{d\chi}{dt} = K_1 \frac{\bar{c} - c_{\rm eq}^{\beta\alpha}}{\bar{c} - c_{\rm eq}^{\alpha\beta}} + K_2 \frac{\bar{c} - c_{\rm eq}^{\beta\gamma}}{\bar{c} - c_{\rm eq}^{\gamma\beta}} \tag{20.34}$$

The concentration $\bar{c}$ is found by equating Eqs. 20.25 and 20.27 to obtain

$$\bar{c} = \frac{K_1 c_{\rm eq}^{\beta\alpha} + K_2 c_{\rm eq}^{\beta\gamma}}{K_1 + K_2} \tag{20.35}$$

Integration of Eq. 20.34 then yields linear growth at a rate that is controlled by the rate constants at the two interfaces. Further analysis (see Exercise 20.3) shows that interface-limited growth also prevails when one ϕ_i is small and the other large. In this case, the β-phase growth rate is controlled by the interface that possesses the larger value of ϕ_i (and the smaller rate constant), as might be expected.

The parameter ϕ_i therefore emerges as the critical parameter determining the mode of growth. The magnitude of ϕ_i, in turn, depends directly upon the magnitude of the rate constant K_i. Determining the magnitude of K_i requires the construction of a detailed model for the source action of the interface based on one (or more) of the mechanisms described in Section 13.4.

Experimental results for the growth of layers has been reviewed [6, 7]. It is recognized that all layer growth should be interface-limited in the early stages when thicknesses are small, since ϕ_i values will then be large [8]. As a layer thickens, the growth may switch at some point from linear interface-limited to parabolic diffusion-limited as ϕ_i decreases. In multilayered systems, such as thin-film electronic devices, this can produce complicated behavior in which some layers first grow, then shrink, and in which some intermediate phases may never appear. Of course, in some systems, the growth may be simpler and remain either linear or parabolic at all observed times.

20.2   GROWTH OF ISOLATED PARTICLES

20.2.1   Diffusion-Limited Growth

The growth of isolated particles is potentially more complex than that of planar layers because any volume changes due to differences in the atomic volume in the particle and matrix phases will be constrained by the surrounding matrix. A realistic solution therefore should include both the rate at which this constraint relaxes and the effect of the stresses generated by the constraint on the diffusion rates. Such a calculation presents a formidable problem and will not be taken up here. Instead, we focus on the relatively simple case where the atomic volumes are assumed to be the same in each phase and there is no change in volume. The composition of the growing particle will be taken to be constant throughout, the interdiffusivity in the matrix will be assumed constant, and the interface will be assumed to be stable against any morphological instability, as described below in Section 20.3. As pointed out by Mullins and Sekerka, this class of problems can be solved conveniently for a variety of particle shapes by using the scaling method in Section 4.2.2 [9].

Spheres.   Consider a B-rich sphere of β phase of radius $R = R(t)$ growing in an infinite α matrix under diffusion-limited conditions as shown in Fig. 20.6. This problem can be solved by using the scaling method with η defined by $\eta \equiv r/(4\widetilde{D}^\alpha t)^{1/2}$. The diffusion equation in the α phase in spherical coordinates in rt-space (see Eq. 5.14) becomes, after transformation into η-space,

$$\frac{d^2 c^\alpha(\eta)}{d\eta^2} = \left(\frac{2}{\eta} + 2\eta \right) \frac{dc^\alpha(\eta)}{d\eta} = 0 \tag{20.36}$$

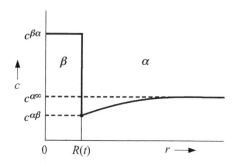

Figure 20.6:   A B-rich sphere (or a cylinder) of β phase of radius $R = R(t)$ growing in an infinite α matrix under diffusion-limited conditions. The solute concentration $c^{\alpha\beta}$ is maintained at the equilibrium value required by the phase diagram.

The boundary and initial conditions in the α phase in rt-space are on the left in Eqs. 20.37–20.39, and the corresponding conditions in η-space are on the right.

$$c^\alpha(r, t = 0) = c^{\alpha\infty} \ (r \geq 0) \quad \text{and} \quad c^\alpha(\infty) = c^{\alpha\infty} \tag{20.37}$$

$$c^\alpha(r = \infty, t) = c^{\alpha\infty} \quad \text{and} \quad c^\alpha(\infty) = c^{\alpha\infty} \tag{20.38}$$

$$c^\alpha(r = R(t), t) = c^{\alpha\beta} \quad \text{and} \quad c^\alpha(\eta_R) = c^{\alpha\beta} \tag{20.39}$$

The entire boundary-value problem is transformed successfully into η-space. $\eta_R = R(t)/(4\widetilde{D}^\alpha t)^{1/2}$ must be constant to satisfy Eq. 20.39. Therefore, the interface will move parabolically according to

$$R(t) = \eta_R \sqrt{4 D^\alpha t} \tag{20.40}$$

Letting $dc^\alpha/d\eta = f(\eta)$ and integrating Eq. 20.36 once yields

$$dc^\alpha = \frac{a_1}{\eta^2} e^{-\eta^2} d\eta \tag{20.41}$$

where $a_1 = $ constant. Integrating again,

$$\int_\eta^\infty dc^\alpha = a_1 \int_\eta^\infty \frac{1}{\xi^2} e^{-\xi^2} d\xi \tag{20.42}$$

Carrying out the integration and determining a_1 by applying the boundary conditions,

$$\frac{c^{\alpha\infty} - c^\alpha}{c^{\alpha\infty} - c^{\alpha\beta}} = \frac{(1/\eta) e^{-\eta^2} - \sqrt{\pi} \, \mathrm{erfc}(\eta)}{(1/\eta_R) e^{-\eta_R^2} - \sqrt{\pi} \, \mathrm{erfc}(\eta_R)} \tag{20.43}$$

Next, η_R is determined in the usual way by invoking the Stefan condition at the interface, which has the form

$$\frac{dR}{dt} \left(c^{\beta\alpha} - c^{\alpha\beta}\right) = \widetilde{D}^\alpha \left(\frac{\partial c^\alpha}{\partial r}\right)_{r=R} \tag{20.44}$$

Use of Eq. 20.43 in Eq. 20.44 then produces the desired relation for η_R:

$$\eta_R^3 = \frac{c^{\alpha\infty} - c^{\alpha\beta}}{2\left(c^{\beta\alpha} - c^{\alpha\beta}\right)} \frac{e^{-\eta_R^2}}{(1/\eta_R) e^{-\eta_R^2} - \sqrt{\pi} \, \mathrm{erfc}(\eta_R)} \tag{20.45}$$

Of special interest is the current of B atoms into the particle, which is given by

$$I = 4\pi R^2 \widetilde{D}^\alpha \left(\frac{\partial c^\alpha}{\partial r}\right)_{r=R} \tag{20.46}$$

Using Eq. 20.43 in Eq. 20.46,

$$I = \frac{4\pi \widetilde{D}^\alpha R \left(c^{\alpha\infty} - c^{\alpha\beta}\right)}{1 - \sqrt{\pi} \, \mathrm{erfc}(\eta_R) \, \eta_R \, e^{\eta_R^2}} \cong 4\pi \widetilde{D}^\alpha R \left(c^{\alpha\infty} - c^{\alpha\beta}\right) \tag{20.47}$$

The last approximation is obtained by an expansion of the denominator to first order and is usually valid since η_R is expected to be small.[3]

---

[3]Note that the approximation given by Eq. 20.47 is identical to Eq. 13.22, which was obtained by an analysis that ignored the effect of the expansion of the sphere during growth. See also Exercise 13.6.

The growth of spherical precipitates under diffusion-limited conditions has been observed in a number of systems, such as Co-rich particles growing in Cu supersaturated with Co (see Chapter 23). In these systems, the particles are coherent with the matrix crystal and the interfaces possess high densities of coherency dislocations, which are essentially steps with small Burgers vectors. The interfaces therefore possess a high density of sites where atoms can be exchanged and the particles operate as highly efficient sources and sinks.

Cylinders, Ellipsoids, and Elliptical Paraboloids. The diffusion-limited growth of particles whose planar intercepts are conic sections can also be analyzed by the scaling method. For example, the scaling function appropriate for a cylinder is $\eta = r/\sqrt{t}$.[4] The solution for the growth of a cylinder is obtained in Exercise 20.5.

20.2.2 Interface Source-Limited Growth

As discussed in Chapter 23, particles produced in phase transformations often nucleate in the form of platelets or needles when they can find special orientations in the matrix where the energies of the platelet faces or needle sides are of relatively low energy. These interfaces will be singular or vicinal and possess low densities of line defects, as described in Section 13.4.1, and will therefore be poor sources for atoms. On the other hand, the platelet edges or needle tips are often general interfaces and act as highly efficient sources. The platelets will then undergo strong edgewise growth (and the needles will undergo lengthwise growth) accompanied by relatively little transverse thickening. The edgewise growth is generally observed to progress linearly with time, in contrast to the parabolic diffusion-limited growth of the layers and particles described above. On the other hand, the transverse growth is episodic and considerably slower, depending upon the infrequent creation of line defects with step and dislocation character which eventually move across the faces as described in Section 13.4.1. No exact solutions describing the simultaneous edgewise and transverse growth kinetics of these particles is available, due to its complexity. However, the edgewise growth can be modeled in an approximate manner which provides some physical insight and accounts for the linear growth kinetics observed.

The geometry of a B-rich β-phase platelet growing edgewise in an α-phase matrix is shown in Fig. 20.7. The growing edge is modeled as a cylindrical interface of radius R where local equilibrium between precipitate and matrix is maintained. Adapting Eq. 15.4 to this cylindrical interface, the concentration in the α phase

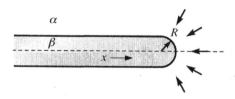

Figure 20.7: Edgewise diffusion-limited growth of β-phase platelet in an α matrix (or, lengthwise growth of a needle-shaped particle).

[4]Scaling functions and examples for other shapes are given by Sekerka et al. [2].

at the cylindrical edge is then given by $c^{\alpha\beta}(R) = c^{\alpha\beta}(\infty)\left[1 + \gamma\Omega/(kTR)\right]$. The concentration in the matrix far from the platelet is $c_B^{\alpha\infty}$, and the concentration in the platelet is fixed throughout at $c^{\beta\alpha}$. Any transverse growth is ignored and it is assumed that the diffusion field reaches a quasi-steady state and is radial, as indicated by the arrows in Fig. 20.7, and advances at the same rate along x as the edge.

As a first approximation, it is assumed that the radial flux entering the edge is constant over the cylindrical interface. The Stefan condition, integrated over the interface, is then

$$2R\frac{d\chi}{dt}[c^{\beta\alpha} - c^{\alpha\beta}] \approx \pi R\tilde{D}^\alpha \left(\frac{\partial c^\alpha}{\partial r}\right)_{r=R} \tag{20.48}$$

where χ is the position of the edge along x and a local cylindrical coordinate system with an origin at the center of curvature of the edge is employed to describe the radial flux. To evaluate the gradient term, it is assumed that the diffusion field in front of the edge is the same as the field that would exist around a cylinder of radius R embedded in a large body of α phase where the concentration at a large distance, R^∞, from the cylinder is c^α and the concentration at the cylinder is $c^{\alpha\beta}$. In the corresponding case of an embedded sphere of radius R, the analyses leading to Eqs. 13.22, 13.52, and 20.47 indicate that the gradient in the α phase at $r = R$ is quite insensitive to the motion of the spherical interface and the exact nature of the boundary condition at R^∞ as long as $R^\infty \gg R$. The present cylindrical system will exhibit the same general behavior [10] and we may therefore evaluate the gradient in Eq. 20.48 by using Eq. 5.11 to obtain

$$\left(\frac{\partial c^\alpha}{\partial r}\right)_{r=R} \approx \frac{c^{\alpha\infty} - c^{\alpha\beta}}{\ln(R^\infty/R)}\frac{1}{R} \tag{20.49}$$

Therefore, combining Eqs. 20.48 and 20.49 yields

$$\frac{d\chi}{dt} \approx \frac{\pi\tilde{D}^\alpha[c^{\alpha\infty} - c^{\alpha\beta}]}{2\ln(R^\infty/R)[c^{\beta\alpha} - c^{\alpha\beta}]R} \tag{20.50}$$

Integration then produces the linear growth expression

$$\chi(t) - \chi(0) \approx \frac{\pi\tilde{D}^\alpha[c^{\alpha\infty} - c^{\alpha\beta}]}{2\ln(R^\infty/R)[c^{\beta\alpha} - c^{\alpha\beta}]R}t \tag{20.51}$$

An expression of the same linear form will be obtained for the growth of a needle if the tip is modeled as a hemisphere. Further results bearing on the diffusion- or interface-limited growth (and shrinkage) of particles have been reviewed by Sutton and Balluffi [6].

20.3 MORPHOLOGICAL STABILITY OF MOVING INTERFACES

In the moving-boundary problems treated above, it was assumed that the interface retained its basic initial shape as it moved. It is important to realize that such problems are a subset of a much wider class of problems known as *free-boundary problems*, in which the boundary is allowed to change its shape as a function of time [2]. A mathematically correct solution for the motion of a boundary of a fixed ideal shape is no guarantee that it is physically realistic.

Moving interfaces can become unstable and break up into complex cellular or dendritic structures, as illustrated in Figs. 20.8 and 20.9. In the cellular case, the surface becomes rough and bumpy, while in the more extreme dendritic case, treelike protrusions, or dendrites, develop. This greatly increases the complexity of the growth process. The origin of this instability has been understood since the 1960s, when Mullins and Sekerka published their celebrated papers on the subject [4, 9]. Further work and reviews have since appeared in many places [5, 11–15]. In the following, we describe qualitatively the sources of the instability in various systems and then present quantitative analysis.

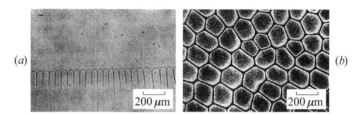

(a)   (b)

Figure 20.8: (a) Cellular interface in a transparent organic material (carbon tetrabromide). From Jackson and Hunt [16]. (b) Cellular interface (observed at normal incidence) in dilute Pb–Sb alloy. From Morris and Winegard [17].

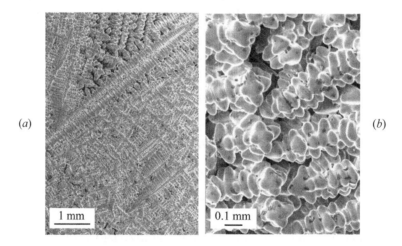

(a)   (b)

Figure 20.9: Dendrites on the surface of a casting of Cu–Ni–Mn alloy made visible when solidification shrinkage causes liquid to retreat into the casting interior. (a) Low magnification. (b) Higher-magnification image showing side-branch evolution near dendrite tips. Micrographs courtesy of J. Feuchtwanger.

20.3.1 Stability of Liquid/Solid Interface during Solidification of a Unary System

Consider first the solidification of a body of superheated pure liquid by removing heat through the wall of its container as in Fig. 20.10. The solidification will begin at the walls, and the solid/liquid interface will move toward the center of the container at a rate dictated by how quickly the latent heat of solidification

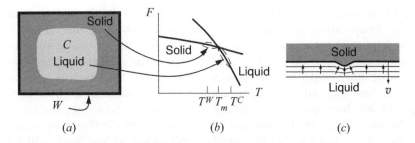

Figure 20.10: **(a)** Solidification of superheated liquid by removing the latent heat through the solid to the walls of its container. **(b)** Corresponding free-energy curves of a solid and liquid as a function of temperature. The interface is maintained at T_m, T^C is the temperature at the center of the liquid, and T^W is the wall temperature. The dashed regions indicate the ranges of temperature that are present in the solid and liquid. **(c)** Detail of the solid/liquid interface. Isotherms in the liquid bunch up at the protrusion, causing increased heat flux (arrows) into the protrusion.

can be conducted out through the freshly solidified solid and the walls. In a local coordinate system with its origin at the interface and pointed toward the liquid, the temperature gradient is positive in both the solid and liquid. In this situation, the interface will be stable. As may be seen from the free-energy diagram, both phases are themselves stable. Any small bump or perturbation on the solid surface that may form randomly on the planar interface will find itself attempting to grow into a hotter region of the liquid and will therefore tend to melt back and disappear. This is illustrated in Fig. 20.10c, where the isotherms bunch up at the protrusion, causing an increased thermal gradient that will increase the flux of heat to the protrusion and melt it back.

Consider now the solidification of a pure liquid that has been supercooled below its melting point without nucleation of the solid, as in Fig. 20.11. If the solid is nucleated by a seed at the center of the container, the solid will grow as the latent heat is conducted to the supercooled liquid and eventually out to the container walls.

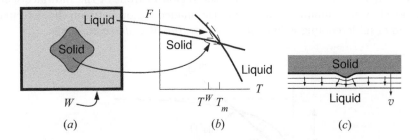

Figure 20.11: **(a)** Solidification of body in a supercooled liquid by removing the latent heat to the liquid. **(b)** Corresponding free-energy curves of solid and liquid as a function of temperature. The interface is maintained at T_m, and T^W is the wall temperature. The dashed regions indicate the ranges of temperature present in the solid and liquid. **(c)** Detail of the solid/liquid interface. The temperature gradient in the liquid is reversed compared to the superheated case (Fig. 20.10c) and the bunched-up isotherms cause increased heat flux out of the protrusion.

In a local coordinate system with its origin at the interface and pointed toward the liquid, the temperature gradient in the liquid is negative. In this situation, the interface is potentially unstable. In contrast to Fig. 20.10, stable solid is now growing into metastable liquid. Any small bump on the solid surface will now find itself growing into a cooler region. As seen in Fig. 20.11c, the bunched-up isotherms increase the heat flux out of the protrusion and so encourage its growth. However, this growth will be opposed by capillary forces that act to reduce the interfacial energy by smoothing out the protrusion. Further analysis is therefore required to answer the question of whether the interface will be stable (see Section 20.3.4).

20.3.2   Stability of α/β Interface during Diffusion-Limited Particle Growth

Growth of a B-rich stable β phase into a metastable α phase in an isothermal, binary solid system is similar to the case directly above since it involves the growth of a stable phase into a metastable phase. If temperature is replaced by concentration in Fig. 20.11c and the direction of the gradients and fluxes are reversed, the diagram will represent the diffusion field expected at a protuberance on the α/β interface. The bunching of the diffusional flux lines then encourages protuberance growth while capillarity discourages it, as described below.

20.3.3   Stability of Liquid/Solid Interface during Binary Alloy Solidification

Consider again the alloy solidification described in Section 20.1.3 and illustrated in Fig. 20.4. Here, the concentration spike in the liquid in front of the interface shown in Fig. 20.4b may cause the liquid in that region to be undercooled. The liquidus temperature for a liquid of composition c^L is given according to the phase diagram in Fig. 20.4a by the relationship

$$T_{\mathrm{liq}}\left(c^L\right) = T_m + m_{\mathrm{liq}}\, c^L \tag{20.52}$$

where m_{liq} is the constant liquidus slope. Since $c^L(x)$ is known in front of the interface, the liquidus temperature may be plotted as a function of x with the aid of Eq. 20.52 to produce the $T_{\mathrm{liq}}(x)$ curve in Fig. 20.12. If the actual temperature in the liquid, $T^L(x)$, is as in Fig. 20.12, the region in front of the interface will be undercooled by the amounts indicated in the shaded region between the two curves. This form of undercooling is known as *constitutional undercooling* [18].

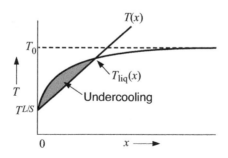

Figure 20.12:   Plot of the liquidus temperature, $T_{\mathrm{liq}}(x)$, vs. distance in the concentration spike in the liquid phase shown in Fig. 20.4b.

The critical condition for the absence of undercooling can then be written

$$\frac{dT^L}{dx} > \left[\frac{dT_{\text{liq}}(x)}{dx}\right]_{x=\chi} = \left[\frac{dT_{\text{liq}}(x)}{dc^L}\frac{dc^L(x)}{dx}\right]_{x=\chi} = m_{\text{liq}}\left[\frac{dc^L(x)}{dx}\right]_{x=\chi} \tag{20.53}$$

or

$$G_T > m_{\text{liq}}G_C \tag{20.54}$$

where G_T and G_C are, respectively, the temperature and composition gradients in the liquid at the interface. Undercooling tends to promote instability. However, stabilizing factors such as capillarity are also present and, again, further analysis is required to settle the question of stability.

20.3.4   Analyses of Interfacial Stability

Interfacial stability has been studied by analytical means [4, 5, 9, 11–15] and also numerically [19–21]. In the following, we present some of the analytical results of most immediate interest. Many other important results may be found in the references cited.

Spherical Particle during Diffusion-Limited Growth in an Isothermal Binary Solid.   This problem was analyzed by Mullins and Sekerka who found expressions for the rate of growth or decay of shape perturbations to a spherical B-rich β-phase particle of fixed composition growing in an α matrix as in Section 20.2.1 [9]. Perturbations are written in the form of spherical harmonics. Steps to solve this problem are:

i. Express the perturbed shape in terms of spherical harmonics.

ii. Write out the boundary conditions, accounting for local variations of curvature (hence concentrations) on the perturbed sphere.

iii. Write out the general solution to the diffusional growth problem. In the limit of low supersaturation, the diffusion field in the α matrix near the sphere is adequately described by Laplace's equation, $\nabla^2 c^\alpha(r, \theta, \phi) = 0$ (an interesting discussion of this *stationary-field approximation* and others encountered in mathematical treatments of diffusional growth problems has been given by Aaron et al. [22]).

iv. Evaluate the solution $c^\alpha(r, \theta, \phi)$ on the interface of the perturbed sphere. This determines the coefficients to the various terms that are present in the general solution.

v. Evaluate the growth velocity of the perturbed sphere from its relation to the flux into the interface. Extract the expression for the rate of change of the amplitude of the perturbation.

Spherical harmonics are derived from solutions of Laplace's equation in spherical coordinates using the method of separation of variables—i.e., a solution of the form

$$c^\alpha = R(r)\,\Theta(\theta)\,\Phi(\phi) \tag{20.55}$$

is assumed (see Arfken [23]). Laplace's equation then yields the relation

$$\frac{1}{\sin\theta}\frac{\partial}{\partial\theta}\left[\sin\theta\frac{\partial(\Theta\Phi)}{\partial\theta}\right] + \frac{1}{\sin^2\theta}\frac{\partial^2(\Theta\Phi)}{\partial\phi^2} - n(n+1)(\Theta\Phi) = 0 \tag{20.56}$$

The $\Phi(\phi)$ function turns out to be an exponential and the $\Theta(\theta)$ function consists of Legendre polynomials. Their product $\Phi(\phi)\Theta(\theta)$ gives the spherical harmonic functions which Arfken writes as $Y_n^m(\theta, \phi)$ and which Mullins and Sekerka write as $Y_{lm}(\theta, \phi)$. Then, from Eq. 20.56,

$$\frac{1}{\sin\theta}\frac{\partial}{\partial\theta}\left(\sin\theta\frac{\partial Y_n^m}{\partial\theta}\right) + \frac{1}{\sin^2\theta}\frac{\partial^2 Y_n^m}{\partial\phi^2} - n(n+1)(\Theta\Phi) = 0 \tag{20.57}$$

The functions $Y_n^m(\theta, \phi)$ are tabulated and can be represented as in Fig. 20.13. A series of spherical harmonics can be used to represent an arbitrary perturbation of a sphere, much the same as a Fourier series can represent an arbitrary function of a single variable.

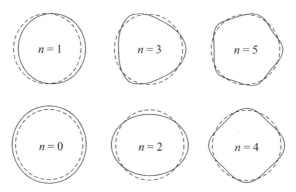

Figure 20.13: Perturbations of a circle by spherical harmonics $Y_n^0(\theta)$. Because $m = 0$, these are functions of θ alone and can be represented by a two-dimensional plot.

The analysis then proceeds:

i. The perturbed shape is given by

$$r_S = R + \delta Y_n^m(\theta, \phi) \tag{20.58}$$

where δ is the amplitude of the perturbation. Because Laplace's equation is linear, each harmonic term can be considered independently. Solutions can be obtained by superposition.

ii. Equation 15.4 can be used to specify the boundary condition on the interface of the perturbed sphere as

$$c^{\alpha\beta}(\kappa) = c^{\alpha\beta}(\kappa = 0)\left[1 + \Gamma\kappa\right] = c^{\alpha\beta}(\kappa = 0)\left[1 + \Gamma\kappa\right] \tag{20.59}$$

where $\Gamma \equiv \gamma\Omega/(kT)$, Ω is the atomic volume, and $\kappa = \kappa_1 + \kappa_2$ is the mean curvature on the interface of the perturbed sphere. Mullins and Sekerka give the angular dependence of κ as

$$\kappa(\theta, \phi) = \frac{2}{r}\left(1 - \frac{\delta Y_n^m}{R}\right) + \frac{n(n+1)\delta Y_n^m}{R^2} \tag{20.60}$$

By substituting Eq. 20.60 into Eq. 20.59, the boundary condition for $c^{\alpha\beta}$ is obtained:

$$c^{\alpha\beta}(\theta, \phi) = c^{\alpha\beta}(\kappa = 0)\left(\frac{1 + 2\Gamma}{R}\right) + \frac{(n-1)(n+2)\Gamma\delta Y_n^m}{R^2} \tag{20.61}$$

iii. The general solution to Laplace's equation in the matrix surrounding the perturbed sphere is

$$c^\alpha (r, \theta, \phi) = \sum_{n=0}^{\infty} \left(A_n r^n + B'_n r^{-n-1} \right) Y_n^m \tag{20.62}$$

Extracting the $n = 0$ term and rearranging,

$$c^\alpha (r, \theta, \phi) = A_0 + \frac{B_0}{r} + \sum_{n=1}^{\infty} \left(A_n r^n + B'_n r^{-n-1} \right) Y_n^m \tag{20.63}$$

c^α approaches $c^{\alpha\infty}$ as r approaches ∞. Thus, all $A_n = 0$ and $A_0 = c^{\alpha\infty}$. The set of constants B'_n can be written as $B_n \delta$, giving

$$c^\alpha (r, \theta, \phi) = c^{\alpha\infty} + \frac{B_0}{r} + \sum_{n=1}^{\infty} \frac{B_n}{r^{n+1}} \delta Y_n^m \tag{20.64}$$

iv. Now evaluate $c^{\alpha\beta} (\theta, \phi)$ on the perturbed interface $R + \delta Y_n^m$;

$$c^{\alpha\beta} (\theta, \phi) = c^{\alpha\infty} + \frac{B_0}{R + \delta Y_n^m} + \sum_{n=1}^{\infty} \frac{B_n}{(R + \delta Y_n^m)^{n+1}} \delta Y_n^m \tag{20.65}$$

Since δ is small, the leading terms of a Taylor expansion give

$$\frac{1}{R + \delta Y_n^m} \approx \frac{1 - \delta Y_n^m / R}{R}$$

and

$$\frac{1}{(R + \delta Y_n^m)^{n+1}} \approx \frac{1 - (n+1)\delta Y_n^m / R}{R^{n+1}}$$

Keeping linear terms in δ and ignoring higher-order terms, the boundary condition on the perturbed interface is

$$c^{\alpha\beta} (\theta, \phi) = c^{\alpha\infty} + \frac{B_0(1 - \delta Y_n^m / R)}{R} + \frac{B_N \delta Y_n^m}{R^{n+1}} \tag{20.66}$$

Equation 20.66 is the general solution for the problem, evaluated at the interface of the perturbed sphere and written for a single harmonic of order n (i.e., the summation has been dropped to simplify comparison with the capillarity condition as given in Eq. 20.61). These two equations can be equated to evaluate the unknown constants B_0 and B_n in Eq. 20.66. From terms not involving δY_n^m,

$$B_0 = \left[c^{\alpha\beta}(\kappa = 0) - c^{\alpha\infty} \right] R + 2\Gamma c^{\alpha\beta}(\kappa = 0) \tag{20.67}$$

From terms involving δY_n^m,

$$B_n = \left[c^{\alpha\beta}(\kappa = 0) - c^{\alpha\infty} \right] R^n + \Gamma c^{\alpha\beta}(\kappa = 0)n (n + 1) R^{n-1} \tag{20.68}$$

These equations for B_0 and B_n are now substituted into Eq. 20.64, giving the general solution which matches the capillarity boundary condition on the interface of the perturbed sphere and which is valid everywhere in the matrix,

$$
c^\alpha (r, \theta, \phi) = c^{\alpha\infty} + \frac{\left(c^{\alpha\beta}(\kappa = 0) - c^{\alpha\infty} \right) R + 2\Gamma c^{\alpha\beta}(\kappa = 0)}{r}
$$
$$
+ \frac{\left[\left(c^{\alpha\beta}(\kappa = 0) - c^{\alpha\infty} \right) R^n + \Gamma c^{\alpha\beta}(\kappa = 0)\, n(n + 1) R^{n-1} \right] \delta Y_n^m}{r^{n+1}}
$$

$$(20.69)$$

Note that the summation over n in this equation has been omitted because we are considering only one spherical harmonic perturbation at a time.

v. The final step is to extract from the growth-velocity expression an expression for the rate of change of the amplitude of the perturbation, $d\delta/dt$. We have from Eq. 20.58 and the Stefan condition at the interface,

$$
v = \frac{dr_S}{dt} = \frac{dR}{dt} + \frac{d\delta}{dt} Y_n^m = \frac{\widetilde{D}^\alpha}{c^\beta - c^{\alpha\beta}} \left(\frac{\partial c^\alpha}{\partial r} \right)_{r=r_S}
$$

$$(20.70)$$

where c^β is the constant composition in the growing precipitate.

Differentiating Eq. 20.69 and evaluating $\partial c^\alpha/\partial r$ at $r = r_S$, then substituting into Eq. 20.70 gives the relation

$$
v = \frac{\widetilde{D}^\alpha}{c^\beta - c^{\alpha\beta}} \left\{ \frac{c^{\alpha\infty} - c^{\alpha\beta}(\kappa = 2/R)}{R} + \left[\frac{\left(c^{\alpha\infty} - c^{\alpha\beta}(\kappa = 0) \right)(n - 1)}{R^2} \right.\right.
$$
$$
\left.\left. - \frac{c^{\alpha\beta}(\kappa = 0)\Gamma[n (n + 1)^2 - 4]}{R^3} \right] \delta Y_n^m \right\}
$$

$$(20.71)$$

where $c^{\alpha\beta}(\kappa = 2/R) = c^{\alpha\beta}(\kappa = 0) (1 + 2\Gamma/R)$. Equating the coefficients of terms in δY_n^m in Eqs. 20.70 and 20.71 gives the relation

$$
\frac{d\delta}{dt} = \frac{\widetilde{D}^\alpha (n - 1)}{(c^\beta - c^{\alpha\beta}(\kappa = 2/R)) R^2}
$$
$$
\times \left[c^{\alpha\infty} - c^{\alpha\beta}(\kappa = 0) \left(1 + \frac{2\Gamma}{R} + \frac{\Gamma}{R}(n + 1)(n + 2) \right) \right] \delta
$$

$$(20.72)$$

$$
\frac{d\delta}{dt} = \frac{\widetilde{D}^\alpha (n - 1)}{(c^\beta - c^{\alpha\beta}(\kappa = 2/R)) R} \left[G - \frac{c^{\alpha\beta}(\kappa = 0)\Gamma (n + 1) (n + 2)}{R^2} \right] \delta \qquad (20.73)
$$

where $G \equiv \left[c^{\alpha\infty} - c^{\alpha\beta}(\kappa = 0) \right]/R$ is the concentration gradient in the matrix at the interface of the unperturbed sphere in the stationary-field approximation. This term favors the growth of the perturbation, as it tends to make $d\delta/dt$ positive. The second term in brackets in Eq. 20.73, which arises from capillarity effects, acts to stabilize the system because it is negative. So, from Eq. 20.72, perturbations for which

$$
(n + 1) (n + 2) + 2 < \frac{R[c^{\alpha\infty} - c^{\alpha\beta}(\kappa = 0)]}{\Gamma c^{\alpha\beta}(\kappa = 0)}
$$

$$(20.74)$$

are unstable. Perturbations for which n is large enough always decay, as expected from interfacial free-energy considerations. For a given n, larger precipitate sizes always show instabilities. It is thus important to know how large R has to be for the onset of instability. A comparison can be made based on the size of the critical nucleus R_c for homogeneous nucleation from a supersaturated solid solution of concentration $c^{\alpha\infty}$ [9]. Since at R_c the precipitate is in (unstable) equilibrium with the matrix at composition $c^{\alpha\infty}$, Eq. 20.59 gives the relation

$$c^{\alpha\infty} = c^{\alpha\beta}(\kappa = 0) \left(\frac{1 + 2\Gamma}{R_c} \right) \qquad (20.75)$$

Thus Eq. 20.74 can be expressed

$$\frac{(n+1)(n+2)+2}{2} < \frac{R}{R_c} \qquad (20.76)$$

If the critical value for R at the stability limit is denoted R^*, then

$$R^* = \frac{[(n+1)(n+2)+2]R_c}{2} \qquad (20.77)$$

Once the particle grows to a size that exceeds R^*, morphological instabilities will set in. The minimum size for instabilities is found from the $n = 2$ spherical harmonic, giving $R^*(n = 2) = 7R_c$.[5] Thus the theory predicts instabilities at very small precipitate sizes.

Another point of interest is how fast the amplitude of the perturbation grows relative to the growth rate of the sphere itself,

$$\frac{(\partial\delta/\partial t)/\delta}{(\partial R/\partial t)/R} = (n-1)\,\frac{1 - R^*/R}{1 - R_c/R} \qquad (20.78)$$

Relative to the growth rate of the precipitate, the growth rate of the perturbation is larger for larger values of n as long as $R \gg R^*$. Equation 20.78 shows that for $R \gg R^*$, when $n = 2$ the growth is *shape-preserving*. That is, an ellipsoidal shape will be maintained during growth under these conditions. This result was also obtained by Ham [10].

Finally, there is a value of n for which the growth rate of perturbations is a maximum; the theory predicts that the "wavelength" $\lambda_{\max}$ of the fastest-growing perturbation is given by

$$\lambda_{\max} \equiv \frac{2\pi R}{n_{\max}} = \pi\sqrt{6RR_c} \qquad (20.79)$$

In spite of these theoretical results, dendritic-type forms for solid-state precipitation processes are the exception rather than the rule. This may happen because the theory is for pure diffusion-limited growth. Interface-limited growth tends to be stabilizing because the composition gradient close to a growing precipitate is less steep when the reaction is partly interface-limited. Thus, G is smaller, and Eq. 20.73 shows that this is a stabilizing effect. This and several other possible explanations for the paucity of observations for unstable growth forms in solid-state

---

[5]Note that $d\delta/dt = 0$ for $n = 1$, because $n = 1$ does not change the shape.

precipitation processes have been suggested [24, 25]. Another possibly important stabilizing effect is the existence of fast interfacial short-circuit diffusion along the advancing interface. If this is rapid enough relative to the crystal diffusion in the matrix, it will be strongly stabilizing. Caroli et al. have shown that elastic stresses arising when the lattice constant is concentration dependent actually destabilize the interface during diffusional growth, due to a reduction in the critical radius for the shape instability [26]. This stability analysis also applies for a solidifying sphere in an undercooled fixed-stoichiometry melt by replacing composition with a temperature field [9].

Planar Liquid/Solid Interface during Alloy Solidification.   Mullins and Sekerka analyzed this problem for the binary system described in Section 20.3.3 by again perturbing the interface and determining whether the perturbation would grow or decay [4]. In this case, sinusoidal perturbations of the flat interface were employed with a coordinate system pointed toward the liquid. Both temperature and solute concentration fields were determined consistent with boundary conditions at the interface, which included the effect of the liquid composition on the liquidus temperature (Eq. 20.52) and the effect of interface curvature on the melting temperature. Also, it was necessary to satisfy the Stefan condition for both the heat flux and the diffusion flux at the interface. The results show that capillarity and positive thermal gradients (pointing toward the liquid) promoted stability as expected. However, a third factor, representing the effect of solute concentration on the liquidus temperature (and therefore related to undercooling), favored growth. Under some conditions, stability criteria were found that agreed fairly closely with the simple constitutional undercooling criterion given by Eq. 20.54.

Bibliography

1. L.I. Rubenšsteĭn. *The Stefan Problem.* Translations of Mathematical Monographs. American Mathematical Society, Providence, RI, 1971. Translated from Russian by A.D. Solomon.

2. R.F. Sekerka, C.L. Jeanfils, and R.W. Heckel. The moving boundary problem. In H.I. Aaronson, editor, *Lectures on the Theory of Phase Transformations*, pages 117–169. AIME, New York, 1975.

3. R.F. Sekerka and S.-L. Wang. Moving phase boundary problems. In H.I. Aaronson, editor, *Lectures on the Theory of Phase Transformations*, pages 231–284. The Minerals, Metals and Materials Society, Warrendale, PA, 1999.

4. W.W. Mullins and R.F. Sekerka. Stability of a planar interface during solidification of a dilute binary alloy. *J. Appl. Phys.*, 35(2):444–451, 1964.

5. J.S. Langer and L.A. Turski. Studies in the theory of interfacial stability—I. Stationary symmetric model. *Acta Metall.*, 25(10):1113–1119, 1977.

6. A.P. Sutton and R.W. Balluffi. *Interfaces in Crystalline Materials.* Oxford University Press, Oxford, 1996.

7. K.N. Tu, J.W. Mayer, and L.C. Feldman. *Electronic Thin Film Science.* Macmillan, New York, 1992.

8. B.G. Guy and H. Oikawa. Calculations of two-phase diffusion in metallic systems including the interfacial reactions. *Trans. AIME*, 245(10):2293–2297, 1969.

9. W.W. Mullins and R.F. Sekerka. Morphological stability of a particle growing by diffusion or heat flow. *J. Appl. Phys.*, 34(2):323–329, 1963.

10. F.S. Ham. Theory of diffusion limited precipitation. *J. Phys. Chem. Solids*, 6(4):335–351, 1958.

11. R.F. Sekerka. Application of the time-dependent theory of interface stability to an isothermal phase transformation. *J. Chem. Phys. Solids*, 28(6):983–994, 1967.

12. R.F. Sekerka. A time-dependent theory of stability of a planar interface during dilute binary alloy solidification. In H.S. Peiser, editor, *Crystal Growth*, pages 691–702. Pergamon Press, Oxford, 1967.

13. R.F. Sekerka. Morphological stability. In P. Hartman, editor, *Crystal Growth, An Introduction*, pages 403–433. North-Holland, Amsterdam, 1973.

14. J.S. Langer. Studies in the theory of interfacial stability—II. Moving symmetric model. *Acta Metall.*, 25(10):1121–1137, 1977.

15. J.S. Langer. Instabilities and pattern formation in crystal growth. *Rev. Mod. Phys.*, 52(1):1–28, 1980.

16. K.A. Jackson and J.D. Hunt. Transparent compounds that freeze like metals. *Acta Metall.*, 13:1212–1215, 1965.

17. L.R. Morris and W.C. Winegard. The development of cells during the solidification of a dilute Pb–Sb alloy. *J. Cryst. Growth*, 5:361–375, 1969.

18. W.A. Tiller, K.A. Jackson, J.W. Rutter, and B. Chalmers. The redistribution of solute atoms during the solidification of metals. *Acta Metall.*, 1:428–437, 1953.

19. S. Kobayashi. Solute redistribution during solidification with diffusion in solid-phase—a theoretical analysis. *J. Cryst. Growth*, 88(1):87–96, 1988.

20. R. Kobayashi. Modeling and numerical simulations of dendritic crystal-growth. *Physica D*, 63(3–4):410–423, 1993.

21. J.A. Warren and W.J. Boettinger. Prediction of dendritic growth and microsegregation patterns in a binary alloy using the phase-field method. *Acta Metall.*, 43(2):689–703, 1995.

22. H.B. Aaron, D. Fanstein, and G.R. Kotler. Diffusion-limited phase transformations: A comparison and critical evaluation of the mathematical approximations. *J. Appl. Phys.*, 41:4404–4410, 1970.

23. G. Arfken. *Mathematical Methods for Physicists*. Academic Press, New York, 1968.

24. P.G. Shewmon. Interfacial stability in solid-solid transformations. *Trans. TMS-AIME*, 233:736–748, 1965.

25. J.W. Cahn. On the morphological stability of growing crystals. In H.S. Peiser, editor, *Crystal Growth (Proceedings of an International Conference on Crystal Growth, Boston 1966)*, pages 681–690. Pergamon Press, New York, 1967.

26. B. Caroli, C. Caroli, B. Roulet, and P.W. Voorhees. Effect of elastic stresses on the morphological stability of a solid sphere growing from a supersaturated melt. *Acta Metall.*, 37:257–268, 1989.

27. D.A. Porter and K.E. Easterling. *Phase Transformations in Metals and Alloys*. Van Nostrand Reinhold, New York, 1981.

28. A.K. Jena and M.C. Chaturvedi. *Phase Transformations in Materials*. Prentice Hall, Englewood Cliffs, NJ, 1992.

29. M. Abramowitz and I.A. Stegun. *Handbook of Mathematical Functions*. U.S. Department of Commerce, National Bureau of Standards, Washington, D.C., 1964.

30. C. Zener. Theory of growth of spherical precipitates from solid solution. *J. Appl. Phys.*, 20:950–953, 1949.

EXERCISES

20.1 Equation 20.51 has sometimes been written in the form

$$\chi(t) - \chi(0) = \frac{\pi \tilde{D}^\alpha}{2R \ln(R^\infty/R)} \frac{c_B^{\alpha\infty} - c_B^{\alpha\beta}(\infty)}{c_B^{\beta\alpha} - c_B^{\alpha\beta}(R)} \left(1 - \frac{R_c}{R}\right) t \qquad (20.80)$$

where R_c is the critical radius of the edge at which the concentration of B in the α phase in equilibrium with the curved edge [i.e., $c_B^{\alpha\beta}(R_c)$], is just equal to $c_B^{\alpha\infty}$ [27, 28]. When R assumes this value, the platelet can no longer grow. Derive this form of the growth expression.

Solution. The critical condition is

$$c_B^{\alpha\infty} = c_B^{\alpha\beta}(R_c) = c_B^{\alpha\beta}(\infty) \left(1 + \frac{\gamma\Omega}{kTR_c}\right) \qquad (20.81)$$

Then

$$c_B^{\alpha\infty} - c_B^{\alpha\beta}(R) = \frac{c_B^{\alpha\beta}(\infty)\gamma\Omega}{kTR_c} \left(1 - \frac{R_c}{R}\right) = \left[c_B^{\alpha\infty} - c_B^{\alpha\beta}(\infty)\right] \left(1 - \frac{R_c}{R}\right) \qquad (20.82)$$

Substitution of Eq. 20.82 into Eq. 20.51 then produces the required expression.

20.2 In all our analyses of diffusion-limited layer growth, we have assumed that the interdiffusivities in the various phases were constants independent of concentration. In each case, the interfaces between phases were found to move parabolically with time. Suppose that assumption is relaxed and the interdiffusivities are allowed to vary with concentration. Will the interfaces still move parabolically?

Solution. Yes. When $\tilde{D}$ varies with concentration we have shown in Section 4.2.2 that the diffusion equation can be scaled (transformed) from xt-space to η-space by using the variable $\eta = x/\sqrt{4Dt}$ (see Eq. 4.19). Also, under diffusion-limited conditions where fixed boundary conditions apply at the interfaces, the boundary conditions can also be transformed to η-space, as we have also seen. Therefore, when $\tilde{D}$ varies with concentration, the entire layer-growth boundary-value problem can be transformed into η-space. Since the fixed boundary conditions at the interfaces require constant values of η at the interfaces, they will move parabolically.

20.3 For the layer system shown in Fig. 20.5, show that the β-phase layer will grow linearly when ϕ_1 is small and ϕ_2 is large (see Eqs. 20.31 and 20.32).

Solution. According to the arguments in Section 20.1.4, when ϕ_1 is small and ϕ_2 is large, $c^{\beta\alpha} \cong c_{eq}^{\beta\alpha}$ and $c^{\beta\gamma} \cong c^{\beta\alpha} \cong c_{eq}^{\beta\alpha}$. Then, by combining Eqs. 20.26 and 20.27,

$$\frac{d\chi_1}{dt} \left(c_{eq}^{\beta\alpha} - c_{eq}^{\alpha\beta}\right) = \kappa_2 \left(c_{eq}^{\beta\alpha} - c_{eq}^{\beta\gamma}\right) \qquad (20.83)$$

and by combining Eqs. 20.27 and 20.28,

$$\frac{d\chi_2}{dt} \left(c_{eq}^{\beta\alpha} - c_{eq}^{\gamma\beta}\right) = \kappa_2 \left(c_{eq}^{\beta\alpha} - c_{eq}^{\beta\gamma}\right) \qquad (20.84)$$

When Eqs. 20.83 and 20.84 are combined to form $d\chi/dt = d(\chi_1 - \chi_2)/dt$, it is seen that $d\chi/dt = $ constant and the growth will therefore be linear.

20.4 Find an expression for the thickness of the growing B-rich β-phase surface layer shown in Fig. 20.3 as a function of time. Details of the growth of this layer have been discussed in Section 20.1.2, and a strategy for determining the growth rate has been outlined. Assume constant diffusivities in both phases.

Solution. Solutions to the diffusion equation in the α and β phases, which match the boundary conditions, are

$$\frac{c_B^\beta - c_B^{\beta 0}}{c_B^{\beta \alpha} - c_B^{\beta 0}} = \frac{\mathrm{erf}(x/\sqrt{4\tilde{D}^\beta t}) - \mathrm{erf}(A_2/\sqrt{4\tilde{D}^\beta})}{\mathrm{erf}(A_1/\sqrt{4\tilde{D}^\beta}) - \mathrm{erf}(A_2/\sqrt{4\tilde{D}^\beta})} \tag{20.85}$$

$$\frac{c_B^\alpha - c_B^{\alpha \infty}}{c_B^{\alpha \beta} - c_B^{\alpha \infty}} = \frac{1 - \mathrm{erf}(x/\sqrt{4\tilde{D}^\alpha t})}{1 - \mathrm{erf}(A_1/\sqrt{4\tilde{D}^\alpha})} \tag{20.86}$$

where

$$A_1 = \frac{\chi_1}{\sqrt{t}} \quad \text{and} \quad A_2 = \frac{\chi_2}{\sqrt{t}} \tag{20.87}$$

The Stefan conditions at interfaces 1 and 2 are given by Eqs. 20.12 and 20.17, respectively. Substituting the appropriate relationships from those given above into these two equations then yields equations that can be solved simultaneously for A_1 and A_2:

$$
A_1 = \frac{Q^\beta \sqrt{4\tilde{D}^\beta}}{\sqrt{\pi}} \frac{c_B^{\beta 0} - c_B^{\beta \alpha}}{c_B^{\beta \alpha} - c_B^{\alpha \beta}} \frac{e^{-A_1^2/(4\tilde{D}^\beta)}}{\mathrm{erf}\left(A_1/\sqrt{4\tilde{D}^\beta}\right) - \mathrm{erf}\left(A_2/\sqrt{4\tilde{D}^\beta}\right)}
$$
$$
- \frac{Q^\alpha \sqrt{4\tilde{D}^\alpha}}{\sqrt{\pi}} \frac{c_B^{\alpha \beta} - c_B^{\alpha \infty}}{c_B^{\beta \alpha} - c_B^{\alpha \beta}} \frac{e^{-A_1^2/(4\tilde{D}^\alpha)}}{1 - \mathrm{erf}\left(A_1/\sqrt{4\tilde{D}^\alpha}\right)} \tag{20.88}
$$

$$
A_2 = \frac{\sqrt{4\tilde{D}^\beta}}{\sqrt{\pi}} \frac{\Omega_B^\beta}{1 - \Omega_B^\beta c_B^{\beta 0}} \frac{\left(c_B^{\beta \alpha} - c_B^{\beta 0}\right) e^{-A_2^2/(4\tilde{D}^\beta)}}{\mathrm{erf}\left(A_1/\sqrt{4\tilde{D}^\beta}\right) - \mathrm{erf}\left(A_2/\sqrt{4\tilde{D}^\beta}\right)} \tag{20.89}
$$

Finally, the layer thickness is given by $\chi = \chi_1 - \chi_2 = (A_1 - A_2)\sqrt{t}$.

20.5 Using the scaling method, find an expression for the diffusion-limited rate of growth of a cylindrical B-rich precipitate growing in an infinite α-phase matrix. Assume the same boundary conditions as in the analysis in Section 20.2.1 (Eqs. 20.37–20.39) for the growth of a spherical particle. Note that Fig. 20.6, which applied to the growth of a spherical particle in Section 20.2.1, will also apply. Use the scaling parameter $\eta = r/(4\tilde{D}^\alpha t)^{1/2}$. You will need the integral

$$E_1(u^2) = \int_{u^2}^\infty e^{-\xi} \frac{d\xi}{\xi} \tag{20.90}$$

which has been tabulated [29].

Solution. Starting with the diffusion equation in cylindrical coordinates (see Eq. 5.8) and using the scaling parameter to change variables, the diffusion equation in η-space becomes

$$\frac{d^2 c_B^\alpha(\eta)}{d\eta^2} + \left(\frac{1}{\eta} + 2\eta\right) \frac{dc_B^\alpha(\eta)}{d\eta} = 0 \tag{20.91}$$

This result, along with the boundary conditions given by Eqs. 20.37–20.39, shows that the particle will grow parabolically according to

$$R(t) = \eta_R \sqrt{4\tilde{D}^\alpha t} \tag{20.92}$$

Integrating Eq. 20.91 once,

$$\frac{dc_B^\alpha}{d\eta} = a_1 \frac{e^{-\eta^2}}{\eta} \tag{20.93}$$

where $a_1 = $ constant. Integrating again yields

$$c_B^{\alpha\infty} - c_B^\alpha = a_1 \int_\eta^\infty e^{-y^2} \frac{dy}{y} \tag{20.94}$$

Determining a_1 from the condition $c_B^\alpha = c_B^{\alpha\beta}$ when $\eta = \eta_R$,

$$\frac{c_B^{\alpha\infty} - c_B^\alpha}{c_B^{\alpha\infty} - c_B^{\alpha\beta}} = \frac{\int_{\eta^2}^\infty e^\eta \frac{d\xi}{\xi}}{\int_{\eta_R^2}^\infty e^{-\eta} \frac{d\xi}{\xi}} = \frac{E_1(\eta^2)}{E_1(\eta_R^2)} \tag{20.95}$$

The Stefan condition at the interface is

$$v\left(c_B^{\beta\alpha} - c_B^{\alpha\beta}\right) = \widetilde{D}^\alpha \left(\frac{dc_B^\alpha}{dr}\right)_{r=R} \tag{20.96}$$

or, in η-space,

$$2\eta_R \left(c_B^{\beta\alpha} - c_B^{\alpha\beta}\right) = \left(\frac{dc_B^\alpha}{d\eta}\right)_{\eta=\eta_R} \tag{20.97}$$

Finally, substituting Eq. 20.95 into Eq. 20.97, we have

$$\eta_R^2 \, e^{\eta_R^2} \, E_1\left(\eta_R^2\right) = \frac{c_B^{\alpha\infty} - c_B^{\alpha\beta}}{c_B^{\beta\alpha} - c_B^{\alpha\beta}} \tag{20.98}$$

for the determination of η_R.

20.6 Find an expression for the rate of thickening of a B-rich β-phase precipitate platelet in an infinite α-phase matrix in a A/B binary system as in Fig. 20.6. Assume diffusion-limited conditions and a constant diffusivity, $\widetilde{D}^\alpha$, in the α-phase matrix. Also, assume that the atomic volume of each species is constant throughout so that there is no overall volume change and that the plate is extensive enough so that edge effects can be neglected. Use the scaling method.

Solution. Let x be the distance coordinate perpendicular to the platelet. The boundary conditions will be the same as Eqs. 20.37–20.39 if $r \to x$ and $\eta \to x/\sqrt{4\widetilde{D}^\alpha t}$. The method is basically the same as that used to obtain the solution for the sphere in Section 20.2.1. In the present case, $\chi(t)$ will be the half-thickness of the plate. η will be constant at the interface at the value $\eta_\chi = \chi(t)/\sqrt{4\widetilde{D}^\alpha t}$, so that $\chi(t)$ will increase parabolically with time according to

$$\chi(t) = \eta_\chi \sqrt{4\widetilde{D}^\alpha t} \tag{20.99}$$

We now determine η_χ by solving for c_B^α and invoking the Stefan condition at the interface. The diffusion equation was scaled and integrated in Cartesian coordinates in Section 4.2.2 with the solution given by Eq. 4.28. When this solution is matched to the present boundary conditions,

$$\frac{c_B^{\alpha\infty} - c_B^\alpha}{c_B^{\alpha\infty} - c_B^{\alpha\beta}} = \frac{\mathrm{erfc}(\eta)}{\mathrm{erfc}(\eta_\chi)} \tag{20.100}$$

The Stefan condition at the interface is

$$\frac{d\chi}{dt}\left(c_B^{\beta\alpha} - c_B^{\alpha\beta}\right) = \widetilde{D}^\alpha \left(\frac{\partial c_B^\alpha}{\partial x}\right)_{x=\chi} \tag{20.101}$$

Use of Eqs. 20.99 and 20.100 in Eq. 20.101 then yields the desired expression for η_χ,

$$\eta_\chi = \frac{\left(c_B^{\alpha\infty} - c_B^{\alpha\beta}\right) e^{-\eta_\chi^2}}{\left(c_B^{\beta\alpha} - c_B^{\alpha\beta}\right) \sqrt{\pi}\, \mathrm{erfc}\,(\eta_\chi)} \tag{20.102}$$

20.7 Consider again the problem posed in Exercise 20.6, whose solution had the form of a transcendental equation. A simple and useful approximate solution can be found by using the linear approximation to the diffusion profile shown in Fig. 20.14. Find the solution based on this approximation.

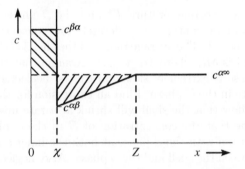

Figure 20.14:   Approximate composition vs. distance profile for determination of the diffusion-limited thickening of a slab of thickness 2χ.

Solution. The Stefan condition at the interface is

$$\widetilde{D}^\alpha \left(\frac{\partial c}{\partial x}\right)_{x=\chi} = \frac{d\chi}{dt}\left(c_B^{\beta\alpha} - c_B^{\alpha\beta}\right) \tag{20.103}$$

Using the linear approximation in Fig. 20.14, $\partial c/\partial x = \left(c_B^{\alpha\infty} - c_B^{\alpha\beta}\right)/(Z - \chi)$. Also, the conservation of B atoms requires that the two shaded areas in the figure be equal. Therefore, $\left(c_B^{\beta\alpha} - c_B^{\alpha\infty}\right)\chi = \left(c_B^{\alpha\infty} - c_B^{\alpha\beta}\right)(Z-\chi)/2$. Putting these relationships into Eq. 20.103 gives

$$\chi\, d\chi = \frac{\widetilde{D}^\alpha \left(c_B^{\alpha\infty} - c_B^{\alpha\beta}\right)^2}{2\left(c_B^{\beta\alpha} - c_B^{\alpha\beta}\right)\left(c_B^{\beta\alpha} - c_B^{\alpha\infty}\right)}\, dt \tag{20.104}$$

Therefore,

$$\chi = \frac{c_B^{\alpha\infty} - c_B^{\alpha\beta}}{\sqrt{c_B^{\beta\alpha} - c_B^{\alpha\beta}}\sqrt{c_B^{\beta\alpha} - c_B^{\alpha\infty}}}\sqrt{\widetilde{D}^\alpha t} \tag{20.105}$$

Zener has compared the approximate solution above with the exact solution found in Exercise 20.6 and finds reasonably good agreement [30].

20.8 Consider a binary system consisting of a thin spherical shell of α phase embedded in an infinite body of β phase. The phase diagram is shown in Fig. 20.15.

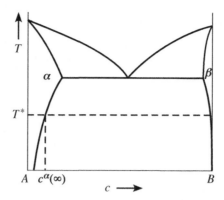

Figure 20.15:   Binary phase diagram.

The system is at the temperature T^* and the β phase is essentially pure B, while the α phase contains a moderately low concentration of B so that Henry's law is obeyed. The average radius of the shell is $\langle R \rangle$ and the thickness of the thin shell is δR, where $\delta R \ll \langle R \rangle$. Assume that the diffusion rate of B in the β phase is extremely slow and can be neglected in comparison to its diffusion rate in the α phase. Find an expression for the shrinkage rate of the shell and show that the shell will shrink at a rate inversely proportional to $\langle R \rangle$. Assume that the concentration of B in the α phase is maintained in local equilibrium with the β phase at both the inner and outer interphase boundaries between the shell and the β phase. Also, neglect any small volume changes that might occur.

Solution. The outer interface is concave and the inner interface is convex with respect to the β phase. The concentrations of B maintained in equilibrium in the α phase at the outer interface, $c_{\text{out}}^{\alpha\beta}$, and at the inner interface, $c_{\text{in}}^{\alpha\beta}$, are given by Eq. 15.4. The concentration difference across the shell is therefore

$$\Delta c^\alpha = c_{\text{out}}^{\alpha\beta} - c_{\text{in}}^{\alpha\beta}$$
$$= c^\alpha(\infty) \left[1 - \frac{2\gamma\Omega}{kT\langle R \rangle} \right] - c^\alpha(\infty) \left[1 + \frac{2\gamma\Omega}{kT\langle R \rangle} \right] = -\frac{4\gamma\Omega c^\alpha(\infty)}{kT\langle R \rangle} \qquad (20.106)$$

The diffusion of B through the shell will be in a quasi-steady state, and since $\delta R \ll \langle R \rangle$, the flux through the shell can be expressed to a good approximation as

$$J = -\widetilde{D}^\alpha \frac{dc^\alpha}{dr} = -\widetilde{D}^\alpha \frac{\Delta c^\alpha}{\delta R} = \frac{4\widetilde{D}^\alpha \gamma \Omega c^\alpha(\infty)}{kT\langle R \rangle (\delta R)} \qquad (20.107)$$

The flux is therefore directed toward the outer interface, causing the spherical shell to shrink toward its center under the driving force supplied by the decrease in interfacial energy that occurs as a result of the shrinkage. (Note that ultimately the shell will shrink to form a solid sphere of α phase at the origin.)

Since $\Delta c^\alpha \ll c^\alpha(\infty)$, the equations of continuity (for Stefan conditions, see Section 20.1.2) at the inner and outer interfaces can be expressed as

$$v_{\text{in}} c_{\text{in}}^{\alpha\beta} = v_{\text{in}} c^\beta + J \qquad (20.108)$$

$$v_{\text{out}} c^\beta + J = v_{\text{out}} c_{\text{out}}^{\alpha\beta} \qquad (20.109)$$

where $v_{\rm in}$ and $v_{\rm out}$ are the velocities of the inner and outer interfaces, respectively. Using these relationships and neglecting small differences between $c_{\rm in}^{\alpha\beta}$, $c_{\rm out}^{\alpha\beta}$, and $c^{\alpha}(\infty)$, the average shell velocity is then

$$\langle v \rangle = \frac{v_{\rm in} + v_{\rm out}}{2} = -\frac{J}{c^{\beta} - c^{\alpha}(\infty)} = -\frac{4\widetilde{D}^{\alpha}\gamma\Omega c^{\alpha}(\infty)}{kT(\delta R)[c^{\beta} - c^{\alpha}(\infty)]}\frac{1}{\langle R \rangle} \qquad (20.110)$$

CHAPTER 21

CONCURRENT NUCLEATION AND GROWTH TRANSFORMATION KINETICS

A discontinuous transformation generally occurs by the concurrent nucleation and growth of the new phase (i.e., by the nucleation of new particles and the growth of previously nucleated ones). In this chapter we present an analysis of the resulting overall rate of transformation. Time–temperature–transformation diagrams, which display the degree of overall transformation as a function of time and temperature, are introduced and interpreted in terms of a nucleation and growth model.

21.1 OVERALL RATE OF DISCONTINUOUS TRANSFORMATION

Consider homogeneous nucleation in a three-dimensional system. In the simplest model, these nuclei form at random locations. The nucleation rate, J, specifies the number of nuclei forming per unit volume per unit time. These nuclei form at locations that have not already been transformed by growth of any previously formed nuclei. Once nucleated, a particle grows at a rate $\dot{R} = dR/dt$ and the untransformed volume decreases. However, no particle can grow indefinitely. A particle nucleated near a surface can grow to impingement with the surface, and the transformation at that location will cease. Similarly, two particles that nucleate near each other grow until they impinge and transformation ceases. Alternatively, growth can be driven by supersaturation and individual nuclei could have their growth limited by the decreasing supersaturation in the untransformed volume.

In previous chapters we have developed models for discontinuous transformations that treat nucleation and growth processes independently. However, when

Kinetics of Materials. By Robert W. Balluffi, Samuel M. Allen, and W. Craig Carter. **533**
Copyright © 2005 John Wiley & Sons, Inc.

these processes occur concurrently, the overall transformation rate (i.e., the volume transformed per unit time) and microstructural characteristics such as particle or grain size depend on the interplay of nucleation and growth processes.

The theory of the kinetics of concurrent nucleation and growth reactions has a rich history that includes work by Kolmogorov [1], Johnson and Mehl [2], Avrami [3–5], Jackson [6], and Cahn [7]. Cahn's *time-cone method* for treating a class of these problems is the most general of these, with the most transparent assumptions, and is presented here. The method of Johnson, Mehl, and Avrami is covered in Section 4 of Christian's text [8].

21.1.1   Time-Cone Analysis of Concurrent Nucleation and Growth

The key to obtaining exact solutions to the transformation kinetics is to make explicit assumptions about the *statistical homogeneity* of nucleation and growth processes in the system. Following Cahn, we denote homogeneous nucleation on randomly dispersed sites in a volume by *volume nucleation*; heterogeneous nucleation on randomly dispersed sites on a surface or interface in a volume by *surface nucleation*, and heterogeneous nucleation on randomly dispersed sites along a linear feature in a volume by *line nucleation*. Exact solutions for transformation kinetics can be obtained when nucleation and growth rates are spatially homogeneous at any instant within the system of interest. The method is applicable to finite samples with a wide range of geometries and can yield position-dependent transformation kinetics.

Statistical homogeneity is necessary for application of the time-cone method. Statistical homogeneity is *not* a valid assumption during precipitation from supersaturated solution because the untransformed regions develop concentration gradients around growing particles and hence the nucleation rate becomes nonuniform. Also, in a material with thermal gradients undergoing a discontinuous transformation (e.g., during continuous cooling), the nucleation rate will be nonuniform. However, the theory *is* applicable to discontinuous transformations in which the parent and product phases have the same composition and in which the temperature is essentially uniform at any instant. Examples include recrystallization, first-order order–disorder transformations, massive transformations, and crystallization during vapor-deposition processes.

Probability that a Point $\vec{r}$ will not be Transformed at Time t.   The probability that a point $\vec{r}$ in the sample will be untransformed at the time t is obtained by computing the probability that no nuclei had formed at any location $\vec{r}\,'$ and any previous time τ that could have grown and led to prior transformation at $\vec{r}$ and t. This is accomplished in three steps:

  i. The set of points that could possibly affect a given point grows with time; therefore, to specify this set of points, a time coordinate is required in addition to the spatial coordinates of the sample. For nucleation and growth in a sample of dimensionality 3, the augmented space has Cartesian coordinates x, y, z, and t; more generally, coordinates $\vec{r}$ and t for any spatial dimension. Emanating from the point $(\vec{r}, t)$ and extending to earlier times is a domain that is the set of all points in the augmented space that would have caused transformation at $(\vec{r}, t)$ if nucleation had occurred at a nearby point and earlier time $(\vec{r}', \tau)$. This subset of points is called V_c. For the $d = 2$ case, the domain V_c is a cone of height t; Cahn refers to this domain as the *time cone*. The cross section of the time cone at any time depends on the growth rate function $\dot{R}(t)$. Figure 21.1 is a representation of the transformation by random nucleation along a one-dimensional sample for the constant-growth-rate case.

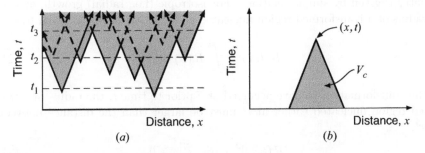

(a) Distance, x (b) Distance, x

Figure 21.1: Nucleation and growth along a one-dimensional specimen. (*a*) Growth cones. The apex of each cone coincides with the time and location of a nucleation event. Once nucleated, constant linear growth transforms the region surrounding a nucleus. Two nuclei have formed at t_1; at t_2 the sample is approximately half transformed; at t_3 it is fully transformed. Transforming regions impinge where growth cones intersect. (*b*) Time cone for the point (x, t). For the point (x, t) to be untransformed at time t, no prior nucleation can have occurred within the time cone volume, V_c.

ii. The nucleation rate can be integrated over the time cone to obtain the number of nuclei expected in V_c, denoted by $\langle N \rangle_c$.

iii. The untransformed fraction $1 - \zeta$ is obtained from the *stochastic independence* of the nucleation events—that is, any particular nucleation event in untransformed material is not influenced by any other nucleation event. Under these conditions, the theory can be formulated using the *Poisson probability equation* [9], which states that if $\mu =$ the mean rate at which events occur, then the probability, $p(k)$, that exactly k events occur in time t is

$$p(k) = e^{-\mu t} \frac{(\mu t)^k}{k!} \tag{21.1}$$

The probability that *exactly zero* events occurs is given by Eq. 21.1 as

$$p(0) = e^{-\mu t} \tag{21.2}$$

and the probability that *at least one* event that leads to some transformation is

$$p(k \geq 1) = 1 - P(0) = 1 - e^{-\mu t} \tag{21.3}$$

In the context of nucleation and growth kinetics, the quantity μ corresponds to the nucleation rate, J; the product μt corresponds to the number of nucleation events, $\langle N \rangle_c$, in the time cone; and the probability that exactly zero nucleation will have occurred within the time cone is equal to the untransformed volume fraction at time t, $1 - \zeta$, giving the relation

$$1 - \zeta = e^{-\langle N \rangle_c} \tag{21.4}$$

Equation 21.4 forms the basis of the theory for the kinetics of concurrent nucleation and growth transformations. Specific cases can be formulated by deriving appropriate expressions for the quantity $\langle N \rangle_c$.

Time Cone V_c for Isotropic, Time-Dependent Growth Rate $\dot{R}(t)$. The time cone's geometry is given by simple relations. For isotropic (i.e., radial) growth, at time t the radius of a transformed region nucleated at an earlier time τ is given by

$$R(t, \tau) = \int_{\tau}^{t} \dot{R}(t') \, dt' \tag{21.5}$$

For the transformation to have occurred at $\vec{r}$ prior to time t, the radius of a transformed region nucleated earlier at $\vec{r'}$ must be larger than the distance between $\vec{r}$ and $\vec{r'}$,

$$[R(t, \tau)]^2 - \left| \vec{r} - \vec{r'} \right|^2 \geq 0 \tag{21.6}$$

The time cone is the set of points Eq. 21.6.

Number of Nuclei Expected in the Time Cone, $\langle N \rangle_c$. For time-dependent nucleation rates $J(t)$ and isotropic growth rates $\dot{R}(t)$ (such as in nonisothermal transformations under conditions in which thermal gradients can be neglected), the number of nuclei in V_c is given for the $d = 3$ case as

$$\langle N \rangle_c(t) = \frac{4\pi}{3} \int_0^t J(t') \left[R(t, t') \right]^3 dt' \tag{21.7}$$

Expressions for Transformation Rate when Nucleation and Growth Rates are Constant. If the growth velocity $\dot{R}$ is isotropic and constant, Eq. 21.5 can be integrated and the time cone is the set of points $\vec{r}$ that obey

$$\dot{R}^2 (t - \tau)^2 - \left| \vec{r} - \vec{r'} \right|^2 \geq 0 \tag{21.8}$$

The radius of the time cone, $|\vec{r} - \vec{r'}|$, is linear in time, and hence the time cone will be a right circular cone of height $t - \tau$. The volume of the time cone, which Cahn calls the *nucleation volume V_c*, for transformation in a system of dimensionality d is given by

$$V_c = \frac{B(t - \tau)}{d + 1} \tag{21.9}$$

where B is an appropriate measure of the base of the cone. Taking $\tau = 0$ as the earliest time that nucleation becomes possible, the following expressions are obtained for systems of dimensionalities 1, 2, and 3:

$$\begin{array}{lll}
B = 2\dot{R}t & V_c = \dot{R}t^2 & (d = 1) \\
B = \pi\dot{R}^2 t^2 & V_c = \frac{\pi}{3}\dot{R}^2 t^3 & (d = 2) \\
B = \frac{4\pi}{3}\dot{R}^3 t^3 & V_c = \frac{\pi}{3}\dot{R}^3 t^4 & (d = 3)
\end{array} \tag{21.10}$$

Equation 21.7 takes a particularly simple form when the nucleation rate is a constant: $\langle N \rangle_c$ is equal to the product of J and V_c for the $d = 3$ case. This result can be generalized to systems of arbitrary dimensionality, d, by use of Eq. 21.9, giving

$$\langle N \rangle_c = \frac{JBt}{d+1} \tag{21.11}$$

Substituting the appropriate factors from Eq. 21.10 into Eqs. 21.11 and 21.4 gives expressions for the fraction transformed in one, two, and three dimensions for the case of constant nucleation and growth rates J and $\dot{R}$. The resulting expressions for the untransformed volume are

$$\begin{aligned}
1 - \zeta &= e^{-J\dot{R}t^2} & (d=1) \\
1 - \zeta &= e^{-(\pi/3)J\dot{R}^2t^3} & (d=2) \\
1 - \zeta &= e^{-(\pi/3)J\dot{R}^3t^4} & (d=3)
\end{aligned} \tag{21.12}$$

The function $\zeta(t)$ in Eq. 21.12 has a characteristic sigmoidal shape with a maximum rate of transformation at intermediate times. Examples are shown in Fig. 21.2. The $d = 3$ form of Eq. 21.12 is commonly known as the *Johnson–Mehl–Avrami equation*.

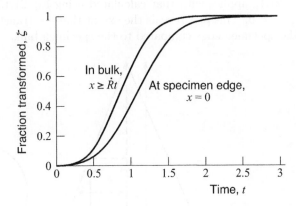

Figure 21.2:   Comparison of volume fraction transformed, ζ, in the interior of a semi-infinite thin-film specimen and at the specimen edge, $s = 0$. Calculations for $J = 1, \dot{R} = 1$.

21.1.2 Transformations near the Edge of a Thin Semi-Infinite Plate

Consider a semi-infinite thin plate that is effectively two-dimensional, lying in the xy-plane, with a single edge along $x = 0$. It is assumed that there is no heterogeneous line nucleation at the edge of the sheet. For constant-volume nucleation and growth rates and isotropic growth, points lying within the near-edge region $x < \dot{R}t$ require special consideration. In the bulk of the plate away from this region, the time cone will be a right circular cone of height t, and the fraction transformed will be given by the $d = 2$ form of Eq. 21.12. Close to the edge, the time cone will be truncated by the plane $x = 0$, its volume will be less than in the bulk, and the number of nuclei $\langle N \rangle_c$ contained in the truncated cone will decrease as $x \to 0$.[1] The transformation rate in the near-edge region will thus be slower than in the

[1]Nuclei cannot form outside the sample and hence cannot influence the transformation anywhere inside it.

bulk, and the grain size after completion of the transformation will be larger near the edge.

To apply Eq. 21.4, an expression for the volume of a right circular time cone of height t having its axis located at (s, y) is required. When $s > \dot{R}t$, the cone volume V_c is given by Eq. 21.10 for the $d = 2$ case, and the edge $s = 0$ has no influence on the transformation kinetics. When $s < \dot{R}t$, the time cone is truncated as in Fig. 21.3a. Its volume may be expressed in terms of quantities shown in Fig. 21.3b and c: its base radius, $\dot{R}t$, height, t, and the distance of the truncated face from the cone axis, s.

$$V_c = \begin{cases} \dfrac{1}{3\dot{R}} \left[\begin{array}{l} \pi \dot{R}^3 t^3 + 2\dot{R}st\sqrt{\dot{R}^2 t^2 - s^2} - \dot{R}^3 t^3 \sec^{-1}\left(\dot{R}t/s\right) \\ + s^3 \ln\left[s/(\dot{R}t + \sqrt{\dot{R}^2 t^2 - s^2}) \right] \end{array} \right] & \text{for } s \le \dot{R}t \\[2em] \dfrac{\pi}{3}\dot{R}^2 t^3 & \text{for } s > \dot{R}t \end{cases}$$

$$(21.13)$$

Figure 21.3a compares the volume fraction transformed, ζ, vs. time for a location $s \le \dot{R}t$ where Eq. 21.12 applies with that calculated using Eq. 21.13 evaluated at the specimen edge, $s = 0$. This plot reveals the extent that the transformation rate is reduced near the specimen edge compared to the specimen bulk.

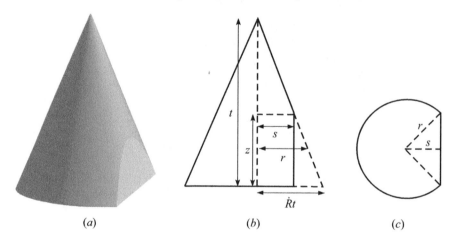

(a)   (b)   (c)

Figure 21.3:   (a) Truncated time cone.   (b) Vertical cross section of time cone.   (c) Horizontal cross section through truncated portion of time cone.

21.2   TIME–TEMPERATURE–TRANSFORMATION (TTT) DIAGRAMS

Useful insights into the kinetics of a phase transformation that proceeds by nucleation and growth can be obtained by observing the fraction transformed, ζ, under isothermal conditions at a series of different temperatures. This is usually done by undercooling rapidly to a fixed temperature and then observing the resulting isothermal transformation. The kinetics generally follows the typical C-shaped behavior described in Exercise 18.4. If a series of such curves is obtained at different temperatures, the time required to achieve, for example, $\zeta = 0.01$, 0.50, and

0.99 fractional transformation at each temperature can be obtained, and a *time–temperature–transformation* (TTT) *plot* of the form illustrated in Fig. 21.4 can then be constructed.

The C shapes of these curves are readily explained in a qualitative way on the basis of nucleation kinetics. As in Section 19.1.5, the overall activation energy for the rate of nucleation generally consists of the free energy to form a critical nucleus, $\Delta \mathcal{G}_c$, plus an activation energy associated with the factor β, G_c^m, which describes the rate at which atoms (or molecules) join a critical nucleus. The activation energy for nucleation is then $[\Delta \mathcal{G}_c + G_c^m]$ and the rate of nucleation is proportional to the factor $\exp[-(\Delta \mathcal{G}_c + G_c^m)/(kT)]$. At $T = T^{\mathrm{eq}}$, there is no undercooling, $\Delta \mathcal{G}_c \to \infty$, and the rate of nucleation is zero. The time thus required for any transformation is infinite at T^{eq}. As the transformation temperature is decreased, the undercooling and the transformation driving force increase, which causes a rapid decrease in $\Delta \mathcal{G}_c$ while G_c^m remains constant. This rapid decrease in $\Delta \mathcal{G}_c$ causes $[\Delta \mathcal{G}_c + G_c^m]$ to decrease more rapidly than kT, which, in turn, causes the factor $\exp[-(\Delta \mathcal{G}_c + G_c^m)/(kT)]$ to increase and the nucleation rate to increase. However, on further cooling, the rate of decrease of $\Delta \mathcal{G}_c$ slows and eventually kT decreases more rapidly than $[\Delta \mathcal{G}_c + G_c^m]$, causing the factor $\exp[-(\Delta \mathcal{G}_c + G_c^m))/(kT)]$ to decrease and the rate of nucleation to decrease. Eventually, the entire system becomes "frozen in" and the time required for any significant transformation again becomes essentially infinite.

An example of such TTT kinetics is solidification. According to Eq. 12.2,

$$\Delta G(T) = \frac{\Delta H(T_m)}{T_m} \Delta T \tag{21.14}$$

if it is assumed that the entropy and enthalpy of solidification are each independent of temperature below T_m. Here, ΔT is the undercooling and ΔH is the enthalpy of solidification. According to Eq. 19.4, $\Delta \mathcal{G}_c$ then has the form $\Delta \mathcal{G}_c = A/(\Delta T)^2$, where A is a constant independent of temperature. The nucleation rate below T_m then has the temperature dependence

$$J(T) \propto e^{-[G_c^m + A/(T_m - T)]/(kT)} \tag{21.15}$$

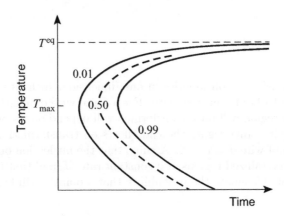

Figure 21.4: TTT diagram for a nucleation and growth phase transformation. At $T = T^{\mathrm{eq}}$ the parent phase is at equilibrium, and no undercooling is present. The three curves indicate the attainment of $\zeta = 0.01, 0.50,$ and 0.99 transformation, respectively.

The nucleation rate will therefore be a maximum at the temperature where the exponent in Eq. 21.15 is a minimum, which occurs at the temperature

$$T_{\max} = T_m \left[1 - \frac{\sqrt{1 + G_c^m T_m/A} - 1}{G_c^m T_m/A} \right] \qquad (21.16)$$

The nose of the C-curve in this case will therefore occur at $T_{\max}$, which typically lies between T_m and $T_m/2$.

Bibliography

1. A.N. Kolmogorov. Statistical theory of metal crystallization (in Russian). *Izv. Akad. Nauk SSSR*, 1:355–359, 1937.

2. W.A. Johnson and R.F. Mehl. Reaction kinetics in processes of nucleation and growth. *Trans. AIME*, 135:416–442, 1939. See also discussion on pp. 442–458.

3. M. Avrami. Kinetics of phase change—I. *J. Chem. Phys.*, 7(12):1103–1112, 1939.

4. M. Avrami. Kinetics of phase change—II. Transformation-time relations for random distribution of nuclei. *J. Chem. Phys.*, 8(2):212–224, 1940.

5. M. Avrami. Kinetics of phase change—III. Granulation, phase change, and microstructure. *J. Chem. Phys.*, 9(2):177–184, 1941.

6. J.L. Jackson. Dynamics of expanding inhibitory fields. *Science*, 183(4123):446–447, 1974.

7. J.W. Cahn. The time cone method for nucleation and growth kinetics on a finite domain. In *Thermodynamics and Kinetics of Phase Transformations*, volume 398 of *Materials Research Society Symposia Proceedings*, pages 425–438, Pittsburgh, PA, 1996. Materials Research Society.

8. J.W. Christian. *The Theory of Transformations in Metals and Alloys*. Pergamon Press, Oxford, 1975.

9. E. Parzen. *Modern Probability Theory and Its Applications*. John Wiley & Sons, New York, 1960.

EXERCISES

21.1 Consider transformation kinetics in one dimension, such as recrystallization (see Section 13.1) in a narrow wire. For a finite wire of length L, the probability that a region will have transformed will depend on its proximity to the end of the wire. Investigate the end effects on transformation kinetics on a finite length of wire $0 < x < L$. Assume that the nucleation occurs uniformly in the unrecrystallized regions at a constant rate, J, and that the growth rate $\dot{R}$ is constant. Calculate the probability that a point x will have transformed at time t.

- The solution should be symmetric around $x = L/2$. There are three separate cases to consider; one of them is $x < \dot{R}t$ and $L - x > \dot{R}t$ (see Fig. 21.5).

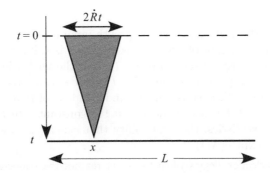

Figure 21.5: Length–time area of influence for a point x.

Solution. Poisson statistics apply when events are random and mutually independent, which is assumed to be the case both in time and along the wire. The probability $p(n, A)$ that n events occur in an "area" (length × time) A with event rate J is given by the Poisson distribution

$$p(n, A) = e^{-JA} \frac{(JA)^n}{n!} \tag{21.17}$$

Therefore, the probability that *no* event occurs is $p(0, A) = e^{-JA}$, and the probability that some (one or more) events occur is $1 - p(0, A)$, which is equal to the probability that a region will have transformed. Therefore, the problem depends only on the area of the time cone illustrated in Fig. 21.5.

Case 1: Very short times or effectively infinite L. There is no interference from the boundaries. The condition for this case is $x > \dot{R}t$. The area of the time cone is $A_1 = \dot{R}t \times t = \dot{R}t^2$. The probability that a point x will have transformed is independent of x when $\dot{R}t < x < L/2$:

$$p(n \geq 1, A_1) = 1 - e^{J\dot{R}t^2} \tag{21.18}$$

Case 2: Near the end of a finite wire. There is interference only from the boundary at $x = 0$. The condition for this case is $L > \dot{R}t$ and $x < \dot{R}t$ (or, in a slightly different form, $x < \dot{R}t$ and $x < L - \dot{R}t$). The area of the time cone, A_2, is A_1 minus the area where $x < 0$:

$$A_2 = A_1 - \frac{\dot{R}t - x}{2} \frac{\dot{R}t - x}{\dot{R}} = \frac{\dot{R}^2t^2 + 2\dot{R}tx - x^2}{2\dot{R}} \tag{21.19}$$

so that $p(n \geq 1, A_2) = 1 - e^{JA_2}$, or

$$p(n \geq 1, A_2) = 1 - [1 - p(n \geq 1, A_1)]e^{J(\dot{R}t - x)^2/(2\dot{R})} \tag{21.20}$$

Case 3: Very short wire or long times. There is interference from both boundaries. The condition for this case is $L < \dot{R}t$ (or $x < \dot{R}t$ and $x < L - \dot{R}t$). The area of the time cone, A_3, is A_1 minus the area where $X > L$:

$$A_3 = A_1 - \frac{\dot{R}t - (L - x)}{2} \frac{\dot{R}t - (L - x)}{\dot{R}} = \frac{2(L\dot{R}t + Lx) - (L^2 + 2x^2)}{2\dot{R}} \tag{21.21}$$

so that $p(n \geq 1, A_3) = 1 - e^{-JA_3}$, or

$$p(n \geq 1, A_3) = 1 - [1 - p(n \geq 1, A_3)] e^{[2J(L\dot{R}t + Lx) - (L^2 + 2x^2)]/(2\dot{R})} \tag{21.22}$$

21.2 Consider recrystallization and grain growth in an infinite thin sheet. Assume that the nucleation rate of recrystallized grains is a linear function of temperature above a critical temperature, T_c, and the nucleation rate is zero for $T < T_c$ [i.e., at temperatures above T_c, $J = \alpha(T - T_c)$]. Also assume that the grain-growth rate, $\dot{R}$, is constant and independent of temperature. Suppose that at time $t = 0$ the sheet is heated at the constant rate $T(t) = T_c/2 + \beta t$. Using Poisson statistics, the probability that exactly zero events occur in a time t is $p_0 = \exp(-\langle N_c \rangle)$.

Solution. Nucleation begins when the sheet reaches the temperature T_c; the time to reach T_c is $T_c/(2\beta)$. The time cone in this two-dimensional sample, for a constant growth rate, is the right circular cone V_c given by

$$V_c = \pi \dot{R}^2 t^3 \tag{21.23}$$

and the expected number of nuclei expected in V_c at time t is

$$\begin{aligned}
\langle N \rangle_c &= \int_{T_c/2\beta}^{t} J(t)\, V_c \, d\tau = \int_{T_c/2\beta}^{t} \alpha \left(\beta\tau - \frac{T_c}{2} \right) \pi \dot{R}^2 t^3 \, d\tau \\
&= \frac{\alpha \pi \dot{R}^2 (T_c - 2\beta t)^2 (T_c + 2\beta t)(T_c^2 + 4\beta^2 t^2)}{128\beta^4}
\end{aligned} \tag{21.24}$$

The volume fraction transformed, ζ, is given by

$$\zeta = 1 - \exp(-\langle N \rangle_c) \tag{21.25}$$

CHAPTER 22

SOLIDIFICATION

The mechanisms of the motion of liquid/crystal interfaces during solidification were discussed in Section 12.3, and aspects of the heat-conduction-controlled motion of liquid/solid interfaces and their morphological stability under various solidification conditions were treated in Chapter 20. This sets the stage for considering the entire process of the solidification of a body of liquid into a solid.

Solidification results in a wide range of structures depending upon the type of material and the conditions under which the solidification occurs [1–3]. Although a huge literature describes and analyzes the many phenomena involved, we restrict ourselves to two cases: when the liquid/solid interface is stable and *plane-front solidification* is achieved, and when the interface is unstable and cellular or dendritic growth occurs. The first mode of solidification has important uses as a method of removing impurities or producing uniform distributions of solute atoms in materials. The second is prevalent in the casting of many alloys and has been the subject of a vast amount of study.

22.1 PLANE-FRONT SOLIDIFICATION IN ONE DIMENSION

22.1.1 Scheil Equation

In Fig. 22.1a one-dimensional solidification is depicted; a liquid binary alloy initially of uniform composition c_0 is placed in a bar-shaped crucible of length L. The bar is progressively cooled from one end, so it solidifies from one end to the other

Kinetics of Materials. By Robert W. Balluffi, Samuel M. Allen, and W. Craig Carter.
Copyright © 2005 John Wiley & Sons, Inc.   **543**

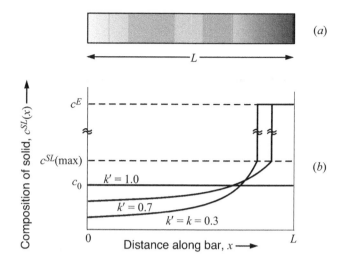

Figure 22.1: (a) Bar-shaped specimen after plane-front solidification. Grayscale intensity indicates solute composition variation. (b) Composition distribution along the bar after plane-front solidification with different effective partition ratios. From Flemings [2].

with a stable and planar liquid/solid interface (see the discussion of stability in Section 20.3). The eutectic phase diagram in Fig. 20.4a shows the liquidus and solidus approximated as straight lines. It is assumed that the interface is a good source or sink (Section 13.4) and that the liquid and solid are essentially in local equilibrium at the interface. As described in Section 20.1.3, the first solid to form is of composition $c_0{}^{SL}$, considerably less than c_0. Excess solute is rejected by the solidifying material and deposited in the liquid because diffusion into the solid is usually negligible. However, for the solute to enter the liquid and become dispersed, it must first diffuse through the liquid boundary layer adjacent to the liquid/solid interface. Because, in this layer, any flow in the liquid is lamellar flow parallel to the interface, solute atoms can pass through it only by diffusional transport. Once the solute has diffused through the layer of thickness δ, it is quickly mixed into the remaining bulk liquid by convection, and it may be assumed that the concentration in the liquid is uniform up to the boundary layer. The boundary layer therefore acts as a "diffusion barrier" and limits the rate at which the rejected solute can be dispersed throughout the liquid. As a result, the solute concentration in the liquid builds up in front of the advancing interface. However, at some time after the onset of solidification, the distribution of solute in the boundary layer reaches a quasi-steady state, as shown in Fig. 20.4b, where a "spike" of concentration has formed. It may be assumed that local equilibrium prevails at the liquid/solid interface, and if the temperature there is $T^{L/S}$, the phase diagram in Fig. 20.4a indicates that the solute concentration in the liquid directly at the interface, c^{LS}, must be related to the concentration of the solid being formed at the interface, c^{SL}, by $c^{SL} = kc^{LS}$. The quantity k, the ratio of the solidus concentration to the corresponding liquidus concentration, is known as the *partition ratio*. When the slopes of the solidus and liquidus curves are constant, as in Fig. 20.4a, the partition ratio is constant.

As solidification continues and solute is continuously rejected into the remaining liquid, the concentration in the bulk liquid increases slowly and the quasi-steady-state solute distribution in the boundary layer evolves. This, in turn, produces corresponding changes in the concentration in the solid being formed. Of primary interest is the final distribution of solute in the solid.

Because the solute diffusivity in the solid is far smaller than in the liquid, any diffusion in the solid will be neglected. In most cases of interest, the transient period required to produce a quasi-steady-state solute distribution at the interface is relatively small.[1] At a relatively short time after the establishment of the quasi-steady-state concentration spike, the flux relative to an origin at the interface moving at velocity v is

$$J^L = -D^L \frac{\partial c^L}{\partial x} - vc^L \tag{22.1}$$

The diffusion equation in the liquid is

$$\frac{\partial c^L}{\partial t} = -\nabla \cdot J^L = D^L \frac{d^2 c^L}{dx^2} + v \frac{dc^L}{dx} = 0 \tag{22.2}$$

which has the general solution

$$c^L(x) = a_1 + a_2 e^{-vx/D^L} \tag{22.3}$$

where x is the distance from the interface. The two constants a_1 and a_2 can be evaluated through use of the boundary conditions[2] $c^L(0) = c^{LS}$ and $c^L(\delta) = c_0$, with the result

$$c^L(x) = \frac{c_0 - c^{LS} e^{-v\delta/(PD^L)} + (c^{LS} - c_0)e^{-vx/D^L}}{1 - e^{v\delta/D^L}} \tag{22.4}$$

Expressions for c^{LS} and c^{SL} can now be found by using the Stefan continuity condition at the interface

$$vc^{LS} + D^L \left[\frac{dc^L}{dx}\right]_{x=0} - vc^{SL} = 0 \tag{22.5}$$

Substituting Eq. 22.4 into Eq. 22.5 and using $c^{SL} = k\, c^{LS}$,

$$c^{SL} = \frac{kc_0 e^{v\delta/D^L}}{1 + k\left(e^{v\delta/D^L} - 1\right)} \tag{22.6}$$

The *effective partition ratio*, k', is defined as the ratio of the concentration in the solid being formed over the concentration in the bulk liquid. Therefore,

$$k' = \frac{c^{SL}}{c_0} = \frac{k}{k + (1 - k)e^{-v\delta/D^L}} \tag{22.7}$$

To obtain the final solute distribution in the solid, the relatively slow changes in the quasi-steady-state system must be considered. If the concentration spike moves forward by dx, the amount of solute that must be rejected into the bulk liquid is $(c^L_\infty - c^{SL})dx$, where the slowly changing concentration in the bulk liquid is

---

[1]When this is not true, the kinetics of the formation of the transient must be taken into account [4].
[2]Note that the amount of solute in the spike is negligible compared to that in the bulk liquid, and therefore the concentration in the bulk liquid remains approximately c_0.

represented by c^L_∞ (rather than its initial value c_0). The change in concentration in the liquid is then $dc^L_\infty = [(c^L_\infty - c^{SL})/(L - x)]\, dx$. Using $c^{SL} = k' c^L_\infty$,

$$\int_0^x \frac{(1 - k')\, d\xi}{L - \xi} = \int_{c_0}^{c^L_\infty} \frac{dy}{y} \tag{22.8}$$

which yields the *Scheil equation*,

$$c^{SL}(x) = k' c_0 \left(1 - \frac{x}{L}\right)^{k'-1} \tag{22.9}$$

Because any diffusion in the solid is neglected, $c^{SL}(x)$ in Eq. 22.9 represents the distribution of solute after the solidification.

Equation 22.9 has two limiting forms. When the parameter $\delta v / D^L \gg 1$, $k' = 1$ according to Eq. 22.7. This situation is encouraged by a lack of convection, a high solidification rate, and a slow rate of diffusion in the liquid. The concentration spike in the liquid is then strong and c^{LS} quickly reaches a level where, according to Eq. 22.9, $c^{SL} = c_0$ and the composition of the solid being formed and the composition of the bulk liquid are the same. On the other hand, when $\delta v / D^L \ll 1$, $k' = k$. There is then rapid mixing in the liquid, the diffusion barrier is nonexistent, and there can be a large difference between the compositions of the solid being formed and the bulk liquid, depending on the factor k. Some typical curves of $c^{SL}(x)$ vs. x under these different conditions are shown in Fig. 22.1*b* for the system in Fig. 20.4. When $k' < 1$, the composition of the solid α phase increases continuously and eventually reaches its maximum value, $c^{SL}(\text{max})$, when the liquid reaches the eutectic composition. After that, the solid forms as a eutectic and its average composition is the eutectic composition, c_E.

As shown by Fig. 22.1*b*, the concentrations of solute atoms are significantly reduced in the material that is solidified early in the solidification process when $k' < 1$. One-dimensional plane-front solidification can therefore be used as a method of purification. However, purification is carried more effectively out by modifying the process and adopting a zone-melting technique.

22.1.2 Zone Melting and Zone Leveling

In *zone melting*, a bar-shaped specimen as shown in Fig. 22.2*a* is first melted at one end to form a melted zone of length l. This zone is then moved along the entire specimen at a constant rate while keeping l constant. As it moves, it picks up solute atoms and eventually deposits them near the other end of the bar, thereby purifying one end. Each iteration of the process leads to increasing purity.

The zone is generally much longer than the width of the liquid boundary layer (i.e., $l \gg \delta$). When the zone moves a distance dx, the amount of solute gained by the zone is $(c_0 - c^{SL})\, dx$, and therefore

$$dc^L = \frac{c_0 - c^{SL}}{l}\, dx \tag{22.10}$$

where c^L is the concentration in the liquid in the zone (effects due to the relatively negligible concentration spike in the zone can be ignored). Now $c^{SL} = k' c^L$ and, by combining this with Eq. 22.10 and integrating,

$$\int_{k' c_0}^{c^{SL}} \frac{dy}{c_0 - y} = \int_0^x \frac{k'}{l}\, dz \tag{22.11}$$

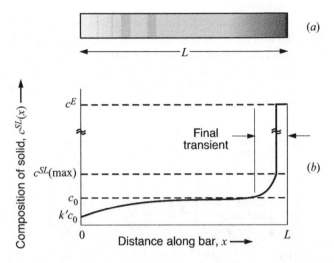

Figure 22.2: (a) Bar-shaped specimen after zone melting. Grayscale intensity indicates solute composition variation. (b) Composition distribution along the bar after a single zone-melting pass. From Flemings [2].

where the lower limit to the first integral occurs because the initial composition of the liquid in the zone is c_0. Integration then yields

$$c^{SL}(x) = c_0[1 - (1 - k')e^{-k'x/l}] \qquad (22.12)$$

A characteristic plot of $c^{SL}(x)$ as a function of x is shown in Fig. 22.2b.

Considerable purification is achieved during zone melting. The final transient at the end begins when the leading end of the zone reaches the end of the specimen. At that point, the solidification becomes very similar to plane-front solidification. Additional passes produce further purification and very small solute concentrations in the first part of the specimen. An asymptotic limit exists, however, as taken up in Exercise 22.2.

When the zone length is relatively short, k' is large, and when the number of passes is small, the bulk of the specimen solidifies at very nearly a uniform composition corresponding to c_0. Zone solidification can be used in this manner to produce compositional uniformity, a technique known as *zone leveling*.

22.2 CELLULAR AND DENDRITIC SOLIDIFICATION

22.2.1 Formation of Cells and Dendrites

When the liquid/solid interface is unstable according to the criteria discussed in Section 20.3.3, a cellular or dendritic structure is developed. When the degree of instability is relatively low, an array of protuberances develops on the interface as shown in Fig. 20.8a. These protuberances, called *cells*, advance perpendicular to the interface. Their shapes vary depending upon the type of material, the orientation of the interface, and other factors. For $\langle 100 \rangle$ liquid/solid interfaces in cubic metals, equiaxed cells form like those in Fig. 20.8b. However, for a $\langle 110 \rangle$ interface, the cells take on a corrugated configuration of long hills and furrows. When the degree of instability is increased by increasing the rate of solidification, fully formed dendrites develop as in Figs. 20.9 and 22.3. As this transition progresses, crystallographic factors play a role and the cell-growth direction deviates toward the preferred growth direction ($\langle 100 \rangle$ for cubic metals) as in Fig. 22.3b. Also, ridges develop along the dendrite sides as in Fig. 22.3c. Finally, secondary dendrite branches appear on the primary dendrites as in Fig. 22.3d.

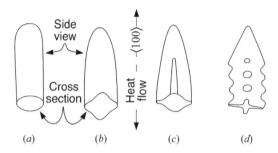

Figure 22.3: Transition from cellular to dendritic growth as the growth velocity is increased. **(a)** Cellular growth at low velocities. **(b)** Cellular growth deviated to the fast growing $\langle 100 \rangle$ direction. **(c)** Appearance of ridges along the primary dendrites. **(d)** Start of secondary dendrite branch formation. After Flemings [2].

22.2.2 Solute Segregation during Dendritic Solidification

During dendritic growth, extensive solute segregation occurs in the interdendritic spaces; this phenomenon is a serious problem in the casting of alloys. The segregation occurs because of the tendency of the solidifying solid to reject excess solute into the remaining liquid and can be understood using the model developed to analyze plane-front solidification. However, the geometry of the dendritic liquid/solid interface and the adjacent diffusion field is complex.

As a reasonable approximation, the dendritic structure may be represented by the diagram in Fig. 22.4. The solidification in the interdendritic space can be described by constructing the cell (shown dashed) and assuming that solidification proceeds in a manner similar to the plane-front solidification of a bar (as discussed in Section 22.1). Under typical casting conditions, $k' \approx k$. Therefore, the segregation

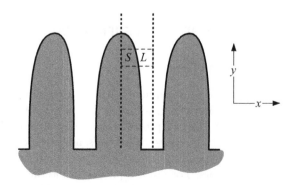

Figure 22.4: Simplified cell model for analyzing interdendritic segregation.

profile along x follows Eq. 22.9 approximately with $k' = k$ and L equal to half the interdendritic spacing. A typical profile is illustrated in Fig. 22.1 for a system with the phase diagram shown in Fig. 20.4a. The formation of the eutectic in the last material to solidify can be expected. When well-defined secondary dendrite branches are present, the solidification cell must be located between the branches rather than between the primary dendrites. Much of the strong segregation present after solidification can be eliminated by *solution treatments* in which castings are annealed at elevated temperatures and the segregates are dispersed by solid-state diffusion [2].

22.3 STRUCTURE OF CASTINGS AND INGOTS

Castings are typically produced by pouring liquid into a relatively cold mold and allowing solidification to take place. Heat is removed from the solidifying material by conduction out through the mold. The classic grain structure obtained after this type of solidification is illustrated in Fig. 22.5.

Three distinct zones are often (but not always) present. The chill zone consists of small equiaxed grains and results from the relatively rapid cooling rate due to the initially rapid outward flow of heat to the cold mold. This produces considerable undercooling, and many small grains with random orientations are nucleated at, or near, the mold surface. These small grains grow until they impinge. Further into the casting, the grains that grow most rapidly in the direction of the thermal gradient will advance most rapidly. Grains that are oriented with slow mobilities in the growth direction will be assimilated by those with faster mobilities and result in a columnar structure. Finally, further into the casting in the liquid ahead of the advancing grains, a third zone consisting of equiaxed grains may form. Here, nuclei for the growth of new grains are provided by small pieces of secondary dendrite branches that have become detached from the stalks of the oncoming primary dendrites. This detachment occurs as a result of temperature fluctuations and convection. For materials of high symmetry (such as cubic metals), these nuclei will grow into approximately equiaxed grains because they are isolated and growing in a very slightly undercooled environment. The latent heat of solidification that is released makes the growing crystals local "hot spots," and the growth therefore

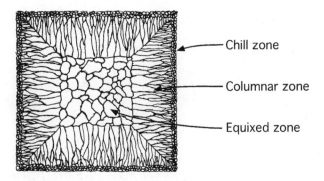

Figure 22.5:   Classic grain structure of casting (or ingot) containing chill zone, columnar zone, and equiaxed zone. From Bower and Flemings [5].

occurs in a direction opposite to the thermal gradient rather than parallel to it as in the case of the dendrites in the columnar zone.

Bibliography

1. W.-G. Pfann. *Zone Melting*. John Wiley & Sons, New York, 1966.

2. M.C. Flemings. *Solidification Processing*. McGraw-Hill, New York, 1974.

3. W. Kurz and D.J. Fisher. *Fundamentals of Solidification*. Trans. Tech. Publ., Switzerland, 1984.

4. V.G. Smith, W.A. Tiller, and J.W. Rutter. A mathematical analysis of solute redistribution during solidification. *Can. J. Phys.*, 33:723–745, 1955.

5. T.F. Bower and M.C. Flemings. Formation of the chill zone in ingot solidification. *Trans. TMS-AIME*, 239:216–219, 1967.

EXERCISES

22.1 An alloy of constant composition can be produced by adding a solute to a bar of pure material by zone melting. This can be done by putting all of the solute into the zone when the trailing end is at $x = 0$ and then moving the zone along the bar. Let $S(x)$ = quantity of solute in the zone when the trailing end of the zone is at x during the pass.

(a) Show that when the length of the zone, l, is kept constant,

$$S(x) = S(0)e^{-k'x/l} \tag{22.13}$$

and therefore that

$$c^{SL}(x) = c^{SL}(0)e^{-k'x/l} \tag{22.14}$$

Note that $c^{SL}(x) \approx c^{SL}(0)$ when k' is small.

(b) Now, show that c^{SL} can be held constant along the bar by continuously reducing the zone-length during the pass according to

$$l(x) = l(0) - k'x \tag{22.15}$$

Solution.

(a) If the zone moves by dx, the amount of solute that is lost from the zone is

$$dS(x) = -c^{SL}(x)\,dx = -k'c^L(x)\,dx = -k'\frac{S(x)}{l}\,dx \tag{22.16}$$

This relation may be integrated in the form

$$\int_{S(0)}^{S(x)} \frac{dy}{y} = -\frac{k'}{l}\int_0^x dx \tag{22.17}$$

so that

$$S(x) = S(0)e^{-k'x/l} \tag{22.18}$$

Also,

$$c^{SL}(x) = k'c^L(x) = \frac{k'S(x)}{l} = k'\frac{S(0)}{l}e^{-k'x/l}$$
$$= k'c^L(0)e^{-k'x/l} = c^{SL}(0)e^{-k'x/l}$$

(b) Now $c^L = S/l$. If c^{SL} is constant, c^L must be constant, and $dl/l = dS/S$. But Eq. 22.16 shows that $dS/S = -(k'/l)\, dx$. Therefore, combining these relationships and integrating,

$$\int_{l(0)}^{l(x)} dl = -k' \int_0^x dx \tag{22.19}$$

so

$$l(x) = l(0) - k'x \tag{22.20}$$

22.2 Suppose that multiple zone-melting passes are made along a bar of length L in an effort to purify it as much as possible. As the number of passes increases, a limiting situation will be reached in which no further purification is obtained by increasing the number of passes. Let $u^{SL}(x)$ be the ultimate solute distribution along the bar after an infinite number of passes.

(a) Show that $u^{SL}(x)$ is given by the solution of the integral equation

$$u^{SL}(x) = \frac{k'}{l} \int_x^{x+l} u^{SL}(\xi)\, d\xi \tag{22.21}$$

where l is the width of the molten zone.

(b) Now show that the solution of this equation is

$$u^{SL}(x) = a_1 e^{a_2 x} \tag{22.22}$$

where a_1 and a_2 are given by

$$a_2 = \frac{k'}{l}(e^{a_2 l} - 1)$$
$$a_1 = \frac{a_2 c_o L}{e^{a_2 L} - 1} \tag{22.23}$$

Solution.

(a) If $u^{SL}(x)$ is not to change during a pass, the total amount of solute in the zone [which is of length l and uniform composition $c^L(x)$] at each stage of the process must be equal to the amount of solute that was in the corresponding length l of the solid bar before it was overrun. Therefore,

$$l\, c^L(x) = \int_x^{x+l} u^{SL}(\xi)\, d\xi \tag{22.24}$$

Because $u^{SL}(x) = k'c^L(x)$,

$$u^{SL}(x) = \frac{k'}{l} \int_x^{x+l} u^{SL}(\xi)\, d\xi \tag{22.25}$$

(b) Trying the solution

$$u^{SL}(x) = a_1 e^{a_2 x} \tag{22.26}$$

a_2 must be

$$a_2 = \frac{k'}{l}(e^{a_2 l} - 1) \tag{22.27}$$

The constant a_1 is determined from the conservation of solute mass:

$$c_o L = \int_0^L u^{SL}(x)\, dx = \int_0^L a_1 e^{a_2 x}\, dx \tag{22.28}$$

Evaluation of the integral then leads to

$$a_1 = \frac{a_2 c_\circ L}{e^{a_2 L} - 1} \tag{22.29}$$

22.3 Consider the growth of an elongated body such as a columnar dendrite under solute-diffusion-controlled conditions. The tip will grow faster than the shank because the solute that must be rejected at the liquid/solid interface can be more rapidly dispersed into the surrounding liquid from the tip region than from the shank. Approximate the shape of the tip by a hemispherical cap and the shank by a cylinder. Now show that the quasi-steady-state velocity with which the dendrite can advance will vary inversely with the radius of the tip. Neglect any effect of capillarity on the solute solubility at the tip and assume that convection in the liquid is low enough so that its effect on mass transport in the liquid can be ignored. Use any reasonable approximations that you need.

Solution. The concentration of solute at the liquid/solid interface at the tip is shown in Fig. 22.6. If the hemispherical tip advances with the velocity v, the rate at which solute must be rejected into the liquid from the tip is

$$I_1 \approx \pi R^2 v (c^{LS} - c^{SL}) \approx \pi R^2 v c^{LS} (1 - k) \tag{22.30}$$

In the quasi-steady state, this must be equal to the rate at which the solute is transported from the tip by diffusion into the liquid. This is given to a reasonable approximation by half of the diffusion rate away from a sphere, which, according to Eq. 13.22, is given by

$$I_2 \approx 2\pi D^L R (c^{LS} - c_\circ) \tag{22.31}$$

The Stefan condition is then $I_1 = I_2$, which yields

$$v \approx \frac{2D^L (c^{LS} - c_\circ)}{c^{LS}(1 - k)} \frac{1}{R} \tag{22.32}$$

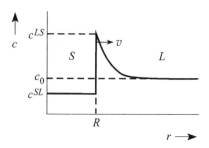

Figure 22.6:   Concentration profile at the liquid/solid interface at the dendrite tip.

22.4 Suppose that the capillarity effect of curvature on solubility is included in Exercise 22.3. Describe qualitatively what happens to the tip growth rate as the tip radius decreases without limit.

Solution. As the hemispherical tip radius becomes smaller (at constant temperature), the equilibrium concentration, c^{LS}, will decrease. This is demonstrated in Fig. 22.7 by employing the common-tangent construction used in Fig. 15.1a. Furthermore, the

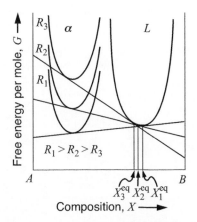

Figure 22.7: Common-tangent construction showing concentration of B, X^{eq}, X_2^{eq}, and X_3^{eq} in liquid in equilibrium with hemispherical tip of α phase of radii $R_1 > R_2 > R_3$.

change in concentration will vary as $1/R$, as in the corresponding case of Eq. 15.4. When c^{LS} is reduced to the point where $c^{LS} = c_o$, equilibrium will prevail at the liquid/solid interface and growth will stop. A calculation of this effect in the solidification of a Al/Cu alloy is plotted in Fig. 22.8.

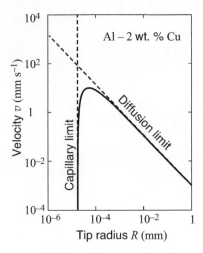

Figure 22.8: Dendrite tip velocity vs. tip radius for an Al/Cu alloy. The diffusion-limit portion of the curve is unaffected by capillarity. The capillarity limit indicates the point where the tip curvature causes the dendrite growth to stop. From Kurz and Fisher [3].

CHAPTER 23

PRECIPITATION

Precipitation occurs when a new phase forms discontinuously within a homogeneous metastable phase to form a two-phase mixture of lower energy. The process occurs by the nucleation and growth of particles (precipitates) of the new phase embedded in the original phase. The form of the precipitation may vary widely depending upon factors such as the degree of coherency between the precipitates and matrix, the degree of supersaturation, and the availability of heterogeneous nucleation sites.

Basic aspects of nucleation and diffusional growth have been described in Chapters 19 and 20. In this chapter we focus on phenomena relevant to precipitation in binary systems such as the morphology and energy of the critical nucleus and precipitate evolution during growth, including loss of coherency. A description of precipitation in two contrasting systems illustrates the wide range of phenomena that can occur in different systems.

23.1 GENERAL FEATURES OF PRECIPITATION

Figure 23.1 is a phase diagram of a system that exhibits precipitation. If cooled along the path indicated, the α phase will become supersaturated with respect to the β phase when it crosses the phase boundary, and if there is no intervening spinodal, the β phase will then precipitate discontinuously in the α phase (matrix phase) as the system attempts to reach equilibrium.

The nucleation and growth processes that produce precipitation can be varied and complex [1–7]. Precipitation will generally involve the formation of critical

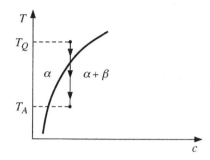

Figure 23.1:   Phase diagram for precipitation.

nuclei possessing a morphology (shape and crystal orientation) of minimum energy and the system will follow the fastest path available. In the classical model, this energy will be the sum of the bulk free-energy change due to the formation of the new phase (a negative quantity) and the interfacial and strain energies (positive quantities that act as nucleation barriers). The interfacial energy is usually the more important barrier and depends upon the nucleus shape and the energies (per unit area) of the interface (or patches of interface) that may be present (see Sections 19.1.1 and 19.1.4). The strain energy depends on shape as well as on the degree of coherence between the nucleus and matrix and the elastic moduli of the nucleus and matrix. To add to the complexity, the nucleation may be homogeneous or inhomogeneous, depending upon the availability of heterogeneous nucleation sites.

23.2   NUCLEUS MORPHOLOGY AND ENERGY

The simplest case occurs when the α-phase and β-phase crystals have different compositions but still match almost exactly in all three dimensions. The critical nucleus can then form with a coherent interface and is therefore of relatively low energy.[1] Also, any strain energy will be small. This condition is met during the precipitation of Ag-rich precipitates in a Al + 4 at. % Ag matrix [8] and Co-rich precipitates in a Cu + 1 at. % Co matrix [9] where the precipitates are coherent and essentially spherical in shape.

When closely matched atomic planes exist in the precipitate and matrix, a low-energy coherent interface can be formed parallel to these planes. A relatively low-energy nucleus can then be produced in the form of a broad-faced disc (or platelet) lying parallel to these planes. As shown by Eqs. 19.26 and 19.27, the strain energy will be relatively small in such a case because the precipitate shape eccentricity $c/a \ll 1$. This common form of precipitation results in platelet structures known as *Widmanstätten structures*. Examples are precipitates in f.c.c./h.c.p. systems with $\{111\}_{\text{fcc}}$ parallel to $\{0001\}_{\text{hcp}}$ as illustrated in Fig. B.6 and the θ'' precipitate and α-matrix phase with their $\{100\}$ planes parallel in the Al–Cu system as illustrated in Fig. 23.2c. In many cases, the narrow platelet edges may be incoherent and the broad faces semicoherent, but this will not change the general picture significantly.

---

[1] As discussed in Section B.6, coherency must always be specified relative to a reference structure. In this case, it can be either the α or β crystal.

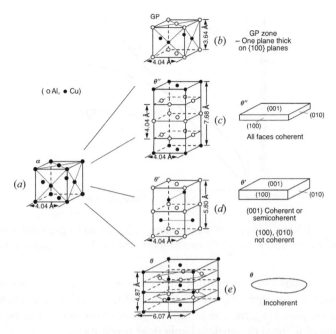

Figure 23.2:   Structures and morphologies of the Cu-rich precipitates which form during precipitation in an Al + 1.7 at. % Cu matrix. **(a)** α-matrix phase. **(b)** Guinier–Preston Zone precipitate. **(c)** θ'' precipitate. **(d)** θ' precipitate. **(e)** Final stable θ precipitate. From Porter and Easterling [4].

When there is no near lattice matching between precipitate and matrix structures in any dimension, the interfacial energy will be relatively high. In such cases, homogeneous nucleation will be slow to occur and the nucleation will be inhomogeneous. When this occurs at grain boundaries, the nuclei are generally more equiaxed because no unusually low-energy interfaces are available. However, the small precipitates formed in this way still possess interfaces which are generally faceted and often contain arrays of line defects [10]. Evidently, these interfaces are semicoherent with respect to bicrystal reference structures (see Section B.6) and are of sufficiently low energy to induce faceting.

23.3 LOSS OF PRECIPITATE COHERENCY DURING GROWTH

Just after nucleation, a small coherent precipitate generally possesses both interfacial energy and strain energy. Its strain energy increases linearly with its size (see Eq. 19.1), which is proportional to $\mathcal{N}$. On the other hand, its total interfacial energy increases linearly with its area, which is proportional to $\mathcal{N}^{2/3}$. Figure 23.3 shows that the interfacial energy dominates at small sizes where the interface-to-volume ratio is large. However, the strain energy eventually becomes dominant as the size increases. Nuclei and small-sized precipitates therefore tend to be coherent because this minimizes the interfacial energy (and the total energy). When this strain energy becomes sufficiently large, it can be reduced if the precipitate is made semicoherent by the introduction of anticoherency dislocations into the interface, as in Figs. 19.8, B.7, and B.8. The anticoherency dislocations, in effect, cancel the

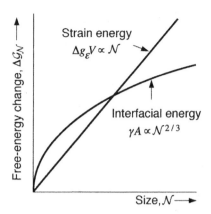

Figure 23.3:   Relative increase of strain energy $(\Delta g_\varepsilon \mathcal{V})$ and interfacial energy (γA) of coherent precipitate as a function of its size as measured by $\mathcal{N}$. $\mathcal{V}$ = precipitate volume.

coherency dislocations (which produce the coherency strains) and thereby reduce the coherency-strain energy. Figure 23.4 illustrates a mechanism, known as prismatic dislocation *punching*, by which anticoherency dislocations can be introduced at the interface of a three dimensional spherical precipitate.

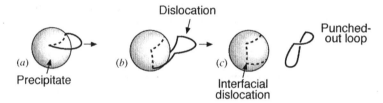

Figure 23.4:   Prismatic dislocation punching at spherical precipitate. **(a)** A dislocation dipole loop is generated in the interface. One side expands into the matrix while the other remains in the interface. **(b)** Segments of the loop in the matrix glide downward to form additional loop length in the interface. **(c)** The loop in (b), which is partially in the interface and partially in the matrix, pinches together at its lowest point and splits into two loops, with one remaining in the interface and the other gliding into the matrix. From Porter and Easterling [4].

23.4   PRECIPITATION IN TWO CONTRASTING SYSTEMS

23.4.1   Cu–Co System

The Cu–Co system is a particularly simple precipitation system in which a Co-rich β phase precipitates in a Cu-rich terminal α phase. The f.c.c. lattices of both phases are well matched in three dimensions, so that the precipitate interfaces are coherent with respect to either lattice as a reference structure and the interfacial energy is sufficiently isotropic so that they are almost spherical, as in Fig. 19.2. Both the interfacial energy and strain energy are therefore relatively low and the nucleation of the β phase is therefore relatively easy and occurs homogeneously. This system has been used to test the applicability of the classical nucleation theory (Section 19.1.1) [11, 12]. In this work, the experimental conditions under which

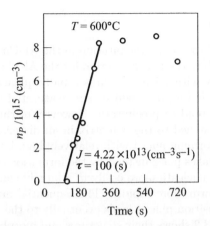

Figure 23.5: Measured density of precipitates, n_P, forming as a function of time in Cu–1 at. % Co alloy during annealing at 600°C. From LeGoues and Aaronson [11].

nucleation might be measurable were first determined by means of calculations. Nucleation rates in this nucleation "window" were then determined experimentally by electron microscopy. Some results shown in Fig. 23.5 were compared with values predicted by the classical model for nucleation using Eq. 19.17. Reasonably good agreement was found, indicating that the classical model is realistic in at least this case. Two nonclassical models for the critical nucleus were also tested; the first used the Cahn–Hilliard gradient-energy continuum formalism described in Section 18.2 and the second employed a detailed discrete-lattice calculation. The results in both cases predicted a somewhat diffuse nucleus/matrix interface and a chemically inhomogeneous critical nucleus. A typical result showing the concentration profile along the nucleus radius is shown in Fig. 23.6; the nucleus has a small inner core of uniform concentration and an interface with a thickness that is about half the nucleus radius. However, the nucleation rates predicted by the three models were in reasonable agreement. The significance of this result is discussed in Section 19.1.7.

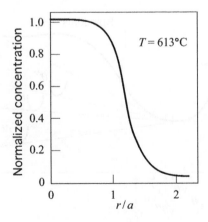

Figure 23.6: Calculated concentration profile (normalized) through a Co-rich critical nucleus in a Cu–1 at. % Co precipitation system. $a =$ lattice constant. From LeGoues and Aaronson [11].

23.4.2   Al–Cu System

In contrast to the Cu–Co system, precipitation in the Al–Cu system is highly complex. Figure 23.1 resembles the phase diagram for the Al–Cu system on the Al-rich side. If an α-phase alloy with 1.7 at. % Cu is rapidly quenched from T_Q to below the solvus, held at T_A, and aged, it will be supersaturated and the equilibrium β phase (i.e., $CuAl_2$) will tend to precipitate. However, in this system, the equilibrium phase is poorly matched to the α matrix in all dimensions and can nucleate only with difficulty by forming nuclei with relatively high interfacial energies. The precipitation therefore takes place instead by the formation of a series of transition copper-rich phases that nucleate more easily in a largely coherent or semicoherent manner and therefore form more rapidly. The sequential nucleation, growth, and dissolution of these transition phases leads eventually to the formation of the equilibrium phase. Figure 23.2 shows their structures and morphologies forming in the sequence

$$\text{Guinier–Preston zones} \rightarrow \theta'' \rightarrow \theta' \rightarrow \theta$$

where the final equilibrium phase ($CuAl_2$) is indicated by θ. The Guinier–Preston zones (*GP zones*)form first and are copper-rich discs of no more than a few atomic layers coherent with respect to a reference structure consisting of either the matrix or precipitate lattice. As time progresses, they serve as preferred sites for the formation of θ'' precipitates that grow in situ on the zones. Eventually, θ' precipitates form on the θ'' precipitates or nucleate heterogeneously on crystal dislocations while the θ'' precipitates dissolve. Finally, stable θ precipitates nucleate heterogeneously on either grain boundaries or θ'/matrix interfaces. As shown in Fig. 23.2, the broad faces of the transition precipitates are either low-energy coherent or semicoherent interfaces which make their nucleation easy. Schematic bulk free energy vs. composition curves for these phases are shown in Fig. 23.7. The free energies of the precipitates decrease in the order: GP zones $\rightarrow \theta'' \rightarrow \theta' \rightarrow \theta$. The common-tangent construction shows that their solubilities in the α phase therefore decrease in the same order. The dissolution of, for example, the θ'' phase and the simultaneous

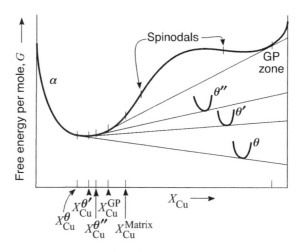

Figure 23.7:   Schematic free energy vs. composition curves during precipitation in the Al–Cu system. From Jena and Chaturvedi [7].

growth of the more stable θ' phase can be understood in terms of diffusion through the matrix from the phase of higher solubility to the one of lower solubility. The phase diagram solvus lines for the transition precipitates in Fig. 23.8 show that the full sequence of transition precipitates will be observed only if annealing is carried out below the GP zone precipitate solvus. If annealing is carried out above the θ' solvus, only the θ precipitate will form and it will nucleate and grow heterogeneously on grain boundaries. Also, if GP zones have formed at a given temperature below the GP solvus and the system is heated to above the GP solvus, the GP zones will dissolve—a phenomenon known as *reversion*.

This complex form of precipitation in the Al–Cu system is of great practical importance. The finely dispersed precipitates act as effective barriers to the glide movement of dislocations during plastic deformation and harden and strengthen the material. This has led to the development of a number of widely used precipitation-hardened Al–Cu alloys [1].

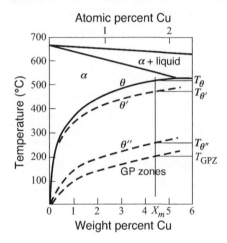

Figure 23.8: Solvus curves for Guinier–Preston zones and θ'', θ', and θ precipitates in the Al–Cu system. From Hornbogen [13] and from Jena and Chaturvedi [7].

Bibliography

1. A. Kelly and R.B. Nicholson. Precipitation hardening. *Prog. Mater. Sci.*, 10(3), 1963.

2. J.W. Christian. *The Theory of Transformations in Metals and Alloys*. Pergamon Press, Oxford, 1975.

3. K.C. Russell and H.I. Aaronson, editors. *Phase Transformations in Materials*. AIME, Warrendale, PA, 1978.

4. D.A. Porter and K.E. Easterling. *Phase Transformations in Metals and Alloys*. Van Nostrand Reinhold, New York, 1981.

5. H.I. Aaronson, D.E. Laughlin, R.F. Sekerka, and C.M. Wayman, editors. *International Conference on Solid→Solid Phase Transformations*, Warrendale, PA, 1982. The Minerals, Metals and Materials Society.

6. G.W. Lorimer. Precipitation in aluminum alloys. In *Precipitation Processes in Solids*, pages 87–119, Warrendale, PA, 1978. AIME.

7. A.K. Jena and M.C. Chaturvedi. *Phase Transformations in Materials*. Prentice Hall, Englewood Cliffs, NJ, 1992.

8. R.B. Nicholson, G. Thomas, and J. Nutting. Electron microscopic studies of precipitation in aluminium alloys. *J Inst. Metals*, 87:429–438, 1959.

9. A.J. Ardell. Precipitate coarsening in solids: Modern theories, chronic disagreement with experiment. In G.W. Lorimer, editor, *Phase Transformations 87*, pages 485–490, London, 1988. Institute of Metals.

10. H.I. Aaronson. Atomic mechanisms of diffusional nucleation and growth and comparisons with their counterparts in shear transformations. *Metall. Trans.*, 24A(2):241–276, 1993.

11. F.K. LeGoues and H.I. Aaronson. Influence of crystallography upon critical nucleus shapes and kinetics of homogeneous f.c.c.-f.c.c. nucleation—IV. Comparisons between theory and experiment in Cu–Co alloys. *Acta Metall.*, 32(10):1855–1864, 1984.

12. H.I. Aaronson and F.K. LeGoues. An assessment of studies on homogeneous diffusional nucleation kinetics in binary metallic alloys. *Metall. Trans. A*, 23(7):1915–1945, 1992.

13. E. Hornbogen. Die elektronenmikroskopise Untersuchung der Ausscheidung in Al–Cu-Mischkristallen. *Aluminium*, 43(1):41–53, 1967.

CHAPTER 24

MARTENSITIC TRANSFORMATIONS

Martensitic transformations are discontinuous transformations that are diffusionless and displacive and occur by the forward glissile motion of the interface between the growing martensite and its parent phase. The theory of the crystallography of these transformations is presented by employing either a pole-figure description or a deformation-tensor formulation. Topics include the prediction of the crystal orientation relationship between the martensite and parent phase, the habit plane of the martensite in the parent phase, and the macroscopic specimen shape change due to the transformation. The glissile nature of the martensite/parent phase interface is explained in terms of a coherency–anticoherency dislocation model. The nucleation of martensite is also considered and found to be heterogeneous in nature. Finally, martensitic transformations in three widely contrasting systems are described to illustrate the wide range of phenomena that can occur.

24.1 GENERAL FEATURES OF MARTENSITIC TRANSFORMATIONS

Martensitic phase transformation is similar in a number of respects to mechanical twinning (as described in Section 13.3.1). Both processes are displacive because they occur by the local transfer of atoms across an advancing interface by a highly organized "military" shuffling of atoms across the interface at conservatively moving dislocations. In both cases, this induces a macroscopic shape change of the specimen. Both processes are conservative; no long-range diffusion is involved, and the martensite and its parent phase must therefore be of the same chemical composition. In both cases, thermal activation is of little importance, and both martensite and twin interfaces often advance at high speeds at relatively low temperatures.

However, important differences exist. Martensite and its parent phase are different phases possessing different crystal structures and densities, whereas a twin and its parent are of the same phase and differ only in their crystal orientation. The macroscopic shape changes induced by a martensitic transformation and twinning differ as shown in Fig. 24.1. In twinning, there is no volume change and the shape change (or *deformation*) consists of a shear parallel to the twin plane. This deformation is classified as an *invariant plane strain* since the twin plane is neither distorted nor rotated and is therefore an *invariant plane* of the deformation.

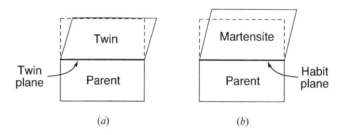

(a)   *(b)*

Figure 24.1: Macroscopic shape changes due to **(a)** twinning and **(b)** martensite formation.

As described below, martensite forms in a manner that produces good dimensional matching along the martensite/parent-phase interface (called the *habit plane*) to reduce any elastic misfit energy between the phases. The habit plane is therefore also an invariant plane to a good approximation. The volume change associated with the martensite phase transformation then appears in the form of a dilation normal to the habit plane, and the total deformation can be factored into a shear parallel to the habit plane and a dilation normal to it as seen in Fig. 24.1*b*. Figure 24.2 shows the typical shape change when the martensite traverses a single crystal of the parent phase. When the martensite develops in the form of inclusions within the parent phase, the inclusions form as thin lenticular platelets with their

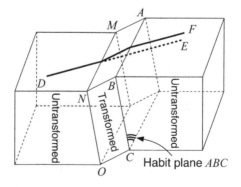

Figure 24.2: Macroscopic shape change produced when a martensite plate traverses a single crystal of the parent phase. A straight fiducial line *DE* in the parent crystal becomes *DF* after the formation of the martensite plate.

broad faces parallel to the habit plane, thereby minimizing the elastic strain energy as discussed in Section 19.1.3.

The driving pressure for a martensitic transformation usually arises from the decrease in bulk free-energy change per unit volume, Δg_B, which occurs in the transformation. However, since the transformation produces a change in specimen shape, an additional pressure can arise if an applied stress is present. This pressure is similar in origin to the pressure exerted by an applied stress on a mechanical twin because of its shape change as discussed in Section 13.3.1. Since the martensite shape change can be factored into an invariant plane strain shear and a dilation normal to the invariant plane, the pressure has two components: the first, due to the shear, can be expressed as $\tilde{\sigma}_s \tilde{\varepsilon}_s$, where $\tilde{\sigma}_s$ is the resolved shear stress along the invariant plane in the direction of the shear and $\tilde{\varepsilon}_s$ is magnitude of the shear; the second, due to the dilation, can be expressed as $\sigma_{nn} \varepsilon_{nn}$, where σ_{nn} is the resolved stress normal to the invariant plane and ε_{nn} is the normal dilation. The total pressure is then

$$\Pi^{\text{phase trans}} = -\Delta g_B + \tilde{\sigma}_s \tilde{\varepsilon}_s + \sigma_{nn} \varepsilon_{nn} \tag{24.1}$$

Martensitic transformations are usually induced by cooling the parent phase below the point where Δg_B turns negative. Martensite then nucleates and grows in conservative glissile fashion under the driving pressure given by the first term on the right side of Eq. 24.1. In some cases, this growth can be very rapid, approaching the speed of sound. However, martensite formation can also be induced by applied stress, as might be expected from the second and third terms in Eq. 24.1. Section 24.3 discusses these aspects of martensite formation. However, it is first necessary to describe how martensite can form with a habit plane that is an invariant plane and an interface that is glissile.

24.2 CRYSTALLOGRAPHY OF MARTENSITIC TRANSFORMATIONS

The crystallography of martensitic transformations has been widely analyzed [1–9]. We shall mainly follow Wayman's description [5] which is largely based on the original work of Wechsler et al. [1]. The famous f.c.c.→b.c.t. (body-centered tetragonal) transformation in iron alloys is the basis for the hardening of steel and will be the focus of the discussion.

24.2.1 Lattice Deformation

Structures of the f.c.c. parent and b.c.t. martensite phases are shown in Fig. 24.3. The f.c.c. parent structure contains an incipient b.c.t. structure with a c/a ratio which is higher than that of the final transformed b.c.t. martensite. The final b.c.t. structure can be formed in a very simple way if the incipient b.c.t. cell in Fig. 24.3a is extended by factors of $\eta_1 = \eta_2 = 1.12$ along x_1' and x_2' and compressed by $\eta_3 = 0.80$ along x_3' to produce the martensite cell in Fig. 24.3b. This deformation, which converts the parent phase homogeneously into the martensite phase, is known in the crystallographic theory as the *lattice deformation*.[1] Unfortunately,

---

[1] In transformations where the parent crystal structure has a basis, the homogeneous lattice deformation may not move every atom within the parent unit cell into its proper position in the martensite unit cell. In such cases, small local shuffles of some of the atoms will be required. However, this is irrelevant with respect to the overall macroscopic shape change.

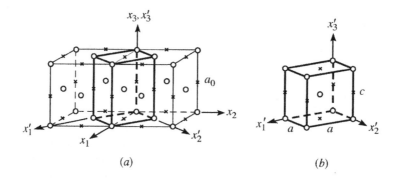

Figure 24.3:   Lattice transformation and lattice correspondence in the f.c.c.→b.c.t. martensitic transformation. **(a)** Initial f.c.c. structure. **(b)** Final b.c.t. structure. Crosses denote interstitial sites partially occupied by C atoms in Fe–C solid solutions.

the lattice deformation in the present case is not an invariant plane strain deformation because there is no plane in the system that remains both undistorted and unrotated. Additional operations are therefore necessary to produce an invariant plane in the transformation.

The absence of an invariant plane in the deformation above can be demonstrated by using some general results for homogeneous deformations. Any homogeneous deformation deforms a unit sphere into an ellipsoid possessing axes corresponding to the three principal deformations as illustrated in Fig. 24.4. Here, the cross section of a unit sphere is assumed to be in the parent phase. It is deformed by the lattice deformation into an ellipsoid in the martensite possessing a shape given by

$$\frac{x_1'^{\,2}}{\eta_1^2} + \frac{x_2'^{\,2}}{\eta_2^2} + \frac{x_3'^{\,2}}{\eta_3^2} = 1 \qquad (24.2)$$

The $\vec{OA'}$, $\vec{OB'}$, $\vec{OC'}$, and $\vec{OD'}$ in the martensite remain unchanged in length in the deformation. The figure has circular symmetry around the x_3' axis, and all of the vectors in the system that satisfy this condition fall on two cones centered on the x_3' axis with their apexes at O and passing through $\vec{OA'}$, $\vec{OB'}$, $\vec{OC'}$, and $\vec{OD'}$, respectively. To find these same vectors in the parent phase, we make use of the

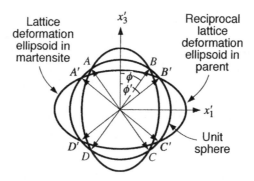

Figure 24.4:   Deformation ellipsoids involved in the lattice deformation in the f.c.c.→b.c.t. martensitic transformation.

reciprocal deformation ellipsoid, which is the particular ellipsoid in the parent phase that is converted into a unit sphere in the martensite by the lattice deformation. The reciprocal deformation ellipsoid therefore has the form

$$\eta_1^2 {x_1'}^2 + \eta_2^2 {x_2'}^2 + \eta_3^2 {x_3'}^2 = 1 \qquad (24.3)$$

The corresponding vectors in the parent phase lie on the two cones that intersect OA and OB and OC and OD, respectively. An invariant plane must contain two noncollinear vectors in the material (defining a plane) which are unchanged in length and unrotated. The vectors on the AOB and COD cones in the parent phase are the only vectors that remain unchanged in length. However, they are all rotated away from the x_3' axis by the deformation through the angle $\phi' - \phi$, and therefore there is no invariant plane in the deformation.

The angle ϕ may be found from the equation for the AOB cone, which can be obtained by setting the equation for the unit sphere, ${x_1'}^2 + {x_2'}^2 + {x_3'}^2 = 1$, equal to Eq. 24.3, to obtain

$$\left(\eta_1^2 - 1\right) {x_1'}^2 + \left(\eta_2^2 - 1\right) {x_2'}^2 + \left(\eta_3^2 - 1\right) {x_3'}^2 = 0 \qquad (24.4)$$

Setting $x_2' = 0$ yields

$$\tan\phi = \frac{x_1'}{x_3'} = \sqrt{\frac{1 - \eta_3^2}{\eta_1^2 - 1}} \qquad (24.5)$$

A corresponding expression for ϕ' may be found by the same general method.

24.2.2 Undistorted Plane by Application of Additional Lattice-Invariant Deformation

Since invariant planes do not exist in the lattice deformation, additional operations are required. A plane that is undistorted can be produced if the parent phase is subjected to two operations: the lattice transformation; and a shear deformation that does not change the crystal structure and is therefore termed a *lattice-invariant deformation*. This lattice-invariant deformation can be accomplished by shearing the crystal by passing perfect dislocations through it (i.e., by dislocation slip). However, even though these operations produce an undistorted plane, they do not generally produce an invariant plane since the undistorted plane they produce is generally rotated away from its inclination in the original parent phase. However, this is easily remedied since the undistorted plane can then be brought back to its original inclination by a final rigid-body rotation. Three operations are therefore required: a lattice deformation, a lattice-invariant deformation (shear), and a rigid-body rotation.

These operations do not occur separately and in any particular sequence but are simply a convenient way to conceptualize the transformation as a series of operations, each of which can be analyzed separately, but which working together produce a martensitic structure containing an invariant plane. As such, they can be imagined to occur in any sequence. For purposes of analysis, it is convenient to imagine that the lattice-invariant deformation occurs first, followed by the lattice deformation, followed finally by the rigid-body rotation. We now show that a lattice-invariant shear by slip followed by the lattice deformation analyzed above can produce an undistorted plane.

The geometrical features of shear deformation are shown in Fig. 24.5. Here, the shear is on the K_1 plane in the direction of d. The initial unit sphere is deformed into an ellipsoid and the K_1 plane is an invariant plane. The K_2 plane is rotated by the shear into the K_2' position and remains undistorted. A reasonable slip system to assume for the lattice-invariant shear deformation is slip in a $\langle 111 \rangle$ direction on a $\{112\}$ plane in the b.c.t. lattice, which corresponds to slip in a $\langle 110 \rangle$ direction on a $\{110\}$ plane in the f.c.c. lattice.

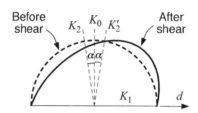

Figure 24.5:   Cross-sectional view of upper half of a unit sphere deformed into an ellipsoid by shear. After Wayman [5].

At this point, it is advantageous to switch to the use of stereographic projections to visualize fully the three-dimensional geometry.[2] The stereographic projection in Fig. 24.6 shows the geometry of the shear illustrated in Fig. 24.5 when the shear

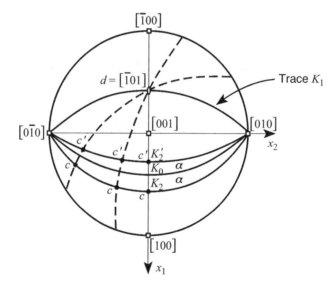

Figure 24.6:   Stereographic representation of lattice-invariant shear deformation in the f.c.c. lattice. Same geometry as in Fig. 24.5. Figure referred to f.c.c. axes. After Wayman [5].

occurs by slip in the f.c.c. structure on the (101) plane in the $[\bar{1}01]$ direction. The (101) K_1 slip plane and $[\bar{1}01]$ slip direction are shown along with the undistorted (but rotated) K_2 plane. Vectors on this plane, such as those at the c positions, will remain undistorted but will rotate on great circles that pass through the slip direction $[\bar{1}01]$. The stereographic projection in Fig. 24.7 shows the geometry of the lattice deformation discussed in Section 24.2.1. The traces of the cones containing

---

[2]Stereographic projections and their uses are described by Barrett and Massalski [10].

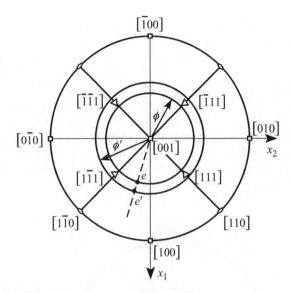

Figure 24.7:   Stereographic representation of lattice deformation in f.c.c.→b.c.t. martensitic transformation. Figure referred to f.c.c. axes. After Wayman [5].

all of the vectors that are unchanged in length appear as circles concentric to the x_3 axis (i.e., [001]). The lattice deformation causes the vector at e to move outward from its initial position on the inner cone to its final rotated position on the outer cone along a path that is radial with respect to the central [001] pole.

The procedure for finding an undistorted plane is shown in Fig. 24.8. If a vector initially at the position a is rotated to the position a' by the lattice-invariant shear

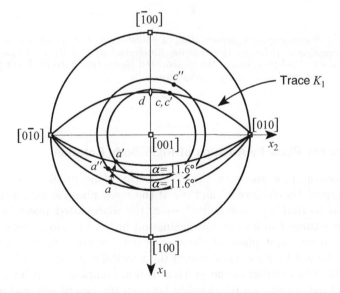

Figure 24.8:   Stereographic representation of lattice-invariant deformation plus lattice deformation to produce undistorted plane in f.c.c.→b.c.t. martensitic transformation. Figure referred to f.c.c. axes. After Wayman [5].

and then subjected to the lattice deformation, it will end up at a'' with its length unchanged. A second vector, initially at c, will remain unchanged at c during the lattice-invariant deformation since it lies in the K_1 shear plane. Then, during the lattice deformation, it will rotate to c'' with its length unchanged. The great circle traces of the two planes containing the vectors at a and c and a'' and c'', respectively, are shown in Fig. 24.9. By adjusting the amount of shear, an arrangement of these vectors can be arrived at in which the included angles between a and c and a'' and c'', respectively, can be made equal. The two traces then represent the initial and final positions of a plane that is undistorted since it contains two noncollinear vectors which have remained unchanged in length and which have maintained a constant included angle with each other.

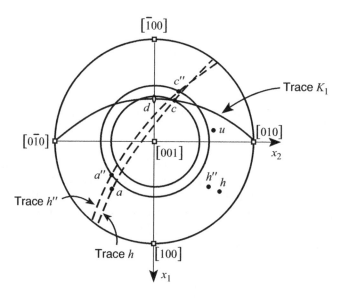

Figure 24.9: Stereographic representation of undistorted plane produced by the lattice-invariant deformation and lattice deformation illustrated in Fig. 24.8. Traces h and h'' represent initial and final positions of the undistorted plane, respectively. Their poles are at h and h''. Figure referred to f.c.c. axes. After Wayman [5].

24.2.3 Invariant Plane by Addition of Rigid-Body Rotation

The plane containing a and c in Fig. 24.9 is the plane in the f.c.c. phase that initially contained the vectors a'' and c''. If the b.c.t. phase is now given a rigid-body rotation so that $a'' \rightarrow a$ and $c'' \rightarrow c$, the undistorted plane in the b.c.t. phase will be returned to its original inclination in the f.c.c.-axis system and will therefore be an invariant plane of the overall deformation. In the present case, this can be achieved by a rotation around the axis indicated by u in Fig. 24.9 (see Exercise 24.3). The solution of the problem is now complete. The invariant plane is known, and the orientation relationship between the two phases and total shape change can be determined from the combined effects of the known lattice-invariant deformation, lattice deformation, and rigid-body rotation.

24.2.4 Tensor Analysis of the Crystallographic Problem

The problem above can also be solved analytically using tensor methods—the preferred technique when higher accuracy is required. In general, any homogeneous deformation can be represented by a second-rank tensor that operates on any vector in the initial material and transforms it into a corresponding vector in the deformed material. For example, in the lattice deformation, each vector, $\vec{v}_{\rm fcc}$, in the initial f.c.c. structure is transformed into a corresponding vector in the b.c.t. structure, $\vec{v}_{\rm bct}$, by

$$\vec{v}_{\rm bct} = B \, \vec{v}_{\rm fcc} \tag{24.6}$$

where B is the lattice deformation tensor given by

$$B = [B_{ij}] = \begin{bmatrix} \eta_1 & 0 & 0 \\ 0 & \eta_2 & 0 \\ 0 & 0 & \eta_3 \end{bmatrix} = \begin{bmatrix} 1.12 & 0 & 0 \\ 0 & 1.12 & 0 \\ 0 & 0 & 0.8 \end{bmatrix} \tag{24.7}$$

If S is the lattice-invariant deformation tensor and R the rigid-body rotation tensor, the total shape deformation tensor, E, producing the invariant plane can be expressed as

$$E = RBS \tag{24.8}$$

Wayman describes in detail how the tensor formalism can be used to solve the crystallographic problem [5]. A simple graphical demonstration, in two dimensions, of how an invariant line (habit plane) can be produced by the deformations B, S, and R is given in Exercise 24.6.

24.2.5 Further Aspects of the Crystallographic Model

The input data for the model consist of the description of the lattice deformation and the choice of the slip system in the lattice-invariant shear. The model has successfully predicted the observed geometrical features of many martensitic transformations. The observed and calculated habit planes generally have high indices that result from the condition that they be macroscopically invariant.

Because of the four-fold symmetry of the [001] pole figures in Figs. 24.6–24.9, additional symmetry-related invariant planes can be produced. Also, further work shows that additional invariant planes can be obtained if a lattice-invariant shear corresponding to $\alpha = 7.3°$ rather than $\alpha = 11.6°$ (see Fig. 24.8) is employed [5]. Multiple habit planes are a common feature of martensitic transformations.

In many cases, the martensite phase is internally twinned and is composed of two types of thin twin-related lamellae, as illustrated in Fig. 24.10. In such cases, the lattice-invariant shear is accomplished by twinning rather than by slip as has been assumed until now (see Fig. 24.10b). The critical amount of shear required to produce the invariant habit plane is then obtained by adjusting the relative thicknesses of the two types of twin-related lamellae shown in Fig. 24.10b.

The crystallographic model for martensite described above is primarily due to Wechsler et al. [1]. A similar model, employing a different formalism but leading to essentially equivalent results, has also been published by Bowles and MacKenzie [2–4]. In both models, a search is made for an invariant (or near-invariant) plane which is then proposed as the habit plane, since the selection of this plane

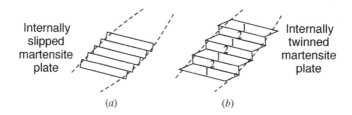

Internally slipped martensite plate   (a)   (b)   Internally twinned martensite plate

Figure 24.10:   (a) Internally slipped martensite. (b) Internally twinned martensite. The two types of twin-related lamellae present are labeled 1 and 2.

by the system minimizes the amount of distortion and corresponding strain energy that accompanies the transformation. More recently, the problem has been approached differently by searching directly for the transformation crystallography that minimizes the energy [11–13]. Not too surprisingly, this direct free-energy-minimization approach leads to habit plane results similar to those predicted by the invariant-plane approach [11, 12].

24.3   GLISSILE INTERFACE

The dislocation structure of the glissile martensite interface can be deduced by following the same sequence of deformation steps as those employed in the crystallographic model above. Consider first the simple case where the martensite forms as a slab with its habit plane extending across the entire specimen as in Fig. 24.2. Figure 24.11a shows the slab-shaped region (dashed lines) of the parent phase, which will be transformed to martensite by this sequence of steps. After the lattice-invariant shear deformation, S, an array of lattice dislocations is deposited at the incipient interface by the dislocation slip process that produced the required lattice-invariant shear. At this point (Fig. 24.11b), the specimen consists entirely of the parent phase, but it contains the array of anticoherency dislocations shown, which produce long-range internal stresses. When the transformation is completed by applying the lattice deformation, B, and rotation, R (Fig. 24.11c), the material in the slab has been transformed to martensite, the interfaces have become invariant planes, and the long-range stresses have been eliminated. In this process, arrays of coherency dislocations have been generated at the interface as shown, which accommodate (coherently) the difference in lattice structure across the interface and also cancel the long-range stress field of the first set of lattice dislocations. The lattice dislocations may therefore be classified as anticoherency dislocations and the Burgers vectors of these anticoherency dislocations and the coherency dislocations cancel, as described in Section B.6. The diagrams in Fig. 24.11 are schematic and show only the total Burgers vector content of the interfaces in an idealized fashion. The detailed way in which this Burgers vector strength is distributed depends upon the atomic structure of the interface [9]. The interface is seen to be glissile, since the coherency dislocations can move conservatively in any direction, and the anticoherency (lattice) dislocations have slip surfaces running into the parent phase.

Consider next the case where the martensite forms in the parent phase as an inclusion. The procedure for obtaining the dislocation structure of the interface is the same as previously. However, the martensite slab is now no longer free to shear

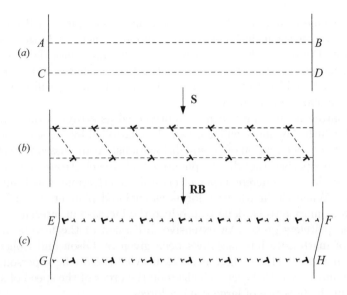

Figure 24.11:   Formation of martensite slab spanning cross section of parent phase. **(a)** Region in parent phase to be transformed. **(b)** Anticoherency dislocations present in interfaces after lattice-invariant deformation, S, by slip. **(c)** Coherency dislocations added as a result of the lattice deformation, B, and the final rotation, R.

at its ends or to expand in a direction normal to the invariant plane. If the shape change is relatively small and is therefore accommodated coherently, everything remains the same as in the previous case except that additional arrays of coherency dislocations will be generated at the ends of the slab, as illustrated in Fig. 24.12a. In this coherent case, these arrays are not canceled by any anticoherency dislocations and they will therefore generate long-range stress fields. In such cases, martensite inclusions generally assume a lenticular shape as in Fig. 24.12b to decrease the elastic strain energy. The added coherency dislocations are then spread out along the two faces of the inclusion in a manner similar to the coherency dislocations in the lenticular twin shown in Fig. 13.2a, where there are only shear displacements parallel to the twin plane that must be accommodated. Again, both interfaces in Fig. 24.12 are glissile.

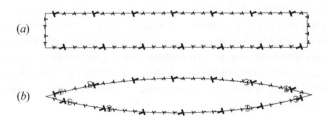

Figure 24.12:   Formation of martensite inclusion. **(a)** Slab-shaped inclusion. Coherency dislocations present at ends that generate long-range stresses. **(b)** Lenticular inclusion. Coherency dislocations present at the ends in (a) are now distributed along the lens-shaped faces.

When the shape change is relatively large, the parent phase will no longer be able to accommodate the inclusion elastically, and anticoherency lattice dislocations will be generated to relieve the long-range stress field and reduce the elastic energy. Plastic deformation will therefore occur in the parent phase, and anticoherency dislocations will be added to the interface. These dislocations will generally tend to reduce the mobility of the interface.

Because martensite interfaces can be represented as arrays of dislocations, the velocity with which they move will generally be controlled by the same factors that control the rate of glide motion of crystal dislocations. As discussed in Section 11.3, these include dissipative drag due to phonons and free electrons and interactions with a large variety of different types of crystal imperfections which hinder their glide motion. When the martensite forms as enclosed platelets as in Fig. 24.12, additional work must also be done to produce the increase in interfacial area that occurs as the platelets grow. An extensive discussion of the factors involved in the motion of martensite interfaces has been given by Olson and Cohen [9]. As pointed out in Section 11.3.4, there is no clear evidence for the supersonic motion of martensite interfaces. However, velocities on the order of the speed of sound can be achieved in the presence of large driving forces.

24.4 NUCLEATION OF MARTENSITE

The homogeneous nucleation of martensite in typical solids is too slow by many orders of magnitude to account for observed results. Calculations of typical values of $\Delta\mathcal{G}_c$ using the classical nucleation model of Section 19.1.4 (see Exercise 19.3) yield values greatly exceeding $76\,kT$. Furthermore, nearly all martensitic transformations commence at very sparsely distributed sites. *Small-particle experiments* [14] have yielded typical nucleation densities on the order of one nucleation event per 50 μm diameter Fe–Ni alloy powder particle.[3] Thus, nucleation of martensite is believed to occur at a small number of especially potent heterogeneous nucleation sites.

The most likely special site for martensitic nucleation is a pre-existing dislocation array, such as a portion of a tilt boundary [9]. The nucleation process involves dissociation of the boundary dislocations, so as to produce periodic faults in the parent crystal and thereby provide a mechanism for the lattice deformation. The process of superimposing lattice-invariant deformation onto the deformation that occurs in the dissociation of the original tilt boundary is used to obtain the equivalent of the lattice deformation $\boldsymbol{RB}$ in the crystallographic model of Section 24.2.4. The rate of initiating such a nucleus is limited by the rate at which the dislocations required to form and then expand the configuration can move under the available driving force. The entire process may be free of any energy barrier under sufficiently high driving forces, or else involve local barriers to certain critical dislocation movements which can be surmounted with the assistance of thermal activation. Details of the specific defects required for the mechanism have been worked out for common structural changes (e.g., f.c.c.$\rightarrow$ h.c.p., f.c.c.$\rightarrow$ b.c.c.) [8, 9].

---

[3]Small-particle experiments are carried out by studying nucleation in small particles of the parent phase and are useful in distinguishing between homogeneous and heterogeneous nucleation. If the nucleation is homogeneous, the nucleation rate is simply proportional to the volume of the particle. On the other hand, if it is heterogeneous, the rate goes essentially to zero when the particle size is lower than $1/\rho$, where ρ is the density of heterogeneous nucleation sites.

24.5   MARTENSITIC TRANSFORMATIONS IN THREE CONTRASTING SYSTEMS

We now describe briefly martensitic transformations in three contrasting systems which illustrate some of the main features of this type of transformation and the range of behavior that is found [15]. The first is the In–Tl system, where the lattice deformation is relatively slight and the shape change is small. The second is the Fe–Ni system, where the lattice deformation and shape change are considerably larger. The third is the Fe–Ni–C system, where the martensitic phase that forms is metastable and undergoes a precipitation transformation if heated.

24.5.1   In–Tl System

Upon cooling, an In–Tl (19% Tl) alloy undergoes an f.c.c. solid solution → f.c.t. solid solution martensitic transformation in which the lattice deformation is relatively slight, corresponding to

$$\boldsymbol{B} = [B_{ij}] = \begin{bmatrix} 0.9881 & 0 & 0 \\ 0 & 0.9881 & 0 \\ 0 & 0 & 1.0238 \end{bmatrix} \tag{24.9}$$

and the shape change is correspondingly small. The lattice-invariant deformation is accomplished by means of twinning in this system, so the low-temperature martensitic phase consists of twin-related lamellae. If a rod-shaped single crystal of the parent f.c.c. phase is carefully cooled in a small temperature gradient from above the transformation temperature, the transformation can be induced so that the martensite first appears at the cooler end of the specimen as a region separated from the parent phase by a single planar interface that spans the entire cross section of the specimen. As cooling continues, the single interface advances along the rod until the entire specimen is transformed. Upon subsequent reverse heating, the transformation is found to be reversible and the original single crystal of the parent phase is recovered with a temperature hysteresis of only about 2°, as shown in Fig. 24.13, where the progress of the transformation is indicated by measurements of the length change of the specimen.

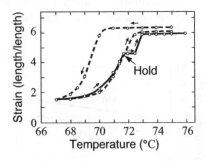

Figure 24.13:   Temperature dependence of the martensitic transformation in In–20.7 at. % Tl. The extent of transformation is revealed by changes of specimen length caused by the transformation. The dashed line shows the reversible transformation resulting from continuous cooling and heating. The solid line shows stabilization of the transformation induced during the heating part of the cycle by a hold of 6 h at constant temperature. From Burkart and Read [16].

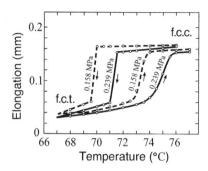

Figure 24.14: Temperature dependence of martensitic transformation in In–20.7 at. % Tl under two different compressive stresses. Phase fraction of martensite is proportional to the permanent strain which can be determined by the stress-free specimen length. From Burkart and Read [16].

The interface motion is jerky on a fine scale and requires a continuous drop in temperature. This indicates that the interface requires a continuous increase in driving pressure (brought about by increased undercooling) to maintain its motion. This may be taken as evidence that the interface must be accumulating defects due to interactions with obstacles in its path which progressively reduce its mobility. If the heating (or cooling) is interrupted by a hold at constant temperature, the interface becomes stabilized as shown in Fig. 24.13. During the holding period, no further transformation occurs, and then a jump in temperature is required to restart the transformation. This is apparently due to an unidentified time-dependent relaxation at the interface that occurs during the hold. The extent of transformation therefore depends primarily on the temperature and not on time. The transformation is therefore considered to be *athermal* to distinguish it from an *isothermal* transformation, which progresses with increasing time at constant temperature.

The transformation can be influenced by an applied stress. As seen in Fig. 24.13, the stress-free transformation to martensite results in a decrease in specimen length. Data in Figs. 24.14 and 24.15 were obtained by applying a series of constant uniaxial stresses at constant ambient pressure, P. The data show that the transformation temperature increases approximately linearly with applied uniaxial compressive stress. This dependence of transformation temperature on stress state follows from minimization of the appropriate thermodynamic function. For a material under

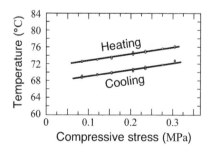

Figure 24.15: Martensite transformation temperature in In–20.7 at. % Tl as a function of applied compressive stress. From Burkart and Read [16].

uniaxial stress, this function takes the form

$$G^{\text{uni}} = U^{\text{uni}} - TS + PV - V_\circ\, \sigma^{\text{app,uni}}(1 + \epsilon^{\text{elas,uni}}) \qquad (24.10)$$

where U^{uni} is the reversible adiabatic work to take a system from a reference state to a state of uniaxial stress.[4] $\sigma^{\text{app,uni}}$ is the applied uniaxial stress above the *gauge hydrostatic stress*, $-P$, and $\epsilon^{\text{elas,uni}}$ is the *elastic* strain in the axial direction. $V_\circ$ is a reference molar volume, which can be taken to be the molar volume of the parent phase at one atmosphere (i.e., $V_\circ = V_{\text{par}}$).

Let the uniaxial strain associated with the martensite transformation be $\delta\epsilon^{\text{uni}}_{\text{mart}}$, and let $\epsilon^{\text{meas,uni}}_{\text{mar}}$ and $\epsilon^{\text{meas,uni}}_{\text{par}}$ be the *measured* uniaxial strains in the martensite and parent phases, respectively. It is not necessary that $\epsilon^{\text{meas,uni}}_{\text{mar}} = \epsilon^{\text{meas,uni}}_{\text{par}}$. The elastic parts of the uniaxial strains in the two phases will be related through their respective elastic constants because the normal components of stress must be equal at the interface.

The differential forms of the molar free energy for the parent and martensite phases are

$$
\begin{aligned}
dG^{\text{uni}}_{\text{mart}} &= -S_{\text{mart}}\, dT + V_{\text{mart}}\, dP - V_\circ(1 + \epsilon^{\text{meas,uni}}_{\text{mart}} - \delta\epsilon^{\text{uni}}_{\text{mart}})\, d\sigma^{\text{app,uni}} \\
dG^{\text{uni}}_{\text{par}} &= -S_{\text{par}}\, dT + V_{\text{par}}\, dP - V_\circ(1 + \epsilon^{\text{meas,uni}}_{\text{par}})\, d\sigma^{\text{app,uni}}
\end{aligned}
\qquad (24.11)
$$

This analysis shows that a compressive load decreases the molar free energy— and that a positive $\delta\epsilon^{\text{uni}}_{\text{mart}}$ reduces the magnitude of the decrease for the martensite phase thereby resulting in an increased transformation temperature, consistent with Fig. 24.16. Further analysis shows that the observed shift in transformation temperature results from differences in the Young's moduli of the two phases (see Exercise 24.5). This result is consistent with LeChatelier's principle.

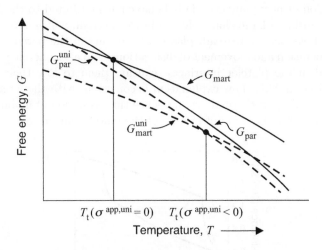

Figure 24.16:   Free energy of parent and martensite phases as a function of temperature, illustrating the effect of compressive uniaxial stress on martensite transformation temperature in In–Tl crystals.

[4]U has the differential $dU^{\text{uni}} = T\, dS - P\, dV + V_\circ \sigma^{\text{app,uni}}\, d\epsilon^{\text{elas,uni}}$. Considering that this energy change must reduce to the fluidlike $P\, dV$ work under pure hydrostatic loading, the $(1 + \epsilon_{ii})$-terms must appear because $\sum \epsilon_{ii} = \Delta V/V_\circ = V/V_\circ - 1$.

Further work found that the transformation in In–Tl alloys could be induced isothermally (i.e., without any cooling whatsoever) by the application and removal of a sufficiently large compressive load [16]. This is consistent with the data in Fig. 24.15, which show that there are conditions where the transformation temperature on cooling of the stressed specimen is above the transformation temperature of the unstressed specimen on heating, as would be required.

24.5.2   Fe–Ni System

Upon cooling, an Fe–Ni (29.3 wt. % Ni) alloy undergoes an f.c.c. solid solution $\rightarrow$ b.c.c. solid solution martensitic transformation in which the lattice deformation is an order of magnitude larger than in the In–Tl transformation and is

$$\boldsymbol{B} = [B_{ij}] = \begin{bmatrix} 1.13 & 0 & 0 \\ 0 & 1.13 & 0 \\ 0 & 0 & 0.80 \end{bmatrix} \tag{24.12}$$

The transformation is again found to be reversible and to exhibit hysteresis as shown in Fig. 24.17, which shows a cooling and heating cycle, detected by means of electrical resistivity measurements. However, the hysteresis, corresponding to about 450°C, is much larger than in the In–Tl system, indicating that a much larger pressure is required to drive the transformation. Examination of the morphology of the transformation shows that it is quite different than in the In–Tl case. The martensite now forms as small lenticular platelets embedded in the parent phase, with their habit planes parallel to variants of the invariant plane, as shown in Fig. 24.18. The manner in which the transformation progresses during cooling is also quite different. After forming, each platelet grows very rapidly to a final size and then remains static. As cooling continues, the transformation then progresses by the formation of new platelets. This behavior is attributed to the large lattice deformation, causing a large shape change in this system, which is too large to be accommodated elastically. Instead, plastic flow occurs in the parent phase in the form of the generation and movement of dislocations, and anticoherency dislocations are introduced in the platelet interfaces, causing them to lose their mobility as described in Section 24.3. This explanation is consistent with the large amount of hysteresis observed upon thermal cycling, since this reduction of mobility makes it difficult to reverse the direction of motion of the platelet interfaces.

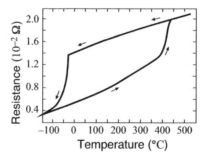

Figure 24.17:   Temperature dependence of the martensitic transformation in the Fe–Ni (29.3 wt. %) system during thermal cycle. Extent of transformation revealed by change of specimen electrical resistivity. From Kaufman and Cohen [17].

Figure 24.18:   Martensite platelets formed in the f.c.c. → b.c.c. transformation in an Fe–32 wt. % Ni alloy. From the ASM *Metals Handbook,* Vol. 8, p. 198.

The phenomenon of stabilization is also observed in this system if the cooling is interrupted and the specimen is held isothermally before cooling is resumed. In this case, the transformation resumes only after the driving force is incrementally increased by a significant drop in temperature. Again, the transformation is primarily athermal, depending upon decreases of temperature which provide corresponding increases in the driving pressure for the formation of more platelets. Also, a relatively small amount of isothermal formation of martensite is observed if the specimen is rapidly quenched into the temperature range where martensite forms and is then held isothermally [18]. However, the isothermal transformation occurs by the formation of new platelets and not by the growth of existing ones.

In general, the result that the platelets form very rapidly (at speeds of the order of the speed of sound) at relatively low temperatures, at rates that are not significantly temperature-dependent, indicates that the platelet growth is not thermally-activated and occurs only when a sufficiently high driving pressure is available.

24.5.3   Fe–Ni–C System

The crystallography of the f.c.c.→ b.c.t. martensitic transformation in the Fe–Ni–C system (with 22 wt. %Ni and 0.8 wt. %C) has been described in Section 24.2. In this system, the high-temperature f.c.c. solid-solution parent phase transforms upon cooling to a b.c.t. martensite rather than a b.c.c. martensite as in the Fe–Ni system. Furthermore, this transformation is achieved only if the f.c.c. parent phase is rapidly quenched. The difference in behavior is due to the presence of the carbon in the Fe–Ni–C alloy. In the Fe–Ni alloy, the b.c.c. martensite that forms as the temperature is lowered is the equilibrium state of the system. However, in the Fe–Ni–C alloy, the equilibrium state of the system in the low-temperature range is a two-phase mixture of a b.c.c. Fe–Ni–C solid solution and a C-rich carbide phase.[5] This difference in behavior is due to a much lower solubility of C in the low-temperature b.c.c. Fe–Ni–C phase than in the high-temperature f.c.c. Fe–Ni–C phase. If the high-temperature

---

[5]The true equilibrium state is the Fe–Ni–C phase plus graphite. However, the carbide phase is so strongly metastable that it can be regarded as an "equilibrium" phase.

f.c.c. Fe–Ni–C parent phase were to be slowly cooled under quasi-equilibrium conditions, it would undergo diffusional phase changes resulting in the ultimate formation of the two-phase mixture. However, if the parent phase is rapidly quenched, these phase changes are bypassed and it transforms martensitically to the solid-solution b.c.t. phase, which is therefore a nonequilibrium phase that is metastable to the formation of the equilibrium two-phase mixture. During the quench, the C atoms are trapped in the interstitial positions they occupied in the parent phase, as shown in Fig. 24.3. By comparing these positions with Fig. 8.8a, it may be seen that they are a subset of the complete set of lattice-equivalent interstitial sites that carbon atoms can occupy in the b.c.c. structure.[6] Carbon atoms occupying interstitial sites generally act as positive centers of dilation that push most strongly against their nearest-neighbors. The carbon atoms that randomly occupy the sites in Fig. 24.3 push most strongly along the z axis and so produce the observed tetragonality. The b.c.t. phase can be considered as a b.c.c. structure that has been forced into tetragonality by quenched-in C atoms that occupy positions inherited from the parent f.c.c. phase.

Once the system is cooled to a low enough temperature to preclude any carbide formation due to diffusion, further martensite can be produced by further drops in temperature. The overall transformation on cooling then has many of the features of the transformation in the Fe–Ni alloy described above. The shape change is large, the martensite forms as embedded lenticular platelets, and the formation is athermal and requires continuously decreasing temperatures to proceed significantly. However, the transformation is not reversible as in theFe–Ni system. When the Fe–Ni–C martensite is heated, it decomposes by precipitating the more stable carbide phase before it is able to transform back to the high-temperature f.c.c. parent phase.

This behavior is typical of steels that are alloys composed mainly of iron and carbon and, in many cases, additional alloying elements such as nickel, chromium, or manganese. The martensite formed directly after quenching is exceedingly hard but quite brittle. However, it can then be toughened by subsequent heating (tempering), which allows some controlled carbide precipitation. Extraordinary mechanical properties can be obtained by this combination of quenching and tempering, and it forms the basis for the heat treatment of steel [15].

Bibliography

1. W.S. Wechsler, D.S. Lieberman, and T.A. Read. On the theory of the formation of martensite. *Trans. AIME*, 197(11):1503–1515, 1953.

2. J.S. Bowles and J.K. MacKenzie. The crystallography of martensite transformations I. *Acta Metall.*, 2(1):129–137, 1954.

3. J.S. Bowles and J.K. MacKenzie. The crystallography of martensite transformations II. *Acta Metall.*, 2(1):138–147, 1954.

4. J.S. Bowles and J.K. MacKenzie. The crystallography of martensite transformations. III. Face-centered cubic to body-centered tetragonal transformations. *Acta Metall.*, 2(2):224–234, 1954.

5. C.M. Wayman. *Introduction to the Crystallography of Martensitic Transformations.* Macmillan, New York, 1964.

[6]Note that the number of carbon atoms occupying these sites is considerably smaller than the number of sites and that the sites are therefore sparsely populated.

6. J.W. Christian. Martensitic transformations. In R.W. Cahn, editor, *Physical Metallurgy*, pages 552–587. North-Holland, New York, 1970.

7. M. Cohen and C.M. Wayman. Fundamentals of martensitic reactions. In J.K. Tien and J.F. Elliott, editors, *Metallurgical Treatises*, pages 455–468. The Metallurgical Society of AIME, Warrendale, PA, 1981.

8. G.B. Olson and M. Cohen. Theory of martensitic nucleation: A current assessment. In *Proceedings of an International Conference on Solid→Solid Phase Transformations*, pages 1145–1164, Warrendale, PA, 1982. The Metallurgical Society of AIME.

9. G.B. Olson and M. Cohen. Dislocation theory of martensitic transformations. In F.R.N. Nabarro, editor, *Dislocations in Solids, Vol. 7*, pages 295–407. North-Holland, New York, 1986.

10. C.S. Barrett and T.B. Massalski. *Structure of Metals: Crystallographic Methods, Principles and Data*. Pergamon Press, New York, 3rd edition, 1980.

11. J.M. Ball and R.D. James. Fine phase mixtures as minimizers of energy. *Arch. Rat. Mech. Anal.*, 100:13–52, 1987.

12. J.M. Ball. The calculus of variations and materials science. *Quart. Appl. Math.*, 56(4):719–740, 1998.

13. J.M. Ball and R.D. James. Theory for the microstructure of martensite and applications. In *Proceedings of the International Conference on Martensitic Transformations*, pages 65–76, Monterey, CA, 1993. Monterey Institute for Advanced Studies.

14. R.E. Cech and D. Turnbull. Heterogeneous nucleation of the martensite transformation. *Trans. AIME*, 206:124–132, 1956.

15. R.E. Reed-Hill and R. Abbaschian. *Physical Metallurgy Principles*. PWS-Kent, Boston, 1992.

16. M.W. Burkart and T.A. Read. Diffusionless phase change in the indium–thallium system. *Trans. AIME*, 197:1516–1524, 1953.

17. L. Kaufman and M. Cohen. The martensitic transformation in the iron–nickel system. *Trans. AIME*, 206:1393–1400, 1956.

18. E.S. Machlin and M. Cohen. Isothermal mode of the martensitic transformation. *Trans. AIME*, 194:489–500, 1952.

19. D.S. Lieberman. Martensitic transformations and determination of the inhomogeneous deformation. *Acta Metall.*, 6:680–693, 1958.

20. J.F. Nye. *Physical Properties of Crystals*. Oxford University Press, Oxford, 1985.

EXERCISES

24.1 It has been stated that "a martensitic phase transformation can be considered as the spontaneous plastic deformation of a crystalline solid in response to internal chemical forces" [9]. Give a critique of this statement.

Solution. According to Eq. 24.1, forward and reverse martensitic transformations can be driven either by internal chemical forces derived from the bulk "chemical" free-energy change, Δg_B, or by forces due to applied stress. In all cases, the transformation causes a shape change that corresponds to plastic deformation. If we regard transformations that occur due to heating or cooling in the absence of applied stress as *spontaneous* and transformations that occur due to applied stress as *driven* then the statement is true. A more inclusive statement might be: "a martensitic phase transformation can be considered as the plastic deformation of a crystalline solid in response to internal chemical forces and/or applied mechanical forces."

24.2 Find an expression for the cone angle, ϕ', in Fig. 24.4 in terms of η_1 and η_3.

Solution. First find the equation for the $A'O'B'$ cone in Fig. 24.4 by setting the equation for the unit sphere, $x_1'^2 + x_2'^2 + x_3'^2 = 1$, equal to Eq. 24.2 to obtain

$$\left(1 - \frac{1}{\eta_1^2}\right)x_1'^2 + \left(1 - \frac{1}{\eta_2^2}\right)x_2'^2 + \left(1 - \frac{1}{\eta_3^2}\right)x_3'^2 = 0 \qquad (24.13)$$

Then, setting $x_2' = 0$ yields

$$\tan\phi' = \frac{x_1'}{x_3'} = \frac{\eta_1}{\eta_3}\sqrt{\frac{1 - \eta_3^2}{\eta_1^2 - 1}} \qquad (24.14)$$

24.3 Section 24.2.3 claims that the rotation axis in the final rigid-body rotation, $\boldsymbol{R}$, which rotates $\vec{a}'' \to \vec{a}$ and $\vec{c}'' \to \vec{c}$ in Fig. 24.9 is located at the position $\vec{u}$. By using the stereographic method, show (within the recognized rather low accuracy of the method) that this is indeed the case.

- The axis of rotation required to bring $\vec{a}'' \to \vec{a}$ by a rigid-body rotation must lie somewhere on a plane normal to the vector $(\vec{a}'' - \vec{a})$.

- Similarly, the axis of rotation required to bring $\vec{c}'' \to \vec{c}$ must lie somewhere on a plane normal to $(\vec{c}'' - \vec{c})$.

- These two rotations can therefore be accomplished simultaneously by a single rotation around a common axis lying along the intersection of these two planes. This axis will therefore be parallel to

$$\vec{u} = (\vec{a}'' - \vec{a}) \times (\vec{c}'' - \vec{c}) \qquad (24.15)$$

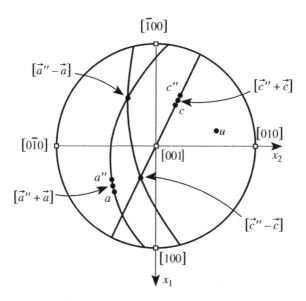

Figure 24.19:   Stereogram showing the method for locating the rotation axis, $\vec{u}$, for the rigid-body rotation, $\boldsymbol{R}$, in Section 24.2.3. From Lieberman [19].

Solution. First find the poles of the vectors $(\vec{a}''' - \vec{a})$ and $(\vec{c}'' - \vec{c})$. The rotation axis, $\vec{u}$, will be the pole of the plane containing these vectors. On a stereogram, this will be the pole of the great circle containing both $(\vec{a}''' - \vec{a})$ and $(\vec{c}'' - \vec{c})$. The vector $(\vec{a}''' - \vec{a})$ is perpendicular to the vector $(\vec{a}''' + \vec{a})$, and they both lie in the same plane. The vector $(\vec{a}''' + \vec{a})$ lies on a great circle going through both $\vec{a}'''$ and $\vec{a}$ and lies midway between them as indicated in Fig. 24.19. Therefore, $(\vec{a}''' - \vec{a})$ lies on this same great circle 90° away from $(\vec{a}''' + \vec{a})$. A similar procedure yields the pole of $(\vec{c}'' - \vec{c})$. The final step is to locate $\vec{u}$ at the pole of the great circle going through both $(\vec{a}''' - \vec{a})$ and $(\vec{c}'' - \vec{c})$.

24.4 In Section 24.3 we pointed out that martensite platelets (Fig. 24.12) can be accommodated elastically in the parent phase when the lattice deformation and shape change are small. Consider such platelets in a polycrystalline parent phase where the platelets have grown across the grains and are stopped at the grain boundaries as in Fig. 24.20. Upon thermal cycling, such a plate will reversibly thicken during cooling and thin during heating due to a "thermoelastic" equilibrium that is reached between changes in its bulk free energy, Δg_B, and the elastic strain energy in the system. Approximate the platelet shape by a thin disclike ellipsoid of aspect ratio c/a as in Section 19.1.3 (Eq. 19.23) and show that the platelet thickness, c, and Δg_B are related by

$$c = -\frac{a}{2A}\Delta g_B \qquad (24.16)$$

where A = constant. Assume an invariant plane strain habit plane and use the elastic-energy expression for an invariant plane strain described in Section 19.1.3.

Figure 24.20: Martensite platelet stopped at grain boundaries in polycrystalline parent phase.

Solution. According to Section 19.1.3, the elastic strain energy (per unit volume of platelet) is proportional to c/a. The free energy associated with the platelet can then be written in the usual way as the sum of a bulk term, an elastic energy term, and an interfacial energy term,

$$\Delta \mathcal{G} = \frac{4}{3}\pi a^2 c\,\Delta g_B + \frac{4}{3}\pi a^2 c\, A\frac{c}{a} + 2\pi a^2 \gamma \qquad (24.17)$$

Here, the interfacial area has been approximated by that of a thin disc. Because a is held constant, the thermoelastic equilibrium requires that $\partial \Delta \mathcal{G}/\partial c = 0$, and this leads directly to the condition

$$c = -\frac{a}{2A}\Delta g_B \qquad (24.18)$$

24.5 Figure 24.15 shows that the martensitic transformation temperature in the In–Tl system is raised by applying a constant uniaxial compressive stress. Using the thermodynamic formalism leading to Eq. 24.11, develop a Clausius–Clapeyron relationship that relates the observed effect of applied stress on transformation temperature to thermodynamic quantities.

Solution. Taking $\Delta G^{\rm uni}$, ΔS, and ΔV as the molar changes for the transformation *parent→martensite*, then

$$d\Delta G^{\rm uni} = -\Delta S\, dT + \Delta V\, dP - V_\circ(\epsilon_{\rm mart}^{\rm meas,uni} - \epsilon_{\rm par}^{\rm meas,uni} - \delta\epsilon_{\rm mart}^{\rm uni})\, d\sigma^{\rm app,uni} \qquad (24.19)$$

At equilibrium, $\Delta G^{\rm uni} = 0$ and

$$\sigma^{\rm app,uni} = (\epsilon_{\rm mart}^{\rm meas,uni} - \delta\epsilon_{\rm mart}^{\rm uni})E_{\rm mart} = \epsilon_{\rm par}^{\rm meas,uni} E_{\rm par} \qquad (24.20)$$

if the applied stress is below the elastic limit for each phase and $E_{\rm mart}$ and $E_{\rm par}$ are the Young's moduli for each phase.[7] At thermodynamic equilibrium subject to linear elasticity, the Gibbs–Duhem equation is

$$0 = -\Delta S\, dT + \Delta V\, dP - \frac{V_\circ}{2}\left(\frac{1}{E_{\rm mart}} - \frac{1}{E_{\rm par}}\right)d(\sigma^{\rm app,uni})^2 \qquad (24.21)$$

At fixed (ambient) pressure, a Clausius–Clapeyron equation relates the change in transformation temperature with applied uniaxial load:

$$\frac{dT}{d(\sigma^{\rm app,uni})^2} = -\frac{V_\circ}{2\,\Delta S}\left(\frac{1}{E_{\rm mart}} - \frac{1}{E_{\rm par}}\right) = -\frac{V_\circ T_\circ (E_{\rm par} - E_{\rm mart})}{2 E_{\rm par} E_{\rm mart}\,\Delta H} \qquad (24.22)$$

where ΔH is the heat absorbed during transformation under no load at the reference temperature $T_\circ$.

24.6 Figure 24.21 shows a two-dimensional martensitic transformation in which a parent phase, P, is transformed into a martensitic phase, M, by a lattice deformation, $\boldsymbol{B}$. Note that there is no invariant line in this two-dimensional transformation. Find a lattice-invariant deformation, $\boldsymbol{S}$, and a rigid rotation, $\boldsymbol{R}$, that together with the lattice deformation, $\boldsymbol{B}$, produce an overall deformation given by

$$\boldsymbol{E} = \boldsymbol{R}\boldsymbol{S}\boldsymbol{B} \qquad (24.23)$$

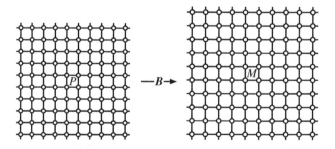

Figure 24.21:   Two-dimensional transformation of parent phase, P, to martensitic phase, M, by the lattice deformation, $\boldsymbol{B}$.

[7]It is assumed that the interface is normal to the applied load. If either phase has anisotropic elastic coefficients, the generalized Young's modulus should be calculated as described by Nye [20].

which produces an invariant line which could then serve as the habit line of the transformation. Accomplish the lattice invariant deformation by means of slip.

- There are many possible solutions to this exercise. Find any one of them.

Solution. One solution is:

(1) Select the proposed interface between the parent phase and the region of the parent phase that will transform to martensite. This lies between AB and $A'B'$ in Fig. 24.22a.

(2) Detach the portion on the right and transform it to martensite as shown in Fig. 24.22b by imposing the lattice deformation, B, illustrated in Fig. 24.21.

(3) Next, as shown in Fig. 24.22c, impose a lattice invariant deformation, S, on the martensite by means of slip on planes of the type indicated so that $|AB| = |A'B'|$.

(4) Finally, rotate the martensite by R as shown in Fig. 24.22d to produce an invariant line along AB. The interface is shown in the unrelaxed state.

Similar procedures can be used to find alternate solutions.

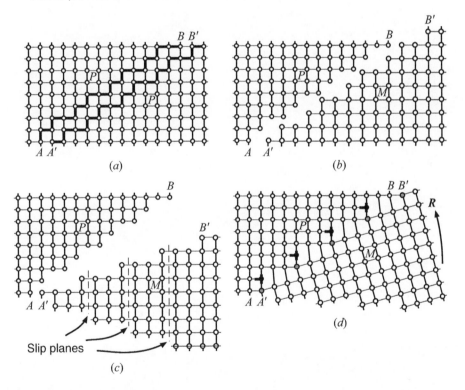

Figure 24.22:   Production of an invariant line (habit line) along AB in a two-dimensional transformation of a parent phase, P, to a martensitic phase, M. The degree of matching of phases is indicated in **(d)** by shading shared sites in the interface.

APPENDIX A

DENSITIES, FRACTIONS, AND ATOMIC

VOLUMES OF COMPONENTS

A.1 CONCENTRATION VARIABLES

Care is required in defining concentration variables for materials. In the following, consider a material comprised of N_i atoms or molecules of type i in a system of N_c components which together occupy a volume V^{tot}. The atomic or molecular weight of each component is M_i°.

Crystalline materials have distinct structures with sites distinguished by their symmetry, and it may be important to specify occupancies of particular types of sites. Vacant sites must be considered as well.

A.1.1 Mass Density

The mass density of material, ρ, is the amount of mass of the material per unit volume (i.e., kg m^{-3}). For component i, the mass density, ρ_i, is therefore

$$\rho_i = \frac{N_i M_i^\circ}{V^{\text{tot}}} \qquad \rho = \sum_{i=1}^{N_c} \rho_i \qquad (A.1)$$

where components $(1, 2, \ldots, N_c)$ include all of the species that make up the material possessing total density ρ. For example, an alloy of copper and zinc has five stoichiometric phases—α (pure Cu), β (Cu$_5$Zn$_8$), γ (CuZn), ϵ (CuZn$_3$), and η (pure Zn)—but only two of the five are independent in a closed system.

Note that for vacancies in crystalline phases, $\rho_V = 0$ because $M_V^\circ = 0$.

A.1.2   Mass Fraction

The mass fraction, ξ_i, is the fraction of the total mass of the material associated with component i:

$$\xi_i = \frac{\rho_i}{\rho} \qquad \sum_{i=1}^{N_c} \xi_i = 1 \tag{A.2}$$

A.1.3   Number Density or Concentration

The number density or concentration, c_i, is the number of atoms, molecules, moles, or other entities of component i per unit volume. Therefore,

$$c_i = \frac{N_i}{V^{\text{tot}}} \qquad \sum_{i=1}^{N_c} c_i = c^{\text{tot}} \tag{A.3}$$

Note that for vacancies in crystalline phases, $c_V \geq 0$.

A.1.4   Number, Mole, or Atom Fraction

The number fraction of component i is

$$X_i = \frac{N_i}{N^{\text{tot}}} = \frac{c_i}{c^{\text{tot}}} \qquad \sum_{i=1}^{N_c} X_i = 1 \tag{A.4}$$

A set of independent number fractions $(X_1, X_2, \ldots, X_{N-1})$ specifies a composition.

A.1.5   Site Fraction

The site fraction is the number of species of a particular component that occupy a particular site divided by the total number of sites of that type. For example, in sodium chloride (NaCl) there is a distinction between cation and anion sites. Impurity species and vacancies may also be present. If there is a total of s distinct types of sites ($s = 2$ in NaCl) and there is a total number, N_j^{tot}, of sites of type j on which are distributed N_i^j atoms (molecules) of component i, the fraction of sites of type j occupied by component i is

$$X_i^j = \frac{N_i^j}{N_j^{\text{tot}}} \qquad \sum_{i=1}^{N_c} X_i^j = 1 \tag{A.5}$$

A.2   ATOMIC VOLUME

The atomic volume of component i, Ω_i, is the volume associated with one atom, molecule, or other entity. The total volume, V^{tot}, is comprised of contributions from each component:

$$dV^{\text{tot}} = \sum_{i=1}^{N_c} \frac{\partial V^{\text{tot}}}{\partial N_i} \, dN_i \tag{A.6}$$

Therefore, upon an Euler-type integration,

$$V^{\text{tot}} = \sum_{i=1}^{N_c} \Omega_i N_i \tag{A.7}$$

where $\Omega_i \equiv \partial V^{\text{tot}}/\partial N_i$ is the *atomic volume* of component i.[1]

Dividing Eq. A.7 through by V^{tot} yields the relation

$$\sum_{i=1}^{N_c} \Omega_i c_i = 1 \tag{A.8}$$

Two differential relationships between the Ω_i and c_i can be derived as follows:

$$\sum_{i=1}^{N_c} c_i \, d\Omega_i = \sum_{i=1}^{N_c} \frac{N_i}{N^{\text{tot}}} \, d\Omega_i \quad \text{and} \quad dV^{\text{tot}} = \sum_{i=1}^{N_c} \left(N_i \, d\Omega_i + \Omega_i \, dN_i \right)$$

$$\sum_{i=1}^{N_c} c_i \, d\Omega_i = \frac{1}{N^{\text{tot}}} \left(dV^{\text{tot}} - \sum_{i=1}^{N_c} \Omega_i \, dN_i \right) = \frac{1}{N^{\text{tot}}} \left(dV^{\text{tot}} - \sum_{i=1}^{N_c} \frac{\partial V^{\text{tot}}}{\partial N_i} \, dN_i \right) = 0 \tag{A.9}$$

and because the total differential of $1 = \sum \Omega_i c_i$ must vanish,

$$\sum_{i=1}^{N_c} \Omega_i \, dc_i = 0 \tag{A.10}$$

The average atomic volume, $\langle \Omega \rangle$, is

$$\langle \Omega \rangle = \frac{V^{\text{tot}}}{N^{\text{tot}}} = \frac{\sum_{i=1}^{N_c} \Omega_i N_i}{N^{\text{tot}}} = \sum_{i=1}^{N_c} \Omega_i X_i \tag{A.11}$$

Also,

$$\langle \Omega \rangle = \frac{V^{\text{tot}}}{N^{\text{tot}}} = \frac{V^{\text{tot}}/N_i}{N^{\text{tot}}/N_i} = \frac{1/c_i}{1/X_i} = \frac{X_i}{c_i} \tag{A.12}$$

---

[1] As defined here, Ω_i is the *partial* atomic volume; for simplicity, we will refer to it as the atomic volume.

APPENDIX B

STRUCTURE OF CRYSTALLINE

INTERFACES

The interfaces of importance in kinetic processes possess a wide range of structures and properties. In this appendix we classify and describe concisely the different types of crystalline materials' interfaces relevant to kinetic processes. The different types of point and line defects that may exist in these interfaces are also described.[1]

B.1 CRYSTALLINE INTERFACES AND THEIR GEOMETRICAL DEGREES OF FREEDOM

Interfaces that involve a crystalline material may be classified in different ways. The broadest system of classification is based on the state of matter abutting the crystal:

- Crystal/vapor interfaces

- Crystal/liquid interfaces

- Internal interfaces in solid and/or crystalline materials

[1]Further information and references may be found in several references [1–3].

These interface types are listed in order of increasing complexity. Crystal/vapor and crystal/liquid interfaces both possess two macroscopic geometrical degrees of freedom corresponding to the parameters required to specify the inclination of the interface plane with respect to the crystal axes.[2] (A convenient choice is the two direction cosines necessary to define a unit vector normal to the interface.) However, the structure of crystal/liquid interfaces is generally more complicated because the first few atomic layers on the liquid side of the interface are significantly affected by the presence of the interface and therefore act as part of the interface. A crystal/crystal interface possesses five macroscopic geometrical degrees of freedom corresponding to the three parameters that specify the misorientation of the two crystals which abut the interface and the two parameters that specify the inclination of the interface plane which separates them. (If the misorientation is described as a rotation of one crystal with respect to the other about a specified axis, the three parameters are then the two direction cosines necessary to specify the rotation axis as a unit vector and the magnitude of the rotation angle.)

B.2 SHARP AND DIFFUSE INTERFACES

Interfaces may be *sharp* or *diffuse*. A sharp interface possesses a relatively narrow core structure with a width close to an atomic nearest-neighbor separation distance. Examples of sharp crystal/vapor and crystal/crystal interfaces are shown in Figs. B.1 and B.2.

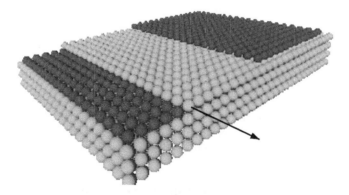

Figure B.1:   Ledged surface in a b.c.c. structure that is vicinal to the {100} singular interface. Body-centered positions are darkened for contrast only.

On the other hand, a diffuse interface possesses a significantly wider core that extends over a number of atomic distances. A diffuse crystalline/amorphous phase interface is shown in Fig. B.3. Similar structures exist in crystal/liquid interfaces [5].

Diffuse crystal/crystal interfaces often appear in systems subject to incipient chemical or structural instabilities associated with phase separation, long-range ordering, or displacive phase transformations [2]. Examples of interfaces associated with the first two types are shown in Fig. 18.7.

---

[2]The number of geometrical degrees of freedom is the number of geometrical parameters that must be specified in order to define the interface.

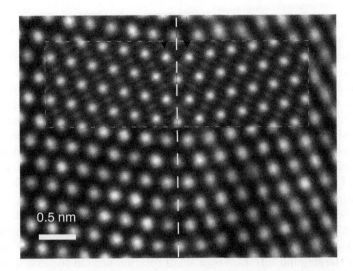

Figure B.2:   Symmetric large-angle $(113)[\bar{1}10]$ tilt boundary in Al viewed along the $[\bar{1}10]$ tilt axis by high-resolution electron microscopy. The tilt angle is $50.48°$. The inset shows a simulated image [4]. Reprinted, by permission, from K.L. Merkle, L.J. Thompson, and F. Phillipp, "Thermally activated step motion observed by high-resolution electron microscopy at a (113) symmetric tilt grain-boundary in aluminum," *Philosophical Magazine Letters*, vol. 82, pp. 589–597. Copyright © 2002 by Taylor and Francis Ltd., http://www.tandf.co.uk/journals.

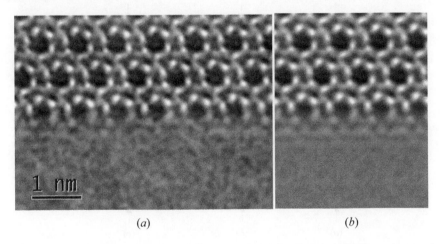

(a)   (b)

Figure B.3:   (a) High-resolution TEM image of interface between Si_3N_4 and amorphous yttrium silicate. (b) Digitally averaged according to the 0.76 nm periodicity along the interface, revealing a gradual loss of order in the interfacial region. Micrographs courtesy Markus Döblinger.

B.3 SINGULAR, VICINAL, AND GENERAL INTERFACES

Interfaces can be further classified as singular interfaces, vicinal interfaces, and general interfaces. An interface is regarded as singular with respect to a degree of freedom if it is at a local minimum of energy with respect to changes in that degree of freedom. It is therefore of relatively low energy and is stable against changes in that degree of freedom. Singular crystal/vapor and crystal/liquid interfaces tend to have dense, relatively close-packed atomic planes in the crystalline phase lying parallel to the interface plane [3]. Singular crystal/crystal interfaces have dense planes parallel to the interface, and their structures have short two-dimensional periodicity in the interface plane [2]. An example is shown in Fig. B.2.

A vicinal interface possesses an interfacial free energy near a local minimum with respect to a macroscopic degree of freedom. The structure of such an interface generally consists of the singular interface at the local minimum containing a superimposed array of discrete line defects, which may be ledges, dislocations, or line defects possessing both ledge and dislocation character. The superimposed array of line defects accommodates the difference between the misorientation and/or inclination of the vicinal interface and that of the nearby singular interface. Vicinal interfaces adopt this type of structure because most of the interface area corresponds to the minimum-energy structure of the nearby singular interface. In the example of a vicinal crystal/vapor interface shown in Fig. B.1, the inclination of the interface is almost parallel to the nearby (100) singular interface and differs from that of the singular interface by a small rotation around the axis shown.[3] The vicinal interface therefore consists of the nearby singular interface with a superimposed array of ledges which accommodates the difference between the inclination of the interface and the inclination of the nearby singular interface.

Examples of vicinal crystal/crystal interfaces are shown in Figs. B.4c, B.5, and B.6. The vicinal interface therefore consists of the singular interface containing a

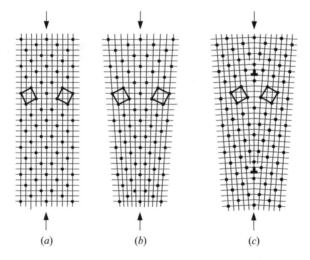

(a) (b) (c)

Figure B.4:   (a) Singular large-angle symmetrical tilt boundary in f.c.c. structure viewed along ⟨100⟩ tilt axis. The tilt angle is 53.1°. The grid is the DSC-lattice of the bicrystal. (b) Establishment of a slightly increased tilt angle [relative to (a)] while maintaining coherence across the boundary. (c) Introduction of dislocations to eliminate the long-range stresses generated in (b). The added dislocation array results in a boundary free of long-range stress and vicinal to the boundary in (a).

---

[3] Although no vapor phase is present in the figure, the surface is interpreted as being in equilibrium with its vapor phase. For many materials, the equilibrium vapor pressure is very small—nevertheless, the differences of surface structure in a vacuum environment compared to the structure in low vapor pressures can be significant.

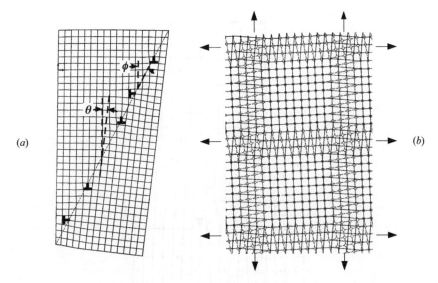

(a)   *(b)*

Figure B.5:   **(a)** Small-angle asymmetric tilt boundary in a primitive cubic lattice viewed along the [100] tilt axis. **(b)** Small-angle twist boundary in a primitive cubic lattice viewed along the [100] twist axis. The open circles represent atoms just above the boundary midplane, and the solid circles are atoms just below. Arrows indicate screw dislocations in the interface structure. From Read [6].

superimposed array of dislocations that accommodates this difference in misorientation angle. In this example, the Burgers vectors of the dislocations are translation vectors of the DSC-lattice (see Fig. B.4a) which is associated with the bicrystal containing the singular interface.[4] In Fig. B.5, interfaces of small crystal misorientation are vicinal to corresponding singular "interfaces" possessing zero degrees of crystal misorientation. In these instances, the perfect crystal is the limiting case of a bicrystal with zero crystal misorientation.

A *general interface* is far from any singular interface with respect to its macroscopic geometric degrees of freedom. It is therefore far from any local energy minimum. General interfaces tend to have high-index planes of the adjoining crystal or crystals running parallel to the interface and possess either very long-period or quasi-periodic structures.

B.4   HOMOPHASE AND HETEROPHASE INTERFACES

Interfaces may also be classified broadly into homophase interfaces and heterophase interfaces. A *homophase interface* separates two regions of the same phase, whereas a *heterophase interface* separates two dissimilar phases. Crystal/vapor and crystal/liquid interfaces are heterophase interfaces. Crystal/crystal interfaces can be either homophase or heterophase. Examples of crystal/crystal homophase interfaces are illustrated in Figs. B.2, B.4, and B.5. Examples of heterophase crystal/crystal interfaces are shown in Figs. B.6 and B.7. Figure B.6a shows an interface between f.c.c. and h.c.p. crystals where the small mismatch between close-packed $\{111\}_{\text{fcc}}$

[4]A full description of the DSC-lattice is given by Sutton and Balluffi [2]. Note that the DSC-lattice of a single crystal is the crystal lattice itself.

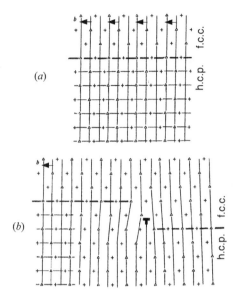

Figure B.6: (a) Singular heterophase interface between an f.c.c. and h.c.p. structures viewed along the $[110]_{fcc}$. Close-packed $\{111\}_{fcc}$ and $\{0001\}_{hcp}$ planes match along the interface. The grid is the DSC-lattice corresponding to the bicrystal. (b) Same as (a) except that the interface is now rotated into a slightly different inclination about an axis normal to the paper. This interface has adopted a stepped structure and is vicinal to the one in (a). From *Interfaces in Crystalline Materials* by A.P. Sutton and R.W. Balluffi (1995). Reprinted by permission of Oxford University Press [2].

and $\{0001\}_{hcp}$ planes is accommodated by elastic strains. If the interface plane is rotated slightly around an axis normal to the plane of the paper while keeping the crystal misorientation constant, the new interfacial structure will consist of the original interface containing an array of superimposed line defects of the type shown in Fig. B.6b. These line defects possess both ledge and dislocation character. Such an interface is therefore vicinal to the singular interface in Fig. B.6a.

B.5 GRAIN BOUNDARIES

Homophase crystal/crystal interfaces are often called *grain boundaries*. It is customary to classify such boundaries as either small-angle grain boundaries or large-angle grain boundaries.

Small-angle grain boundaries, which are interfaces for which the angle of crystal misorientation is less than about 15°, consist of arrays of discrete dislocations as illustrated in Fig. B.5. The dislocations possess Burgers vectors that are translation vectors of the crystal lattice, and the dislocations accommodate the crystal misorientations of the boundaries. These boundaries are vicinal to corresponding *singular boundaries* possessing no crystal misorientation in the fictive perfect-crystal lattice. As the crystal misorientation increases, more dislocations must be added to compensate for the increased misorientation, and the dislocation spacings therefore decrease. When the misorientation reaches about 15°, the dislocation spacing becomes sufficiently small so that the cores of the dislocations begin to overlap. At

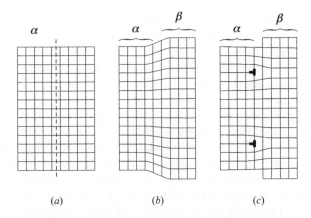

Figure B.7:   Construction of a heterophase interface. **(a)** Reference crystal taken to be the α phase. **(b)** Transformation of the region on the right of the desired interface into the β phase while maintaining coherence. **(c)** Elimination of the long-range stresses present in (b) by the introduction of an array of dislocations in the α/β interface.

this point, the boundaries become, in essence, continuous slabs of dislocation core material that can no longer be described as arrays of discrete lattice dislocations. Boundaries of this misorientation, or larger, are termed *large-angle boundaries*.

Grain boundaries can also be classified as tilt boundaries, twist boundaries, and mixed boundaries. A *tilt boundary*'s plane is parallel to the rotation axis used to define its crystal misorientation, as in Fig. B.4c. The crystals adjoining the boundary are related by a simple tilt around this axis. A *twist boundary*, as in Fig. B.5b, is a boundary whose plane is perpendicular to the rotation axis. The two crystals adjoining the boundary are then related by a simple twist around this axis. All other types of boundaries are considered to be *mixed*.

B.6 COHERENT, SEMICOHERENT, AND INCOHERENT INTERFACES

All sharp crystal/crystal homophase and heterophase interfaces can be classified as coherent, semicoherent, and incoherent. The structural features of these interfaces can be revealed by constructing them using a series of operations which always starts with a reference structure.

The construction of the heterophase interface between α and β phases in Fig. B.7c starts with a reference structure, which is taken to be the single crystal of α phase in Fig. B.7a. The interface is to be located along the plane indicated by the dashed line. In the first operation, the portion of the α crystal on the right of the desired interface plane is transformed into the β phase while maintaining registry along the interface as illustrated in Fig. B.7b. The resulting interface is *coherent* because the two crystals adjoining it are maintained in registry. Long-range coherency stresses are required to maintain the interface registry.

In a further operation, these stresses can be eliminated by introducing an array of dislocations in the interface as in Fig. B.7c. The resulting interface consists of patches of coherent interface separated by dislocations. The cuts and displacements necessary to introduce the dislocations destroy the overall coherence of the interface, which is therefore considered to be *semicoherent* with respect to the reference structure in Fig. B.7a. Because of the good atomic matching across coherent interfaces, the energetic contribution from mismatch is generally small. The energy of semicoherent interfaces is minimized when most of interfacial area consists of patches of the coherent reference structure. This reduces the core width of the line defects that delineate the coherent regions of the interface, and the result is well-defined fit–misfit structures containing line defects with localized cores.

Semicoherent interfaces can also be constructed by employing a bicrystal containing a periodic interface as a reference structure. The initial reference structure is the bicrystal in Fig. B.4a. A new boundary of increased misorientation can be produced by increasing the misorientation angle while maintaining coherence everywhere as in Fig. B.4b. Long-range stresses are required to maintain coherency, but they may again be relieved by introducing an array of dislocations as in Fig. B.4c. The result is a semicoherent interface consisting of patches of the coherent interface of the reference structure separated by dislocations that have destroyed the overall boundary coherence. In this example, the Burgers vectors of the dislocations are translation vectors of the DSC-lattice of the reference bicrystal.

The coherence attributed to a semicoherent interface is the coherence of the reference structure, which in different situations can be either a single crystal or a bicrystal containing a periodic interface. (A single crystal is the limiting case of a bicrystal containing an interface of zero misorientation.) The reference structure must be specified in any meaningful description of interface coherence. The Burgers vectors of the dislocations in a semicoherent interface will generally be translation vectors of the DSC-lattice of the reference structure. The bicrystal reference structures, which are of most physical relevance, will generally contain interfaces of relatively low energy.

It is often useful to describe the dislocation content of coherent and semicoherent interfaces in terms of another framework which employs *coherency dislocations* and *anticoherency dislocations*. The basic idea is illustrated in Fig. B.8, which shows the same two boundaries shown previously in Fig. B.7b and c. The coherency dislocations possess a stress field equivalent to the long-range coherency stresses associated with the coherent interface. They are not "real" dislocations in the

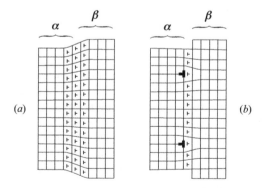

Figure B.8:   (a) Same structure as in Fig. B.7b. However, the presence of an array of coherency dislocations is indicated. (b) Same structure as in Fig. B.7c. The coherency dislocations shown in (a) are again present (in a more localized distribution), and an array of anticoherency dislocations has been added. From *Interfaces in Crystalline Materials,* by A.P. Sutton and R.W. Balluffi (1995). Reprinted by permission of Oxford University Press [2].

conventional sense—they are line defects, but they do not contain bad material in their cores. However, they serve as constructs to model the displacement and stress fields associated with the coherent interface. These long-range stresses are then eliminated by adding *anticoherency dislocations* as shown in Fig. B.8b. These dislocations destroy the boundary coherence, and the result is the same semicoherent interface free of long-range stress as in Fig. B.7c. However, the interface is now considered to contain two sets of dislocations—coherency dislocations and anticoherency dislocations—whose long-range stress fields (and Burgers vectors) cancel. Coherency and anticoherency dislocations are often useful in modeling interfaces in cases where there is incomplete cancellation of the coherency and anticoherency dislocations and residual long-range stresses are therefore present.

Finally, *incoherent interfaces* can be regarded as the limiting case of semicoherent interfaces for which the density of dislocations is so great that their cores overlap and that essentially all of the coherence characteristic of the reference structure has been destroyed. The cores of incoherent interfaces are therefore continuous slabs of bad material, and consequently the interfaces lack long-range order.

B.7   CLASSIFICATION OF LINE DEFECTS IN CRYSTAL/CRYSTAL INTERFACES

The line defects that can exist in crystal/crystal interfaces can be classified as pure dislocations, dislocation/ledges (i.e., line defects with both dislocation and ledge character), and pure ledges. Examples of *pure dislocations* are shown in Figs. B.4c and B.7c. In these cases, there is no ledge in the boundary at the dislocation. An example of a *dislocation/ledge* is shown in Fig. B.6b, and a *pure ledge* without any dislocation content is shown in Fig. B.9.

The line defects which are either dislocations or dislocation/ledges may be further classified as intrinsic or extrinsic. So far, only intrinsic line defects have been considered. These line defects are arranged in uniform arrays and accommodate deviations of interface misorientation and/or inclination from certain reference structures. As part of the minimum-energy equilibrium structure of the interfaces, they are termed *intrinsic*. On the other hand, similar line defects can be present in interfaces in a more or less random fashion, so that their Burgers vectors cancel. In

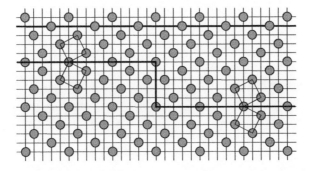

Figure B.9:   Example of a pure ledge in the boundary shown previously in Fig. B.4a. The ledge has zero dislocation character. A detailed discussion of the topological basis of these different types of line defects is given by Sutton and Balluffi [2].

such cases they do not systematically accommodate deviations from any reference structures. Such defects are not part of the minimum-energy equilibrium structure of the interface. They are in a sense "extra" line defects and are therefore termed *extrinsic*. Such line defects could, for example, be present in an interface as a result of the impingement of lattice dislocations from one of the adjoining crystals during plastic deformation or annealing.

Bibliography

1. D. Wolf and S. Yip. *Materials Interfaces*. Chapman & Hall, London, 1992.

2. A.P. Sutton and R.W. Balluffi. *Interfaces in Crystalline Materials*. Oxford University Press, Oxford, 1996.

3. J.M. Howe. *Interfaces in Materials*. John Wiley & Sons, New York, 1997.

4. K. L. Merkle, L. J. Thompson, and F. Phillipp. Thermally activated step motion observed by high-resolution electron microscopy at a (113) symmetric tilt grain-boundary in aluminium. *Phil. Mag. Lett.*, 82:589–597, 2002.

5. J.Q. Broughton, A. Bonissent, and F.F. Abraham. The FCC (111) and (100) crystal-melt interfaces—a comparison by molecular-dynamics simulation. *J. Chem. Phys.*, 74(7):4029–4039, 1981.

6. W.T. Read. *Dislocations in Crystals*. McGraw-Hill, New York, 1953.

APPENDIX C

CAPILLARITY AND MATHEMATICS OF

SPACE CURVES AND INTERFACES

Reviews of the theory of capillarity and its application to solid-state processes have been written by Herring [1], Mullins [2], and Blakely [3]. Adam wrote a classic text on fluid surfaces [4]. For modern mathematical treatments of capillarity, consult Finn's book [5]. For a mathematical treatment of curvature and anisotropic interfaces written for materials scientists, see Taylor's review article [6].[1] There are useful analogies between interfaces and phase diagrams which are particularly instructive for materials scientists [7]. Anybody with a milligram of curiosity and a sense of humor must read C.V. Boys's book on soap bubbles; although written for children, the book is full of useful insights about the nature of interfaces [8].

C.1 SPECIFICATION OF SPACE CURVES AND INTERFACES

C.1.1 Space Curves

A *space curve* is a trajectory of points in three dimensions and can be described mathematically in terms of a position vector $\vec{r}$ that depends on a parameter u (see

---

[1]For convenience in this appendix, we refer to all interfaces in materials, including free surfaces and internal interfaces, simply as interfaces.

Kinetics of Materials. By Robert W. Balluffi, Samuel M. Allen, and W. Craig Carter.   **601**
Copyright © 2005 John Wiley & Sons, Inc.

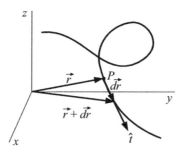

Figure C.1: Space curve $\vec{r}(u)$ and unit tangent $\hat{t}$ at point P.

Fig. C.1); for example,

$$\vec{r} = \vec{r}(u) \tag{C.1}$$

In Cartesian coordinates, u may be the parameter that specifies position in terms of coordinates x, y, and z, giving the functional form

$$\vec{r} = \vec{r}[\vec{x}(u)] = \vec{r}[x(u), y(u), z(u)] \tag{C.2}$$

The *unit tangent vector* at a point P on the curve $\vec{r}(u)$ is defined as

$$\hat{t} = \frac{d\vec{r}(u)}{ds} \tag{C.3}$$

where ds is an element of arc length of the curve $\vec{r}(u)$. It is often convenient to specify the curve using the arc length, s, as a parameter, $\vec{r} = \vec{r}(s)$. Because this vector is between points infinitesimally separated on the space curve at any point, it is tangent to the curve at that point, consistent with Eq. C.3.

The curvature of a space curve, κ^c, is equal to the rate at which the tangent vector changes as the curve is traversed and is therefore given by the relation

$$\frac{d\hat{t}}{ds} = \kappa^c \hat{n} \tag{C.4}$$

where $\hat{n}$ is called the *principal normal vector*. It is a unit vector normal to the curve "in the plane of the curve." For the time being, κ^c is considered to be a positive quantity, and thus $\hat{n}$ points toward the center of curvature of the curve. The radius of curvature R^c is the reciprocal of κ^c, which may be interpreted as the radius of a circle (the *osculating circle*) constructed so that it touches the curve at P and has the same unit tangent vector and derivative $d\hat{t}/ds$ as the curve at P (i.e., a circle that most closely matches the curve in the immediate neighborhood of P). In two dimensions, for a plane curve described by the function $y = y(x)$, the curvature is given by

$$\kappa^c = \frac{1}{R^c} = -\frac{\frac{d^2 y}{dx^2}}{[1 + (\frac{dy}{dx})^2]^{3/2}} \tag{C.5}$$

The minus sign in Eq. C.5 is a matter of convention; it can be either plus or minus, and one should always *question or verify* which convention is being used.

For the more general case of a curve in three dimensions, note that

$$\hat{t} \cdot \hat{t} = \frac{d\vec{r}(s)}{ds} \cdot \frac{d\vec{r}(s)}{ds} = 1 \tag{C.6}$$

so that, by taking the derivative of Eq. C.6 with respect to arc length,

$$\hat{t} \cdot \frac{d\hat{t}(s)}{ds} = 0 \tag{C.7}$$

This defines a new vector, $\vec{\kappa}^c = d\hat{t}/ds$, which is perpendicular to $\hat{t}$ and has magnitude

$$|\vec{\kappa}^c| \equiv \kappa^c = \pm \sqrt{\frac{d\hat{t}(s)}{ds} \cdot \frac{d\hat{t}(s)}{ds}} = \pm \sqrt{\left(\frac{d^2x}{ds^2}\right)^2 + \left(\frac{d^2y}{ds^2}\right)^2 + \left(\frac{d^2z}{ds^2}\right)^2} \tag{C.8}$$

where the sign again depends on the convention used (e.g., the curvature vector may be considered to be positive if it points in the same direction as the curve is bending). Note that Eq. C.5 can be obtained by using Eq. C.8 and the identity $ds^2 = dx^2 + dy^2 + dz^2$.

C.1.2   Interfaces

Interfaces are two-dimensional objects embedded in three dimensions. There is, therefore, one relationship between the three space variables, (x_1, x_2, x_3). If that relation is F, then

$$F(x_1, x_2, x_3) = \text{constant} \tag{C.9}$$

expresses an interface (sometimes called a *level set*) for any chosen constant. A vector ∇F, which is normal to the interface, can be obtained by considering nearby level sets (i.e., two slightly different constants). The unit normal must be

$$\hat{n} = \frac{\nabla F}{|\nabla F|} \tag{C.10}$$

where the derivative is evaluated at a particular point P and on a particular interface (a given constant in Eq. C.9).

There are two different varieties of the curvature of an interface which are convenient to use in capillarity studies: *mean curvature*, denoted by κ, and the *weighted mean curvature*, denoted by κ_γ.

Mean Curvature of an Interface.   The mean curvature is simply the sum of the curvatures of two curves on the interface that intersect at right angles. Any two such curves, c_1 and c_2, can be obtained by the intersection of orthogonal planes with the interface, as illustrated in Fig. C.2. The planes are chosen so that the interface normal $\hat{n}$ lies completely in each plane; the line of intersection between the planes is parallel to $\hat{n}$ (a crystallographer might think of $\hat{n}$ as a *zone axis*). There are an infinite number of choices for these planes, but all are related by rotation around the axis $\hat{n}$.

Let the coordinate system have x_3 parallel to $\hat{n}$, x_1 normal to one plane, and x_2 normal to the other (a local orthonormal coordinate system). Then x_3 can be

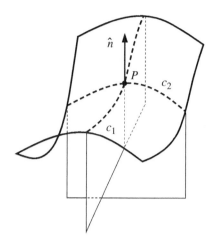

Figure C.2:   Orthogonal planes intersecting interface with plane intersection parallel to interface normal $\hat{n}$.

expanded in the Taylor series,[2]

$$x_3 = f(x_1, x_2) = \frac{a}{2}x_1^2 + bx_1x_2 + \frac{c}{2}x_2^2 + \cdots \tag{C.11}$$

There is one particular rotation of the planes about x_3 where the cross term vanishes,[3]

$$f(\tilde{x}_2, \tilde{x}_3) = \frac{\kappa_1}{2}\tilde{x}_1^2 + \frac{\kappa_2}{2}\tilde{x}_2^2 + \cdots \tag{C.12}$$

where κ_1 and κ_2 are the eigenvalues of a Hessian matrix and $\tilde{x}_1$ and $\tilde{x}_2$ are the axes of the principal coordinate system. Choosing the sum of the eigenvalues is useful because that sum is independent of the rotation and therefore is an invariant definition of the curvature,

$$\kappa = \kappa_1 + \kappa_2 = \frac{1}{R_1^c} + \frac{1}{R_2^c} \quad \text{for any } c_1 \perp c_2 \text{ at } P \tag{C.13}$$

Equations C.12 and C.13 correspond to the curvatures along space curves in independent directions when the derivative in Eq. C.8 is applied. The curvature formula corresponding to the choice of coordinates in Eq. C.11 is

$$\kappa = \frac{(1 + f_2^2)f_{11} - 2f_1f_2f_{12} + (1 + f_1^2)f_{22}}{(1 + f_1^2 + f_2^2)^{3/2}} \tag{C.14}$$

where

$$f_i \equiv \frac{\partial f}{\partial x_i} \quad f_{ij} \equiv \frac{\partial^2 f}{\partial x_i \partial x_j} \tag{C.15}$$

[2]We assume that the interface is twice differentiable. Very often, this is *not* the case for material interfaces, and in this case, anisotropy and the anisotropic equivalent to curvature must be considered [6].

[3]The expansion in the local coordinate system is

$$f(x_1, x_2) = f(0,0) + \begin{bmatrix} f_1 \\ f_2 \end{bmatrix} \cdot [x_1, x_2] + \frac{1}{2}[x_1, x_2] \cdot \begin{pmatrix} f_{11} & f_{12} \\ f_{21} & f_{22} \end{pmatrix} \cdot \begin{bmatrix} x_1 \\ x_2 \end{bmatrix} + \cdots$$

The first two terms vanish because of the particular choice of local coordinate system. In the principal coordinate system, the Hessian (i.e, the matrix of second derivatives) is diagonal.

For the particular case of an axisymmetric surface $r(z)$, the curvature is the sum of the radius of the osculating circle (in the plane shared by the surface normal) and the curvature of $r(z)$ in two dimensions:

$$\kappa(z) = \frac{1}{r(z)\sqrt{1 + \left(\frac{\partial r}{\partial z}\right)^2}} - \frac{\frac{\partial^2 r}{\partial z^2}}{\left[1 + \left(\frac{\partial r}{\partial z}\right)^2\right]^{3/2}} \qquad (C.16)$$

The convention that a convex interface of a solid body has positive mean curvature and a concave interface has negative mean curvature is adopted throughout this book (see Section 14.1). A table of surface formulae is provided in Table C.1.

As will become evident below, mean interface curvature is useful when the interfacial energy is isotropic (not dependent on interface inclination).

Weighted Mean Curvature of an Interface.   The weighted mean curvature, κ_γ, has exactly the same geometrical properties as the mean curvature except that it is weighted by the possibly orientation-dependent magnitude of the interfacial tension. It is particularly useful for addressing capillarity problems when the interfacial energy is anisotropic, that is, dependent upon the interface orientation (Section C.3).

C.2   ISOTROPIC INTERFACES AND MEAN CURVATURE

The curvature of an interface generally introduces a driving force for mass transport. Because this quantity relates to the potential for an interface to move in such a way that interfacial energy decreases, the curvature is very important when considering how the shape or morphology of a material changes. As seen in Chapters 3, 14, 15, and 16, it is the essential driving force for many types of morphological evolution, such as surface smoothing, particle coarsening, grain growth, and sintering.

C.2.1   Implications of Mean Curvature

There are several geometrical aspects of interface mean curvature that are particularly important when the interfacial energy is isotropic and the curvature becomes a driving force for mass transport. We present several equivalent *cursory* statements regarding mean curvature that have rigorous counterparts in differential geometry [6].

- *The mean curvature is the local rate of interface area change with a local addition of volume.*[4] This is perhaps the most important aspect of curvature, especially when combined with γ, which is the work required to create an interface per unit area, A. Imagine that in a pure material the addition of a small volume makes an interface develop a localized small "bump." The statement above implies that $\kappa = \Delta A/\Delta V$ in the limit of small volumes; therefore, the work to create the bump is $\gamma \Delta A = \gamma \kappa \Delta V$, where γ is the interfacial energy per unit area (see Eq. 3.73). Equating this work to the work done by the system $P \Delta V$, where P is the net pressure on the interface

---

[4]See Exercise 3.11 for a demonstration of this.

Table C.1:   Geometrical Formulae in Various Surface Representations

<u>**Level Set Surfaces**</u>**: Tangent Plane, Surface Normal, and Curvature**

$$F(x, y, z) = \text{const}$$

Tangent Plane ($\vec{x} = (x, y, z)$, $\vec{\xi} = (\xi, \eta, \zeta)$)

$$\nabla F \cdot (\vec{\xi} - \vec{x}) \text{ or } \frac{\partial F}{\partial x}(\xi - x) + \frac{\partial F}{\partial y}(\eta - y) + \frac{\partial F}{\partial z}(\zeta - z)$$

Normal

$$\frac{\xi - x}{\frac{\partial F}{\partial x}} = \frac{\eta - y}{\frac{\partial F}{\partial y}} = \frac{\zeta - z}{\frac{\partial F}{\partial z}}$$

Mean Curvature

$$\nabla \cdot \left(\frac{\nabla F}{|\nabla F|} \right) \text{ or }$$

$$\frac{\left[\left(\frac{\partial^2 F}{\partial y^2} + \frac{\partial^2 F}{\partial z^2} \right) \left(\frac{\partial F}{\partial x} \right)^2 + \left(\frac{\partial^2 F}{\partial z^2} + \frac{\partial^2 F}{\partial x^2} \right) \left(\frac{\partial F}{\partial y} \right)^2 + \left(\frac{\partial^2 F}{\partial x^2} + \frac{\partial^2 F}{\partial y^2} \right) \left(\frac{\partial F}{\partial z} \right)^2 \atop -2 \left(\frac{\partial F}{\partial x} \frac{\partial F}{\partial y} \frac{\partial^2 F}{\partial x \partial y} + \frac{\partial F}{\partial y} \frac{\partial F}{\partial z} \frac{\partial^2 F}{\partial y \partial z} + \frac{\partial F}{\partial z} \frac{\partial F}{\partial x} \frac{\partial^2 F}{\partial z \partial x} \right) \right]}{\left(\frac{\partial F}{\partial x}^2 + \frac{\partial F}{\partial y}^2 + \frac{\partial F}{\partial z}^2 \right)^{3/2}}$$

<u>**Parametric Surfaces**</u>**: Tangent Plane, Surface Normal, and Curvature**

$$\vec{x} = (p(u, v), q(u, v), s(u, v)) \text{ or } x = p(u, v) y = q(u, v) z = s(u, v)$$

Tangent Plane ($\vec{x} = (x, y, z)$, $\vec{\xi} = (\xi, \eta, \zeta)$)

$$(\vec{\xi} - \vec{x}) \cdot \left(\frac{d\vec{x}}{du} \times \frac{d\vec{x}}{dv} \right) \det \begin{pmatrix} \xi - x & \eta - y & \zeta - z \\ \frac{\partial p}{\partial u} & \frac{\partial q}{\partial u} & \frac{\partial s}{\partial u} \\ \frac{\partial p}{\partial v} & \frac{\partial q}{\partial v} & \frac{\partial s}{\partial v} \end{pmatrix} = 0$$

Normal

$$\frac{\xi - x}{\frac{\partial(q, s)}{\partial(u, v)}} = \frac{\eta - y}{\frac{\partial(s, p)}{\partial(u, v)}} = \frac{\zeta - z}{\frac{\partial(p, q)}{\partial(u, v)}}$$

Mean Curvature

$$\frac{\left(\frac{d\vec{x}}{du} \cdot \frac{d\vec{x}}{du} \right) \left(\frac{d\vec{x}}{du} \times \frac{d\vec{x}}{dv} \cdot \frac{d^2\vec{x}}{dv^2} \right) - 2 \left(\frac{d\vec{x}}{du} \cdot \frac{d\vec{x}}{dv} \right) \left(\frac{d\vec{x}}{du} \times \frac{d\vec{x}}{dv} \cdot \frac{d^2\vec{x}}{dudv} \right) + \left(\frac{d\vec{x}}{dv} \cdot \frac{d\vec{x}}{dv} \right) \left(\frac{d\vec{x}}{du} \times \frac{d\vec{x}}{dv} \cdot \frac{d^2\vec{x}}{du^2} \right)}{\left(\frac{d\vec{x}}{du} \times \frac{d\vec{x}}{dv} \cdot \frac{d\vec{x}}{du} \times \frac{d\vec{x}}{dv} \right)^{3/2}}$$

<u>**Graph Surfaces**</u>**: Tangent Plane, Surface Normal, and Curvature**

$$z = f(x, y)$$

Tangent Plane ($\vec{x} = (x, y, z)$, $\vec{\xi} = (\xi, \eta, \zeta)$)

$$\frac{\partial f}{\partial x}(\xi - x) + \frac{\partial f}{\partial y}(\eta - y) = (\zeta - z)$$

Normal

$$\frac{\xi - x}{\frac{\partial f}{\partial x}} = \frac{\eta - y}{\frac{\partial f}{\partial y}} = \frac{\zeta - z}{-1}$$

Mean Curvature

$$\frac{(1 + \frac{\partial f}{\partial x}^2) \frac{\partial^2 f}{\partial y^2} - 2 \frac{\partial f}{\partial x} \frac{\partial f}{\partial y} \frac{\partial^2 f}{\partial x \partial y} + (1 + \frac{\partial f}{\partial y}^2) \frac{\partial^2 f}{\partial x^2}}{\sqrt{1 + \frac{\partial f}{\partial x}^2 + \frac{\partial f}{\partial y}^2}}$$

due to its curvature, establishes the *Gibbs–Thomson equation,*

$$P = \gamma\kappa = \gamma \left(\frac{1}{R_1^c} + \frac{1}{R_2^c} \right) \tag{C.17}$$

The quantity $\gamma\kappa$ may be regarded as the local potential due to the interface curvature to add a chemical species per unit volume of the species. On interfaces where the mean curvature is constant everywhere (such as on a sphere where $\kappa = 2/R^c$, on a cylinder where $\kappa = 1/R^c$, and on a plane and a catenoid where $\kappa = 0$), this potential is uniform and thus these are equilibrium interfaces. There is an infinite number of equilibrium interfaces; a three-parameter family of minimal interfaces has been described [9].

- *Of all local motions occurring with the velocity $\vec{v}(\vec{r})$ of an interface which pass the same amount of volume from one side to the other, the motion that is normal to the interface with magnitude proportional to the curvature [i.e., $\vec{v}(\vec{r}) \propto \kappa\hat{n}$], increases the area the most quickly.* This provides a variational statement which is useful for calculating the evolution of interfaces of nonuniform curvature.[5]

- *The mean curvature is the interface divergence of the normal vectors.* The *interface divergence* is the divergence in two dimensions, calculated with the local coordinates in the interface as in Eq. C.11 (see Fig. C.3a). It can be tedious to calculate using the interface divergence as an operator because of its explicit dependence on the interface geometry. The divergence operator indicates how rapidly vectors in a vector field in the bulk vary through the volume; the interface divergence of the interface normals indicates how rapidly the normals are varying along the interface. On a flat interface, the normals are all the same and the interface divergence is zero. On an interface that has large curvature, the normals vary rapidly, so the interface divergence is large.

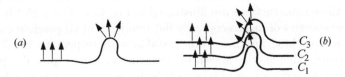

Figure C.3: The mean curvature of an interface is related to **(a)** the interface divergence of the interface normal vectors and **(b)** the divergence of the normal vectors to the level sets.

- *The mean curvature is the divergence of the normals to the level sets.* This is usually a much easier calculation than the one employing the interface divergence. The interface is represented as a level set of $F(x_1, x_2, x_3)$ as in Eq. C.9. The normals to the level sets are given by Eq. C.10 and are defined on all (x_1, x_2, x_3) (see Fig. C.3b). The mean curvature is then

$$\kappa = \nabla \cdot \hat{n} = \nabla \cdot \left(\frac{\nabla F}{|\nabla F|} \right) \tag{C.18}$$

[5]This particular statement lacks rigor in the comparison between interfaces with the *same amount of volume.* A more precise statement involves those interfaces where the integral of the squared difference are equivalent. This is called the *L2 norm on functions.*

This extension to all of space is used in the derivation of the Cahn–Hoffman ξ-vectors—a convenient way to study capillarity when the interfacial energy depends on the inclination, $\hat{n}$ [7, 10].

Curvature relates to the local change in interface area when an interface moves. The energy change per unit volume swept out by the interface is equal to the product of κ and the interfacial energy per unit area γ. Normally, for fluids, γ is independent of the interface inclination $\hat{n}$; in this case, the interface is isotropic. For example, a soap bubble has isotropic interface tension. If perturbed, a floating individual soap bubble will quickly re-establish its equilibrium form—a sphere of fixed volume. Such a soap bubble will also shrink slowly—the gas will diffuse out of the bubble because of a pressure difference across the soap film ($\Delta P = \gamma \kappa = 2\gamma/R^c$). Thus, there are two kinds of equilibrium being established: at short times, the interface changes so that its curvature becomes uniform; at long times, that same curvature drives gas through the interface and the bubble slowly shrinks.

A fixed amount of condensed phase enclosed by an interface will undergo essentially the same process, except that the time scales may differ greatly. For solid phases, the interfaces will reduce gradients in curvature by diffusional processes such as interface diffusion, crystal diffusion, and vapor transport. At similar time scales (in the case of crystal diffusion) interfaces will move because atoms will experience differences in diffusion potential across an interface arising from differences in the curvature according to Eq. 3.76.

C.3 ANISOTROPIC INTERFACES AND WEIGHTED MEAN CURVATURE

C.3.1 Geometric Constructions for Anisotropic Surface Energies

The γ-plot and a variety of geometric constructions based upon it are highly useful in the treatment of anisotropic interfaces (see Section 14.2.1). Several important examples of these constructions are illustrated in Fig. C.4. The γ-plot in Fig. C.4a represents the energies of macroscopically flat interfaces of all possible inclinations. However, some of these interfaces are unstable with respect to breakup into a faceted structure. This may be shown by constructing the reciprocal γ-plot shown in Fig. C.4b. Here, inclinations of relatively high energy appear in the depressions, and the common-tangent construction on the plot indicates that the regions of inclination between segments such as between B and C are unstable against the breakdown of the surface into facets with inclinations corresponding to the points of common tangency at B and C [11]. As discussed by Cahn and Carter, there is a close resemblance between this construction and the common-tangent construction applied to free-energy curves of binary systems (shown in Fig. 17.6), which reveals the region in composition space where the single phases are unstable with respect to the formation of two phases with compositions corresponding to the points of common tangency [12]. Because the reciprocal γ-plot of a spherical γ-plot that intersects the origin at one point is a plane, the common-tangent construction on the reciprocal γ-plot in Fig. C.4b is equivalent to the tangent-sphere construction on the γ-plot shown in Fig. C.4c. Those inclinations for which a tangent sphere with one point at the origin lies completely within a γ-plot are stable. The tangent-sphere construction in Fig. C.4c therefore shows that the regions of inclination, such as between A and B, are stable, whereas regions such as between B and C

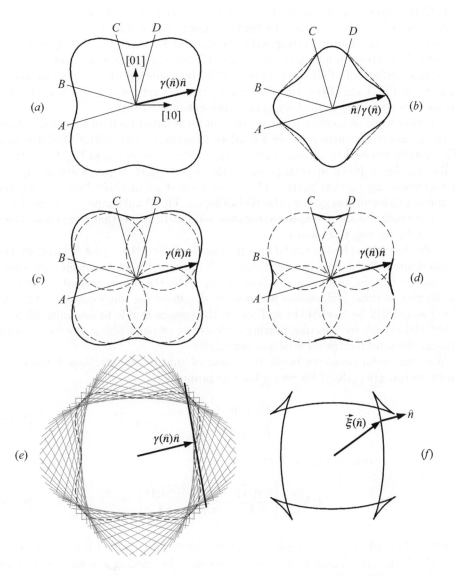

Figure C.4:   Constructions in two dimensions used to determine equilibrium shapes of anisotropic surfaces. **(a)** Plot of $\gamma(\hat{n})$ as a function of $\hat{n}$ for macroscopically flat nonfaceted surfaces, some of which are unstable. **(b)** Plot of $1/\gamma(\hat{n})$ as a function of $\hat{n}$. Plot is reciprocal to the plot in (a). **(c)** Same as (a) but including tangent circles (dashed). **(d)** Plot of $\gamma(\hat{n})$ as a function of $\hat{n}$ for macroscopically flat but stable surfaces. Segments such as between A and B are smooth interfaces inherited from the stable regions of the plots in (a) and (b). Segments such as between B and C are faceted with the facets corresponding to the surface inclinations at their endpoints. The segment between B and C is circular and coincides with the portion of the tangent circle shown between B and C in (c). **(e)** Wulff shape obtained by Wulff construction applied to the γ-plot in (a). **(f)** Plot of capillarity vector $\vec{\xi}(\hat{n})$ as a function of $\hat{n}$.

are unstable, in agreement with the results of the common-tangent construction in Fig. C.4b. Surfaces in the unstable region between B and C delineated in Figs. C.4b and c can reduce their energies by breaking down into stable structures consisting of facets with inclinations corresponding to those at B and C. The energies of these faceted structures will fall on the tangent circles shown in Fig. C.4c, and the γ-plot for stable surfaces of minimum energy at all inclinations will then appear as dashed tangent in Fig. C.4d. Here, the solid curve segments represent smooth nonfaceted surfaces, whereas the dashed segments represent faceted surfaces.

Another topic of interest is the shape that an isolated body of constant volume with an anisotropic surface energy will adopt to minimize its total interfacial energy. This can be resolved by means of the *Wulff construction* shown in Fig. C.4e. Here, a line has been drawn at each point on the γ-plot which is perpendicular to the $\hat{n}$ corresponding to that point. The interior envelope of these lines is then the shape of minimum energy (i.e., the Wulff shape). The Wulff shape for the γ-plot in Fig. C.4a contains sharp edges and contains only inclinations that have been shown to be stable in Fig. C.4b and c.

Note that when the interfacial energy is isotropic and the γ-plot is a sphere, the Wulff shape will also be a sphere. However, if the γ-plot possesses deep depressions or cusps at certain inclinations such as in Fig. C.4a, the planes normal to the radii of the plot at these inclinations will tend to dominate the inner envelope, and the Wulff shape will be faceted. In such cases, the system is able to minimize its total interfacial energy by selecting patches of interface of particularly low energy even though the total interfacial area increases.

Another useful construct in the treatment of anisotropic interfaces is the capillarity vector, $\vec{\xi}(\hat{n})$ [13]. This vector has the properties

$$\vec{\xi}(\hat{n}) \cdot \hat{n} = \gamma(\hat{n})$$
$$\hat{n} \cdot d\vec{\xi}(\hat{n}) = 0$$

(C.19)

It can also be expressed in the form

$$\vec{\xi}(\hat{n}) = \left[\frac{\partial \gamma(\vec{A})}{\partial A_x}, \frac{\partial \gamma(\vec{A})}{\partial A_y}, \frac{\partial \gamma(\vec{A})}{\partial A_x} \right]$$

(C.20)

where $\gamma(\vec{A}) = |\vec{A}|\gamma(\hat{n})$ is the surface energy of the surface of area $|\vec{A}|$ and inclination $\hat{n} = \vec{A}/|\vec{A}|$ [6, 14]. A $\vec{\xi}$-plot can also be produced by making a polar plot of $\vec{\xi}(\hat{n})$ as a function of $\hat{n}$. According to Eqs. C.19, the projection of $\vec{\xi}(\hat{n})$ on $\hat{n}$ must be equal to $\gamma(\hat{n})$, and the inclination of the surface of the plot at $\vec{\xi}(\hat{n})$ must be normal to $\hat{n}$. The $\vec{\xi}$-plot corresponding to the γ-plot in Fig. C.4a is shown in Fig. C.4f. It has the same form as the Wulff shape, as may be seen by comparing Figs. C.4e and f. The swallowtail-shaped "ears" at the four corners of the plot correspond to nonequilibrium inclinations of the γ-plot in Fig. C.4a and may be ignored because they are not part of the Wulff shape.

C.3.2 Implications of Weighted Mean Curvature

The capillarity vector, $\vec{\xi}(\hat{n})$, plays the same role as the interfacial energy multiplied by $\hat{n}$ [i.e., $\gamma(\hat{n})\hat{n}$]. Just as the mean curvature was related to derivatives on the vector field $\hat{n}$, the weighted mean curvature is related to derivatives on $\vec{\xi}$ [6]. We now make cursory statements about κ_γ, along the same lines as our previous remarks about κ. These statements do not require the interface to be everywhere differentiable.

- *The weighted mean curvature is the local rate of interfacial energy change with a local addition of volume.* This establishes the connection to the work, δW, to pass a small volume of material, δV, through an interface. $\delta W/\delta V = \kappa_\gamma(\vec{r})$, in the limit of small volumes.

- *Of all local motions, $v(\vec{r})$, of an interface that pass the same amount of volume from one side to the other, the motion that is normal to the interface with magnitude proportional to the weighted mean curvature, $v(\vec{r}) \propto \kappa_\gamma \hat{n}$, increases the interfacial energy the fastest.* However, "fastest" depends on how distance is measured. How this distance metric alters the variational principles that generate the kinetic equations is discussed elsewhere [14].

- *The weighted mean curvature is the interface divergence of the $\vec{\xi}$ evaluated on the unit sphere.* The interface divergence is defined within the interface, and if the interface is not differentiable, subgradients must be used. The convex portion of $\vec{\xi}$ is equivalent to the the Wulff shape, so the interface divergence is operating from one interface onto another. This form can get very complicated.

- *The weighted mean curvature is the divergence of $\vec{\xi}$ (see Eq. 14.40).* This is a simple form. $\vec{\xi}$ is a vector defined on all space, so there is no confusion in the application: $\kappa_\gamma = \nabla \cdot \vec{\xi}$.

C.4 EQUILIBRIUM AT A CURVED INTERFACE

C.4.1 Gibbs–Thomson Equation

Consider a two-phase system of fixed total volume, with constant T and μ (an open system with respect to matter flow), as illustrated in Fig. C.5. Under these conditions, the function $\Omega = E - TS - \mu_1 N_1 - \mu_2 N_2$ is the appropriate thermodynamic potential. For any small variation at equilibrium, such as an infinitesimal variation

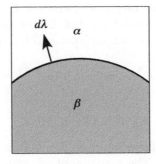

Figure C.5: An open isothermal system that allows for reversible motion of a curved α/β interface.

$d\lambda$ in position of the α/β interface, $d\Omega = 0$. The Ω function is appropriate for such changes at an interface because, in general, the phases involved are so extensive relative to the volume of the interfacial region that they behave essentially like chemical reservoirs.

For the variation being considered,

$$d\Omega = -P^\alpha \, dV^\alpha - P^\beta \, dV^\beta + \gamma \, dA = 0 \tag{C.21}$$

Since the total volume is constant, $dV^\alpha = -dV^\beta$. For an interface of area A moving a distance $d\lambda$, as illustrated in Fig. C.5, $dV^\beta = A \, d\lambda$ and the interfacial area change according to Section C.2.1 is $dA = \kappa \, dV$, where $dV = dV^\beta$. So the free-energy change is

$$d\Omega = P^\alpha \, dV^\beta - P^\beta \, dV^\beta + \gamma\kappa \, dV^\beta = 0 \tag{C.22}$$

or

$$P^\beta - P^\alpha = \gamma\kappa = \gamma(K_1 + K_2) \tag{C.23}$$

Equation C.23 is the form of the Gibbs–Thomson equation introduced in Eq. C.17. It is a condition for mechanical equilibrium in a two-phase system with a curved interface. The phase located on the side of the interface toward its center of curvature (e.g. the β phase in Fig. C.5), has the higher pressure. Note also that for a flat interface, Eq. C.23 gives $P^\alpha = P^\beta$, as expected.

C.4.2   Equilibrium Solubilities of Small Dispersed-Phase Particles

Consider a two-component, two-phase fluid system. Let β be a spherical fluid droplet of radius r surrounded by α, the second fluid phase. The equilibrium conditions are

$$\begin{aligned}
\mu_A^\alpha &= \mu_A^\beta \\
\mu_B^\alpha &= \mu_B^\beta \\
T^\alpha &= T^\beta \\
P^\beta - P^\alpha = \gamma(K_1 + K_2) &= \frac{2\gamma}{r}
\end{aligned} \tag{C.24}$$

To find the equilibrium concentration of component B in α as a function of r, the change of μ_2 with r is considered first. Let X^{eq} be the equilibrium solubility of component B in α for a system with a planar α/β interface. Then

$$\mu_B^\alpha(P^\alpha, X^{\text{eq}}) = \mu_B^\beta(P^\alpha) \tag{C.25}$$

and

$$\mu_B^\alpha(P^\alpha, X) = \mu_B^\beta(P^\beta) \tag{C.26}$$

where Eq. C.25 holds for a planar interface, Eq. C.26 holds for a curved interface, and X is to be determined. The variation of μ_B^β with P is given by the Gibbs–Duhem equation,

$$V^\beta \, dP = S^\beta \, dT + N_A^\beta \, d\mu_A + N_B^\beta \, d\mu_B \tag{C.27}$$

By assuming that β is pure component B (i.e., $N_A^\beta = 0$), at constant temperature Eq. C.27 yields

$$\Omega_B\, dP = d\mu_B \quad \text{or} \quad \left.\frac{\partial \mu_B}{\partial P}\right|_{T\text{ constant}} = \Omega_B \qquad (C.28)$$

where Ω_B is the atomic volume in the β phase. The change of chemical potential is then

$$\mu_B^\beta(P^\beta) - \mu_B^\beta(P^\alpha) = \int_{P^\alpha}^{P^\beta} \left.\frac{\partial \mu_B}{\partial P}\right|_{T\text{ constant}} dP = \int_{P^\alpha}^{P^\beta} \Omega_B\, dP \qquad (C.29)$$

Assuming that β is incompressible and that Ω_B is independent of P,

$$\mu_B^\beta(P^\beta) - \mu_B^\beta(P^\alpha) = \Omega_B(P^\beta - P^\alpha) = \Omega_B\gamma(K_1 + K_2) \qquad (C.30)$$

For a spherical particle of radius r, the chemical potential difference is

$$\mu_B^\beta(P^\beta) - \mu_B^\beta(P^\alpha) = \frac{2\Omega_B\gamma}{r} \qquad (C.31)$$

Therefore, in a system with a small β particle that is rich in component B, the value of μ_2 characteristic of equilibrium is raised compared with a system in which the mean curvature of the α/β interface is zero.

This finding is illustrated on a free energy vs. composition diagram in Fig. C.6. The free-energy "curve" for the β phase is a vertical line at $X = 1$, because the β phase has been assumed to be pure component B. Note especially that the change in μ_2 with P (or r) causes a change in the equilibrium solubility of component B in α.

If the temperature dependence of the free-energy curves is considered in addition to the β particle size, a temperature–composition diagram for the system can be plotted. This is illustrated in Fig. C.7, which shows clearly the way in which the solubility of component B in α increases with decreasing β particle size.

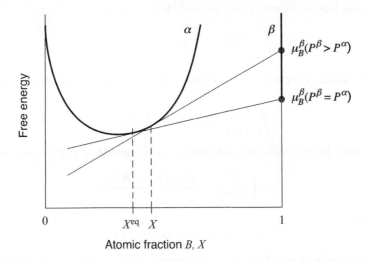

Figure C.6: Free energy vs. mole fraction diagram, showing a shift of chemical potentials with pressure in the β phase.

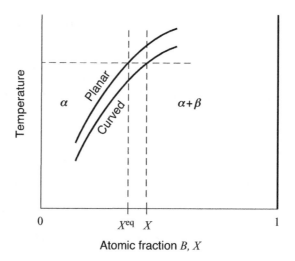

Figure C.7:   Temperature vs. mole fraction diagram, illustrating shift of an α/β coexistence curve with the radius of the β phase and the resulting change of solubility of component B in α.

The effect of β particle size on the solubility of component B in α can be quantified. For the α phase,

$$\mu_B^\alpha(P^\alpha, X) = \mu_B^\alpha(P^\alpha, X^{\text{eq}}) + \int_{X^{\text{eq}}}^{X} \frac{d\mu}{dX} \, dX \tag{C.32}$$

Substituting Eqs. C.25 and C.26 into Eq. C.32 gives

$$\mu_B^\beta(P^\beta) - \mu_B^\beta(P^\alpha) = \int_{X^{\text{eq}}}^{X} \frac{d\mu}{dX} \, dX \tag{C.33}$$

By combining Eqs. C.30 and C.33, we find that

$$\int_{X^{\text{eq}}}^{X} \frac{d\mu}{dX} \, dX = \Omega_B \gamma (K_1 + K_2) \tag{C.34}$$

For a dilute solution (in which the activity coefficient is independent of concentration),

$$\int_{X^{\text{eq}}}^{X} \frac{d\mu}{dX} \, dX = kT \ln\left(\frac{X}{X^{\text{eq}}}\right) \tag{C.35}$$

The final result for the shift of equilibrium solubility with β particle size is

$$\ln\left(\frac{X}{X^{\text{eq}}}\right) = \frac{\Omega_B \gamma (K_1 + K_2)}{kT} \tag{C.36}$$

Equivalently,

$$X = X^{\text{eq}} e^{\Omega_B \gamma (K_1 + K_2)/(kT)} \tag{C.37}$$

which is valid subject to all of our assumptions. Note in Eqs. C.36 and C.37 that $X > X^{\text{eq}}$, so that the solubility is always enhanced as the curvature of the β particle increases, or as the β particle size decreases.

The main assumptions in the preceding derivation are:

- The phase β is pure component B.

- The phase β is incompressible.

- The matrix phase α is a dilute solution.

Each of these assumptions has been made only for the sake of algebraic simplicity. In principle, the same derivation could be repeated with fewer assumptions, but the result would be more complicated.

Bibliography

1. C. Herring. Surface tension as a motivation for sintering. In W.E. Kingston, editor, *The Physics of Powder Metallurgy*, pages 143–179, New York, 1951. McGraw-Hill.

2. W.W. Mullins. Solid surface morphologies governed by capillarity. In N.A. Gjostein, editor, *Metal Surfaces: Structure, Energetics and Kinetics*, pages 17–66, Metals Park, OH, 1962. American Society for Metals.

3. J.M. Blakely. Thermodynamics of surfaces and interfaces. In M.B. Bever, editor, *Encyclopedia of Materials Science and Engineering*, pages 4962–4967. Pergamon Press, New York, 1986.

4. N.K. Adam. *The Physics and Chemistry of Surfaces*. Oxford University Press, London, 1941.

5. R. Finn. *Equilibrium Capillary Surfaces*. Springer-Verlag, New York, 1986.

6. J.E. Taylor. Overview No. 98. II—Mean curvature and weighted mean curvature. *Acta Metall.*, 40(7):1475–1485, 1992.

7. J.W. Cahn and W.C. Carter. Crystal shapes and phase equilibria: A common mathematical basis. *Metall. Mater. Trans. A*, 27(6):1431–1440, 1996.

8. C.V. Boys. *Soap Bubbles and the Forces Which Mould Them*. Society for Promoting Christian Knowledge, London, 1902. Reprinted, 1959, Doubleday and Co., New York.

9. W.C. Carter. The forces and behavior of fluids constrained by solids. *Acta Metall.*, 36(8):2283–2292, 1988.

10. J.W. Cahn and D.W. Hoffman. A vector thermodynamics for anisotropic surfaces. II. Curved and facetted surfaces. *Acta Metall.*, 22(10):1205–1214, 1974.

11. F.C. Frank. The geometrical thermodynamics of surfaces. In W.D. Robertson and N.A. Gjostein, editors, *Metal Surfaces*, pages 1–15. American Society for Metals, Metals Park, OH, 1963.

12. J.W. Cahn and W.C. Carter. Crystal shapes and phase equilibria: A common mathematical basis. *Metall. Trans.*, 27A(6):1431–1440, 1996.

13. D.W. Hoffman and J.W. Cahn. A vector thermodynamics for anisotropic surfaces. I. Fundamentals and applications to plane surface junctions. *Surf. Sci.*, 31:368–388, 1972.

14. W.C. Carter, J.E. Taylor, and J.W. Cahn. Variational methods for microstructural-evolution theories. *JOM*, 49:30–36, 1997.

ILLUSTRATION CREDITS

We are grateful to have received permission to use figures and content from a variety of sources. To the best of our abilities, we have attempted to identify and attribute all nonoriginal material and intellectual property. We regret any omissions and will make corrections in later editions.

Fig. 3.11 Reprinted, by permission, from L.S. Darken, "Diffusion of carbon in austenite with a discontinuity in composition," *Trans. AIME*, Vol. 180, pp. 430–438. Copyright © 1949 by TMS (The Minerals, Metals, and Materials Society). **Exercise 3.4** Adapted, by permission, from P.G. Shewmon, *Diffusion in Solids*. Copyright © 1989 by the MMM Society. **Fig. 4.8** Reprinted, by permission, from L.C. Correa da Silva and R.F. Mehl, "Interface and marker movements in diffusion in solid solutions of metals," *Trans. AIME*, Vol. 191, pp. 155–173. Copyright © by TMS (The Minerals, Metals, and Materials Society). **Fig. 6.2** and **Fig. 6.3** Reprinted, by permission, from A. Vignes and J.P. Sabatier, "Ternary diffusion in Fe–Co–Ni alloys," *Trans. AIME*, Vol. 245, pp. 1795–1802. Copyright © 1969 by TMS (The Minerals, Metals, and Materials Society). **Fig. 6.2** Reprinted, by permission, from J.S. Kirkaldy, *Diffusion in the Condensed State*. Copyright © 1987 The Institute of Metals (Maney Publishing). **Fig. 9.1** Reprinted, by permission, from N.A. Gjostein, "Short circuit diffusion," in *Diffusion*. Copyright © 1973 by The American Society for Metals (ASM International). **Fig. 9.2**, **Fig. B.6**, and **Fig. B.8** From *Interfaces in Crystalline Materials* by A.P. Sutton and R.W. Balluffi (1995). Reprinted by permission of Oxford University Press. **Fig. 9.2** Reprinted, by permission, from I. Herbeuval and M. Biscondi, "Diffusion du zinc dans les joints de flexion symmetriques de l'aluminium," *Canadian Metallurgical Quarterly*, Vol. 13, pp. 171–175. Copyright © 1974 by the Canadian Institute of Mining and Metallurgy. **Fig. 9.3** Reprinted, by permission, from A. Atkinson, "Diffusion in Ceramics," in *Materials Science and Technology—A Comprehensive Treatment*, Vol. 11: Structure and Properties of Ceramics, R.W. Cahn, P. Haasen, and E. Kramer, eds. Copyright © 1994 by VCH Publishers. **Fig. 9.5** Reprinted, by permission, from D. Turnbull, *Grain Boundary and Surface Diffusion*. Copyright © 1951 by American Society for Metals (ASM International). **Fig. 9.6** Reprinted from *Scripta Metallurgica*, Vol. 13, J.W. Cahn and R.W. Balluffi, "Diffusional mass-transport in polycrystals containing stationary or migrating grain boundaries," 499–502, Copyright 1979, with permission from Elsevier. **Fig. 9.9** Reprinted from *Scripta Metallurgica*, Vol. 8, R.W. Balluffi, T. Kwok, P.D. Bristowe, A. Brokman, P.S. Ho, and S. Yip, "Determination of the vacancy mechanism for grain-boundary self-diffusion by computer simulation," 951–956, Copyright 1981, with permission from Elsevier. **Fig. 10.2** Reprinted from "Tracer Diffusion of Fe-59 in Amorphous $Fe_{40}Ni_{40}B_{20}$," J. Horvath and H. Mehrer, 1986, *Crystal Lattice Defects and Amorphous Materials*, Taylor and Francis, http://www.tandf.co.uk/journals/. **Fig. 10.5** Reprinted from *Progress in Materials Science*, Vol. 32, R. Kirchheim, "Hydrogen solubility and diffusivity in defective and amorphous metals," 261–325, Copyright 1989, with permission from Elsevier. **Fig. 10.6** Reprinted, by permission, from H. Hahn and R.S. Averback, "Dependence of tracer diffusion on atomic size in amorphous Ni–Zr," *Phys. Rev. B*, Vol. 37, p. 6534. Copyright © 1988 by the American Physical Society. **Fig. 10.7a** From *Introduction to Ceramics*, Kingery et al., 2nd ed., Copyright 1976 John Wiley and Sons, Inc. This material is used by permission of John Wiley & Sons, Inc. **Fig. 10.7b** Reprinted from *Journal of Non-Crystalline Solids*, Vol. 71, G.N. Greaves, "EXAFS and the structure of glass," 203–217, 1985, with permission from Elsevier. **Fig. 10.8** Reprinted, by permission, from G.H. Frischat, *Ionic Diffusion in Oxide Glasses*. Copyright © 1975 by Trans Tech Publications. **Fig. 10.10** and **Fig. 10.11** From *Advances in Chemical Physics*, Vol. 79, Prigogine and Rice (eds.), Copyright 1990 John Wiley and Sons, Inc. This material is used by permission of John Wiley & Sons, Inc. **Fig. 11.7** Reprinted with permission from W.G. Johnston and J.J. Gilman, "Dislocation velocities, dislocation densities, and plastic flow in lithium fluoride," *Journal of Applied Physics*, Vol. 30, pp. 129–144. Copyright 1959, American Institute of Physics. **Fig. 12.6** From *Interfaces in Materials* by James M. Howe, Copyright © 1997 John Wiley & Sons. Reprinted with permission of John Wiley & Sons, Inc. Original figure is reprinted, by permission, from S.T. Peteves and R. Abbaschian, "Growth kinetics of solid-liquid Ga interfaces: Part 1 experimental," *Met. Trans. A*, Vol. 19, pp. 1259–1271. Copyright © 1991 by TMS (The Minerals, Metals, and Materials Society). **Fig. 13.5a** Reprinted from *Acta Metallurgica*, Vol. 17, H. Gleiter, "The mechanism of grain boundary migration," 565–573, Copyright 1969, with permission from Elsevier. **Fig. 13.5b** Reprinted from *Acta Metallurgica*, Vol. 27, D.J. Dingley and R.C. Pond, "On the interaction of crystal dislocations with grain boundaries," 667–682, Copyright 1979, with permission from Elsevier. **Fig. 13.6** Reprinted from Acta Metallurgica, Vol. 10, J.W. Cahn, "The impurity-drag effects in grain boundary motion," 789–798, Copyright 1962, with permission from Elsevier. **Fig. 13.7** Reprinted from *Acta Metallurgica*, vol. 33, E. Nes, N. Ryum, and O. Hunderi, "On the Zener drag," pp. 11–22, Copyright 1985, with permission from Elsevier. **Fig. 13.8** Reprinted, by permission, from E.M. Fridman, C.V. Kopesky, and L.S. Shvindlerman, "Effects of orientation and concentration factors on migration of individual grain boundaries in aluminium," *Z. Metallk.*, Vol. 66, pp. 533–539. Copyright © 1975 by Carl-Hanser-Verlag. **Fig. 13.9** Reprinted, by permission, from D.A. Molodov, C.V. Kopetskii, and L.S. Shvindlerman, "Detachment of a special ($\Sigma = 19$, $\langle 111 \rangle$) tilt boundary from an impurity in iron-doped aluminum bicrystals," *Soviet Physics, Solid State*, Vol. 23, pp. 1718–1721. Copyright © 1981 by American Institute of Physics. **Fig. 13.10** Reprinted from *Acta Metall. Mater.*, vol. 28, R.W. Siegel, S.M. Chang, and R.W. Balluffi, "Vacancy loss at grain-boundaries in quenched polycrystalline gold," 249–257, 1980, with permission from Elsevier. **Fig. 13.11** Reprinted from "Direct observation of grain boundary dislocation climb in ion-irradiated gold bicrystals," Y. Komen, P. Petroff, and R.W. Balluffi, 1972, *Philosophical Magazine*, Taylor and Francis, http://www.tandf.co.uk/journals/titles/14786435.html. **Fig. 13.12** Reprinted, by permission, from A.H. King, unpublished research. **Fig. 13.13a** Reprinted from *Acta Metall. Mater.*, Vol. 37, K.E. Rajab and R.D. Doherty, "Kinetics of growth and coarsening of faceted hexagonal precipitates in an fcc matrix. 1. Experimental observations," 2709–2722, Copyright 1989, with permission from Elsevier. **Fig. 13.13b** Reprinted from *Acta Metallurgica*, Vol. 19, Weatherly, "The structure of ledges at plate-shaped precipitates," 181–192, Copyright 1971, with permission from Elsevier. **Fig. 14.9** Reprinted, by permission, from J.M. Dynys, *Sintering Mechanisms and Surface Diffusion for Aluminum Oxide*, PhD thesis, Department of Materials Science and Engineering, Massachusetts Institute of Technology, 1982. **Fig. 14.18** Reprinted, by permission, from J.W. Cahn, J.E. Taylor, and C. Handwerker, "Evolving crystal forms: Frank's characteristics revisited," in *Sir Charles Frank OBE, FRS: An Eightieth Birthday Tribute.* Copyright © 1991 by the Institute of Physics. **Fig. 15.6**, **Fig. 15.8**, and **Fig. 15.9** Reprinted, by permission, from S.C. Hardy and P.W. Voorhees, "Ostwald ripening in a system with a high volume fraction of coarsening phase," *Metallurgical Transactions A*, Vol. 19A, pp. 2713-2721. Copyright © 1988 by TMS (The Minerals, Metals, and Materials Society). **Fig. 15.10** and **Fig. 16.8** Reprinted, by permission, from E.J. Siem, unpublished research. Copyright © 2004 by Ellen J. Siem. **Fig. 15.11** Reprinted from *Solid State Physics*, Vol. 55, C.V. Thompson, "Grain growth and evolution of other cellular structures," 269–314, Copyright 2001, with permission from Elsevier. **Fig. 15.14** Reprinted, by permission, from M. Marder, "Soap bubble growth," *Phys. Rev. A*, 36, 438–440 1987. Copyright © 1987 by the American Physical Society. **Fig. 15.14** Reprinted, by permission, from J.A. Glazier, S.P. Gross, and J. Stavans, "Dynamics of two-dimensional soap froths," *Phys. Rev. A*, Vol. 36, p. 306-312. Copyright © 1987 by the American Physical Society. **Fig. 16.3** Reprinted from *Proceedings of the Fifth International Conference on Sintering and Related Phenomena, 1980*, p. 146, "Initial stage solid state sintering models. A critical analyisis and assessment," W. Coblenz et al., © Plenum Press, with kind permission of Springer Science and Business Media. **Fig. 16.4** Reprinted, by permission, from W. Beeré, "Stresses and deformations at grain boundaries," *Philosophical Transactions of the Royal Society*, Vol. 288, 177–195. Copyright © 1978 by Royal Society Publishers. **Fig. 16.5** Reprinted from *Acta Metallurgica*, M.F. Ashby, "A first report on deformation mechanism maps," Vol. 20, 887–797, Copyright 1972, with permission from Elsevier. **Fig. 16.6** With permission from Prof. Hans-Eckart Exner. **Fig. 16.9** Reprinted from *Acta Metallurgica*, Vol. 5, B.H. Alexander and R.W. Balluffi, "Mechanism of sintering of copper," 666–677, Copyright 1957, with permission from Elsevier. **Fig. 16.10** Reprinted from *Acta Metallurgica*, Vol. 22, M.F. Ashby, "A first report on sintering diagrams," 275–289, Copyright 1974, with permission from Elsevier. **Fig. 18.11** Reprinted from *Acta Metallurgica*, Vol. 32, S.S. Brenner, P.P. Camus, M.K. Miller, W.A. Soffa, "Phase separation and coarsening in Fe–Cr–Co alloys," 1217–1227, Copyright 1984, with permission from Elsevier. **Fig. 18.12** Reprinted, by permission, from *ASM Handbook: Alloy Phase Diagrams*, Vol. 3, H. Baker (Ed.), p. 2•56. Copyright © 1992 by ASM International. **Fig. 18.13a** Reprinted from "Phase separation of Fe-Al alloys with Fe₃Al order," S.M. Allen, 1977, *Philosophical Magazine*, Taylor and Francis, http://www.tandf.co.uk/journals/titles/14786435.html. **Fig. 18.13b** Reprinted from *Acta Metallurgica*, Vol. 28, T.Miyazaki, S. Takagishi, H. Mori, T. Kozakai, "The phase decomposition of iron-molybdenum binary alloys by spinodal mechanism," 1143–1153, Copyright 1980, with permission from Elsevier. **Fig. 19.2** Reprinted, by permission, from F.K. LeGoues, H.I. Aaronson, Y.W. Lee, and G.J. Fix, "Influence of crystallography upon critical nucleus shapes and kinetics of homogeneous f.c.c.→f.c.c. nucleation. I. The classical theory regime," *Proceedings of an International Conference on Solid to Solid Phase Transformations (1981)*, pp. 427-431. Copyright © 1982 by the Metallurgical Society of AIME. **Fig. 19.9** Reprinted from *Scripta Metallurgica*, Vol. 8, D.M. Barnett, J.K. Lee, H.I. Aaronson, and K.C. Russell, "The strain energy of coherent ellipsoidal precipitates," 951–956, 1974, with permission from Elsevier. **Fig. 19.10** Reprinted, by permission, from J.K. Lee, D.M. Barnett, and H.I. Aaronson, "The elastic strain energy of coherent ellipsoidal precipitates in anisotropic crystalline solids," *Met. Trans. A*, Vol. 8, pp. 963–970. Copyright © 1977 by TMS (The Minerals, Metals, and Materials Society). **Fig. 19.13** and **Fig. 19.14** Reprinted from *Acta Metallurgica*, Vol. 4, J.W. Cahn, "The kinetics of grain boundary nucleated reactions," 449–459, Copyright 1956, with permission from Elsevier. **Fig. 19.15** and **Fig. 19.17** Reprinted from *Acta Metallurgica*, Vol. 5, J.W. Cahn, "Nucleation on dislocations," 169–171, 1957, with permission from Elsevier. **Fig. 20.8a** Reprinted from *Acta Metallurgica*, Vol. 13, K.A. Jackson and K.D. Hunt, "Transparent compounds that freeze like metals," 1212–1215, Copyright 1965, with permission from Elsevier. **Fig. 20.8b** Reprinted from *Journal of Crystal Growth*, Vol. 5, L.R. Morris and W.C. Winegard, "The development of cells during the solidification of a dilute Pb–Sb alloy," 361–375, Copyright 1969, with permission from Elsevier. **Fig. 22.1, Fig. 22.2,** and **Fig. 22.3** Reprinted, by permission, from M.C. Flemings, *Solidification Processing*. Copyright © 1974 by McGraw-Hill. **Fig. 22.5** Reprinted, by permission, from T.F. Bower and M.C. Flemings, "Formation of the chill zone in ingot solidification," *Trans. AIME*, Vol. 239, pp. 216–219. Copyright © 1967 by TMS (The Minerals, Metals, and Materials Society). **Fig. 22.8** Reprinted, by permission, from W. Kurz and D.J. Fisher, *Fundamentals of Solidification.* Copyright © 1984 by Trans Tech Publications. **Fig. 23.2** and **Fig. 23.4** Reprinted, by permission, from D.A. Porter and K.E. Easterling, *Phase Transformations in Metals and Alloys*, 2nd ed. Copyright © 1992 by Chapman and Hall. **Fig. 23.5** and **Fig. 23.6** Reprinted from *Acta Metallurgica*, Vol. 32, F.K. LeGoues and H.I. Aaronson, "Influence of crystallography upon critical nucleus shapes and kinetics of homogeneous f.c.c.-f.c.c. nucleations— IV. Comparisons between theory and experiment in Cu–Co alloys," 1855–1870, Copyright 1984, with permission from Elsevier. **Fig. 23.7** Reprinted, by permission, from A.K. Jena and M.C. Chaturvedi, *Phase Transformations in Materials.* Copyright © 1992 by Prentice Hall. **Fig. 23.8** Reprinted, by permission, from R.H. Beton and E.C. Rollason, "Hardness reversion of dilute aluminum–copper and aluminum–copper–magnesium alloys," *Journal of the Institute of Metals*, Vol. 86, pp. 77–117. Copyright © 1957 by the Institute of Metals (Maney Publishing). **Fig. 23.8** Reprinted, by permission, from E. Hornbogen, "Die electronenmikroskopisce untersuchung der ausscheidung in Al-Cu-mischkristallen," *Aluminium*, Vol. 47, pp. 4147. Copyright © 1967 by Aluminium-Verlag. **Fig. 24.5, Fig. 24.6,** and **Fig. 24.7** From *Introduction to the Crystallography of Martensitic Transformations* by C.M. Wayman. Many of Wayman's stereographs are drawn from "Martensitic Transformations and Determination of the Inhomogeneous Deformation" by D.S. Lieberman, *Acta Metallurgica*, Nov. 1958. **Fig. 24.13, Fig. 24.14,** and **Fig. 24.15** Reprinted, by permission, from M.W. Burkart and T.A. Read, "Diffusionless phase change in the indium–thallium system," *Trans. AIME*, Vol. 197, pp. 1516–1524. Copyright © 1953 by TMS (The Minerals, Metals, and Materials Society). **Fig. 24.17** Reprinted, by permission, from L. Kaufman and M. Cohen, "The martensitic transformation in the iron–nickel system," *Trans. AIME*, Vol. 206, pp. 1393–1400. Copyright © 1956 by TMS (The Minerals, Metals, and Materials Society). **Fig. 24.18** Reprinted, by permission, from the *ASM Metals Handbook*, 8th ed. Vol. 8, p. 198. Copyright © 1973 by ASM International. Reprinted with permission from the estate of William C. Leslie. **Fig. 24.19** Reprinted from *Acta Metallurgica*, Vol. 6, D.S. Lieberman, "Martensitic transformations and determination of the inhomogeneous deformation," 680–693, Copyright 1958, with permission from Elsevier. **Fig. B.2** Reprinted, by permission, from K.L. Merkle, L.J. Thompson, and F. Phillipp, "Thermally activated step motion observed by high-resolution electron microscopy at a (113) symmetric tilt grain-boundary in aluminum," *Philosophical Magazine Letters*, Vol. 82, p. 591. Copyright © 2002 by Taylor and Francis Ltd., http://www.tandf.co.uk/journals. **Fig. B.3** Reprinted, by permission, from M. Döblinger, unpublished research. Copyright © 2005 by Markus Döblinger. **Fig. B.5** Reprinted from W.T. Read, *Dislocations in Crystals*, McGraw-Hill, 1953.

CITED AUTHOR INDEX

FIGURE INDEX

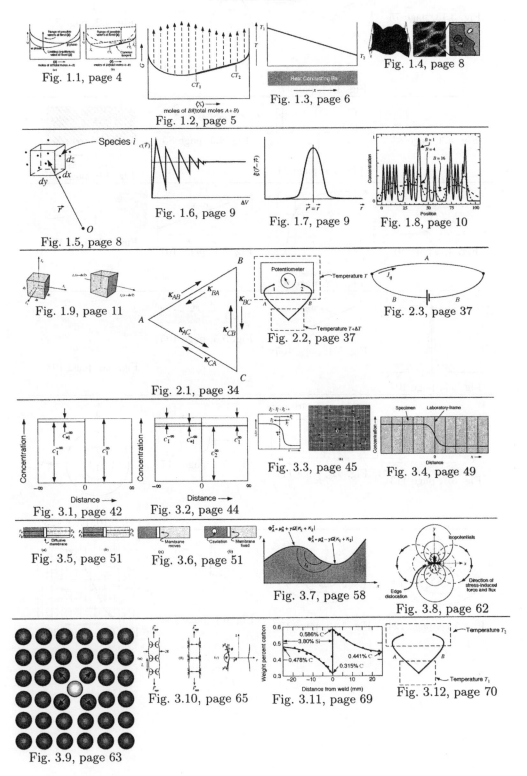

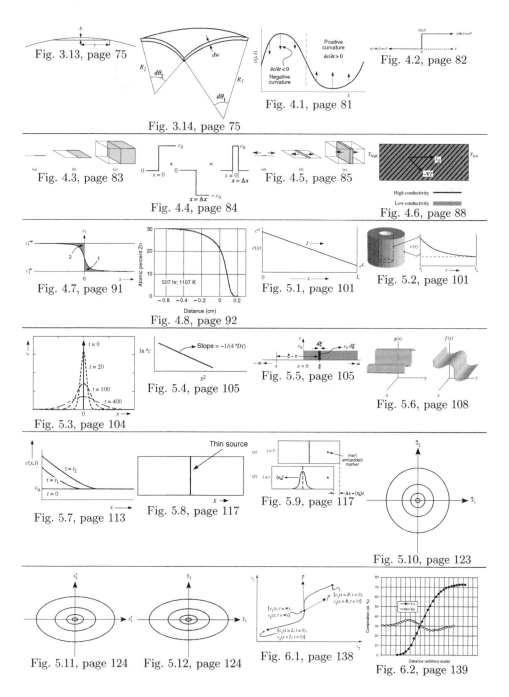

Fig. 3.13, page 75

Fig. 3.14, page 75

Fig. 4.1, page 81

Fig. 4.2, page 82

Fig. 4.3, page 83

Fig. 4.4, page 84

Fig. 4.5, page 85

Fig. 4.6, page 88

Fig. 4.7, page 91

Fig. 4.8, page 92

Fig. 5.1, page 101

Fig. 5.2, page 101

Fig. 5.3, page 104

Fig. 5.4, page 105

Fig. 5.5, page 105

Fig. 5.6, page 108

Fig. 5.7, page 113

Fig. 5.8, page 117

Fig. 5.9, page 117

Fig. 5.10, page 123

Fig. 5.11, page 124

Fig. 5.12, page 124

Fig. 6.1, page 138

Fig. 6.2, page 139

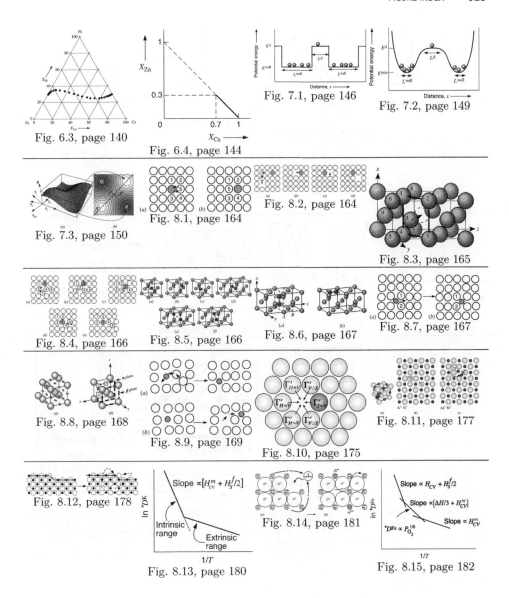

Fig. 6.3, page 140

Fig. 6.4, page 144

Fig. 7.1, page 146

Fig. 7.2, page 149

Fig. 7.3, page 150

Fig. 8.1, page 164

Fig. 8.2, page 164

Fig. 8.3, page 165

Fig. 8.4, page 166

Fig. 8.5, page 166

Fig. 8.6, page 167

Fig. 8.7, page 167

Fig. 8.8, page 168

Fig. 8.9, page 169

Fig. 8.10, page 175

Fig. 8.11, page 177

Fig. 8.12, page 178

Fig. 8.13, page 180

Fig. 8.14, page 181

Fig. 8.15, page 182

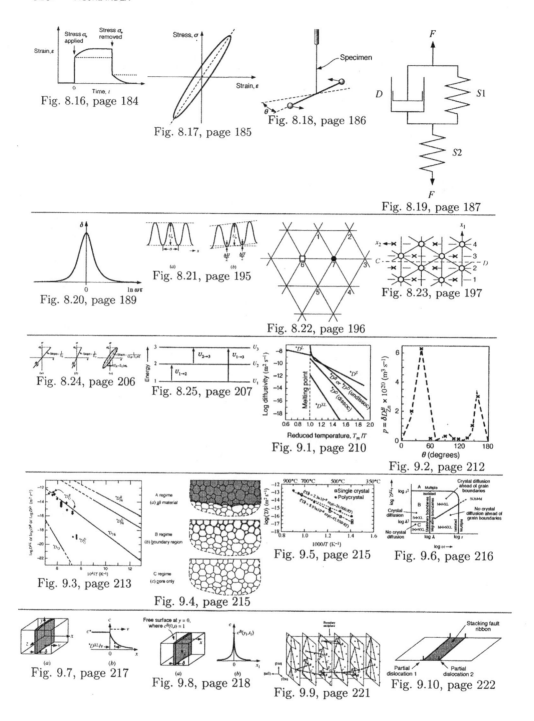

Fig. 8.16, page 184

Fig. 8.17, page 185

Fig. 8.18, page 186

Fig. 8.19, page 187

Fig. 8.20, page 189

Fig. 8.21, page 195

Fig. 8.22, page 196

Fig. 8.23, page 197

Fig. 8.24, page 206

Fig. 8.25, page 207

Fig. 9.1, page 210

Fig. 9.2, page 212

Fig. 9.3, page 213

Fig. 9.4, page 215

Fig. 9.5, page 215

Fig. 9.6, page 216

Fig. 9.7, page 217

Fig. 9.8, page 218

Fig. 9.9, page 221

Fig. 9.10, page 222

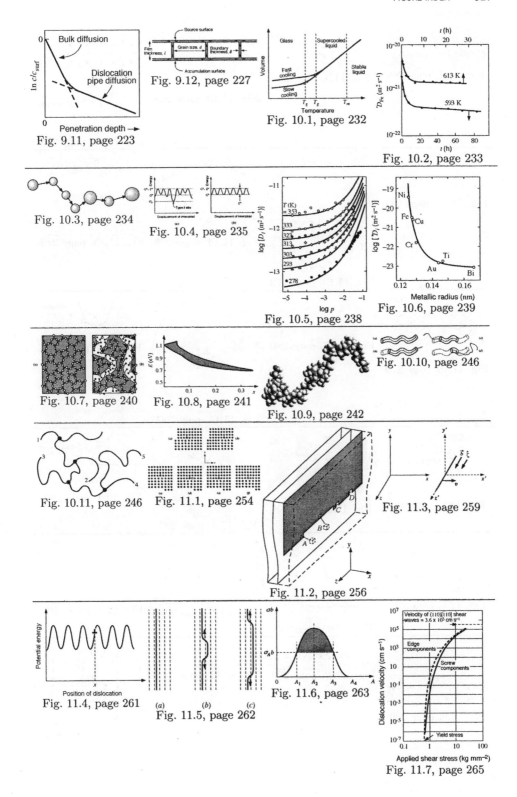

Fig. 9.11, page 223

Fig. 9.12, page 227

Fig. 10.1, page 232

Fig. 10.2, page 233

Fig. 10.3, page 234

Fig. 10.4, page 235

Fig. 10.5, page 238

Fig. 10.6, page 239

Fig. 10.7, page 240

Fig. 10.8, page 241

Fig. 10.9, page 242

Fig. 10.10, page 246

Fig. 10.11, page 246

Fig. 11.1, page 254

Fig. 11.2, page 256

Fig. 11.3, page 259

Fig. 11.4, page 261

Fig. 11.5, page 262

Fig. 11.6, page 263

Fig. 11.7, page 265

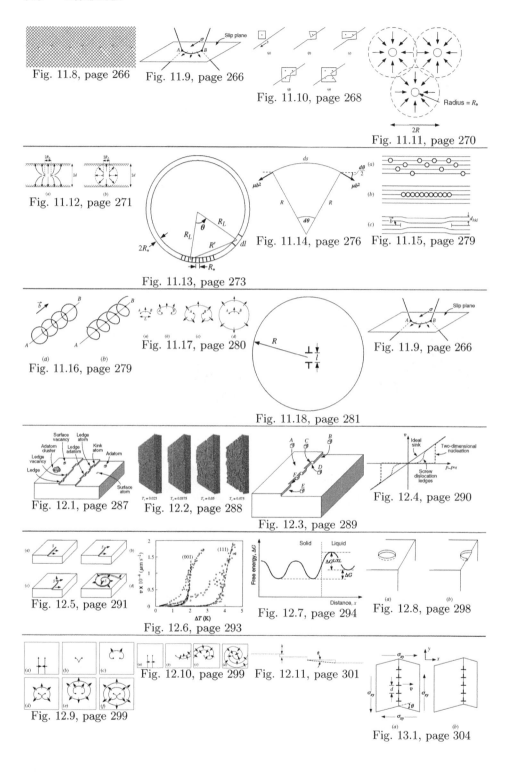

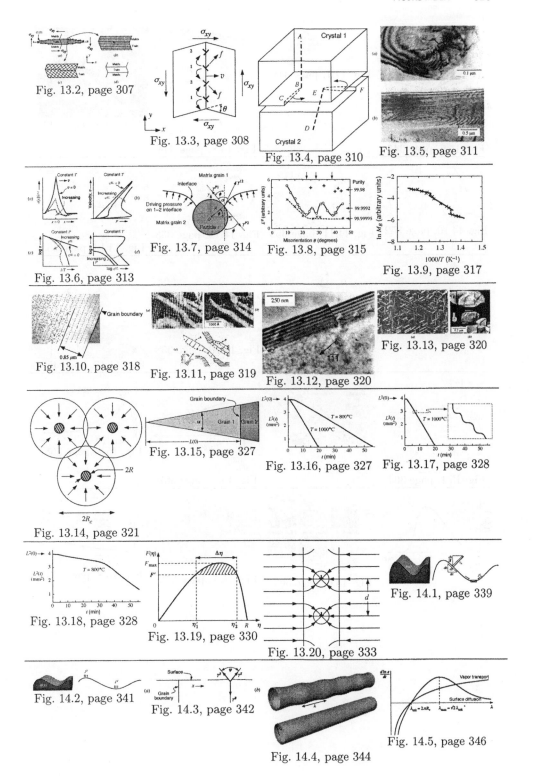

Fig. 13.2, page 307

Fig. 13.3, page 308

Fig. 13.4, page 310

Fig. 13.5, page 311

Fig. 13.6, page 313

Fig. 13.7, page 314

Fig. 13.8, page 315

Fig. 13.9, page 317

Fig. 13.10, page 318

Fig. 13.11, page 319

Fig. 13.12, page 320

Fig. 13.13, page 320

Fig. 13.14, page 321

Fig. 13.15, page 327

Fig. 13.16, page 327

Fig. 13.17, page 328

Fig. 13.18, page 328

Fig. 13.19, page 330

Fig. 13.20, page 333

Fig. 14.1, page 339

Fig. 14.2, page 341

Fig. 14.3, page 342

Fig. 14.4, page 344

Fig. 14.5, page 346

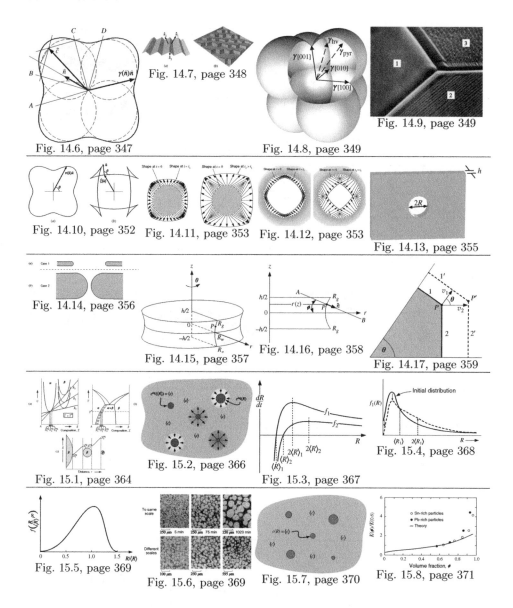

Fig. 14.6, page 347

Fig. 14.7, page 348

Fig. 14.8, page 349

Fig. 14.9, page 349

Fig. 14.10, page 352

Fig. 14.11, page 353

Fig. 14.12, page 353

Fig. 14.13, page 355

Fig. 14.14, page 356

Fig. 14.15, page 357

Fig. 14.16, page 358

Fig. 14.17, page 359

Fig. 15.1, page 364

Fig. 15.2, page 366

Fig. 15.3, page 367

Fig. 15.4, page 368

Fig. 15.5, page 369

Fig. 15.6, page 369

Fig. 15.7, page 370

Fig. 15.8, page 371

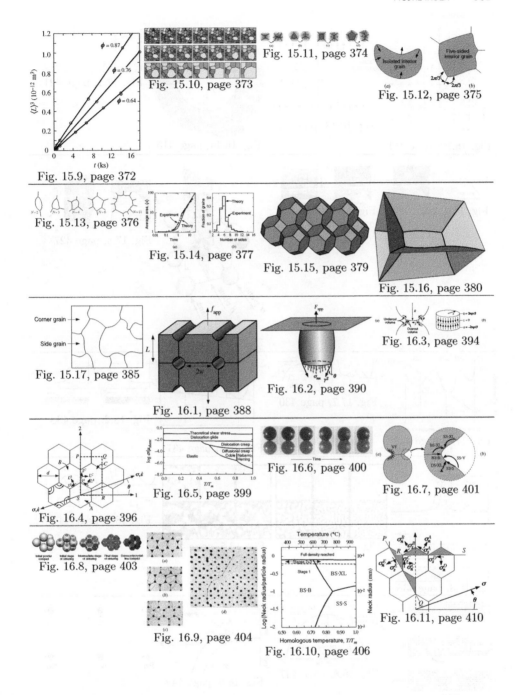

Fig. 15.9, page 372

Fig. 15.10, page 373

Fig. 15.11, page 374

Fig. 15.12, page 375

Fig. 15.13, page 376

Fig. 15.14, page 377

Fig. 15.15, page 379

Fig. 15.16, page 380

Fig. 15.17, page 385

Fig. 16.1, page 388

Fig. 16.2, page 390

Fig. 16.3, page 394

Fig. 16.4, page 396

Fig. 16.5, page 399

Fig. 16.6, page 400

Fig. 16.7, page 401

Fig. 16.8, page 403

Fig. 16.9, page 404

Fig. 16.10, page 406

Fig. 16.11, page 410

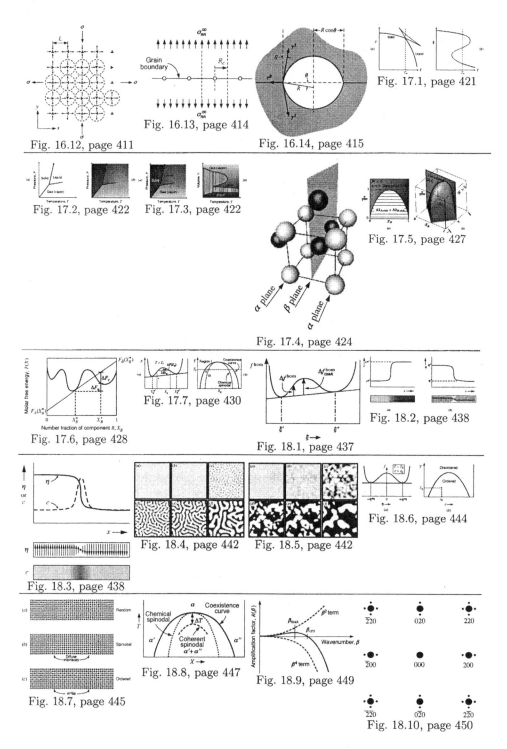

Fig. 16.12, page 411

Fig. 16.13, page 414

Fig. 16.14, page 415

Fig. 17.1, page 421

Fig. 17.2, page 422

Fig. 17.3, page 422

Fig. 17.5, page 427

Fig. 17.4, page 424

Fig. 17.6, page 428

Fig. 17.7, page 430

Fig. 18.1, page 437

Fig. 18.2, page 438

Fig. 18.3, page 438

Fig. 18.4, page 442

Fig. 18.5, page 442

Fig. 18.6, page 444

Fig. 18.7, page 445

Fig. 18.8, page 447

Fig. 18.9, page 449

Fig. 18.10, page 450

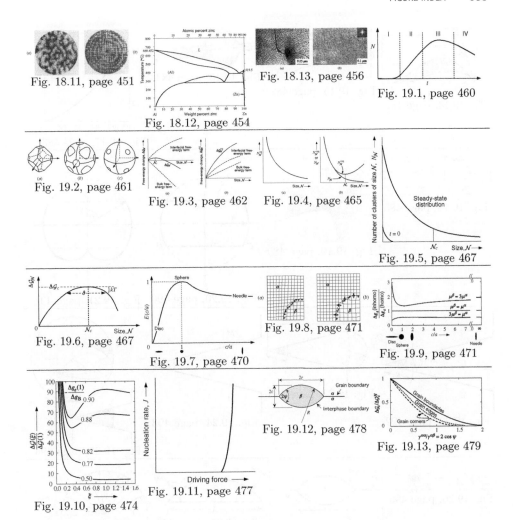

Fig. 18.11, page 451

Fig. 18.12, page 454

Fig. 18.13, page 456

Fig. 19.1, page 460

Fig. 19.2, page 461

Fig. 19.3, page 462

Fig. 19.4, page 465

Fig. 19.5, page 467

Fig. 19.6, page 467

Fig. 19.7, page 470

Fig. 19.8, page 471

Fig. 19.9, page 471

Fig. 19.10, page 474

Fig. 19.11, page 477

Fig. 19.12, page 478

Fig. 19.13, page 479

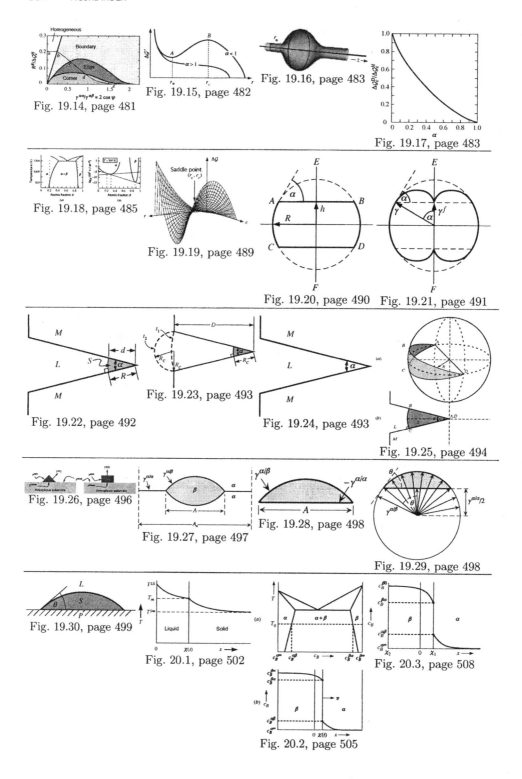

Fig. 19.14, page 481

Fig. 19.15, page 482

Fig. 19.16, page 483

Fig. 19.17, page 483

Fig. 19.18, page 485

Fig. 19.19, page 489

Fig. 19.20, page 490

Fig. 19.21, page 491

Fig. 19.22, page 492

Fig. 19.23, page 493

Fig. 19.24, page 493

Fig. 19.25, page 494

Fig. 19.26, page 496

Fig. 19.27, page 497

Fig. 19.28, page 498

Fig. 19.29, page 498

Fig. 19.30, page 499

Fig. 20.1, page 502

Fig. 20.2, page 505

Fig. 20.3, page 508

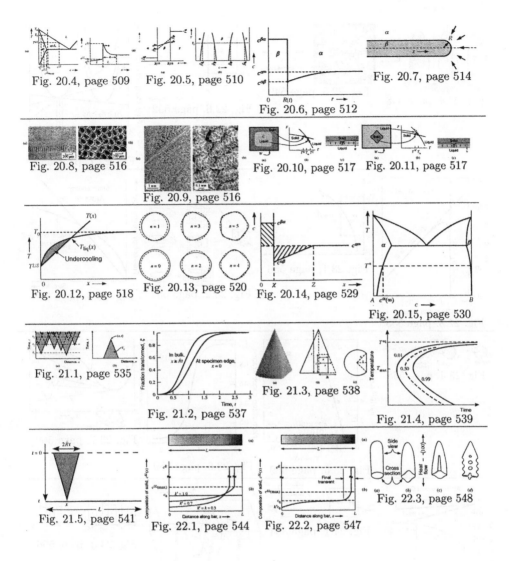

Fig. 20.4, page 509   Fig. 20.5, page 510

Fig. 20.7, page 514

Fig. 20.6, page 512

Fig. 20.8, page 516

Fig. 20.10, page 517   Fig. 20.11, page 517

Fig. 20.9, page 516

Fig. 20.12, page 518   Fig. 20.13, page 520   Fig. 20.14, page 529   Fig. 20.15, page 530

Fig. 21.1, page 535

Fig. 21.3, page 538

Fig. 21.2, page 537

Fig. 21.4, page 539

Fig. 21.5, page 541   Fig. 22.1, page 544   Fig. 22.2, page 547   Fig. 22.3, page 548

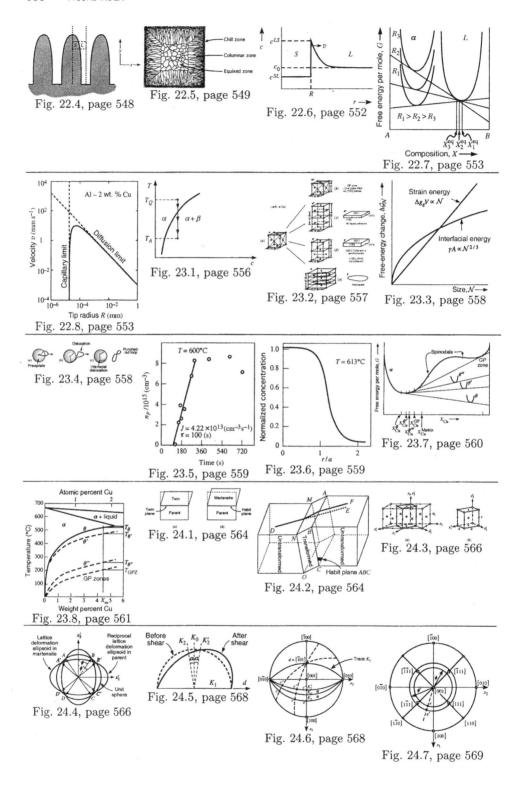

Fig. 22.4, page 548

Fig. 22.5, page 549

Fig. 22.6, page 552

Fig. 22.7, page 553

Fig. 22.8, page 553

Fig. 23.1, page 556

Fig. 23.2, page 557

Fig. 23.3, page 558

Fig. 23.4, page 558

Fig. 23.5, page 559

Fig. 23.6, page 559

Fig. 23.7, page 560

Fig. 23.8, page 561

Fig. 24.1, page 564

Fig. 24.2, page 564

Fig. 24.3, page 566

Fig. 24.4, page 566

Fig. 24.5, page 568

Fig. 24.6, page 568

Fig. 24.7, page 569

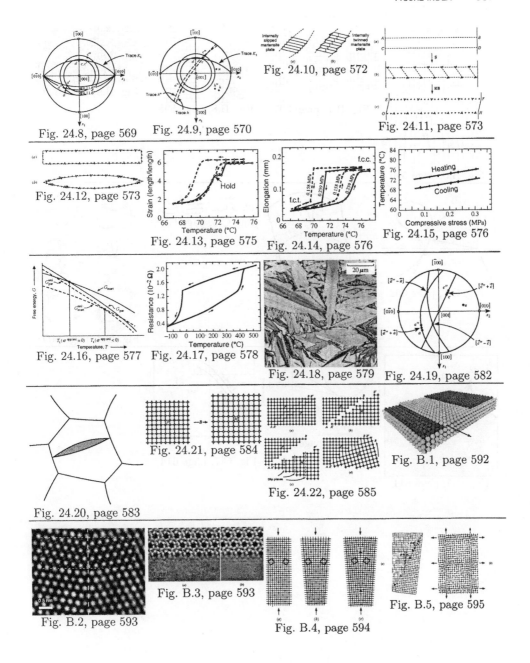

Fig. 24.8, page 569

Fig. 24.9, page 570

Fig. 24.10, page 572

Fig. 24.11, page 573

Fig. 24.12, page 573

Fig. 24.13, page 575

Fig. 24.14, page 576

Fig. 24.15, page 576

Fig. 24.16, page 577

Fig. 24.17, page 578

Fig. 24.18, page 579

Fig. 24.19, page 582

Fig. 24.20, page 583

Fig. 24.21, page 584

Fig. 24.22, page 585

Fig. B.1, page 592

Fig. B.2, page 593

Fig. B.3, page 593

Fig. B.4, page 594

Fig. B.5, page 595

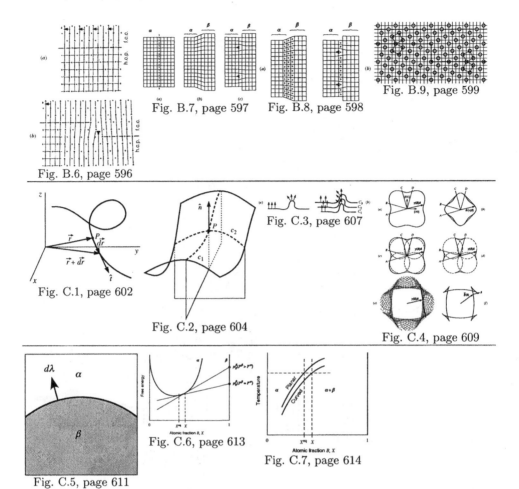

Fig. B.6, page 596

Fig. B.7, page 597

Fig. B.8, page 598

Fig. B.9, page 599

Fig. C.1, page 602

Fig. C.2, page 604

Fig. C.3, page 607

Fig. C.4, page 609

Fig. C.5, page 611

Fig. C.6, page 613

Fig. C.7, page 614

TOPIC INDEX

Printed and bound by CPI Group (UK) Ltd, Croydon, CR0 4YY

16/04/2025

14658587-0005